LAKE KARIBA:
A Man-Made Tropical Ecosystem in Central Africa

MONOGRAPHIAE BIOLOGICAE

Editor

J. ILLIES

Schlitz

VOLUME 24

DR. W. JUNK b.v. PUBLISHERS THE HAGUE 1974

LAKE KARIBA:
A Man-Made Tropical Ecosystem in Central Africa

Edited by

E. K. BALON & A. G. COCHE

DR. W. JUNK b.v. PUBLISHERS THE HAGUE 1974

ISBN-13: 978-94-010-2336-8   e-ISBN-13: 978-94-010-2334-4
DOI: 10.1007/978-94-010-2334-4
© 1974 by Dr. W. Junk b.v., Publishers, The Hague
Softcover reprint of the hardcover 1st edition
Cover design M. Velthuijs, The Hague
Zuid-Nederlandsche Drukkerij N.V., 's-Hertogenbosch

GENERAL CONTENTS

	Preface	VII
	Abstract	IX
Part I	**Limnological Study of a Tropical Reservoir** by A. G. Coche	1
	Contents of Part I	3
	Introduction and Acknowledgements	7
Section I	*The Zambezi catchment above the Kariba Dam: general physical background*	11
1.	Physiography	13
2.	Geology and soils	18
3.	Climate	25
4.	Flora, fauna and human population	41
Section II	*The rivers and their characteristics*	49
5.	The Zambezi River	51
6.	Secondary rivers in the lake catchment	65
Section III	*Lake Kariba physico-chemical characteristics*	75
7.	Hydrology	77
8.	Morphometry and morphology	84
9.	Sampling methodology	102
10.	Optical properties	108
11.	Thermal properties	131
12.	Dissolved gases	164
13.	Mineral content	183
Section IV	*Conclusions*	231
14.	General trophic status of Lake Kariba with particular reference to fish production	233
	Literature cited	236
	Annex I	244
	Annex II	246

Part II		Fish Production of a Tropical Ecosystem by E. K. BALON[1] . .	249
		Contents of Part II	253
		A Parable	255
	1.	Introduction and acknowledgements	257
	2.	Methods.	265
	3.	Age and growth studies by E. K. BALON, J. HOLČIK, S. FRANK, I. BASTL, K. ČERNY, K. CHITRAVADIVELU, A. KIRKA, I. KRUPKA, E. D. MUYANGA & K. PIVNIČKA	280
	4.	Total production, available production and yield of major fish taxa from Lake Kariba	428
	5.	The eels	446
	6.	Fish production of the drainage area and the influence of ecosystem changes on fish distribution (with a section by J. M. KAPETSKY)	459
	7.	The success and failure of the clupeid introduction (with a section by J. G. WOODWARD)	524
		Concluding discussion	542
		Epilogue.	554
		Literature cited	558
Appendix A.		List of symbols used	574
Appendix B.		Efficiency of rotenone-cove samples by G. P. BAZIGOS	575
Appendix C.		Tables of mean sizes of individual species, life intervals and growth intensity	595
Appendix D.		Tables of individual species and single sample production computations	603
Appendix E.		Morphometry of sampling sites and standing crop tables . . .	625
Appendix F.		Time of annulus inception: a pond experiment by E. K. BALON & E. M. CHADWICK	643
Appendix G.		Lepidological study: Key scales of Lake Kariba fishes	647
Part III		Plates: An Annotated Photographic Summary of the Lake Kariba Ecosystem by E. K. BALON	677
		General Index	749

---

[1] Sections written by other authors bear their names after each title (see also Contents of Part II, p. 253–254).

PREFACE

In 1964 the Lake Kariba Fisheries Research Institute (LKFRI) was created in Kariba, Rhodesia as a United Nations Development Program Project, and executed by the Food and Agriculture Organization (FAO) in cooperation with the Governments of Rhodesia and Zambia. Dr. A. G. COCHE took charge of the Limnological Section and conducted research on the entire lake between January 1965 and January 1966.

In 1966 the Central Fisheries Research Institute (CFRI) was created in Chilanga, Zambia by the Department of Wildlife, Fisheries, and National Parks. It was also supported by a UNDP Project executed by FAO. Between 1967 and 1971 Dr. E. K. BALON & Dr. A. G. COCHE were in charge of the Sections of Ichthyobiology and of Limnology respectively.

The results of their FAO research activities on Lake Kariba are united in this volume. In the first part A. G. COCHE presents a limnological synthesis. In the second part E. K. BALON studies in detail the fish production and succession.

The views expressed are those of the authors and do not necessarily coincide with those of the Food and Agriculture Organization of the United Nations.

EUGENE K. BALON
Department of Zoology
University of Guelph
Guelph, Ontario, Canada

ANDRÉ G. COCHE
Lake Kossou Fishery Development
Project (FAO)
Kossou, Ivory Coast, W. Africa

The associate authors I. BASTL (Bratislava), G. P. BAZIGOS (Rome), K. ČERNY (Prague), E. M. CHADWICK (Guelph), K. CHITRAVADIVELU (Ceylon), S. FRANK (Prague), J. HOLČIK (Bratislava), J. M. KAPETSKY (Ann Arbor), A. KIRKA (Bratislava), I. KRUPKA (Bratislava), E. D. MUYANGA (Zambia), K. PIVNIČKA (Prague) and J. WOODWARD (England) are not responsible for the arrangement of their contributions or for the final format of the book.

ABSTRACT

Kariba Dam was built to harness the Zambezi River for the production of hydro-electric power. Its closure in December 1958 created Lake Kariba, which to-day is one of the largest reservoirs in the world. Background information is presented about the physiography of the Zambezi catchment above Kariba, its geology and soils, and the climate and the biology of the lake area. The rivers are studied, in particular the Zambezi River. In the chapter about hydrology, the water budget components are discussed. It is concluded that the Zambezi River contributes on an average 77 percent of the inflow, other rivers 16 percent, and rainfall 7 percent. Evaporation accounts for 14 percent of the water losses. The theoretical renewal time of the water mass is close to three years.

Lake Kariba is naturally subdivided into four distinct basins. Within the lower one several sub-basins can be defined mainly on the basis of the local influence of affluents and submerged topography. Morphometric parameters which characterize in detail the planimetry and the bathymetry of the lake are given with reference to the normal operating water level of 485 m.

The physico-chemical limnology of the water mass was studied in 1965 and in 1968/69. It is compared to data compiled for other African lakes and reservoirs. The changes which have taken place along the longitudinal axis of the reservoir between its upper and lower ends were especially put into evidence in the optical, thermal, and chemical properties.

The depth of visibility generally ranged from 50 to 1060 cm, averaging 405 cm. It is equivalent to the depth at which about 22.7 percent of the incident solar light was found. The depth of the euphotic zone could be estimated as 3.54 DV. The penetration of the visible light changes markedly with time. The average vertical attenuation coefficient for total light varied from 0.20 to 2.20 according to the basin considered. The study of the penetration of spectral light blocks (445, 530, and 630 nm) showed that the green light was generally the most penetrating. The depth of the euphotic zone was relatively small, averaging 10 m in the upper basins and 16 m in the lower ones. It could be estimated as equal to 4.2/AVAC min.

From the thermal point of view, the two lower thirds of the reservoir presented all the characteristics of a warm monomictic lake. Upstream, riverine conditions prevailed. Total circulation occurred in the lacustrine part during the cool dry season around 22° C. During stratification, a tilted thermocline existed mostly around the 20–25 m depth.

The annual heat budget approximated 15000 cal. $cm^{-2}yr^{-1}$. It represented only 13 percent of the global radiation delivered to the surface of the lake. Residual heat was about 40000 cal. cm^{-2}. To better compare lakes among themselves new indices were defined relating heat contents to mean depth. On such a basis, the Annual Heat Index for L. Kariba was greater than that of any other lake studied. The Tropicality Index (1400 cal. $cm^{-2} m^{-1}$) characterized the lake as truly tropical in nature. Estimated energy contents in the lower basin pointed to a relatively high work of the wind with a very low efficiency and to a stability representing 37 percent of the total work.

The lake surface waters were well supplied with dissolved oxygen. The latter's depth distribution was typically clinograde outside the total circulation period. The oxycycle was closely correlated to the thermal cycle, a strong oxycline developing at the thermocline and the DO content drastically dropping below. Relatively small hypolimnetic area deficits pointed to low productivity in tropical waters. The average rate of hypolimnetic oxygen depletion varied around 0.100 to 0.150 $mg.cm^{-2} day^{-1}$ during the stagnation period. Considering that 2 mg. L^{-1} of dissolved oxygen represented the lower average limit for fish occurrence in tropical waters, the seasonal depth variation of this index showed that unfavourable oxygenation conditions existed in the eastern part of the lake for at least five months of the year.

The total mineral content was small. In 1965 the lake's annual average for total solids was 58 mg. L^{-1}, for salinity 42.4 mg. L^{-1}, for conductivity (20°C) 73 micromhos. Waters belonged to the calcico-carbonate type. The average ionic composition closely resembled that most currently encountered in the world. Water was very soft, total alkalinity varying between 0.42 and 0.88 meq. L^{-1}. Mineral replenishment mainly took place through river inflows but thermal stratification and density currents contributed to the delay in recycling processes of minerals brought into the lake's hypolimnion with the river floods. Quantitative evaluation of the annual chemical budget showed that the Zambezi River imported more than two million tons of minerals per year. The secondary rivers contributed at least 400 thousand tons and rain 42 thousand tons. Most chemical properties point to a low productivity potential. In particular the average quantity of non-carbonated salts available for biological production was very low.

Both the morphoedaphic index and its newly proposed form point to a very low potential of fish production. This conclusion was reinforced by other limnological characteristics which may adversely affect the fish production such as timing of water level fluctuations, shallow depth of the trophogenic zone, seasonally unfavourable oxygenation conditions, and general chemical characteristics of the water mass.

The biological evolution of Lake Kariba appears to have followed a sequence similar to that experienced under temperate climates. The first phase of maturation characterized by maximum biological productions probably lasted until 1964. The second phase of depressed productivity may last until 1974. A slight increase in productivity should normally take place then. But in view of its inherent characteristics of low productivity it is believed that in the future Lake Kariba, relatively to other tropical water bodies, will retain its general oligotrophic status.

The second part of this monograph is devoted entirely to fish fauna, its succession, densities and ecological production, as these relate to limnological, social and environmental issues. Exact usage and definition of abundance, density, standing crop and stock, biomass and production are given; values of total and available production are introduced and terms such as harvest or catch are used to designate that part of production which is above the minimum harvestable size and used by man.

Age, absolute and relative growth values were examined by ten authors on 21 species of Lake Kariba fishes. The average density of the fish population was estimated as 97,000 per hectare. For the entire area of the lake inhabited by fish (determined by echo-sounding), it amounted to 19.6×10^9 specimens. The initial biomass is 2,830 kg/ha, which for the whole lake gives a value 5.7×10^8 kg; the mean biomass was estimated as 3,855 kg/ha or 7.8×10^8 kg. The total production is 3,468 kg/ha/yr, or over 700 thousand metric tons for the whole lake. The estimated available production is 682 kg/ha/yr or 140 thousand metric tons. Yield, the harvestable part of production, was determined according to the minimum harvestable sizes interpolated graphically. The total yield is divided between natural mortality, actual catch and the available yield (i.e. the remainder). It amounted to 599 kg/ha/yr, while the assessed actual catch was 14.8 kg/ha/yr. Since the available yield was estimated to be 189 kg/ha/yr, 395.2 kg/ha/yr of the total yield is attributable to natural mortality. Fishery could increase its exploitation by 32% of the total yield, while 66% is lost through natural mortality and 2.5% forms the actual catch. The 'maximum sustained yield' is thus in the vicinity of 40 thousand metric tons per year, of which actual fishing has been exploiting only 3 thousand tons:

	in kg/ha/yr	in %
Total production ($A = G\bar{B}$)	3,468	100
Natural mortality (M)	$\simeq$ 2,786	80.3
Final production ($\bar{P} = P' + E'$)	$\simeq$ 697	20.1
Available production ($P' = N'_i(\bar{w} - \bar{w}_{i-1})$)	682	19.7
Total yield ($Y_A = A$ above minimum harvestable size)	599	17.3
Natural mortality of harvestable part (M')	$\simeq$ 395	11.4
Sustained yield ($Y_{P'} + E'$)	$\simeq$ 204	5.9
Available yield ($Y_{P'} = P'$ above minimum harvestable size)	189	5.4
Actual catch (E')	15	0.4

However a substantial stock of the catadromous eel *Anguilla nebulosa labiata*, subsequently discovered and evaluated as well as the introduced Tanganyikan anchoveta *Limnothrissa miodon* have to be added to the above values.

It has been predicted that eels will disappear from Lake Kariba because the juveniles will be unable to surmount the dam. This prediction has been proved wrong. An abundant population of eels was discovered in the lake at a depth of 25 to 40 m.

The age structure of the eel cohorts suggests that juveniles surmount the dam in their second year of life and then spend approximately seven years in streams of the lake drainage. The faster maturing individuals which are in better condition emigrate into the ocean, the slower maturing eels remain in the lake longer. The oldest eel in the samples was 18 years old.

The catch per unit of effort for hoopnets was 2.35 kg in the upper part of the lake and 0.41 kg for the lower part of the lake. The density at Namazambwe was assessed at 46 eels per 1 ha. The unexploited eel population of Lake Kariba could form a valuable resource.

Several rivers of the Lake Kariba drainage were investigated as sources of comparative pre-impoudment production values, as well as habitats enabling sustainment of original riverine fish fauna, and as sources from which fish invaded the new lake. The fishes of the Zambezi River and drainage streams fed the local people for many centuries. The whole system formed a valuable source of sustained fish protein harvest.

The original 28 fish species of Lake Kariba increased from 1963 to 1971 to 41. Most of these were from the Upper Zambezi. Samples of fishes from the edge of Victoria Falls were utilized to explain the history of exchanges between two fish faunas and the invasion of Lake Kariba. Fishes of the pre-Upper Zambezi River are of a different origin from those in the pre-Middle and Lower Zambezi. When the two separate river systems which now make up the Zambezi were united as a result of a tectonic upwarp, the point of unification became the Victoria Falls. The rushing waters of the river cut steep gorges along the fault lines in the basalt lava. The Falls were considered a physical barrier separating fishes of the Upper and Middle Zambezi. Recent environmental changes which accompanied the creation of man-made Lake Kariba, together with the invasion of the Lake by Upper Zambezi fishes demonstrate the ineffectiveness of the Falls as a downstream barrier. A theory of an ecological barrier caused by differences in habitat and by differences in number of available niches is presented here. This ecological barrier was functional for 500,000 years, when the separate rivers first united. Creation of Lake Kariba changed the character of this ecological barrier so that lentic fishes flushed over the edge of Victoria Falls could invade the lake.

The successful introduction into Lake Kariba of the Lake Tanganyika clupeid, when evaluated from its density point of view reveals yet another useless deed of fishery practitioners. Catch survey as well as echo-sounding records indicate that this fish may not attain densities worthwhile for commercial exploitation and that it will have to share the available nutrients with other indigenous fishes probably more desirable for human consumption.

If the fish production values of Lake Kariba are recalculated for the whole lake area instead for the 38% of the total area inhabited by fish it certainly will reveal a picture of low fishery potential. Fishery potential presented in this manner (E' 5.59 kg/ha/yr) will support the conclusions that Lake Kariba is an oligotrophic impoundment of low productivity; it may, however, have little value in terms of energy distribution and utilization. The relationship of nutrients locked in inshore waters to those of open waters is an unknown quantity. In order to draw valid conclusions on the ecological production potential of Lake Kariba these phenomena will have to be understood much better than they are now.

The data obtained are examined with regard to the fish population stability and to the nutrient contents. It is concluded that in terms of production the population as a whole is already stable, but that the single taxa are still in the process of adaptation. Due to the fact that one third of the lake water is replaced each year and some nutrients are brought in with inflowing river waters, the loss of nutrients, as a consequence of removal of the sustained yield of fish, could be negligible. Other variables, in which the turnover of nutrients is reflected, are discussed as well as some estimates of the lake's fish potential predicted in the past. Finally, some conclusions concerning fishery policies are given and an attempt is made to debunk man's justification for such environmental changes.

PART I

LIMNOLOGICAL STUDY OF A TROPICAL RESERVOIR

by

ANDRÉ G. COCHE

CONTENTS OF PART I

		Introduction and acknowledgements	7
Section I		**The Zambezi catchment above the Kariba dam: general physical background**	11
	1.	Physiography	13
	1.1.	*The Zambezi River catchment*	13
	1.2.	*The Lake Kariba drainage basin*	13
	1.3.	*The Gwembe Valley*	16
	2.	Geology and soils	18
	2.1.	*The Zambezi River catchment*	18
	2.2.	*Geology of the Zambezi catchment above the Kariba dam*	18
	2.3.	*Geology of the Gwembe Valley and L. Kariba floor*	18
	2.4.	*The valley soils and lake bottom deposits*	21
	2.5.	*Seismography of the Lake Kariba region*	24
	3.	Climate	25
	3.1.	*The normal climate*	25
	3.1.1.	*General climatology of the Zambezi catchment*	25
	3.1.2.	*The normal climate in the vicinity of Lake Kariba*	27
	3.2.	*The local climate during the years of study*	35
	4.	Flora, fauna and human population	41
	4.1.	*The flora of the Gwembe Valley*	41
	4.1.1.	*Terrestrial vegetation*	41
	4.1.2.	*Semi-aquatic and aquatic vegetation*	43
	4.2.	*The fauna of the Gwembe Valley*	46
	4.2.1.	*Vertebrate fauna*	46
	4.2.2.	*Invertebrate aquatic fauna*	46
	4.3.	*The human population of the Gwembe Valley*	47
Section II		**The rivers and their characteristics**	49
	5.	The Zambezi River	51
	5.1.	*The Zambezi flow regime upstream from L. Kariba*	51
	5.2.	*Water quality in the Zambezi River above L. Kariba*	54
	5.2.1.	*Water temperature*	54
	5.2.2.	*Salinity*	54
	5.2.3.	*Silt load and depth of visibility*	58
	5.2.4.	*Chemical composition*	58
	5.2.5.	*General comparison with some other large African rivers*	61
	5.3.	*The Zambezi River below the Kariba dam*	61
	5.3.1.	*The Zambezi flow below the dam*	61
	5.3.2.	*Water quality of the Zambezi River below the dam*	63
	6.	The secondary rivers in the lake catchment	65
	6.1.	*General characters of the secondary rivers*	65

6.2.	Total discharge in the secondary rivers	65
6.3.	Water quality in some secondary rivers	67
6.3.1.	Sanyati River	67
6.3.2.	Lufua River	70
6.3.3.	Miscellaneous rivers	70
6.3.4.	Conclusions	73
Section III	**Lake Kariba physico-chemical characteristics**	75
7.	Hydrology of Lake Kariba	77
7.1.	The hydrological control of the lake level	77
7.2.	The average water budget of the lake	77
7.3.	Fluctuations of the water level	80
7.3.1.	The filling phase	80
7.3.2.	The post-filling phase	82
7.4.	Sedimentation in Lake Kariba	82
7.5.	Water movements: surface seiche	83
8.	Morphometry and morphology of Lake Kariba	84
8.1.	General considerations	84
8.2.	Planimetry of Lake Kariba	86
8.2.1.	Planimetry at the 485-m water level	86
8.2.2.	Planimetry at various water levels	87
8.3.	Bathymetry of Lake Kariba	88
8.3.1.	Bathymetry at the 485-m water level	88
8.3.2.	Bathymetry at various water levels	93
8.3.3.	Areas and volumes of depth zones	98
8.4.	Lake Kariba islands	98
8.5.	Bush-cleared fishing grounds	99
8.6.	Morphology of Lake Kariba	99
8.7.	Comparative morphometry	99
9.	Sampling methodology	102
9.1.	Sampling scheme	102
9.2.	Sampling the water optical properties	102
9.2.1.	Apparent colour of the lake	102
9.2.2.	Depth of visibility	103
9.2.3.	Light attenuation underwater	103
9.3.	Sampling water thermal properties	105
9.4.	Sampling water chemical properties	107
9.4.1.	Water sampling	107
9.4.2.	Routine chemical measurements	107
9.4.3.	Detailed chemical analyses	107
10.	Optical properties of the lake water	108
10.1.	Apparent colour of the lake	108
10.2.	Depth of visibility and its relationship to incident light	108
10.3.	Penetration of visible light	111
10.3.1.	Definitions and methodology	112
10.3.2.	Total light penetration	114

10.3.3.	*Penetration of spectral light blocks*	116
10.3.4.	*Green light penetration*	121
10.3.5.	*Penetration of light into L. Kariba*	123
10.3.6.	*Comparative notes on light penetration in some African lakes*	127
10.3.7.	*Applicability of the standard attenuation concept to Lake Kariba*	127
11.	Thermal properties of Lake Kariba	131
11.1.	*Thermal cycle in the pelagic region*	131
11.1.1.	*Terminology and definitions*	131
11.1.2.	*Longitudinal thermal profiles of the pelagic region*	132
11.1.3.	*Mean vertical temperature distribution*	132
11.1.4.	*The annual thermal cycle in L. Kariba basins*	133
11.1.5.	*The annual thermal cycle in some African reservoirs*	146
11.2.	*Energy content of Lake Kariba*	146
11.2.1.	*The birgean Annual Heat Budget and the Annual Heat Index*	147
11.2.2.	*Residual Heat and Tropicality Index*	151
11.2.3.	*Variation of the Heat Content throughout the annual cycle*	154
11.2.4.	*Maximum Heat Content and Maximum Heat Index*	156
11.2.5.	*The work of the wind in warming L. Kariba*	157
11.2.6.	*Stability in L. Kariba*	160
11.3.	*Conclusions*	161
12.	Dissolved gases in Lake Kariba	164
12.1.	*Dissolved oxygen*	164
12.1.1.	*Oxygen content of surface waters*	164
12.1.2.	*Distribution of oxygen within the water mass*	165
12.1.3.	*Oxygen deficits of Basins III and IV*	169
12.1.4.	*Evolution of hypolimnetic oxygen depletion during the oxycycle*	170
12.1.5.	*Evolution of the oxycycle*	174
12.1.6.	*Oxygen distribution and the fish stocks*	174
12.1.7.	*Comparison with other African reservoirs*	179
12.2.	*Hydrogen sulphide*	180
12.3.	*Conclusions about dissolved gases*	181
13.	Mineral content of lake waters	183
13.1.	*Hydrogen-ion concentration*	183
13.2.	*Total mineral content of pelagic surface waters*	183
13.2.1.	*Total solids*	184
13.2.2.	*Salinity*	187
13.2.3.	*Conductivity*	188
13.2.4.	*Conclusions*	191
13.3.	*Inorganic ions in pelagic surface waters*	192
13.3.1.	*General ionic composition*	192
13.3.2.	*Alkali metals: sodium and potassium*	197
13.3.3.	*Total hardness: calcium and magnesium*	198
13.3.4.	*Cationic composition: conclusions*	198
13.3.5.	*Total alkalinity: carbonate and bicarbonate*	199
13.3.6.	*Chloride and sulphate*	200

13.3.7.	*Silica*	201
13.3.8.	*Nitrate and phosphate*	201
13.3.9.	*Anionic composition: conclusions*	202
13.3.10.	*Main salts in surface waters*	203
13.3.11.	*Seasonal variations of the mineral content*	205
13.4.	*Mineral content of deep waters*	209
13.5.	*Relation between the chemical composition of lake and inflow waters*	210
13.5.1.	*The Zambezi River and the lake*	210
13.5.2.	*The secondary rivers and the lake*	210
13.5.3.	*The supply of ions by rain*	211
13.6.	*The chemical budget of Lake Kariba*	213
13.6.1.	*Qualitative assessment of the chemical budget*	213
13.6.2.	*Quantitative assessment of the chemical budget*	215
13.7.	*The chemical evolution since the closure of the dam*	219
13.7.1.	*Evolution of the total mineral content*	219
13.7.2.	*Evolution of the water chemical composition*	222
13.8.	*The chemical status of L. Kariba among African lakes and reservoirs*	224
13.8.1.	*Total mineral content of surface waters*	224
13.8.2.	*Chemical composition of surface waters*	224
13.8.3.	*Conclusions*	227
13.9.	*General conclusions about the mineral content in Lake Kariba*	227
Section IV	**Conclusions**	231
14.	The general trophic status of Lake Kariba with particular reference to fish production	233
14.1.	*Relationship of trophic status to environmental factors*	233
14.2.	*Past, actual, and future trophic status of Lake Kariba*	235
Literature cited		236
Annex I	General characteristics of the large African reservoirs	244
Annex II	List of abbrevations and symbols	246

INTRODUCTION AND ACKNOWLEDGEMENTS

Tropical regions of the world are now being investigated more closely than ever for potential hydro-electric development. Huge projects are proposed to harness the energy of their major rivers in South America, Asia (Mekong Valley), and in Africa. This last continent in particular is extremely rich in hydro-electric potential energy, one third of the world's prime capability being part of its vast natural resources (SMITH, 1968).

It is only recently, since about 1958 that the development of this African energy potential has really begun (Ann. I). In December 1958 the first of a series of large dams was closed in Kariba, on the Middle Zambezi River. In 1964 both the Akosombo Dam (Volta R., Ghana) and the Aswan High Dam (Nile R., Egypt) were put into operation. Four years later the Kainji scheme (Niger R., Nigeria) was completed while work was being started in Ivory Coast on the Kossou Dam (Bandama R., closed in February 1971). To-day dam construction rapidly progresses as for example downstream from Kariba (Cabora Bassa, Mozambique), in the Orange River Valley (Rep. South Africa), and on the Congo River (Inga, Zaïre). It is considered that in West Africa the basic and most economic source of electric energy for the future lies in its waterways and plans have been made for building dams on the Niger, Benue, Volta, Senegal, Bandama, Cavally, Konkoure, Sassandra, and Comoe Rivers (SMITH, 1968).

Although primarily built for the generation of electricity, these African dams and the associated man-made lakes created upstream have in fact multi-purpose uses. Among these uses river regulation, flood control, water supply, and irrigation are outstanding. Particularly in developing countries, they may help to promote agriculture, industry, water transport, tourism, and fisheries. On the other hand, dams bring with them new problems, related to social development (resettlement), public health (schistosomiasis), economic development (disruption of communications), and ecological balance (downstream areas) for examples.

Within the inundated area a lake is artificially created. Practically overnight a new aquatic environment replaces the more or less stabilized ecosystem of land and river, and a new type of biological evolution is started, which is of prime interest to all concerned with future developments. The civil engineer should be prepared to face and prevent possible damages due to the presence of hydrogen sulphide in the lake water (e.g. corrosion of the turbines and clogging of the cooling system by algae). Water transports might have to cope with the explosive development of aquatic plants (e.g. *Salvinia, Eichornia, Pistia*). Drawdown agriculture will depend on the seasonality of water level fluctuations. Irrigation schemes and industries will require or prefer a water supply with qualities suitable for their purpose. Most of all, the enlarged fishing industry will depend on the limnological characteristics of the new reservoir.

The possibilities of a hydro-electric scheme on the Zambezi River in South-Central Africa were first investigated by the Rhodesian government in 1925. But only in

February 1955 was the decision taken by the government of the Central African Federation to implement the Kariba development project. On December 2, 1958 the dam was closed, the double-curvature concrete arch – 128 m high and 617 m long – taming the Zambezi River for the first time in history. Lake Kariba started forming, the average operating water level being reached for the first time four-and-a-half years later. Today it is one of the world's largest man-made lakes, politically shared by Zambia and Rhodesia. The wealth of information accumulated on this part of Africa which was so poorly known before the Kariba Project, was recently brought together and indexed (COCHE, 1971).

Hydrobiological research was initiated in 1956 during the pre-impoundment period by the Joint Fisheries Research Organization, particular emphasis being placed on seasonal fishery surveys (BOWMAKER, 1960). In 1959 the Kariba Lake Co-ordinating Committee provided additional man-power and limnological data started to be collected on a routine basis in the forming lake (HARDING, 1962; 1964a, 1965). In late 1963 the United Nations Development Programme (UNDP) declared the Lake Kariba Project operational. Because of the changing political scene, it took another year before the establishment of the Lake Kariba Fishery Research Institute (LKFRI) in Kariba, staffed by the Food and Agriculture Organization of the UN (FAO) and by the governments of Rhodesia and Zambia. The first overall lake surveys for defining the physico-chemical characteristics were conducted between January 1965 and January 1966 (COCHE, 1965, 1968, 1969). For the second time, changes in the political situation interrupted limnological research on Lake Kariba in 1966 when the international LKFRI staff had to move from Kariba to Chilanga, Zambia. Fortunately overall surveys started again in 1967 (BEGG, 1971) and were continuing since, in the Rhodesian territory of the lake (VAN DER LINGEN, 1973). At the Zambian Central Fisheries Research Institute (CFRI) created in 1966 with the assistance of UNDP and FAO, I studied Lake Kariba again since late 1967. Overall limnological surveys were made from January 1968 to February 1969, additional data being gathered on a less intensive basis until 1971.

It is the purpose of this study to present a synthesis of the observations collected and of the results obtained since the creation of Lake Kariba. On the basis of my personal studies however, particular emphasis is given to the years 1965 and 1968. From the limnological point of view, the thermal properties (Chapter 11) and the occurrence of dissolved oxygen (Chapt. 12) have received an increased attention because of their direct effect on the distribution of fishes. In a final chapter, the junction between physico-chemical considerations and potential biological productivity of fish in particular is realized, the pure ichthyological aspects being later detailed by Dr. E. K. BALON in Part II.

Gratitude is expressed to Mr. L. S. JOERIS, FAO Project Manager in Rhodesia and Zambia, for his invaluable help and encouragement during the course of these studies. All staff members of LKFRI (1965) and CFRI (1967–1971) contributed in one way or another to the success of the limnological surveys. I wish to remember thankfully the long days of hard field work shared by the Rhodesian and Zambian junior staff, in particular assistant L. M. MWELAISHA, coxswain M. KAMWAYA, and helper P. MALILAH. In 1965 Dr. D. S. MITCHELL, Lecturer at the Botany Department of the

University College of Rhodesia, supervised the chemical analyses of the water samples collected during lake cruises. He kindly provided the results. Further data were also obtained from G. BEGG, LKFRI actual limnologist, to whom I am most grateful.

I am also indebted to the following scientists for providing constructive criticism while reviewing the manuscript: Dr. W. C. BECKMAN, Aquatic Resources Improvement and Environment Service, FAO, Rome; Dr. H. F. HENDERSON, Aquatic Resources Survey Evaluation Service, FAO, Rome; Dr. K. F. LAGLER, School of Natural Resources, University of Michigan, USA (Chapt. 1 to 10); Dr. J. TALLING, Freshwater Biological Association, Ambleside, United Kingdom (Chapt. 10 & 11).

SECTION I

THE ZAMBEZI CATCHMENT ABOVE THE KARIBA DAM:
General Physical Background

1. PHYSIOGRAPHY

1.1. The Zambezi River catchment

The Zambezi River, southern Africa's largest river, rises from a swamp at 1400 m altitude on the southern slopes of the South Equatorial Divide. Through an easterly course it flows over nearly 2500 km to the Indian Ocean. Its vast catchment area (1 193 500 sq.km) can be sub-divided into the following physiographic entities (Fig. 1);
(i) Catchment above Kariba Dam 663 820 sq.km
Northern Highlands: 220 670 sq km
Central Plains: 286 970 sq km
Rhodesian Highlands: 156 180 sq km
(ii) Catchment below Kariba Dam 529 680 sq.km
Kafue, Luangwa and Malawi catchments
Tete Basin and Mozambique Plain

From the ecological point of view as reflected for example in the fish fauna (JUBB, 1967), the Zambezi River is divided into three reaches:
(i)   Upper Zambezi R.: from source to the Victoria Falls (1078 km)
(ii) Middle Zambezi R.: between Victoria Falls and Cabora Bassa Rapids (853 km)
(iii) Lower Zambezi R.: from Cabora Bassa to the Ocean (563 km)

The barriers separating each of these sections are well illustrated in a longitudinal profile of the river bed (Fig. 1).

1.2. The Lake Kariba drainage basin

The Zambezi catchment above the Kariba Dam (Fig. 2) is bounded on the north by the South Equatorial Divide, a well-defined boundary separating the Zambezi Basin from the Congo Basin. On the west and south the watershed divide between the Zambezi and the Okavango Basins is low and during high floods they may become connected. To the east there is a distinct drainage divide between the Zambezi River and the Kafue River.

Physiographic features (Fig. 2) also point to the three distinct ecological areas referred to above:
(i) Northern Highlands, a belt of high ground (1000–2000 m alt.) including the Lungwebungu and Kabompo River catchments.
(ii) Central Plains, a relatively flat plateau (1000–1500 m alt.) characterized by large swampy areas, the Barotse Flood Plain (Lukulu to Senanga – 209 km) and the Chobe Swamps which exert a marked controlling effect on the Zambezi River discharge downstream.
(iii) Rhodesian Highlands (alt. from 650 m in the north to 1300 m in the south) which comprise a peneplain eroded subsequent to uplift (see Geology). Here the Zambezi

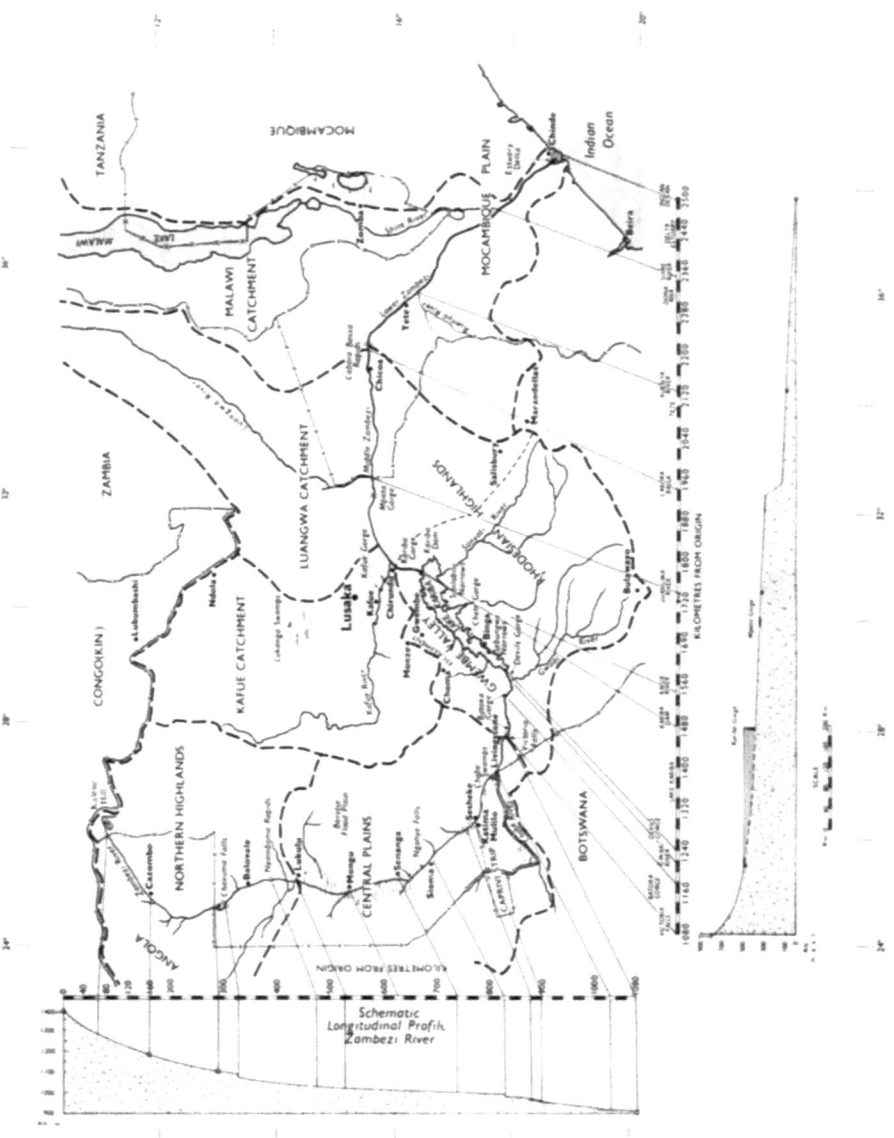

Fig. 1. General physiography of the Zambezi River catchment and longitudinal profile of the river bed.

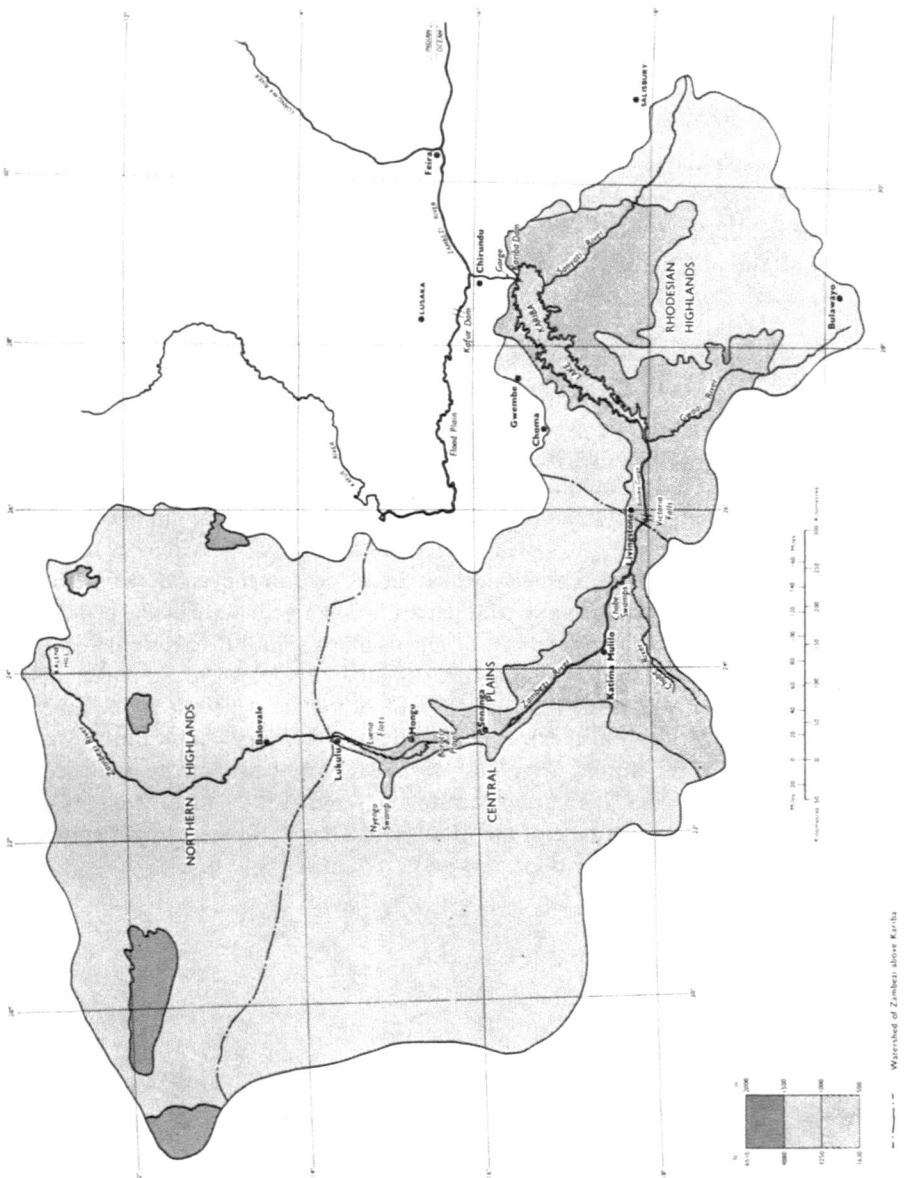

Fig. 2. Physical features of the Zambezi River catchment above the Kariba Dam (acc. Federal Power Board, unpubl. data).

River follows a large rift valley – the Gwembe Valley – 65 to 80 km wide on average and overlooked on either side by escarpments up to 650 m high (Fig. 3)[1]

1.3. The Gwembe Valley

Throughout most of the Gwembe Valley and transversely on either side of the Zambezi River channel, several topographical zones could be recognized before the creation of Lake Kariba. In Zambia COLSON (1960) described four zones between the River (alt. 400 to 450 m) and the Highlands (alt. 900 to 1250 m) as follows:

Zone 1. Zambezi Plain, flat land with small isolated hills and including banks of alluvium.

Zone 2. Zone of Lower Hills whose axis generally runs parallel to the river. Rivers break out between the hills at right angle.

Zone 3. Zone of High Upland Valleys (max. alt. 610 m) where rivers flow parallel to the Zambezi until suddenly swinging at a right angle to cross the preceeding zone.

Zone 4. Escarpment Region, a zone several kilometers broad formed by a series of rough sandstone ridges deeply carved by tributary rivers. It makes up the High Plateau margin (Fig. 3).

On the southern side of the Zambezi River the descent to the Valley from the Rhodesian Plateau follows the same general pattern as the foregoing, but the plateau margin is not so well defined and the belt of hills is much wider. It follows that the drainage from the North (Zambia) is relatively small and confined to a number of short rivers. The main watershed lies to the South (Rhodesia) where it follows the crest of a 1220-m high ridge which extends between Salisbury, Gwelo, and Bulawayo.

After the closure of the Kariba Dam most of the Gwembe Valley became flooded and the actual shores of the lake (alt. 485 m) follow some of the hills of Zone 2. A main exception is found at the base of the Matusadona Range where a broad apron of flat land is now a game reserve (CHILD, 1968).

---

[1] The Gwembe Valley has been defined by SCUDDER (1962) as 'the area from the Zambezi-Gwaai confluence ... to the Zambezi/Kafue confluence 230 miles (370 km) further downriver.' Gwembe is the Matabele name for the Zambezi River (MACRAE, 1938).

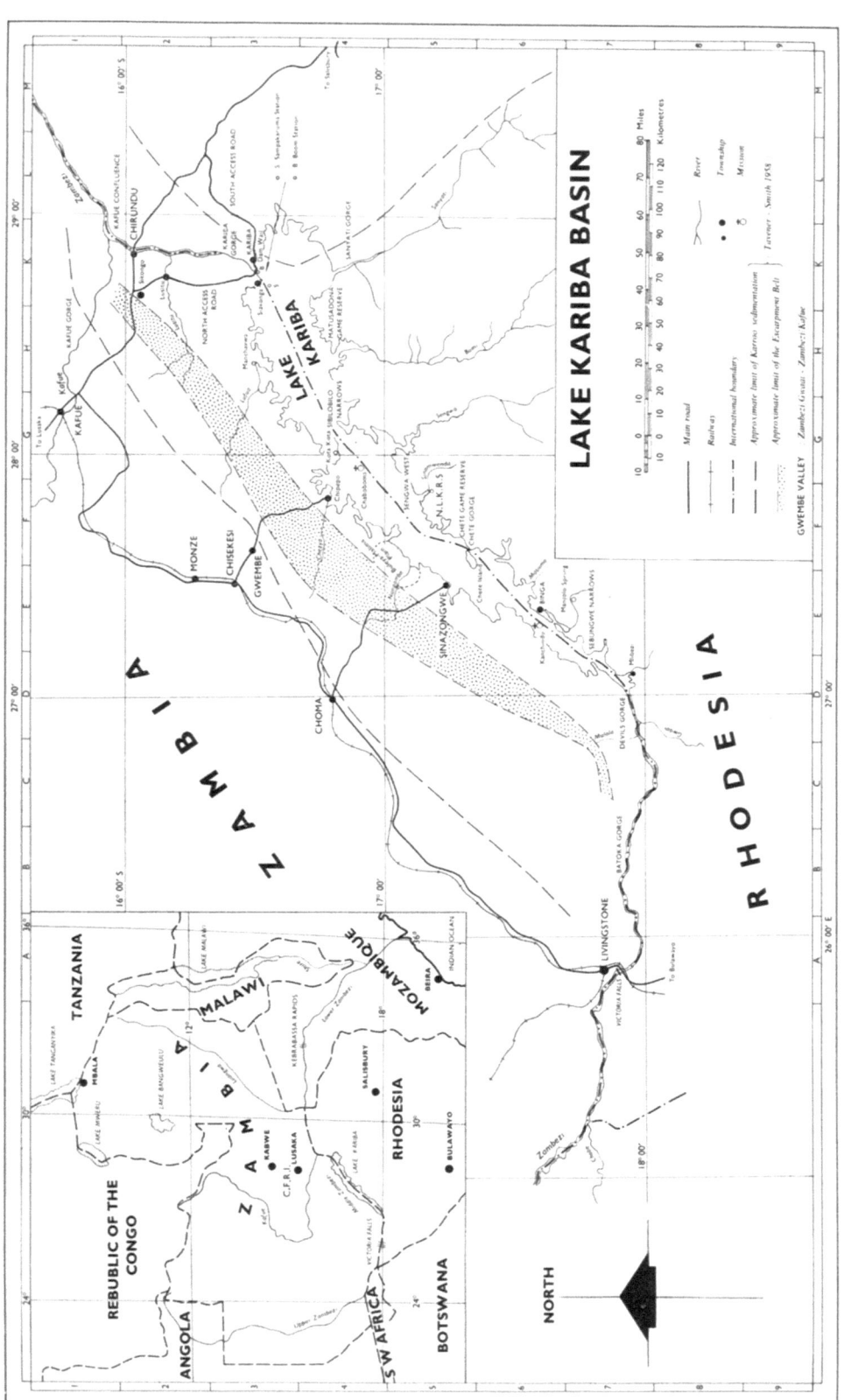

Fig. 3. Lake Kariba Basin: reference map for locations and geographical features.

2. GEOLOGY AND SOILS

2.1. The Zambezi River Catchment

On the basis of its geological features, WELLINGTON (1946) divided the Zambezi River Catchment (Fig. 1) into three regions:
1. Plateau Tract dominated by Kalahari Sediments (Barotse Plain and northern edge of the Ngami Depression). In its southeastern part between the Zambezi/Chobe confluence and the end of the Batoka Gorge (Fig. 4), the Zambezi Valley is cut in basalts (basic lavas) part of the Batoka Basalt of the Stormberg Age (Upper Karroo). To the east the basalt outcrop is terminated by a north-easterly trending fault downthrowing basalt against the much softer Upper Karroo sandstones. It is thought that the Victoria Falls might well have been initiated at this point (ANON., 1962).
2. Trough Tract, a structural feature in which Karroo sediments have been deposited between rocks of the Basement Complex (Fig. 4). It includes the Gwembe Valley and terminates eastwards at the Cabora Bassa rapids.
3. Lowland Tract, composed of the Manica Platform (Cabora Bassa to Luputa Gorge) and of the Mozambique Plain (marine deposits, Cretaceous to Recent).

2.2. Geology of the Zambezi Catchment above the Kariba Dam

Most of the catchment west of the 25°E meridian is mantled by unconsolidated, wind-blown sands associated with consolidated sands and gravels, belonging to the Kalahari System (REEVE, 1963). Here and there older sediments are surfacing generally as isolated features (Fig. 4).

East of the Zambezi/Chobe confluence, Karroo sediments and rocks of the Basement Complex predominate. The presence also of Karroo Basalt has been stressed earlier.

2.3. Geology of the Gwembe Valley and L. Kariba floor[2]

Fairly recent geological research has been conducted in the mid-Zambezi Valley. In Zambia, GAIR (1959) and TAVENER-SMITH (1960) mapped the Karroo rocks in detail between 1951 and 1955; NEWTON (1963) worked on the Zambian Plateau and part of the Escarpment; DE SWARDT & DRYSDALL (1964) dealt with the pre-Cambrian rocks (Basement Complex and Katanga System) of the area, with MATHESON (1969) concentrating on the southwestern part of it; and HITCHON (1958) investigated pre-Karroo rocks in the vicinity of the Kariba Dam. In Rhodesia, BOND (1965) studied the lake area, while LONEY (1966) concentrated on the geology of the Kariba District.

[2] I am grateful to Dr. G. MATHESON, geologist at the Geological Survey Department of Zambia, for his cooperation while compending the geological map (Fig. 5).

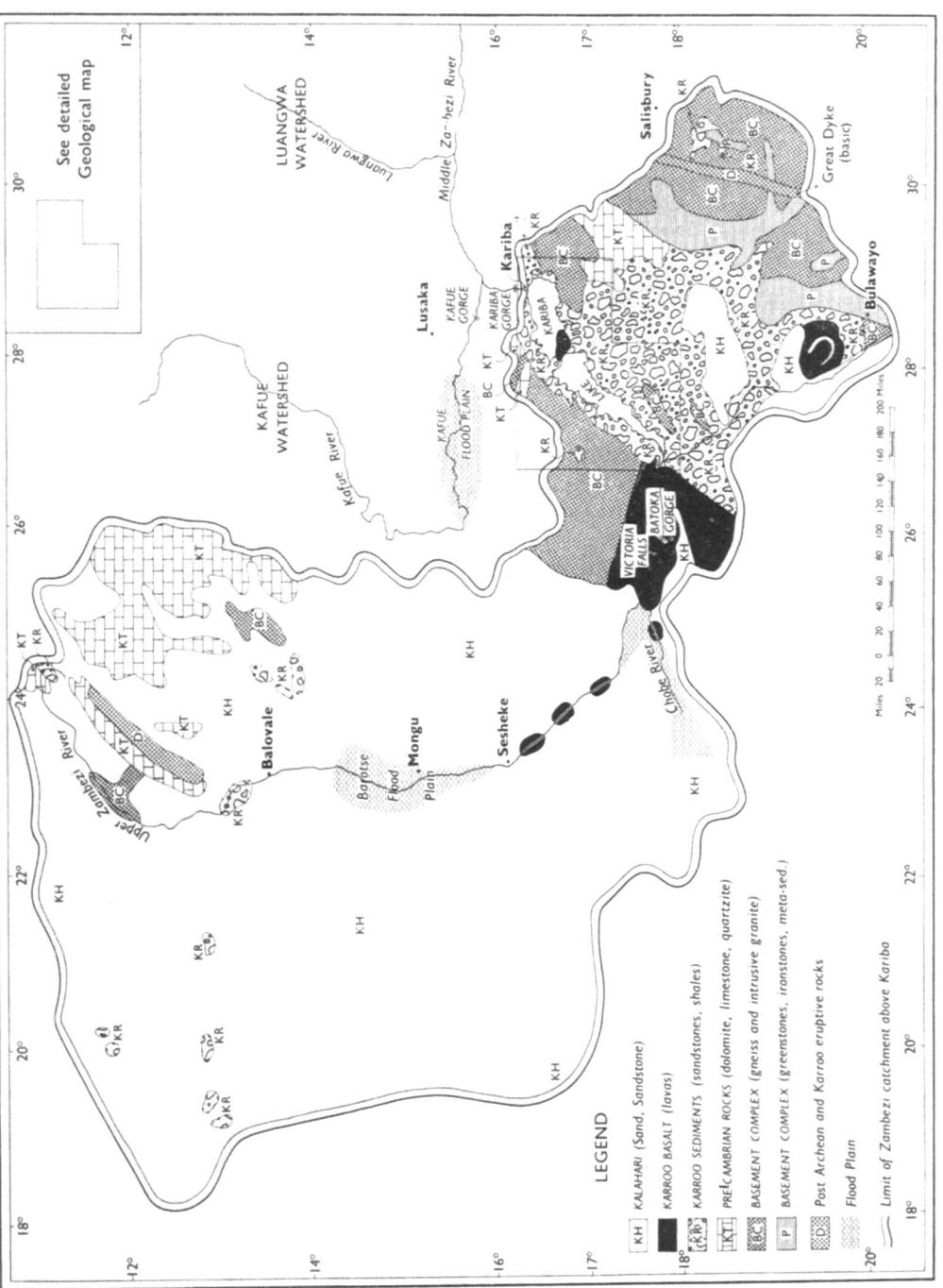

Fig. 4. Schematic geology of the Zambezi catchment above the Kariba Dam (acc. Federal Power Board, unpubl. data).

The Mid-Zambezi Valley constitutes the southern-most extremity of the East African rift system. Its geological development and that of adjacent areas is now fairly well established and was summarized by NEWTON (1963 p. 4) as follows:

'The pre-Karroo floor was probably a peneplained surface of low relief upon which the major basins of Karroo deposition were initiated by downwarping. As the basins deepened the downwarping was supplemented by faulting, although the faults probably did not at that time extend to the surface. A stage of approximate equilibrium was probably reached immediately before new uplift in the positive areas surrounding the Karroo basins led to the deposition of the coarse Escarpment Grit. This formation was not confined to the Karroo basins, since remnants are scattered over the plateau surface. The final phase in the history of the Karroo was a volcanic episode during which vast areas were covered by basalt. Following this there was renewed activity along the Karroo fault-lines, letting down the Karroo rocks and forming a structural depression along the line of the Zambezi Valley. A new period of erosion followed, during which almost all the cover of Karroo rocks was stripped off the plateau areas, eventually resulting in a generally peneplained surface (usually inferred to be mid-Tertiary). Downwarping and downfaulting of the Karroo basin again took place and the Zambezi Valley took on roughly its present appearance. In this downwarped surface was initiated the vigorous river system draining into the Zambezi Valley'.

The Gwembe Valley can be structurally considered as a huge asymmetrical faulted syncline with the steep limb on the Zambian side (GAIR, 1959). The present form of the Valley floor was generated by erosion adjusted to the underlying structure (see Physiography 1.1–1.3).

Soft Karroo sediments (Sandstones and Escarpment Grit) dominate most of the Mid-Zambezi Valley floor and the Lake Kariba area (Fig. 5). This is further emphasized by rough estimates of the percentage composition of the Rhodesian shore line (excl. islands and long inlets of tributaries) as proposed by BOND (1965):

	Percentage composition	
Molteno Series (grits, sandstones . . .)	41	
Forest Sandstone	17	
Lower Karroo Sandstone	8	
Fine Red Marly Sandstone	2	
Basalt and interbedded sandstones	15	
KARROO ROCKS	—	83%
Gneisses	14	
Pebbly Arkose	2	
Sandstone (? Kalahari)	1	
PRE- AND POST KARROO ROCKS	—	17%

Important exceptions are the following (Fig. 5):
(i) Batoka basalts (hard basic lava) and interbedded sandstones occur in the Sibilobilo region (Kota Kota hill cap, string of islands, Bumi area) and upstream from the Mulolo/Zambezi confluence (Batoka Gorge).
(ii) Madumabisa mudstones are found east of the Sanyati River within the Charara and Naodza River valleys.

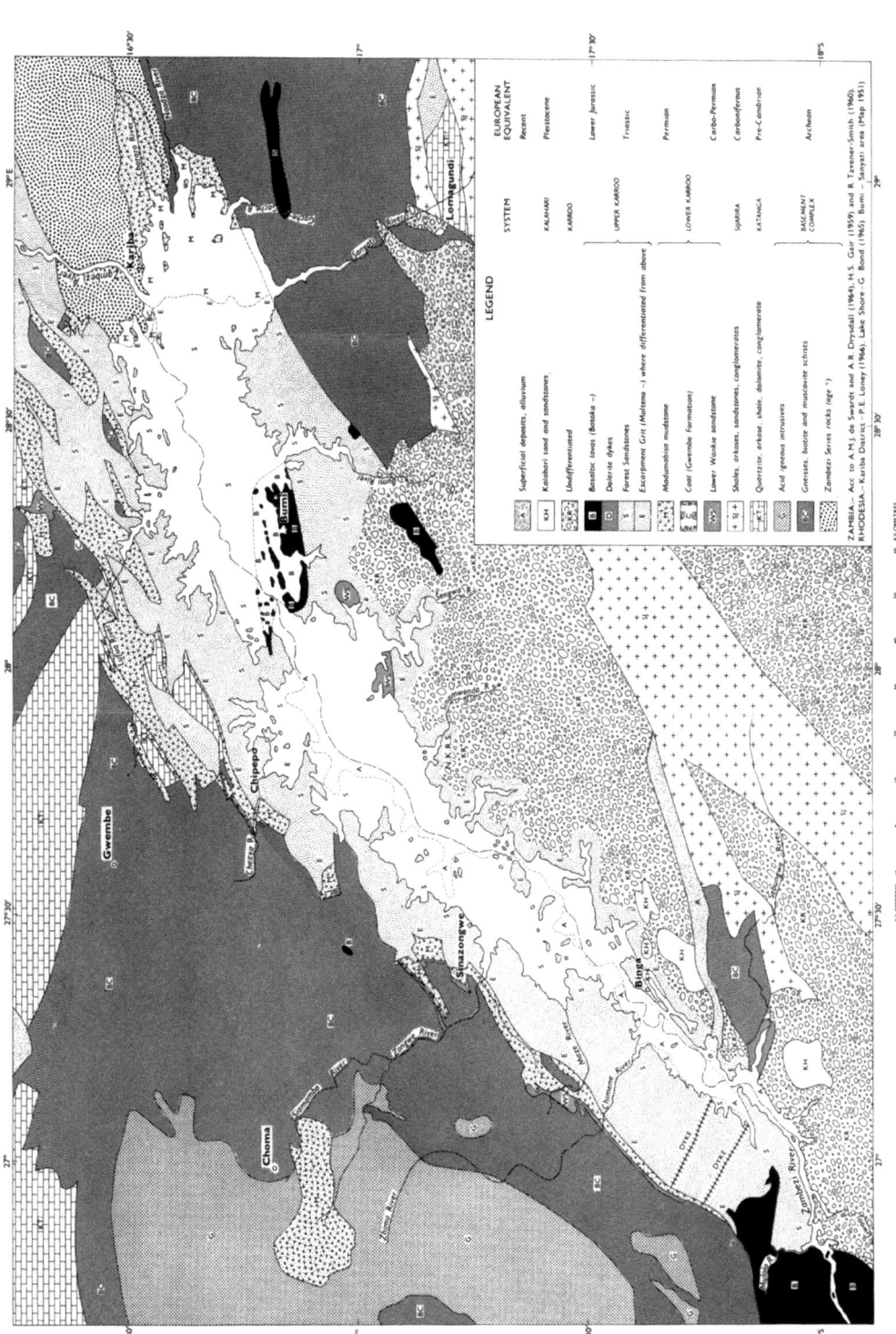

Fig. 5. Geological map of the Lake Kariba area.

(iii) Gneisses appear on either side of the Sanyati Gorge, at the Gache-Gache R., north of the Naodza R., and around the Kariba Gorge.[3]

The Karroo sequence in the Mid-Zambezi Valley consists of several thousand metres of sediments (REEVE, 1963). The thickness of these deposits taper off in depth toward the outer valley margins (Figure 6).

Thermal and mineral springs are present in the Valley (MAUFFE, 1933) and are closely associated with Karroo formations. BOND (1953) pointed out that these springs lay east of the Gwaai R. in two belts: one on the southern side of the boundary faults of the Sijarira Horst structure; another along faults (mainly NE-SW) near the course of the Zambezi River (Fig. 6). He concluded that they were most likely of meteoric rather than juvenile origin. Chemical analyses results were given by GAIR (1959), MAUFFE (1933), and REEVE (1963).

Quaternary deposits consist mainly of gravel terraces and beds of alluvium (BOND & CLARKE, 1954). The most extensive deposits of fertile alluvia were to be found in the upper part of the Valley between the Masumo and Sebungwe Rivers, where the Zambezi meandered through a plain up to eight kilometers wide (Fig. 5). Further downriver alluvia were found along the edge of the Zambezi River channel as a narrow belt generally less than 800-m wide especially on the southern bank (SCUDDER, 1962). According to BOND (cited in SCUDDER, 1962) these alluvia could be broadly classified into two groups:

(i) Alluvium II older upper-terrace alluvia,
(ii) Alluvium III younger lower-terrace alluvia.

Tributary rivers after breaking out of Zone 2 across the Valley floor also formed alluvial deposits at first narrow but widening later near their confluence with the Zambezi River (TRAPNELL & CLOTHIER, 1957).

2.4. The Valley soils and lake bottom deposits

The Gwembe Valley carries its own characteristic range of soils akin to those of the Luangwa Valley, formed under hot and relatively dry conditions. Except for the immature alluvial soils, their types are closely related to underlying geological formations (Fig. 5). GAIR (1959) recognizes three general soil groups:

1. Mature soils on pre-Karroo rocks: derived from schist and gneiss they are relatively thin and infertile; derived from limestone, they are fertile and sometimes relatively deep on gentle slopes. Rarely found in the Zambezi Plain (Zone 1).

2. Mature soils on Karroo rocks: derived from coarse-grained sandstones and other underlying sedimentary rocks of the Karroo formation. These are soils of pedocal (lime-accumulating) tendencies where depth and derivation permit (TRAPNELL & CLOTHIER, 1957). They comprise a wide range from residual sands to lower lying clay-sands and clays. They are the most extensive and SCUDDER (1962) mentions three types of such mature soils according to their derivation:

---

[3] Recent investigations (LONEY, 1966; MATHESON, 1970) have cast some doubts about the age of the gneisses (psammite group) found in the Kariba area. They have therefore been left 'open' in Fig. 5.

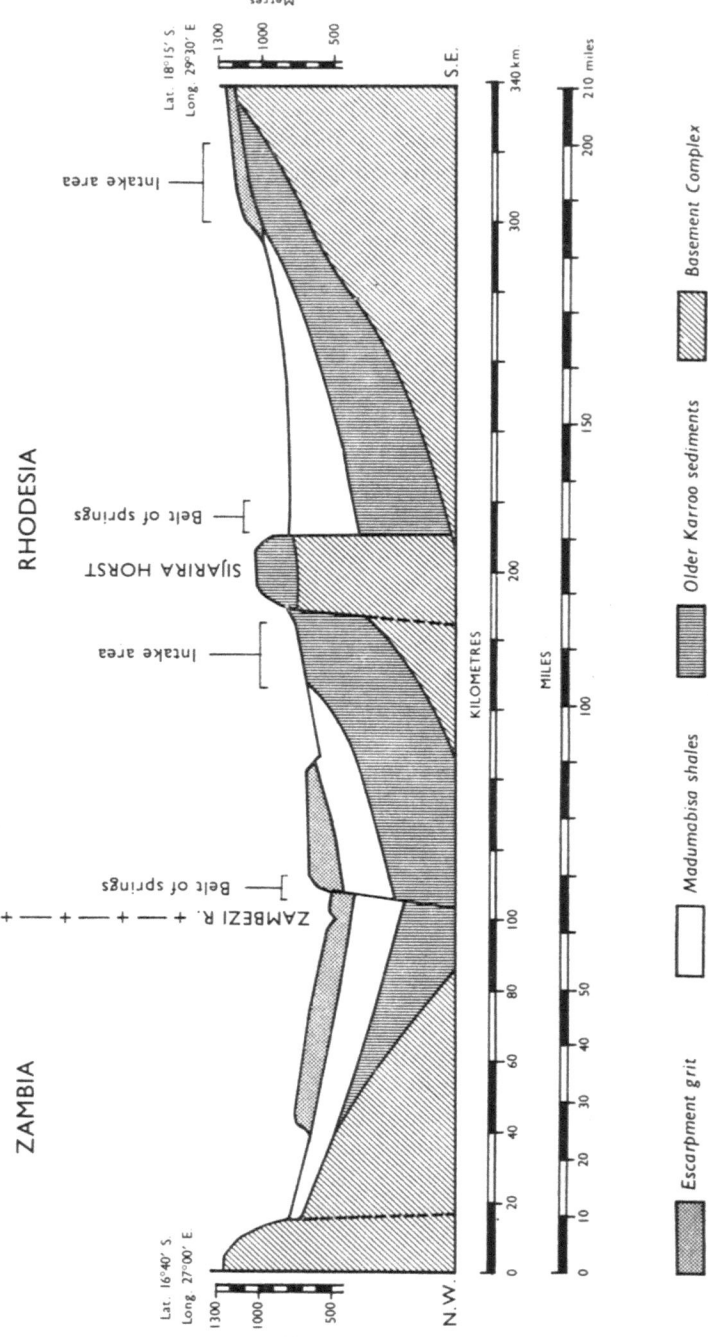

Fig. 6. Generalised geological section across the Gwembe Valley (adapted from BOND, 1953).

(i) from Madumabisa Mudstones: fine textured, clay-type soils, usually highly calcareous; fertile where sufficiently deep.

(ii) from Escarpment Grit: sandy, well-drained soils, very liable to drought; the fertility of some of them seems to have been underestimated in the past.

(iii) from Fine Red Marly Sandstone: very localised (Sinakatenge – Sengwa area), they appear to be quite fertile soils.

3. Immature alluvial soils: derived from pre-Karroo (pebble and grit grades) and Karroo (clay and sand grade) rocks. The annually inundated lower-terrace alluvia, of high clay content, are the most fertile soils in the Valley (SCUDDER, 1962). The non-annually inundated alluvia vary in fertility. The banks of recent alluvium deposited by tributary rivers across the Zambezi Plain are derived from gneisses and mica-schists of the Escarpment. They vary from loose, coarse loamy sands and soft micaceous sandy loams to fine silty soils.

Following the need to resettle the local population displaced by the creation of Lake Kariba, a general survey of the amount of cultivable land available in the Zambian part of the Gwembe Valley was carried out (SCUDDER, 1970). It revealed that out of a total area of 102 850 ha, less than 40 percent could support semi-permanent cultivation (5 to 10 years of culture followed by a fallow of equal length). These some 40 000 ha consisted of deep woodland soils mostly sandy clays, probably derived from non-Karroo rocks and transported into the Valley from the adjacent Escarpment and High Plateau. They were found in three regions (Fig. 3): Lusitu (SW of Chirundu), Buleya-Malima Plain (NE of Sinazongwe), and Mpendele-Mutulanganga (between Kariba Hills and Lusitu).

Lake bottom deposits have been recently studied by S. M. MCLACHLAN (1965). Of particular interest are her prelimary results concerning the ion-exchange properties of the various sediments. Since these properties depend greatly on the type and quantity of clay minerals, she classified the major rocks found in L. Kariba (Fig. 5) into two groups as follows:

Group 1. Rocks likely to yield low ion-exchange clays: Escarpment Grit (Molteno Series) Sandstones; Forest Sandstones; Gneisses.

Group 2. Rocks likely to yield high base ion-exchange clays: Basalts and interbedded Sandstones; Fine Red Marly Sandstones; Madumabisa Mudstones.

In the mouth of the Sinamwenda River the dominant clay is kaolinite (Group 1). Sediments from an exposed coast (with little organic matter) had the low exchange value of about 40 meq. per 100 g clay; from a sheltered bay (more organic matter) the value doubled to 80–100 meq. Sediments from the Sibilobilo area (basalts – Group 2) had a value of 176 meq per 100 g clay or nearly as much as the exchange value for clays of high organic content.

Since the majority of the L. Kariba sediments having a high mineral content belong to Group 1, it is most likely that low ion-exchange properties prevail. Only a high organic content might compensate for such a deficiency. It might also be expected that areas of L. Kariba where rocks of the Group 2 occur (Sengwa, Sibilobilo, and Sanyati Sub-Basin) will be potentially more productive.

2.5. Seismography of the L. Kariba region

Prior to the filling of Lake Kariba, the Middle Zambezi region was seismically inert. But as the reservoir started to fill, the increased load of the water mass on the earth's crust reactivated the existing faults. The first seismic activity was measured in 1959 when the water depth at the dam was about 60 m. It culminated in 1963–1964 when the reservoir became full for the first time. According to the actual observations the reservoir is approaching the final phase of its seismological evolution with occasional bursts of low level seismic activity (ARCHER, 1969 and GOUGH & GOUGH, 1970 cited in VAN DER LINGEN, 1973).

3. CLIMATE

3.1. *The normal climate*

3.1.1. General climatology of the zambezi catchment

The climate of the region including the Zambezi River catchment above Kariba depends entirely on the relative position of the Inter-tropical Convergence Zone. The latter is rather complicated in structure as it provides the meeting ground for three distinct air currents:
(i) The Southeast Trade Winds (Indian Ocean) which normally cover the southern part of Rhodesia,
(ii) The Northeast Monsoon (East African coast), and
(iii) 'Congo Air', a monsoonal indraught across the west coast of Africa approaching the region from the north-west.

Generally, the Intertropical Convergence Zone (ICZ) stays in the area between November and March, and accounts for the main rainy season. During the rest of the annual cycle, the ICZ having moved north, the region becomes influenced by a zone of high pressure. Then dry and sunny conditions prevail.

Combined into the general physiography of the Zambezi catchment above Kariba (Fig. 2), the ICZ latitudinal displacement generates two prevalent types of climate best defined according to KÖPPEN's (1931, 1936) ecological classification as follows:
1. BSwh: Steppe climate dry in winter (w), the latter being relatively mild. Mean annual temperature of the air is greater than 18° C (h) or even than 22° C (h' = additional subdivision proposed by the Rhodesian Meteorological Department).
2. Cwah: humid mesothermal climate with a mild and dry winter (w) and a warm summer (a = mean temperature of the warmest month higher than 22° C). Mean annual temperature greater than 18° C (h).

The generalized distribution of these types of climate in Zambia and Rhodesia (Fig. 7) shows that:
(i) The Middle Zambezi region, including the Gwembe Valley and L. Kariba, lies entirely within the BSwh' Zone.
(ii) The Central Plains, Northern Highlands and Zambian High Plateau practically belong to Cwah.
(iii) The Rhodesian Highlands have both BSwh and Cwah climates which share the eastern and western halves respectively – a notable exception being the headwaters of the Sanyati River which rises in a Cwbh zone, a cooler variant of the Cw climate (mean annual temperature below 22° C).

Rain belts intersect the Zambezi River channel at right angles. Along the course of the latter the following 'belts' (mean annual rainfall) can be recognised:

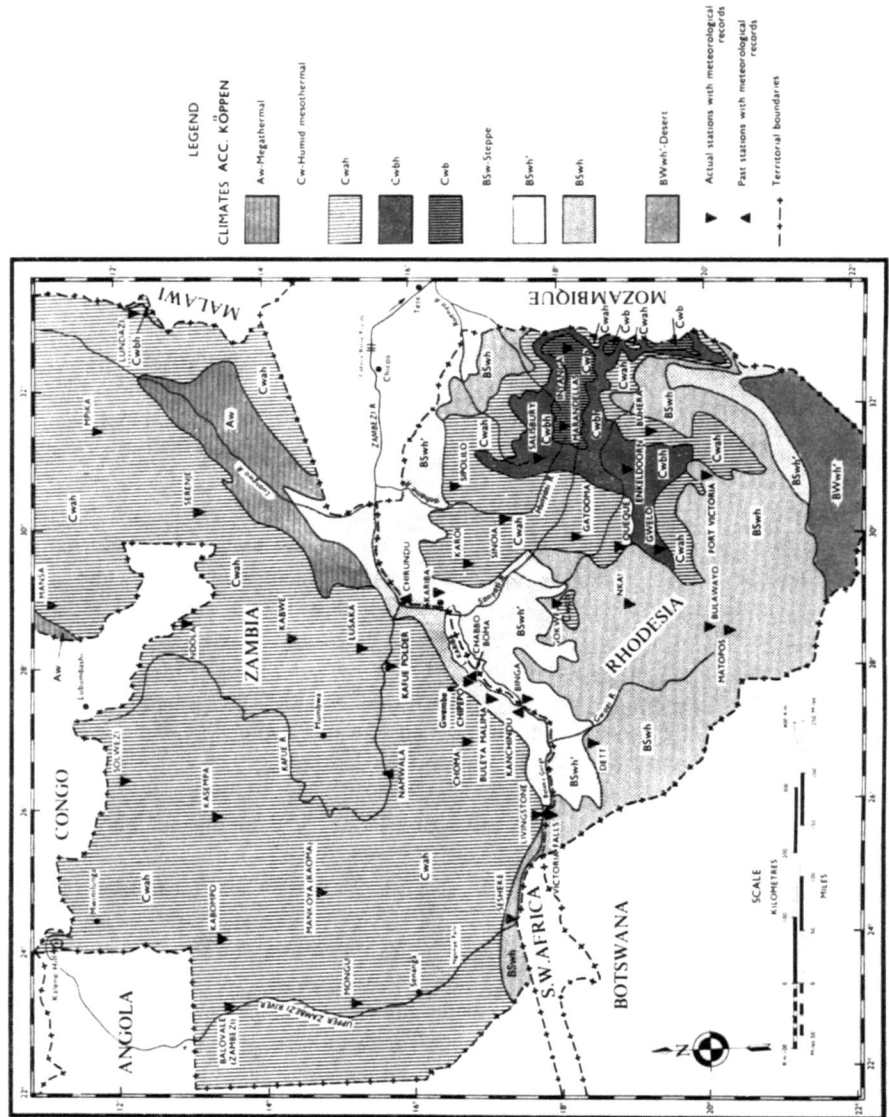

Fig. 7. Distribution of Köppen climates in Rhodesia and Zambia (acc. Meteorological Dept., Rhodesia).

Zambezi R. headwaters	1422–1219 mm
Angola belt	1219–1016 mm
Angola/Zambia border–Mongu	1016– 813 mm
Mongu–Batoka Gorge	813– 610 mm
GWEMBE VALLEY above the Kariba Gorge:	
– Zambezi Plain	610– 406 mm
– Lower Hills, Upland Valleys	610– 813 mm
– Escarpment	813–1016 mm
Kariba Gorge–Cabora Bassa	610– 813 mm
Cabora Bassa–Tete–Tambara	406– 610 mm
Chemba–Indian Ocean	above 610 mm

One third of the Zambezi catchment above Kariba lies in the rain belt 610–813 mm, more than half of the remaining area having an annual rainfall between 813 mm and 1219 mm (Table 1). Less than seven per cent of the area, mostly now inundated under Kariba waters, has less than 610 mm of rain annually.

Table 1. Percentage of the Zambezi catchment area covered by intervals between isohyets (After SHAND, 1960).

Rainfall millimetres	Percent of Zambezi catchment from source to:		
	Ngonye Falls	Kariba Dam	Luangwa confl.
Over 1219	32.4	19.4	15.0
1219–1016	28.1	16.8	18.5
1016– 813	39.5	23.6	32.5
813– 610	—	33.3	29.7
under 610	—	6.9	4.3

Additional details of the climate in the Gwembe Valley prior to flooding are given by SCUDDER (1962).

3.1.2. THE NORMAL CLIMATE IN THE VICINITY OF LAKE KARIBA

The Normal Climate for the Gwembe Valley can be most clearly defined by long-term average values calculated for the various meteorological factors recorded (Fig. 7 and Table 2)[4].

The Normal Climate varies at different stations situated along the actual shore of Lake Kariba (Tables 3 to 5). The available data describing these variations relate to: air temperature measured in a standard meteorological shed; total rainfall; mean relative humidity of the air (generally as reductions from hydrographs); mean wind speed (15-cm cup anemometer in Rhodesia; monthly run of wind in Zambia); mean

---

[4] I am most grateful to the Central African Power Corporation in Salisbury and to the Rhodesian and Zambian Meteorological Departments for supplying regularly the basic data.

Table 2. Meteorological stations in the Gwembe Valley.

Station	Latitude S. Equator	Longitude E. Greenwich	Altitude m a s l	General Situation	Period of records
Kanchindu Mission[1]	17°37'	27°16'	460	High on banks of Zambezi R.	Jan. 1938–Dec. 1939 / Sep. 1954–July 1957
Binga Township	17°38'	27°20'	617	On top of Lower Hill facing lake	Since Jan. 1960
Buleya-Malima	17°04'	27°32'	494	In Buleya-Malima Plain	Since Mar. 1969
Chabbo Boma Mission[1]	16°55'	27°55'	418	On banks of Zambezi R., very dry and subject to low minimum winter temperatures	Oct. 1951– Jul. 1957
Chipepo Harbour	16°47'	27°52'	493	On hill facing the lake	Since Jan. 1965
Kariba Airport	16°31'	28°53'	518	Along the lake shore	Since May 1962
Chirundu Estate[1]	16°00'	28°54'	395	Open exposure near sugar field adjacent to Zambezi R.	June 1951– June 1961

[1] Not functional any more

sunshine from daily totals (Campbell-Stokes recorder and WMO-type reduction); mean global radiation from daily totals (Gunn-Bellani spherical pyranometer, shielded). On the basis of the monthly values for mean air temperature and rainfall, climatographs for Binga, Kariba and Chipepo stations have been drawn (Fig. 8)

The normal climate in the Gwembe Valley in the vicinity of Lake Kariba may be characterized as follows (Fig. 8 and Tables 3 to 5):

a. Air temperature. Although the air temperature is consistently high (annual mean over 24 C) it varies in an annual cycle. Highest in October-November (mean max. 33 to 38 C), it gradually drops as the rainy season progresses to become lowest in June to July (absolute min. 2.8 C in the Kariba area). Then it rapidly increases as the dry season intensifies.

b. Rainfall. An outstanding feature of the climate is the marked seasonality of the rainfall as indicated previously. Precipitation normally starts sporadically in October. In November rainfall slightly increases to reach its peak, generally between December and February. Marked decrease in rainfall follows until April when the dry season practically starts. Very rarely does it rain between May and early October. Total precipitation averages a little more than 600 mm but varies greatly from year to year especially in the Kariba area (range from 350 to 1015 mm). Such large variation exists also for the rainy months as shown by the range of the mean monthly totals (Tables 3–5). Rainfall generally occurs as heavy showers often associated with thunderstorms (SCUDDER, 1962) and they are sometimes highly localized.

c. Relative humidity (RH). Except for the period from December to March (mean RH near or over 70 per cent) and the early hours of the morning in other months, relative humidity of the air is moderate to low. As a result evapo-transpiration (see below) takes place freely.

d. Wind. Three main categories of wind exist in the vicinity of and on Lake Kariba (LAW, 1965):
(i) Land-sea breezes and katabatic winds: both are local and probably occur around many parts of the lake shore.
(ii) Local winds and squalls in the vicinity of thunderstorms: coming from any direction such winds may have peak speeds of 50 to 75 km per hour or more.
(iii) Winds covering a large area controlled by the prevalent synoptic situation: with speeds over 40 km/hr they have been reported sometimes to blow as continuous gales for over 30 hours. These gales usually come from west-southwest.

But in general wind speeds are low and on many days they do not reach 16 km/hr. The windiest period extends from September to November when warm, strong, and very gusty winds generally blow from the northern quadrant (Tables 5 and 6). The strongest winds occur in October blowing during most of the day. When occurring at night they start about 2100 hr and are most commonly from the southern quadrant. During the following rainy season they are replaced by thunderstorm squalls which may suddenly rise from any direction. Between April and August the latter are replaced by cold dry, southeast winds.

e. Evaporation and potential evapo-transpiration. Evaporation from the free water surface of the lake is high. It has been estimated to range between 2 500 and 3 600 mm annually (CAPCOR, unpubl. data). SHAND (1960) proposes the average value of 8 mm

Table 3. Normal climate of Binga Township. Mean values over 10+ years (January 1960–June 1970).

	Jul	Aug.	Sept.	Oct.	Nov.	Dec.	Jan.	Feb.	March	April	May	June	Annual
Air temperature, C													
24-hr mean	20.0	22.6	26.2	29.2	27.7	25.9	25.8	25.4	25.7	24.7	22.7	20.2	24.7
Range of mean	18.8–21.1	21.5–23.6	25.1–27.2	28.1–30.7	26.1–29.2	24.3–28.1	24.1–27.3	24.0–26.4	24.2–27.9	23.5–25.9	20.6–24.3	18.7–21.1	24.2–25.1
Mean maximum	25.7	27.9	31.1	34.0	33.0	30.7	30.8	30.5	30.9	30.3	28.2	25.7	30.0
Mean minimum	14.3	17.0	20.4	23.6	23.5	22.1	22.1	21.8	21.6	20.4	17.4	14.8	19.9
Abs. maximum	30.6	34.4	37.2	40.0	40.0	38.9	38.3	35.0	36.7	35.0	35.0	31.1	40.0
Abs. minimum	10.0	11.7	15.6	16.7	15.6	18.3	18.9	19.4	18.3	16.1	12.2	7.8	7.8
Rainfall, mm													
Total	N	~0	~0	10.9	42.3	157.0	149.6	126.0	90.7	14.2	4.6	0	608
Range of total	N–N	N–3.3	N–3.6	N–51.1	1.3–103	9–333	77–358	27–405	9.9–408	N–53	N–26	N–6.9	432–1001
Relative humidity, percent													
Daily mean	50	44	37	38	53	69	74	76	68	62	56	53	57
Wind speed, km hr^{-1}													
mean	8.8	8.6	9.4	10.1	8.6	7.9	6.8	6.8	7.2	8.3	8.3	8.0	8.6
Range of mean	4.7– 9.0	6.1– 9.7	6.5– 1.9	7.6–12.2	6.1–11.2	4.7– 9.7	4.0– 9.4	3.6–10.1	4.7–9.4	4.7–10.8	4.3–11.9	4.3–11.9	5.0–10.1
[1]Sunshine, hr.day^{-1}													
mean	9.6	9.8	10.1	9.0	8.2	4.1	8.1	7.9	8.1	9.3	9.5	8.3	8.0
Range of mean	8.6–10.4	9.3–10.5	10.0–10.3	7.9– 9.9	6.1– 9.4	2.4– 5.7	7.2–8.9	7.0– 8.6	4.9– 9.7	8.8– 9.9	9.1– 9.9	6.4– 9.6	7.5– 8.5
[2]Global radiation, cal.cm^{-2}.day^{-1}													
mean	484	534	597	610	599	530	593	613	596	582	517	441	558
minimum	167	167	216	102	119	171	164	195	181	254	260	150	102
maximum	581	670	708	784	797	797	804	766	753	701	598	556	804

[1] Sunshine for the last 2+ years; [2] Global radiation since July 1966, N = nil.

Table 4. Normal Climate at Chipepo Harbour. Mean values for period Jan. 1965–Dec. 1969 (from Hydrological Dept., Lusaka).

Variable	Jul.	Aug.	Sept.	Oct.	Nov.	Dec.	Jan.	Feb.	March	April	May	June	Annual
Air temperature, C													
Mean (Max.+min/2)	19.4	22.2	25.2	28.4	28.1	27.2	26.7	26.1	25.8	25.2	22.4	19.9	24.6
Range of mean	18.8–20.0	21.7–22.8	23.9–26.1	27.2–29.4	25.6–29.4	21.9–28.3	25.0–28.3	25.0–26.7	24.4–26.7	24.4–26.1	22.1–22.8	17.8–21.1	23.9–25.3
Mean maximum	26.0	28.7	31.4	34.5	33.3	30.2	30.8	30.0	30.2	30.6	28.4	25.9	30.1
Mean minimum	12.9	15.5	18.8	22.2	22.9	21.8	22.5	22.2	21.4	19.6	16.3	14.1	19.2
Rainfall, mm													
Total	N	N	~0	6.4	39	154	138	128	106	27.5	10.1	1.7	611
Range of total	N–N	N–N	N–0.8	N–20	10–76	28–228	54–228	53–304	22–282	6–77	N–28	N–7.6	542–795
Relative humidity, per cent													
Daily mean	60	55	47	47	56	68	75	76	72	72	63	59	62
Evaporation, mm													
Daily mean	6.5	7.5	8.5	10.7	9.8	7.9	6.9	7.1	6.8	6.7	6.5	6.9	7.0
Wind speed, m. sec^{-1}													
Daily mean	1.3	1.4	2.0	2.2	1.8	1.7	1.4	1.6	1.2	1.3	1.4	1.3	1.5
Range of mean	1.1– 1.5	1.3– 1.6	1.4– 2.1	2.1– 2.3	1.0– 2.0	1.4– 1.8	1.3– 1.4	1.2– 2.3	1.0–1.3	1.2– 1.4	1.0– 1.9	1.1– 1.4	—
Sunshine, hrs. day^{-1}	9.8	9.9	10.0	9.3	8.2	6.2	6.4	6.2	7.9	8.9	9.6	8.5	7.3

Table 5. Normal climate at Kariba Airport – Mean values over 7+ years (May 1962–June 1970).

	Jul.	Aug.	Sept.	Oct.	Nov.	Dec.	Jan.	Feb.	March	April	May	June	Annual
Air temperature, C													
24-hr. mean	18.5	21.7	25.9	29.5	28.6	26.1	25.6	25.5	25.5	24.4	21.1	18.6	24.3
range of mean	17.0–20.0	19.9–22.9	25.0–26.6	26.9–30.8	26.4–29.9	24.8–27.6	23.9–26.7	24.6–26.3	24.4–26.7	23.2–25.7	19.2–23.2	17.1–20.7	23.4–25.0
mean maximum	26.2	28.8	31.6	35.3	38.2	31.0	30.7	30.9	31.5	30.9	28.4	26.1	30.5
mean minimum	10.5	13.7	18.2	23.1	23.5	21.9	21.6	21.2	20.2	18.2	13.8	11.1	18.1
abs. maximum	32.8	35.6	38.9	40.6	40.6	38.3	37.8	35.6	36.1	35.0	35.0	32.2	40.6
abs. minimum	5.6	8.9	11.1	15.6	15.6	17.2	18.3	17.8	15.0	11.1	6.7	2.8	2.8
Rainfall, mm													
Total	Nil	~0	~0	8.4	49.8	143.8	157.6	147.0	69.2	18.0	1.9	~0	609
Total range	N–N	N–1.5	N–3.3	N–251	12–79	52–282	76–280	52–232	2.3–228	1–29	N–7	N–4.1	350–1015
Relative humidity, per cent													
daily mean	54	47	39	37	49	69	76	77	70	64	62	60	59
at 0700 hrs[1]	58	52	49	47	60	77	81	91	84	72	68	63	67
at 1400 hrs[1]	33	28	28	28	38	55	58	67	53	42	37	37	42
Wind speed, km. hr⁻¹													
mean	6.5	7.6	10.4	10.1	8.3	6.8	5.0	5.0	5.4	5.8	6.1	6.5	7.2
range of mean	4.7–7.6	5.0–10.1	6.1–11.2	6.5–11.5	5.4–10.8	4.7–9.4	3.6–6.5	3.6–6.1	4.0–7.2	4.3–7.6	3.6–9.4	4.0–9.0	5.0–9.4
[2] **Sunshine, hr.day⁻¹**													
mean	10.0	9.8	10.1	9.6	7.5	5.9	7.0	7.8	7.9	9.1	9.9	8.7	8.5
range of mean	10–10	9.1–10.6	10.1–10.2	8.5–10.9	5.6–9.3	3.4–7.9	4.7–8.4	7.3–8.5	5.1–9.9	9.0–9.3	9.3–10.3	7.7–9.3	8.2–8.7
[3] **Glob. rad., cal.cm⁻² day⁻¹**													
mean	506	536	622	632	615	558	576	597	565	561	518	444	561
minimum	242	140	242	155	128	121	83	265	147	246	201	197	83
maximum	570	679	755	801	831	869	820	816	778	729	638	558	869

[1] Four-year average; [2] For last 2+ years; [3] Since July 1966.

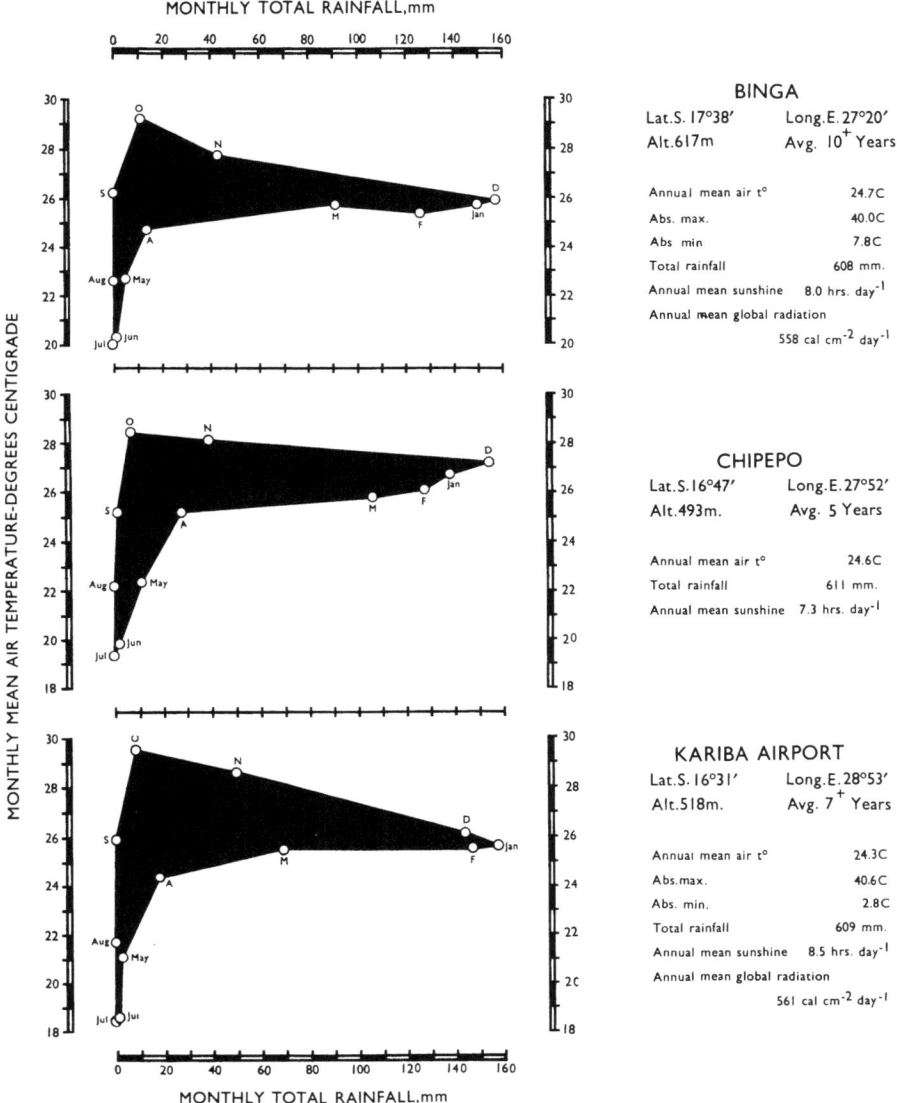

Fig. 8. Normal climatographs for Binga, Chipepo, and Kariba stations.

per day (2900 mm per year). He estimates the total annual loss of water by evaporation from the lake to be 4190 million cubic metres. In Chipepo evaporation averages 7 mm per day in evaporation pans (Table 4).

Potential evapo-transpiration (PET), the quantity of water that would be evaporated from the surface of the soil and transpired from plants if the soil contained the

Table 6. Wind characteristics at Kariba (mean values over 1 to 3 years, acc. to ANON., 1963).

Wind	July	Aug.	Sept.	Oct.	Nov.	Dec.	Jan.	Feb.	Mar.	Apr.	May	June
Strongest ten- min. value, km.hr^{-1}	22.5	29.0	33.7	24.2	33.7	30.6	41.8	32.2	22.5	17.7	19.3	16.1
Maximum gust, km.hr^{-1}	41.8	54.7	56.3	66.0	61.2	86.9	67.6	69.2	64.4	67.6	56.3	40.3
No. of days with gusts of 51.5 km.hr^{-1} or over	1	2	5	5	5	7	5	7	2	2	1	0

optimum content of moisture, has been computed according to the THORNTHWAITE's method by HOWE (1953). He concludes that the annual PET in the Middle Zambezi Valley is greater than 1400 mm, reaching a record value at Chirundu (1485 mm).

After accounting for the annual rainfall the average water deficiency along the Valley would be greater than 760 mm. In the rest of the Zambezi catchment above Kariba, PET decreases as one travels towards the Plateau or within the Upper Zambezi. The average water deficiency follows the same trend to reach a minimum value (ab. 12 mm) in the most elevated areas (Zambezi R. and Sanyati R. headwaters).

f. Conclusions. On the basis of the above characteristics the climate of the Gwembe Valley may be defined as a true tropical semi-arid climate with three or four distinct seasons (Fig. 9).

3.2. The local climate during the years of study

The climate during the first year of study (Jan. 1965 – Jan. 1966) has been defined earlier on the basis of data for the Binga and the Kariba Airport stations (COCHE, 1968).

Basic meteorological data from Binga, Chipepo, and Kariba key stations pertinent to the second annual cycle (Feb. 1968 – Feb. 1969) are graphed in Figures 10 to 12.

Some of the previous data have been related to the 'normal' values (Sec. 3.1) in an attempt to better define the particular climatic deviations during the years of study from the long-term average values. This has been possible only for the stations at Binga and Kariba Airport. It has involved mean monthly values for the air temperature (24-hr mean), total rainfall, and global radiation.

From the results (Fig. 13) it can be concluded that during the two particular years under study:

a. Air temperatures followed an average pattern very close to normal. The observed deviations always remained within $\pm$ 10 per cent of the latter but most usually within $\pm$ 5 percent. During 1965 there was a slight tendency for the mean maximum and the mean minimum temperatures to be lower than normal. The reverse was true during the second year of the study.

b. Rainfalls deviated rather greatly from the long-term average as expected in tropical climates. There were also considerable differences between Binga and Kariba. At this last station the 1967/68 season broke all records (1015 mm).

c. Global radiation, except in a few cases, varied as expected from the 'normal' curves.

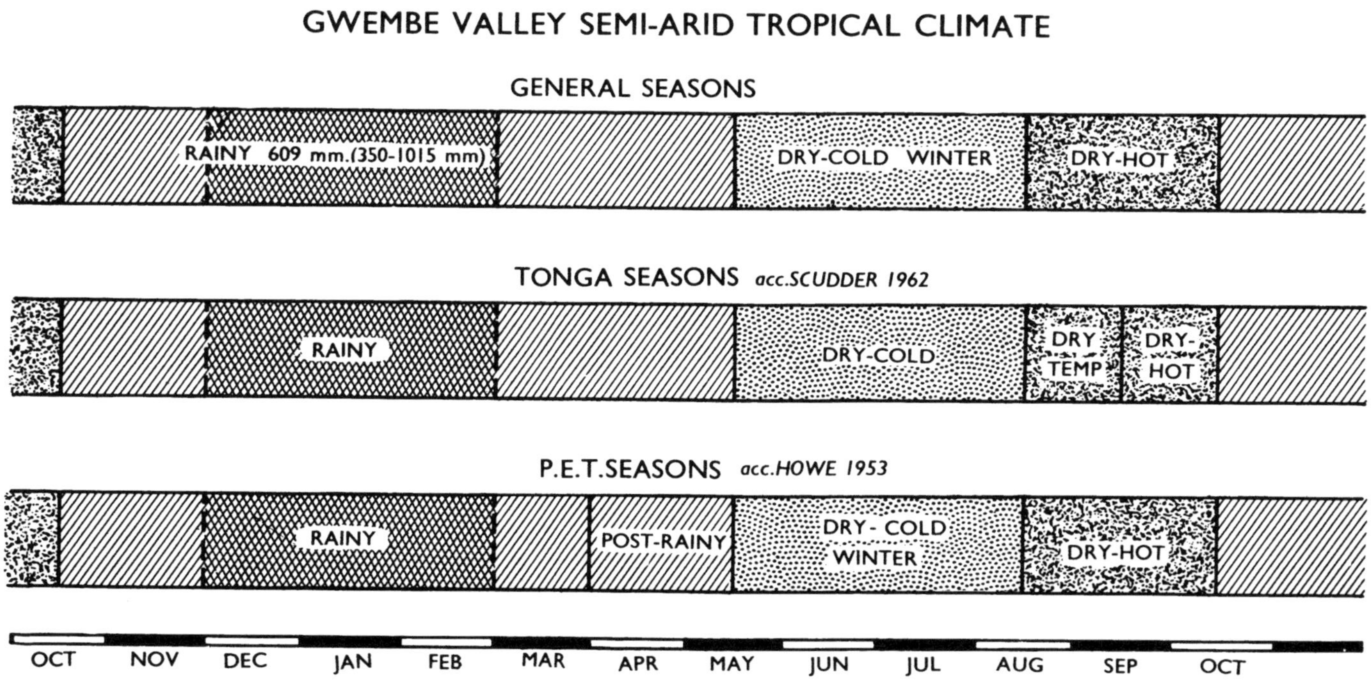

Fig. 9. Schematic climatology of the Gwemble Valley.

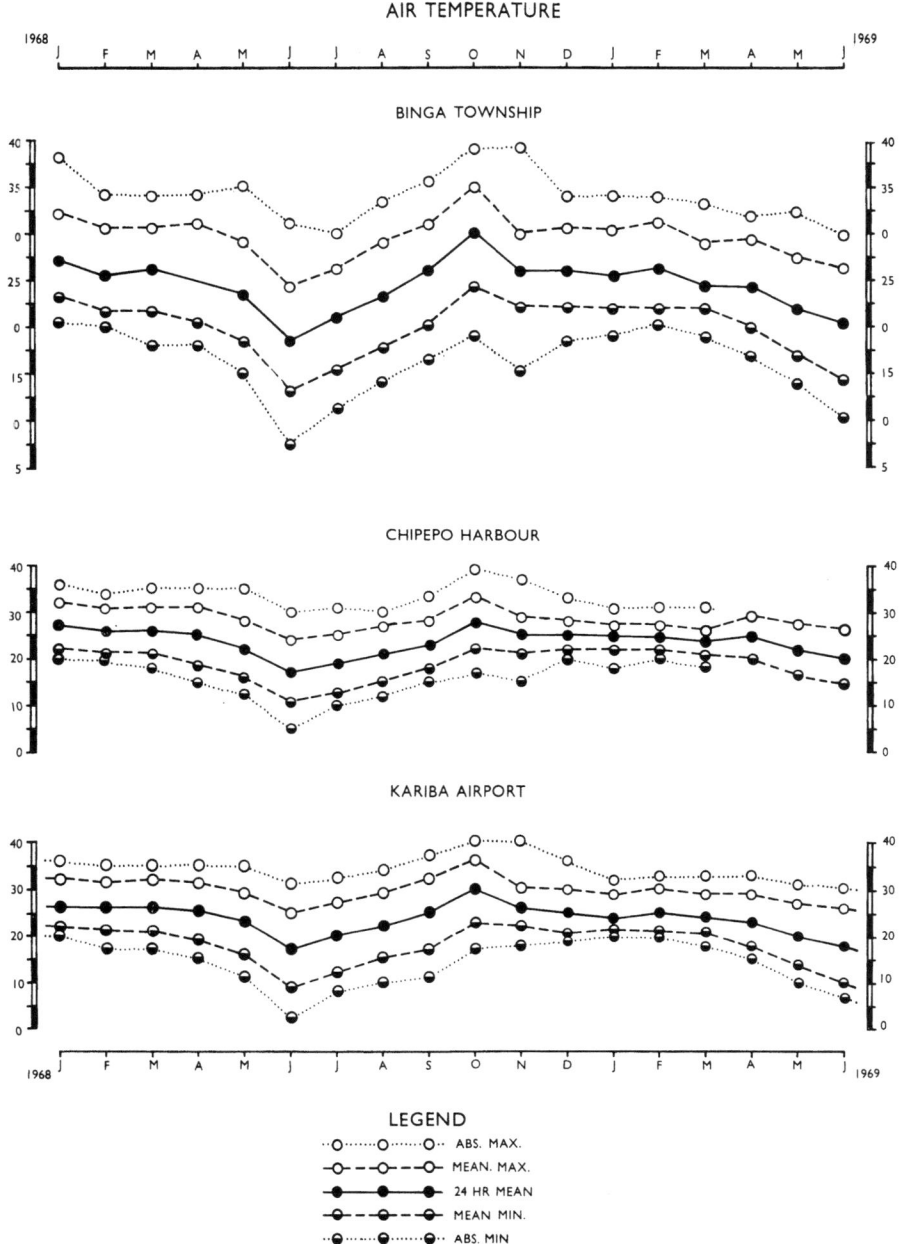

Fig. 10. Variations of the air temperature in 1968–1969 (centigrades).

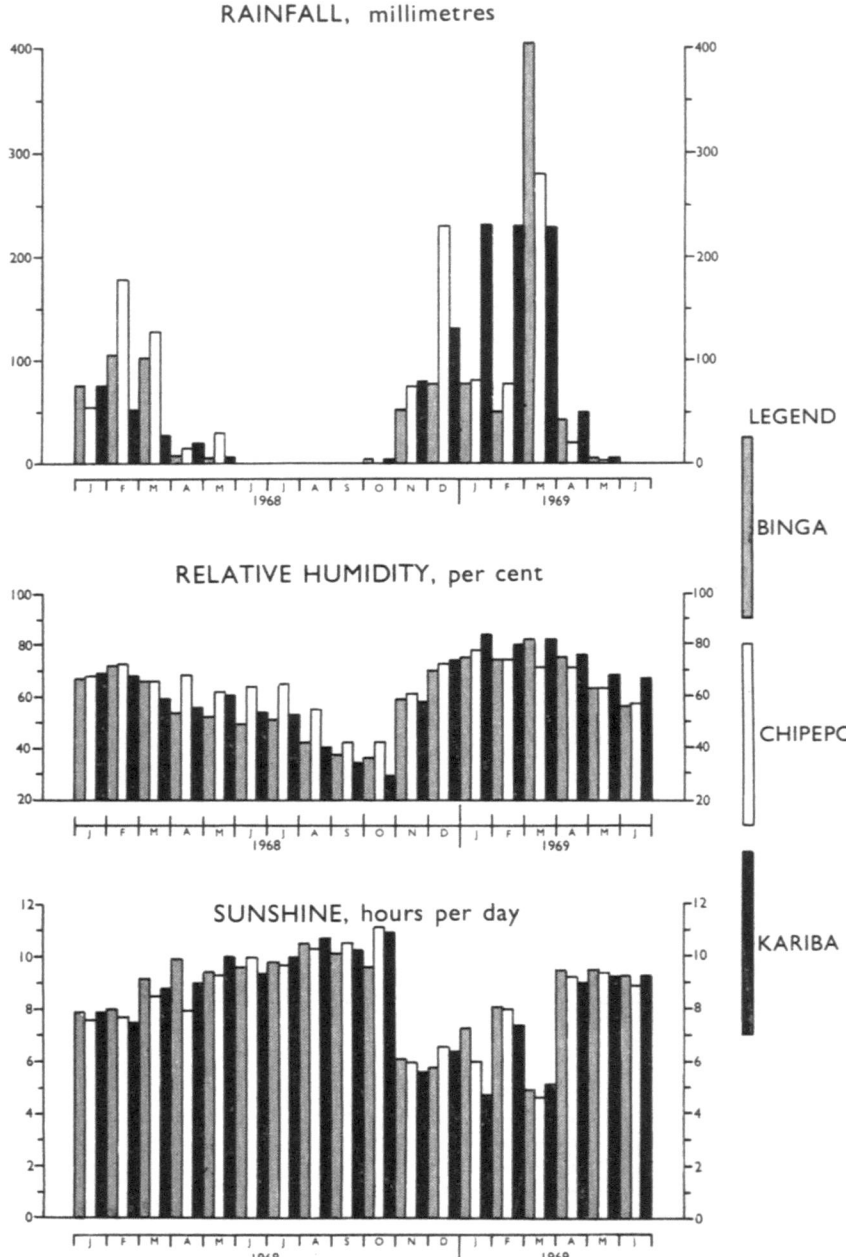

Fig. 11. Rainfall, relative humidity, and sunshine in 1968–1969.

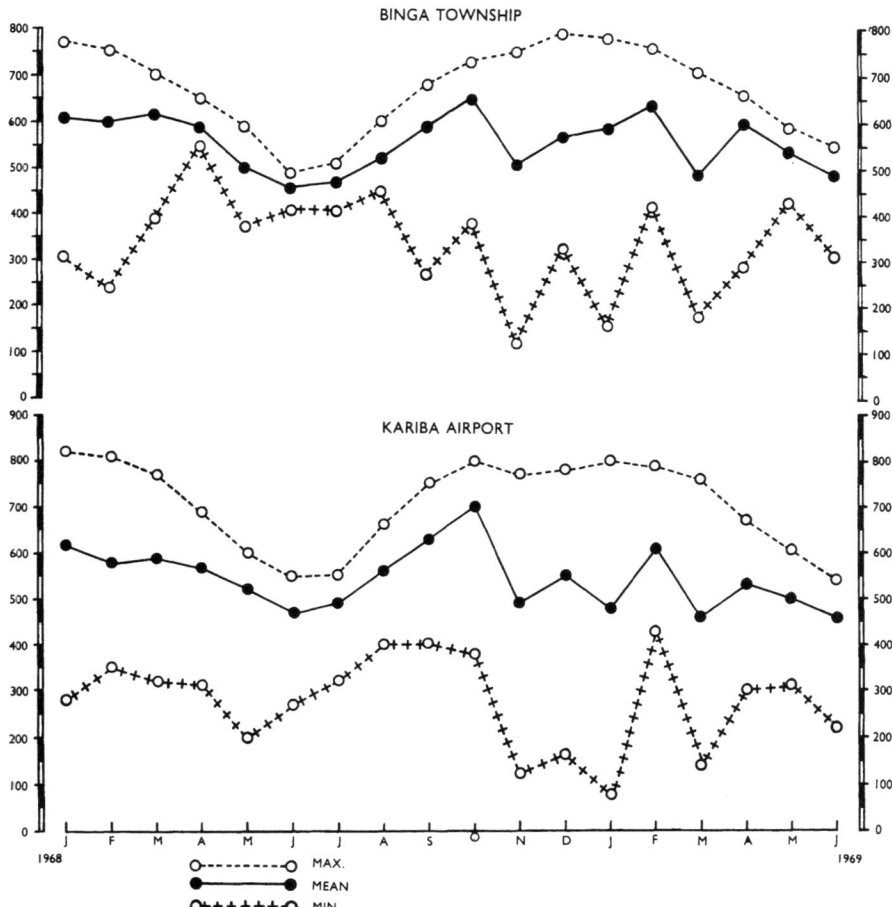

Fig. 12. Global radiation in 1968–1969 (calories per sq.cm per day).

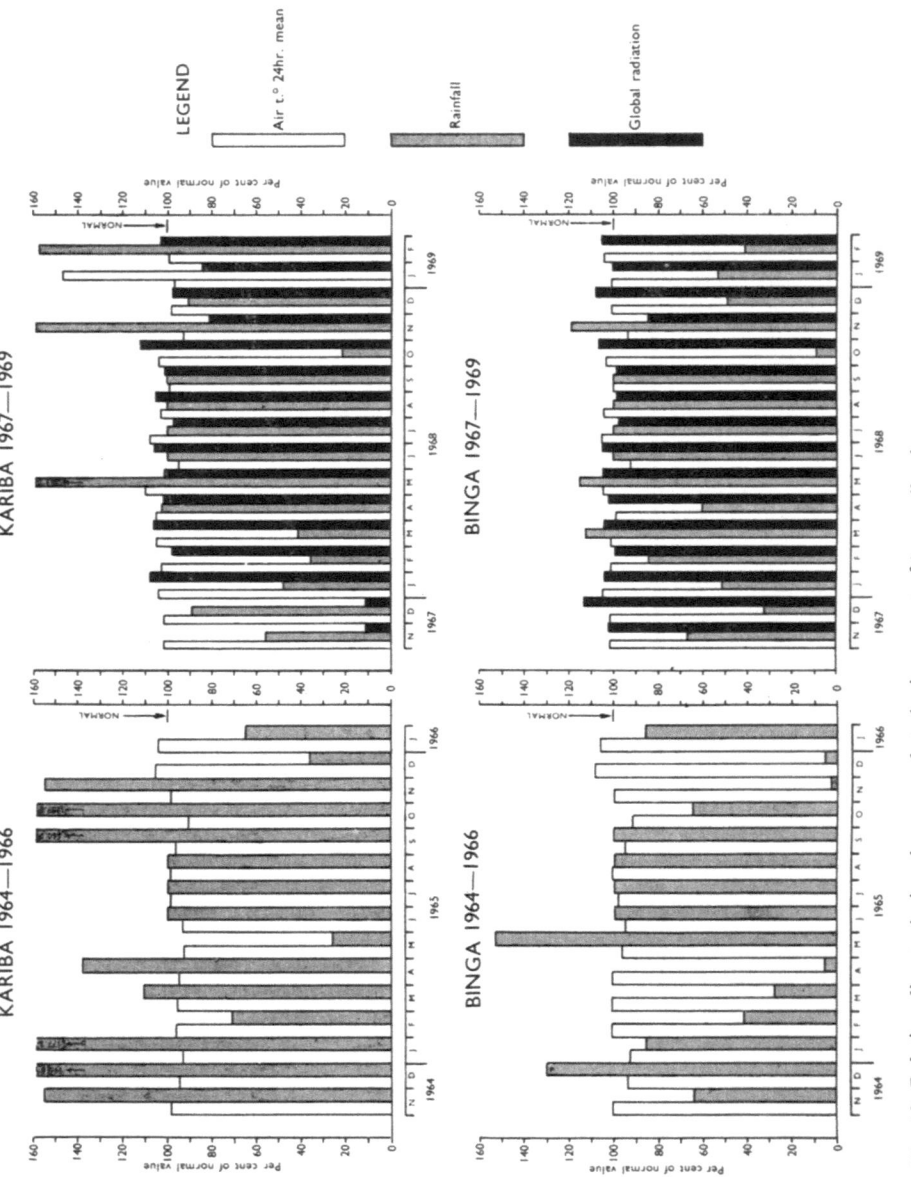

Fig. 13. Relative climate during the years of study, in percent of 'normal' values.

4. FLORA, FAUNA, AND HUMAN POPULATION

4.1. The flora of the Gwembe Valley

4.1.1. Terrestrial vegetation

The terrestrial vegetation of the Gwembe Valley now flooded was characteristically deciduous except for a narrow fringe of evergreen trees on fertile alluvia along the rivers. It was described as predominantly savannah woodland dominated by mopane, *Colophospermum mopane* (CHILD, 1968) with widely distributed vegetation of a thicket type (TRAPNELL & CLOTHIER, 1957). Angularly branched, often spiny, shrubs and other small trees of an early-deciduous type (notably *Commiphora* spp) combined with a similar variety of species to form impenetrable belts. Three types of thicket *(lusaka, londe,* and *luumpa)* were described by SCUDDER (1962).

As may be expected, a close relationship existed between phytosociological and pedological characteristics (SCUDDER, 1962). The most frequent complex was the Mopane woodland (classified acc. Lovemore) which occupied up to half of the flatter areas (Table 7). On rocky ridges a typical association of *Euphorbia* spp., *Sterculia* spp. and *Adansonia digitata* (baobab) was found.

Today most of the riverine vegetation and much of the mopane woodland are underwater. Submerged trees have become a conspicuous feature of Lake Kariba particularly in shallower areas. Between 1964 and 1966 A. J. MCLACHLAN (1970a) studied the importance of this biota from a biological point of view, as substrate for benthic fauna. He discovered that submerged woodland supported a distinctive fauna of which *Chironomidae* formed the major part numerically. The faunal biomass reached up to 97 mg per sq. m after four months of immersion. On the average, littoral trees supported a faunal weight of 60 mg per sq. m. Taking into account the considerable surface area available to such colonization it compared favourably to faunal weights present in the mud (206 mg/sq.m) and in aquatic vegetation (1064 mg/sq.m). He concluded that trees 'provide a valuable additional habitat when vegetation is present, and substitute to some extent for (aquatic) vegetation before it appears'. This is specially true in man-made lakes where the development of the shoreline (BOND, 1965) and of its vegetation (BOUGHEY, 1965), the appearance of rooted aquatic vegetation (MITCHELL, 1969) and the accumulation of sediments (A. J. MACLACHLAN, 1970b) are locally hampered by water level fluctuations.

Bush clearing. In a plan to benefit fishing and navigation, a total of 954 sq.km (about 18 percent of the lake area) was cleared of woodland vegetation before inundation started (Table 8; CLEMENTS, 1959). The location of the areas to be cleared was chosen so as to be within the 20-m contour below maximum water level for possible advantage to future fishing (Fig. 23). Since the cutting of the vegetation had to be done some time in advance of the rise of the lake water and since such a vast area was involved, regeneration of the vegetation occurred in various intensities as studied by CHILD (1968 – Table 9).

Table 7. Vegetation types in the Gwembe Valley prior to inundation (after SCUDDER, 1962).

Vegetation Type	Soil Type	Human Interference
1. *Acacia* woodland – *A. albida* and dense thicket and grass – *Acacia spp* in dense, riverine forest	alluvial soils – lower-terrace alluvia – upper-terrace alluvia	cleared for permanent cultivation into a parkland complex
2. Mopane woodland – Pure Mopane – Mopane with low shrub – Dry shrub Mopane with other genera – opane – other genera woodland Mopane dominant (*luwani*) Other genera dominant (*lusaka*)	poorly drained, clayey soils better drained, but shallow soils on Karroo rocks pebbly, barren lower slopes of hills or hummocks throughout Zambezi plain relatively heavy, clay soils of alluvial or colluvial origin residual soils (Karroo) comparatively deep, sandy and clay loams	not cultivated not cultivated not cultivated infrequently cultivated cultivated
3. Mixed woodland (*Commiphora, Pterocarpus, Sterculia, Kirkia, Terminalia*)	broken hill country near Zambezi (now esp. on ridges forming lake shore)	
4. *Brachystegia* spp. woodland	higher hills and escarpment	

Table 8. Bush – cleared fishing grounds in Lake Kariba.

Basin	Denomination	Surface area at the 485-m water level square km	
		Zambia	Rhodesia
IV	Sanyati A (Kariba)	—	30.3
	Sanyati East B	—	9.7
	Sanyati C	—	33.0
	Sanyati D	—	70.0
	Siavonga E	10.0	—
	Loteri F	3.0	—
	Sanyati West G	—	48.0
	Bumi H	—	5.2
	Lufua J	8.0	—
	Sibilobilo K	—	104.2
III	Chipepo East M	72.0	—
	Sengwa West N	—	36.0
	Sengwa West O	—	50.0
	Chipepo West P	26.0	—
	Sinazongwe East Q	120.0	—
	Sinazongwe West R	161.0	—
II	Masumo S	—	8.4
	Kanchinotu – Binga T	110.0	49.0
Lake	Total cleared area 953.8 sq. km	510.0 53.5 per cent	443.8 46.5 per cent

4.1.2. SEMI-AQUATIC AND AQUATIC VEGETATION

The development of biological communities within the newly formed L. Kariba occurred in phases mainly related to the water level. The first phase started at the closure of the dam (1959) with explosive development (BALINSKY & JAMES, 1960) and it ended when the water reached its maximum level in September 1963. It lasted nearly five years. The second phase was then initiated with the annual fluctuations in water level becoming one of the most influential factors of vegetation development.

MITCHELL (1969) has recently reviewed the main events in the development of the vascular hydrophytes of Lake Kariba.

a. Free-floating hydrophytes

a.a. *Salvinia auriculata* AUBL. (water fern) was first recorded from above the Victoria Falls in 1948. It very soon appeared in L. Kariba and underwent an explosive growth which peaked in 1962 with some 21.5 per cent of the lake's surface colonized. A rapid drop followed in 1963 due to the decrease in sheltered areas and to the increase of the effects of wave action. In 1964 a sudden drop of the water level brought yet another reduction in the amount of *Salvinia*. Since 1964 the coverage appears to have reached

Table 9. Regeneration time of vegetation in bush-cleared areas (from Child, 1968).

Transect	No. of points	Season	Interval since clearing, months	Per cent bare ground	Per cent litter	Per cent grass	Per cent grass canopy	Per cent woody canopy	Max. ht. woody canopy ft.	Mean ht. woody canopy ft.	Habitat
1.	100	growing	5	86	12	12	29	0	—	—	Mopane veld.
2.	100	dry	12	97	3	3	3	3	3.0	2.2	Infertile alluvium with mopane.
3.	100	dry	12	97	14	2	7	4	2.5	1.9	Infertile alluvium mixed species.
4.	100	dry	12	92	22	6	19	2	3.0	3.0	As above, but soil more fertile
5.	100	dry	12	94	10	11	8	2	3.0	2.5	As No. 4, but burnt over.
6.	200	dry	12	98	8.5	0.2	3.5	4.5	6.0	2.9	Fertile alluvium with *Acacia*.
7.	200	dry	12	99.5	1	0	0.5	0	—	—	Down-graded alluvium old cultivation.
8.	100	dry	12	92	22	6	19	2	3.0	3.0	Down-graded alluvium.
9.	100	dry	12	94	10	11	8	2	3.0	2.5	Fertile alluvium burnt over old cultivation.
10.	50	growing	12	94	22	6	6	4	2.0	1.5	Fertile alluvium with *Acacia*.
11.	100	growing	12	91	3	6	4	9	7.0	4.9	Fertile alluvium with *Acacia*.
12.	100	growing	12	96	10	4	10	3	7.0	4.5	Fertile alluvium with *Acacia*.
13.	100	growing	12	88	3	8	49	2	7.0	4.5	Fertile alluvium with *Acacia*.
14.	100	growing	30	97	21	0	1	32	9.0	4.8	Mopane with *H. spinescens* heavily over-populated with game.
15.	100	growing	30	98	20	1	16	15	10.0	6.6	As above, but jesse with the mopane.
16.	100	growing	30	97	46	3	15	39	10.0	4.5	As No. 14 but *T. brachystemma* with mopane.

a measure of stability at about 15 percent of the total lake area, confined mainly to creeks and protected inlets. The possibility to further reduce its spread wherever desirable is being investigated using *Paulinia acuminata* (MITCHELL cited in VAN DER LINGEN, 1973). The autecology of this water fern was studied in detail by MITCHELL (1970) while its relative contribution towards increased biological productivity was shown by BOWMAKER (1969). Early in the lake's life *Salvinia* has retained within the lake basin the nutrients released from the flooded land during the filling phase and, in the absence of submerged vegetation, it has provided cover and food organisms to the juvenile fish. This aquatic fern now constitutes an important reservoir of nutrients on one hand but on the other hand it greatly delays the cycling of these nutrients within the lake ecosystem (MITCHELL, 1970).

a.b. *Pistia stratiotes* L.: although recorded from large areas of calm water until 1961, it later proved unable to compete successfully with *Salvinia*.

b. Emergent hydrophytes

b.a. The *sudd community* is a floating colony of plants resulting from the colonisation of a stable mat (e.g. *Salvinia* mat) by other vascular plants (BOUGHEY, 1963). Such colonization started very early in the life of the lake. By 1961 the list of sudd plants comprised 40 species belonging to two groups, semi-aquatics and ruderals of open ground.

b.b. *The shoreline vegetation.* Through 1970 this group of hydrophytes had not succeeded in forming permanent colonies due probably to an interaction of such factors as: the wave action effect on the shore line; the inhibitory effect of the dry *Salvinia* on seed germination; steep shore gradients; the considerable annual fluctuations in water level. Two species *(Panicum repens* and *Ludwigia stolonifera)* might be the first to be successful in the future.

c. The attached hydrophytes with floating leaves

Although uncommon in the lake probably because of lake level fluctuations, very recently *Potamogeton sweinfurthii* has been recorded.

d. The submerged hydrophytes

Most of the development of submerged aquatic vegetation has taken place in areas where *Salvinia* does not form permanent mats, including areas previously cleared of trees and shrubs. *Ceratophyllum demersum* has always been present in the lake from its earliest stage of filling. In late 1964 *Potamogeton pusillus* appeared, followed by *Lagarosiphon ilicifolius* two years later. Recently additional species have been collected (*Vallisneria aethiopica*, *Najas* sp.). The vital role of submerged plants on the development of the invertebrate fauna (BOWMAKER, 1968; A. J. MCLACHLAN, 1969) and of the fish fauna (e.g. DONNELLY, 1969) has been well documented.

e. The phytoplankton

Samples collected with a net No. 12 by HARDING in May 1959 a few months after the closure of the dam, were studied by THOMASSON (1965). He identified 170 species of algae among which the following species were found to be dominant for at least one station: *Microcystis flos-aquae, Eunotia garusica, E. pectinalis, Melosira granulata, Eudorina elegans, Closterium kuetzingii,* and *Xanthidium subtrilobum.* He concluded to the existence of a close relationship between the algal vegetation of the Kariba and Bangweulu areas. According to BEGG (cited in VAN DER LINGEN, 1973), in 1967 only

31 species of phytoplankters were present (39 sp. in 1970) among which 15 sp. Chlorophyta (8 dominant), 10 sp. Chrysophyta (3d), 4 sp. Cyanophyta (2d), and 2 sp. Pyrrophyta (2d). The overall dominant species belonged to the genus *Anacystis* Meneghini 1837, Cyanophyceae, which greatly increased in numbers a short time after the annual turnover of the water mass.

4.2. The fauna of the Gwembe Valley

4.2.1. VERTEBRATE FAUNA

The vertebrate fauna of the Gwembe Valley gained notoriety between 1957 and 1962 when Operation Noah, the rescue of animals trapped on islands by rising water, received much publicity (LAGUS, 1959). It provided a rare opportunity for gathering scientific information on Central African mammals (e.g. CHILD, 1968).

By focusing the world public attention on the existence of such varied and colourful animals, it most probably served well the cause of the African wildlife and, among other results, lead to the creation of several sanctuaries in the Mid-Zambezi Valley: Mana-Pools (flood plain downstream from Chirundu); Kariba, Matusadona, and Chete Game Reserves (along the Rhodesian lake shore). In these protected areas research has been carried out on the ecology of the vertebrate fauna (e.g. ATTWELL, 1970; JARMAN, 1965).

Faunal lists have become available. Many of the mammals found in the Valley were related to the subsistence economy of the human inhabitants (SCUDDER, 1962). JUBB (1961, 1967) dealt with fish taxonomy while BALON (1971a, b) and JACKSON (1960a, b; 1961) reported on fish biology.

Fish catches were studied by HARDING (1964b, 1966) and others as the fish catch rose to ab. 4800 tons (1963) and then declined (See BALON's Part II). Exposed trees in flooded areas were reported to be heavily attacked by terrestrial wood-borers *Xyloborus torquatus* (VAN DER LINGEN, 1973).

4.2.2. INVERTEBRATE AQUATIC FAUNA

a. Benthos. In Lake Kariba explosive development of the invertebrate fauna occurred during the filling phase between 1959 and late 1963 (e.g. BALINSKY & JAMES, 1960). The following phase of development was characterized by shore-line development, appearance and expansion of rooted hydrophytes, accumulation of sediments, annual water level fluctuations, and reduced nutrient concentrations. Under these new conditions, A. J. MCLACHLAN (1969, 1970b) conducted his detailed studies on benthic invertebrates, particularly on Chironomidae (Diptera) the dominant constituent of the benthos. In presence of a static water level maximum biomass occurred in 2-m deep water (23 species, 500 mg/sq. m dry weight), the fauna disappearing at water depths of 8 to 12 m. Reviewing the phases of evolution of the benthos he concluded that mesotrophy succeeded to eutrophy as the water level stabilised (1963). But with the establishment of the aquatic vegetation, since 1967 especially, the trend toward eutrophy was progressing in such areas.

Extensive benthos surveys (pers. data) conducted in 1969–70 in the Zambian cleared areas of Lake Kariba showed how relatively low the abundance of the benthic community still was, in areas devoid of submerged vegetation. The collected organisms belonged to three groups – Oligochaeta, Mollusca, and Diptera (Chironomidae). In 1970 BEGG (cited in VAN DER LINGEN, 1973) determined 11 species of mollusks from the Sanyati Sub-Basin comprising two lamellibranchs and 9 gastropods.

b. Zooplankton. From samples collected in May 1959 THOMASSON (1965) identified 46 species Rotifera, 12 sp. Cladocera, and 4 sp. Protozoa. The dominant species were *Brachionus falcatus* and *Bosmina longirostris*. In 1967 BEGG (cited in VAN DER LINGEN, 1973) found only 20 species present in zooplankton samples, among which 9 Rotifera, 9 Cladocera *(Bosmina longirostris)*, and 2 Copepoda *(Tropodiaptomus kraepelini)*. In 1971, he observed that this last copepod species dominated the zooplankton. After the annual turnover of the water mass, *Limnocnida rhodesiae* (Coelenterata) occurs in swarms in Lake Kariba.

4.3. The human population of the Gwembe Valley

Archaeological evidence suggests that the Gwembe Valley has been inhabited by man since fairly early Acheulian (Stone Age) time (BOND & CLARK, 1954). Such occupation although at first sporadic, increased during Sangoan times but even by then the population probably amounted to less than a hundred hunter-food gatherers.

The origin of the present major inhabitants, the Valley Tonga, and the length of time of their residence in the Gwembe Valley are unknown although archaeological, physical, anthropological, and ecological facts suggest a long time of residence (SCUDDER, 1962).

By the mid-fifties the total population of the Valley approximated 86 000 people (Zambia 55 000 and Rhodesia 31 000), the majority of them being Valley Tonga (COLSON, 1960). They were subsistence cultivators in permanent or semi-permanent villages clustered in small groups around the more fertile alluvial deposits. In the actual lake area the highest density of population was found in the area situated between the Sebungwe River and the Masumo River, the lowest density prevailing between Kota-Kota and Kariba (Fig. 3).

Detailed studies of the Valley Tonga were made between 1956 and 1957 prior to their evacuation from the flooding area and their resettlement on higher grounds: social organisation (COLSON, 1960); agriculture and ecology (SCUDDER, 1962); and material culture (REYNOLDS, 1968). Follow-up studies took place a few years later (1962/63 and 1967) pointing out the vast problems faced by the resettled populations and the responsible authorities (SCUDDER, 1968, 1970).

Resettlement took place between 1956 and 1959. In Zambia all Valley inhabitants (except 3 000) were allocated new homes within the Gwembe Valley in the Upland Valleys, and in the Lusitu area. A Fisheries Training Centre was created in Sinazongwe to help in readjustment of the people to the emerging fishery resources. A pilot irrigation scheme was started in the Buleya-Malima Plain to take advantage of the new water regime. For results of these schemes refer to the Concluding discussion and Epilogue of Part II.

SECTION II

THE RIVERS AND THEIR CHARACTERISTICS

5. THE ZAMBEZI RIVER

Since the Zambezi River contributes in average 77 percent of the total inflow into L. Kariba (Sec. 7.2) it is very important to know the basic characteristics of the water quality in order to understand the lake's limnology and fish production.

As pointed out earlier (Sec. 1 & Fig. 1) the course of the Zambezi River is composed of three ecologically distinct zones, the Upper, Middle, and Lower Zambezi. Within the Gwembe Valley, the Middle Zambezi was surveyed by the Joint Fisheries Research Organisation before the creation of Lake Kariba. This region was defined (JACKSON 1961) as a 'sand-bank' river (such as the Luangwa and the Limpopo Rivers) as opposed to 'reservoir-rivers' (such as the Kafue, Luapula, and Shire Rivers). The latter have their flow better regulated either by a natural lake or by extensive swamps, and they display marginal aquatic vegetation at all times of the year. On the contrary, 'sand-bank rivers ... have very deep beds with well-defined steep banks cut in the alluvial earth ... through which the rivers meander at low level, often little more than a connection between a series of pools fringed with rocks and sand-banks, with little aquatic vegetation, and still less marginal vegetation ... on the banks, while the flood is violent but of comparatively short duration, with the floodplains inundated for a relatively short space of time' (JACKSON, 1961: p. 3).

The natural river situation has been changed drastically by the creation of Lake Kariba both within the flooded area and downstream from the dam. The actual relationship between river and lake will be examined in the next sections. Further material will be presented later, when after having discussed the flow regime of the lake tributaries, the general hydrology of the lake will be considered (Chap. 7).

5.1. *Zambezi flow regime upstream from L. Kariba*

The annual cycle of rainfall (Section 3.1.2) gives rise to a corresponding clearly marked cycle of runoff. The double flood peak recorded upstream from the Barotse Plain is evened out by the latter. The resulting single flood peak is then further flattened out by the Chobe Swamps and appears as the characteristic flood peak observed at Livingstone. The latter forms the base flood flow of the total flood wave reaching Lake Kariba. Normally it takes approximately three to four weeks for the base-flood peak to travel from Balovale to Kariba (Fig. 1).

The Zambezi flow has been recorded at the Livingstone Pump Station since October 1924 providing an unusually long-term set of monthly and annual averages to characterize the annual cycle (Fig. 15)[5].

Flow starts evenly increasing in December to reach its peak value (ab. 3 500 cum per second) in April. This peak is followed by an equally even fall-off until a minimum

---

[5] I am most grateful to Mr. W. G. WANNELL, Chief Hydrological Engineer, Hydrological Branch, Salisbury, for providing me with the basic data.

Table 10. Period and frequency of occurrence at Kariba of minimum and maximum discharge for the Zambezi River (period 1925–1966).

Minimum discharge		Maximum discharge	
Month	Frequency	Month	Frequency
September	9	February	2
October	26	March	11
November	20	April	23
December	1	May	6
—	—	June	1

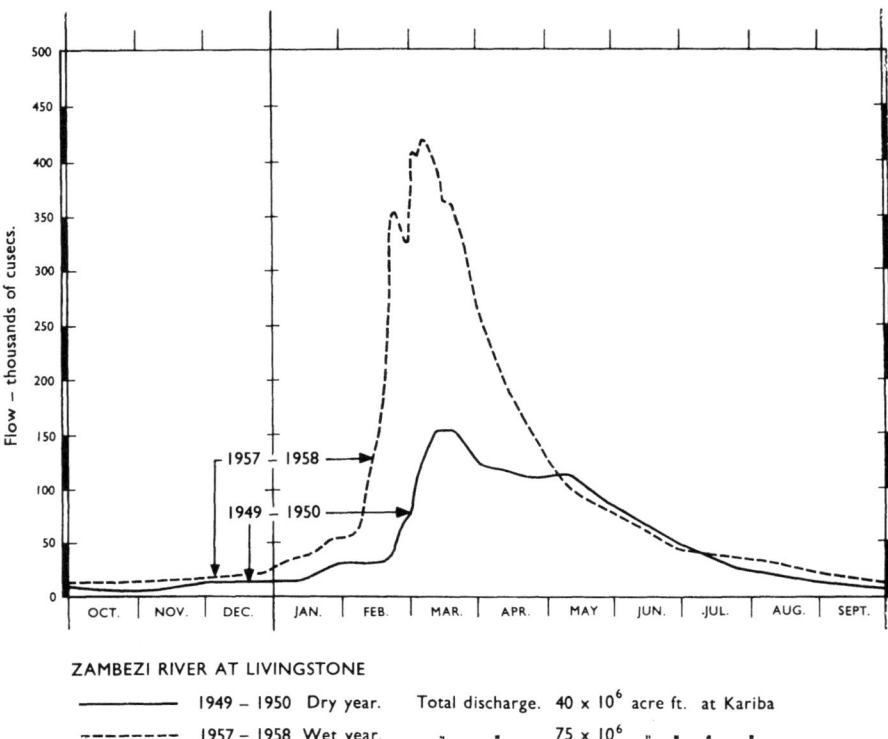

ZAMBEZI RIVER AT LIVINGSTONE
————— 1949 – 1950 Dry year.   Total discharge. 40 x 10⁶ acre ft. at Kariba
- - - - - - - 1957 – 1958 Wet year.   "   "   75 x 10⁶ "   "   "

Fig. 14. Hydrographs of the Zambezi River at Livingstone Pump Station for a dry year (1949–50) and a wet year (1957–58) acc. to the Federal Power Board (unpubl. data).

flow (ab. 350 cum/sec) is reached in October to November. But because of the erratic rainfall pattern which characterizes the region, the Zambezi flow greatly varies from year to year. In the past, maximum discharge has occurred between February and June while minimum discharge has taken place between September and December (Table 10).

Table 11. Annual regime (cum. sec^{-1}) of the Zambezi River at Kariba Dam site during record dry (1948-49) and wet (1957-58) years compared to the long-term averages (period 1925-66).

	Oct.	Nov.	Dec.	Jan.	Feb.	Mar.	Apr.	May	June	July	Aug.	Sep.	Annual
1948-49	472	472	519	566	1180	944	1038	1086	661	472	378	283	667
1957-58	378	425	802	3021	6561	10762	6042	3304	1746	1038	755	566	2925
1925-66	331	345	765	1482	1982	3092	3483	2794	1756	1001	604	425	1493

Variations exist not only in time but also in quantity (Table 17). At Livingstone, minimum total annual flows (less than 700 cum/sec) were recorded in 1928–29 (665 cum/sec), 1932–33 (688 cum/sec), and 1948–49 (566 cum/sec). Maximum flows (more than 2000 cum/sec) happened only in 1957–58 (2516 cum/sec) and in 1962–63 (2232 cum/sec). The long-term average (1925–66) reaches 1200 cum per second. Data at hand seem to indicate that the Zambezi River flow has increased in the last two decades, the average flow for 1959–66 being 1517 cum/sec. Peak flood also tends to occur slightly earlier than before.

The pattern of the annual regime at the Kariba Dam site in extreme years is compared to long-term average values in Table 11 and graphed in Figure 14. Average monthly discharges calculated for the period 1925–1954 for the Livingstone Station (Table 18) show that the flood season contributes 81.2 percent of the total Zambezi inflow into L. Kariba, only 18.8 percent being fed during the dry season.

The Zambezi River catchment above L. Kariba includes (Sec. 1.1) the Northern Highlands (220 670 sq.km) and the Central Plains (286 970 sq.km). To these areas situated upstream from the Victoria Falls (Fig. 1) should be added the Gwaai R. catchment (47 140 sq.km – Sec. 6.1) which raises the total estimate (A_c) to about 555 000 sq.km. If the actual annual average flow (Q for 1959/66) is equivalent to 1517 cum.sec^{-1}, the specific discharge of the Zambezi River into Lake Kariba is obtained as:

$$q = Q.A_c^{-1} = 2.73 \text{ l. sec}^{-1}. \text{km}^{-2}$$

Such specific discharge is very close to the value calculated by SYMOENS (1968) for the Luvua R. at Pweto ($q = 2.60$ l.sec^{-1}. km^{-2}), at the exit of L. Mweru.

5.2. Water quality in the Zambezi River above L. Kariba

On the basis of all available published and unpublished data the variable physical and chemical characteristics of the Zambezi R. water during one annual hydrological cycle have been estimated (Table 12 and Fig. 15).

5.2.1. WATER TEMPERATURE

Maximum temperature (30 C) is reached in January as the flow increases. Then temperature decreases at first slowly but later rather rapidly to reach a minimum (17 C) in late June. The temperature rises between July and the following January. This thermal cycle is closely related to the flow pattern, to the climatic conditions existing in the upper catchment, and to the time it takes for the water to travel down to the Gwembe Valley.

5.2.2. SALINITY

The variation in salinity (as defined in Sec. 13.2) shows a general inverse relationship with the flow regime. It remains about constant (50–70 mg. L^{-1}) between October and December when relatively concentrated water arrives from the upper and middle catchments. Later there is a drop in salinity until a lower level (35–45 mg. L^{-1})

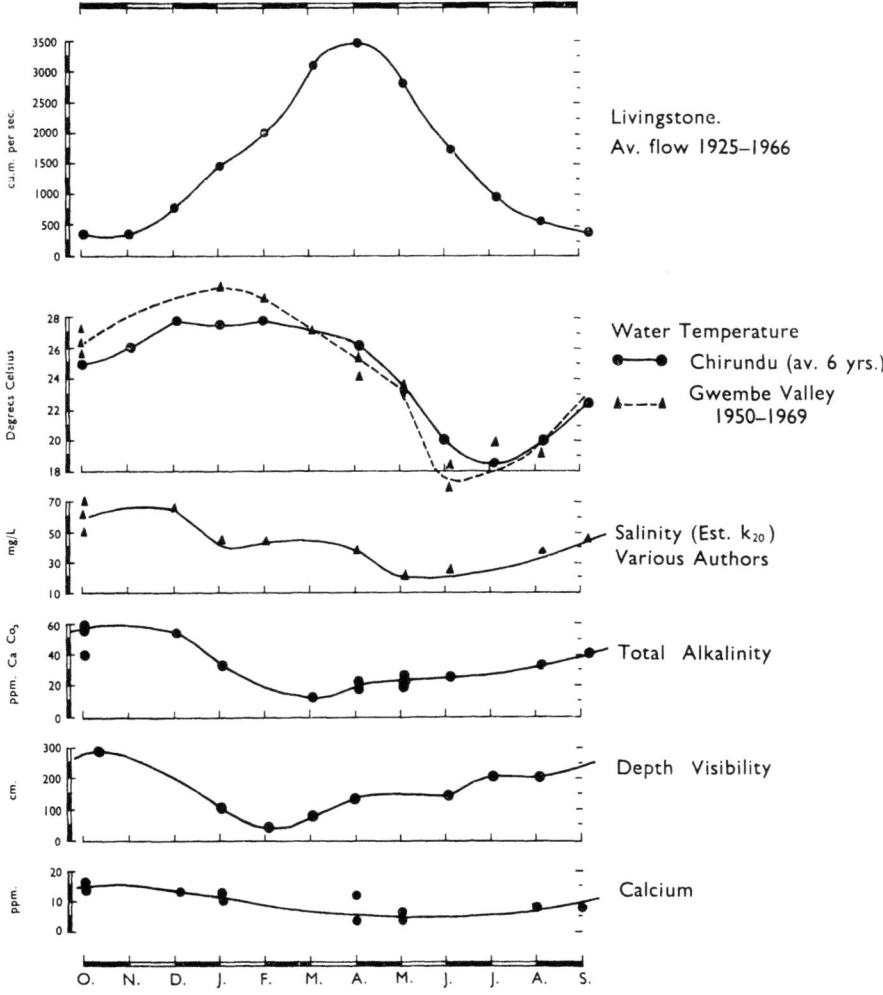

Fig. 15. Zambezi River: average flow regime (1925–1966) at the Kariba Dam site and annual variation of the water quality (acc. Table 12).

which characterizes the river during its phase of increasing flow and which lasts until the interval between January and April. A minimum salinity (ab. 20 mg. L^{-1}) is usually reached in May when the Livingstone peak flood (April) reaches Lake Kariba. As the flow slowly decreases, a gradual rise of the salinity brings it back to its maximum level.

Table 12. Water quality of the Zambezi River above Lake Kariba.

Reference	Harding 1964a			Jackson 1961			Harding 1962			Bow-maker 1960	Unpubl. LKFRI data		
Date	210650	011255	210156	250456	261056	110557	241058	151258	small[2]	May 1959	070164	190564	070964
Flow character	flood	—	—	flood	v. small	flood	—	increas. (closure of dam)		(flood)	(in-creas.)	(flood)	—
Temperature, C	—	—	—	25.4	27.3	23.8	—	—		23.3	—	23.3	—
pH	—	—	—	7.0	7.9	7.6	7.3	6.9		7.2	—	6.9	7.4
Conductivity, k_{20}	46	—	—	68	104	38	121	116		38	—	42	82
[1] Salinity, mg. L^{-1}	26	92	—	(39)	(60)	(21)	69	66		(21)	—	(24)	(47)
Tot. solids, mg.L^{-1}	60	—	70	—	—	—	103	91		—	—	—	—
Totl. alk, ppm $CaCO_3$	—	—	—	25	57	22	59	56		28	—	25	44
Tot. hardness, ppm $CaCO_3$	—	—	—	—	—	—	49	50		—	40	17	38
Calcium, mg. L^{-1}	—	—	—	13	16.9	7.2	14	14		—	12.8	5.2	8.4
Magnesium, mg. L^{-1}	—	—	—	2	7.3	4	3.4	3.7		—	1.9	0.9	3.9
Sodium, mg. L^{-1}	—	—	—	1	—	—	8	7		—	4.7	1.4	3.3
Potassium, mg. L^{-1}	—	—	—	—	—	—	0	0		—	2.0	T	0.9
Sulphate, mg. L^{-1}	—	—	—	—	—	—	3	3		—	—	—	—
Chloride, mg. L^{-1}	—	—	—	—	—	—	3	3		—	—	—	—

Table 12. (concluded)

Reference	Symoens 1968		Coche 1968				Begg 1970		Pers. data, CFRI				General ranges
Date	140465	080165	270465	270665	080865	311065	160166	May 1967	300368	130768	161068	120269	
Flow character	(flood)	(increas.)	flood[2]	decreas.	—	small	increas.	peak flow	flood	decreas.	small	increas.	
Temperature, C	—	—	24.3	18.5	19.3	25.7	30.0	—	27.1	20.0	26.4	29.4	17 – 30
pH	6.9	—	7.0	—	7.5	7.3	8.0	—	7.0	6.4	7.5	7.1	6.4– 8.0
Conductivity, k_{20}	38	—	40	50	66	87	78	39	36	58	104	79	36 –121
[1] Salinity, mg.L^{-1}	—	—	(23)	(28)	(38)	(50)	(45)	(22)	(20)	(33)	(60)	(46)	20 – 69
Tot. solids, mg.L^{-1}	34.4	—	—	—	—	—	—	—	—	—	—	—	34 –103
Tot. alk, ppm CaCO$_3$	24	—	20	28	35	41	35	—	15	—	—	—	15 – 59
Tot. hardness, ppm CaCO$_3$	—	40	18	—	31	—	—	—	—	—	—	—	17 – 50
Calcium, mg.L^{-1}	4.71	10.1	4.7	—	10.0	—	—	—	—	—	—	—	4.7– 12.8
Magnesium, mg.L^{-1}	1.5	3.5	1.5	—	1.4	—	—	—	—	—	—	—	0.9– 7.3
Sodium, mg.L^{-1}	—	4.4	1.7	—	2.9	—	—	—	—	—	—	—	1 – 8
Potassium, mg.L^{-1}	—	1.6	T	—	0.4	—	—	—	—	—	—	—	T – 2
Sulphate, mg.L^{-1}	3.8	—	—	—	—	—	—	—	—	—	—	—	3 – 3.8
Chloride, mg.L^{-1}	<0.1	—	—	—	—	—	—	—	—	—	—	—	T – 3
Depth visib, cm	—	—	150	160	220	—	110	—	90	220	300	50	50 –300
App. colour	—	—	brown	brown	gr/br	—	brown	—	brown	brownish	gr/br	brown	br–gr/br

[1] When in parentheses, estimated value from k_{20} – see Fig. 51.
[2] Quality considered as characteristic of a particular flood regime (see Table 13).

Table 13. Chemical composition of Zambezi River water above Lake Kariba during low and high flood.

Element	Low water conditions (24 Oct. 1958)			High water conditions (27 April 1965)		
	mg/L	meq/L	percent	mg/L	meq/L	percent
Ca	14	0.6986	52.0	4.7	0.2345	51.8
Mg	3.4	0.2796	20.8	1.5	0.1234	27.3
Na	8	0.3480	25.9	1.7	0.0740	16.4
K	(0.7)	(0.0169)	1.3	(0.8)	(0.0205)	4.5
Total cations	26.1	1.3431	100.0	8.7	0.4524	100.0
$HCO_3 + CO_3$	35.4	1.1801	88.9	12.0	0.4000	75.5
SO_4	(3)	0.0846	6.4	3.8	0.0791	23.8
Cl	(3)	0.0625	4.7	0.09	0.0020	0.7
Total anions	41.4	1.3272	100.0	15.89	0.4811	100.0
Salinity	67	2.67	—	24	0.93	—
pH		(7.5)			7.0	
k_{20} micromhos		121			40	
Reference	Harding 1962			Coche 1968		

5.2.3. Silt load and depth of visibility

Because of the presence in the Zambezi catchment above L. Kariba of the Barotse Plain and of the Chobe Swamps which act as sediment traps (Fig. 1), the silt load of the river water as it reaches the lake is relatively light. During the flood season the silt load in the Zambezi River is about 1:1 600 by volume (1:3 200 by weight) but it drops to only about one tenth of these values when the flow becomes small (Wellington's report cited by Scudder, 1962). Between the last sedimentation area (Chobe) and the lake, the Zambezi River water picks up mainly fine sand (Bond, 1965) due to the characteristics of the geological formations encountered (mainly hard Batoka Basalt as shown in Fig. 4). The preceding trends are reflected in the variation of the depth of visibility (Fig. 15) which in general ranges between 100 and 300 cm in the Devil's Gorge. Maximum values coincide with minimum flow. Minimum Secchi disc values were observed in February at a time when upstream tributary rivers (Gwaai R., Mulolo R.) were in full flood.

5.2.4. Chemical composition

Two sets of analyses have been chosen to characterize the chemical composition of the Zambezi River water according to the flow conditions (Table 13). Under high flood conditions and reduced salinity the chemical composition of the water changes with the percentages (by meq.) of magnesium and potassium increasing to the detriment of sodium. The predominance of fixed carbonate decreases to the profit of sulphate which greatly reinforces their relative proportion.

There is no doubt that during the flood season (ab. Feb-June), rapid flow and large volume greatly inhibit the influence which local conditions might have on the river water chemical composition. Salinity is reduced to less than 50 mg. L^{-1}, total alkalinity is low (0.3 to 0.6 meq. L^{-1}), pH is close to neutrality and, in general, concentrations of major ions are lowered to one third of their annual maximum values. In contrast during the dry season, salinity rises above 50 mg. L^{-1}, total alkalinity averages 1.0 meq. L^{-1}, and the mineral contents increase, except for potassium.

The annual average quality of the Zambezi River water feeding Lake Kariba may be obtained as weighted averages (0.8 high flood and 0.2 low flood water) from the data presented in Table 13 (Table 14). The average ionic composition by equivalents indicates the overall importance of fixed carbonate and calcium (calcico-carbonate water). It is very similar in composition to the world average (Fig. 52). But its lower mineral content reflects the mineral poverty and the relative insolubility of the rocks and soils present in the catchment above L. Kariba. The salinity averages only 1.283 meq. L^{-1} (32.2 mg. L^{-1}). Among the cations, calcium represents (by equivalents) 52.2 percent, the magnesium and sodium proportions being similar but potassium being proportionally very little represented. Among the anions, the fixed carbonate accounts for about 85 percent while sulphate (11.6 percent) and chloride (4 percent) are much less important.

Similar calculations were made for 1965/66 on the basis of chemical analyses performed on water samples collected in January 1965 and late April 1965 (high flood) and in August 1965 and September 1964 (low flood) as given in Table 12. Results show that during this particular year of study the Zambezi R. water was slightly richer than average, salinity being about 1.5 meq. L^{-1}. The ionic proportions were very similar except for magnesium which raised up to 27.6 percent of the total ionic content by equivalents, sodium decreasing to 17.8 percent.

Cationic ratios (by equivalents) were calculated from Tables 13 and 14 giving the following results:

Ratio	Low water	High water	Ann.wgt.avg.
K/NA	0.05	0.28	0.16
Ca/Mg	2.50	1.90	2.11
Ca + Mg/Na + K	2.68	3.79	3.23

These ratios point to the variability of the water composition relative to the two main flow patterns. The difference is particularly pronounced for the ratio of monovalent cations, greatly affected by the changes in sodium content which shows an important increase during the dry season. Compared to similar data from Central and East African waters (TALLING & TALLING 1965 – Fig. 5), both the average ratios of calcium to magnesium and of divalents to monovalents are relatively high. The third average ratio (K/Na) classifies among the middle values.

Table 14. Average composition of the water of the Zambezi River above and below Lake Kariba, the Luapula R. and the Kafue R.

Element	Zambezi river above Lake Kariba			Zambezi river below Lake Lariba			Luapula river Kasenga		Kafue river road/rail bridge	
	mg/L	meq/L	percent	mg/L	meq/L	percent	meq/L	percent	meq/L	percent
Ca	6.56	0.327	52.2	9.74	0.486	55.8	0.395	38	1.05	51
Mg	1.88	0.155	24.7	2.19	0.180	20.7	0.414	39	0.75	36
Na	2.96	0.129	20.6	3.97	0.173	19.9	0.208	20	0.22	11
K	0.77	0.020	2.5	1.20	0.031	3.6	0.036	3	0.04	2
(Na + K)			(23.1)			(23.5)		(23)		(13)
Total cations	12.17	0.631	100	17.10	0.870	100	1.053	100	2.06	100
$HACO_3 + CO_3$	16.70	0.557	85.4	24.9	0.83	(90.2)	0.827	86	1.92	97
SO_4	3.64	0.076	11.6	(3.0)	(0.06)	(6.5)	0.039	10	(T)	(1)
Cl	0.67	0.019	4.0	(1.0)	(0.03)	(3.3)	0.095	4	0.04	2
Total anions	21.01	0.652	100	(28.9)	(0.92)	100	0.961	100	—	100
Salinity	33.18	1.283	—	46.6	1.79	—	2.0	—	4.1	—
Reference	Weighted avg. – Tables 11 & 13			COCHE 1968			DE KIMPE 1964		TAIT 1967	

5.2.5. GENERAL COMPARISON WITH SOME OTHER LARGE AFRICAN RIVERS

The general quality of the Zambezi River water is compared to that of some other large African rivers which constitute major inflows for actual or future lakes and reservoirs (Table 15).

It shows again the similarity existing between the Zambezi and the Luapula River (L. Mweru) waters. The Volta and the Niger Rivers feeding L. Volta and L. Kainji respectively appear to be slightly poorer in minerals. Obviously the Kafue River and the three Niles (L. Nubia/Nasser) are much more endowed with nutrients and their content in fixed carbonate may reach much higher values. The Bandama River actually feeding L. Kossou (Ivory Coast) has an intermediate position while the lower Congo River exhibits greatly dilute concentrations except for chloride and sulphate.

5.3. The Zambezi River below the Kariba Dam

The damming of the Zambezi River at the entrance of the Kariba Gorge in December 1958 has radically changed the flow regime of the river downstream. The impact on agriculture in the region between the dam and the Kafue confluence has been shown to be drastic (SCUDDER, 1970). Effects on the ecology of the Mana flood plain (below Chirundu) have also been important (ATTWELL, 1970).

5.3.1. THE ZAMBEZI FLOW BELOW THE DAM

The Zambezi flow or discharge below the Kariba dam is made of two components whose importance greatly varies according to the engineering needs during the year:
a. The turbine flow released through the tail races is closely related to electric power requirements and its associated financial considerations. When all six actual turbines are operating, the turbine flow ranges between 600 and 900 cum per second. Over the period 1963/66 the annual average turbine flow has probably been close to 650 cum per second (BEGG, 1970) or 20.5 cu km per year. The centres of the lake water intakes (7 m high) are situated at about 462.5 and 447.5 m asl respectively. At the average operating water level of the reservoir (485 m asl) the water therefore is drawn from around the 20-m depth at least. During most of the year such depth corresponds with the cooler and deoxygenated hypolimnion (Chapt. 11 & 12). Most probably downstream aquatic life is then affected by this poor quality of the discharged water for a certain length of the Zambezi River below the dam.
b. The spilling flow (spillage) is released through one (or more) of the six sluice gates built into the dam (457–466 m asl) for controlling the lake water level. At the 485-m average operating level each fully opened gate discharges a little more than 1500 cum per second. In a normal year spilling would probably occur only through one gate throughout May to December. But if the Zambezi inflows were higher than normal, some spilling would be required between January and May. Conversely with a lower-than-normal inflow, gradual spilling during the dry season would be reduced accordingly (ALLISON, 1965).

The resulting water discharges below the Kariba Dam (Table 16) have fluctuated

Table 15. Water quality of some African rivers, major inflows for lakes and reservoirs.

River	Range temp. °C	Spec. Cond. at 20 C micromhos	Salinity mg.L⁻¹	Cations, mg. L⁻¹				Anions, mg. L⁻¹			Reference
				Ca	Mg	Na	K	HCO₃+CO₃	Cl	SO₄	
Zambezi (L. Kariba)	17–30	36–121	20–70	4.7– 16.9	0.9– 7.3	1–8	0 –2	12– 36	(3)	(3)	Original data
Blue Nile (Khartoum)	—	140–390	—	19.6– 28.1	5.0– 6.5	4.5– 9.0	1.7–2.9	49– 80	2.0–7.3	—	Hammerton 1972
White Nile (Khartoum)	—	220–500	—	14.2– 19.0	7.5–11.7	25 –41	6.9–9.2	69– 99	6.0–9.5	—	
Nile (L. Nubia)	(17–28)	(190)	—	25	7	17	4	134	4.5	9	Entz (pers. comm.)
Niger (L. Kainji)	(25–30)	34– 80	—	4.11	2.60	3.49	2.41	12– 19	1.36	T	Imevbore 1970
Volta (Ajena, L. Volta)	(25–30)	(60)	—	6.67	2.9	(2.4– 3.5)	(1.4–2.6)	40.3	4.6	12.7	Entz 1969 (Biswas 1968)
Bandama (L. Kossou)	(25–29)	90–200	—	4.4– 5.0	—	5.8– 6.4	1.6–2.5	24– 36	1 –3.5	0.2–1.0	CTFT 1972 / Welcomme 1972
Kafue (Flood Plain)	15–32	135–350	Avg.202	18 –102	7 –57	1.7–15	0.8–4	25–114	1.6–10	(T)	Tait 1967
Luapula (L. Mweru)	21–27	44–108	32–72	4.3– 12.8	1.8– 8.6	3.7– 6.2	0.6–1.7	15–38	1.8–5.8	1 –3	De Kimpe 1964 / Symoens 1968
Congo (Matadi, April 1966)	—	40	30.7	2.71	1.81	3.15	1.72	9.46	6.48	5.32	Symoens 1968

wildly in amount especially since 1963 when the lake reached its operating water level. From 1963 to 1966 monthly averages ranged between 4767 cum sec^{-1} and 425 cum sec^{-1}, a ten-fold range. Annual averages for the same period varied between 1573 and 591 cum sec^{-1}, the overall annual average being 1374 cum sec^{-1} or 43.3 cu km year^{-1}.

5.3.2. WATER QUALITY OF THE ZAMBEZI R. BELOW THE DAM

The chemical composition of the water discharged into the Zambezi bed below the Kariba Dam was studied in 1965 (Table 15 in COCHE, 1968). From these data the average ionic composition has been estimated (Table 14). Electrical conductivity (k_{20}) averaged 80 micromhos, pH varied around 7.1 and salinity about 1.8 meq. L^{-1}. Compared with the average ionic composition of the Zambezi River water entering L. Kariba (Table 14), the average outflowing water is proportionally (by equivalents) richer in calcium, potassium, and fixed carbonate but poorer in magnesium. The total result is an overall slight enrichment of the original river water within the lake basin. The effect of the latter on the river water composition seems mainly to be the exchange of ions rather than concentration or dilution. Similar observations have been made from Lake Kainji and L. Volta (H. F. HENDERSON, pers. comm.).

As pointed out in Sec. 5.3.1 (a) the quality of the water discharged from Lake Kariba also greatly varies according to the limnological conditions prevailing in the vicinity of the dam (Sta. 6348B), to the reservoir's water level, and to the turbine intake being in use.

Table 16. Lake Kariba monthly discharges in cubic metres per second.

Year	Oct.	Nov.	Dec.	Jan.	Feb.	March	April	May	June	July	Aug.	Sept.	Annual
1958/59	472	472	33	24	14	19	19	19	24	194	288	278	153
1959/60	283	283	142	142	142	142	142	189	189	236	283	283	203
1960/61	330	283	330	330	283	330	330	330	330	330	378	378	293
1961/62	378	378	378	378	330	378	378	378	425	425	425	378	383
1962/63	378	378	378	1,982	2,218	3,398	3,870	2,360	425	425	425	425	1,381
1963/64	2,690	4,767	4,626	4,578	3,162	802	472	472	472	519	519	472	1,252
1964/65	472	472	472	472	1,227	519	472	897	519	519	566	519	591
1965/66	519	472	3,257	3,446	519	519	472	2,643	519	3,446	2,171	991	1,573

6. THE SECONDARY RIVERS IN THE LAKE CATCHMENT

6.1. General characters of the secondary rivers

The Lake Catchment which coincides with the Rhodesian Highlands region except for the lake itself (5364 sq km) has an area of about 150 816 sq km representing 22.7 per cent only of the total Zambezi catchment above the Kariba Dam (Sec. 1.1). It has been described earlier how the juvenile rivers of the area surrounding Lake Kariba run from the Plateau down the Escarpment, meander through the Upland Valleys, and break out at right angles between the Lower Hills before joining the lake.

It has been pointed out how the profile of the Gwembe Valley is asymmetrical in cross-section with the steep side to the north (Fig. 6). The drainage reflects this asymmetry (Fig. 2). Rhodesian rivers draining into the Zambezi River above the lake and in the lake itself take their source on the divide between Salisbury and Bulawayo, some 200 to 300 km away. On the contrary Zambian rivers have their sources on the Lusaka-Choma ridge only 45 to 75 km away.

The result of such peculiar physiography and in particular of the relative steepness especially in Zambia, is that in the Lake Catchment the runoff comes down in spate and gives rise to flash floods. The hydrographs of the secondary rivers are characterized by many high-peak floods of short duration. They form the 'local floods' and superimpose themselves on the 'base flood' which travels down the Victoria Falls and the subsequent gorges before reaching the lake (Sec. 5.1).

6.2. Total discharge in the secondary rivers

In order to arrive at an estimate for the annual average discharge of these secondary rivers (Table 17) two methods have been used:
(i) Before the existence of L. Kariba the estimate has been obtained by difference between the annual discharges recorded at the Livingstone Pump Station (base flood) and at the Kariba Dam site. The former values have been estimated for the period 1925/26–1955/56 from a diagram (ANON., 1969); the later values (until 1965/66) were kindly supplied by the Hydrological Branch in Salisbury. Estimated values earlier than 1946/47 for the Kariba station have also been provided by this last agency.
(ii) Since the presence of L. Kariba (1959/60) the net evaporation losses (Table 17) from the lake's surface (CAPCOR – unpub. data) have been taken into account and added to the Kariba station discharges before subtracting the Livingstone discharge.

Results (Table 17) show that in the secondary rivers: a. The flow greatly varies from year to year. Absolute minimum and maximum average annual discharges were respectively 75 cum. sec^{-1} in 1946/47 and 852 cum. sec^{-1} in 1938/39. b. Overall average discharge was 318 cum sec^{-1}. It has not changed since the creation of Lake Kariba although the average Zambezi inflow has increased. On a long-term basis the

Table 17. Average annual discharges (cum.sec^{-1}) at Livingstone Pump Station, Kariba Dam site (Zambezi River) and in lake catchment rivers.

Hydrological year	Kariba Dam site	Livingstone Pump Station	Lake catchment (secondary rivers)
1925/26	1,439	1,110	329
1926/27	1,271	1,026	245
1927/28	1,080	909	171
1928/29	1,131	665	466
1929/30	1,131	831	300
1930/31	1,154	977	177
1931/32	1,498	1,095	403
1932/33	893	688	205
1933/34	1,427	1,212	215
1934/35	1,326	997	329
1935/36	1,162	880	282
1936/37	1,431	1,144	287
1937/38	1,303	909	394
1938/39	1,790	938	852
1939/40	1,763	1,344	419
1940/41	1,326	1,002	324
1941/42	885	743	142
1942/43	1,178	763	415
1943/44	1,474	967	507
1944/45	1,232	1,036	196
1945/46	1,193	812	381
1946/47	1,307	1,232	75
1947/48	1,880	1,642	238
1948/49	667	566	101
1949/50	1,556	1,309	247
1950/51	1,412	1,217	195
1951/52	2,114	1,584	530
1952/53	2,040	1,525	515
1953/54	1,388	1,144	244
1954/55	1,849	1,271	578
1955/56	1,931	1,681	250
1956/57	1,794	1,592	202
1957/58	2,925	2,516	409
1958/59	1,420	1,206	214

Closure of Kariba Dam		Net Evaporation from lake			Sec. rivers' inflow percent of total inflow
1959/60	1,314	(+ 90)	1,204	200	16.6
1960/61	1,751	(+121)	1,611	261	16.1
1961/62	2,059	(+172)	1,858	373	20.5
1962/63	2,726	(+129)	2,232	623	27.9
1963/64	1,268	(+200)	1,212	256	21.1
1964/65	1,197	(+168)	1,136	229	20.5
1965/66	1,474	(+168)	1,365	277	20.3
Average 1925–1966	1,493	—	1,199	318	26.5
Average 1959–1966	1,679	(+150)	1,517	317	21.0

secondary rivers have contributed 26.5 percent of the total inflow. c. Since October 1959 the relative contribution of the secondary rivers to the general water inflow has decreased to about 21 percent. This stresses how comparatively small is the volume of water they contribute to the general Lake Kariba water budget (Sec. 7.2).

Because of the marked seasonality of the rainfall (Sec. 3.1.) the discharges of the secondary rivers vary according to a cyclic pattern. A general estimate of this pattern is obtained from monthly data presented graphically by SCUDDER (1962 – Fig. 2) and reported in Table 18. The resulting estimates for the monthly discharges in the secondary rivers show that:

a. They greatly vary from month to month, the main flood season taking place between January and March (about 56 percent of the annual discharge). Therefore the 'local floods' in the Lake Catchment generally reach Lake Kariba before the Zambezi River 'base flood' (March-May) as cited previously (Sec. 5.1).

b. During the dry season (esp. July-Nov) the flow becomes much reduced and most river beds are dry at least in their lower reaches. A more permanent flow exists in tributary rivers upstream from the Chete Gorge especially on the southern shore. But throughout the Gwembe Valley other tributaries maintain a continual flow in the Upland Valley where they are fed by springs whose waters sink below the sands before they reach the Lake (SCUDDER, 1962).

On the basis of their individual catchment areas the main secondary rivers are in order of importance: 1. Gwaai River: 47 140 sq km; 2. Sanyati River: 43 500 sq km; 3. Sengwa River: 5 200 sq km.

All together these three rivers, situated in Rhodesia, drain 95 840 sq km or 63.6 percent of the Lake Catchment area. As pointed earlier (Sec. 6.1) the Zambian rivers have relatively small catchments. According to Dr. J. BALEK (pers. comm.) main river drainage areas in Zambia are as follows: Chimene (Zhimu) River: 2567 sq km; Zongwe River: 2290 sq km; Lufua River: 2274 sq km; Chezya River: 850 sq km.

6.3. Water quality in some secondary rivers

Results of chemical analyses performed on several secondary rivers have been given elsewhere (COCHE, 1968). It is intended in this particular section to reach general conclusions about the relative importance of Lake Kariba affluents. Previous geological (Figs. 4 & 5) and climatological (Fig. 7) considerations should be referred to when interpreting the following results.

6.3.1. SANYATI RIVER[6]

Before the creation of Lake Kariba, the Sanyati River joined the Zambezi near the actual dam wall. Now it enters the lake through the deep narrow Sanyati Gorge, 30 km south of Kariba (Fig. 23).

The Sanyati River is to be considered as the most important secondary river for several reasons:

---

[6] Sometimes referred to as the Umniati River

Table 18. Average monthly discharges (cum. sec^{-1}; period 1925–1954) in the Zambezi River (Livingstone and Kariba Stations) and in secondary rivers of the lake catchment.

Month	Oct.	Nov.	Dec.	Jan.	Feb.	March	April	May	June	July	Aug.	Sep.
Kariba[1]	320	392	672	1,288	1,904	2,604	3,080	2,576	1,680	952	504	392
Livingstone[1]	230	232	406	644	1,050	1,820	2,632	2,240	1,456	812	448	364
Secondary rivers	90	160	266	644	854	784	448	336	224	140	56	28

[1] Estimated from Figure 2, SCUDDER, 1962.

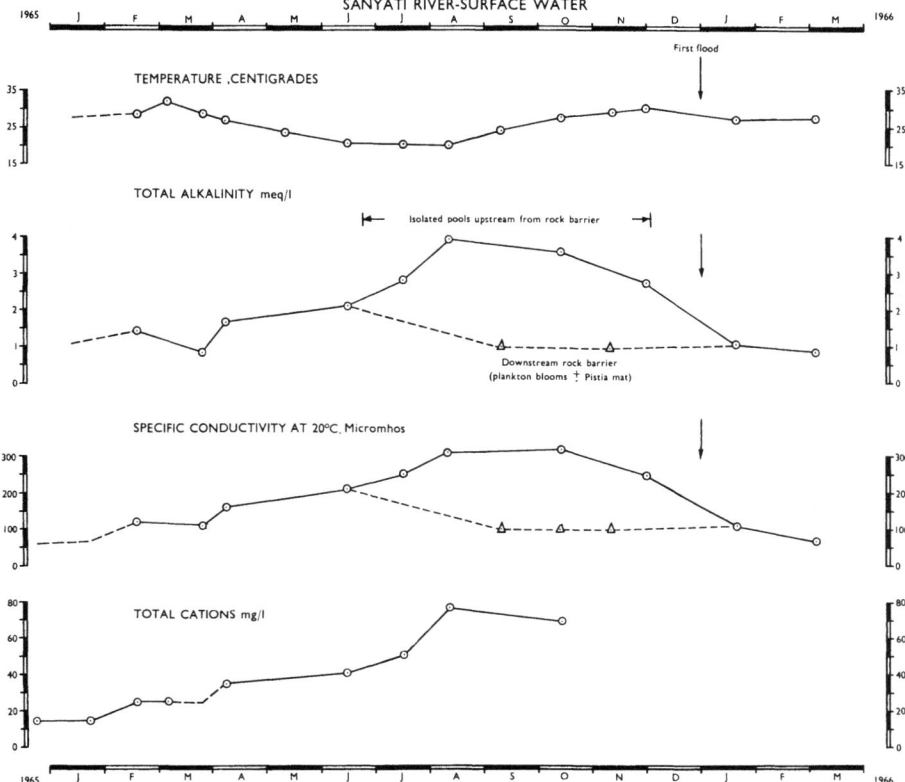

Fig. 16. Variation of the quality of the Sanyati River water in 1965–1966.

(i) It is one of the largest relatively permanent rivers although in certain years (e.g. 1965) its bed dries out in its lower part at least in surface. JACKSON (1961) has aptly described it to be 'in October ... very low, consisting mainly of large and sometimes shallow pools of water ... connected by a comparatively small stream of running water with little aquatic vegetation ...'

(ii) It drains the upland Rhodesian farmlands and brings down to Lake Kariba highly mineralized water (see further).

(iii) It serves as migration route for important commercial fish species (e.g. *Hydrocynus vittatus* and *Labeo* spp.) which vigorously swim upstream to spawn.

The water quality and its variations during 1965/66 are illustrated in Figure 16. During that particular year the flood lasted until April. Between August and December there was no apparent flow. The first new flood did not appear until January 1966 because of abnormally late rains in the catchment. The subsurface water temperature ranged between 32 C (March) and 20 C (August). Highly mineralized water was found during the dry season in isolated pools situated upstream from the rock barrier (total alkalinity up to 4 meq. L^{-1}). The first flood flushing this nutrient-rich water into the lake resulted in a sharp drop of the mineral content of the river water.

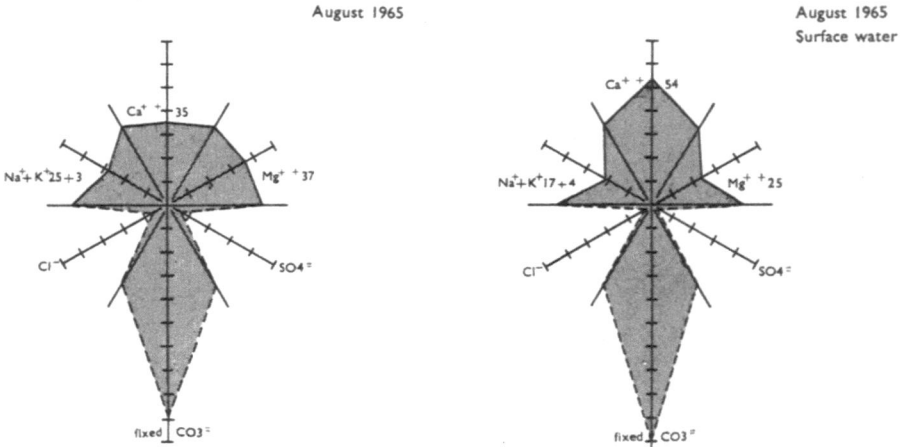

Fig. 17. Average percentage ionic composition (1965) of the Sanyati River and Lufua River waters.

Only the cationic composition of the Sanyati water was fully determined (Fig. 17 and Table 19). Calcium and magnesium were the prevalent cations by equivalents (ab. 36 percent each) but sodium was also well represented. There is no doubt that bicarbonate largely dominate among the anions. The high average conductivity (155 micromhos at 20 C) is characteristic of relatively well mineralized water.

6.3.2. Lufua river

The Lufua River is a typical secondary river of the northern shore characterized by a relatively small catchment area and by flash-floods. A few kilometers upstream from its mouth it is permanently choked by a thick *Salvinia* mat.

There seems to be a direct relationship between flow and mineral content of the water, the latter being definitely higher during the flood season (Table 20 and Fig. 17). Most probably this is due to stronger erosion in the Escarpment Region and in the Upland Valleys, and to the effect during the dry season of the presence of a permanent *Salvinia* mat. But the average percentage ionic composition closely resembles that of the Zambezi River water with calcium predominance among the cations. The average conductivity is moderate (114 micromhos at 20 C) while pH varies around neutrality.

6.3.3. Miscellaneous rivers

Further chemical data about various other secondary rivers in the Gwembe Valley are available from Bowmaker (1960), Coche (1968), Jackson (1961), and Mitchell (1970). Some of the personal data collected in 1965 from the Sebungwe R., Chimene R., and Zongwe R. are summarized in Table 21.

Table 19. Chemistry of Sanyati River water, December 1964 to March 1966 (Station 6437).[1]

Date 1964–66	Temp. C	pH	k_{20} μmhos	TA meq/L	Ca^{++} mg/L	Mg^{++} mg/L	Na^+ mg/L	K^+ mg/L	Nitrite ppb	Nitrate ppb	Phosphate ppb	Total cations mg/L
23 December	—	—	(100)	(0.98)	7.8	0.39	4.40	2.08	7.52	252.5	52.5	14.67
23 January	—	—	(100)	(0.98)	4.8	1.94	5.64	1.78	7.08	99.4	81.8	14.16
19 February	28.6	8.5	120	1.44	9.6	6.34	7.40	1.88	7.08	82.2	42.5	25.22
5 March	32.0	—	—	—	10.1	5.98	7.30	1.98	1.60	67.8	—	25.36
24 March	28.8	7.6	110	0.82	—	—	—	—	—	—	—	—
8 April	26.7	8.2	160	1.86	15.9	8.26	8.54	2.08	2.80	64.4	74.3	34.78
11 May	23.9	7.6	(180)	(1.95)	—	—	9.38	1.91	3.56	55.1	17.0	—
14 June	20.4	7.6	210	2.10	16.8	12.10	10.20	2.31	Nil	29.5	—	41.41
16 July	20.3	8.2	248	2.80	18.0	15.76	13.78	2.82	6.36	72.4	18.8	50.36
12 August	20.0	8.1	310	3.92	23.8	18.30	33.90	0.54	1.28	12.7	69.0	76.54
10 September	24.0	6.8	100	0.98	11.3	2.47	4.60	1.51	T	25.3	18.9	19.88
12 October	27.5	9.0	—	3.60	23.4	18.20	23.00	4.01	15.00	—	—	68.61
11 November	28.7	8.9	99	0.96	—	—	—	—	—	—	—	—
1 December	30.0	9.0	248	2.72	—	—	—	—	—	—	—	—
20 January	27.0	8.1	110	1.04	—	—	—	—	—	—	—	—
3 March	27.5	—	63	0.82	—	—	—	—	—	—	—	—
Average 1965, mg/L			(155)	(57.9)	14.1	9.0	11.6	2.1				
Average 1965, meq/L			—	1.93	0.70	0.74	0.50	0.05				
Avg. Ionic Comp., per cent			—	—	35	37	25	3				

[1] Complete chemical analyses performed at the University College of Rhodesia, Salisbury, under the supervision of D. S. MITCHELL.

Table 20. Chemistry of Lufua River water in 1965 (Station 4948)[1].

Date 1965	Temp. C	pH	k_{20} μmhos	TA meq/L	Ca^{++} mg/L	Mg^{++} mg/L	Na^+ mg/L	K^+ mg/L	Nitrite ppb	Nitrate ppb	Phosphate ppb
5 January	—	—	—	—	11.15	2.43	5.23	3.68	6.60	61.6	330.1
11 March	29.2	7.2	193	—	20.71	7.76	7.30	2.98	10.00	42.0	—
23 April	26.5	7.0	150	1.60	20.50	7.75	6.75	2.45	1.44	22.6	20.3
25 May	24.3	7.0	108	1.16	13.50	2.40	5.07	1.60	4.76	36.3	—
2 July	21.0	7.1	98	1.04	11.76	2.80	4.65	1.50	13.52	26.3	38.5
4 August	17.5	7.0	82	0.88	13.00	1.20	4.24	1.27	N	15.1	26.0
30 August	23.0	7.0	80	0.88	10.00	2.60	2.19	1.12	T	13.4	17.6
7 October	23.6	7.2	87	1.00	10.86	3.44	4.30	1.12	2.54	14.7	31.5
3 November	27.0	7.4	82	—	—	—	—	—	—	—	—
Average composition in 1965											
mg/L				(37.5)	13.9	3.8	5.0	2.0			
meq/L				1.25	0.694	0.313	0.218	0.051			
per cent				(98)	54	25	17	4			

[1] Complete chemical analyses performed at the University College of Rhodesia, Salisbury, under the supervision of D. S. MITCHELL.

Table 21. Water quality in the Sebungwe River, Chimene R., and Zongwe R. in 1965.

1965	k_{20} μmhos	TA meq/L	pH	Ca^{++} mg/L	Mg^{++} mg/L	Na^+ mg/L	K^+ mg/L	NO_3N μg/L	PO_4P μg/L
			Sebungwe River – Basin I to II						
10 Jan	—	—	—	10.8	3.34	4.08	1.28	20	31
26 Apr	60	0.50	—	7.9	N	2.32	0.22	59	16
8 Aug	52	0.56	—	6.9	1.88	2.54	T	39	27
			Chimene River – Basin II						
7 Jan	—	—	—	7.3	N	8.34	2.48	64	246
28 Apr	65	0.72	6.8	11.2	0.49	3.78	0.82	76	29
7 Aug	54	0.58	6.6	6.9	1.88	2.53	0.24	91	35
			Zongwe River – Basin III						
12 Jan	—	—	—	8.8	3.29	3.78	1.28	36	48
28 Apr	100	0.92	7.0	11.9	2.43	5.02	1.03	23	26
9 Aug	82	0.84	6.7	13.8	0.47	4.70	3.84	130	348

6.3.4. Conclusions

The water of the secondary rivers is relatively richer in minerals than that of the Zambezi River. The high mineral content of the Sanyati River at certain times of the year is associated with the size of its flood. The ionic composition of the secondary rivers is in general similar to that of the Zambezi water, calcium being dominant among the cations whereas bicarbonate dominate the anions. The Sanyati River however makes exception, both calcium and magnesium being prevalent among the cations in similar proportions.

There is a general trend for the river tributaries to exhibit an increased fertility from the Zambezi River inflow northeastward up to the Sibilobilo Narrows (Fig. 23).

SECTION III

LAKE KARIBA

PHYSICO-CHEMICAL CHARACTERISTICS

7. HYDROLOGY OF LAKE KARIBA

7.1. The hydrological control of the lake level

The hydrological control of Lake Kariba (ALLISON, 1965, 1969) has been until now the main interest of the Central African Power Corporation (CAPCOR) whose main function is to produce electric power from the Kariba hydro-electric scheme for Rhodesia and Zambia[7].

Between December and May (Zambezi flood season) it operates a flood forecasting system based on: (i) daily water flow data from the Zambezi catchment above Kariba obtained from automatic telemetering stations or direct measurements (telegrams, radios); and (ii) rainfall data. With this information and under fairly normal conditions, CAPCOR can assess inflow conditions at Livingstone some thirty days in advance. A major improvement was expected from the determination of the rainfall/runoff relationships for the various sections of the catchment. When the North Bank scheme comes into operation it is planned to keep the retention level high. Spilling will only be necessary at intervals of three to five years (ALLISON, 1970 cited in VAN DER LINGEN, 1973).

The lake water discharges at the dam from the following levels (a.s.l.):
a. Turbine intakes: a.a. Between 459 and 466 m (avg. 462.5 m); a.b. Between 444 and 451 m (avg. 447.5 m).
b. Spillway: (6 gates) between 457 and 466 m. The average operating water level is 485 m asl.

7.2. The average water budget of the lake

Monthly water gains and losses for Lake Kariba are available from CAPCOR (unpubl. data). Examples of average values are presented in Table 22 which describes the average cyclic pattern of the water budget components throughout the year. To the credit of this balance are the rainfall on the lake surface and the runoff from the Zambezi River catchment above the Kariba Dam. The latter results in:
(i) Zambezi River discharge as measured at Livingstone Station and
(ii) the discharge of the secondary rivers, all rivers flowing into the Zambezi R. and the lake between the Victoria Falls and Kariba Dam[8].

---

[7] CAPCOR a joint organisation for Rhodesia and Zambia was created in 1964 to succeed the Federal Power Board (May 1956–Dec. 1963). In late 1970 the Zambian Parliament took action to allow for a newly constituted CAPCOR to handle the Kariba North Bank (Zambia) scheme being started.
[8] The possible contribution of the water springs existing in the lake area has been neglected as well as the evapo-transpiration of the vegetation which has not been estimated.

Table 22. Examples of monthly water gains and losses for Lake Kariba, in cubic kilometres.

Component	Oct.	Nov.	Dec.	Jan.	Feb.	March	April	May	June	July	Aug.	Sept.	Annual
1. Gains[1]													
1.1. Rainfall on the lake area	0.04	0.30	0.86	0.89	0.88	0.33	0.06	—	—	—	—	—	3.36
1.2. Inflow (run-off):													
1.2.1. Zambezi R. at Livingstone	1.05	0.98	1.56	2.45	3.80	7.47	10.10	8.44	5.05	3.05	1.85	1.30	47.10
1.2.2. Secondary rivers	—	0.04	1.02	1.35	2.01	0.98	0.23	0.01	—	0.01	0.01	—	5.66
1.3. Total gains	1.09	1.32	3.44	4.69	6.69	8.78	10.39	8.45	5.05	3.06	1.86	1.30	56.12
2. Losses													
2.1. Gross evaporation from lake[2]	1.04	0.91	0.95	0.73	0.55	0.80	0.78	0.68	0.58	1.30	0.67	0.85	9.84
2.2. Outflow[3]:													
2.2.1. Turbine flow	1.53	1.52	1.52	1.43	1.36	1.54	1.48	1.64	1.63	1.67	1.68	1.49	18.49
2.2.2. Spillage	—	—	—	—	1.01	3.70	3.44	2.33	—	—	—	—	10.48
2.3. Total losses	2.57	2.43	2.47	2.16	2.92	6.04	5.70	4.65	2.21	2.97	2.35	2.34	38.81

[1] Mean values for period July 1961–June 1967 (after MITCHELL, 1970).
[2] Mean values for 1963–64 (CAPCOR, unpublished data).
[3] Values for 1967 (BEGG, 1970).

Table 23. Average yearly water budget for Lake Kariba (1963–66).

Component	Water Volume	
	cu.km	per cent
1. Gains		
1.1. Rainfall on the lake area	3.5	7.0
1.2. Inflow (run-off):		
1.2.1. Zambezi River at Livingstone	39.0	77.2
1.2.2. Secondary rivers (Liv.–Kariba)	8.0	15.8
1.3. Total gains	50.5	100.0
2. Losses		
2.1. Gross evaporation from the lake surface	(7.2)	14.3
2.2. Outflows:		
2.2.1. Turbine flow	20.5	40.6
2.2.2. Spillage	22.8	45.1
2.3. Total losses	50.5	100.0

The cyclic variations reflect previous conclusions. To the debit of the water balance are the gross evaporation from the lake surface[8], the turbine flow (relatively constant), and seasonal spillage losses (variable from year to year).

Estimates of the yearly water budget for Lake Kariba from averages of available data for the period 1963–1966 lead to the following remarks (Table 23):

a. Rainfall: the average rainfall for the period under consideration was 646 mm, this amount of rain falling on the lake surface at the average operating level of 485 m (5364 sq km).

b. Runoff: averages have been obtained using data for recent years (Table 17). Because of the high evaporation rate prevailing over the Zambezi River catchment, it is estimated that only about 6.5 percent of the mean annual rainfall appears as runoff (long-term average). However under conditions of intense, localised, and continous rains, the runoff yield has risen to almost 13 percent (CAPCOR unpubl. data).

c. Evaporation: gross evaporation from a free-water surface has been estimated (SHAND, 1960) to average 8 mm per day (2920 mm/year) under the Gwembe Valley conditions which would amount to 15.66 cu km per year from the lake surface at 485 m asl. Data from Chipepo evaporation pans (Table 4) point to a slightly smaller average value (7 mm per day or 13.7 cu km per year). Monthly estimates from CAPCOR (Table 22) add up to still a smaller value (9.84 cu km per year or 5 mm per day). Therefore it would appear that the gross evaporation value retained in Table 23 and simply estimated as the complement of the outflow to balance gains and losses, has been underestimated. It should be increased to a minimum value of 9 cu km per year which comes in line with the general estimate given by BEGG (1970). Moreover such value would better agree with the average net evaporation estimate (gross evapora-

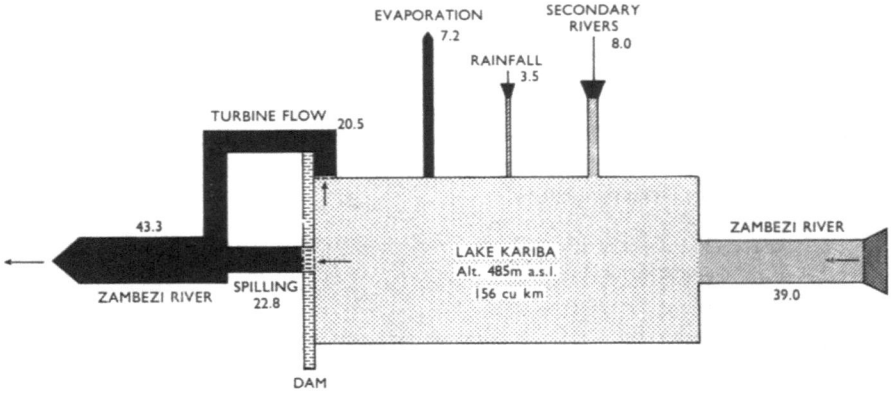

AVERAGE WATER BALANCE – CUBIC KM PER YEAR – 1963/66

Fig. 18. Relative contribution of the various components to the water budget for Lake Kariba.

tion minus rainfall) for the period 1963–66 given by CAPCOR as 5.6 cu km per year.
d. Turbine flow: the mean value given by BEGG (1970) has been used (Sec. 5.3.1); it is slightly higher than the 1967 estimate shown in Table 22 (18.5 cu km).
e. Spillage: it has been calculated by difference the estimated total discharge of the Zambezi River below the dam being 43.3 cu km (Sec. 5.3.1).

From these considerations the relative importance of each of the budget components can be judged for Lake Kariba from the percentage contributions which appear in Table 23 and the graphical analysis shown in Figure 18. There is a net loss by evaporation of at least 4 cu km per year.

The greatest gain contribution is provided by the Zambezi River (at least 77 percent). The annual water gains representing 32.3 percent of the lake volume it theoretically takes a little more than three years to 'renew' the whole lake.

7.3. *Fluctuations of the water level*

Since the closure of the dam the water level in Lake Kariba has fluctuated in two general patterns (Fig. 19) which coincide with phases of the lake's development: its filling phase (1959–1963) and its post-filling phase. Detailed values of the water level have been tabulated elsewhere (COCHE, 1968).

7.3.1. THE FILLING PHASE

The closure of the dam took place on 2 December 1958 when the lake started to fill from its river-bed base of 391 m above mean sea level. At first the level increase was very rapid (about 60 m during the first year) but gradually slowed down thereafter

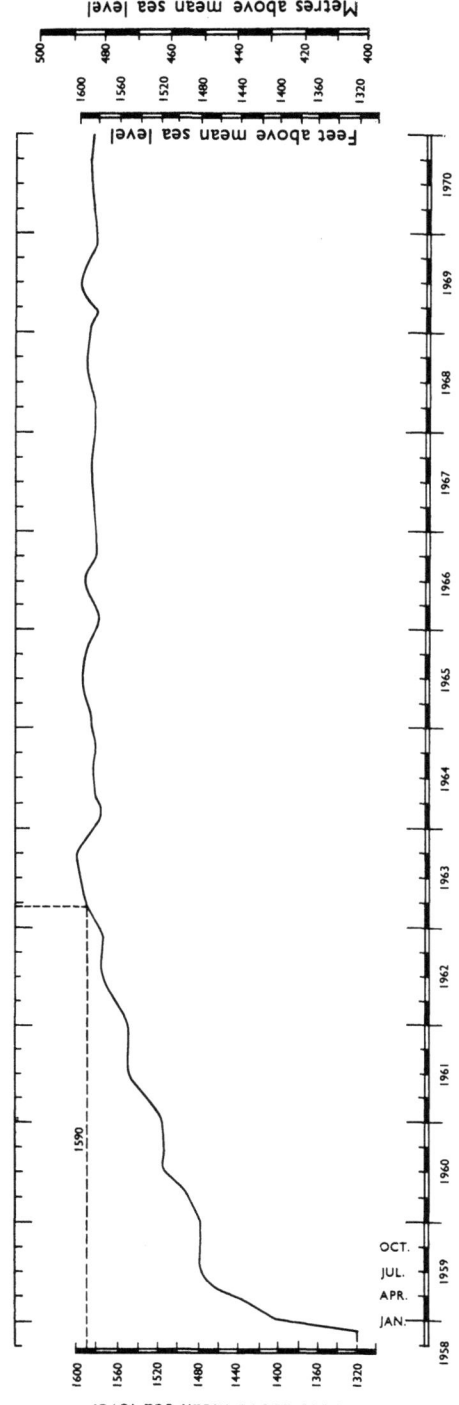

Fig. 19. Water level in Lake Kariba on the first day of each month (period November 1958–December 1970; data from CAPCOR).

(12 to 9 m per year). The average operating water level (485 m a.s.l.) was reached for the first time in April 1963 during the fifth flood season.

7.3.2. THE POST-FILLING PHASE

Once the normal operating water level had been reached the fluctuations became cyclic being mainly related to the variations of the runoff in the catchment above the dam.

For electric power production there must be compliance with the following critical levels: – Maximum flood storage level: 489 m asl; – Minimum drawdown level: 475 m asl. CAPCOR's basic policy is to draw the lake down to a level of about 482 m (1580 ft) by the end of December each year, the objective being 'to provide a steady medium level flow downstream simulating conditions which will exist when full development of Kariba is complete and turbine discharge may be two and a half times the present volume' (ALLISON, 1965). Under 'normal' conditions, there would therefore be two distinct periods within each annual cycle:
(i) May–December: falling water level down to about 482 m (limited spilling);
(ii) January–May: rising water level up to about 487 m (no spilling). Unfortunately a 'normal' year is seldom experienced and the level control has to be adjusted accordingly.

This is reflected by the actual fluctuations of the lake water level experienced since 1963 (Fig. 19):
(i) Very high maximum level in 1963 and 1969 (ab. 487.5 m)
(ii) Very low minimum level in 1964 and 1966 (ab. 480 m)
Although diel fluctuations of the water level are expected to be negligible, the seasonal variations have greatly varied from year to year:
(i) Very great amplitude (6–8 m) in 1963/64 and 1969,
(ii) Great amplitude (4–5 m) in 1964/65, 1965/66 and 1968/69,
(iii) Small amplitude (ab. 2 m) in 1967/68 and 1970.

There is no doubt that the biological production of the lake margin can be greatly affected, positively or negatively, by both levels of inundation and seasonal amplitudes. From the above data it might be inferred that 1967 and 1970 were years with a high potential fish production. On the contrary 1963 and 1969 were most probably responsible for a great decline in the potential fish production following the destruction of numerous cichlid nests during the relatively rapid drop of the water level in a particularly active nesting season (Sept. to Dec.). Areas of lake floor exposed following water level drawdown can be estimated from Table 28.

It is expected that when the North Bank Scheme participates to the hydro-electric exploitation of Lake Kariba, the retention level will be kept high, the water level fluctuations being much reduced.

7.4. Sedimentation in Lake Kariba

As pointed out earlier the Zambezi River has a low silt load constituted mainly by sand (Sec.5.2.3). Such transported materials are deposited at the head of the lake in the Devil's Gorge where a sudden change of flow velocity takes place. Moreover, the steep gradients of the tributary rivers being suddenly arrested far up their long, submerged inlets, sedimentation happens before their water reaches the main lake. Wave action on exposed shorelines of islands and lake shore is responsible for erosion-sedimentation processes taking place in the inshore littoral zone. All forementioned phenomena lead to the conclusion that sedimentation in the open lake must be very small indeed (BOND, 1965).

Provision for the study of the deposition rate of materials in the river mouths has been made by the Hydrological Branch, Rhodesia (MUNCASTER, 1965). The planned method is by means of closely controlled echosounding runs over survey profile lines. Three stations were ground-surveyed prior to inundation (Mlibizi, Sebungwe, and Binga) and two stations were air-surveyed (Sengwa and Sanyati). A series of lines have also been established across the Zambezi River upstream from the Devil's Gorge.

7.5. *Water movements: surface seiche*

The inertial oscillations of the Kariba water surface (seiche) have been recorded for several years (MUNCASTER, 1965). Stations have been placed along the Rhodesian lake shore at the entrance of the Kariba Gorge, in the Sanyati Gorge, in the Sinamwenda R. mouth, in the Binga pump well, at Sebungwe, and Mbilizi. In Zambia they have functioned at Sikolwinzola (Lukunzu R.) and in the Chimene R. mouth. The limnographs consist of Leupold Stevens level recorders mounted on top of modified windmill towers erected in 10 m of water (avg. normal operating lake level).

Preliminary results suggest: (i) a persistent twelve hour period and therefore evidence for a pronounced tidal component; (ii) a maximum amplitude of 25 cm; (iii) an oscillation of the lake over its entire length. Kariba and Sanyati oscillations were in phase with each other; the same was observed from Sinamwenda and Sebungwe records. But at these last two stations the seiche was half a period 'out of phase' with the former ones (MUNCASTER, 1965).

8. MORPHOMETRY AND MORPHOLOGY OF LAKE KARIBA

8.1. General considerations

It is generally agreed that Lake Kariba can be defined as the expanse of water artificially created on the Middle Zambezi River between the Kariba Dam and the Deka/Zambezi R. confluence. Extreme geographical coordinates are therefore: latitudes 16°28′–18°04′ South and longitudes 26°42′–29°03′ East. Its longitudinal axis roughly coincides with the political boundary between Zambia (North) and Rhodesia, and shows a general SW-NE orientation.

The appended bathymetric map (Fig. 20) shows the morphometric features of the lake. It was compiled from the topo-hydrographic set of three maps (scale 1:100 000) published by the Surveyor General of Rhodesia in 1965 (Sheets No. 17, 18 and 19). All morphometric data which follow have been calculated from them.

Lake Kariba is naturally subdivided into relatively distinctive basins. Although I have earlier functionally defined five Basins (COCHE, 1968) as shown in Figures 20 and 23, I consider the lake-proper to include Basins I to IV only. MITCHELL (1970) followed by BEGG (1970) also consider five Basins but with different definitions: their Basin I combines my Basins O and I, whereas my Basin IV combines their Basins IV and V. The latter is rather considered by me as formed by two Sub-Basins of Basin IV (S. B. Sanyati and S. B. Kariba – Fig. 23) which are each relatively homogeneous entities. This is based on the fact that all Kariba Basins are separated from each other either by narrows between promontories or by chains of islands, topographical features. Sub-Basin delimitation in contrast depends mainly on the local influence of particular factors. In the eastern part of the lake for example, the Sanyati River water inflow combines with the drowned physiography to create special limnological conditions within a defined area of Basin IV.

Water level variations are characteristic of reservoirs such as Lake Kariba (Sec. 7.3) and morphometric measurements vary accordingly. Therefore it becomes necessary to define a 'reference water level' for purposes of generalization. The normal operating water level (485 m asl) of Lake Kariba has been chosen for this reference purpose. In past years the surface water level at Kariba Dam has fluctuated between 490 m and 480 m (Fig. 19) and it is planned to keep it so in the future.

All morphometric parameters are defined following HUTCHINSON (1957) except when stated otherwise. Calculations have been made separately for the four functional Basins and for the entire lake. But in this last instance Basin O (from the Deka R. confluence to the end of the Devil's Gorge) has always been excluded because of its major riverine character along most of its length throughout the year. Thus Lake Kariba-proper is considered only and selected parameters are grouped in Table 24.

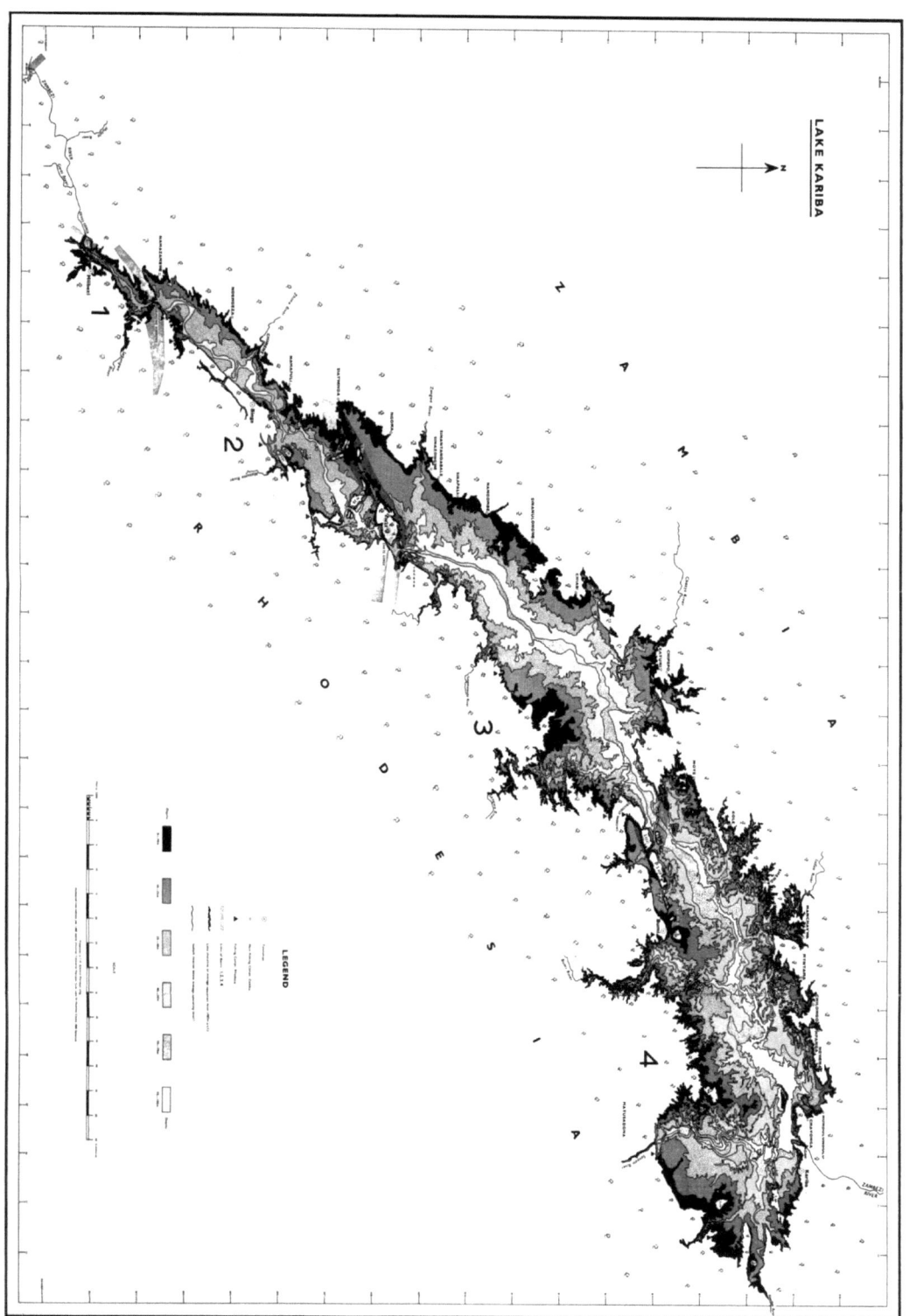

Fig. 20. Bathymetric map of Lake Kariba.

Table 24. Morphometry of Lake Kariba and its Basins at the normal operating water level (485 m).

Parameter	Symbol	Unit	0	I	II	III	IV	Lake Kariba[1]
Length	l	km	42	23	56	96	102	277
Mean breadth	$\bar{b} = A_o l^{-1}$	km	(0.4)	4.0	12.1	21.2	25.1	19.4
Surface area	A_o	sq. km	(18)	91	677	2033	2563	5364
Shore line exposed[2]	—	per cent	—	0.0	36.7	40.0	31.1	32.6
Shore line sheltered[2]	—	per cent	—	68.2	31.5	30.3	38.7	36.9
Shore line estuarine[2]	—	per cent	—	31.8	31.9	29.7	30.2	30.5
Shore line total	L_o	km	—	141	355	704	964	2164
Developm. of shore line	$D_L = L_o(2\sqrt{\pi A_o})^{-1}$	—	—	4.1	3.9	4.4	5.4	8.3
Bush-cleared areas	Table 8	sq.km	—	—	167.4	465.0	321.4	953.8
Islands, number	—	—	—	21	40	87	145	293
Islands, shore line[2]	L_I	km	—	38	121	159	286	604
Islands, surf. area	A_I	sq. km	—	5.53	46.46	29.34	65.54	146.87
Insulosity	A_I	per cent	—	6.1	6.9	1.4	2.6	2.7
Volume	V_o	cu.km	(0.4)	1.145	16.254	53.958	85.142	156.5
Mean depth	$\bar{z} = V_o A_o^{-1}$	m	(22)	12.58	24.01	26.54	33.22	29.18
Max. depth[3]	z_m	m	40	37	52	66	93	93
Depth ratio	$\bar{z} z_m^{-1}$	—	—	0.34	0.46	0.40	0.36	0.31
Relative depth	$z_r = 50 z_m \sqrt{\pi(\sqrt{A_o})^{-1}}$	—	—	0.34	0.18	0.13	0.16	0.11
Development of volume	$D_V = 3\bar{z}(z_m)^{-1}$	—	—	1.02	1.38	1.20	1.08	0.93
Mean slope of bottom	$\bar{S}$	per cent	—	2.8	1.7	1.4	2.3	2.1
	ϵ	degree	—	1°33′	0°58′	0°49′	1°19′	1°13′

[1] Basin 0 excluded.
[2] Data from Mitchell (1970) for slightly different Basin limits.
[3] Occasional deeper 'holes' not considered.

Table 25. Estimated lengths of shore line at water level 485 m, in kilometres.

Shoreline	Basin				
	I	II	III	IV	Lake
Zambia	46	142	370	425	983
Rhodesia	95	213	334	539	1,181
Lake Kariba	141	355	704	964	2,164

8.2. Planimetry of Lake Kariba

8.2.1. PLANIMETRY AT THE 485-M WATER LEVEL

Parameters defining the planimetry of the lake have been first calculated for its average operating water level (Table 24).

a. The lengths (l), shortest distances on the water surface between most distant points on the lake shore, were measured along the longitudinal axis of the lake. It totals 319 km between the Deka/Zambezi confluence and the most eastern shore. The lake itself is 277 km long. The individual Basins increase in length from west to east, maximum value being observed in Basin IV (102 km).

b. The mean breadths ($\bar{b}$) given as ratios of surface area to length, increase from 4 km in Basin I to 25.1 km in Basin IV. The lake mean breadth equals 19.4 km.

c. The surface areas (A_0) were determined by planimetry and include the islands (Sec. 8.4). The area of Basin I is relatively small (91 sq km) but the size of the Basins greatly increases as one progresses towards the dam. Basins III and IV are greater than 2000 sq km each. The total area of the lake-proper is 5 364 sq km at the 485 m asl reference level.

d. The lengths of the shoreline (L_0) were measured on the maps by means of a curvimeter, two independent measurements being made and their mean value recorded. The perimeter totals 2 164 km of which 77 percent are found around Basins III and IV, with different fractions in Zambia and in Rhodesia (Table 25).

MITCHELL (1970) after qualifying the shore lines as exposed, sheltered or estuarine, has estimated their relative importance in each Basin (Table 24). In Basin I most of the shore is sheltered. In the other Basins the percentages are more similar to each other: exposed shores are slightly more common in Basins II and III than sheltered ones, whereas in Basin IV the reverse occurs. On the whole estuarine shores make up about 30 percent of the total shore line, 37 percent are sheltered shores, and 33 percent are exposed.

e. The developments of the shore line (D_L) or indices of sinuosity (TONOLLI, 1969) have been calculated as the ratios of the shoreline lengths to the lengths of the circumferences of circles of areas equal to those of the regions considered as follows:

$$D_L = L_0(2\sqrt{\pi A_0})^{-1}$$

The relatively high values obtained (Table 25) are characteristic of dendritic lakes

Table 26. Length of depth contours for the 485 m lake level, in kilometres.

Altitude a.s.l., metres	485	476	460	445	430	415
Depth contour, metres	0	10	25	40	55	70
Basin I	141	106	55	—	—	—
Basin II	355	289	290	241	—	—
Basin III	704	690	554	502	190	—
Basin IV	964	939	920	877	422	359
Lake Kariba	2164	2024	1819	1620	612	359

Table 27. L. Kariba. Estimates for surface areas and capacities from hypsographic curves.

Water level, a.s.l.		Lake surface area	Lake volume	Hydroelectric scheme
m	ft	sq km	cu km	
489	1605	5820	178.0	Max. flood storage
488	1600	5630	170.4	—
485	1590	5364	156.5	Normal operating level
482	1580	5000	141.3	—
479	1570	4660	124.1	—
476	1560	4325	108.2	Lowest drawdown

with great irregularity of shore line, numerous embayments and shore projections. The Lake Kariba value ($D_L = 8.3$) is greater than any of the values mentioned by HUTCHINSON (1957) for large natural lakes, the maximum D_L reported being 5.5 for fjord lake Salsvatn. No doubt that elongation combines in Lake Kariba with sinuosity of shore line to give such high values.

8.2.2. PLANIMETRY AT VARIOUS WATER LEVELS

Parameters defining the lake planimetry at various water depths (z) below the reference surface level of 485 m asl have been estimated.
a. The lengths of the major depth contours (L_z) shown on the bathymetric map (Fig. 20) have been measured with a curvimeter (Table 26). These measurements have been used to calculate mean slopes (Sec. 8.3.1 f).
b. The areas of the major depth contours (A_z) have been estimated using a polar planimeter. On the basis of the results the hypsographic curve for the lake surface area has been plotted (Fig. 21) from which the lake surface area at various water levels has been obtained (Table 27). Although at maximum storage level (489 m) Lake Kariba covers 5 820 sq km, at the lowest drawdown level (476 m) its waters extend only over 4 325 sq km, a difference of 1 495 sq km (26 percent).

Such considerations have led to the calculation of the areas of lake floor exposed as the water level fluctuates around the normal operating level (Table 28). As the

Table 28. Area of lake floor exposed as the water level varies around 485 m.

Water level asl		Level difference	Change in area of lake floor
m	ft	m	sq km
489	1 605	+4	+456
488	1 600	+3	+266
485	1 590	0	
482	1 580	−3	−364
479	1 570	−6	−704
476	1 560	−9	−1 039

lake level drops a considerable area of lake floor is exposed with all the resulting consequences (rapid oxidation of the organic matter, destruction of the fish nests, drying up of the aquatic vegetation, etc). As much as 10 percent of the total lake area becomes affected by a drop of only five metres under the normal operating water level.
c. In order still more efficiently to represent the shape of the various Basins of Lake Kariba, the areas of the successive isobaths have been expressed in percent of the surface area at the 485 m level (A_0). The relative hypsographic curve for surface area (TONOLLI, 1969) has then been plotted for each separate Basin and for the lake as a whole (Fig. 22). The flatter the curve on the abscisse, the smaller the mean slope of the lake floor. Comparing the different Basins it can be seen that the floor slope becomes steeper in average as one progresses from Basin I to the dam.
d. Relative areas of depth contours ($RA_z = A_z \cdot A_0^{-1}$) have been estimated from Figure 22 for 5-m depth intervals (Table 29). Differences between successive values (ΔRA) give an idea of the average floor slope within each water stratum.
e. Areas of depth contours (A_z) at 5-m intervals have been obtained by simple multiplication of RA_z by A_0 (Table 30).

8.3. *Bathymetry of Lake Kariba*

8.3.1. Bathymetry at the 485-m water level (Table 24)

a. The volumes (V_0) have been evaluated by summing the volumes of strata 10 to 15-m thick based on the areal data available from the bathymetric map (Fig. 20). Strata volumes between isobaths 1 and 2 were estimated using Simpson's formula:

$$V_{1-2} = \frac{z_2 - z_1}{3}(A_1 + A_2 + \sqrt{A_1 A_2})$$

The total water volume of the lake is about 156 cu km. The volumes of Basins I (in particular) and II are relatively small while in Basins III and IV the volume reaches 54 and 85 cu km respectively. Volumes between the water surface and horizontal planes of varying depth are given in Table 31.
b. The mean depths ($\bar{z}$) obtained as the ratios of volumes to corresponding areas vary

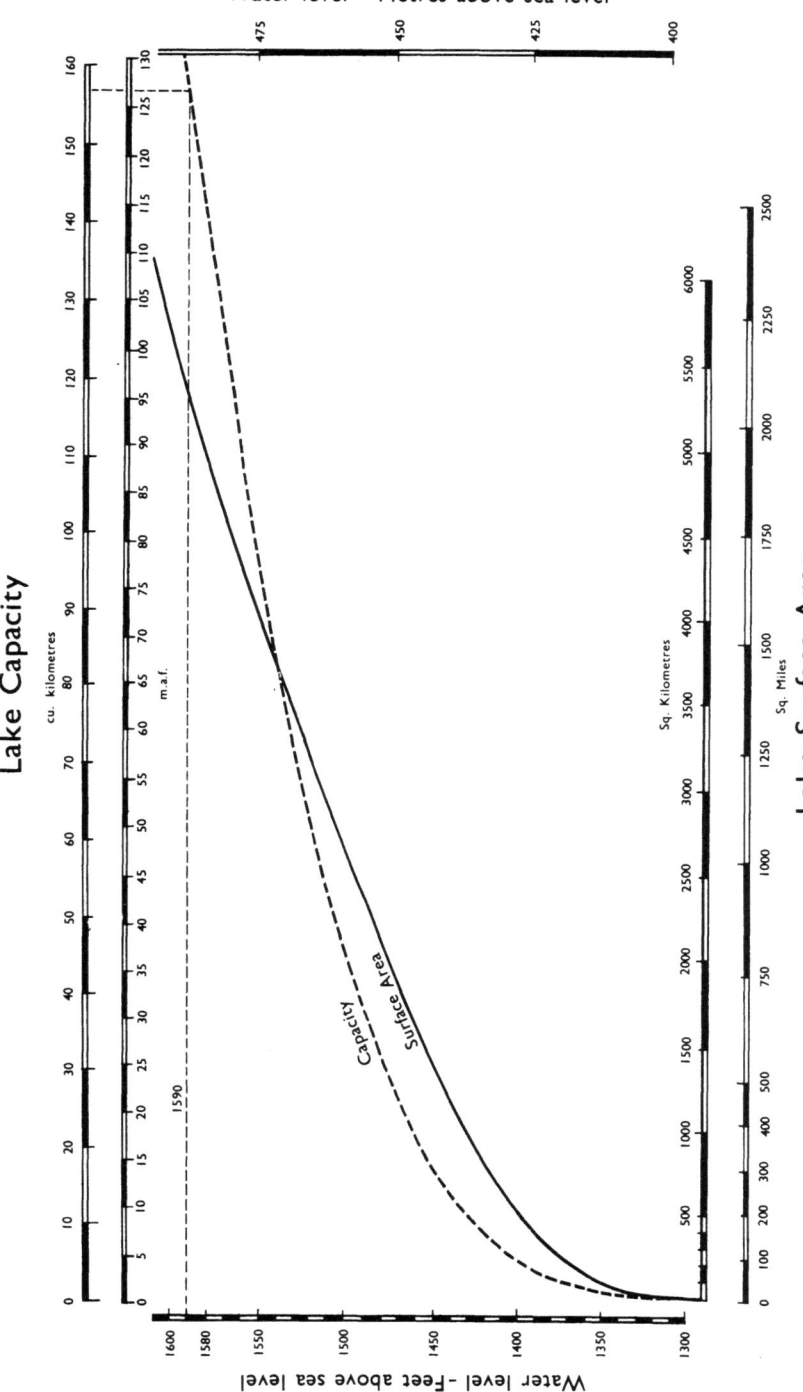

Fig. 21. Hypsographic curves for the lake surface area and volume.

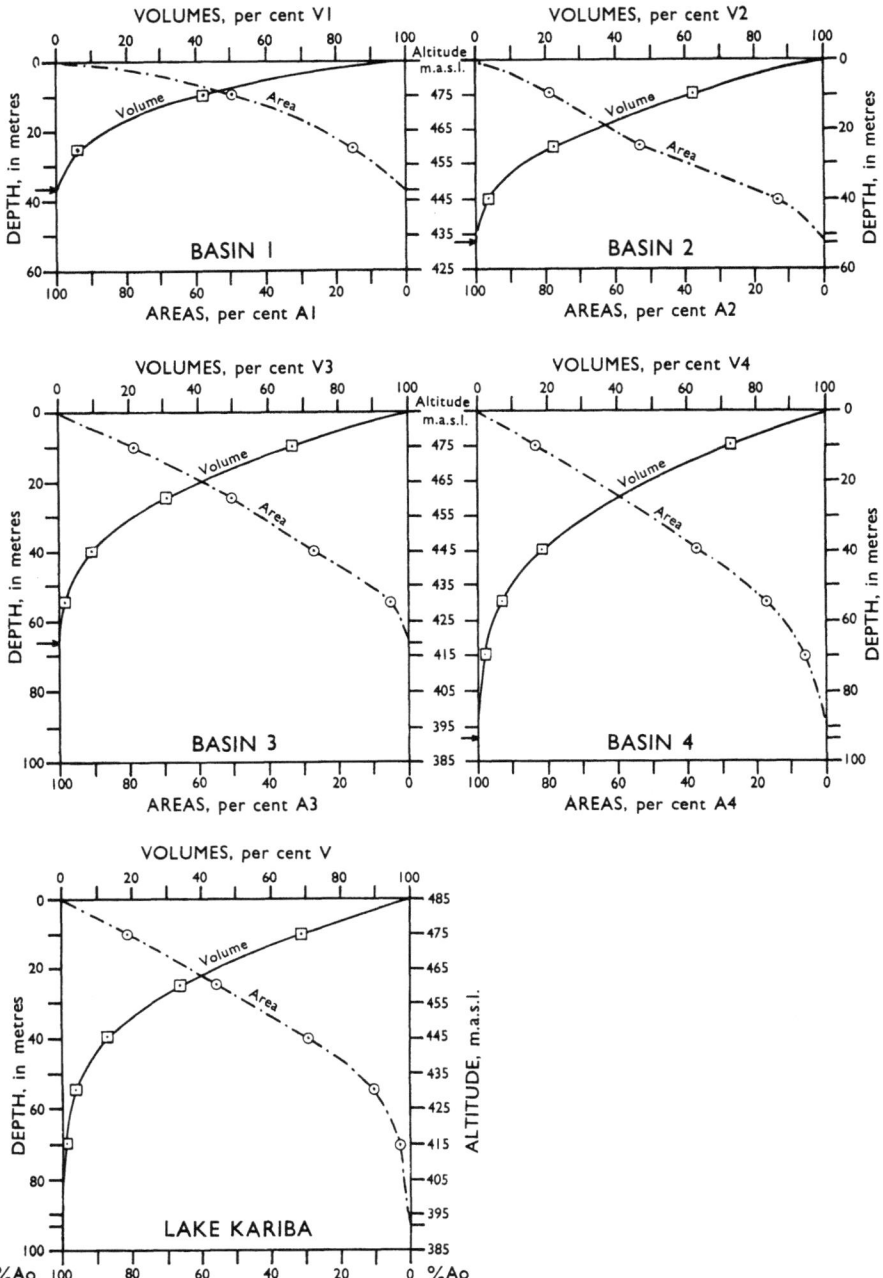

Fig. 22. Relative hypsographic curves for surface area and volume of Lake Kariba and its four Basins.

Table 29. Relative areas of depth contours (RA) for $A_o = 485$ m asl.

Depth m	Basin I RA	ΔRA	Basin II RA	ΔRA	Basin III RA	ΔRA	Basin IV RA	ΔRA
0	1.000		1.000		1.000		1.000	
		0.350		0.123		0.121		0.086
5	0.650		0.877		0.879		0.914	
		0.166		0.084		0.095		0.077
10	0.484		0.793		0.784		0.837	
		0.141		0.105		0.112		0.087
15	0.345		0.688		0.672		0.750	
		0.105		0.088		0.082		0.081
20	0.240		0.600		0.590		0.669	
		0.086		0.076		0.084		0.072
25	0.154		0.524		0.506		0.597	
		0.054		0.124		0.079		0.077
30	0.100		0.400		0.427		0.520	
		0.086		0.153		0.084		0.080
35	0.014		0.247		0.343		0.440	
		0.014		0.116		0.074		0.075
40	0.000		0.131		0.269		0.365	
				0.065		0.077		0.075
45			0.068		0.192		0.290	
				0.053		0.074		0.058
50			0.015		0.118		0.232	
				0.015		0.073		0.063
55			0.000		0.045		0.169	
						0.030		0.039
60					0.015		0.130	
						0.011		0.035
65					0.004		0.095	
						0.004		0.033
70					0.000		0.062	
								0.021
75							0.041	
								0.015
80							0.026	
								0.010
85							0.016	
								0.010
90							0.006	
								0.006
93							0.000	
Total	1.000		1.000		1.000		1.000	
Max. depth, m	37		52		66		93	
Mean depth, m	12.58		24.01		26.54		33.22	

Table 30. Areas of depth contours at lake level 485 m, in sq. km.

Depth contour m	Basin				Lake
	I	II	III	IV	
0	91.00	677.00	2,033.00	2,563.00	5,364.00
5	59.15	593.73	1,787.01	2,342.58	4,782.47
10	44.00	537.00	1,539.00	2,146.00	4,320.00
15	31.40	465.78	1,366.18	1,922.25	3,785.61
20	21.84	406.20	1,199.47	1,714.65	3,342.16
25	14.00	355.00	1,028.00	1,530.00	2,927.00
30	9.10	270.80	868.09	1,332.76	2,480.75
35	1.27	167.22	697.32	1,127.72	1,993.53
40	0.00	89.00	546.00	935.00	1,570.00
45		46.04	390.34	743.27	1,179.65
50		10.16	239.89	594.62	844.67
55		0.00	91.00	433.00	524.00
60			30.50	333.19	363.70
65			8.13	243.49	251.62
70			0.00	159.00	159.00
75				105.08	105.08
80				66.64	66.64
85				41.01	41.01
90				15.38	15.38
93				0.00	0.00

from 12.6 m to 33.2 m, increasing towards the dam. The mean depth of the entire lake is 29.2 m.

c. The maximum depths (z_m) increase from 37 m in Basin I to 52, 66, and 93 m in the following Basins. There are of course a few deeper 'holes' in each Basin.

d. The developments of volume (D_v) express the form of the Basins. According to one definition D_v relates the volume of the Basin to that of a cone of basal area A_0 and height z_m such as:

$$D_v = 3\,\bar{z}\,.\,z_m^{-1}$$

Therefore it is thought (HUTCHINSON, 1957) that the ratio of mean to maximum depth provides a similarly good measure of D_v. NEUMANN (1959) reviewing the values for these two last depths in numerous lakes, concluded that on the average their ratio was 0.467. The ideal general shape of a lake basin therefore was an elliptic sinusoid ($\bar{z}\,.\,z_m^{-1} = 0.464$). In contrast ratios for Lake Kariba suggest that in general the Basin depressions have a conical shape ($\bar{z}\,.\,z_m^{-1} = 0.33$) except in Basin II.

e. The relative depths (z_r) defined on a percentage basis as $z_r = 50.\,z_m.\,\sqrt{\pi}.(\sqrt{A_0})^{-1}$ vary in most of the lake around 0.15.

f. The mean slopes ($\bar{S}$) expressing the proximity of the depth contours to one another (REID, 1961) have been estimated in percent as:

Table 31. Volumes between the water surface (level 485 m) and horizontal planes of varying depth, in cu. km.

Depth, z	10 m	25 m	40 m	55 m	70 m	100 m
Basin I	0.661	1.075	1.145	—	—	—
II	6.057	12.700	15.809	16.254	—	—
III	18.085	37.588	49.204	53.503	53.958	—
IV	23.514	50.954	69.259	79.280	83.552	85.142
Lake Kariba	48.317	102.317	135.417	150.182	154.909	156.499

$$\bar{S} = \frac{1}{n} \cdot \left(\frac{1}{2} L_0 + L_1 + L_2 \ldots + L_{n-1} + \frac{1}{2} L_n \right) \frac{z_m}{A_0}$$

where n is the number of depth contours.

The corresponding angle of the mean slope with the horizontal (ε) is obtained from $\bar{S} = $ tang ε. Values for $\bar{S}$ vary between 1.4 and 2.8 percent for Lake Kariba as a whole with the mean slope being the steepest in Basin I.

8.3.2. BATHYMETRY AT VARIOUS WATER LEVELS

a. Strata volumes between major isobaths (Sec. 8.3.1 a) have been used to calculate water volumes present in Lake Kariba at various lake surface levels. These results have been plotted to provide the hypsographic curve for volume (capacity – Fig. 21) from which the lake volume at major water levels has been obtained (Table 27). At maximum storage level Lake Kariba contains 178 cu km of water, whereas at lowest drawdown its capacity decreases to 108 cu km, a difference of 70 cu km.

b. A similar treatment of basic volume data for the various Basins followed by their expression in percentage of the volumes at 485 m level (V_0), have provided the bases for plotting the relative hypsographic curves for volumes (TONOLLI, 1969) for each separate Basin and for the lake as a whole (Fig. 22).

c. Relative volumes (RV = $V_z \cdot V_0^{-1}$) have been estimated from Figure 22 at 5-m depth intervals (Table 32). The differences between each two successive values provide the Partial Volume Ratios (PVR) for water strata 5-m thick at the 485 m surface level, to be used in later calculations.

d. Volumes of water at depth contours (V_z)[9] have been calculated by simple multiplication of RV_z by V_0 (Table 33).

e. Reduced Thickness (RT cm) was introduced by BIRGE (1916) as 'the thickness of any given stratum if its area is made equal to that of the lake and its sides are vertical.' It is computed as follows:

$$RT = (Vol. stratum) \cdot (A_0)^{-1} = PVR \cdot \bar{z}$$

---

[9] V_z in our mind represents the volume of water present between the horizontal plane of depth z and the lake floor. HUTCHINSON's V_z (1957) refers to the volumes of water between the surface and the plane of depth z.

Table 32. Relative Volumes (RV) and Partial Volume Ratios (PVR) at 485-m lake level.

Depth m	Basin I RV	Basin I PVR	Basin II RV	Basin II PVR	Basin III RV	Basin III PVR	Basin IV RV	Basin IV PVR
0	1.000		1.000		1.000		1.000	
5	0.645	0.355	0.788	0.212	0.811	0.189	0.856	0.144
10	0.423	0.222	0.627	0.161	0.665	0.146	0.724	0.132
15	0.260	0.163	0.486	0.141	0.532	0.133	0.600	0.124
20	0.147	0.113	0.347	0.139	0.410	0.122	0.500	0.100
25	0.061	0.086	0.219	0.128	0.303	0.107	0.402	0.098
30	0.028	0.033	0.138	0.081	0.225	0.078	0.325	0.077
35	0.008	0.020	0.072	0.066	0.150	0.075	0.250	0.075
40	0.000	0.008	0.027	0.045	0.088	0.062	0.187	0.063
45			0.012	0.015	0.048	0.040	0.139	0.048
50			0.004	0.008	0.024	0.024	0.100	0.039
55			0.000	0.004	0.008	0.016	0.069	0.031
60					0.005	0.003	0.046	0.023
65					0.002	0.003	0.030	0.016
70					0.000	0.002	0.019	0.011
75							0.010	0.009
80							0.007	0.003
85							0.004	0.003
90							0.001	0.003
93							0.000	0.001
Total		1.000		1.000		1.000		1.000
Max. depth, m	37		52		66		93	
Mean depth, m	12.58		24.01		26.54		33.22	

Table 33. Volumes of water (cu.km) at depth contours for the 485-m lake level.

Depth m	Basin I	Basin II	Basin III	Basin IV	Total lake
0	1.145	16.254	53.958	85.142	156.499
5	0.739	12.808	43.760	72.882	130.189
10	0.484	10.197	35.873	61.628	108.182
15	0.298	7.899	28.706	51.085	87.988
20	0.168	5.640	22.123	42.571	70.502
25	0.070	3.554	16.370	34.188	54.182
30	0.032	2.243	12.141	27.671	42.087
35	0.009	1.170	8.094	21.286	30.559
40	0.000	0.445	4.754	15.883	21.082
45		0.195	2.590	11.835	14.620
50		0.065	1.295	8.514	9.874
55		0.000	0.455	5.862	6.317
60			0.270	3.917	4.187
65			0.108	2.554	2.662
70			0.000	1.590	1.590
75				0.851	0.851
80				0.596	0.596
85				0.341	0.341
90				0.085	0.085
93				0.000	0.000

Table 34. Reduced thickness (RT) and work constant (WC)[1] for the 485-m lake level, in each Basin considered as a separate entity.

Stratum m	Basin I		Basin II		Basin III		Basin IV	
	RT cm	WC cm^2	RT cm	WC cm^2	RT cm	WC cm^2	RT cm	WC cm^2
0– 5	446.59	111648	509.01	127253	501.61	125402	478.37	119593
5–10	279.28	209460	386.56	289920	387.48	290610	438.50	328875
10–15	205.05	256313	338.54	423175	352.98	441225	411.93	514913
15–20	142.15	248763	333.74	584045	323.79	566633	332.20	581350
20–25	108.19	243428	307.33	691493	283.98	638955	325.56	732510
25–30	41.51	114153	194.48	534820	207.01	569278	255.79	703423
30–35	25.16	81770	158.47	515028	199.05	646913	249.15	809738
35–40	10.06	36216	108.05	405188	164.55	617063	209.29	784838
40–45			36.02	153085	106.16	451180	159.46	677705
45–50			19.21	91248	63.70	302575	129.56	615410
50–55			9.60	48960	42.46	222915	102.98	540645
55–60					7.96	45770	76.41	439358
60–65					7.96	49750	53.15	332188
65–70					5.31	34780	36.54	246645
70–75							29.90	216775
75–80							9.97	77268
80–85							9.97	82253
85–90							9.97	87238
90–93							3.32	30378

[1] WC = RT multiplied by the distance in cm from the lake surface to the middle of the stratum.

Table 35. Areas of depth zones in Lake Kariba at normal operating water level (485 m), Basin 0 excluded.

Basin	0–10 m sq.km	0–10 m %	10–25 m sq.km	10–25 m %	25–40 m sq.km	25–40 m %	40–55 m sq.km	40–55 m %	55–70 m sq.km	55–70 m %	70 m + over sq.km	70 m + over %
I	47	4.5	30	2.2	14	1.0	—	0.0	—	0.0	—	0.0
II	140	13.4	182	13.0	266	19.6	89	8.5	—	0.0	—	0.0
III	440	42.2	565	40.6	482	35.5	455	43.5	91	24.9	—	0.0
IV	417	39.9	616	44.2	595	43.9	502	48.0	274	75.1	159	100
Lake	1044		1393		1357		1046		365		159	

Depth zone	Basin I sq.km	Basin I %	Basin II sq.km	Basin II %	Basin III sq.km	Basin III %	Basin IV sq.km	Basin IV %	Lake sq.km	Lake %
0–10 m	47	51.6	140	20.7	440	21.6	417	16.3	1044	19.5
10–25 m	30	33.0	182	26.9	565	27.8	616	24.0	1393	26.0
25–40 m	14	15.4	266	39.3	482	23.7	595	23.2	1357	25.2
40–55 m	—	0.0	89	13.1	455	22.4	502	19.6	1046	19.5
55–70 m	—	0.0	—	0.0	91	4.5	274	10.7	365	6.8
70 m + over	—	0.0	—	0.0	—	0.0	159	6.2	159	3.0

Table 36. Volumes of depth zones in Lake Kariba[1] at normal operating water level (485 m asl), Basin 0 excluded.

Depth zone	0–10 m		10–25 m		25–40 m		40–55 m		55–70 m		70 m + over	
Basin	cu.km	%	cu.km	%	cu.km	%	cu.km	%	cu.km	%	cu.km	%
I	0.661	1.4	0.414	0.8	0.070	0.2	—	—	—	—	—	—
II	6.057	12.5	6.643	12.3	3.109	9.4	0.445	3.0	—	—	—	—
III	18.085	37.4	19.503	36.1	11.616	35.1	4.299	29.1	0.455	9.6	—	—
IV	23.514	48.7	27.440	50.8	18.305	55.3	10.021	67.7	4.272	90.4	1.590	100.0
Lake	48.317		54.000		33.100		14.765		4.727		1.590	

Basin	I		II		III		IV		Lake	
Depth zone	cu.km	%	cu.km	%	cu.km	%	cu.km	%	cu.km	%
0–10 m	0.661	57.7	6.057	37.3	18.085	33.6	23.514	27.6	48.317	30.9
10–25 m	0.414	36.2	6.643	40.9	19.503	36.1	27.440	32.2	54.000	34.5
25–40 m	0.070	6.1	3.109	19.1	11.616	21.5	18.305	21.5	33.100	21.2
40–55 m	—	—	0.445	2.7	4.299	8.0	10.021	11.8	14.765	9.4
55–70 m	—	—	—	—	0.455	0.8	4.272	5.0	4.727	3.0
70 m + over	—	—	—	—	—	—	1.590	1.9	1.590	1.0
Total volumes	1.145		16.254		53.958		85.142		156.499	

[1] Acc. Formula (3) in HUTCHINSON, 1957, p. 166.

Table 37. Islands in L. Kariba, surface areas in sq. km.

Basin	Zambia	Rhodesia	Total
I	2.77	2.76	5.53
II	37.82	8.64	46.46
III	14.57	14.77	29.34
IV	17.46	48.08	65.54
Total	72.62	74.25	146.87

Values have been calculated for strata 5-m thick and for each of the four Basins (Table 34) to be used later in the computation of the work of the wind (Sec. 11.2.5). For the same purpose the constant (WC) which 'states the work that would be done in warming the stratum by mixture if the water density were reduced to zero' (BIRGE, 1916) has been calculated for each 5-m thick stratum as the product of the stratum's RT by the distance from the lake's surface to the middle of that stratum (Table 34).

8.3.3. AREAS AND VOLUMES OF DEPTH ZONES

On the basis of the data described previously the areas and the volumes of the major depth zones (Fig. 20) have been computed because of their usefulness in assessing biological productions. The normal operating water level (485 m asl) has been taken again as the reference level (Tables 35 and 36). The potentially most productive zone (0–10 m) represents 19.5 percent (1044 sq km) of the surface area and 30.9 percent of the total lake volume. Regions deeper than 55 m are very little represented while the best represented zone is the one comprised between the 10 and 25-m isobaths.

8.4. Lake Kariba islands

The islands created in Lake Kariba have their origin mainly in geological faulting (Sec. 2.3). As a result they are found in chains or lines especially south-west of the Chete Gorge and in the Sibilobilo area (Fig. 23). Their shores are generally steep. The majority are covered with sandstone boulders and solid outcrops of sandstones. Sandy Karroo soils predominate with some Kalahari sands in upper reaches and clayed soils from shales near Kariba. In the Sibilobilo area basalt is predominant (Sec. 2.3). ROBERTS, MULLINS & BARNETT (1960) have studied in detail the islands, including size and development potential. A map was drawn (drawing KCC/160 – April 1960) with each island being given a number in separate series for Zambia (N-) and Rhodesia (S-), starting at the Devil's Gorge end. The most important islands are shown in Fig. 20.

There are 103 islands in Zambia and 190 in Rhodesia (Table 24). Their total surface area is some 147 sq km which corresponds to an insulosity of 2.7 percent. MITCHELL (1970) has estimated that their shore line totals 604 km. The largest island is Chete Island (N21) 2637 hectares in size.

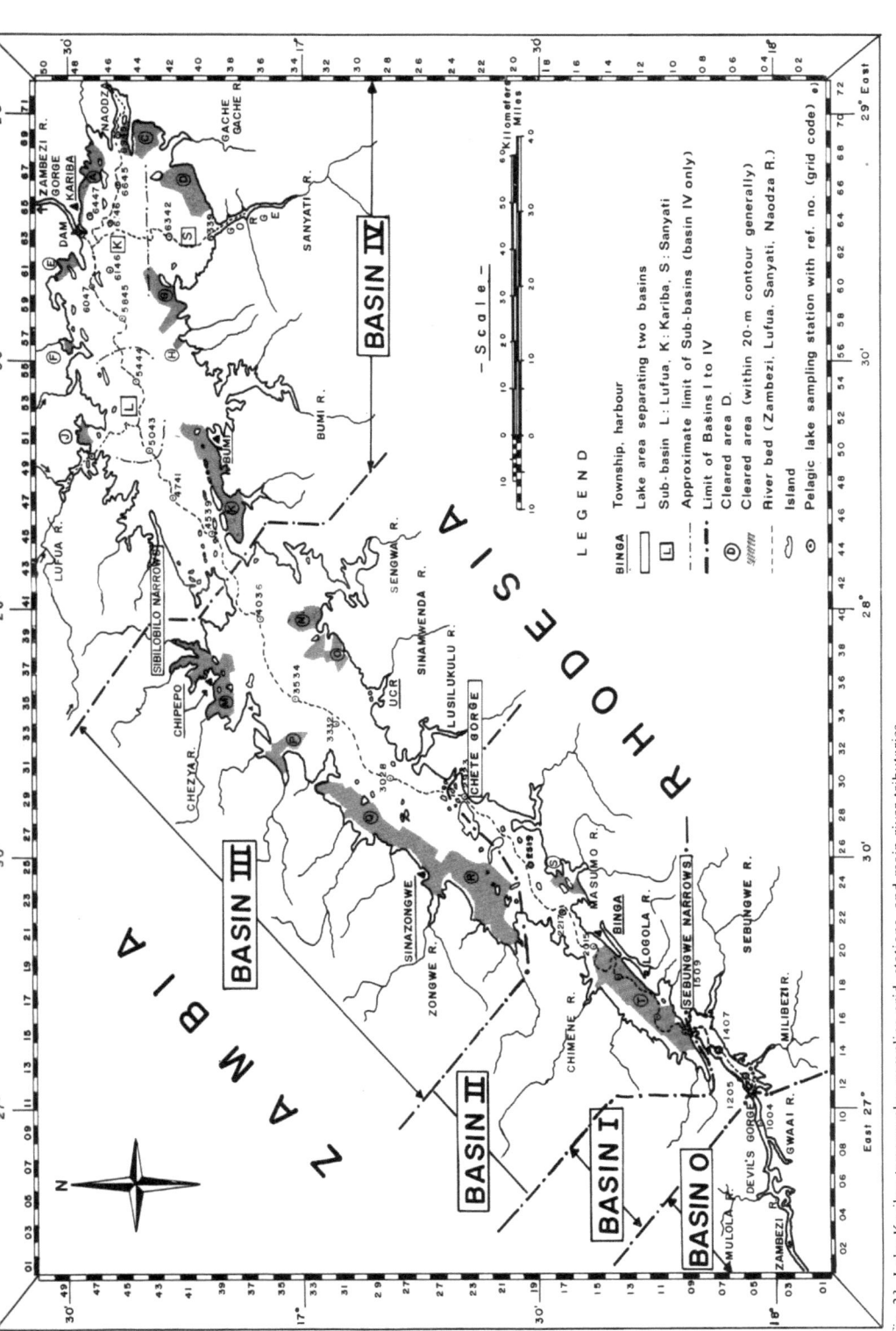

Fig. 23. Lake Kariba: topography, sampling grid, stations, and main river tributaries.

Insulosity differs by Basin considered (Tables 24 and 37) with the greatest in Basin II (6.9 per cent) where the surface area of the islands adds up to 46.5 sq km. Most of these islands are grouped along a line forming the limit between Basins II and III.

8.5. Bush-cleared fishing grounds

As previously stated, for fishing purposes a total of about 954 sq km (17.8 percent A_0) was cleared of woody vegetation before inundation (Table 8). Between the time of clearing and inundation some regeneration of the vegetation took place (Table 9). These cleared areas (Table 38) are distributed all around the lake (Fig. 23) where they lie within the 20-m isobath below maximum water level. The largest cleared area (Sinazongwe Q and R) covers 281 sq km, the second largest (Sibilobilo K) being 104 sq km in size.

There has existed some doubts about the wiseness of bush-clearing such large lake floor areas for fishery development following the poor fish catches experienced in these areas. But recently DONNELLY (1969) surveying *Tilapia* and *Sarotherodon* (Pisces) nurseries in Lake Kariba has observed that the cleared areas provide a greater diversity of aquatic plant life than uncleared areas. *Salvinia* tends to form broad mats in the latter, finding anchorage among the emergent trees and shrubs. Submerged aquatic growth is inhibited by shading which exclude such regions as potential fish nurseries. On the contrary in the bush-cleared areas well colonized by submerged vegetation juvenile *Tilapia* are plentiful as well as benthic organisms.

There has been the tendency to rather recommend strip-clearing both for navigation and fishing purposes, oriented from shallow to deep waters. In Lake Kossou, Ivory Coast where this approach has been taken (Centre Technique Forestier Tropical, 1970) the future will show whether such method has decisive advantages over the prior one and whether it could still be improved by varying the orientation of the cleared strips and changing their wideness actually fixed at 100 m.

8.6. Morphology of Lake Kariba

Morphological observations were started in February 1964 by G. BOND (1965) during a cruise on Lake Kariba along the Rhodesian shore. Main points of interest were erosion and deposition phenomena, shaping of the shore line and beach formation being related to geological features.

In Rhodesia stony beaches were much more common than sandy beaches. The latter were only developing extensively on the soft forest sandstones, in particular between the Sanyati and Bumi rivers. Sandy beaches were also developing on interbedded basalts and sandstones, and dip slopes on Molteno sandstones. In contrast armored shores resulted from wave action on Molteno (scarp slopes) and Basement Complex (gneiss) rocks, with very little further erosion taking place after the washing away of the top soil.

8.7. Comparative morphometry

Morphometric data for major African reservoirs are summarized in Table 38. For

Table 38. Morphometry of major reservoirs and of some natural lakes in Africa[1].

Lake	Alt. m asl	Type of basin	l km	$\bar{b}$ km	A_o sq.km	V_o cu.km	z_m m	$\bar{z}$ m	z_r %	L_o km	D_L	D_V $3\bar{z}/z_m$	D_V $\bar{z}/z_m$
Kariba	485	Artificial dam	277	19.4	5 364	156	93	29.2	0.11	2.164	8.3	0.93	0.31
Kainji	142	Artificial dam	137	9.3	1 280	15.8	50	12.3	0.12	720	5.6	0.74	0.23
Nasser-Nubia	(182)	Artificial dam	482	12.9	6 222	157	90	25.2	0.10	8.803	31.5	0.84	0.28
Volta	85	Artificial dam	1 296	6.8	8 845	165	75	18.6	0.07	5.271	15.8	0.74	0.24
Cabora Bassa	?	Artificial dam	250	11.0	2 739	69	157	25.2	0.27	2.000	10.7	0.48	0.16
Kossou	204	Artificial dam	180	8.9	1 600	29.5	54	18.4	0.12	3.500	24.7	1.02	0.34
Tanganyika	773	tectonic, graben	650	52.3	34 000	18 940	1 470	572	0.70	1.900	2.9	1.17	0.39
Malawi	472	tectonic, graben	560	55.0	30 800	8 400	706	273	0.36	1.500	2.4	1.47	0.49
Victoria	1 135	tectonic, epirogen	403	170.7	68 800	2 700	79	40	0.027	3.440	3.7	1.53	0.51
Chad	283	? tectonic ?	270	61.1	16 500	40	12	4	0.008	700	1.6	1.00	0.33

[1] For additional data relative to the African reservoirs, see Annex I.

comparison purposes data are also given on four natural water bodies: Lake Tanganyika, L. Malawi, L. Victoria, and L. Chad. Major African reservoirs are more completely characterized in Annex I against their general background.

On the basis of surface area and volume, Lake Volta is by far the largest of the African reservoirs. L. Kariba comes third although its water volume is practically the same as the volume of L. Nasser/Nubia will be, when full.

The three other reservoirs (L. Kainji, Cabora Bassa, and Kossou) are much smaller. The deepest of all will be L. Cabora Bassa ($z_m = 157$ m) although its mean depth will be similar to L. Nasser's and shallower than L. Kariba's. Mean breadth is generally small compared to natural lakes, L. Kariba and L. Volta being relatively the widest in average among the African reservoirs.

The length of the shore line (L_0) as well as its index of development (D_L) differ greatly among the reservoirs. The differences reflect mainly the elongation of the lake (e.g. L. Nasser/Nubia) as well as the highly dendritic nature of their shores (e.g. L. Volta and L. Kossou). There is a marked difference with the D_L from natural lakes which present much less tormented shorelines in general.

9. SAMPLING METHODOLOGY

9.1. Sampling scheme

Two annual cycles have been studied during nine lake cruises which took place according to the following schedule:

A. Cycle 1965–1966
    K4/65 : 22 April to 1 May 1965
    K6/65 : 22 June to 2 July 1965
    K8/65 : 3 to 12 August 1965
    K10/65: 27 October to 4 November 1965
    K1/66 : 12 to 20 January 1966

B. Cycle 1968–1969
    K3/68 : 28 March to 4 April 1968
    K7/68 : 12 to 19 July 1968
    K10/68: 15 to 22 October 1968
    K2/69 : 11 to 17 February 1969

During these cruises one of the research objectives was to sample the longitudinal axis of Lake Kariba within as short a time as possible. In 1965/66 sampling was done at stations of 10-km intervals, between the eastern end (Naodza region) and the Devil's Gorge which took up to four days. On the basis of the results obtained initially (COCHE, 1968) the sampling frequency along the axis was reduced subsequently to 11 key stations in 1968/69 which then permitted a run that could be completed in one or two days.

Further data were obtained in 1965–66 during monthly cruises within Basin IV and at routine stations (e.g. Boom Sta. 6348B; Sampakaruma Sta. 6047). In later years (1970 and 1971) additional data were taken on occasional lake cruises within the Zambian territory.

Data from these numerous stations have been classified on an area-and-time basis. For this purpose the lake surface has been subdivided into square areas (2' lat. by 2' long.) using a numbered grid (Fig. 23). Each sampling station is thus identified by a four-digit code number, the first pair of digits referring to the longitudinal position (grid abscissa) and the second one to the latitudinal position (grid ordinate). For example key stations in 1968/69 were: 1004 (Zambezi R.), 1407, 1611, 2015, 2519, 2926, 3332, 4036, 4741, 5344 and 5845.

9.2. Sampling the water optical properties

9.2.1. APPARENT COLOUR OF THE LAKE

The apparent colour of the lake was determined comparing a FOREL-ULE standard scale (HUTCHINSON, 1957, Table 45) with the colour of the Secchi-disc white quadrants

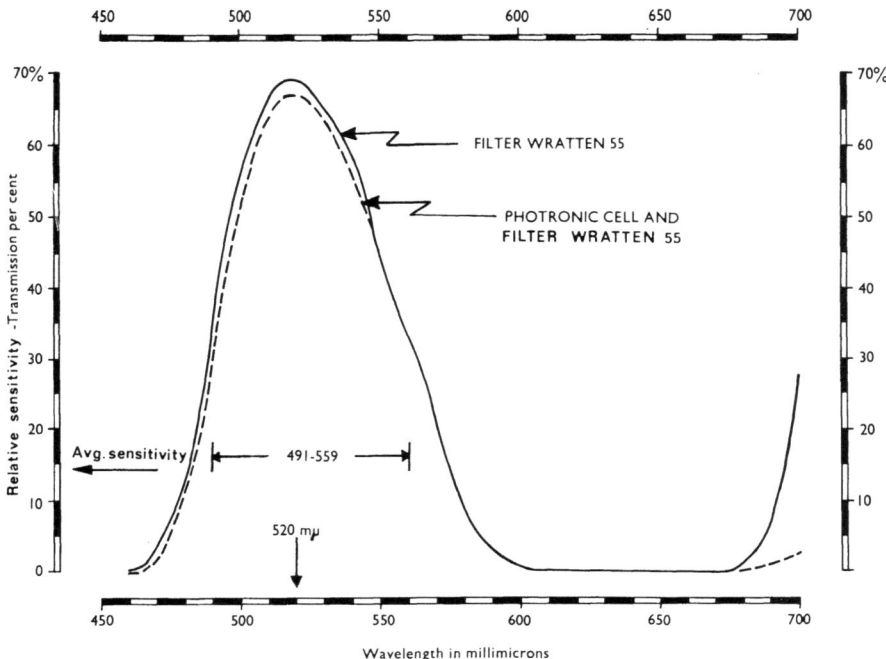

Fig. 24. Optical properties of the photometer used in 1965–1966.

held one meter under the water surface. This empirical method has the following correspondences:

I–II	III–IV	V–VII	VIII–X	XI–XV	XVI–XIX	XX–XXII
Blue	Greenish blue	Bluish green	Green	Greenish yellow	Yellow	Brown

9.2.2. Depth of visibility

The depth of visibility (DV) was measured at each sampling station with a Secchi disc (black and white quadrants) of 30-cm diameter under standard conditions as far as possible (WELCH, 1948). The size of the disc was chosen according to the estimated values of DV (mostly less than 17 m) following JOSEPH (cited in SAUBERER, 1962).

9.2.3. Light attenuation underwater

In 1965/66 the relative light intensity (T_z^{520}) at various depths, expressed in percent of the surface light, was directly measured with a Photometer (Ocean Research Equipment USA), equipped with a barrier-layer photoelectric cell (Photronic, Weston) and a green filter (Wratten 55). The transmission curve of the filter and the spectral sensitivity of the photocell (manufacturer's data) have combined to determine the

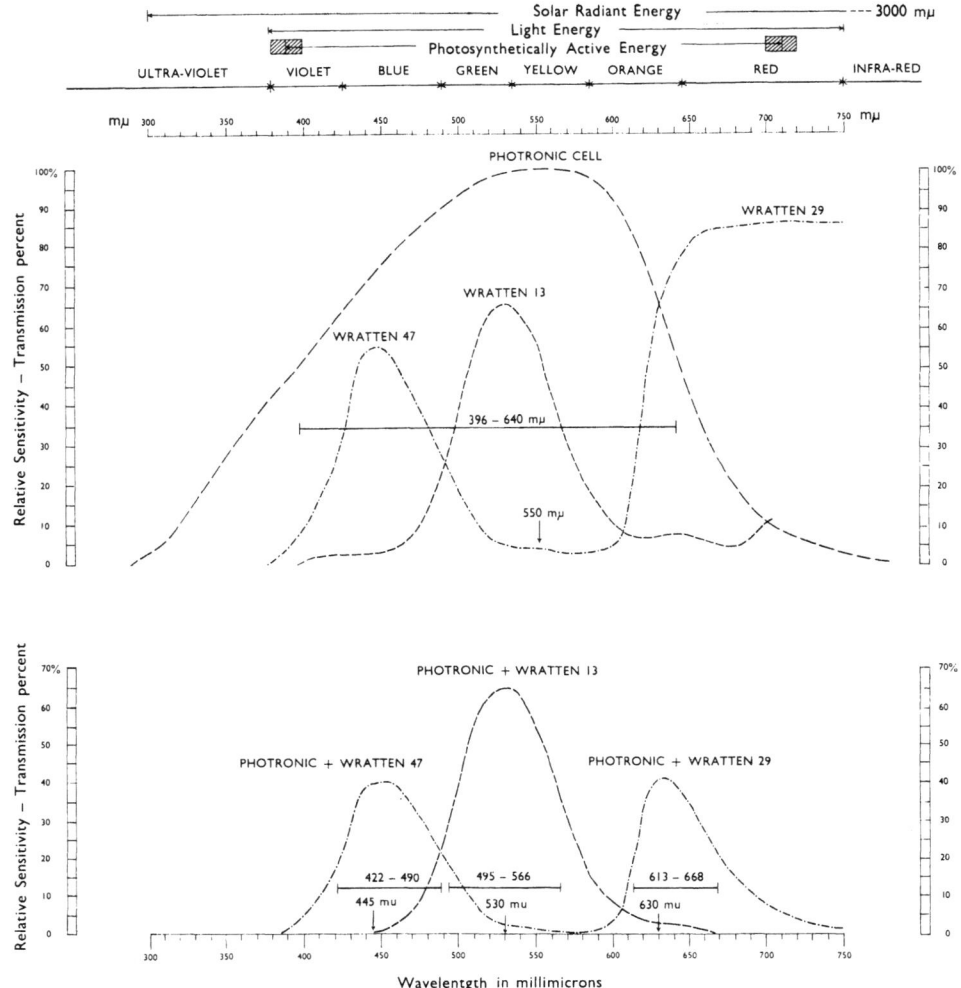

Fig. 25. Optical properties of the photometer and filters used since 1968.

spectral sensitivity of the photometer in sun-and-sky light (Fig. 24). Optical parameters (WESTLAKE, 1965) have thus been obtained (Table 39) without taking into account the colour filter action of the overlying water.

Since 1968 light intensity at various depths was measured with a submarine photometer (mod. 15MO4, G.M. Mfg. U.S.A.) equipped with a barrier-layer photoelectric cell (Photronic, Weston) situated under a cosine filter and a pressure plate. Blue, green and red filters (Wratten 47, 13 and 29) were added in turn on some occasions to assess the spectral composition of the light at various depths. Transmission curves of the filter and spectral sensitivity of the Photronic cell were combined to determine the spectral sensitivity of the equipment in each particular case (Fig. 25). The appropriate optical parameters have been calculated (Table 39) as above.

During the first year of study it was decided to make only single series of measurements and, following STRICKLAND's (1958) advice, a green filter with maximum transmission around 520 millimicrons was chosen. Later when more time became available for light intensity measurements, they were made in quadruple series. The total amount of radiant energy (no filter) was first measured (STRICKLAND, 1958) and then followed immediately by successive determinations with the three spectral wavebands (filters) recommended by RUTTNER (1963; p. 19) as 'sufficient to provide an indication of the optical properties of a lake'. Assuming that the transmission of the glass window and of the opal diffusing disc is non-selective and can be ignored (WESTLAKE, 1965), the relative sensitivity curve for the Photronic cell (Fig. 24) shows it to respond to light of wave length between 290 and 775 nm with a maximum response around 550 nm. Its current output is known to be linear when the light intensity exceeds about 10^{-4}ly.min^{-1}. By adding filters, its sensitivity centre was moved to 520 nm (1965/66) and 445, 530 and 630 nm (since 1968) respectively (Table 39).

Measurements were made at selected stations starting either above the water surface (T_0^{520} adjusted to 100 percent) or at 10 cm below the surface to avoid the 'surface effect' (VOLLENWEIDER, 1969). Readings were made at 1 m of depth and deeper at 1-m intervals until the level of one-percent relative light intensity was reached. Sometimes further readings were taken below that level. As far as the cruise schedule would permit the measurements were taken when the sun was high (between 0900 and 1500 hr SAST) and when there was no cloud interference.

Using a similar methodology comparable data were obtained in November 1970 for Lake Mweru-Luapula and Lake Bangweulu.

9.3. *Sampling water thermal properties*

Nearly all thermal profiles from the lake's water surface down to a depth of 70 metres at the most, were recorded using a bathythermograph (Wallace and Tiernan) and metalized slides. Because of the presence of numerous trees underwater this instrument could only be used from a stationary boat after having been adapted for vertical retrieving. Only until June 1965 were the thermal profiles made with a calibrated thermistor.

In 1965/66 about 320 profiles were recorded throughout Lake Kariba. An additional 261 profiles were made during the study of the second annual cycle. Each of them was read on the spot and critical temperatures noted down. Bathythermograph slides were annotated and stored for later printing on photographic paper as a permanent record.

Water temperatures were read to 0.1 C at 1-m depth intervals. They were frequently checked and corrected if necessary against temperatures measured in the insulated water sampling bottle with a calibrated laboratory thermometer on samples obtained from known depths.

Table 39. Optical parameters for photocell and filters.

Photocell-filters	Sensitivity range nm	50 per cent band[1] nm	Sensitivity centre nm	Optical centre of gravity[2] nm	Average sensitivity per cent
A. Photronic cell	290–775	396–640	550	—	—
B. 1965–1966 Photronic cell combined with filter Wratten 55	460–600	491–559	520	525	14.3
C. Since 1968 Photronic cell combined with					
(a) Filter Wratten 47	385–565	422–490	445	465	10
(b) Filter Wratten 13	400–700	495–566	530	532	8.4
(c) Filter Wratten 29	500–750	613–668	630	643	4.9

[1] Fifty-per-cent band: waveband width at half the max. sensitivity.
[2] Optical centre of gravity: wavelength which divides the area under the spectral sensitivity curve in half (WESTLAKE, 1965).

9.4. Sampling water chemical properties

9.4.1. WATER SAMPLING

In 1965/66 all water samples were obtained with a 2-litre Ruttner bottle containing a thermometer (RUTTNER, 1963), a hand winch equipped with a cable marked at 1-m intervals, and a metre-wheel. Later a 1-litre insulated Rigosha bottle was used.

Surface water was sampled from practically all stations. Deeper layers were sampled at stations selected according to either the particular thermal characteristics or the water sampling designed in collaboration with the University College of Rhodesia in January 1965 (COCHE, 1965). In the latter case, the water sample was treated with saturated mercuric chloride (1 ml/L) as preservative.

9.4.2. ROUTINE CHEMICAL MEASUREMENTS

Routine chemical measurements were made on all water samples immediately after securing them. The measurements involved:

a. Hydrogen-ion concentration (pH) by colorimetry (Lovibond comparator) until September 1965 and by electrometry (Radiometer pH-meter) thereafter.

b. Total alkalinity (TA) by titration of 100 ml of water with N/50 sulphuric acid until pH 5.1 using BDH 4.5 or brom-cresol green as indicator.

c. Conductivity (k_t) with a Dionic Tester (k_{20}) until August 1965 and with a Wheatstone-Bridge meter (k_{25}) thereafter.

d. Hydrogen sulphide (H_2S) was estimated in deoxygenated water samples using the Hach test (Iowa, USA) which consists in comparing the colouration developed after exposure to the released gas of chemically treated paper discs to a standard chart (0.1 to 5 mg/L H_2S).

e. Dissolved oxygen content (DO) was also routinely measured in most instances on 250-ml water samples using the unmodified Winkler method.

9.4.3. DETAILED CHEMICAL ANALYSES

Between February 1965 and January 1966, 412 water samples were collected throughout Lake Kariba at quarterly intervals. They were fixed with mercuric chloride (1 mg. L^{-1}) and sent to Dr. D. S. MITCHELL for detailed analyses at the University College of Rhodesia in Salisbury. Standard chemical methods were used to determine contents for calcium, magnesium, sodium, potassium, nitrite, nitrate, and phosphate (COCHE, 1965, 1968).

Since 1970 further water samples were collected to be analysed under the supervision of Dr. R. ARMSTRONG (FAO) in the chemical laboratory of the Central Fisheries Research Institute in Chilanga, Zambia. Standard chemical methods were used to determine contents for major cations (Ca, Mg, Na, K) and major anions (fixed CO_3, Cl, SO_4).

10. OPTICAL PROPERTIES OF THE LAKE WATER

10.1 Apparent colour of the lake

The apparent colour of the lake or colour of the light of the water as it emerges from the lake surface (HUTCHINSON, 1957) varies through the year. Each of the previously defined Basins (river mouth excluded) has its own variation pattern (Table 40).

In general the water in Basins 0 and I appears yellow to slightly brown. As one progresses towards the dam the water often appears more and more greenish at first and bluish later, being fully green during phytoplankton blooms only.

This seasonal pattern is related to the importance of the Zambezi flow (BI-II) and to water thermal properties (circulation phenomena) to be discussed later. Most probably the yellow apparent colour is due mainly to the presence of colloidal clay (Sec. 10.3).

10.2. Depth of visibility and its relationship to incident light

The depth of visibility (DV = Secchi-disc transparency) although determined by a very simple technique still retains its value, a high correlation being observed between light penetration and DV in a single lake or in a homogeneous group of lakes (HUTCHINSON, 1957). Variations in seston account for most of the seasonal DV variations. The depth of the euphotic zone, the daily photosynthesis per unit of the crop, the standing crop of plankton per unit of volume, all have been related by various authors to the depth of visibility (STRICKLAND, 1958).

In the limnetic region of Lake Kariba the depth of visibility has been found to vary between 50 cm and 1060 cm (Table 41). This is a much wider range than observed in other African reservoirs where 410 cm has been reported as the upper limit (Annex I).

As a rule the depth of visibility increases from Basin I to the dam. Exceptions exist from October to December when the Zambezi inflow is relatively transparent (DV 280 cm) which increases the transparency in Basin I while plankton blooms develop in Basin IV.

Average DV in the limnetic region of Lake Kariba (Basin 0 excluded) have also greatly varied from Basin to Basin and according to the season considered (Table 42). The annual mean DV increases towards the dam, doubling between Basin I and IV. The Kariba general average (405 cm) is smaller than averages calculated for the North American Great Lakes: L. Superior 1000 cm; L. Michigan 600 cm; and L. Erie 450 cm (BEETON, 1965). But on the basis of the data available (Annex I) the water transparency in L. Kariba is most probably greater than in other African reservoirs. The relatively high values observed in Basins III and IV away from most of the influence of colloidal clay and dissolved organic matter, could point out to oligomesotrophic conditions such as in certain Canadian lakes (RAWSON, 1960).

Table 40. Apparent colour of the lake water in L. Kariba Basins.

Month	0	I	II	III	IV
January	yl	yl/gr-gr	gr	bl/gr	gr/bl
February	yl	yl	gr/yl-gr	bl/gr-gr/bl	gr/bl-bl
March	yl	yl	—	—	—
April	⎧ yl ⎨ yl	yl yl	yl-gr/yl yl	gr/yl-gr gr-bl/gr	bl/gr gr/bl
June	yl	yl	gr/yl	bl/gr	gr/bl-bl
July	yl	gr	gr	gr-bl/gr	gr/bl-bl
August	yl	yl	gr/yl-gr	bl/gr-bl	gr/bl-bl
October	⎧ gr/yl ⎨ yl	gr yl	gr(bloom) gr/yl	gr bl/gr	gr gr
Year	yl	yl	yl-gr/yl-gr	gr/yl-gr/bl	gr-bl

yl = yellow (XVI–XIX); gr/yl = greenish/yellow (XI–XV); gr = green (VIII–X); bl/gr = bluish/green (V–VII); gr/bl = greenish/blue (III–IV); bl = blue (I–II).

Table 41. Depths of visibility in the limnetic regions of L. Kariba, ranges in cm.

Month	0	I	Basin II	III	IV
January	90–110	230–300	300–400	540–640	340– 640
February	50	60	200–300	300–850	500– 900
March	(40)[1]	60– 80	140–160	320	320– 340
April	⎧ 90 ⎨ 160	130–170 120–200	220–240 160–240	310–500 300–560	660– 780 520– 740
May[1]	140–160	140–180	220–360	420–700	500– 880
June	160	160–260	240–280	380–480	400– 650
July	220	320	300–360	460–600	600– 880
August	200	300–320	340–480	450–760	600–1060
September[1]	300	340	340–400	460–760	300– 720
October	⎧ 300 ⎨ —	340–500 340–400	230–320 260–440	380–480 400–440	480– 620 220– 320
November[1]	280	380–400	360–600	540–580	340– 640
December	—	—	—	—	—
Year	50–300	60–500	140–600	300–760	220–1060

[1] BEGG, 1970.

Twenty five Secchi disc readings have been systematically combined with simultaneous photometer measurements of total light[10] penetration. The relationships between the depth of visibility and relative total light intensity (T_z^T in percent T_0^T) have been calculated (Table 43). Results show that on average DV equals the depth

[10] Total light (T^T): part of the solar spectrum as measured since 1968 with a selective photocell alone. For details see Sec. 9.2.3.

Table 42. Average depths of visibility in the limnetic region of Lake Kariba, in centimetres.

Month	Basin				
	0	I	II	III	IV
January	100	277	350	556	472
February	50	60	330	670	800
March	(40)[1]	—	—	—	—
April	90	150	235	405	720
	140	155	203	386	625
May[1]	150	132	240	476	517
June	160	253	268	440	523
July	220	320	333	533	727
August	200	307	425	565	713
September[1]	300	320	323	427	459
October	300	420	247	430	527
	—	375	318	397	253
November[1]	280	317	462	486	496
December	—	—	—	—	—
Year	169	257	311	481	569

[1] BEGG, 1970.

at which 22.7 percent of the incident solar light are found. Most usually the equivalent T_z^T varies between 15 and 27 percent. The general order of magnitude is 15 percent (VOLLENWEIDER, 1969). The values recorded from Lake Kariba are higher than the average value (14.7 ± 3.95) calculated by BEETON (1958) for Lake Huron. It is also higher than the range of 12 to 15 percent obtained by KIKUCHI (cited in HUTCHINSON, 1957) using a similar method.

On the basis of such measurements the depth of visibility can be used to estimate the depth of the euphotic zone for a particular body of water (TYLER, 1968). Being a common practice to limit the euphotic (trophogenic) zone to the depth at which only one percent of the incident light is left $(T_z^T\ 1\%)$[11] this particular level has been related to the corresponding DV (Table 43). On average it has been calculated that in Lake Kariba the depth of visibility must be multiplied by 3.54 to estimate the depth of the euphotic zone. The general range of the factor (2.05 to 5.24) very closely fits to the extreme values quoted by STRICKLAND (1958) from various authors for the oceans (2.5) and for lakes (5). VERDUIN (1956) obtained this last value for Lake Erie. In

---

[11] There is variability and confusion in the literature about the definition of the euphotic zone limit. Dr. TALLING prefers it as one percent of the photosynthetically available radiation. Others prefer to use either one percent of the most penetrating spectral component or one percent of photometer response. It is this last approach which is used here. Because of the photocell characteristics (Sec. 9.2.3) results should not greatly differ from those based on Dr. TALLING's point of view. See also 10.3.5 c.

Table 43. Relationship between Depth of Visibility (DV) and Relative Total Light Intensity ($T_z{}^T$).

Basin	Station	Date	DV cm	$T_z{}^T$ 1% at-cm	Equiv. $T_z{}^T$ per cent	Ratio $T_z{}^T$ 1% to DV
I	1205	050371	60	260	33	4.33
	1407	0269	60	200	25	3.33
	1508	0368	170	420	14	2.47
II	1611	0368	220	450	9	2.05
	1611	0269	190	600	21	3.16
	1915	0368	240	565	14	2.35
	2015	030371	155	630	27	4.06
	2016	0269	280	845	27	3.02
	2620	0269	520	1540	19	2.96
III	2826	141169	540	1360	11	2.52
	2926	310170	360	1320	9.5	3.67
	2926	030470	300	780	19	2.60
	3332	0269	500	1960	30	3.92
	3332	061270	280	1240	24	4.43
	3332	100371	320	1260	22	3.94
	4036	070371	320	1610	31	5.03
IV	4741	0269	880	2400	18	2.73
	4741	071270	300	1320	46	4.40
	4741	090371	325	1700	27	5.23
	5344	071270	300	1000	20	3.33
	5344	080371	340	1780	40	5.24
	5444	0269	800	2100	16	2.63
	5845	081270	300	1140	19	3.80
	5845	080371	340	1480	26	4.35
	5847	0468	660	2000	21	3.03

Great Slave Lake RAWSON (1950) calculated a factor of 4.3, while a factor of three was estimated by RILEY (1941) for Georges Bank in the Western North Atlantic.

10.3. Penetration of visible light

The importance of light in the aquatic environment has been stressed on many occasions. Of particular importance from our point of view is the manner in which knowledge of subaquatic light conditions in lakes can contribute to a better understanding of the problems related to heat budgets, thermal stratification, and related subjects (VOLLENWEIDER, 1961). Such data are also of great value to estimate potential primary productivity to be usefully compared with actual values (LEVRING & FISH, 1956) and to understand the dynamics of the latter (e.g. TALLING, 1965, 1966). Light provides the most useful sensory information for many fish, while also coordinating patterns of their behaviour and their physiology through diel and seasonal changes (WOODHEAD, 1966). The schooling of fish for example has recently been shown to stop at particular light intensities (WHITNEY, 1969). Intensities of illumination at which major phenomena take place have been summarized by BLAXTER (1965) and STRICKLAND (1958).

10.3.1. DEFINITIONS AND METHODOLOGY

As described earlier (Sec. 9.2.3) the penetration of solar light (visible part of the radiation) was measured with a photocell whose spectral sensitivity coincided reasonably well with the part of the solar spectrum that is active photosynthetically (390–710 nm ± 10 nm) or 46–48 percent of the total available energy (VOLLENWEIDER, 1969).[12]

The fate of the solar radiation as it reaches the water surface of the lake and travels underwater is described in several limnological treatises (e.g. HUTCHINSON, 1957; REID, 1961; RUTTNER, 1963). But nevertheless there is still a confused usage of the terms 'absorption', 'extinction' and 'attenuation'. WESTLAKE (1965) very aptly attempted to stop the existing confusion. I have chosen to follow his recommendations by avoiding the use of extinction and by replacing it by attenuation which is defined as the diminution with depth of radiant energy by absorption and scattering.

The variations in vertical illumination due to the height of the sun have been neglected following POOLE & ATKINS (cited in HUTCHINSON, 1957). Primary light data have been treated as suggested by VOLLENWEIDER (1969). The microammeter readings from various depths have been plotted on semilogarithmic paper. Straight line sections have been fitted and extrapolated to the zero depth to provide the estimated incident light (T_0^λ). Taking the latter as 100 percent, all values were recalculated on this basis as Relative Light Intensities (T_z^λ) and they were replotted on semi-logarithmic paper. Optical properties were then characterized by:
(i) Percentile vertical transmissions:

$$T_p^\lambda \text{ percent} = 100\, T_{z_2}^\lambda (T_{z_1}^\lambda)^{-1}$$

where $z_2 - z_1 = 1$ metre.
(ii) Vertical attenuation coefficients:

$$\text{VAC}^\lambda = (\ln T_{z_1}^\lambda - \ln T_{z_2}^\lambda)$$

with $z_2 - z_1 = 1$ metre, being the rate per metre at which the visible radiation decreases with depth. These coefficients have been estimated from SAUBERER's (1962) nomograph (Fig. 26) based on the relationship existing between percentile transmission and attenuation coefficients as follows:

$$T_p^\lambda = e^{-\text{VAC}^\lambda}$$

---

[12] In fact photosynthesis can proceed at least down to 350 nm. Conventionally the spectral limits applicable to photosynthetically available radiation are taken as 400 to 700 nm. As there is not much energy available below 400 nm, in practice it does not make much difference to ignore the fraction below 400 nm (Dr. J. TALLING, pers. comm.).

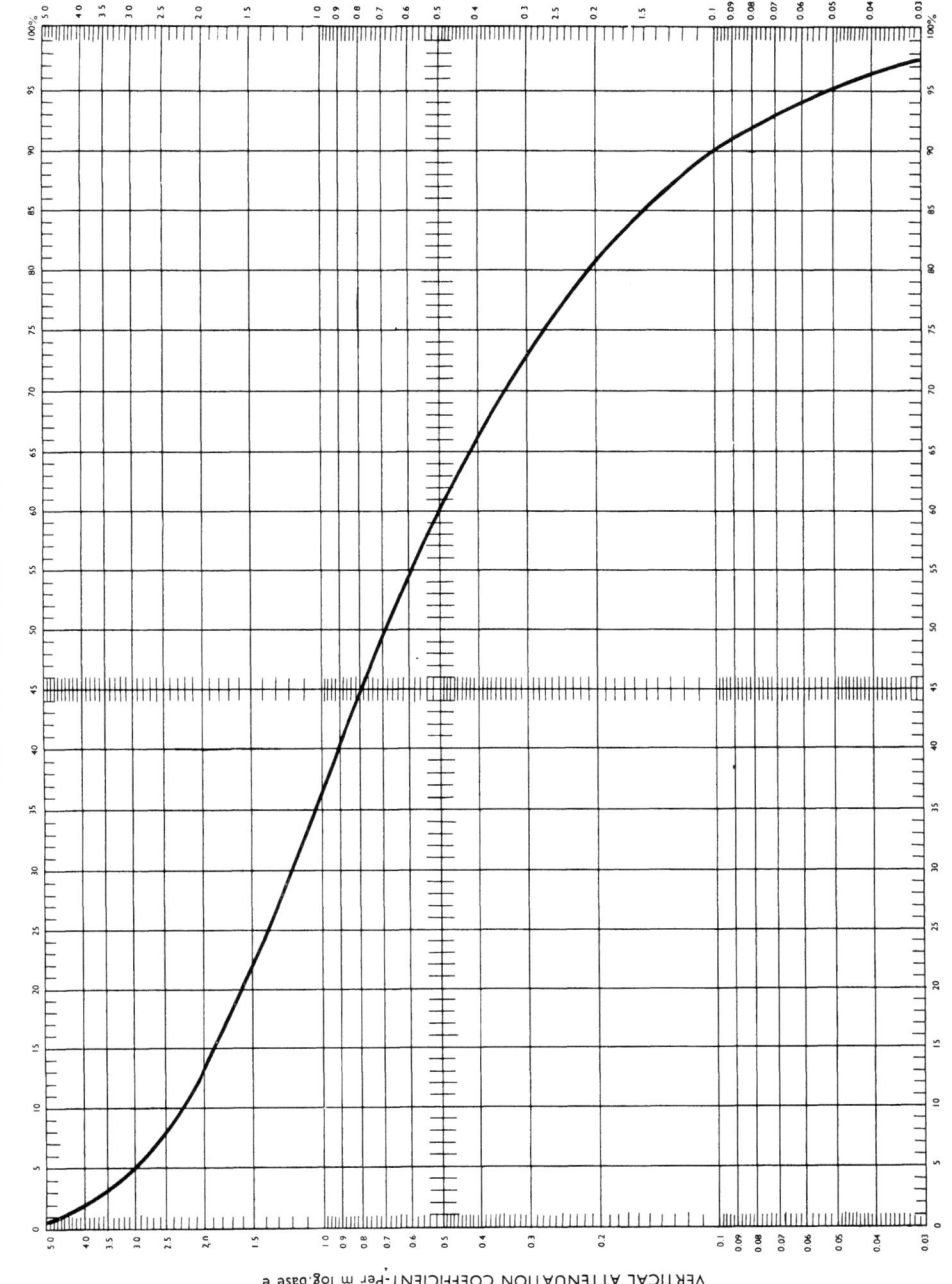

Fig. 26. Improved Sauberer's nomograph.

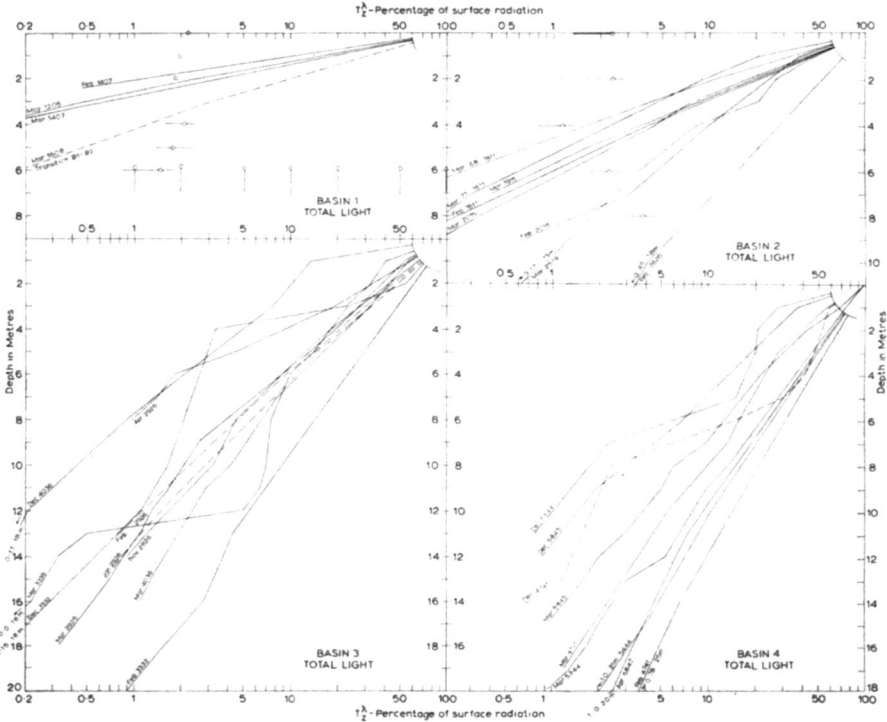

Fig. 27. Penetration of total light in Lake Kariba.

(iii) Spectral transmission curve, T_p^λ being plotted against the sensitivity centre (λ nm) of the three filter-photocell combinations to provide an estimate of the light spectral composition at various depths.

If necessary the relative light intensities could be approximately converted into the absolute values (cal. cm^{-2}. min^{-1}) with the incident solar energy present at the water surface being estimated from available meteorological data (global radiation – Tables 3 and 5) or from other published data for the tropics and subtropics (e.g. ORTH, 1939).

Due to technical difficulties light data could not be collected according to a definite sampling scheme. Therefore all similar data have been grouped (1965–66 and 1968–1971) in an attempt to include various seasonal conditions and all Lake Kariba Basins. Three groups have resulted from such treatment, to be successively examined in the following sections: total-light data (no filter), spectral data (three filters), and green-light data (green filter).

10.3.2. TOTAL LIGHT PENETRATION

Measurements for total light within the limits expressed in Sec. 9.2.3 were made whenever possible between 1968 and 1971 in the Zambian lake territory. In such

Table 44. Total light penetration characteristics in Lake Kariba.

Basin	Station	Date	Z_E	Vert. Perc. Transm. per cent per m		VAC^T ln. m^{-1}	
			cm	Range	VAC^T	Range	VAC^T
I	1205	Mar. 71	260	—	15.4	—	1.85
	1407	Feb. 69	200	—	10.3	—	2.20
		Mar. 71	275	17.5–19.5	18.8	1.7 –1.6	1.65
Trans.	1508	Mar. 68	420	31.4–38.8	35.1	1.1 –0.94	1.02
II	1611	Feb. 69	600	45.0–49.4	47.2	0.80–0.70	0.74
		Mar. 68	450	32.7–41.8	37.4	0.87–1.10	0.97
		Mar. 71	525		54.8		0.59
	2016	Feb. 69	845	37.1–68.2	65.9	0.98–0.37	0.39
	1915	Mar. 68	565	43.7–50.0	45.0	0.83–0.68	0.80
	2015	Mar. 71	630	41.4–51.8	48.8	0.88–0.62	0.71
	2620	Feb. 69	1540	72.7–84.2	74.4	0.30–0.16	0.28
	2519	Mar. 71	980	65.2–69.2	67.6	0.41–0.35	0.38
III	2926	Jan. 70	1340	68.8–78.6	70.7	0.36–0.23	0.33
		Feb. 70	1240	67.7–70.5	68.9	0.38–0.33	0.36
		Mar. 71	1340	71.8–79.1	72.9	0.32–0.21	0.30
		Apr. 70	780	45.5–57.8	55.4	0.78–0.54	0.58
		Nov. 69	1360	65.2–84.3	71.6	0.41–0.16	0.32
	3332	Feb. 69	1960	75.3–86.7	79.1	0.26–0.13	0.22
		Mar. 71	1260	41.3–90.4	79.5	0.88–0.08	0.21
		Dec. 70	1240	14.7–89.0	77.4	1.90–0.11	0.24
	4036	Mar. 71	1610	56.5–82.0	75.6	0.56–0.18	0.26
IV	4741	Feb. 69	2400	78.6–85.4	82.4	0.22–0.14	0.18
	4741	Mar. 71	1700	74.7–82.9	79.5	0.27–0.17	0.21
		Dec. 70	1320	43.0–85.0	70.6	0.84–0.15	0.33
	5444	Feb. 69	2100	79.4–82.6	80.3	0.21–0.17	0.20
	5344	Mar. 71	1780	56.6–79.9	78.3	0.56–0.20	0.23
		Dec. 70	1000	25.0–85.8	68.0	1.40–0.14	0.37
	5845	Mar. 71	1480	66.9–77.7	73.3	0.39–0.23	0.29
	5847	Apr. 68	2000	49.6–85.2	80.0	0.70–0.15	0.20
	5845	Dec. 70	1140	34.5–76.8	68.7	1.10–0.25	0.36

cases the resulting output of the photocell at any depth is considered to be related to the light energy present at that particular depth (STRICKLAND, 1958).

Resulting curves for relative light intensities present at various depths (T_z^T) are grouped by Basin in Figure 27. At a glance differences between Basins can be easily recognized, with the slope of the curves generally increasing towards Basin IV. Homogeneous water masses can also be identified with the plots in such cases resulting in more or less straight lines. Homogeneity occurred in surface waters of Basin I and, in some instances, in other Basins. But most of the time in the latter changes in the slope most probably testify to a heterogeneous water mass at the time of the sampling. The increase of the slope within the first metre corresponds to the total removal of

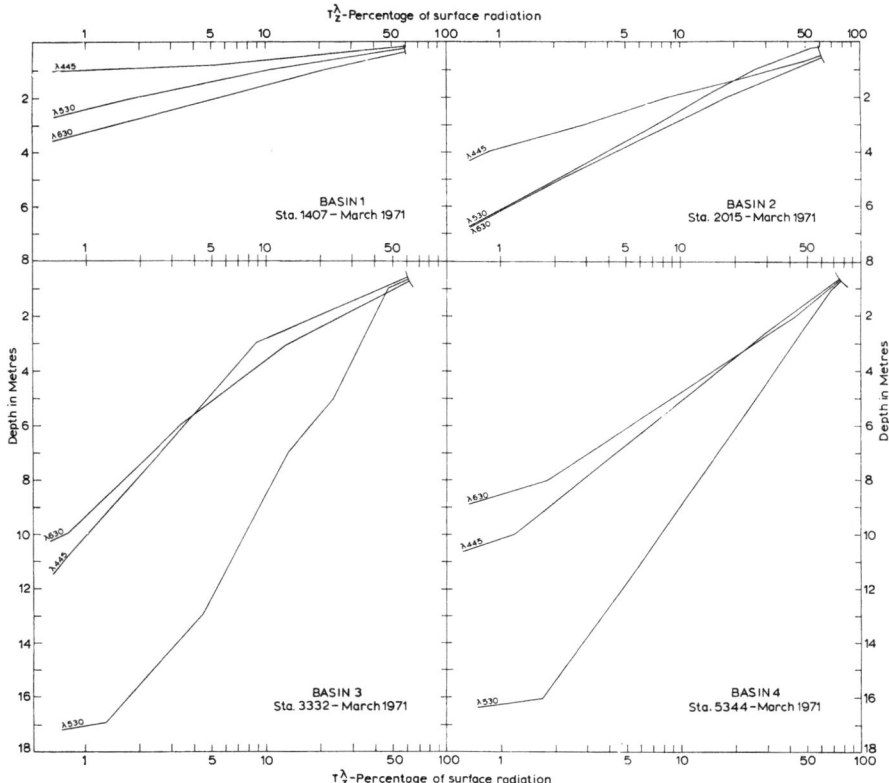

Fig. 28. Penetration of spectral light blocks in Lake Kariba.

the ultra-violet and infra-red parts of the spectrum (within the photocell sensitivity).

At greater depths it indicates the presence of highly coloured water (colloidal clay, organic matter dissolved or in suspension eg. seston or plankton).

Vertical percentile transmissions and corresponding vertical attenuation coefficients (VAC) have been calculated on the basis of Figure 27 (Table 44). Averages over the depth of the euphotic zone (z where $T_z^T = 1$ percent) have been obtained to characterize the light penetration at each station ($\overline{VAC^T}$).

$\overline{VAC^T}$ varied in general between 2.20 (Basin I) and 0.20 (Basin IV). It decreases as one progresses from the Zambezi River towards the dam. Except in Basins I and II when floods pass through, $\overline{VAC^T}$ are much lower than in Lake Mweru-Luapula (1.37) and L. Bangweulu (1.48) at a time when they were not yet influenced by inflowing floods.

10.3.3. PENETRATION OF SPECTRAL LIGHT BLOCKS

Light measurements with a set of coloured filters started in November 1969, three spectral blocks being studied centered respectively on wavelengths 445, 530 and 630 nanometers (Table 39).

Table 45. Penetration characteristics of three major wavelengths in L. Kariba. Average values for T_P^λ and VAC^λ over Z_P, indices of light penetration (IP), and optical depths (OD).

Basin	Station	Date	Most penetrating ray		AT_P^λ per cent m^{-1}			$AVAC^\lambda$			IP	OD
			λ nm	$Z_E = Z_P$ m	445 nm	530 nm	630 nm	445 nm	530 nm	630 nm		
I	1205	Mar. 71	630	3.0	3	22	24	3.50	1.50	1.40	0.7	6.1
	1407	Mar. 71	630	3.3	12	18	24	2.10	1.70	1.40	0.7	6.7
II	1611	Mar. 71	630	5.7	25	47	49	1.45	0.75	0.71	1.4	5.8
	2015	Mar. 71	530–630	6.2	35	52	47	1.02	0.64	0.75	1.6	5.7
	2519	Mar. 71	530	11.1	53	71	59	0.62	0.33	0.52	3.0	5.3
III	2926	Jan. 70	530	14.6	66	73	63	0.40	0.30	0.45	3.3	6.3
		Feb. 70	530	12.2	52	69	56	0.64	0.36	0.56	2.8	6.3
		Mar. 71	530	15.1	64	74	64	0.43	0.23	0.43	4.3	5.0
		Apr. 70	(530)	7.0	43	52	52	0.84	0.64	0.64	1.6	6.5
		Nov. 69	530	13.5	56	71	63	0.56	0.33	0.45	3.0	6.4
		Mar. 71	530	17.1	67	80	61	0.39	0.20	0.48	5.0	4.9
	3332	Dec. 70	445	10.3	71	64	69	0.33	0.43	0.36	3.0	4.9
	4036	Mar. 71	530	18.7	66	79	63	0.40	0.22	0.45	4.5	5.9
		Dec. 70	630	12.6	72	70	84	0.31	0.34	0.16		
IV	4741	Mar. 71	530	16.2	67	76	28	0.39	0.27	1.23	3.7	6.3
		Dec. 70	530	10.4	63	67	66	0.45	0.39	0.40	2.6	5.9
	5344	Mar. 71	530	16.2	48	78	47	0.73	0.23	0.75	4.3	5.4
		Dec. 70	445	10.8	66	65	62	0.40	0.42	0.47	2.5	6.2
	5845	Mar. 71	530	14.2	44	76	23	0.82	0.26	1.45	3.8	5.3
		Dec. 70	445	12.8	72	62	59	0.31	0.47	0.52	3.2	5.7

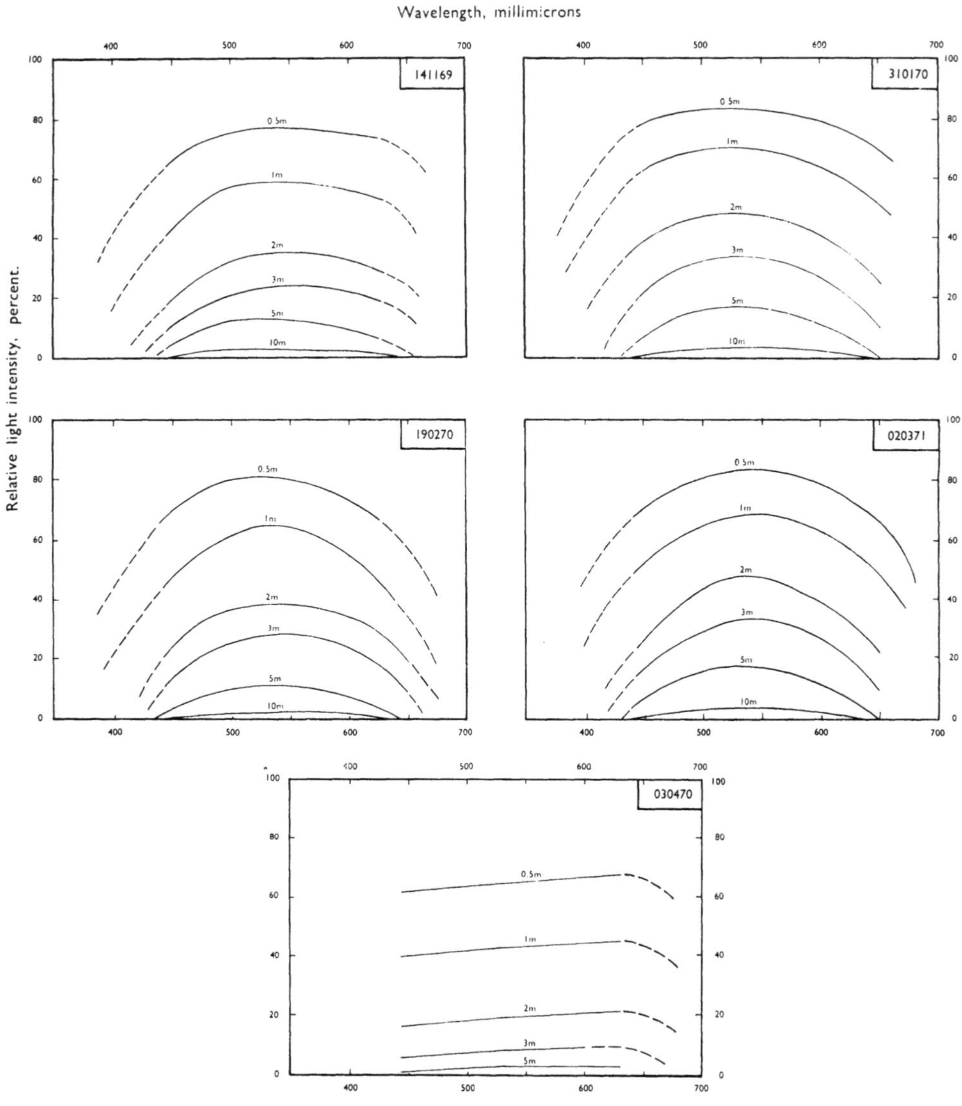

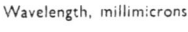

Fig. 29. Intensity and spectral composition of light at various depths in Lake Kariba, Station 2926.

Examples of the type of light penetration curves obtained in each of the Lake Kariba Basins are illustrated in Figure 28. Data are summarized in Table 45. The changes recorded at Station 2926 on various dates (Fig. 29) further illustrate the intensity and spectral composition of the light at various depths in L. Kariba.

Light attenuations have been characterized by AVAC minimum, the average vertical attenuation coefficient (AVAC) calculated over the depth (Z_p) where the relative intensity of the light of the wavelengths least strongly attenuated equals one

Table 46. RUTTNER's Optical Code (RUTTNER 1963)[1].

Lake	Optical Code	Reference
Como, Italy	745	TONELLI, 1969
Achensee (Tyrol), Austria	697	in RUTTNER, 1963
Victoria, E. Africa	687	TALLING, 1965
Maggiore, Italy		
Spring	587	TONELLI, 1969
Winter	798	,,
Albert, E. Africa	465	TALLING, 1965
Lunzer Untersee, Austria	376	in RUTTNER, 1963
Lugano, Switzerland	366	TONELLI, 1969
Lunzer Obersee, Austria	145	in RUTTNER, 1963
Volta, Ghana (1966)	144	VINER, 1970
Edward, E. Africa	023	LEVRING & FISH, 1956
Mweru-Luapula, Zaire/Zambia	023	Original data
Bangweulu, Zambia	022	,,
Skärshultsjön, S. Sweden	012	in RUTTNER, 1963
Lammen, S. Sweden	002	,,
Kariba (year averages)		
Basin I	122 (flood)–(365)	
Basin II	355 (flood)–(475)	Original data
Basin III–IV	675	

[1] Numbers refer to tens of percentile transmission for wavelengths 400, 500 and 600 nanometers respectively, from curves in Figure 30.

percent.[13] AVAC min. corresponds to the average percentile transmission (AT_p). Resulting values differ from basin to basin throughout the year (Table 45).

Whereas in Basins I and II, AVAC min. varied between 0.33 and 1.40 ln units per metre, in Basins III and IV it ranged from 0.16 to 0.64. These last values if compared with slightly differently defined estimates for Lake Victoria (min VAC 0.16 to 0.33 in TALLING, 1965) show that in general the light attenuation is probably greater in Lake Kariba. It is much greater in the upper part of the lake under the direct influence of the inflowing Zambezi flood which progressively travels downstream within the euphotic zone.

One point of particular interest is the wavelength of the most penetrating spectral block. For L. Kariba AVAC min. lies in the green spectral region most generally as recorded by TALLING (1965) in L. Victoria (min. VAC 0.165–0.33), L. Albert (min. VAC 0.35–0.72), and L. Edward (min. VAC 1.30). VINER (1970) also found that 530 nm was generally the most penetrating wavelength in L. Volta (min. VAC 0.7–1.2). Occasionally it is found in L. Kariba that red light penetrates deeper (Table 45). It is

[13] Dr. TALLING (pers. comm.) prefers to use the minimum value, over the spectrum, of the vertical attenuation coefficient without water depth specification and without mathematical average. To compare to our estimate (AVAC min.) his point of view will be referred to as min. VAC.

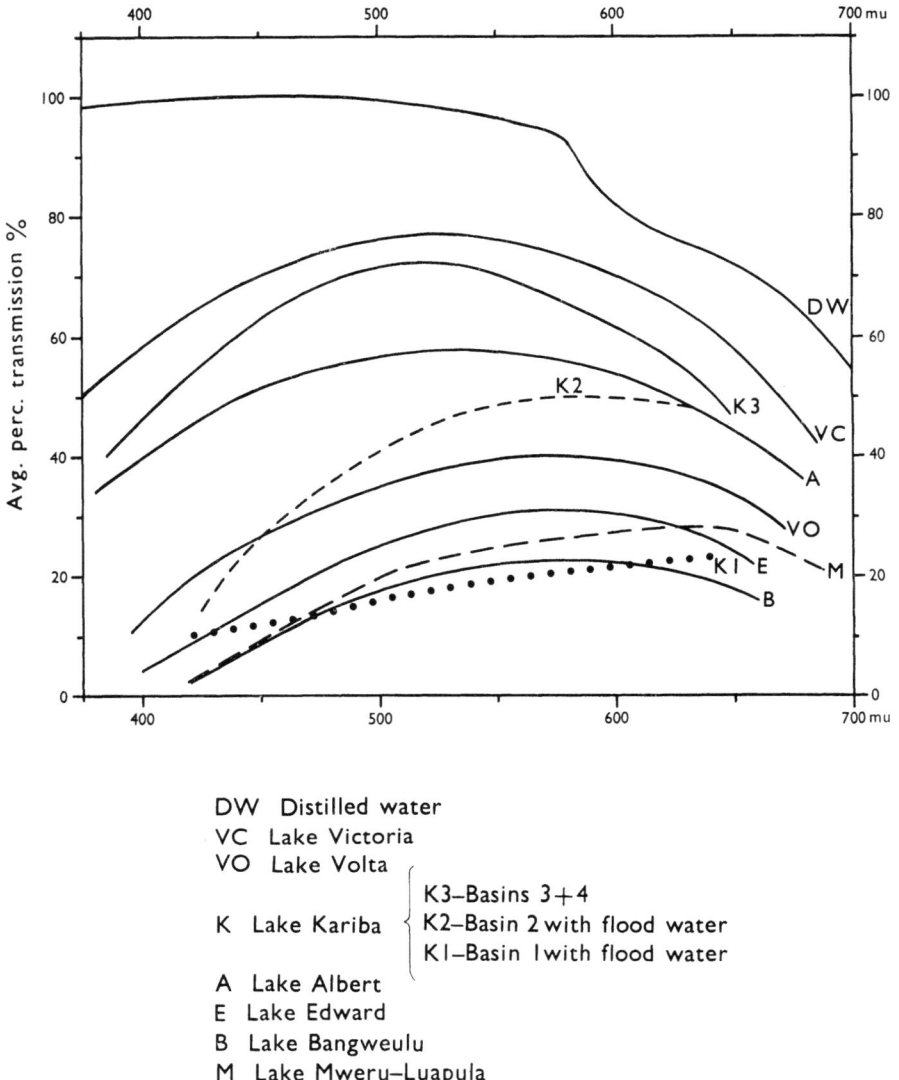

DW Distilled water
VC Lake Victoria
VO Lake Volta
K Lake Kariba { K3–Basins 3+4
K2–Basin 2 with flood water
K1–Basin 1 with flood water
A Lake Albert
E Lake Edward
B Lake Bangweulu
M Lake Mweru–Luapula

Fig. 30. Spectral transmission of light in various African lakes.

known that suspended material has little effect on the wavelength of the transmitted light and that transmission differences may be largely due to colouring agents (RUTTNER, 1963). As the content in humus or in colloïdal iron increases, the absorption for short wavelengths strongly increases, attenuation increases, and its minimum value shifts towards the red part of the spectrum. In L. Volta during the two first years of impoundment, the relatively high transmission of red light was due to colloïdal iron (G. BRETSCHKO, pers. comm.). In L. Kainji it has also been observed that the red component of the spectrum is less affected than green and blue by the colloïdal clay of the white flood (F. HENDERSON, pers. comm.). Recorded shifts of the minimum VAC are almost certainly due to colloïdal scattering rather than humic colouring. This is probably what happens in Basins I and II under the direct influence of the Zambezi River. A similar minimal attenuation around 630 nm was generalised in November 1970 in Lake Mweru-Luapula (AVAC min = 1.22) and in Lake Bangweulu (AVAC min = 1.52) where humic colouring might be rather responsible because of the presence of vast papyrus swamps within the catchment. TALLING (1965) has recorded the same pattern in L. George (min. VAC = 4.6–6.3).

In pure water, the blue spectral block penetrates deepest. As already inferred from the apparent colour of the lake (Sec. 10.1) and now confirmed by actual measurements, such relative water purity exists in Basin IV especially at certain times of the year (Table 40 and 45). The attenuation however remains practically the same (AVAC min 0.3 to 0.4) being still much greater than in oceanic conditions.

Light penetration characteristics can change markedly with time (Fig. 29). In April the attenuation was suddenly increased and greater penetration shifted from within the green spectral block towards longer wavelengths. This shift most probably reflects the arrival in the area of flood water loaded with colloidal clay (Sec. 11). In Fig. 28 this progressive change in water quality is illustrated as one passes from Basin I (presence of colloidal clay in flood water) into Basin II (proportion of clay smaller), and into Basins III and IV where the green light penetrates relatively deep, followed by the blue light.

RUTTNER (1963) has proposed to characterize the optical properties of lakes by a code number based on percentile transmissions at 400, 500, and 600 nanometers rounded off to tens. The idea has been applied to a few lakes of the world for comparative purposes (Table 46). It is worthwhile to note how spectral transmissions are similar in a few couples of lakes as also seen from Fig. 30: L. Victoria and European subalpine lakes; L. Volta and Lunzer Obersee in Austria; Lakes Edward, Mweru, and Bangweulu, all fringed by extensive papyrus swamps. The optical properties of L. Kariba might be partly compared (Fig. 30) to those found in Lake Volta (Basin I during flood season), to L. Lugano and L. Albert (Basin II), and to L. Victoria (Basins III and IV). But on a more general basis the comparison to L. Victoria should be retained (Fig. 30).

10.3.4. GREEN LIGHT PENETRATION

During the first year (1965–66) a series of measurements were made with a permanently mounted green filter which sensitivity centered on 520 nanometers (Sec. 9.2.3). As seen above, green light usually penetrated the deepest in Basins III and IV. Therefore early measurements provide additional data on AVAC values, the most important parameter to characterize the optical properties of the lake water for biological processes. In Basins I and II however AVAC min. based on green light measurements might be overestimated when the red part of the visible spectrum was in fact the least strongly absorbed by colloidal clay during the flood season especially.

All available penetration curves for green light (λ 520 and λ 530 nm) have been

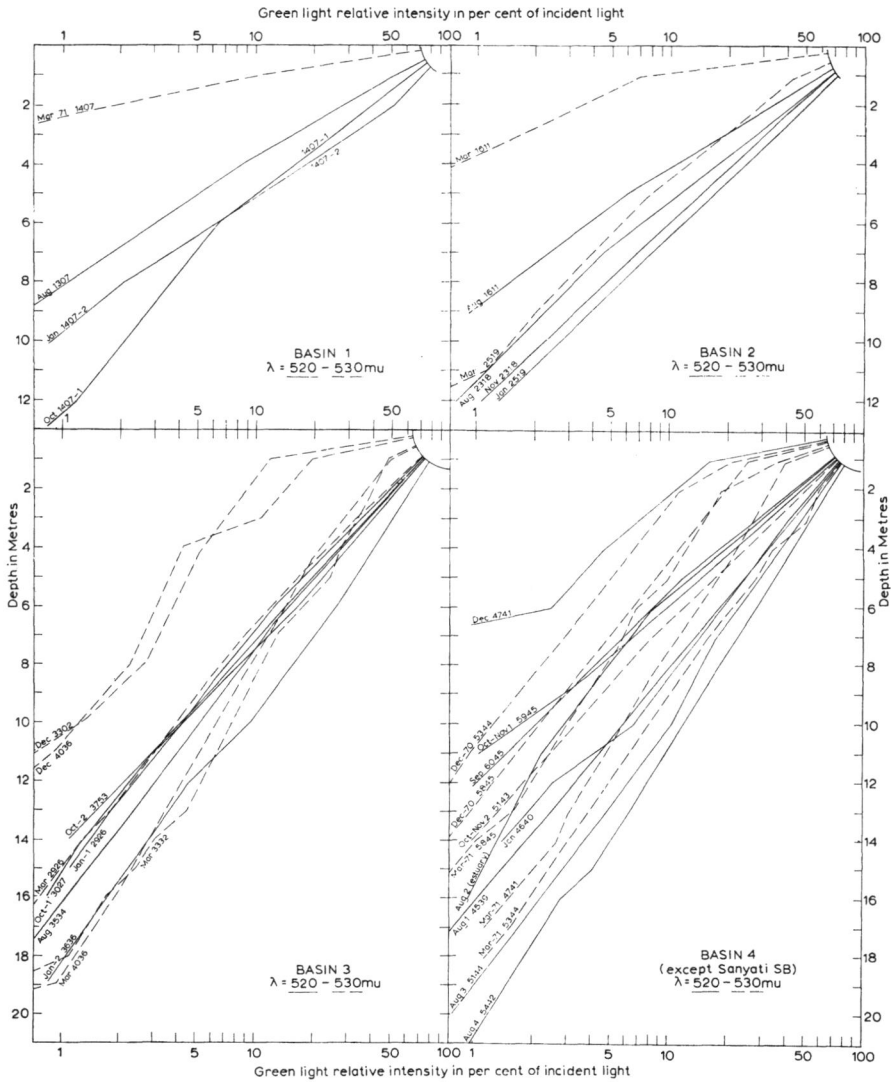

Fig. 31. Green light penetration in Lake Kariba Basins.

assembled by basin (Fig. 31) with data relevant to the eastern sub-basins (Kariba and Sanyati) being presented separately (Fig. 32). There is little doubt that the green light penetration measured in Basins I and II in August, October and January at times when the Zambezi River was not in flood (Fig. 15), closely represents the most penetrating part of the light spectrum. During flood time (e.g. March) the situation was completely changed as can be seen at Station 1407 in Basin I and at Sta. 1611 in Basin II (Fig. 31). But even by that time the Zambezi flood water had not yet reached Station 2519 (eastern BII). Therefore most of the present data characterizing green light penetration

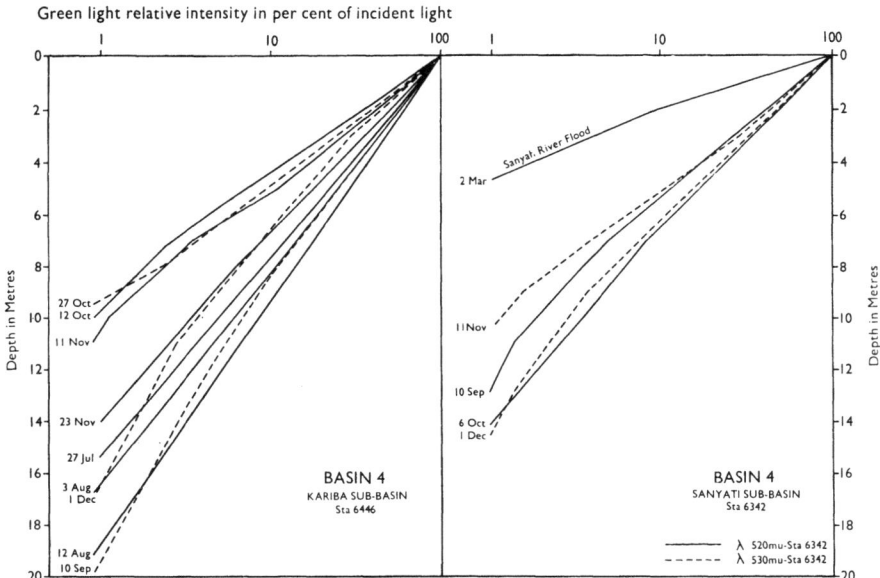

Fig. 32. Green light penetration in the Kariba and Sanyati sub-basins of Basin IV.

can confidently be used to characterize light penetration in general by the method applied in the previous section to the most penetrating wavelengths.

In summary (Table 47) $AVAC^{520}$ ranged throughout the year from 0.20 to 0.56 with one exception at Station 6342 in March (AVAC = 0.92) when there was much reducing green light penetration in the Sanyati flood water. All other measurements having been taken during the dry season or in early rainy season, flood conditions are not characterized here in general and these data will have to be combined with previous ones to provide a picture of the variations during the entire annual cycle.

10.3.5. Penetration of light into Lake Kariba

a. General average characteristics. It has been shown in the previous section how most of the data pertinent to green light penetration in 1965–66 (Table 47) could be considered as reflecting the attenuation characteristics of the most penetrating wavelengths. Therefore they have been combined with the more sophisticated data obtained later (Table 45) in an attempt to better define in general the optical properties of Lake Kariba water in the limnetic region.

Average characteristics (percentile transmissions and average attenuation coefficients over Z_P) have been calculated from available data by Basins, taking into consideration how the Zambezi River directly influences Basin I (especially) and Basin II (progressively) during the annual cycle. Within Basins III and IV it has been considered that the green light is the most penetrating light throughout the year. But it must be kept in mind that under particular conditions such a state of affair can be changed

Table 47. Green light penetration characteristics in Lake Kariba.

Basin	Station	Date 1965–66	Min Z_E m	Tp^{520} per cent.m^{-1} Range	ATp^{520}	VAC^{520} ln.m^{-1} Range	AVAC	IP min.	OD
I	1407	Aug.	8.15	53.9–59.9	56.9	0.50–0.61	0.56	1.8	6.6
	—	Oct.	12.5	63.4–75.5	69.5	0.26–0.44	0.35	2.8	6.3
	—	Jan.	9.8	58.6–72.1	62.7	0.31–0.52	0.45	2.2	6.4
II	1611	Aug.	8.8	56.6–63.3	59.6	0.44–0.56	0.51	2.0	6.5
	2318	Aug.	11.25	64.0–70.8	66.5	0.33–0.43	0.40	2.5	6.5
	—	Nov.	12.2	64.8–69.2	68.5	0.36–0.42	0.36	2.8	6.3
	2519	Jan.	12.7	—	69.7	—	0.35	2.8	6.4
III	3534	Aug.	16.3	72.5–77.1	75.4	0.24–0.31	0.26	3.8	6.1
	3027	Oct.	15.2	71.9–80.9	73.9	0.20–0.36	0.29	3.4	6.4
	3735	Oct.	14.3	71.2–74.2	72.5	0.29–0.33	0.31	3.2	6.4
	2926	Jan.	15.4	70.8–78.2	74.1	0.23–0.34	0.28	3.6	6.2
	3636	Jan.	18.3	70.0–79.9	77.9	0.21–0.35	0.23	4.3	6.1
IV	4539	Aug.	16.1	—	75.1	—	0.27	3.7	6.3
	5144	Aug.	19.0	72.2–84.1	78.5	0.16–0.25	0.22	4.5	6.0
	5442	Aug.	20.7	67.5–81.0	80.1	0.19–0.38	0.20	5.0	6.0
	6045	Sep.	12.0	64.7–70.7	68.7	0.34–0.43	0.36	2.8	6.2
	5945	Oct.–Nov.	11.1	61.8–67.7	65.6	0.38–0.47	0.41	2.4	6.6
	5143	Oct.–Nov.	13.8	69.7–73.7	72.0	0.29–0.35	0.31	3.2	6.2
	4640	Jan.	15.0	61.4–77.2	68.9	0.25–0.48	0.36	2.8	7.8
	6446	Jul.	15.3	—	74.0	—	0.29	3.4	6.4
	—	Aug.	16.4–18.8	75.6–78.3	77.0	0.23–0.27	0.24	4.2	5.7–6.5
	—	Sep.	19.6	75.6–82.9	79.3	0.17–0.27	0.21	4.8	5.9
	—	Oct.	9.5	53.1–71.0	61.8	0.33–0.62	0.47	2.1	6.5
	—	Nov.	10.50–14.0	57.0–83.6	69.0	0.17–0.55	0.36	2.8	5.5–7.3
	—	Dec.	16.6	66.2–83.6	76.8	0.17–0.40	0.24	4.2	5.8
	6945	Aug.	15.6	73.6–76.7	74.6	0.25–0.29	0.28	3.6	6.3
	6645	Jan.	16.4	72.8–78.9	75.7	0.22–0.30	0.26	3.8	6.2
	6342	Sep.	12.5	63.5–84.6	70.7	0.15–0.44	0.33	3.0	6.0
	—	Oct.	14.0	69.8–74.2	72.3	0.28–0.35	0.31	3.2	6.3
	—	Nov.	10.3	60.7–73.3	64.6	0.29–0.49	0.42	2.4	6.3
	—	Dec.	14.5	68.9–83.8	73.7	0.16–0.36	0.29	3.4	6.1
	—	Mar.	4.6	30.7–42.6	39.6	0.86–1.20	(0.92)	(1.1)	6.1

Table 48. Average penetration of light into L. Kariba.

Basin	No flood				Flood			
	λ nm	ATp %	AVAC min	IP	λ nm	ATp %	AVAC min	IP
I	530	63	0.45	2.2	630	24	1.40	0.70
	July?–January				February–June?			
II	530	66	0.40	2.5	630	49	0.71	1.40
	July (Aug)?–January (Mar)				February (Apr.)–July (Aug)?			
III	$\lambda = 530$ nm $-$ ATp $= 72\%$ $-$ AVAC min $= 0.31$ $-$ IP $= 3.2$ Range AVAC min 0.20 $-$ 0.64 (IP 1.6 $-$ 5.0)							
IV	$\lambda = 530$ (445) nm $-$ ATp $= 72\%$ $-$ AVAC min $= 0.31$ $-$ IP $= 3.2$ Range AVAC min 0.20 $-$ 0.47 (IP 2.1 $-$ 5.0)							

momentarily; after a heavy rain water might reach the limnetic region of the lake within the euphotic zone and favour the penetration of red light; or the water might become so pure to favour penetration of blue light. Indices of light penetration (IP = 1/AVAC min) have also been calculated (Tables 45 and 47) to compare the various basins (TALLING, 1965). Summarized data are grouped in Table 48.

Within the two Upper Basins one must take into account whether or not the Zambezi flood water is flowing through them. In the first instance greatest penetration occurs in the red part of the spectrum and absorption of other wavelengths by colloidal clay is relatively important (AVAC min 0.7–1.4). Later when the Zambezi discharge has become much reduced upstream and the river has returned to its channel, the water becomes more transparent (AVAC min 0.4–0.5 on average). Because of the lack of complete data the length of each period could only be estimated.

Within the two lower basins flood water is found in the euphotic zone only occasionally. In most cases greatest penetration takes place in the green part of the visible spectrum and light attenuation is relatively small (AVAC min less than 0.5). Although Basins III and IV have similar average values the range of AVAC min differs from one basin to the other, being wider within Basin III.

b. Optical depth. The notion of 'optical depth' (OD) has been introduced by TALLING (1965) when trying to eliminate among different lakes the effects of varying light penetration characteristics on the form of the profiles of photosynthetic rates against depth. It has been defined as:

$$OD = z(\min VAC)(\ln 2)^{-1}$$

where z is the depth in metres and ln 2 = 0.692 so that each OD unit corresponds to a halving of the intensity of light of the wavelengths least strongly absorbed. A similar approach has been taken here replacing min VAC by AVAC min.

Because of the relationship (see next) which seems to exist between optical depth and depth of the euphotic zone (z_E) the former has been evaluated for L. Kariba with $z = z_E$. The observed OD values derived from measurements made with various filters (Table 45) average 5.8 whereas OD values based on green light data (Table 47) average 6.3. The overall average is 6.1 which agrees well to TALLING's (1965) value for East African lakes (OD = 5.4) based on a slightly different definition.

c. Depth of the euphotic zone. The depth of the 'euphotic zone' (z_E) has been defined in the past on the basis of various light penetration criteria.[14] STRICKLAND (1958) considered that the depth where the total incident light (photometer response) is attenuated to one percent of its original value is 'a not unreasonable rough approximation of the depth of the euphotic zone'. This last depth has been approximated earlier for Lake Kariba (Sec. 10.2) as about 3.54 times the depth of visibility. Others consider that z_E being primarily determined by the vertical attenuation of particular wavelengths its downward limit extends until the most penetrating part of the visible spectrum is attenuated to one percent of its surface.

The depth of the euphotic zone (z_E) in Lake Kariba has been estimated from all photometer's data available following both above definitions (Tables 44, 45 and 47). The following ranges within the basins and yearly averages at central-basin stations have been observed:

	Range, cm	Ann. avg., cm
Basin I	200 to 1250	844
Basin II	420 to 1540	998
Basin III	700 to 1960	1650
Basin IV	950 to 2400	1538

It appears that the depth of the euphotic zone does not exceed 24 m in the limnetic region of Lake Kariba. But in general it is much shallower averaging about 10 m in Basins I and II and 16 m in the rest of the lake, river mouths being excluded. In Lake Victoria it has been recorded as ranging between 15 and 20 m (TALLING, 1965). In Lake Volta it probably reaches about 10 m now (BRETSCHKO, pers. comm.). In L. Nasser z_E varies greatly with the season, ranging from 1 m to 3 m during the flood season (Aug.–Oct.) and from 4 to 10 m in winter (ENTZ & RAMSEY, 1973). In L. Mweru-Luapula and in L. Bangweulu I estimated the depth of the euphotic zone to reach 3.5 and 3 m respectively, towards the end of the dry season (November 1970).

Because of the difficulty involved in physically defining the response of a photocell without filters, J. TALLING (pers. comm.) prefers to relate the depth of the euphotic zone to the optical depth (OD). In East African lakes (TALLING, 1965) he has established that:

$$z_E(\text{metres}) = 5.4 \pm 0.6 \text{ units OD} = 5.4 (\ln 2)(\min \text{VAC})^{-1}$$

and consequently that:

$$z_E (\text{metres}) = 3.7 (\min \text{VAC})^{-1}.$$

From a slightly different point of view it has been seen in the previous section that the depth of the euphotic zone in L. Kariba corresponds on the average to 6.1 units OD, which means that a useful estimate might be:

$$z_E (\text{metres}) = 4.2 (\text{AVAC min})^{-1}$$

[14] See footnote 11, Sec. 10.2.

10.3.6. COMPARATIVE NOTES ON LIGHT PENETRATION IN SOME AFRICAN LAKES

The spectral quality of the visible light in African lakes has been summarized from available data (Table 49 – Fig. 30). If these data are compared it is readily seen how light penetrates in the greatest part of Lake Kariba. Penetration in Basins III and IV is similar to that in Lake Victoria although in the latter attenuation is less pronounced. In contrast, Basin I in flood approximates conditions found in Lake Mweru-Luapula toward the end of the dry season. Basin II in flood holds an intermediate position.

For comparing various lakes among themselves, TALLING (1965) has proposed to use the reciprocal of min VAC as an index of light penetration (IP). Such index has been computed for Lake Kariba (Tables 45, 47 and 48) and for other African lakes (Table 50). In Lake Kariba the corresponding index of light penetration ranged from 0.7 to 5.0 because of the great basin heterogeneity. In Basin IV the range was slightly narrower than in Lake Victoria while in Basin III it became wider. Basin I had a range very similar to the one calculated from very patchy data for L. Volta during its first years of impoundment (VINER, 1970).

10.3.7. APPLICABILITY OF THE STANDARD ATTENUATION CONCEPT TO LAKE KARIBA

A decade ago VOLLENWEIDER (1961) proposed the concept of standard attenuation curves in analogy with RODHE's (1949) concept of standard water composition of the major chemical constituents. He related his standard attenuation curves to the mean vertical attenuation coefficient (MVAC), a measure derived from the total light attenuation and defined, in our case as:

$$MVAC = \tfrac{1}{3}(VAC^{430} + VAC^{530} + VAC^{630}).$$

This concept was reduced to three equations of linear regression as follows: (i) $VAC^{430} = 1.57$ MVAC $- 0.18$; (ii) $VAC^{530} = 0.83$ MVAC $- 0.06$; (iii) $VAC^{630} = 0.74$ MVAC $+ 0.21$. Whenever such standard concept is applicable it becomes possible to determine MVAC from one series of light intensity measurements (e.g. at $\lambda = 530$ nm) within the euphotic zone as:

$$MVAC = 1.20\,(VAC^{530}) + 0.07$$

and, using this MVAC value within the equations (i) and (iii) above to estimate VAC^{430} and VAC^{630}.

Although the standard transmission concept seems to have a wide applicability, it applies only to lake waters where light attenuation is primarily a function of light scattering and not of light absorption. The concept would therefore apply neither to Lake Mweru-Luapula nor to Lake Bangweulu, but it might apply to L. Kariba and save on future field work in particular in Basins III and IV. If not, it would point out to the most probable presence of dissolved coloured substances in the water.

Mean vertical attenuation coefficients (MVAC) have been calculated for the data of Section 10.3.2 and plotted against their corresponding vertical attenuation coefficients

Table 49. Average percentile transmissions (ATp%) and corresponding vertical attenuation coefficient (VAC) in some African lakes.

Lake	Geogr. position of station	λab. 430 nm		λab. 530 nm		λab. 630 nm		Reference
		ATp%	VAC	ATp%	VAC	ATp%	VAC	
L. Victoria	0°05′S –33°03′E	66	0.40	77	0.24	64	0.43	Talling 1965
L. Albert	1°30′N–30°52′E	48	0.73	58	0.53	49	0.71	Talling 1965
L. Volta (1966)	general average	18	1.69	38	0.96	39	0.93	Viner 1970
L. Edward	0°14′S –29°50′E	10	2.24	28	1.23	28	1.23	Talling 1965
L. Mweru-Luapula	9°20′S –28°34′E	6	2.75	23	1.45	28	1.23	Pers. data
L. Bangweulu	11°08′S –29°46′E	5	2.95	20	1.60	21	1.54	Pers. data
L. Kariba	Basin I flood	11	2.15	18	1.70	24	1.40	Pers. data
	no flood	(30)	(1.17)	63	0.45	(50)	0.68	
	Basin II flood	17	1.75	47	0.75	49	0.71	
	no flood	(40)	(1.00)	66	0.40	(50)	0.68	
	Basin III and IV	58	0.53	72	0.31	54	0.59	
Distilled water		99	—	98	—	75	—	James & Birge 1938 in Hutchinson 1957

Table 50. Minimum average vertical attenuation coefficient (AVAC min) and light penetration index (IP) in some African lakes.

Lake	Vertical attenuation Range	IP Range	Usual most penetrating light
Kariba (offshore)	AVAC min		
Basin I–II	0.33–1.40	0.7 –3.0	red-green
Basin III	0.16–0.64	1.6 –6.3	green
Basin IV	0.20–0.47	2.1 –5.0	green (blue)
	min. VAC		
Victoria (offshore)[1]	0.16–0.33	3.0 –6.3	green
Albert[1]	0.35–0.72	1.4 –2.9	green
Volta[2]	0.71–1.2	0.8 –1.4	green
Edward[1]	1.30	0.8	green (red)
Mweru-Luapula	1.15–1.30	0.8 –0.9	red (green)
Bangweulu	1.40–1.67	0.6 –0.7	red
Jebel-Aulia[3]	1.5 –3.5	0.67–0.29	red (green)
George[1]	4.6 –6.3	0.16–0.20	red

[1] Basic data from TALLING, 1965; [2] VINER, 1970; [3] TALLING, 1957.

at 445, 530, and 630 nanometers (Fig. 33). On the same graph the three standard regression lines (incl. data for distilled water as origins) have been added. It results that: (i) values relative to $\lambda = 445$ nm although fitting very well in some cases, are generally shifted above the corresponding standard line; (ii) red values do not fit well and in most cases appear below their standard line; (iii) data relative to $\lambda = 530$ nm fit reasonably well to the corresponding standard distribution.

It is therefore to be concluded that light attenuation in Lake Kariba is most probably and most often a function of light absorption by dissolved coloured substances. Although these are probably present in very small amounts in the water of the lower basins, nevertheless they modify the attenuation of longer wavelengths enough to preclude the possibility of applying VOLLENWEIDER's standard attenuation concept to Lake Kariba.

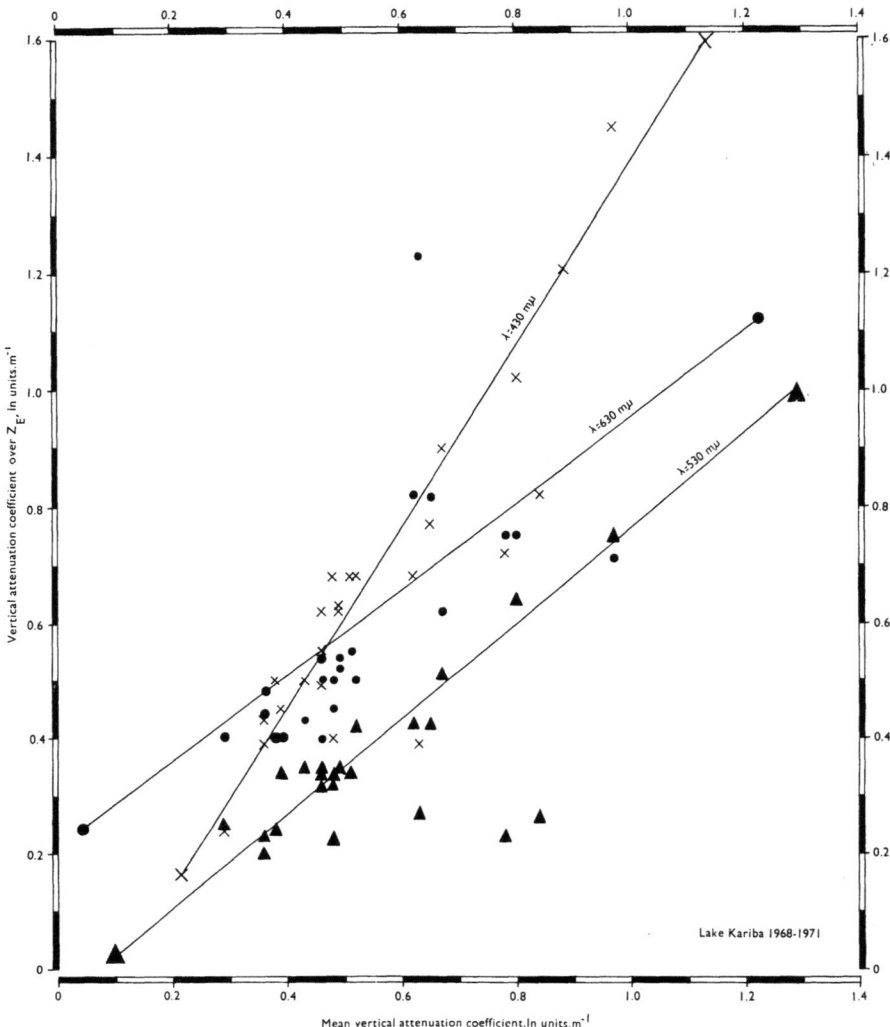

Fig. 33. Comparison of the light attenuation properties of Lake Kariba water with Vollenweider's standard attenuation curves.

11. THERMAL PROPERTIES OF LAKE KARIBA

11.1. Thermal cycle in the pelagic region

11.1.1. Terminology and definitions

On the basis of its thermal properties, Lake Kariba has been defined as a warm monomictic reservoir (COCHE, 1968; BEGG, 1970). Its thermal cycle is characteristically composed of a period of total circulation followed by a period of partial circulation. During the latter, the water mass is typically stratified into three layers: a surface – or mixed layer (epilimnion), a bottom – or stagnant layer (hypolimnion), and an intermediate region (metalimnion) which includes the plane of maximum rate of decrease in temperature known as thermocline (HUTCHINSON, 1957).

From the above considerations, it results that during stratification:
(i) the depth of the thermocline corresponds to the depth at which the point of maximum thermal gradient (TG) is found;
(ii) the depth of the mixed layer equals the vertical distance from the water surface to the point where the temperature gradient exceeds a predefined value for the first time (SWEERS, 1968).

Since the introduction of the term 'thermocline' by BIRGE in 1897 as the thermal boundary between two distinct water masses, it has become customary to adopt the original gradient of $TG = 1$ C per metre as this predefined value. For the study of temperate waters such a value is most probably satisfactory. But in the case of tropical lakes, because of the greater relative thermal resistance at high temperatures (VALLENTYNE, 1957), the value of the limiting thermal gradient must be decreased accordingly. DUNCAN (1964) studying the thermal structure of the inshore waters off the south-west Cape coast (Rep. South Africa) adopted the predefined limit of $TG = 0.1$ C per metre. Personally I have adopted $TG = 0.2$ C per m as the thermal gradient to be exceeded to leave the epilimnion and enter the metalimnion.

Secondary thermoclines which develop in the epilimnion during warm, windless days, have been observed in L. Kariba on many occasions. But in general they will be omitted from later considerations.

Multiple thermoclines have also been often recorded on bathythermograph's slides. They are known (HUTCHINSON, 1957) to occur following alternatively warm and cooler weather. In areas under the direct influence of river inflows, in particular in Basin I (Zambezi R.) and in the Sanyati Sub-Basin, a density current limited on either side by a metalimnion has been observed at certain times of the year to reach the pelagic region. Only this last phenomenon, because of its influence on the dynamics of the upper third of the lake, will be considered later.

11.1.2. LONGITUDINAL THERMAL PROFILES OF THE PELAGIC REGION

Water temperatures measured during fast longitudinal transects have been plotted to show the relative position of 0.5 C – interval isotherms in the pelagic region of Lake Kariba. Results of the five earlier cruises (1965–66) are summarized in Figure 46 (COCHE, 1968). Observations made during 1968–69 cruises (Fig. 34) confirmed and completed these past results.

In mid-July (Fig. 34), total circulation existed in Basin IV and to a lesser degree in Basin III. In the upper basins where Zambezi water flowed at an intermediate level as a density current, partial circulation was being established. In mid-October heterothermy had started to appear throughout the lake. Warmer Zambezi water was flowing into Basin I closer to the lake's surface. Four months later (K2/69), a strong thermocline existed in Basins III and IV whereas the thermal gradient in the metalimnion of Basin II showed a tendency to decrease. The Zambezi R. had started to progress through the entire depth of Basin I throwing a surface tongue in front of itself. By the end of March (K3/68), it had nearly completely flushed out the lake water from Basin I, simultaneously pushing the thermocline deeper in Basin II. A strong thermocline was present in the lower basins.

11.1.3. MEAN VERTICAL TEMPERATURE DISTRIBUTION

a. Methodology. Two methods of reducing large numbers of vertical thermal profiles to series of average values were recently described by SWEERS (1968). These methods being complementary, they have both been used to obtain the mean vertical temperature distribution in each basin at the time of each cruise.

The first method consists in averaging the water temperatures from the various thermal profiles considered at a number of standard fixed depths. It provides 'mean temperature' curves, $\bar{T}(z)$. In the present case standard depths have been taken at 5-m intervals starting at the lake's surface.

In the second method mainly applicable to a well-stratified lake, the depth of the isotherms (here at 0.5 C intervals) are averaged and 'mean depth' curves, $\bar{Z}(t)$ are obtained. It leads to a better description of the thermal profile in the thermocline region and gives the possibility to calculate average values for the maximum thermal gradient ($\overline{TG}$) and the depth of the thermocline ($\bar{z}_T$).

One further point of interest in these two averaging methods is that the two traces $\bar{Z}(t)$ and $\bar{T}(z)$ intersect near the point of maximum gradient of $\bar{Z}(t)$, thus defining the average depth of the thermocline ($\bar{z}_T$).

To be able to apply these two methods certain requirements are to be fulfilled (SWEERS, 1968):
(i) all observed thermal profiles must extend well into the hypolimnion;
(ii) they must show a clear thermocline and a mixed layer near the surface;
(iii) they must be monotone, decreasing from surface to bottom;
(iv) the observations should form a regular, either temporal or spatial, series.

In the case of Lake Kariba most of the observations conformed to these requirements except during periods of total circulation and, in Basin I, when a density current

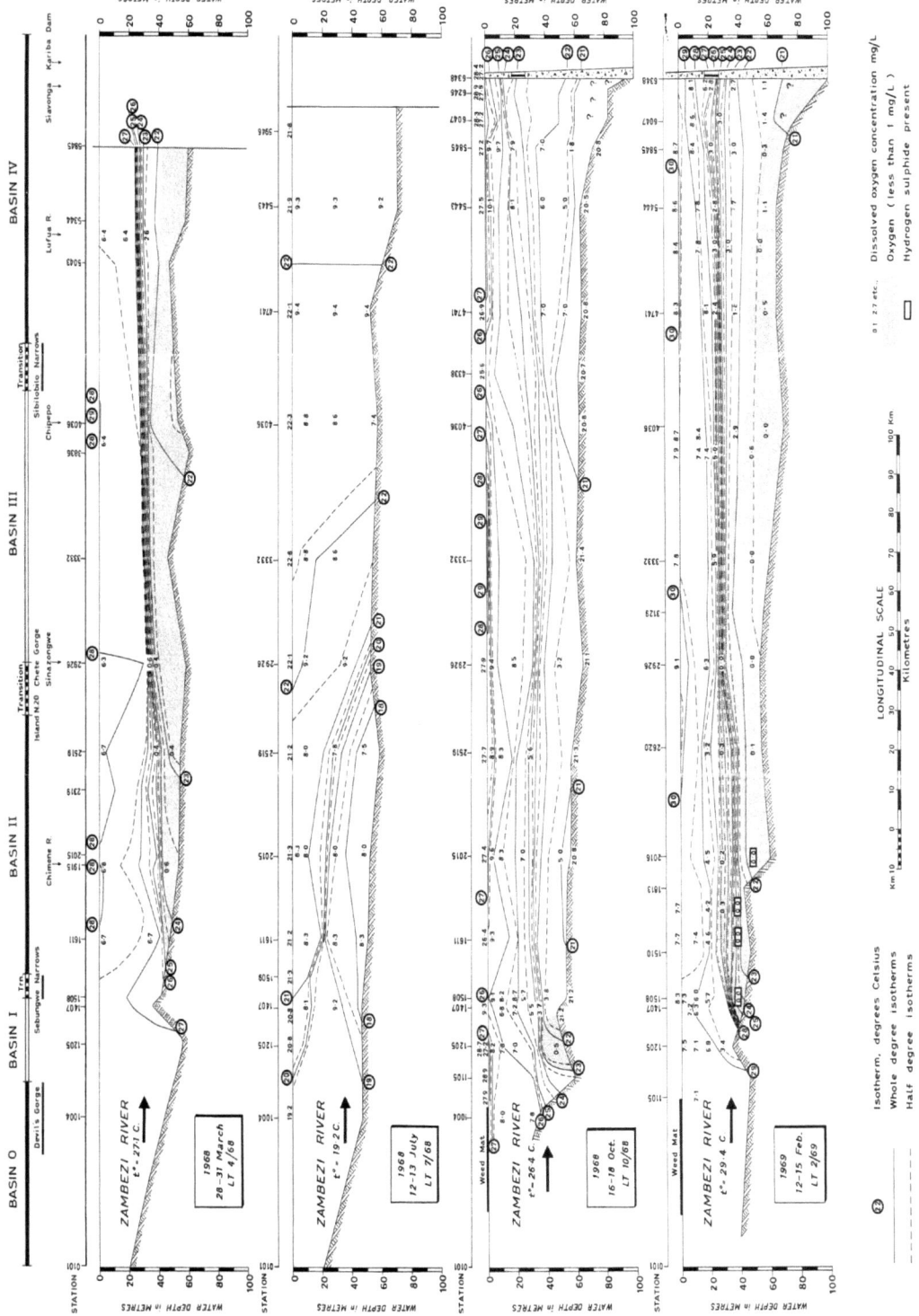

Fig. 34. Longitudinal profiles of Lake Kariba for 1968/69: thermal and dissolved-oxygen conditions.

was present. The measurements were taken from stations situated along the longitudinal axis at even distances (spatial series) and it is considered that each of these stations were characteristic of the pelagic region in that particular area. Whenever possible, the two methods have been used to characterize the mean vertical temperature distribution in each basin separately.

b. Mean temperature curves. The mean temperature curves $\bar{T}(z)$ depict clearly the progressive change of the thermal structure in each of the Lake Kariba basins as observed during two annual cycles (Fig. 35).

The two alternating phases of warming and cooling are well marked in Basins III and IV, where they lead to complete vertical mixing in June–July. Such total circulation does not last very long and by October–November it becomes already hampered by thermal differences between surface and bottom water masses. In Basin I average homothermy takes place in May – when the Zambezi flood flushes out the 'old' lake water into Basin II.

c. Thermal gradients and depth of thermocline. As said above (see a.) average values for the maximum thermal gradient ($\bar{T}\bar{G}$) and for the depth of the thermocline ($\bar{z}_T$) have been obtained from the traces of $\bar{Z}(t)$ and $\bar{T}(z)$ shown in Figures 37, 38 and 39. Results (Table 51 – Figure 42) picture the establishment of the thermocline and its evolution as the cycle progresses, in each basin.

The overall maximum $\bar{T}\bar{G}$ was 5.52 C. m^{-1}, calculated for Basin III, January 1966. Values of 5.0 C. m^{-1} were estimated twice in Basin III also (March 1968 and April 1965) but only once in Basin IV (February 1969). Such high $\bar{T}\bar{G}$ correspond to very great thermal resistances in the tropical conditions under consideration. From that point of view, Basin III stands out as the region of L. Kariba where the barrier effect of the thermocline is the most pronounced.

The mean depth of the thermocline in general increases as the year progresses, a normal pattern. But in Basins II and III the existence of a density current (Zambezi R. water-of-the-year) complicates matters until November. The deepest discontinuity water layer (T_2) is being displaced upwards following its volume increase while the true thermal discontinuity (T_1) is pushed downwards. Observations made in October (1965, 1968) still showed the presence of two thermoclines while in January (1966, 1968) only one was present, attesting of their possible junction in mid-water (20–25 m).

11.1.4. THE ANNUAL THERMAL CYCLE IN L. KARIBA BASINS

a. Basin I. The thermal cycle in Basin I depends entirely upon the Zambezi River, its water temperature (see 5.2) and its flow regime (see 5.1). On the basis of the data observed at Station 1205 in the upper part of the basin, the water dynamics in Basin I has been tentatively schematized (Fig. 36).

In mid-April the Zambezi peak flood reaches Basin I and progressively displaces downstream the lake water present. Minimum homothermy occurs in May, the temperature of the whole water mass dropping rapidly from about 24 C to 18 C. BEGG (1970) observed a minimum temperature of 17 C. In early June the Zambezi flow having decreased does not occupy the basin from surface to bottom any more, and

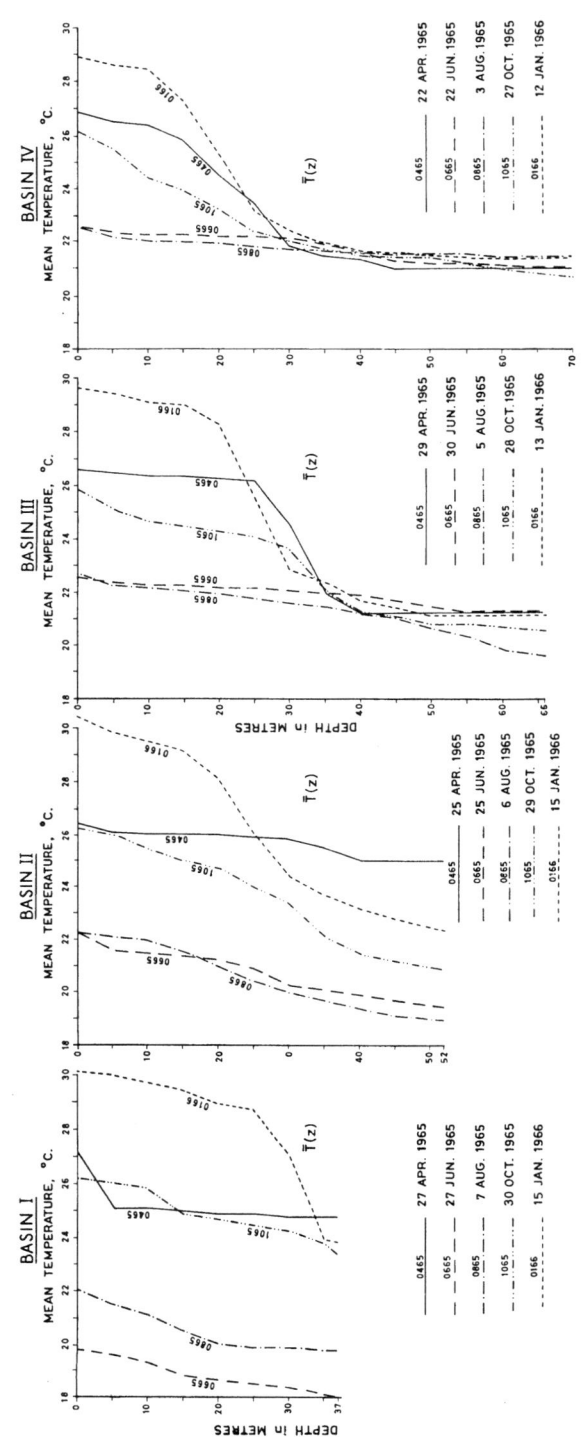

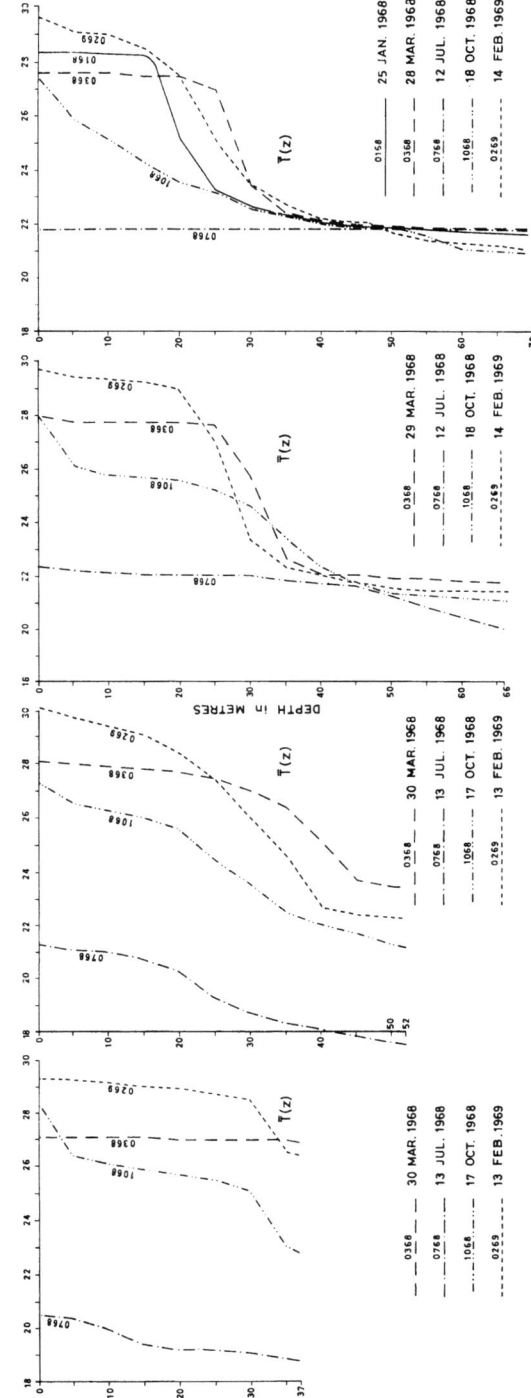

Fig. 35. Mean-temperature curves for Basins I to IV during two annual cycles (1965/66: top row; 1968/69: bottom row). Number on each curve indicates date of sampling as per individual legends.

Table 51. Average values for the maximum thermal gradients ($\overline{TG}$) and for the depths of the thermocline ($\bar{z}_T$).

Lake Cruise	6/65	7/68	8/65	10/65	10/68	1/66	1/68	2/69	3/68	4/65
Month	June	July	August	October	October	January	January	Febr.	March	April
Basin II (ig. 37)										
$\overline{TG}$, C/m	incompl. circul.	(0.34)	incompl. circul.	{ 0.30 { 0.30	{ 0.54 { 0.34	0.56	—	0.54	0.52	incompl. circul.
$\bar{z}_T$, m	— (Zambezi from 30m)	19.7 (Zambezi from 23m)	— (Zambezi from 25m)	{10 {32	{ 1.8 {30.0	24	—	35	38	— (Zambezi from 40m)
Basin III (Fig. 38)										
$\overline{TG}$, C/m	incompl. circul.	incompl. circul.	incompl. circul.	{ 0.28 { 0.28	{ 1.52 { 0.50	5.52	—	1.78	5.0	5.0
$\bar{z}_T$, m	(homothermy)	(Zambezi from 45m)	(Zambezi from 52m)	{ 2.0 {32.5	{ 1.0 {32.0	25	—	27.5	31	31.5
Basin IV (Fig. 39)										
$\overline{TG}$, C/m	circulation incompl.	Homothermy	reduced homothermy	0.28	0.26	0.96	0.44	5.00	1.42	0.82
$\bar{z}_T$, m				18.5	15.5	17.3	18.5	24	28	22.5

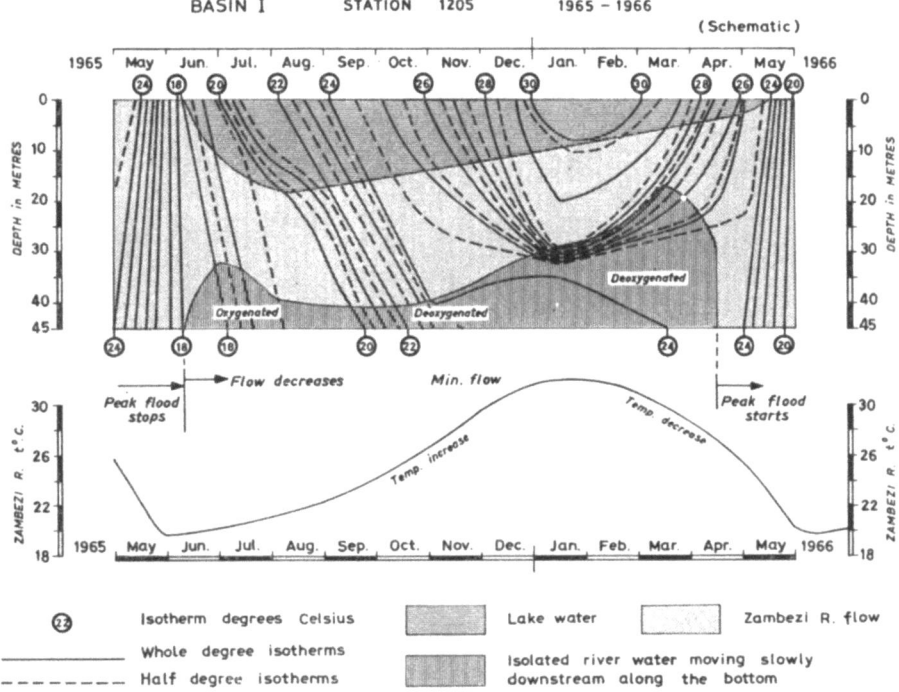

Fig. 36. The limnological cycle in Basin I (schematized).

the Zambezi water takes an intermediate level determined by its temperature-related density. A cold bottom layer becomes isolated from the rest of the water mass and flows independently along the basin's floor. As the Zambezi flow decreases and its temperature increases, the 'new' river water takes a higher level within the water mass. Maximum temperature is reached in January as the flow starts to increase. Butting against the bottom layer, a strong thermocline is created for a short while (late January – early February). As the flow increases and the temperature decreases, this thermocline is progressively weakened, and in April the peak flood invades the basin restarting the cycle.

b. Basin II. The thermal cycle in Basin II greatly depends upon the events taking place upstream, in Basin I, as described above. Mean temperature curves $\bar{T}(z)$ together with their intersection with $\bar{Z}(t)$ curves have been used to define the water masses present and their average characteristics within Basin II at various times of the annual cycle (Fig. 37 and Table 51).

The 'new' Zambezi water progressively leaves the lake's floor, first to take an intermediate level (Sep.–Oct.), then to reach the lake's surface (Jan.–Feb.). As in Basin I, this displacement isolates a bottom layer by a thermocline. In October, warm climatic conditions lead to the formation of a second thermocline which deepening later,

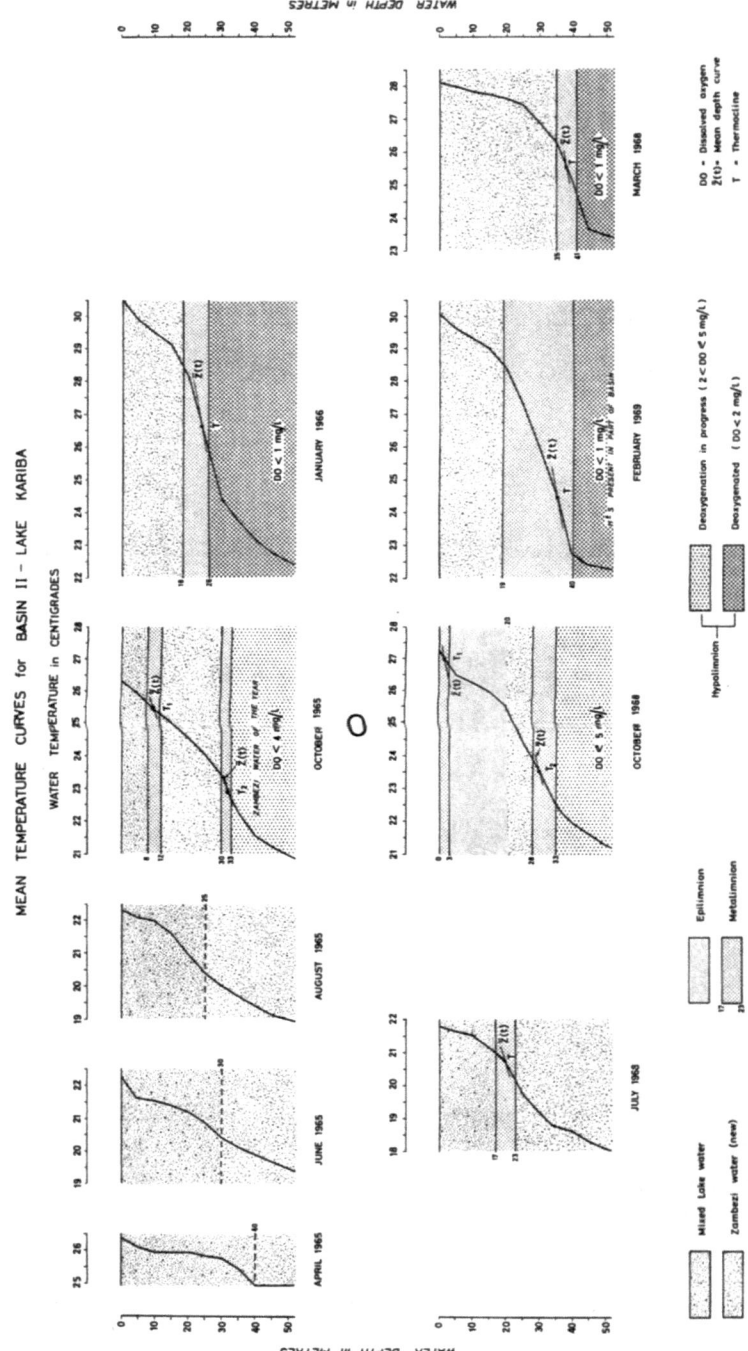

Fig. 37. The limnological cycle in Basin II. Mean-temperature curves $\bar{T}(z)$ and part of mean-depth curves $\bar{Z}(t)$, defining the position of the thermoclines (T) and their thermal gradient [slope of $\bar{Z}(t)$].

probably joins the first one towards the end of the calendar year. The $\bar{TG}$ generally approximates 0.50 C.m^{-1}. The joint influence of the Zambezi River and of the climate results in the absence of true 'average' homothermy in Basin II.

c. Basin III. A similar treatment of the cruise data has been adopted to define the average thermal cycle and its characteristics in Basin III (Fig. 38 and 42; Table 51). Average homothermy occurs in late June–early July just before cool Zambezi water-of-the-year reaches the basin in July, slowly flowing along the bottom of the old Zambezi channel. A thermocline soon establishes itself at the limit between the two water masses, 'old' and 'new' water (August). As the importance of the bottom layer increases, this thermocline ($\bar{TG}$ 0.28–0.50 C.m^{-1} in October) moves upwards from ab. 50 m to ab. 30 m (October), while simultaneously the increasing solar radiation generates a thermocline in the surface water layer (Oct. $\bar{TG}$ 0.28–1.52 C.m^{-1}). By January, the two metalimnion have joined together at 25 m depth and form a very strong barrier ($\bar{TG}$ 5.0 C.m^{-1}). Later the epilimnion is progressively deepened to ab. 30–35 m depth until the drop in air temperature (May–June) brings about total circulation which lasts only until the arrival of Zambezi water-of-the-year. It may be possible, on certain years, that homothermy only exists in the lower part of Basin III.

As pointed out earlier, it is in Basin III that the thermal resistance in the metalimnion is the greatest, most probably because of the combined effects of the Zambezi density flow and the local climate.

d. Basin IV. Following a similar treatment of the field data for Basin IV (Fig. 39 and 42; Table 51) average homothermy is shown to mainly occur in July at 21–22 C. By October cool Zambezi water-of-the-year has reached the basin, flowing along its floor in the deepest part of it. Sometimes the temperature difference between these two water masses is sufficient for a small metalimnion to be created. Within the surface water mass (upper 50 m), difference in temperature soon leads to the presence of a thermocline (October $\bar{TG} = 0.27$). The latter evolves following the standard pattern ($\bar{z}_T$ from 15 to 30 m) until its disappearance in late May–June, before total circulation (July). In February 1969 a very strong thermal resistance was observed around the 25-m depth. But in most cases the $\bar{TG}$ did not exceed 1.5 C.m^{-1}.

The annual cycle of thermal stratification in Basin IV was closely followed at station 'Sampakaruma' (Sta. 6047) in 1965 (Fig. 40). It shows how the difference between surface and deep water temperatures increased from overturn (July) until January, then started to decrease until homothermy was reached again. A thermal discontinuity reformed soon after; it suddenly disappeared again (high winds), before becoming strong enough (October) to successively establish itself. The metalimnion, as time passed, sunk deeper. The variation of the air temperature shown as monthly 24-hr mean values, was responsible for the thermal cycle observed. There was a steady increase of the thickness of the mixed layer and a deepening of the thermocline as the temperature fell down. As the cooling proceeded the epilimnion finally included the entire water mass. Homothermy occurred and total circulation followed soon later.

e. Conclusions: the annual thermal cycle. L. Kariba basins differ in the average characteristics of their annual thermal cycle. After the study of the 1965–66 cycle the dynamics involved in each case had been tentatively described (COCHE, 1968). But

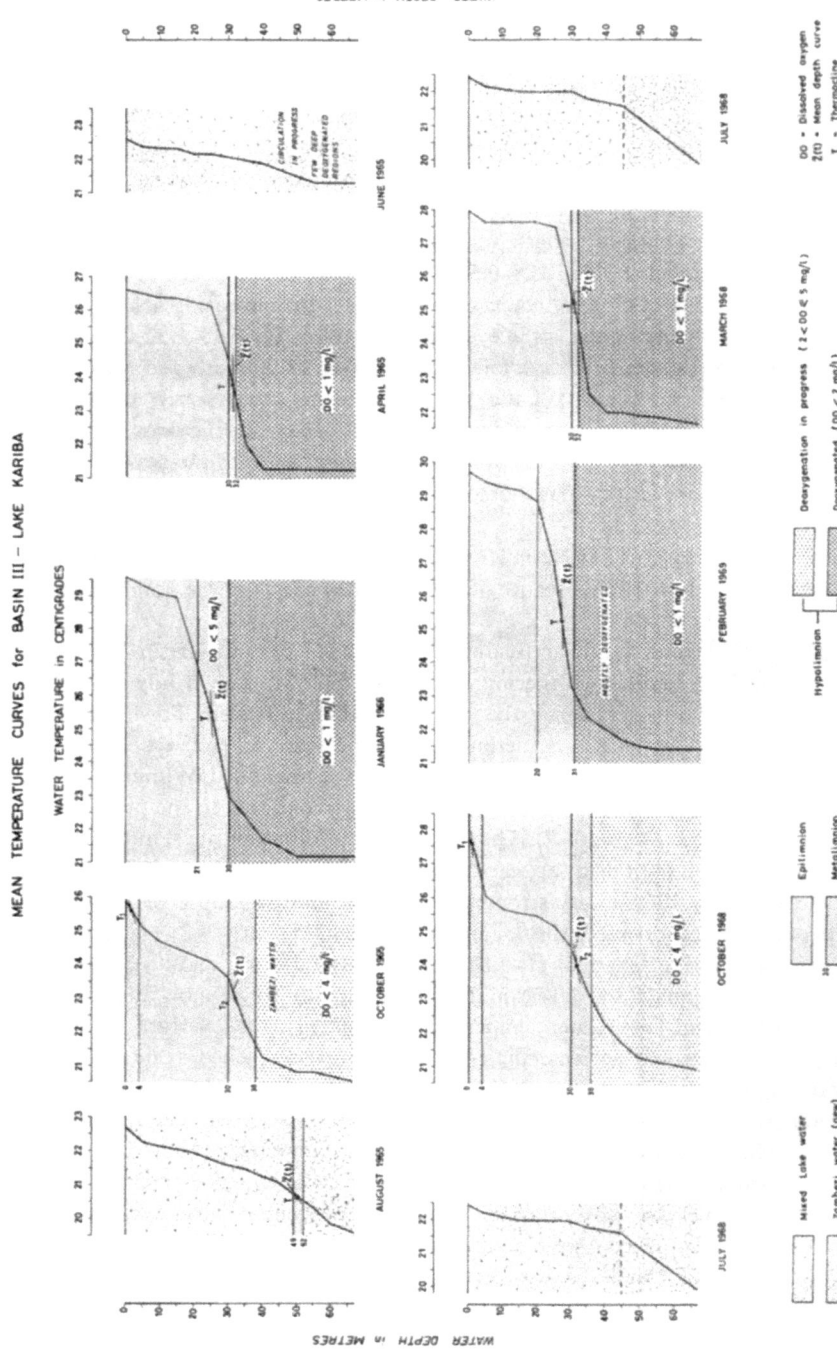

Fig. 38. The limnological cycle in Basin III [ff(z), Z(t), and T as defined for Figure 37].

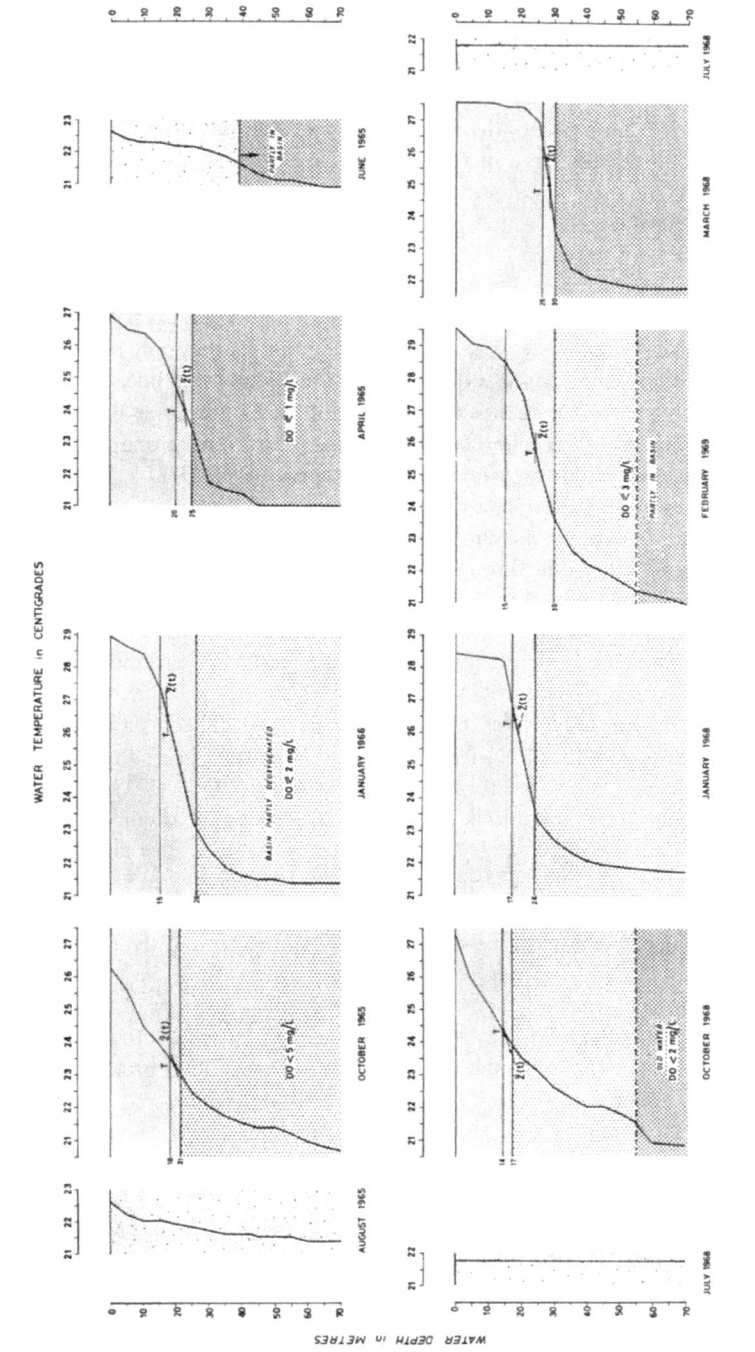

Fig. 39. The limnological cycle in Basin IV [$\bar{T}(z)$, $\bar{z}(t)$, and T as defined for Figure 37].

to-day additional data (incl. BEGG, 1970) from later years provide a sounder basis for generalization (Table 52 and Fig. 41).

As also pointed out by BEGG (1970), Basin I is mostly riverine in nature and under the direct influence of the Zambezi River. It partly loses this character as the flow of the latter drops to its minimum and takes an intermediate level within the water mass (density current) for several months. A strong stratification exists only towards January as the Zambezi discharge increases and pushes against the bottom layer of cooler water. In April–May homothermy occurs following the arrival of the Zambezi peak flood which flushes out Basin I, a total water replacement process termed by BEGG (1970, p. 782) entirety shift.

Basin II also possesses riverine characteristics but to a much lesser degree. Because here the influence of both the Zambezi River and the climate meet, average circulation never seems to be really complete. The above flushing-out process only takes place in the upper part of the basin whereas downstream from Binga longitude, the Zambezi water occupies only part of the basin's depth. It results that on average Basin II never reaches homothermy and must be considered as a transitional zone of the lake. In some years this zone might extend in the upper part of Basin III.

The lower two basins are more lacustrine, in particular Basin IV where the thermal cycle is typically dependent on the climatic conditions. In both basins however there is a riverine element left in the through-flow along the bottom of the old Zambezi channel of cool Zambezi water-of-the-year. Total circulation occurs in June (B III) and July (B IV) around 22° C. Stratification reappears soon after (September) and leads to a strong thermocline being present between November and May, around the 20–30 m depth. Resistance to mixing is particularly high in Basin III in December–January.

On the basis of the above considerations, Lake Kariba is defined as a warm monomictic reservoir (acc. classif. HUTCHINSON, 1957). The pattern of the thermal cycle appears well established since the formation of the lake in 1959 (HARDING, 1966, COCHE, 1968). Only slight variations in timing have probably occurred reflecting variations in the Zambezi R. water characteristics (esp. temperature and flow) and in local meteorological conditions. Density current, tilt of the isotherms, and changing water quality along the longitudinal axis point out to the river–lake nature of Lake Kariba, dammed at one end and receiving its major inflow through the opposite end. Such hybrid nature leads to a relatively complex thermal cycle, best studied and described on the basis of relatively homogeneous longitudinal sections of the reservoir.

The average depth of the mixed layer (taken as the depth at which the thermocline is found $= \bar{z}_T$) is generally greater than 15 m in Basin IV and 20 m in Basins III and II (Tables 51, 52). Therefore (Fig. 21), at least 40, 60 and 65 percent of the water volume in Basin IV, III and II respectively are at all times available to biological production.

But, for primary production, MORTIMER (1969) has recently stressed the importance of the ratio of stirred depth (SD) to illuminated depth (ID). The illuminated depth (euphotic zone) in Lake Kariba has been shown earlier to average 15.4 m, 16.5 m, and 10 m in Basins IV, III and II respectively (Sec. 10.3.5c). The SD/ID ratio therefore has the following minimum values: Basin IV: $15/15.4 = 0.97$; Basin III: $20/16.5 = 1.21$; Basin II: $20/10 = 2.00$.

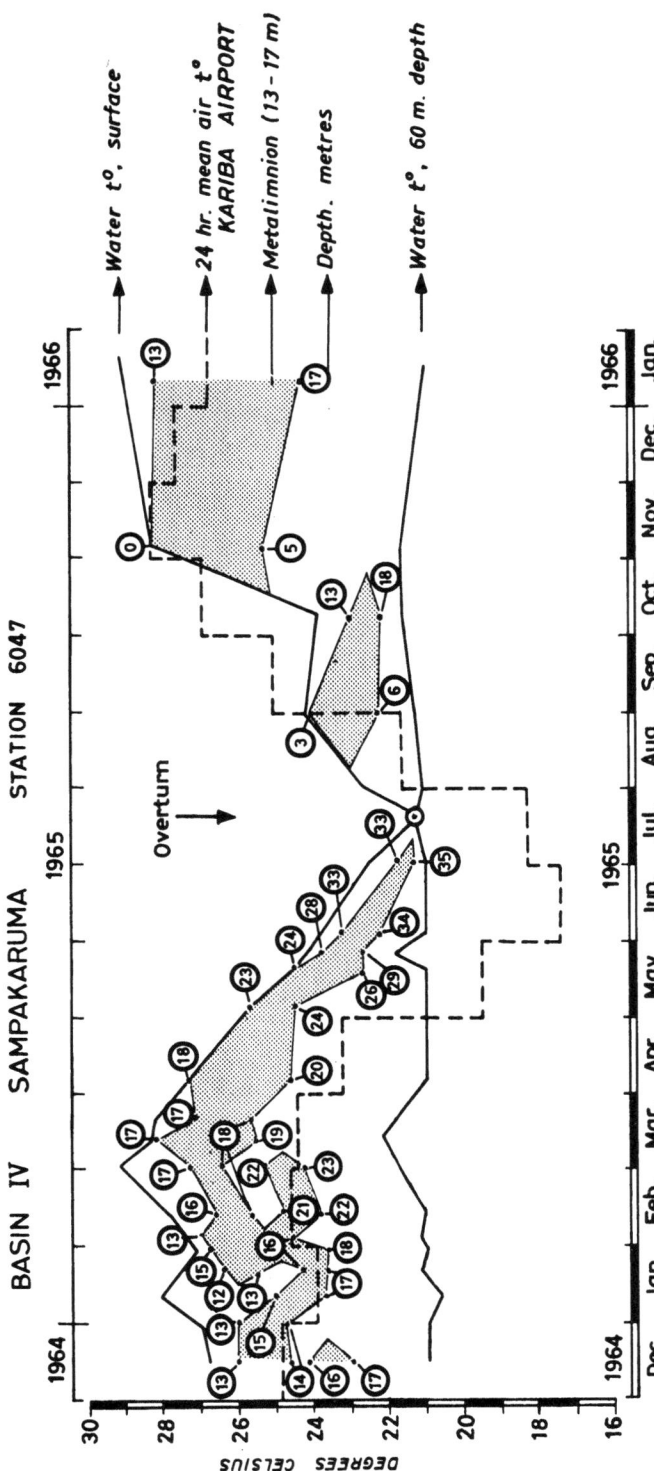

Fig. 40. Annual variation of the temperature differences between surface and deep water, and stratification characteristics as related to the air-temperature cycle at Station 6047 in 1965.

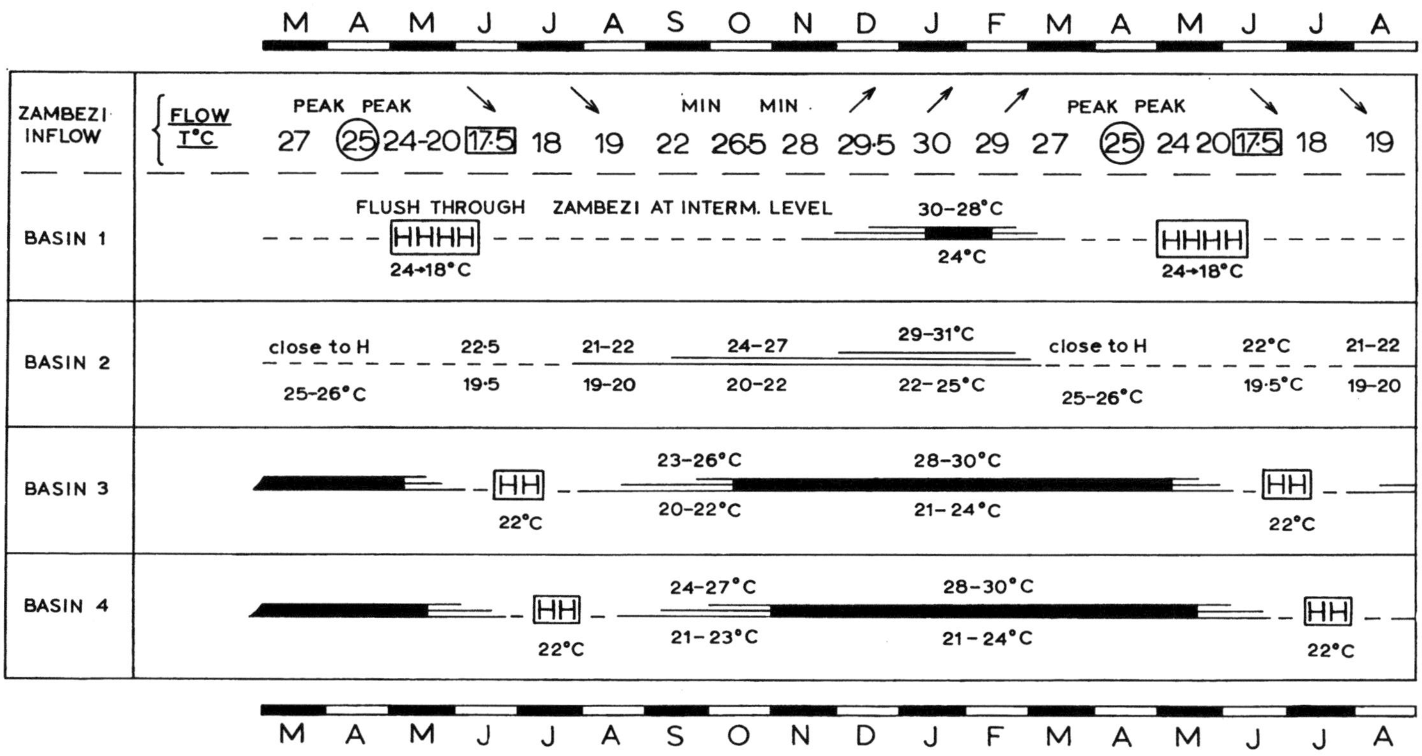

Fig. 41. Schematized thermal cycles in the four basins of Lake Kariba. General values of water temperatures in °C; homothermy (H); thickness of the line is proportional to the strength of the stratification.

Table 52. Average characteristics of the thermal cycles in Lake Kariba.

Basin II – Fig. 36							Basin III – Fig. 37							Basin IV – Fig. 38						
Month	Thermocline		Metalimnion, m			Dynamics	Month	Thermocline		Metalimnion, m			Dynamics	Month	Thermocline		Metalimnion, m			Dynamics
	Depth, m	Slope °C/m	From	To	Thickness			Depth, m	Slope °C/m	From	To	Thickness			Depth, m	Slope °C/m	From	To	Thickness	
Apr.						Fresh Zambezi water invades the basin and spreads along floor.														
May																				
Jun.						Volume of Zambezi water increased; great decrease in water t°.														
Jul.	20	0.25	19	23	4	Same trend accentuated, water t° reaching its minimum.	Jul.						Total circulation. New Zambezi water reaches basin.	Jul.						Total circulation and homothermy. Circulation; surface heating increases. Stratification starts as surface
Aug.						Zambezi water flows at intermediate level;	Aug.	51	(0.17)	49	52	3	Top layer in circulation. Zambezi water over bottom.	Aug.						
Sep.							Sep.							Sep.						
Oct.	30	0.25	28	35	7	cool water-of-the-year becomes isolated along bottom.	Oct.	33	0.35	30	36	6	Strong surface heating. Zambezi water over bottom.	Oct.	16	0.20	14	17	3	heating increases. New Zambezi water along bottom.
	32	0.20	30	33	3			35	0.40	30	38	8			19	0.25	18	21	3	
Nov.							Nov.						Rains start.	Nov.						Rains start.
Dec.							Dec.						Two metalimnion join together.	Dec.						
Jan.	25	0.50	18	27	9	Fresh Zambezi water at surface.	Jan.	24	1.50	21	30	9		Jan.	17	1.90	15	26	9	
															18	0.70	17	24	7	
Feb.	37	0.50	24	40	16	Increasing volume of Zambezi water.	Feb.	27	1.80	20	31	11	Mixed layer thickness increases.	Feb.	24	1.60	15	30	15	Mixed layer thickness increases.
Mar.	39	0.35	35	41	6	Rapidly decreasing water t° (cf. Basin I).	Mar.	31	2.00	30	32	2		Mar.	28	1.10	26	30	4	
							Apr.							Apr.						
							May	31	3.00	30	32	2	Rains stop. Air temp. drops.	May	23	0.60	20	25	5	Rains stop. Air temperature drops.
							Jun.						Circulation increases to become total at end of month.	Jun.						Circulation increases and thermocline breaks down.

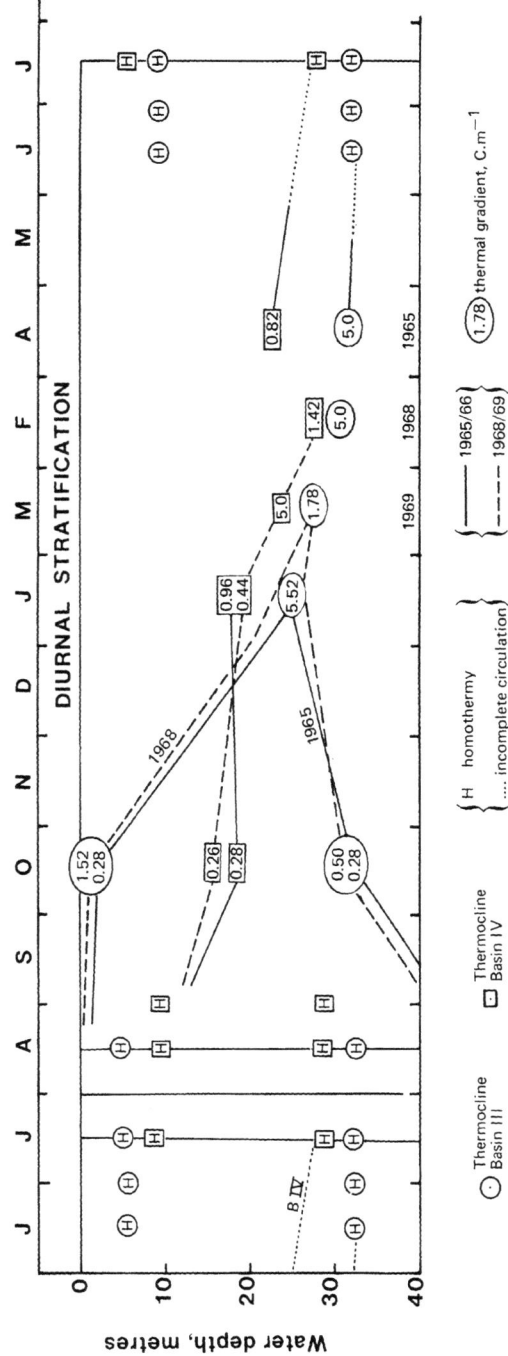

Fig. 42. Average maximum thermal gradients and mean depths of the thermocline in Basins III and IV throughout the annual cycle.

On the basis of TALLING's (1971) research such ratios are not high enough to impose a condition of morphometric oligotrophy on the primary production. But these values are minimal and, particularly in Basin II, it might happen during the year that the SD/ID ratio becomes greater than five.

11.1.5. THE ANNUAL THERMAL CYCLE IN SOME AFRICAN RESERVOIRS

The annual thermal cycle has been recently studied in some of the newly created reservoirs in Africa (Annex I) which enable comparisons with results obtained in L. Kariba. VINER (1970a) and CZERNIN-CHUDENITZ (FAO/UNDP 1971) studied thermal conditions in L. Volta, Ghana: IMEVBORE (1970) and HENDERSON (1973) reported on L. Kainji, Nigeria; ENTZ & RAMSEY (1973) worked on L. Nasser, Egypt. Limnological research has also started on L. Kossou, Ivory Coast.

Only L. Kariba and L. Nasser are truly monomictic reservoirs, total circulation being in both related partly to the incoming flood water and partly to the cooling of the air temperature in the winter season. In L. Kainji the situation is more complex because of the strong influence of drawdown and inflowing flood. Full circulation occurs in December–January each year. A strong thermal stratification follows (Feb.–mid May) which disappears following an important drawdown of the water level (9–10 m) to accommodate the incoming floods. During the refilling period vertical mixing is reduced but without evidence of stability. Rapid cooling in November–December (flood and air) leads to strong vertical mixing and complete circulation in December. L. Volta may be defined as a warm polymictic reservoir (G. BRETSCHKO, pers. comm.). One circulation period is onset by heavy rains and the arrival of the flood, the second homothermal conditions being caused by the cooling brought about by the harmattan, a cool dry wind.

Temperature at homothermy is related to the previous 'winter' temperature and varies according to the climate. It is lowest in L. Nasser (18° C) and highest in L. Volta (28.5° C). The period of total circulation is relatively short in all these reservoirs (three months max). Stratification reappears quickly and a metalimnion is formed. The latter is found on average at a relatively shallow depth throughout L. Kainji (15–20 m), L. Kariba (15–25 m), and L. Nasser (13–20 m) whereas it occurs much deeper in L. Volta (25–35 m).

11.2. Energy content of Lake Kariba

The energy content of the lake imparted by temperature and density stratification, should be considered to provide a better understanding of the hydrographic and biological phenomena which are observed in Lake Kariba (RUTTNER, 1963).

The energy entering a lake thermally stratified exists under two distinct forms:
(i) the thermal energy or stored heat which the lake can impart to its surroundings on cooling. In a tropical monomictic lake both the annual heat budget (11.2.1) and the amount of residual heat (11.2.2) present at homothermy must be studied successively. The variation of the heat content throughout the annual cycle (11.2.3) and the maximum heat content (11.2.4) will then be considered.

(ii) the mechanical energy. Not considering the kinetic energy incorporated in water masses in motion, the potential energy present in a stratified lake should be estimated using BIRGE's concept of the work performed by the wind (11.2.5) and SCHMIDT's approach by way of computation of the stability (11.2.6) of the stratification (BIRGE, 1916).

In order to be able to compare L. Kariba results with those obtained from various other monomictic lakes, three new indices are introduced which not only express the estimated values on a surface unit basis but also relate them to the mean depth of the lake under consideration.

11.2.1. THE BIRGEAN ANNUAL HEAT BUDGET AND THE ANNUAL HEAT INDEX

a. Birgean Annual Heat Budget

BIRGE (1915) has defined the annual heat budget of a lake (BAHB) as the amount of heat necessary to raise a column of water of unit base (sq.cm) and a height equal to the mean depth of the lake (cm) from the minimum temperature of winter to the maximum summer temperature.

On the basis of this definition, the BAHB is estimated for Lake Kariba as follows:
(i) The lake is considered as made of four distinct basins. Although interrelated by the existing flow-through from the Zambezi River down to the dam and back into the downstream channel, these four basins are thought to be sufficiently distinct to justify computations of BAHB for each of them separately. A general BAHB for Lake Kariba as a whole would not have much meaning.
(ii) Each basin is divided into strata 5-m thick, down to the maximum depth. Partial Volume Ratios (PVR – Table 32) and Birge's Reduced Thickness (RT – Table 34) are calculated for each stratum.
(iii) BAHB is obtained from PVR values and mean temperatures (T_{zs}, T_{zw}) of the strata, and mean depth of the lake ($\bar{z}$) as shown as an example in Table 53.
(iv) To provide most representative results in each basin, sets of data have been chosen from the following cruises:

Cycle	1965–1966	1968–1969
Basin I	K6/65 vs. 1/66	K7/68 vs. 2/69
Basins II–III–IV	K8/65 vs. 1/66	K7/68 vs. 2/69

Results (Table 55) show Birgean Annual Heat Budgets varying between 10 800 and 19 300 cal. cm^{-2}. yr^{-1}. Taking into account possible errors introduced by out-of-date sampling, these values have been rounded off. The highest BAHB is found in Basin II (20 000 cal. cm^{-2}. yr^{-1}) while in most of Lake Kariba the BAHB approximates 15 000 cal. cm^{-2}. yr^{-1}.

For Lake Victoria (homothermy at 23° C), TALLING (1966) has estimated a BAHB between 9 000 and 11 000 cal. cm^{-2}. yr^{-1}, a considerable depth (30–60 m) being

Table 53. Birgean Annual Heat Budget for Basin III – An example of computations.

Depth Stratum m (1)	PVR (2)	August 1965		January 1966		July 1968		February 1969	
		Mean Temp.C (3)	(2)×(3) (4)	Mean Temp.C (5)	(2)×(5) (6)	Mean Temp.C (7)	(2)×(7) (8)	Mean Temp.C (9)	(2)×(9) (10)
0– 5	0.189	22.50	4.25	29.50	5.58	22.30	4.21	29.55	5.58
5–10	0.146	22.25	3.25	29.25	4.27	22.15	3.23	29.35	4.29
10–15	0.133	22.15	2.95	29.05	3.86	22.05	2.93	29.25	3.89
15–20	0.122	22.05	2.69	28.65	3.50	22.00	2.68	29.05	3.54
20–25	0.107	21.90	2.34	26.95	2.88	22.00	2.35	27.95	2.99
25–30	0.078	21.70	1.69	24.25	1.89	22.00	1.72	25.15	1.96
30–35	0.075	21.55	1.62	22.65	1.70	21.90	1.64	22.80	1.71
35–40	0.062	21.40	1.33	22.05	1.37	21.75	1.35	22.15	1.37
40–45	0.040	21.20	0.85	21.60	0.86	21.65	0.87	21.85	0.87
45–50	0.024	20.90	0.50	21.35	0.51	21.40	0.51	21.60	0.52
50–55	0.016	20.55	0.33	21.20	0.34	21.00	0.34	21.45	0.34
55–60	0.003	20.15	0.06	21.20	0.06	20.80	0.06	21.40	0.06
60–65	0.003	19.80	0.06	21.20	0.06	20.25	0.06	21.40	0.06
65–66	0.002	19.70	0.04	21.20	0.04	20.00	0.04	21.40	0.04
Total	1.000	—	21.96	—	26.92	—	21.99	—	27.22

Temp. difference (Δt)   4.96   5.23
BAHB cal. cm^{-2} ($\Delta t.\bar{z}$)   13164   13880

reached by the upper mixed layer. Under somewhat warmer conditions (global radiation: 550 vs. 400 cal. cm^{-2} day^{-1}), Lake Kariba presents a slightly greater BAHB (15 000 cal. cm^{-2} yr^{-1}). As pointed out by TALLING (1966), these values tend to show that monomictic lakes sited in tropical climates might have a BAHB only one half to one third of the most common values, observed in temperate climates for lakes with similar thermal cycle (Table 56). But as seen later (see b), this does not seem to be the case if the mean depth is taken into consideration.

In temperate conditions the warming of lakes generally takes between 40 and 60 percent of the total radiation delivered to their surface (HUTCHINSON, 1957). On the contrary, in Lake Kariba and probably in other tropical areas, the warming itself from the minimum to the maximum heat content (e.g. BAHB) takes much less of the total radiation available. Based on climatological data available for Kariba and Binga townships (Tables 3 and 5), the mean global radiation delivered to the surface of the lake during the warming phase (22 July–15 February) was 114 000 cal. cm^{-2} (avg. 550 cal. cm^{-2} day^{-1}). From these, 15 000 cal. cm^{-2} were used as BAHB or about 13 percent only. This low percentage with which the available radiation is used in L. Kariba most probably relates to a high Residual Heat (Sec. 11.2.2). But it might also be correlated with the great evaporation rate (avg. 7 mm/day) present in subtropical conditions throughout the year (Table 4).

b. Annual Heat Index

The above BAHB are based on the mean depth ($\bar{z}$) of the lake following the original method proposed by BIRGE (1916) and the synthesis presented by HUTCHINSON (1957). To better compare L. Kariba results among themselves and with those estimated for other monomictic lakes, I have computed the BAHB per metre of mean water depth to obtain the Annual Heat Index (cal. cm^{-2} m^{-1} yr^{-1}) as follows:

$$AHI = BAHB \cdot \bar{z}^{-1}$$

the mean depth ($\bar{z}$) being expressed in metres. Such AHI have been computed from the world data available (Table 56).

Results show that on such basis of comparison, L. Kariba basins have a decreasing AHI as one progresses towards its deepest areas (dam). Basin I has the highest index most probably because of the direct influence of the Zambezi River (T_w around 19° C and entirety shift of the water mass). Basin II is intermediate, as from other points of view. In the lower two basins, AHI ranges from 420 to 540 cal. cm^{-2} m^{-1} yr^{-1}.

Compared to the values observed in other parts of the world (Table 56), the Lake Kariba minimum AHI (420 cal.) is higher than any other AHI. Only L. Mead (410 cal.) and L. Bolsena (407 cal.) have a comparable average heat budget. On the other hand, L. Victoria index (224–275 cal.) is quite comparable to L. Lucerne and L. Geneva AHI, but greater than those for L. Como and L. Thun.

Therefore on a mean depth basis monomictic tropical lakes do not necessarily exhibit an annual heat budget smaller than those observed in temperate climates, as had appeared from the comparison of BAHB (see a).

Table 54. Basin IV. Variation of the heat content above 4° C during the 1968/69 cycle (HC, cal. cm^{-2}).

Depth Stratum	RT cm	January 1968		March 1968		July 1968		October 1968		February 1969	
		T-4C	HC	T-4C	HC	T-4C	HC	T-4C	HC	T-4C	HC
0– 5	478.37	24.35	11648	23.60	11290	17.80	8515	22.65	10835	25.35	12127
5–10	438.50	24.35	10677	23.60	10349	17.80	7805	21.55	9450	25.05	10984
10–15	411.93	24.30	10010	23.55	9701	17.80	7332	20.75	8548	24.75	10195
15–20	332.20	22.80	7574	23.50	7807	17.80	5913	19.95	6627	24.00	7973
20–25	325.56	20.35	6625	23.25	7569	17.80	5795	19.40	6316	22.30	7260
25–30	255.79	19.00	4860	21.25	5436	17.80	4553	18.90	4834	20.30	5193
30–35	249.15	18.45	4597	18.95	4721	17.80	4435	18.40	4584	19.05	4746
35–40	209.29	18.20	3809	18.25	3820	17.80	3725	18.15	3799	18.35	3840
40–45	159.46	18.00	2870	18.05	2878	17.80	2838	18.10	2886	18.05	2878
45–50	129.56	17.90	2319	17.95	2326	17.80	2306	18.00	2332	17.85	2313
50–55	102.98	17.85	1838	17.85	1838	17.80	1833	17.75	1828	17.55	1807
55–60	76.41	17.80	1360	17.80	1360	17.80	1360	17.35	1326	17.35	1326
60–65	53.15	17.75	943	17.80	946	17.80	946	17.05	906	17.25	917
65–70	36.54	17.65	645	17.80	650	17.75	649	16.95	619	17.10	625
70–75	29.90	17.55	525	17.80	532	17.70	529	16.90	505	17.00	508
75–80	9.97	17.50	174	17.80	177	17.70	176	16.90	168	16.90	168
80–85	9.97	17.50	174	17.80	177	17.70	176	16.90	168	16.90	168
85–90	9.97	17.50	174	17.80	177	17.70	176	16.90	168	16.90	168
90–93	3.32	17.50	58	17.80	59	17.70	59	16.90	56	16.90	56
Total heat content			70880		71813	RH =	59121		65955	MHC =	73252

RH: Residual Heat; MHC: Maximum Heat Content; RT: Reduced Thickness: T: mean temperature of stratum.

11.2.2. RESIDUAL HEAT AND TROPICALITY INDEX

a. Residual Heat

For warm monomictic lakes, the minimum 'winter' value of the heat content gives a measure of the tropical against temperate nature of the lake considered (BIRGE, 1915). This permanent stock of heat is here termed Residual Heat (RH). It is equivalent to the negative Winter Heat Incomes given by HUTCHINSON (1957 – Table 53) for monomictic lakes.

By definition Residual Heat is estimated as the amount of heat above 4° C present in a column of water of unit base (sq.cm) and a height equal to the mean depth of the lake (cm), at the time of its lowest heat content (e.g. circulation). The bases for the computations have been BIRGE's Reduced Thickness (Table 34) and mean temperatures of 5-m thick strata as shown in Table 54 for the July 1968 data.

Results (Table 55) show that Residual Heat varies in L. Kariba from basin to basin, between 18 000 (B I) and 59 000 (B IV) cal. cm^{-2}. It roughly increases as the average depth of the basin. It also points out that the tropical nature of the Basins increases towards the dam.

This tropicality of Lake Kariba (in general higher than 40 000 cal. cm^{-2}) originates from the relatively high water temperature at which homothermy occurs (ab. 20° C). In Lake Victoria where isothermal conditions happen around 23° C (TALLING, 1966), the estimated RH is even greater (ab. 76 000 cal. cm^{-2}). But strangely enough on similar bases most of Lake Kariba is less 'tropical' than L. Como (European Alps) which stands out among other lakes under a temperate climate (Table 56). From the same comparative data it also appears that Basin I is less 'tropical' than several other warm monomictic lakes sited in much cooler climates.

b. Tropicality Index

In view to better assess the relative tropical nature of lakes from the data available, the above estimates for Residual Heat (RH) have been reduced to a common metre-of-depth basis leading to the definition of the Tropicality Index (TI cal. cm^{-2} m^{-1}). The mean depth ($\bar{z}$) being expressed in metres:

$$TI = RH \cdot \bar{z}^{-1}$$

In fact this index equals 100 ($T_w - 4$), T_w being the mean isothermal temperature (in °C) of the water mass during the full circulation period. On the other hand, the latter can be estimated as:

$$T_w = (0.01 \, TI) + 4$$

The Tropicality Index of known monomictic lakes has been calculated for further comparisons (Table 56).

Within L. Kariba, TI ranges from about 1400 to 1850 cal. cm^{-2}. m^{-1}. Its value increases towards the dam but the maximum average tropicality is observed in Basin III. It has been seen before (11.1.4c) that homothermy does not last very long in this

Table 55. Birgean Annual Heat Budgets (BAHB cal. cm^{-2} yr^{-1}), Residual Heats (RH cal. cm^{-2}), and Maximum Heat Contents (MHC cal. cm^{-2} ab. 4C) in Lake Kariba basins.

Basin	I		II		III		IV	
Year	1965/66	1968/69	1965/66	1968/69	1965/66	1968/69	1965/66	1968/69
BAHB	12907	11498	16231	19256	13164	13880	10863	14119
RH	19176	20063	41498	39097	47650	47785	59408	59121
MHC	32083	31561	57729	58353	60814	61665	70271	73240
BAHB	15000		20000		14000		14000	
RH	18000		39000		48000		59000	
MHC	33000		59000		62000		73000	
z̄ metres	12.6		24.0		26.5		33.2	

Table 56. Birgean Annual Heat Budget (BAHB), Annual Heat Index (AHI), Residual Heat (RH), Tropicality Index (TI), Maximum Heat Content (MHC), and Maximum Heat Index (MHI) of adequately known monomictic lakes of the world.

Lake	z m	BAHB[1] cal.cm^{-2} yr^{-1}	AHI cal.cm^{-2}m^{-1}yr^{-1}	RH[1] cal.cm^{-2}	TI cal.cm^{-2} m^{-1}	MHC[1] cal.cm^{-2}	MHI cal.cm^{-2}m^{-1}
Kariba, Africa							
Basin I	13	15,000	1150	18,000	1390	33,000	2540
Basin II	24	20,000	834	39,000	1625	59,000	2459
Basin III	26	14,000	539	48,000	1846	62,000	2385
Basin IV	33	14,000	423	59,000	1787	73,000	2210
Victoria, Africa	40	9–11,000	224–275	ab. 76,000	1900	ab. 85,000	ab. 2124
Como, Eur. Alps	185	32,800	177	50,000	270	82,800	447
Thun, Eur. Alps	135	24,200(?)	179	9,200	68	33,200(?)	247
Lucerne, Eur. Alps	104	24,500(?)	236	14,500	139	39,000(?)	375
Geneva, Eur. Alps	154	36,600(?)	238	23,200	150	59,800	388
Ness, Britain	133	37,200	280	14,600	110	51,800	390
Lugano, Eur. Alps	130	40,000(?)	307	17,000	131	57,000(?)	438
Morar, Britain	87	29,400	338	13,000	150	42,400	488
Orta, Eur. Alps	71	25,400	358	6,400	90	31,800	448
Bourget, Eur. Alps	81	29,200	361	2,800	35	32,000	396
Bolsena, Eur. Alps	78	31,600	407	29,000	372	60,600	779
Mead, N. Amer.	59	24,200	410	22,000	373	46,200	783

[1] Acc. BIRGE, 1915 and HUTCHINSON, 1957 modified.

particular basin, taking place earlier at a slightly higher temperature than in Basin IV. L. Victoria has a slightly greater TI (1900 cal), again homothermy taking place at a higher water temperature (ab. 23° C, TALLING, 1966) than in L. Kariba.

Better comparisons (cf. a above) can now be made for known monomictic lakes, among themselves as well as between lakes sited in different climatic regions (Table 56). For example the outstanding tropicality of L. Como (RH 50 000 cal.) now drops to average level (TI 270 cal.) lower than L. Bolsena and L. Mead. Also L. Geneva and L. Morar become similar from TI point of view. It also shows how much more 'tropical' are the lakes sited in tropical climates. Their Tropicality Index is at least five times greater.

11.2.3. VARIATIONS OF THE HEAT CONTENT THROUGHOUT THE ANNUAL CYCLE

Heat contents above 4° C (HC) present on different dates of the annual cycle have been obtained from computations similar to those used to estimate the Residual Heat (see Table 54 for example). In fact, the latter corresponds to HC minimum. The Maximum Heat Content (MHC) is of particular interest as it coincides with the sum of Birgean Annual Heat Budget and Residual Heat (Table 55). This MHC also corresponds in warm monomictic lakes to the Summer Heat Income[15] of a temperate dimictic lake as tabulated for the latter by HUTCHINSON (1957 – Table 53).[16]

The variation of the heat content in the four basins is graphically presented in Figure 43 which groups results obtained during the two observed annual cycles (1965/66 and 1968/69), in the pelagic region. More detailed data are given in Table 54 for Basin IV.

Basin I has to be considered separately (Fig. 43): its heat content is minimum in early June when cold Zambezi water invades the basin. This heat content then increases until mid-January (max. 33 000 cal. cm^{-2}). The loss of heat starts, increases gradually, and reaches its maximum rate in May, as the temperature of the inflowing Zambezi River drops while the flood reaches its peak.

In Basins III and IV, the HC variation follows a different pattern: minimum HC coincides with homothermy in July (circulation). The warming phase lasts from late July until mid-February, the rate of warming being most consistent in Basin IV (ab. 2165 cal. cm^{-2} month^{-1}). In Basin III the rate of warming gradually decreases as the warming progresses. MHC is observed in mid-to late February (62–73 000 cal. cm^{-2}), four months after global radiation has reached its maximum. From April to June, the stored heat is then lost at a fast rate, similar in the two basins (ab. 4 200 cal. cm^{-2} mth^{-1}). The warming phase thus lasts six months, the cooling phase being

---

[15] 'The amount of heat needed to raise the lake from an isothermal condition at 4° C up to the highest observed summer heat content' (HUTCHINSON, 1957).

[16] A different terminology is preferred here to distinguish temperate dimictic lakes from warm monomictic lakes. It leads to Summer Heat Incomes different from those presented as such (θ_{bs}) by HUTCHINSON for monomictic lakes. Comparisons then become possible (same ref. temperature 4 °C) with temperate dimictic lakes.

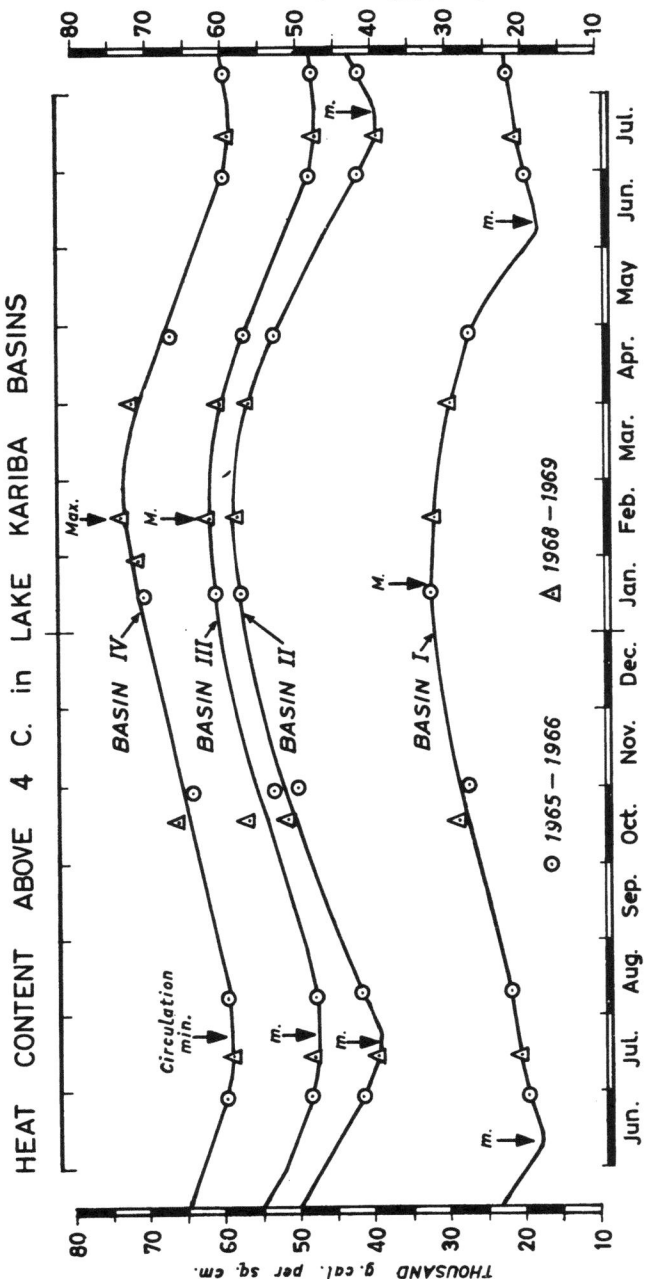

Fig. 43. Seasonal variation of the heat content above 4 °C in Lake Kariba basins.

one and a half times shorter. To the contrary of what has been observed in temperate climates (HUTCHINSON, 1957), the rate of heating is not directly related to the rate of delivery of solar radiation.

Basin II shows a similar but more accentuated pattern of variation during the time of minimum HC. This is due as seen earlier, to the existence in July of a double cooling effect: as the air temperature reaches its minimum, cool Zambezi water invades the upper half of Basin II accelerating the cooling rate and depressing the minimum HC.

These seasonal variations of Lake Kariba heat content (Fig. 43 exc. Basin I) are closely related to those of the global radiation (Tables 3-5) combined with those of the wind regime. The start of the cooling phase (April) coincides with the drop of the mean air temperature and with the appearance of the cold, dry, south-east winds which contribute to accelerate the lake water cooling in May–June. Later, the start of the warming phase is related to the increase of the sun radiation and to the switch (September) to warm northerly winds. It does not appear that the HC variations are mainly related to those of the evaporation rate as observed in Lake Victoria where the seasonal pattern is much less pronounced (TALLING, 1966).

Changes in heat content in 5-m thick strata are typically exemplified by data calculated for Basin IV in 1968/1969 (Table 54). Most of the heating takes place in the surface 5-m layer. Throughout the year, between one-fourth to one-third of the total HC is contributed by the top 10-m water layer.

Moreover the behaviour of the HC variations is different above and below the 50-m depth. This is related to the flow of cold Zambezi water along the bottom of the lake. As warming starts in the top half of the water mass, cooling still progresses below the 50-m depth (October). In the top layer maximum heat content is reached earlier (February) than in the deep layer (ab. March).

11.2.4. MAXIMUM HEAT CONTENT AND MAXIMUM HEAT INDEX

a. Maximum Heat Content

Maximum Heat Content above $4°$ C (MHC) varies in Lake Kariba between 33 000 and 73 000 cal. cm^{-2} (Table 55). In most of the lake, it varies between 60 000 and 70 000 cal. cm^{-2} being contributed mostly by Residual Heat, in particular in Basins III and IV. Compared to other warm monomictic lakes of the world (Table 56), MHC in L. Kariba are greater than most of those listed. They are similar to the values calculated for L. Geneva and L. Bolsena. But they are definitely smaller than those presented for L. Como and L. Victoria. Basin I is again outstanding from this point of view.

Earlier it has been shown that the Annual Heat Budget (BAHB) of Basin IV represents about 13 percent only of the total radiation available to the water surface. But if a simila rrelationship is calculated for Maximum Heat Content (above $4°$ C as for temperate dimictic lakes), it is found that MHC (Basins III–IV) represents 54–65 percent of the global radiation available during the warming phase. These last percentages are very close to the range of 40 to 60 percent observed for the BAHB of lakes in temperate regions (HUTCHINSON, 1957).

b. Maximum Heat Index

For similar reasons as presented earlier (11.2.2 b), a Maximum Heat Index (MHC per metre of water depth) is defined as:

$$\text{MHI} = \text{MHC} \cdot \bar{z}^{-1} = \text{AHI} + \text{TI}$$

the mean depth ($\bar{z}$) being expressed in metres.

Calculations have been made for the available data (Table 56). Results point out to the clear distinction to be made between true tropical lakes and lakes of a tropical nature, sited in temperate regions. On the other hand, within L. Kariba MHI regularly decreases from the point of major inflow towards the dam. The average MHI equals 2 400 cal. cm^{-2}. m^{-1}, a higher value than L. Victoria's. Most values for temperate monomictic lakes range from 400 to 450 cal., L. Bolsena and L. Mead being outstanding again (ab. 780 cal.).

11.2.5. The work of the wind in warming Lake Kariba

Considering that heating by radiation in a lake is negligible except at the surface, BIRGE (1916) suggested that the warming of the water is mainly done through the transference of warm lighter water from the surface to the deeper strata by the agency of the wind. This involves an expenditure of work part of which, being recorded in the thermal history as a change in temperature distribution, can be measured and calculated as 'the work of the wind'. A work budget can be established to be chiefly of use in aiding to interpret the heat budget.

In monomictic lakes the approach practically consists in computing the amount of work per unit area of the lake surface (B=g–cm. cm^{-2}) needed to distribute the amount of heat received during the warming phase, from the minimum to the maximum heat content such as (BIRGE, 1916):

$$B = \sum_{z=0}^{z_m} RT \cdot z \cdot (d_{zw} - d_{zs})$$

where, for each stratum considered separately: RT = Reduced Thickness, in cm; z = distance from the surface to the centre of the stratum in cm; d_{zw} = water density corresponding to the winter mean temperature, g.cm^{-3}; d_{zs} = water density corresponding to the summer mean temperature, g.cm^{-3}.

The work of the wind in Lake Kariba (Table 57) has been estimated from data collected in Basin IV during the 1968–1969 annual cycle on the following bases:

Top layer 50-m deep subdivided into 5-m thick strata; bottom layer (below 50 m) considered as not affected by the wind, its thermal regime being entirely dependent on the water flow moving through the basin at a temperature generally lower than T_w.

T_{zw} the homothermal water temperature (July 1968).

T_{zs} the mean temperature of each stratum at the time of maximum heat content (February 1969).

WC = (RT.z), work constant for Lake Kariba (Table 34).

Table 57. Work of the wind (B) done in behalf of 5-m strata to distribute heat (BAHB) through Basin IV during the 1968–1969 cycle (July–February).

Stratum	Mean Temp. °C		[1]Water density g.cm^{-3}			Work Constant[2]	B	BAHB
m	T_{zw}	T_{zs}	d_w	d_s	$d_w - d_s$	cm^2	gcm. cm^{-2}	calories
0– 5	21.80	29.35	0.9978444	0.9958705	0.0019739	119593	236	3612
5–10	—	29.05	—	.9959638	.0018806	328875	618	3179
10–15	—	28.75	—	.9960465	.0017979	514913	926	2863
15–20	—	28.00	—	.9962623	.0015821	581350	920	2060
20–25	—	26.30	—	.9967326	.0011118	732510	814	1465
25–30	—	24.30	—	.9972511	.0005933	703423	714	639
30–35	—	23.05	—	.9975556	.0002888	809738	234	311
35–40	—	22.35	—	.9977193	.0001251	784838	98	115
40–45	—	22.05	—	.9977879	.0000565	677705	38	40
45–50	—	21.85	—	.9978332	.0000112	615410	7	6
							4308	14290

[1] Acc. HUTCHINSON, 1957 – Table 7 (p. 204–205).
[2] From Table 34 as WC = RT × z.
[3] BAHB = $(T_{zs} - T_{zw})$.RT.

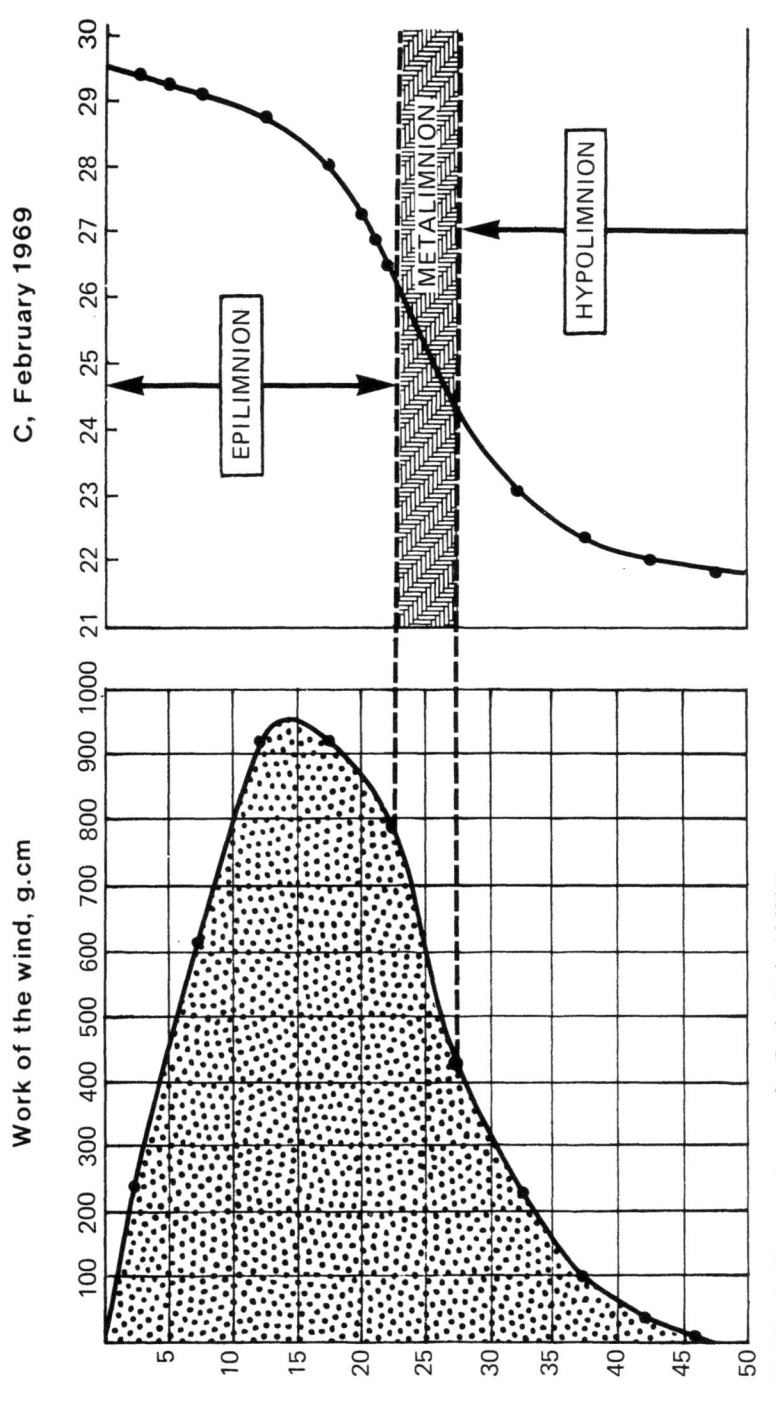

Fig. 44. Birge's direct-work curve for Basin IV in 1968/69.

The results show that the heat gained above 21.8° C by the Basin IV in 1968/69 amounted to 14 300 cal. per square centimetre of surface. To distribute this heat through the basin, it required about 4308 gcm of work. Comparable data given by HUTCHINSON (1957 – Table 58, p. 508) point out that this work budget is relatively high. Only two out of the sixteen lakes cited have a similarly high work budget: L. Pyramid, a desert lake in Nevada (3700–4055 gcm with TI = 125 cal cm^{-2}. m^{-1}) and L. Atitlan in Guatemala (3741 gcm), a truly tropical higher lake (TI ab. 1554 cal. cm^{-2} m^{-1}).

From world data available (HUTCHINSON, 1957), the work required to mix one calorie of heat averages between 0.08 and 0.12 gcm. In Lake Kariba, this efficiency amounts to 0.302 gcm.cal^{-1} if one only involves the BAHB figure. It stands out of the known range and points out to a very low efficiency of the work of the wind in mixing the water mass. This is probably correlated with the presence of a strong diurnal surface thermocline and the consequent high thermal resistance (BIRGE, 1916) existing in Lake Kariba for a great part of the year.

The work of the wind done in behalf of each 5-m water stratum in Basin IV (Table 57) has been plotted as BIRGE's direct-work curve, opposite to the mean temperature profile (Fig. 44). It shows how the strata 10–25 m situated in the lower epilimnion and in the upper metalimnion, contribute more than 60 percent of the total work of the wind performed. Maximum contribution originates from the 15-m level, at a depth about two-third the depth of the epilimnion. It rapidly decreases within the metalimnion and reaches low values as one progresses deeper through the hypolimnion.

11.2.6. STABILITY IN LAKE KARIBA

SCHMIDT's stability of a lake is defined 'as the amount of work needed to mix the entire body of water to uniform temperature without addition or subtraction of heat' (HUTCHINSON, 1957). While BIRGE's work of the wind (B) estimated the amount of work required to produce a certain stable equilibrium (i.e. stability), SCHMIDT's stability (S) estimates the amount of work necessary to continue the distribution of the heat content until homothermy – or indifferent equilibrium – is reached again. The summation of these two separate estimates (B + S) provides a measure of the work (G) needed to maintain hypothetical homothermy in the lake throughout the warming season.

The computation of the stability (S gcm) for a monomictic lake as a whole rests on applying the modified practical form of SCHMIDT's general formula (ECKEL, 1950):

$$S = z_g \sum_{z=0}^{z_m} A_z(d_{zw} - d_{zs}) - \sum_{z=0}^{z_m} A_z - z(d_{zw} - d_{zs})$$

where z_g is the depth of the centre of gravity of the lake and where, for each stratum considered, z is the distance from the surface to the centre of the stratum, A_z the area at depth z, d_{zw} and d_{zs} being the water densities corresponding to the winter and summer temperatures respectively. To express S per unit area of lake surface the result is to be divided by A_0.

But the computation of the stability can be greatly simplified if the work of the wind (B) has been first estimated on the basis of the following relationship (BIRGE, 1916):

$$S = [z_g \cdot \bar{z}(d_w - d_s)] - B = G - B$$

where $(d_w - d_s)$ is the mean value for the lake of the differences in density between winter and summer conditions, considering water strata. It is this last approach which has been adopted hereunder.

The depth of the geometrical centre of gravity (z_g metres) in each Lake Kariba basin has been calculated on the basis of 5 m-strata following ECKEL's (1950) method whereby:

$$z_g = \left(\sum_{z=0}^{z_m} A_z \cdot z \right) \left(\sum_{z=0}^{z_m} A_z \right)^{-1}.$$

Results are as follows:

BASIN:	I	II	III	IV
z_g in m:	8.5	14.5	17.2	22.0

For Basin IV, during the 1968–1969 cycle, further results were as follows:

$$G = z_g \cdot \bar{z}(d_w - d_s) = 2200 \times 3322 \times 0.0009421$$

or $G = 6885$ gcm cm^{-2}, and B being equal to 4308 gcm cm^{-2} (Table 57),

$$S = G - B = 2577 \text{ gcm cm}^{-2}.$$

Compared to data available from other lakes (HUTCHINSON, 1957, Table 59), the stability observed in Lake Kariba is much smaller than the estimates obtained for other relatively large monomictic lakes such as Lake Atitlan, Guatemala ($S = 21\,500$ gcm.cm^{-2}) and Lake Pyramid, Nevada ($S = 7$–$10\,000$ gcm.cm^{-2}).

In Basin IV of Lake Kariba, the work to be done to keep the lake circulating (G), given its heat budget, amounted to 6 885 gcm. Although not to be considered as an excessively high value (cf. L. Pyramid 11–14 000 gcm. cm^{-2}, op.cit.), this work is still much greater than the recorded possible values of the work of the wind (Table 58 in HUTCHINSON, 1957), which leads to the conclusion that the thermal cycle in Basin IV of L. Kariba will most probably remain a permanent feature under present hydrological and climatological conditions.

11.3. Conclusions

For the study of a tropical lake such as Lake Kariba, a thermal gradient of at least 0.2° C per metre has been adopted to define the metalimnic layer. The lake has been subdivided into four basins relatively homogeneous from a limnological point of view. Each of these basins has been studied as a separate entity but keeping in mind the possible relationships existing between each of them or with the upstream conditions.

Table 58. Summary: thermal properties of the water in Lake Kariba.

Basin	→ River → I Western Basin	II Mid-Western Basin	III Mid-Eastern Basin	IV → Dam Eastern Basin
Nature	riverine	transition	mostly lacustrine	lacustrine
Major regulator	Zambezi River	Zambezi R. + climate	climate	climate
Annual cycle	late April: entirety shift	Partial water displacement	—	—
– Circulation	May–June (24–18 C)	incomplete (20–22 C)	late June–early July (22C)	July (22C)
– Stratification	Jan–Feb (30–24 C)	Dec–Feb. (22–31 C)	Sep–Mar (21–30 C)	Oct–June (21–30 C)
– Density current	June–December	July–October	August–June	October–July
Stirred depth	—	min. 20m	min. 20m	min. 15m
Annual Heat Budget, cal.cm^{-2}	15,000	20,000	14,000	14,000
Annual Heat Index, cal.cm^{-2}m^{-1}	1,150	834	539	423
Residual Heat, cal. cm^{-2}	18,000	39,000	48,000	59,000
Tropicality Index, cal.cm^{-2}m^{-1}	1,390	1,625	1,846	1,787
Maximum Heat Content, cal.cm^{-2}	33,000	59,000	62,000	73,000
Maximum Heat Index, cal.cm^{-2}m^{-1}	2,540	2,459	2,385	2,210
Work of the wind, gcm.cm^{-2}	—	—	—	4,308 (63%)
Stability, gcm, cm^{-2}	—	—	—	2,577 (37%)
Total work G, gcm. cm^{-2}	—	—	—	6,885 (100%)

Lake Kariba being a reservoir has a river–lake nature. Its basins differ in the average characteristics of their annual thermal cycle (Table 58). Basin I is typically riverine during most of the year, his thermal characteristics being imposed by the Zambezi River. Total water replacement takes place in late April. On the contrary Basin IV exhibits lacustrine thermal characteristics its annual cycle being most dependent on the local meteorological conditions. Basin II constitutes a transitional zone in the lake between riverine and lacustrine conditions. The latter affirm themselves as one progresses towards the dam, Basin III already being considered as a relatively typical lake.

The two-thirds lower part of the reservoir (B III–IV) present all the thermal characteristics of a warm monomictic lake, except for its deepest layer (below 55-m depth) where cool river water may flow along the old channel. Total circulation occurs in June–July around 22° C, during the cool dry season. As the air temperature increases a thermal stratification develops, a strong thermocline becoming present around the 20–25 m depth (Fig. 42).

Other thermal characteristics are summarized in Table 58. In most of lake Kariba the Annual Heat Budget approximates 15 000 cal. cm^{-2} yr^{-1}, a relatively low value, only one half to one third of the most common values cited for temperate lakes. But on a metre-of-depth basis, the Annual Heat Index has a greater value than in any of the other lakes studied. The Annual Heat Budget represents only 13 percent of the global radiation delivered to the surface of the lake.

Both Residual Heat (above 40 000 cal. cm^{-2}) and Tropicality Index (ab. 1400 cal. cm^{-2} m^{-1}) point out to the tropical nature of Lake Kariba, which increases towards the dam.

Except in Basin I the seasonal variations of the heat content are closely related to those of the global radiation combined with those of the wind regime. But their rate of variation is not related, as in temperate climates, to the rate of delivery of solar radiation.

Maximum Heat Content above 4° C varies generally between 60 000 and 70 000 cal. cm^{-2}, mostly contributed by the Residual Heat. This represents an average of 60 percent of the global radiation available during the warming period. The Maximum Heat Index averages 2400 cal. cm^{-2} m^{-1}, more than five times the value observed in temperate lakes.

Energy contents of Basin IV have been estimated for 1968–1969 (Table 58) considering that they represent fairly well the general conditions to be found in most of Lake Kariba. Of the total work to be done to maintain hypothetical homothermy throughout the warming period (G), 63 percent is used to distribute the heat throughout the water mass (B) while 37 percent belongs to the water mass stability (S). The work of the wind is relatively high, its efficiency being very low (0.302 gcm. cal^{-1}). More than 60 percent of this work is contributed by the lower epilimnion and the upper metalimnion (10–25 m depth). The maximum contribution originates from a depth (15 m) equal to two-thirds that of the epilimnion. The stability measured in Lake Kariba is relatively small if compared to the values known from large monomictic lakes of the world.

12. DISSOLVED GASES IN LAKE KARIBA

12.1 Dissolved oxygen

Oxygen is one of the most significant of all the chemical substances present in natural waters. It is essential, and in many instances even the limiting factor, for maintaining aquatic life. It regulates metabolic processes of communities and organisms, in particular fish, and it may be taken as an indicator of lake conditions for their potential production.

Dissolved oxygen in lakes originates from the atmosphere mainly through wind action on the surface waters and from photosynthesis in the euphotic zone of the water mass. It is mostly lost through respiration processes and in the oxidative breakdown of organic materials, either produced within the trophogenic zone or of allochtonous origin.

In Lake Kariba dissolved oxygen (DO) has been successively studied since the closure of the dam by HARDING (1966), COCHE (1968), and BEGG (1970). Personally I have been able to conduct intensive samplings covering the whole lake between April 1965 and January 1966 (Fig. 46). From March 1968 to March 1969 only the Zambian territory of the lake could be studied (Fig. 34). Further data have been given in COCHE (1968). The general sampling methodology has been previously described (Sec. 9.4). Further calculations have been made as follows:

(i) Saturation values: based on the solubility of oxygen in water in equilibrium with wet air at 760 mm Hg pressure, as determined by MONTGOMERY et al. (1964).

(ii) Conversion of observed concentrations into equivalent saturation values using a nomogram adapted to high water temperatures (GOLTERMAN, 1969, p. 133).

(iii) Correction factor for altitude: all above calculated values have been corrected for 485 m asl equivalent to 716 mm Hg average atmospheric pressure (DUSSART & FRANCIS-BOEUF, 1949, cited in GOLTERMAN, 1969), such as:

$$S_{716} = 0.914 \, S_{760} \quad \text{and} \quad P_{716} = 1.06 \, P_{760} ,$$

S being the oxygen solubility and P the percentage saturation, at chosen Hg pressures.

12.1.1. OXYGEN CONTENT OF SURFACE WATERS

All available data show that L. Kariba surface waters (top 5 m) are generally well supplied with dissolved oxygen. Rarely has it been observed that the DO concentration had decreased below 70 percent of the saturation value. In fact most of the time the surface waters were either nearly saturated or even supersaturated with oxygen (Table 59).

Maximum DO saturations were observed towards the end of the circulation period. Later the saturation percentages decreased to reach their minimum at the end of the stagnation period. In the particular situation of the Kariba Gorge (Boom station),

Table 59. Dissolved oxygen concentrations (percent saturation) in surface waters of the pelagic region, 1968-69.

Date	BI Sta 1407	BII Sta 2015	BIII Sta 3332	BIV	
				Sta 4741	Sta 5845
Late March	93	91	90	—	88
Mid-July	95	99	106	112	111
Mid-October	122	127	126	133	127
Mid-February	103	106	120	116	120

a record 168 percent DO saturation was noted in February and March 1965. In each lake basin the variations are related to those of the thermal conditions.

The DO contents of the surface waters have ranged mostly between 90 and 130 percent saturation. Such variations around saturation, in particular between March and August, might point out to the unproductive character of L. Kariba during more than half of the year (HUTCHINSON, 1957).

12.1.2. DISTRIBUTION OF OXYGEN WITHIN THE WATER MASS

The distribution of DO in L. Kariba is typically of the clinograde type as shown by selected data, characteristic of each of the three lower basins (Fig. 45). The distribution is uniform only during the circulation period (ab. July). Shortly later oxidative processes occurring in the hypolimnion contribute to the removal of oxygen (October). As the thermal stratification intensifies, the DO content decreases towards zero (February). In Basin IV below the 55-m depth cold river water and increased deoxygenation are found along the old Zambezi channel. In general, as pointed out also by BEGG (1970), the sharpness of the DO gradient increases near the bottom of the lake indicating a particularly high degree of biochemical activity at the water-mud interface (HUTCHINSON, 1957).

The above data already exhibit a definite pattern of variation between one circulation period and the next one. Further data gathered along the lake axis (Fig. 34 and 46) and at selected stations (Figs. 5 to 10 in COCHE, 1968) have permitted a better definition of the oxycycle in the various basins of Lake Kariba.

The oxycycle has always shown a very good correlation with the thermal cycle. A similar correlation between water temperatures and DO concentrations has also been described for L. Volta (ENTZ, 1969 a), confirming the general statement that in tropical lakes the oxygen curve is determined chiefly by temperature (RUTTNER, 1953). The thermal cycle in Lake Kariba having been studied in detail (Sec. 11.1), only a few additional remarks will be needed now to help define the oxycycle in the various basins.

During the period of homothermy mixing takes place and the more oxygenated surface water moves to greater depth. Fortunately the circulation phenomenon involves a relatively small anaerobic hypolimnion only[17] and deoxygenation does not

---

[17] Ratios of volumes (epi: hypolimnion) are relatively large: Basin III = 3.4; Basin IV = 1.0.

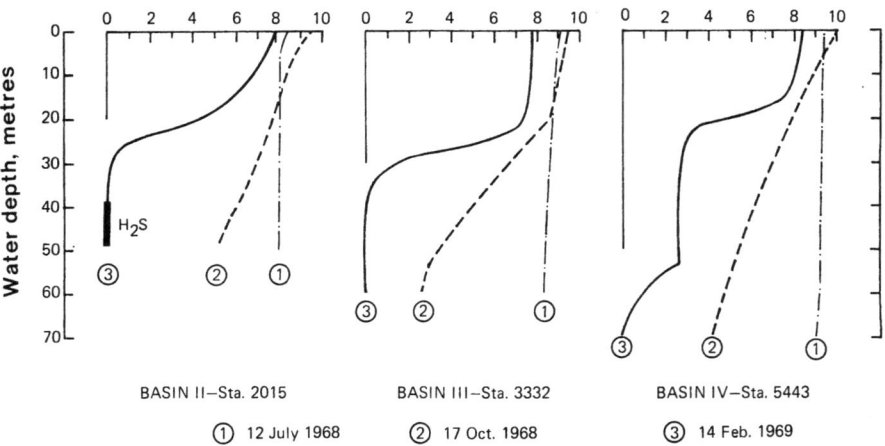

Fig. 45. Dissolved oxygen distribution in Lake Kariba throughout the annual cycle.

spread throughout the water mass. The latter normally becomes completely reoxygenated from surface to bottom without drastic decrease of overall DO content. But it might happen as observed in monomictic subtropical lakes (HUTCHINSON, 1957) that only incomplete circulation takes place during a particular year and/or in particular zones. For example a very mild winter might be able to induce partial circulation only, the deep waters not being fully aerated. Such a possibility has been envisaged for the protected Kariba Gorge region for example (BEGG, 1970). Or full circulation might occur so late in the annual cycle that it would last for such a short time that the water mass would not become completely saturated with oxygen.

As thermal stratification establishes itself and as the temperature gradient increases, oxygen depletion progresses in the hypolimnion. A sharp oxycline generally develops at the thermocline and the DO content drastically drops to very low values. This is particularly observed in the lacustrine part of the lake (Figs. 34 and 46). In the two upper Basins the lotic element helps improving the oxygen conditions.

Towards the end of the stratification period, as a result of the sinking thermocline and of its increasing weakness the oxygen content of the water masses increases. In Basin IV this phenomenon starts at least two and half months (mid-April to early May) before total circulation occurs.

As described for the thermal stratification, the deoxygenation of the hypolimnion processes in time from Basin II towards Basin IV. The rate at which the DO locked in the hypolimnion is consumed also increases as one progresses towards the dam, as will be discussed later (Sec. 12.1.4). In Basin I on the contrary, the constant inflow of oxygenated Zambezi water helps keep the upper part of the lake well oxygenated throughout most of the year.

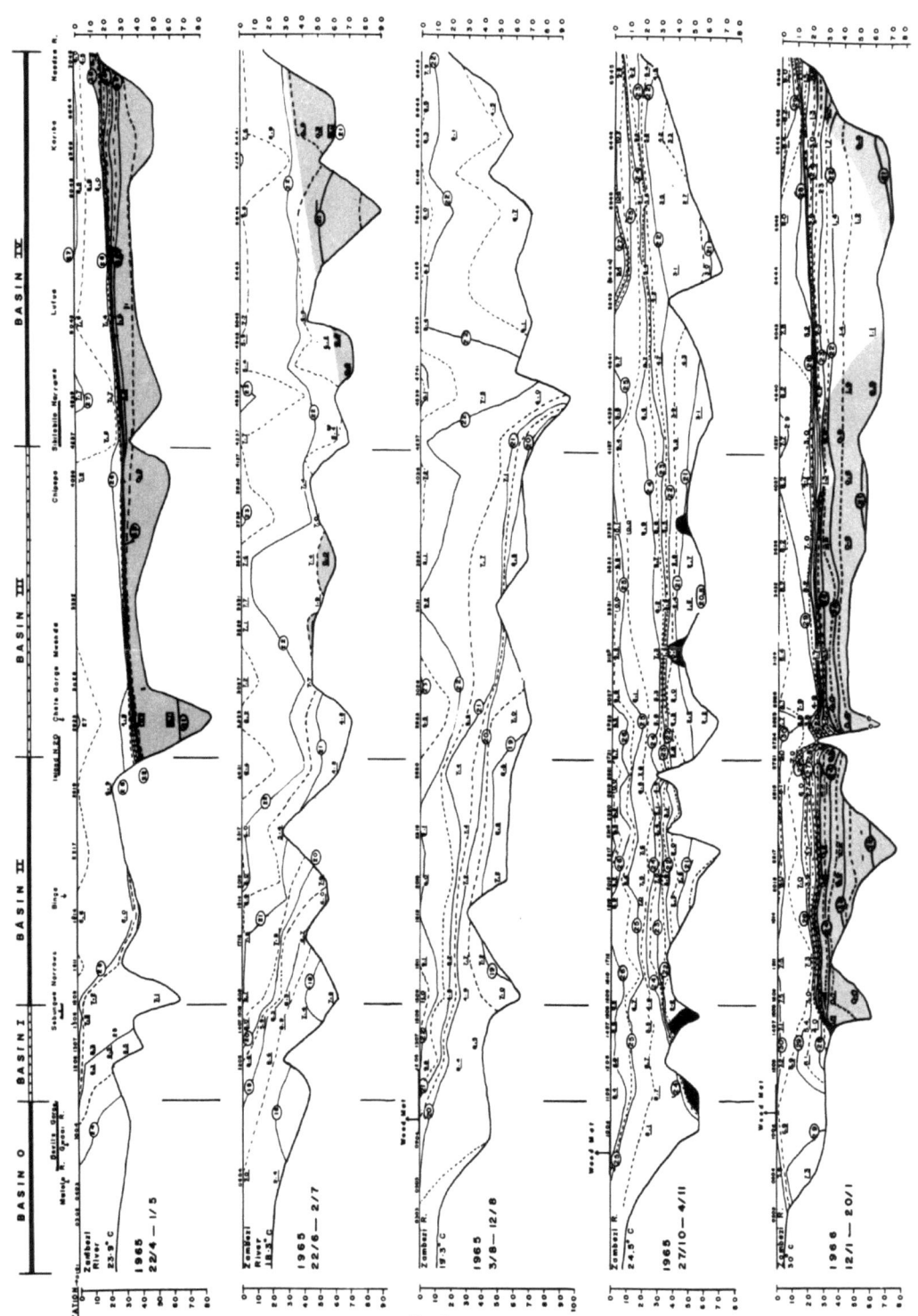

Fig. 46. Longitudinal profiles of L. Kariba in 1965–1966: isotherms and dissolved oxygen concentrations.

Table 61. Oxygen deficit in 1964-65, Lake Kariba Basin IV.

z m	A_z 10^4 m²	t°C	22 April 1965				DO deficit		$V_{z_1-z_2}$ 10^6 m³	DO total on 220465 10^6 g	DO actual deficit		DO relative deficit	
			DO sat mg.L⁻¹	DO mg.L⁻¹	Mean DO mg.L⁻¹	% DO Sat	Actual mg.L⁻¹	Relat.[1] mg.L⁻¹			10^6 g	mg.cm⁻² A_o	10^6 g	mg.cm⁻² A_o
0	256 300	26.5	7.60	7.8		104	−0.20	0.6						
					7.7				23 514	181 058	−2 704	−0.02	16 460	1.31
10		26.3	7.63	7.6		100	−0.03	0.8						
					7.5				19 057	142 927	2 096		17 151	
20	171 465	26.1	7.65	7.4		95	0.25	1.0						
					4.6				14 900	68 540	50 064		56 620	
30		21.8	8.28	1.8		42	6.48	6.6						
					1.4				11 788	16 503	81 455		82 516	
40		21.4	8.34	1.0		12	7.34	7.4						
					0.5				7 369	3 685	57 847		58 215	
50		21.3	8.36	0.0		0	8.36	8.4						
					0.0				4 597	0	38 477		38 615	
60		21.2	8.38	0.0		0	8.38	8.4				10.17		10.49
					0.0				2 327	0	19 500		19 547	(∴ A_H = 15.68)
70		21.2	8.38	0.0		0	8.38	8.4						
					0.0				994	0	8 330		8 350	
80		21.2	8.38	0.0		0	8.38	8.4						
					0.0				511	0	4 285		4 292	
90		21.1	8.39	0.0		0	8.39	8.4						
					0.0				85	0	713		714	
93		21.1	8.39	0.0		0	8.39	8.4						
Tot.					21.7				Tot. hypol. Tot. gen.	88 728 412 713	260 671 260 063	10.14	268 869 302 480	— 11.80

[1] DO Primary Constant: 8.4 mg. L⁻¹.

Table 60. Oxygen deficit in 1964–65, Lake Kariba Basin III.

z m	A_z 10^4 m²	23 April 1965					DO deficit		$V_{z_1-z_2}$ 10^6 m³	DO total on 230465 10^6 g	DO actual deficit		DO relative deficit	
		t°C	DO sat mg.L⁻¹	DO mg.L⁻¹	Mean DO mg.L⁻¹	%DO Sat	Actual mg.L⁻¹	Relat.[1] mg.L⁻¹			10^6 g	mg.cm⁻² A_o	10^6 g	mg.cm⁻² A_o
0	203 300	26.3	7.63	7.8		106	−0.17	0.8						
					7.8				18 085	141 063	−2 984		14 468	
10	—	26.2	7.64	7.8		104	−0.16	0.8						2.44
					7.7				13 750	105 875	−756	0.62	12 375	
20	—	26.1	7.65	7.6		99	0.05	1.0						
					6.3				9 982	62 887	16 470		22 958	
30	86 809	22.0	8.25	5.0		60	3.25	3.6						
					2.9				7 387	21 422	39 853		42 106	
40	—	21.4	8.34	0.8		10	7.54	7.8						4.01
					0.4				3 459	1 384	27 499		28 364	(÷ A_H = 9.40)
50	—	21.3	8.36	0.0		0	8.36	8.6						
					0.0				1 025	0	8 584	3.84	8 815	
60	—	21.1	8.39	0.0		0	8.39	8.6						
					0.0				270	0	2 265		2 322	
70	—	21.1	8.39	0.0		0	8.39	8.6						
Tot.					25.1				Tot. hypol.	22 806	78 201	—	81 607	—
									Tot. gen.	332 631	90 931	4.47	131 408	6.46

[1] DO Primary Constant: 8.6 mg.L⁻¹.

Table 62. Oxygen actual and relative deficits, 1964–1965.

Element	Basin III	Basin IV
1. Ratio vol. (epi: hypolimnion)	3.4	1.0
2. Period considered:		
– end circulation	15 Sept.	15 Oct.
– late stagnation	23 Apr.	22 Apr.
– length, days	220	188
3. Final mean DO, mg L^{-1}:		
– epilimnion	7.27	7.60
– hypolimnion	0.82	0.89
– total column	3.58	2.33
4. Final ratio DO content (epi: hypolimnion)	13.58	3.56
5. Oxygen actual deficit, mg.cm^{-2} of lake surface		
– epilimnetic	0.62	−0.02
– hypolimnetic	3.84	10.17
– total column	4.47	10.14
6. Oxygen relative deficit, mg. cm^{-2} of lake surface		
– epilimnetic	2.44	1.31
– hypolimnetic	4.01	10.49
– total column	6.46	11.80
7. Oxygen relative deficit, mg. cm^{-2} hypolimnion surface		
– hypolimnetic	9.40	15.68
8. Areal relative deficit, mg. cm^{-2}. day^{-1}		
– of lake surface, hypolimnetic	0.018	0.055
– of lake surface, total column	0.029	0.062
– of hypolimnion surface, hypolimnetic	0.042	0.083

12.1.3. Oxygen deficits of basins III and IV

Several ways of expressing the oxygen deficits in lakes have been proposed in the past as reviewed by Hutchinson (1957). Two of them will be briefly considered here:
(i) the actual deficit (OAD) defined as the difference between the DO concentration observed and the equivalent saturation concentration.
(ii) the relative deficit (ORD) or difference between the DO contents observed successively at circulation and at later dates, preferably at the time of maximum stagnation.

From the ORD the average rate of loss of oxygen in the hypolimnion during the stagnation period (hypolimnetic areal deficit) has been estimated, the conditions of applicability for water colour (less than 10 Pt units) and maximum depth (20 to 75 m) being satisfactorily fulfilled in the two basins considered.

Basic data and computations according to DUSSART (1966 – Table 26) are given in Tables 60 (Basin III) and 61 (Basin IV) for 1964–1965. Results are summarized in Table 62.

Within Lake Kariba oxygen deficits confirm previous observations about the relative magnitude of organic productivity in Basins III and IV. Although the volume ratio of epilimnion to hypolimnion is much higher in Basin III and therefore might be favourable to a greater elaboration of organic matter through photosynthesis, the smaller areal relative deficits point out to a relatively smaller productivity.

It has been stressed by RUTTNER (1953) that because of much higher water temperatures and much faster chemical reactions, a marked oxygen deficiency exists in the hypolimnion of most tropical lakes ... 'regardless of whether they are eutrophic or oligotrophic in productivity'. Therefore it is only tentatively that we can suggest data comparing the warm monomictic L. Kariba with temperate dimictic lakes. For the latter limits between oligotrophy, mesotrophy, and eutrophy have been at best estimated on the basis of the hypolimnetic areal deficit (ref. hypolimnion surface – HAD) at 0.025 and 0.055 mg. cm^{-2}. day^{-1} respectively (MORTIMER in HUTCHINSON, 1957). RUTTNER (1953) has estimated the rate of use of oxygen to be at least four times as high in a tropical lake. On the basis of such conservative estimates: HAD Lake Kariba, range 0.042 – 0.083 mg. cm^{-2}. day^{-1}; HAD max. temperate oligotrophy < 0.025 mg. cm^{-2}. day^{-1}; est. HAD max. tropical oligotrophy < 0.100 mg. cm^{-2}. day^{-1}.

Such reasoning would classify L. Kariba as oligotrophic in nature. Evidently such rough approximation does not take into account that the stagnation period lasts more than twice as long in L. Kariba than in temperate lakes in general!

12.1.4. EVOLUTION OF HYPOLIMNETIC OXYGEN DEPLETION DURING THE OXYCYCLE

In an attempt to follow the evolution of oxygen depletion in Basins III and IV during the annual cycle, data on DO concentrations have been considered at three representative stations in each of the two years of observations, as follows:

Year	Area	Station	Period, months
1965–66	Basin IV		
	S.B. Loteri	5845	Aug–Oct; Oct–Nov; Nov–Jan.
	S.B. Kariba	6446	Jul–Oct; Oct–Dec; Dec–Jan; Jan–March
	S.B. Sanyati	6342	
1968–69	Basin III	3332	
	Basin IV		Jul–Oct and Oct–Feb.
	S.B. Lufua	5443	
	S.B. Loteri	5845	

At each station on the sampling date, the DO content (mg. L^{-1}) was obtained at 10-m depth interval from surface to bottom. The average between two consecutive values was considered to represent the mean DO content (mg. cm^{-2}) present in a

Table 63. Hypolimnetic oxygen depletion (HOD) in Basin III, Station 3332 in 1968-69.

z m	12 Jul 1968 DO mg. L^{-1}	12 Jul 1968 Mean DO mg. cm^{-2}	17 Oct 1968 DO mg. L^{-1}	17 Oct 1968 Mean DO mg. cm^{-2}	14 Feb 1969 DO mg. L^{-1}	14 Feb 1969 Mean DO mg. cm^{-2}
0	9.2		9.4		7.8	
		9.0		9.2		7.8
10	8.8		9.0		7.8	
		8.8		8.7		7.6
20	8.7		8.5		7.4	
		8.65		7.8	5.9	3.3
30	8.6	—	7.0	—	1.6	1.9
		8.55		6.0		0.8
40	8.5		5.0		0.0	
		8.45		4.1		0.0
50	8.4		3.2		0.0	
		8.35		2.9		0.0
60	8.3		2.6		0.0	

Epilimnion (DOE)	26.4		25.7		18.7	
Hypolimnion (DOH)	25.4		13.0		2.7	
Ratio (DOE: DOH)	1.039		1.977		6.926	
HOD, mg.cm^{-2}			−12.4		−10.3	
Days			97		120	
RHOD, mg.cm^{-2}, day^{-1}			0.128		0.086	

water column 10 m high and one square-centimetre in area. Total mean contents were then obtained separately for epilimnion (DOE) and hypolimnion (DOH), from which the ratio (DOE/DOH) was calculated. The hypolimnetic oxygen depletion observed between one sampling date and the next one was divided by the number of days elapsed to give an estimate of the average rate of hypolimnetic oxygen depletion (mg.cm^{-2}. day^{-1}) at this particular station. An example of computations is presented in Table 63. Results are grouped in Table 64.

The ratio of epilimnetic to hypolimnetic oxygen content generally decreases as one moves down towards the dam, reflecting of the similar relationship which exists between the depths of the two strata. In July 1968 for example the ratio was 1.039 at station 3332, 0.564 at station 5443, and 0.450 only at station 5845. But as the oxycycle progresses the difference between Basin III (3332) and Basin IV greatly increases attesting to a greater oxygen depletion towards the western end of the lake. A similar but less marked increase is observed as one moves from the Kariba Sub-Basin (6446) into the Sanyati S.B (6342). The ratios vary also according to the year considered. At Station 5845 for example in 1968-69, it varied between 0.45 and 1.50 (a 3.3-fold variation) while three years earlier it had varied at least between 0.18 and 2.0 (an 11-fold variation), during the same period.

Table 64. Dissolved oxygen ratios and average rates of depletion in Basins III and IV.

A. Ratios of epilimnetic to hypolimnetic oxygen contents

	J	A	S	O	N	D	J	F	M	
III-3332	1.039			1.977				6.926		
IV-5443	0.564			0.870				1.851		
IV-5845	0.450			0.693				1.486		1968–69
	J	A	S	O	N	D	J	F	M	
IV-5845		0.183	0.233		0.355		1.914			
IV-6446	0.295		0.689			0.575		1.154	4.380	
IV-6342	0.212		0.258			0.825		1.178	4.600	1965–66

B. Average rates of hypolimnetic oxygen depletion, mg. cm^{-2} day^{-1}

	J	A	S	O	N	D	J	F	M	Tot
III-3332		0.128				0.086				0.104
IV-5443		0.163				0.138				0.150
IV-5845		0.149				0.134				0.141
										1968–69
IV-5845		0.092		0.517	0.314	0.200				0.227
IV-6446		0.252		0.134		0.124	0.114	0.090		0.167
IV-6342		0.212		0.242		0.158	0.071	0.078		0.157
										1965–66

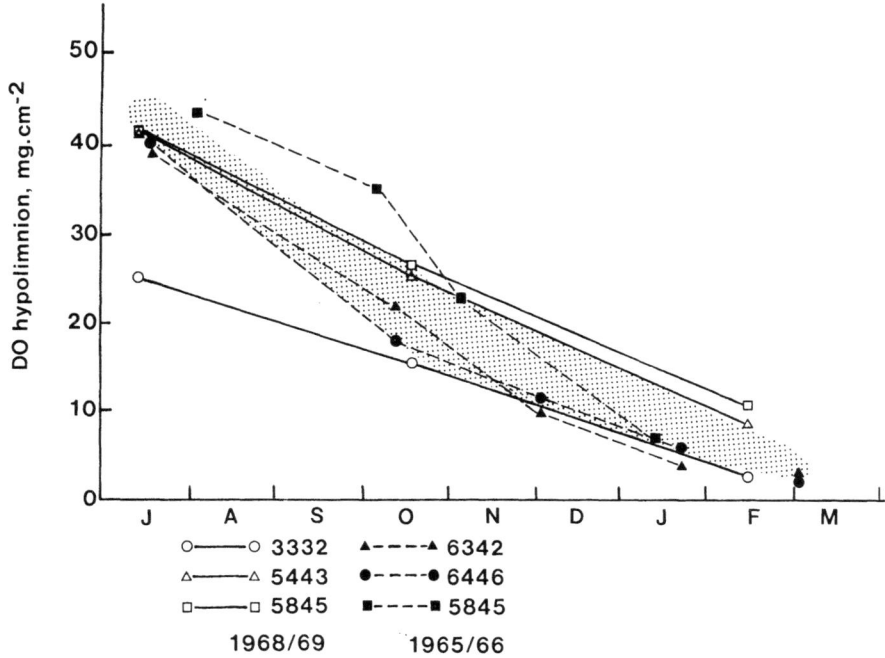

Fig. 47. Variation of the average rate of hypolimnetic oxygen depletion in Basins III and IV.

The average rate of hypolimnetic oxygen depletion (RHOD) has on the whole varied around 0.150 mg. cm^{-2} day^{-1} during a complete oxycycle at stations situated in Basin IV (Table 64). In Basin III the average rate is definitely smaller and about 0.100 mg. cm^{-2} day^{-1}, confirming the difference in rate of oxygen depletion between the two basins. When compared to the areal relative deficits-hypolimnetic (ARDH) much closer to reality (Table 62), the following ratios might be of practical use to provide a rapid estimate of the latter:

Ratio	Basin III	Basin IV
RHOD: ARDH of water surface	5.5	2.7
RHOD: ARDH of hypolimnion surface	2.4	1.8

During the oxycycle the average rate of hypolimnetic oxygen depletion generally decreases from the beginning of the cycle to the end (Table 64 – B). Exceptions to this observation (e.g. Sta. 5845 in 1965) are due to 'anormalities' such as a sudden storm reoxygenating the whole water mass after the start of restratification.

In Basin IV (Fig. 47) the average rate decreases from about 0.220 in July–November to about 0.100 mg. cm^{-2} day^{-1} later. During the first part of the cycle the DO content of the hypolimnion drops from 40 mg to 10–15 mg. By the end of the cycle there are only a few milligrams of oxygen left under the thermocline. In Basin III on the contrary, the rate is much more constant oscillating around 0.100 mg. cm^{-2}day^{-1}. This originates from the fact that less oxygen is present in the hypolimnion (ab. 25 mg) at the beginning of the cycle, a direct consequence of a lower depth ratio of hypo- to epilimnion.

12.1.5. EVOLUTION OF THE OXYCYCLE

Since the closure of the dam in December 1958 and the beginning of the flooding of the Gwembe Valley, the oxygen content of the hypolimnion has varied according to the early established pattern of the thermal cycle. But the length of the period of deoxygenation (DO less than one mg. L^{-1}) has gradually decreased until stabilization (Fig. 48).

At the Station 6348B in November 1959 after one year of filling the new reservoir, DO was absent below the 10-m depth (HARDING, 1966). The same situation was still encountered one year later (HARDING, 1961). Only in 1961/62 during the third year was oxygen present in the hypolimnion until late December at least (HARDING, 1964a). The situation still improved during the next cycle (HARDING, 1966). The general pattern of oxygenation established itself after the complete filling of the lake (April 1963), most probably in 1963/64. From 7 to 8 months of complete deoxygenation of the hypolimnion during the first evolutionary phase (intense mineralization of organic matter from flooded land vegetation), the deoxygenated period progressively shortened and finally stabilized itself around four months per year. In fact the length of this period now varies depending on locality within the lake and seasonal conditions (COCHE, 1968; BEGG, 1970).

Whereas the thermal regime became established immediately after the formation of the lake in 1959, it took several years before the stabilization point was reached for the oxycycle. Deoxygenation was widespread during the first two years. It later improved and stabilization closely followed the complete filling of the lake. A similar evolution has been described for L. Volta (Sec. 12.1.7).

12.1.6. OXYGEN DISTRIBUTION AND THE FISH STOCKS

Particular emphasis has been given to the identification of water masses with a low dissolved oxygen content because of their major impact on aquatic organisms in general, and on the fish in particular. Effects of oxygen depletion in water on freshwater species of fish have been extensively studied. This work has recently been critically reviewed by DOUDOROFF & SHUMWAY (1970) and by a Working Party of the European Inland Fisheries Advisory Commission (EIFAC, 1973).

Sensitivity to low dissolved oxygen concentrations differs between fish species, between various life stages, and between the different life processes such as feeding, growth, and reproduction. The effect of dissolved oxygen on fish is also influenced by several other factors, in particular water temperature. Tentative minimum sustained oxygen content for normal survival of juvenile and adult European fish has been set at 3 mg. L^{-1}. For normal fecundity, hatching of eggs, larval survival, and growth, this minimum DO has been raised to 5 mg. L^{-1}. On the other hand in a

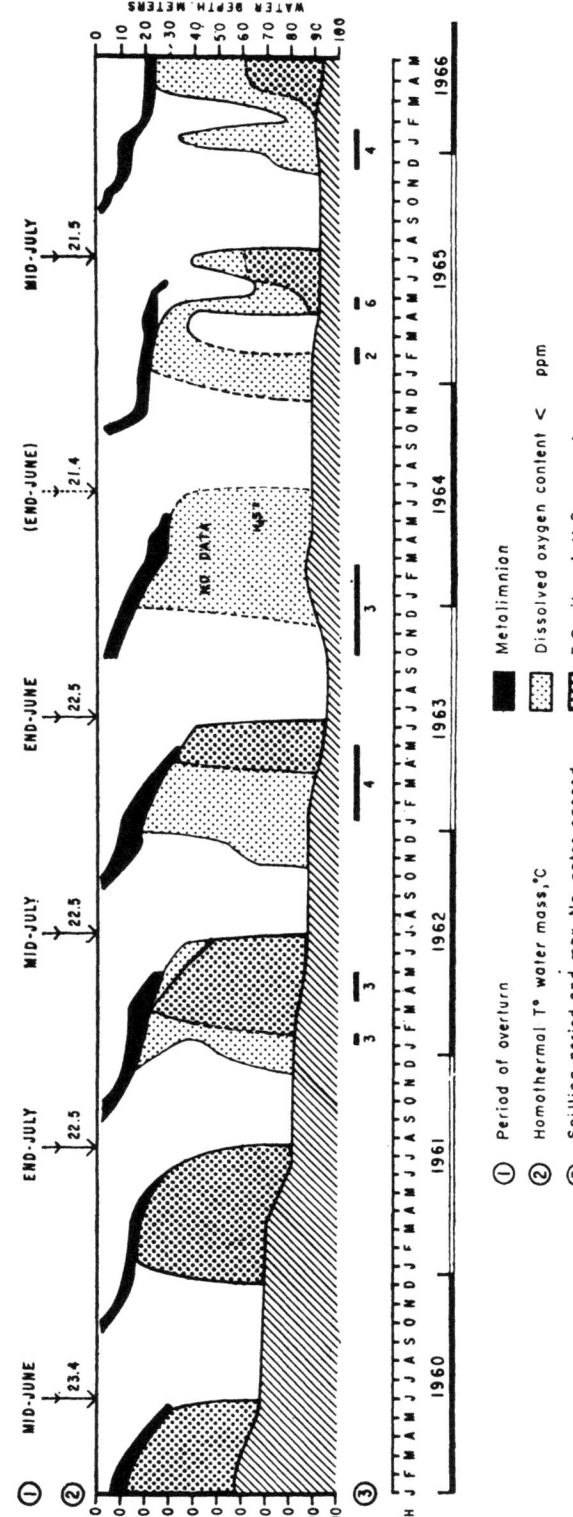

Fig. 48. Evolution of the oxygen and hydrogen sulphide conditions at the Boom Station (6348B) between 1960 and 1966 (acc. to HARDING's and personal data).

North American reservoir it has been observed that oxygen contents lower than 2 mg. L^{-1} limited the fish distribution (ELEY et al., 1967). Therefore as far as the fishes are concerned, it seems reasonable in tropical waters to consider that most fishes will avoid water masses with less than 2 mg. L^{-1} DO content. This particular DO concentration also usually marks the layer in which there is a rapid approach to anoxic conditions (TALLING, 1969).

On these bases and following the proposal made by TALLING the oxycycle in Lake Kariba has been redefined from the fishery point of view on the basis of the 'variation with time of the depth at which the concentration of oxygen is reduced to 2 mg. L^{-1}' (op. cit.). Concurrently the variation of the depth of the euphotic zone (Sec. 10.3.5) has been examined (Table 65 – Fig. 49).

The average length of the period when limiting dissolved oxygen concentrations are present in the water mass (Fig. 49A) increases from four months in Basin I and II up to seven months in Basin III and 8–9 months in Basin IV. This period roughly varies as follows:
  (i) Basin I: October to February
  (ii) Basin II: November to March
  (iii) Basin III: November to June
  (iv) Basin IV: October to June (July).

But most important from the fishery aspect is the extent to which these limiting conditions are present. In Basin I the deoxygenated zone only represents a relatively small volume of the water mass throughout the year. The riverine character of the basin predominates. In Basin II depths below 30 m seem to be possibly affected around February only. Limiting DO concentrations of some importance are therefore uncommon in this basin. In Basin III although the period of deoxygenation is much longer, its potential harmful effect on the fishery is also relatively well restricted in time (Jan. to March), January being particularly affected. But on the whole these three upper basins present a well oxygenated trophogenic zone throughout the year[18].

The situation in Basin IV is radically different as seen from Figure 49(A) where the observed range for the depth of the euphotic zone (z_E) is emphasized. From the start of the oxycycle the depth of the 2 mg. L^{-1} oxygen isopleth quickly decreases, at least in the eastern half of the basin. By January it coincides there with the bottom limit of the euphotic zone at 15 m. As the latter increases the depth of hospitable waters goes on decreasing until April when the worst situation is encountered. The depth at which the concentration of oxygen is reduced to 2 mg. L^{-1} may be as shallow as 10 m thus reducing the inhabitable water mass considerably. It can therefore be stated that in the eastern half of Basin IV DO conditions may unfavourably affect the fish distribution for at least five months of the year (Jan. to May). As far as general organic production is concerned the volume of the potential trophogenic zone (down to about 25 m) is greatly reduced because of the lack of oxygen in its bottom half. This reduction in food elaboration most probably contributes to further depress the fish production, even for a while after oxygen conditions have ameliorated.

---

[18] The maximum depth of the euphotic zone in Basins I, II and III was about 13, 16, and 20 m respectively (Sec. 10.3.5).

Table 65. Mean depth (metres) at which the D.O. concentration is reduced to 2 mg. L^{-1} in L. Kariba Basins. (A = absent from the water mass. Parentheses indicate part-time value).

Basin – Year	Month											
	A	S	O	N	D	J	F	M	A	M	J	J	A	S

Basin – Year	A	S	O	N	D	J	F	M	A	M	J	J	A	S
Basin I 1965/66	A	A	45	45	—	40	—	—	A	A	A	A	A	A
1968/69	—	—	34	—	—	—	35	A	A	—	—	—	—	—
Basin II 1965/66	—	A	—	45	—	35	—	—	(A)	A	A	A	A	A
1968/69	A	A	A	—	—	—	30	(55)	A	—	—	—	A	A
Basin III 1965/66	A	A	A	(40)	—	25	35	32	35	(40)	(A)	A	A	A
1968/69	—	—	A	—	—	—	—	—	—	—	—	A	—	—
Basin IV 1965/66	A	(A)	(A)	(30)	—	15–30	—	—	—	10–27	27–60	(A)	A	(A)
1968/69	—	—	60	—	—	—	45	30	—	—	—	A	—	—

E. Basin IV 1965/1966 (See COCHE, 1968 Fig. 5, 9, and 10)[2].

	A	S	O	N	D	J	F	M	A	M	J
Loteri SB (Sta. 5845)	A	37	(75)[1]	60	40	25	—	16	20	30	A
Kariba SB (Sta. 6446)	A	45	(75)[1]	45	30	18	16	16	15	19	29
Sanyati SB (Sta. 6342)	A	A	(70)	54	25	16	20	23	28	35	45

	J	A	S
Loteri SB (Sta. 5845)	A	A	37
Kariba SB (Sta. 6446)	A	A	45
Sanyati SB (Sta. 6342)	A	A	A

[1] Following a storm in late September.
[2] Data for the middle of each month.

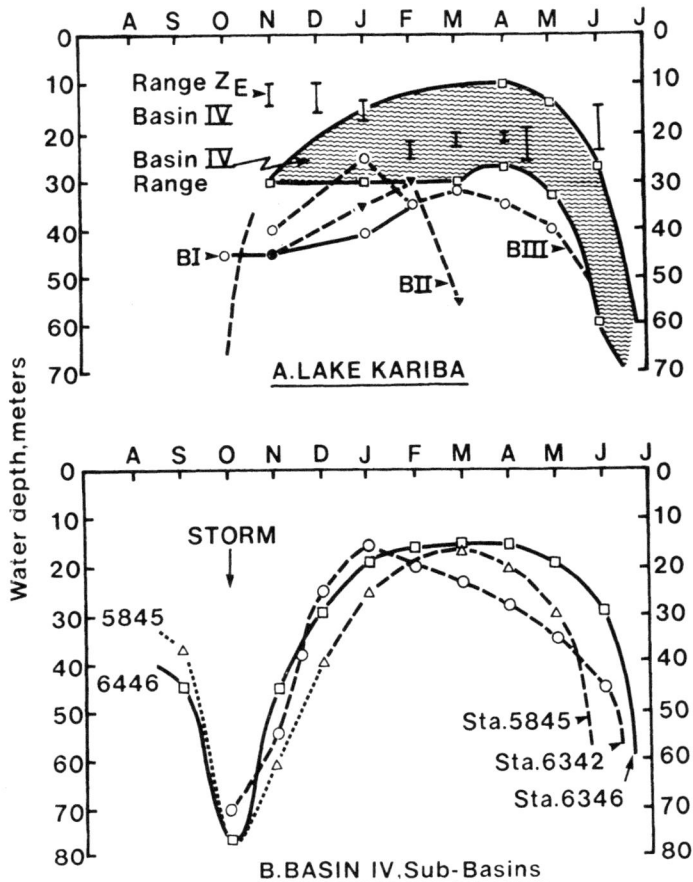

Fig. 49. Variation of the mean depth at which the dissolved oxygen concentration was reduced to 2 mg.L^{-1} in L. Kariba Basins.

Data for 1965/66 (Table 65) throw additional details concerning the oxycycle in Eastern Basin IV by permitting comparisons between stations representative of three major sub-basins (fig. 49 B): Loteri SB (Sta. 5845), Kariba SB (Sta. 6446), and Sanyati SB (Sta. 6342). A violent storm in late September reoxygenated the water mass from the surface down to 60 m after an early start of the oxycycle in September. Comparatively deoxygenation was less pronounced in the most western station 5845. But from January, as the Sanyati River flow started to influence station 6342, reoxygenation gradually increased in that particular sub-basin. On the contrary, oxygen depletion further extended into the other two sub-basins for another two (west) to three (east) months. Reoxygenation then took place, the two stations however showing a one-month displacement in time. Such observations contribute to demonstrate the importance of the "Sub-Basin" limnology in the Kariba reservoir.

12.1.7. COMPARISON WITH OTHER AFRICAN RESERVOIRS

Earlier it has been described how during the first years of the life of L. Kariba the absolute quantity of dissolved oxygen in the total water mass became low and practically nil as soon as thermal stratification established itself. Pronounced deoxygenation of the hypolimnion lasted until the next circulation period. As time passed the situation gradually ameliorated itself and stabilization was reached one year after filling of the reservoir became completed (Sec. 12.1.5).

A similar evolution occurred in Lake Volta, Ghana (ENTZ, 1969a; VINER, 1970b). The Akosombo dam was closed in May 1964 and the reservoir started filling. For the next two years very low dissolved oxygen concentrations were observed, the well-oxygenated zones being limited to the top 5–10 m layer. No marked seasonal variations took place. But in 1967 and during the following two years, a great increase in the oxygen concentrations was recorded, the well oxygenated zones (over 50 percent saturation) extending down to 20–30 m. The dimictic oxycycle came into effect with two periods of maximal and two periods of minimal DO contents: a first maximum in Jan. to Feb. (harmattan season) was followed by a first minimum (March–April); a second maximum in July and Sep.–Oct. (rainy season) was followed by a second minimum (Nov. to Dec.). Complete circulation was observed (ENTZ, 1969a). It appears that, as happened in L. Kariba, the oxycycle most probably reached stabilization in 1968 with the complete filling of the lake.

A similar pattern of evolution has been started following the construction in Ivory Coast of the Kossou dam and the subsequent filling since February 1971 of Lake Kossou. Regular monitoring in 1972/73 of dissolved oxygen in the forming lake has shown the oxygenated zone to be restricted to less than the top 10-m water layer for most of the year (KRZELJ, pers. comm.).

In L. Nasser/Nubia the evolution has been completely different because of the absence of land vegetation in the flooded area. From actual data (ENTZ & RAMSEY, 1973), it appears as if the oxycycle has already established itself before the complete filling of the reservoir. The oxycycle is closely related to the thermal cycle and complete circulation (DO 70–100 percent saturation) occurs in December each year. Deoxygenation of the hypolimnion starts in June and intensifies rapidly. Progressive reoxygenation takes place from South to North under the main influences first of the Nile flood (Aug. to Sept.) and later of the cooling of the water mass.

In Lake Kainji, Nigeria, during initial flooding the oxygen content never fell below 20 percent saturation (IMEVBORE cited in HENDERSON, 1973) because of the fast rate of water replacement which characterizes this particular reservoir. The filling occurred the same year that the dam was closed and the oxycycle established itself very early in the life of this lake. The DO content in surface waters never falls below 70 percent saturation and for most of the year it remains higher than 4 mg. L^{-1} in deep waters (HENDERSON, 1973). But two periods of strong deoxygenation (less than 1 mg. L^{-1}) exist. As soon as thermal stratification has established itself the dissolved oxygen rapidly disappears (March to mid May) until an important drawdown from the hypolimnion eliminates the thermocline. A less-marked DO deficiency below

30 m takes place as flood water invades the lake (Oct. to Nov.). Total circulation starts in December and reoxygenates the water mass completely.

12.2. Hydrogen sulphide

Under anaerobic conditions the sulphates are reduced and hydrogen sulphide (H_2S) is formed as a decomposition product, colloidal sulphur being precipitated out and utilized in chemosynthetic processes of sulphur bacteria (REID, 1961). The chemical dissociation of H_2S greatly varies according to the pH of the water, the ratio $[H_2S]$: $[HS^-]$ being close to 1 at pH 7 but increasing about 10 times for each unit decrease of the pH (DUSSART, 1966). Undissociated H_2S being much more toxic to the living organisms than the HS^- ion, acid waters loaded with hydrogen sulphide are most toxic. In lakes such waters are devoid of life which depresses their potential organic production, in particular fish production.

During the L. Kariba surveys deoxygenated water masses have been routinely checked for the presence of H_2S (Sec. 9.4.2). In 1965/66 this gas could be detected in the lake at pH 6.3–6.8 from three sites only (COCHE, 1968):

(i) Chete Gorge (Sta. 2923): in late April 0.1 mg. L^{-1} H_2S was measured from 40 m down.

(ii) Sanyati Gorge (Sta. 6339): in May 0.3–0.5 mg. L^{-1} H_2S from 30 m down and in January 0.3 mg. L^{-1} H_2S from 35 m down.

(iii) Kariba Gorge (Sta. 6348B): up to 1 mg. L^{-1} in April to July 1965 below the 60-m depth.

In February 1969, hydrogen sulphide was observed (in acid conditions) in the Zambian lake territory below the 40-m depth in Basin I (Sta. 1508) and in Basin II (Sta. 1611, 1712 and 2016). BEGG (1970) reported the brief appearance of H_2S around Nov. to Dec. 1967 in these two upper Basins, believing that it originated from neighbouring rivers infested with *Salvinia* weed. During the same year, H_2S was present at Station 6348B (boom in front of the dam) for seven months being however totally absent from that station in 1968. He related this last particular phenomenon to the poor flow conditions of the Sanyati River during the previous rainy season: H_2S did not penetrate the Sanyati sub-basin further than south of the Long Island-Redcliffe ridge (ab. Sta. 6346). In 1970, hydrogen-sulphide scarcely protruded further than the river mouths because of poor rainfall. On the contrary in 1972 the Sanyati R. flowed so strongly that its waters traversed the Sub-Basins in a few days and the H_2S gas reached the Kariba Gorge very quickly. It was dissipated within a short time due to strong discharges from the floodgates (G. BEGG, pers. comm.).

The above observations concur to show that a definite improvement in the L. Kariba water quality has taken place as far as the presence of toxic hydrogen sulphide is concerned since the early years of the reservoir's existence (Fig. 48). From the closure of the dam (December 1958) until three years later, this gas was most probably present in the whole deoxygenated hypolimnion for at least seven months (Dec. to July). Only in 1963 as the flooded vegetation was disappearing did the situation greatly improved, H_2S appearing during the second half of the stagnation period only (HARDING, 1966). Later further improvement took place until 1965 when H_2S laden water could only be found in isolated deep troughs (COCHE, 1968). As seen for dissolved oxygen it appears as if stabilization for hydrogen sulphide also coincided with the complete filling of the reservoir over a few years.

Nowadays, as judiciously pointed out by BEGG (1970), the primary source of hydrogen-sulphide has shifted from the decay of flooded land vegetation to the rotting of *Salvinia* weed present in old riverbeds. In river localities heavily infested by such floating vegetation mass, the water rapidly deoxygenates and H_2S appears at the bottom of the channel over the organic debris. The majority of the river waters contain this toxic gas while the lake water mass is still well oxygenated. As the rainy season starts and rivers begin to flow, H_2S-laden water is flushed into the lake following the old river channels.

It may be concluded that the hydrogen-sulphide phase at any locality 'has become a variable phenomenon' (BEGG, 1970). It varies depending mainly on the bottom topography, on the timing of the rivers flow, and on the magnitude of their flood.

In young Lake Kossou, Ivory Coast, hydrogen-sulphide is commonly present (up to 1 mg. L^{-1}) in the deoxygenated hypolimnion during most of the stagnation periods (S. KRZELJ, pers. comm.). In Lake Nasser H_2S has been reported to occur in the hypolimnion of the northern sections of the lake towards the end of the stagnation period only (ENTZ & RAMSEY, 1973). In Lake Kainji concentrations in H_2S may reach 0.6 mg. L^{-1} in the deepest waters but generally they are much lower (HENDERSON, 1973). In L. Volta (G. BRETSCHKO, pers. comm.) free H_2S was observed during the first years of impoundment, even in the oxygenated surface layers, attesting of very fast organic decomposition processes. To-day its presence in the lacustrine part of the lake is correlated with the oxygen cycle and with the mixing pattern, being therefore restricted to the hypolimnion. In the riverine northern lake, the hydrogen-sulphide content depends also on the decomposition of allochtonous organic matter.

12.3. Conclusions about dissolved gases

Lake Kariba surface waters are generally well supplied with dissolved oxygen, the concentration varying around the saturation value. The distribution of this gas within the water mass is typically clinograde outside the total circulation period. The oxycycle shows a very good correlation with the thermal cycle, a strong oxycline developing at the thermocline and the dissolved oxygen content drastically dropping below. Because of the relatively small hypolimnion involved, circulation normally reoxygenates the water completely. This reoxygenation starts at least two months before total circulation happens.

The hypolimnetic areal deficit within the two lacustrine basins ranges from 0.042 to 0.083 mg. cm^{-2} day^{-1}, pointing in tropical waters to a relatively low productivity. The average rate of hypolimnetic oxygen depletion generally varies around 0.100 (B III) and 0.150 (B IV) mg. cm^{-2} day^{-1}. It decreases from the beginning to the end of the oxycycle.

Since the closure of the dam the oxygen content of the hypolimnion has varied according to the yearly pattern established for the thermal cycle. But the length of the deoxygenated period has gradually decreased from seven to eight months after the closure of the dam to about four months per year. Today this period essentially depends on locality and seasonal conditions. Stabilization has closely followed the complete filling of the lake.

It seems reasonable to consider the 2 mg. L^{-1} content in dissolved oxygen as the average limit for the fish distribution in inland tropical waters. On the basis of this index, the upper three lake basins present a well oxygenated trophogenic zone throughout the year. But in the eastern half of Basin IV the oxygenation conditions may unfavourably affect the fish stocks for at least five months of the year.

Both the oxygen content of surface waters and the hypolimnetic areal deficit tend to point to a relatively low organic productivity in L. Kariba. But here again basin individuality is clearly marked, the potential improving from the upper end of the lake towards the dam area.

Hydrogen sulphide has practically disappeared from the lake water mass since 1967. Its actual presence and development mainly depends on the bottom topography, on the timing of the rivers flow, and on the magnitude of their flood.

13. MINERAL CONTENT OF LAKE WATERS

13.1. Hydrogen-ion concentration

Lake Kariba surface waters are normally alkaline. The hydrogen – ion concentration expressed in pH units varies between 7.5 and 8.5. Values above pH 8.1 up to 9.0 are mostly measured in Basins III and IV during mid-day periods, when photosynthesis is particularly active.

Generally pH increases from Basin I to Basin IV, particularly when passing from Basin II into Basin III as also observed by BEGG (1970). Furthermore in the upper two basins greater ranges of variations exist due to the influence of the Zambezi River. As soon as its flow decreases (June) there is a gradual increase in the pH of the epilimnetic waters of Basins I and II.

At greater depth the above-mentioned values gradually start decreasing as soon as the thermocline establishes itself. At the beginning of the stagnation period, the hypolimnetic pH averages from 7.3 to 7.0. As stratification intensifies and deoxygenation increases the water becomes acid (pH around 6.8). BEGG (1970) reports pH as low as 6.3 in extreme cases.

It may be generalized that the epilimnetic waters of L. Kariba are slightly alkaline. They are therefore favourable to biological processes in general and to fish production in particular. Slightly acid conditions occur in the hypolimnion only but, even in extreme cases, the hydrogen – ion never reaches concentrations endangering fish life. Highly alkaline pH (9–10) which might be dangerous if prolonged, exist only momentarily.

13.2. Total mineral content of pelagic surface waters

The total mineral content of water may be expressed in three different ways, the following terminology and methods being adopted here:

(i) Total solids (TS). Total inorganic material dissolved in water. In the present case TS represents the dry residue on evaporation at 180° C of a filtrated aliquot of water.

(ii) Salinity (SL). Concentration of all the ionic constituents present estimated as the sum of the concentration of Na, K, Ca, Mg, CO_3, SO_4 and halide (Cl), all bicarbonate being converted to carbonate (HUTCHINSON, 1957). The last conversion has been made by multiplying bicarbonate concentration (mg. L^{-1}) by 0.4917 (HEM, 1959) or by multiplying the total alkalinity (meq. L^{-1} fixed CO_3) by 30.

(iii) Conductivity (k_{20}). The specific electrical conductance of a cube of water one centimetre on a side, referred to 20° C and expressed in reciprocal micromhos (μmhos).

Because of the availability of numerous measurements of conductivity from L. Kariba as well as from other African waters, the observed relationships between k_{20} on one hand and total solids, salinity, total ionic concentration, and total alkalinity

Table 66. Relationship of conductivity (k_{20}) to total solids (TS) at Station 6348 B (acc. basic data from HARDING).

Date	k_{20} μmhos	TS mg. L^{-1}	Coefficient a[1]
1958	116	91	0.784
1959	117	88	0.752
	116	87	0.750
	115	84	0.730
1959	100	74	0.740
	91	71	0.780
1960	90	72	0.800
	88	70	0.795
	93	69	0.742
1961	87	65	0.747
	84	64	0.762
	83	66	0.795
	82	66	0.805
1962	81	68	0.840
	79	68	0.861
	80	69	0.863
	79	68	0.861
	85	64	0.753
	82	67	0.817
	82	68	0.829
	79	68	0.861
	73	60	0.822
1963	75	60	0.800
	76	66	0.868
	77	64	0.831
AVG.	88.40	70.48	0.799

[1] $a = TS. (k_{20})^{-1}$.

on the other hand have been calculated in an attempt to facilitate future estimations.

13.2.1. TOTAL SOLIDS

Unfortunately no recent determination of total solids has been made from L. Kariba. The only data available are those given by HARDING (1962, 1964a, 1965 and 1966). They refer to surface water samples taken at Station 6348B in the Kariba Gorge between December 1958 and November 1963.

Although dating from Phase I of the lake evolution (as will be discussed later), the range of the recorded concentrations (from 60 to 91 mg. L^{-1}) seems to justify the determination of the possible relationship (coefficient a) existing between these total solids data and concurrently measured conductivities such as:

184

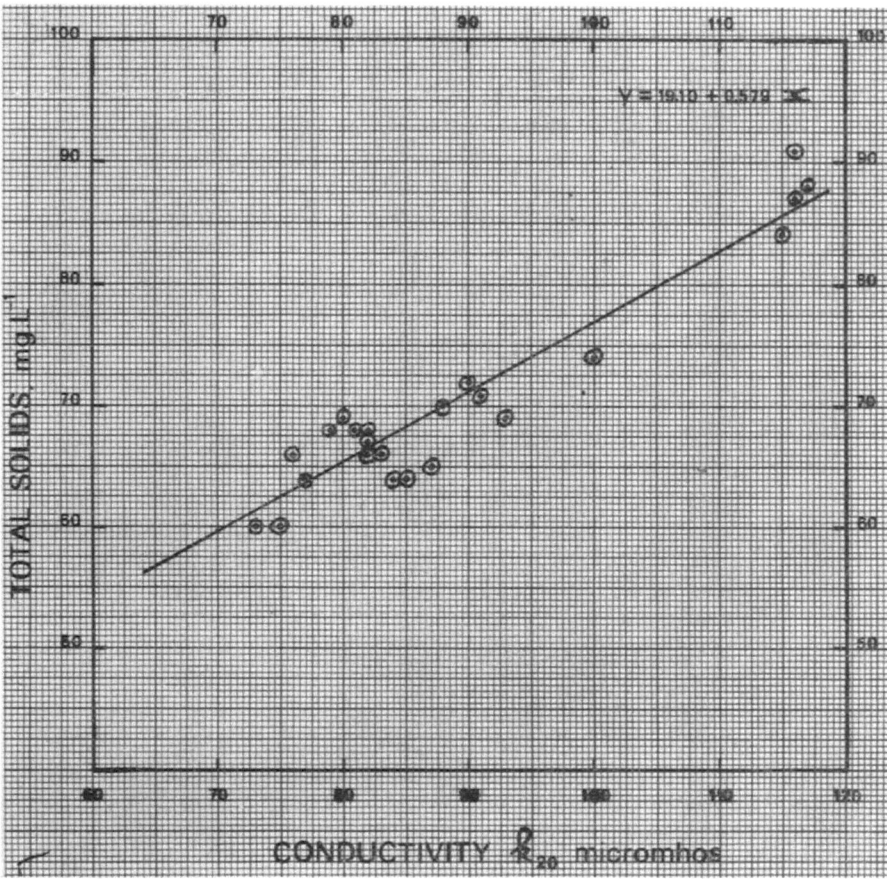

Fig. 50. Relationship between total solids and conductivity in L. Kariba water (based on data from Table 66).

$$TS(mg.L^{-1}) = a.k_{20}$$

This correlation if found sufficiently strong, could be useful for estimating TS from available k_{20} values. Such estimation is required for example to calculate RYDER's morphoedaphic index related to fish harvest (Chap. 14).

In simple dilute solutions of a salt, it is common knowledge that a linear relation exists between total solids concentration and conductivity. On the contrary this has not been found in solutions of several electrolytes such as natural water. However if it is dilute enough (low TS), this relation does not differ greatly from linearity.

On this basis the correlation existing between 25 paired measurements (Table 66) has been statistically studied using the standard method of least squares. An excellent correlation exists (r = 0.947) and the regression line (Fig. 50) can be used with con-

Table 67. Variation in 1965 of the mean annual salinity (SL) and conductivity (k_{20}) in Lake Kariba basins.[1]

Basin	Station	No Samples	Mean. Salinity		Mean k_{20} μmhos	b_2[2] for meq.L^{-1}	b_1[2] for mg.L^{-1}
			meq.L^{-1}	mg.L^{-1}			
I	1205	5	1.384	34.60	59	0.0234	0.586
II	1509	5	1.316	33.17	55	0.0239	0.603
	2015	3	1.332	34.93	57	0.0233	0.613
	2519	3	1.600	40.87	73	0.0210	0.538
	Avg.	11	1.416	36.32	62	0.0224	0.586
III	3028	5	1.694	44.67	78	0.0217	0.573
	3332	11	1.732	45.80	79	0.0219	0.580
	4036	6	1.756	45.71	80	0.0219	0.571
	Avg.	22	1.728	45.40	79	0.0215	0.575
IV	5043	12	1.796	46.13	79	0.0227	0.584
	5845	6	1.784	46.55	81	0.0220	0.575
	6446	9	1.812	46.84	80	0.0226	0.586
	6342	9	1.822	47.38	79	0.0230	0.600
	Avg.	36	1.806	46.74	80	0.0225	0.587
Lake	Gen. avg.	74	1.639	42.44	73	0.0224	0.581

[1] For details see Tables 71 and 72.
[2] b = SL. $(k_{20})^{-1}$ per μmhos.

fidence to estimate total solids from conductivity, particularly within the range of 70 to 100 micromhos:

$$TS = 19.10 + 0.579 \, k_{20}.$$

For most natural waters the simpler relationship $TS = a.k_{20}$ has been used, a varying generally between 0.50 and 0.67 (HEM, 1959). For the waters of the Bangweulu-Luapula Basin SYMOENS (1968) has found that a varies between 0.555 and 0.769. In L. Kariba near the dam (Table 66) a has ranged from 0.730 to 0.868. Such higher values probably correlate with the higher content in organic matter (SYMOENS, 1968) which prevailed at the time of these measurements. Thus in average $a = 0.799 \pm 0.058$ at the 95 percent confidence level.

On the basis of the above data, total solids in Lake Kariba have most probably varied in 1965–1966 between 35 and 80 mg. L^{-1}, the mineral content increasing from Basin I to Basin IV in general. For the lake as a whole the total solids averaged 58 mg. L^{-1}. It should be considered as a relatively low mineral content comparable to the L. Mweru value (DE KIMPE, 1964).

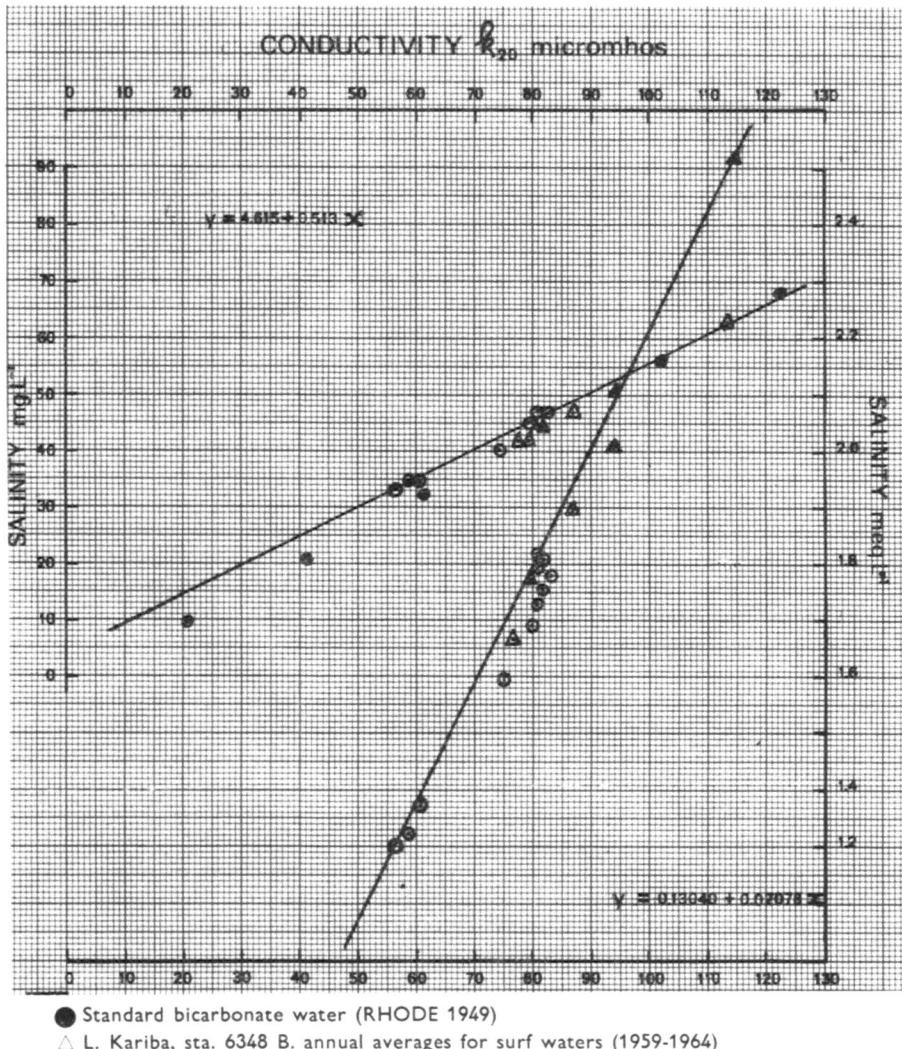

● Standard bicarbonate water (RHODE 1949)
△ L. Kariba, sta. 6348 B. annual averages for surf waters (1959-1964)
○ L. Kariba, misc. stations, annual averages for pelagic surf. water (1965)

Fig. 51. Relationship between salinity and conductivity in L. Kariba water.

13.2.2. SALINITY

The average annual salinity (SL) of the pelagic surface waters has been estimated from the data available for eleven stations situated along the major lake axis and covering the whole year 1965 (Tables 4 to 10 in COCHE, 1968). Data are summarized in Table 67.

The average salinity generally increases as one progresses from the upper end of the lake (Sta. 1205) towards the Kariba area (Sta. 6446 and 6342). In 1965, it ranged from 1.316 meq. L^{-1} (33.17 mg. L^{-1}) to 1.882 meq.L^{-1} (47.38 mg. L^{-1}), the general average for the lake as a whole being estimated at 1.639 meq.L^{-1} (42.44 mg. L^{-1}). Such relatively low salinity again ranks L. Kariba close to L. Mweru below other major African lakes as will be discussed later. According to CONWAY (cited by HUTCHINSON, 1957), salinities below 50 mg.L^{-1} are generally characteristic of igneous drainages.

The particular relationship existing between salinity (SL) and conductivity (k_{20}) has been statistically determined using the standard method of least squares. Two sets of data have been chosen: (i) paired measurements from samples taken at Sta. 6348B between 1959 and 1964 (HARDING, 1962, 1964a, 1965 and 1966; pers. data); (ii) paired measurements from 73 samples collected along the lake longitudinal axis in 1965 from 11 stations. In both cases annual averages have been calculated, providing 17 paired measurements to which the regression lines were fitted (Fig.51) as follows:

$$SL\ (meq.\ L^{-1}) = 0.13040 + 0.02078\ k_{20}$$
$$SL\ (mg.\ L^{-1}) = 4.615 + 0.513\ k_{20}$$

In both cases the correlation is strong (r equals 0.991 and 0.985 respectively). RODHE (1949) has determined this relationship for standard bicarbonate waters with salinities ranging from 10.5 mg. L^{-1} (k_{20} = 20 µmhos). Pertinent data are plotted in Fig. 51 also and results point to the striking similarity existing between L. Kariba waters and RODHE's bicarbonate waters from the point of view of the SL/k_{20} relationship.

The relationship between salinity and conductivity is more usually expressed in a simpler form such as:

$$SL = b. k_{20}$$

where b is a coefficient specific to the water studied. For Lake Kariba, two coefficients are defined according to the unit of measurement, b_1 for mg. L^{-1} and b_2 for meq. L^{-1} (Table 67). Their ranges of variation have been:

$$0.538 \leqslant b_1 \leqslant 0.613\ mg.\ L^{-1}\ per\ micromhos,$$
$$0.0210 \leqslant b_2 \leqslant 0.0239\ meq.\ L^{-1}\ per\ micromhos.$$

The general mean values and their 95-percent confidence limits have been as follows:

$$b_1 = 0.5741 \pm 0.0101\ (mg.\ L^{-1}/\mu mhos),$$
$$b_2 = 0.0225 \pm 0.00037\ (meq.\ L^{-1}/\mu mhos).$$

13.2.3. CONDUCTIVITY

The electrical conductivity has always been used by limnologists as a valuable method to estimate the degree of mineralization of waters and as a guide for studying its global variations in rivers and lakes. The amount of dissolved ionizable salts in fresh waters being generally considered as related to their potential biological productivity, conductivity measurements have also been widely taken to provide the basis for the estimation of fish production.

Table 68. Range of the surface water conductivity, k_{20} micromhos.

Period 1965–1966	Basin			
	I	II	III	IV
22 Apr.–1 May	40–45	60–80	80	80–99
22 June–2 July	45–52	45–53	73–80	75–81
3–12 August	50–52	55–64	75–80	77–80
27 Oct.–4 Nov.	71–75	71–76	76–78	77–82
12–20 January	83–86	80–81	76–83	78–81
Year	40–86	45–80	73–83	75–99
Oct. 1968	58–75	61–73	83–85	87–98

Conductivity increases with water temperature and must therefore be referred to a standard temperature to become comparable. The conductivity coefficient for temperature correction (K_t) from observed (t_0) to standard (t_s) temperature is calculated as:

$$K_t = \frac{1}{1 + q\,(t_0 - t_s)},$$

where q is a coefficient increasing from 0.021 to 0.027 with the increase of water temperature from 0 to 30° C (SMITH, 1962). For African waters in general, TALLING & TALLING (1965) have estimated and used q = 0.026 corresponding to a 2.3 percent change in conductivity per centigrade. A similar rate has been adopted in the present study to convert observed conductivities to the standard temperature of 20° C.

In 1965 the mean annual conductivity of pelagic surface waters varied between 55 and 81 μmhos (Table 67). Its value regularly increased as one moved from Basin I towards Basin IV. The general mean k_{20} for the lake was 73 μmhos. Lower conductivities were mostly observed in the upper two basins (Table 68) except during the period when the Zambezi River had a reduced discharge (September-December). Higher conductivities were always observed in the lower two basins particularly in eastern Basin IV. The overall range in 1965 for k_{20} was from 40 to 100 μmhos. Greater conductivities are only present locally in particular situations such as the Sanyati River pools during the dry season (more than 300 μmhos in August 1965).

The total ionic concentration (TIC) is known to be closely related to conductivity. The equivalent conductivity (c) is defined as

$$c = k_{20} \cdot (TIC)^{-1},$$

the latter being expressed in mg-equivalents per litre. The total ionic concentration is adequately estimated as the sum of the concentrations of the principal cations Ca, Mg, Na, and K (TALLING & TALLING, 1965). This method has been applied here

Table 69. Relationship of mean annual total ionic concentration (TIC) and total alkalinity (TA) to electrical conductivity.[1]

Basin	Station	No Samples	Mean k_{20} μmhos	Mean TIC meq.L^{-1}	c^2 Eq.cond.	Mean TA meq.L^{-1}	d^3
I	1205	5	59	0.692	85.26	0.584	101
II	1509	5	55	0.658	83.59	0.556	99
	2015	3	57	0.666	85.59	0.607	94
	2519	3	73	0.800	91.25	0.727	100
	Avg.	11	62	0.708	87.57	0.630	98
III	3028	5	78	0.847	92.09	0.800	98
	3332	11	79	0.866	91.22	0.833	95
	4036	6	80	0.878	91.12	0.817	98
	Avg.	22	79	0.864	91.44	0.817	97
IV	5043	12	79	0.898	87.97	0.837	94
	5845	6	81	0.892	90.81	0.847	96
	6446	9	80	0.906	88.30	0.847	94
	6342	9	79	0.911	86.72	0.860	92
	Avg.	36	80	0.903	88.59	0.848	94
Lake	Gen. avg.	74	73	0.820	88.59	0.756	96

[1] For details see Table 72.
[2] $c = k_{20}.\,(TIC)^{-1}$ μmhos per meq. L^{-1}.
[3] $d = k_{20}.\,(TA)^{-1}$ μmhos per meq. L^{-1}.

to the mean annual ionic composition of pelagic surface waters in 1965 (Table 69). Similar calculation has been applied to data obtained by HARDING and myself from Sta. 6348B between 1959 and 1965 (Table 87).

The general average equivalent conductivity for the lake as a whole in 1965 was 88.6 μmhos per meq. L^{-1}. In the various basins it ranged from 84 to 92 μmhos. At the Boom Station, the range of the annual mean c extended from 87 to 91 μmhos, the general average 1959–1965 being also 88.6 μmhos per meq. L^{-1}. Therefore this estimate for c appears to be reasonably adequate for estimating the total ionic concentration from k_{20} when no detailed chemical analysis has been performed. This Kariba equivalent conductivity lies in between the general average (85 μmhos) determined for African lakes (TALLING & TALLING, 1965) and the value applying to RODHE's (1949) standard bicarbonate water with similar conductivity ($k_{20} = 80$; $c = 94.6$ μmhos).

When the carbonate and bicarbonate ions are predominant, there exists a strong correlation between conductivity and total alkalinity (TA) so that:

$$k_{20} = d \cdot TA,$$

TA being expressed in meq. L^{-1} total fixed CO_3 (HEM, 1959). Applying this principle to L. Kariba data available from Sta. 6348B for the period 1959–1965 (HARDING, misc. and pers. data) and transformed into annual averages, d is equivalent to 94 μmhos per meq. L^{-1}. Applied to the 1965 averages (Table 69) it is found that d slightly varies from basin to basin (94–100 μmhos). It decreases towards Basin IV where the average d equals 94 μmhos, confirming the previous estimate for Sta. 6348B. The overall lake average is 96 μmhos per meq. L^{-1}. In the Bangweulu-Luapula basin d mostly varies between 80 and 110 (SYMOENS, 1968). For African waters in general TALLING & TALLING (1965) have proposed d = 100 which closely agrees with the estimate proposed for L. Kariba.

13.2.4. CONCLUSIONS

In Lake Kariba, the total mineral content of pelagic surface waters is relatively small as attested by the low values measured for total solids, salinity, and conductivity (Table 70). This mineral poverty finds its origin mostly in the character of the Zambezi River water but also in the character of the lake sediments and of the thermal properties, as will be further discussed later (Sec. 13.6.1).

Strong relationships have been found existing between conductivity on one hand and on the other hand total solids content, salinity, total ionic concentration and total alkalinity. The general average coefficients have been determined (Table 70) and they should be very helpful for further rapid estimations of general trends of evolution.

Table 70. Total mineral content of pelagic surface waters in 1965 and basic relationships to conductivity.

Item	Lake	Basin			
		I	II	III	IV
Total solids (TS), mg. L^{-1}	(58)	(47)	(49)	(63)	(64)
Salinity (SL), mg. L^{-1}	42.4	34.6	36.3	45.4	46.7
Salinity (SL), meq. L^{-1}	1.639	1.384	1.416	1.728	1.806
Conductivity, k_{20} μmhos	73	59	62	79	80
Coefficients					
a = TS. $(k_{20})^{-1}$	0.799	—	—	—	—
b_1 = SL. $(k_{20})^{-1}$ for mg.L^{-1}	0.574	0.586	0.586	0.575	0.587
b_2 = SL. $(k_{20})^{-1}$ for meq.L^{-1}	0.0225	0.0234	0.0224	0.0215	0.0225
c = k_{20}.(TIC)$^{-1}$	88.59	85.26	87.57	91.44	88.59
d = k_{20}.(TA)$^{-1}$	96	101	98	97	94

13.3. Inorganic ions in pelagic surface waters

Already when defining 'salinity' (Sec. 13.2) the relative importance on one hand of sodium (Na), potassium (K), calcium (Ca), and magnesium (Mg) as cations and, on the other hand of fixed carbonate (CO_3), sulphate (SO_4), and chloride (Cl) as anions, has been stressed. It is the purpose of this section to present and discuss data on the relative abundance of these major ionic constituents in so far as they determine the overall chemical characteristics of pelagic surface water in Lake Kariba (HUTCHINSON, 1957).

13.3.1. GENERAL IONIC COMPOSITION

The methodology used to determine the general ionic composition of lake water has been dealt with earlier (Sec.9.4). Detailed results of the numerous chemical analyses performed in 1965 have been given elsewhere (Table 4 to 10 in COCHE, 1968). Our actual purpose being to ascertain general rather than detailed trends, individual results at each representative station have been grouped, the mean annual ionic compositions being considered only. From such data basin and lake averages have been computed, the latter being obtained from all eleven stations rather than from the basin's averages to better account for differences arising from surface areas of the various the basins (weighted averages).

Results are presented in milligrams per litre (Table 71) and in milligrams-equivalent per litre (Table 72). From these last data percentage ionic compositions have been determined for further comparisons.

The ionic composition of waters has been graphically represented using the modified MAUCHA's method as proposed by KUFFERATH (1951). This simplified method has the main advantage to enable the graphical representation of water analyses in which neither the two major alkali metals (Na and K) nor the carbonate and bicarbonate anions have been separately determined. Only six major ions (or group of ions) are therefore taken into account: Na + K, Ca, Mg among the cations and HCO_3 + CO_3 (fixed CO_3), SO_4, Cl among the anions.

The ionic concentrations are calculated in equivalents per litre and expressed in percent of the sum for each of the two ionic groups separately. These percentages are transferred onto a six-sided regular polygon subdivided into six triangular sectors, so that in any diagram the length of the line starting from the polygon's centre and bisecting a sector is proportional to the concentration of the given ion (or group of −) as equivalent percent of the total ionic composition. A cation field is provided on the top half of the hexagon, the anion field being below. Both fields are subdivided into three equal sectors each of them being allowed to a specific ion (or group of −). The polygon is built and analyses results are transferred into it, in such a way that the area of the tetragon delimited by two consecutive bisectors is proportional to the concentration of the corresponding ion (or group of −). KUFFERATH (1951) provides a ready-made model with the necessary graduations for this purpose. Examples are given in Figure 52.

The average ionic composition in 1965 of Lake Kariba waters from the surface

Table 71. Mean annual ionic composition of Zambezi River water and lake pelagic surface water in 1965, mg. L^{-1}.

Ion	Zambezi R.	Basin I				Basin II				Basin III				Basin IV				Lake
		1205	1509	2015	2519	Avg.	3028	3332	4036	Avg.	5043	5845	6446	6342	Avg.	Avg.
No. samples		5	5	3	3	11	5	11	6	22	12	6	9	9	36	74
Ca	8.27	7.82	7.89	7.84	8.50	8.08	10.53	10.10	10.55	10.39	9.71	9.70	9.80	10.03	9.81	9.32
Mg	2.13	1.90	1.64	1.63	2.34	1.87	1.44	1.83	1.63	1.63	2.44	2.19	2.20	2.14	2.24	1.95
Na	3.00	2.88	2.53	2.81	3.40	2.91	3.52	3.77	3.96	3.75	3.58	4.05	4.11	4.12	3.97	3.52
K	0.67	0.48	0.43	0.44	0.82	0.56	1.18	1.11	1.06	1.12	1.29	1.20	1.32	1.29	1.28	0.97
CO$_3$	19.09	17.52	16.68	18.21	21.81	18.90	24.00	24.99	24.51	24.51	25.11	25.41	25.41	25.80	25.44	22.68
Cl + SO$_4$ [1]	(4.00)	(4.00)	(4.00)	(4.00)	(4.00)	(4.00)	(4.00)	(4.00)	(4.00)	(4.00)	(4.00)	(4.00)	(4.00)	(4.00)	(4.00)	(4.00)
Salinity	37.16	34.60	33.17	34.93	40.87	36.32	44.67	45.80	45.71	45.40	46.13	46.55	46.84	47.38	46.74	42.44

[1] Estimated mean annual contents: Cl = 1 mg. L^{-1}; SO$_4$ = 3 mg. L^{-1}

Table 72. Mean annual ionic composition of Zambezi River water and lake pelagic surface water in 1965, meq. L⁻¹.

Ion	Zambezi R. 1965 meq. L⁻¹		Basin I		Basin II					Lake Kariba Basin III				Basin IV					Lake
			1205		1509	2015	2519	Avg.		3028	3332	4036	Avg.	5043	5845	6446	6342	Avg.	Avg.
No. samples	3		5		5	3	3	11		5	11	6	22	12	6	9	9	36	74
Ca	.413	55%	.390	56%	.394	.391	.424	.403	57%	.525	.504	.526	.518 60%	.485	.484	.489	.500	.490 55%	.465 57%
Mg	.175	23%	.156	23%	.135	.134	.192	.154	22%	.118	.150	.134	.134 15%	.201	.180	.181	.176	.184 20%	.160 20%
Na	.131	18%	.125	18%	.110	.122	.148	.127	18%	.153	.164	.172	.163 19%	.156	.176	.179	.179	.173 19%	.153 18%
K	.029	4%	.021	3%	.019	.019	.036	.024	3%	.051	.048	.046	.049 6%	.056	.052	.057	.056	.056 6%	.042 5%
Tot. ionic conc.	.748	100%	.692	100%	.658	.666	.800	.708	100%	.847	.866	.878	.864 100%	.898	.892	.906	.911	.903 100%	.820 100%
CO₃	.636	(85%)	.584	(84%)	.556	.607	.727	.630	(89%)	.800	.833	.817	.817 (95%)	.837	.847	.847	.860	.848 (94%)	.756 (89%)

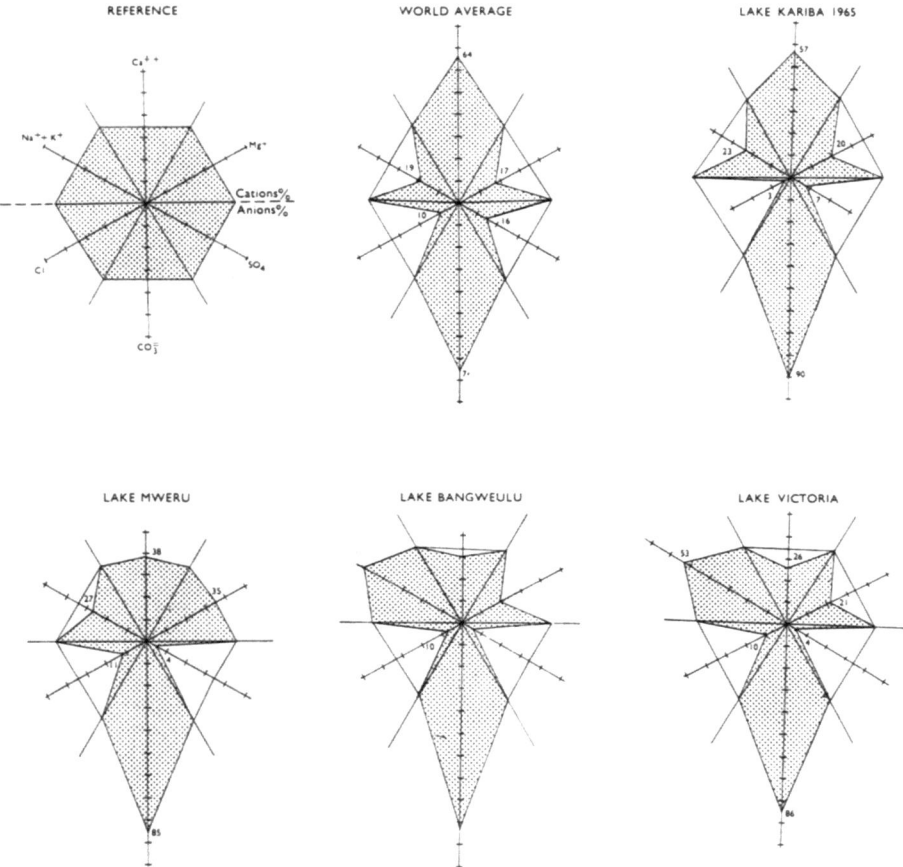

Fig. 52. Ionic composition of L. Kariba water compared to L. Mweru, L. Bangweulu, L. Victoria, and the world average freshwater.

of the pelagic zone (Tables 71 and 72; Fig. 52) reveals a striking similarity with the composition of the world average freshwater as cited by GORHAM (1957). Although the salinity of the latter (4.66 meq.L^{-1}) is nearly three times as high as in L. Kariba, the percentage cationic composition is only slightly different (Table 73). Differences are greater when the anion's relative abundance is compared, the proportion of carbonate and bicarbonate being greater in L. Kariba. A similar disproportion is observed in the Zambezi River water (Table 72). Both graphs however display a strikingly similar general shape (Fig. 52) characteristic of calcico-carbonate waters.

VISSER (cited in IMEVBORE, 1970) has calculated the average composition of East and West African lakes and rivers (Table 73). Comparatively and percentage wise (by equivalents), L. Lariba water appears as relatively enriched in calcium and fixed

Table 73. Average ionic composition of L. Kariba surface water compared to average African and World freshwaters.

Ion	L. Kariba 1965 average		World freshwater average			African freshwater average		East and West African lakes and rivers		
	meq.L^{-1}	percent	meq.L^{-1}	percent		meq.L^{-1}	mg.L^{-1}	meq.L^{-1}	percent	mg.L^{-1}
Calcium	0.465	57	1.49	64		0.624	12.5	0.324	29	6.5
Magnesium	0.160	20	0.42	17		0.313	3.8	0.239	21	2.9
Sodium	0.153	18	0.36	16		0.479	11.0	0.392	34	9.0
Potassium	0.042	5	0.08	3		?	?	0.182	16	7.1
Total cations	0.820	100	2.35	100		—	—	1.137	100	25.5
Fixed CO$_3$	0.756	89	1.71	74		?	?	(0.725)	(64)	(21.8)
SO$_4$	(0.062)	(7)	0.37	16		0.281	13.5	0.229	(20)	11.0
Cl	(0.028)	(3)	0.23	10		0.341	12.1	0.183	(16)	6.5
Salinity	1.67	—	4.66	—		(4.5)	—	(2.27)	(100)	(39.3)
SiO$_2$	12.3 mg/L		17.1 mg/L			—	23.2	—		15.5
Reference	Pers. data		Gorham 1957			Livingstone 1963		Visser cited in Imevbore 1970		

Table 74. Variations of three ratios of the mean annual concentrations of major cations, by equivalents (from Tables 72 and 73).

Region		K/Na	Ca/Mg	Ratio Ca+Mg/Na+K	Ca+Mg/CO$_3$
Zambezi R.	low water	0.05	2.50	2.68	0.83
	high water	0.28	1.90	3.79	0.90
	wgtd. avg.	0.16	2.11	3.23	0.87
Basin I		0.17	2.50	3.74	0.93
Basin II		0.19	2.62	3.69	0.88
Basin III		0.30	3.86	3.08	0.80
Basin IV		0.32	2.66	2.94	0.79
Lake Kariba		0.27	2.91	3.21	0.83
World freshwaters		0.22	3.55	4.34	1.12
African freshwaters		—	1.99	—	—
E. and W. Africa		0.46	1.36	1.02	—

Table 75. Ranges of variation of the concentration in major cations (mg. L^{-1}) in L. Kariba, 1965.

Cation	Basins I–II	Basins III–IV
Calcium	2 –10	7 –13
Magnesium	0.5– 3.0	0.5– 3.0
Sodium	1.8– 3.6	3.2– 4.7
Potassium	T– 1.8	0.6– 1.6

carbonate but poor in monovalent cations, sulphate, and chloride. Salinity is smaller than average.

The spatial variation of the major ionic composition along the longitudinal axis of the lake (Table 72) appears to be small, confirming observations from other African lakes (TALLING & TALLING, 1965). Although the total ionic concentrations progressively increase from Basin I towards Basin IV the percentage equivalent composition does not exhibit a similar trend. Only Basin III might be considered as out-of-line from this point of view, particularly in its cationic relative composition (Sec. 13.3.4).

13.3.2. ALKALI METALS: SODIUM AND POTASSIUM

Among the two major alkali metals, sodium is by far the more abundant (Tables 71 and 72). Its concentration varies around 3.5 mg. L^{-1} representing 18 percent of the total ionic concentration (TIC). Although this percentage does not greatly vary between Basin I and Basin IV, the average content slightly increases as the salinity increases, ranging in the lake from 2.53 to 4.12 mg. L^{-1}. The overall range for 1965 extended from 1.8 to 4.7 mg. L^{-1} (Table 75).

Potassium makes up from 3 to 6 percent only of TIC, the average content remaining under 1 mg. L^{-1} in the two upper Basins under the direct influence of the Zambezi River. It is only in Basins III and IV that the average concentration increases above 1 mg. L^{-1} varying between 1 and 1.3 mg. L^{-1}. In 1965, the overall variation ranged from traces to 1.6 mg. L^{-1} (Table 75).

The ratio K/Na by equivalents averages 0.27 but it varies from basin to basin (Table 74). A sudden increase is noted as one moves from Basins I–II into Basins III–IV.

13.3.3. Total hardness: calcium and magnesium

In Lake Kariba, calcium is the predominant cation its average concentration varying between 7.8 and 10.5 mg. L^{-1} (Table 71). The overall variation ranged from 2 to 13 mg. L^{-1} (Table 75). By equivalents calcium makes up for 55 to 60 percent of TIC (Table 72). This percentage does not seem to have any particular pattern of variation throughout the lake.

Magnesium, on average, ranks second (20 percent TIC) but particularly so in Basins I and II (Table 72). In the lower basins it loses some of its relative importance to the benefit of the monovalent cations. This reduction in magnesium is particularly marked in Basin III where the average content drops to 1.63 mg.L^{-1} while the calcium content reaches maximum values.

The ratio Ca/Mg by equivalents (Table 74) averages 2.9 for 1965, but in most of the lake the ratio is 2.5 to 2.6 with Basin III constituting the exception (3.86). Even the mean value is already relatively high, if compared to other African waters (TALLING & TALLING, 1965; SYMOENS, 1968) where it mostly varies between 0.5 and 2.0.

Definitions for hardness are generally conflicting (HUTCHINSON, 1957; HEM, 1959). For some authors it is defined in terms of calcium and magnesium hardness; for others it includes all substances which react with soap. But the effect of other alkaline earths which may be present being in fact included as equivalent quantities of the reported Ca and Mg, total hardness may be considered for general purposes as equivalent to the sum of these two ions. Total hardness averages 11.27 mg. L^{-1} (0.625 meq. L^{-1}) which equals to about one third only of the world average. Compared to the average African water (Table 73), Kariba hardness still represents only two-thirds of it.

13.3.4. Cationic composition: conclusions

Among the four major cations calcium is largely predominant, the commonest order of relative importance by equivalents being Ca > Mg ⩾ Na > K. It follows the most common tendency in freshwaters of the world (HUTCHINSON, 1957). In East and Central Africa sodium has been shown to be generally predominant except where

Table 76. Ranges of variation of total alkalinity (10^{-2} meq. L^{-1} fixed CO_3) in L. Kariba basins in 1965.

Period 1965-66	Basin				Lake
	I	II	III	IV	
22 Apr.–1 May	42–48	58–72	80–84	80–84	42–84
22 June–2 July	48–52	46–56	76–84	80–86	48–86
3–12 August	56–58	56–64	80–84	82–88	56–88
27 Oct.–4 Nov.	72–74	74–80	80–82	80–86	72–86
12–20 January	80	80–84	76–84	82–86	76–86
Year	42–80	46–84	76–84	80–88	42–88

saline inflows are of little importance (TALLING & TALLING, 1965). Lake Kariba is part of these exceptions its major inflow, the Zambezi River, having a very low total ionic concentration (Sec. 5.2.4).

The sum of the divalent cations (hardness) represents the greatest proportion of the cationic composition (77 percent), the divalent to the monovalent cations ratio averaging 3.2 (Table 74). Among the anions the fixed carbonate being predominant (Sec. 13.3.5); there exists a certain correlation between hardness (Ca + Mg) and total alkalinity (SYMOENS, 1968). In L. Kariba the average ratio Ca + Mg/CO_3 (Table 74) varies between 0.8 and 0.9 (by milliequivalents). Normally the ratio ranges from 1.0 to 2.0 which points out to an unusually low ratio in L. Kariba.

13.3.5. TOTAL ALKALINITY: CARBONATE AND BICARBONATE

The property of alkalinity in water is its ability to neutralize acid. For practical purposes, it is generally considered that in waters with pH less than 8.2 the carbonate alkalinity is nil, total alkalinity (TA) being caused by bicarbonate ions only (SYMOENS, 1968). But it should be kept in mind that in fact the total alkalinity measurement, best expressed in milliequivalent fixed carbonate (CO_3), lumps together the effects of all the anions entering into hydrolysis reactions. Thus it may also include the equivalent of all or part of such ions as silicate, phosphate, borate, and possibly fluoride (HEM, 1959).

In 1965, total alkalinity in Lake Kariba varied between 0.42 and 0.88 meq. L^{-1} fixed CO_3 (Table 76). The greatest annual ranges of variation exist in Basins I and II related to the changing quality of the Zambezi River water throughout the year. As soon as the inflow decreases, the total alkalinity starts to increase gradually. The range of variation is relatively narrow in the lower two basins (0.76–0.88 meq. L^{-1} CO_3). Maximum for total alkalinity is observed in the Sanyati River pools during the dry season when values up to 4 meq. L^{-1} CO_3 have been recorded (Aug. 1965). The absolute minimum value is probably close to 0.38 meq. L^{-1} measured in Basins I and II in May 1967 during peak Zambezi flood (BEGG 1970).

The mean annual total alkalinity (TA) has been computed for eleven stations representing the four basins and the lake as a whole (Table 72). Average TA varied in 1965 between 0.56 and 0.86 meq. L^{-1} CO_3 representing about 84 to 95 percent of the total anions. The Lake average for the year amounts to 0.756 meq. L^{-1} fixed CO_3 (89 percent).

Total alkalinity increases from Basin I to Basin IV, the greatest change occuring between Basins II and III. The relative abundance (by equivalents) among the anions follows a similar trend down to Basin III only. Again the Zambezi River is responsible for the observed horizontal pattern of variation.

The pH in Basins I and II exceeds 8.2 on a very limited number of occasions only (Sec. 13.1), total alkalinity can be there generally considered as representing the bicarbonate (HCO_3) content. This content varied in 1965 between 0.42 and 0.84 meq. L^{-1} (25.6–51.2 mg. L^{-1} HCO_3), a two-fold variation. Annual average concentration in bicarbonate (from Table 72) ranged from about 0.56 to 0.73 meq. L^{-1} (33–44 mg. L^{-1}).

The large predominance of carbonate and bicarbonate anions in L. Kariba waters accounts for the close correlation found between total alkalinity on one hand and on the other hand total minerals (conductivity – Sec. 13.2.3) and hardness (Ca + Mg – Sec. 13.3.3) as discussed earlier. It also suggests that it is the variation in total alkalinity which is mainly responsible for the variations in conductivity (TALLING & TALLING, 1965).

13.3.6. CHLORIDE AND SULPHATE

Chloride (Cl) is present in igneous rocks in small quantities only. These rocks constitute a minor source for this ion in natural waters (HEM, 1959). Commonly, chloride originates either from the atmosphere (oceans) or from ground water (mineral springs).

Situated near the heart of Central Africa at more than 700 km from the Indian Ocean and at more than 1500 km from the Atlantic Ocean, it is most probable that L. Kariba catchment only receives very little chloride of atmospheric origin. One of the main sources of chloride, though still very limited from a volume point of view, might well be the mineral springs described from the Middle Zambezi Valley (MAUFFE, 1933).

Chemical analysis performed in 1970 on twenty water samples have shown chloride concentrations ranging from 0.016 meq. L^{-1} (late stagnation, Basin IV) to 0.071 meq. L^{-1} (after overturn, Basin III). The average chloride content of all analyses was 0.025 meq. L^{-1} (0.89 mg. L^{-1}) a result close to the previous estimate of 1 mg. L^{-1} (HARDING, 1961, 1962, 1964a). It is this last value which has been taken as a basis for further computations (Table 71 and 72). Thus the chloride anion represents about three percent only (by equivalents) of the average ionic composition. According to HUTCHINSON (1957) concentrations of less than 1 mg. L^{-1} are common, although general averages for the world freshwaters (8.16 mg. L^{-1}), for the African freshwaters (12.1 mg L^{-1}), and for Eastern and Western African lakes and rivers (6.5 mg L^{-1}) are much higher (Table 73). There is no doubt that the continentality of L. Kariba's geographic position is determinant in limiting the chloride concentration to a relatively low value.

Sulphate (SO_4). Very few data are available from L. Kariba concerning sulphate. During the filling of the Lake HARDING (1962, 1964a, 1965) estimated concentrations ranging from zero to 4 mg.L^{-1}. From his studies in 1963–1965, MITCHELL (1970). reports 0–0.5 mg. L^{-1} SO_4. In 1970 four analyses made at the Central Fisheries Research Institute resulted in concentrations ranging from less than 1 mg L^{-1} (late stagnation Basin IV) to 4 mg L^{-1} (after overturn, Basin III). TALLING & TALLING (1965) have pointed out to the possible systematic error of the analytical procedure involving the precipitation of the barium salt, unreliable for such small concentrations. They report that 'the smallest concentration of sulphate obtained by the ion-exchange procedure is 1.0 mg/L from L. Bangweulu. The next lowest concentration (2.3 mg/L)... from offshore L. Victoria, is between two and three times higher than the earlier values' (op. cit. p. 440).

On such bases the annual average SO_4 concentration in Lake Kariba has been estimated to be about 3 mg L^{-1} (0.062 meq. L^{-1}) for further computations (Tables 71 and 72). It results that sulphate represents about seven percent only of the ionic composition which corresponds to the general range of proportions found in African waters, 4 to 12 percent (TALLING & TALLING, 1965). But the concentration is definitely small when compared to more general averages (Table 73). It is similar to L. Mweru's average content in SO_4 (DE KIMPE, 1964). This paucity in sulphate relates to the small amounts generally present in soils and non-sedimentary rocks in Africa.

13.3.7. SILICA

Except for oxygen, silicon is the most abundant element in the earth's crust and silicate (SiO_2) minerals are well represented in the composition of igneous and metamorphic rocks (HEM, 1959). These rocks being particularly abundant in Africa and the tropical temperature favoring their weathering, their chemical breakdown, and their solubility into water, high contents (above 10 mg. L^{-1}) of dissolved silica are very common in African waters (TALLING & TALLING, 1965).

Data about silica in L. Kariba have been provided by HARDING (1964a, 1965, 1966) for the period Dec. 1961 to Oct. 1964. They are very consistent, the SiO_2 concentrations ranging from 10 to 15 mg. L^{-1} and averaging 12.3 mg L^{-1}. Compared to more general values (Table 73), this appears as a below-average content for Africa. It is nevertheless relatively high when compared to silica data from L. Mweru (below 5 mg. L^{-1} in DE KIMPE, 1964) which is otherwise similar to L. Kariba for so many other chemical characteristics.

13.3.8. NITRATE AND PHOSPHATE

Nitrate-nitrogen and phosphate-phosphorus are most important in determining the productivity of waters being both indispensable for biological growth processes.

Nitrate-nitrogen (NO_3-N) representing the final stage of nitrogeneous organic matters dissolved in waters or found in bottom muds, has been extensively measured in 1965 (Tables 4 to 10 in COCHE, 1968). The observed ranges in surface waters were as follows: (i) Basins I and II: 3–60 μg. L^{-1}; (ii) Basins III and IV; 10–97 μg. L^{-1}.

But as also pointed out by MITCHELL (1970), it is seldom that the NO_3-N concentration reaches over 20 µg. L^{-1}. Such very low concentrations are in line with observations made in most East African lakes where nitrate usually remains below 30 µg. L^{-1}. In freshwater this is far below the world average content (300 µg. L^{-1} in GORHAM, 1957) which corresponds to VISSER'S average for East and West African lakes and rivers (cited in IMEVBORE, 1970). The mean content for African waters (800 µg. L^{-1}) would even be further away (LIVINGSTONE, 1963).

Recently CAULTON (1970) has suggested that nitrate-nitrogen in L. Kariba has a predominant atmospheric origin (electrical rainstorms), very little originating from ground leaching. He based his theory on the relatively high content in NO_3-N measured in rain water (17.4 µg. L^{-1} on average). This would restrict the replenishment of nitrate to the rainy season (Nov.-April).

Phosphate-phosphorus (PO_4-P) on the contrary would originate from the leaching of the soil following rains. The latter in average contained only 7.4 µg. L^{-1} in January 1970 (CAULTON, 1970). Therefore the geology of the surface strata in the lake catchment would be of vital importance, replenishment in PO_4-P however coinciding also with the rainy season. Future agricultural development based on the addition of fertilizers could greatly influence the PO_4-P content of lake water.

In 1965 phosphate-phosphorus has been determined in samples representing the whole lake (Tables 4 to 10 in COCHE, 1968). The observed ranges in surfaces waters were mostly as follows: (i) Basins I and II: trace to 50 µg. L^{-1}; (ii) Basins III and IV: trace to 72 µg. L^{-1}. Generally the concentration in PO_4-P remained below 25 µg. L^{-1} throughout the year. The limiting concentration for phytoplankton development (10 µg. L^{-1} in TALLING & TALLING, 1965) was reached on few occasions only in the lower basins, being more frequently present in the upper two basins.

Concentrations of inorganic PO_4-P in East and Central African waters are generally high (TALLING & TALLING, 1965), being either slightly below or above the world average content in freshwaters (110 µg. L^{-1} in GORHAM, 1957). But notable exceptions exist particularly in lakes with low conductivity. L. Kariba belongs to this group of lakes poor in inorganic phosphate-phosphorus.

13.3.9. ANIONIC COMPOSITION: CONCLUSIONS

Among the three anions the ranking by equivalents coincides with the most currently encountered situation in freshwaters (HUTCHINSON, 1957): $CO_3 > SO_4 > Cl$. Although in line with the general trend observed elsewhere (Table 73), it contrasts with the data available from East and Central Africa where chloride is generally present in a higher concentration than sulphate (TALLING & TALLING, 1965).

Bicarbonate and carbonate are largely predominant representing in average 89 percent of the total ions. Such large predominance of fixed CO_3 also characterizes most of the world freshwaters (HUTCHINSON, 1957) as well as numerous lakes in Africa (TALLING & TALLING, 1965 and Table 73). It leads to the fairly good relationship existing between total alkalinity and total minerals (k_{20}) or hardness, as described earlier (Sec. 13.2.3 and 13.3.3).

L. Kariba water is very soft, the fixed CO_3 content rarely exceeding 1 meq. L^{-1}

Table 77. Main salts in surface water, station 6348B (after HARDING's data).

Period	Feb–Nov. 1959		Jan–Aug. 1960		1961		1962	
Salts	mg.L^{-1}	percent	mg.L^{-1}	percent	mg.L^{-1}	percent	mg.L^{-1}	percent
Bicarbonate Ca	36	57.1	31	64.6	28	62.2	28	62.3
Bicarbonate Mg	11	17.5	8	16.7	8	17.8	8	17.8
Bicarbonate Na	8	12.7	7	14.6	7	15.6	5	11.1
Bicarbonate K	—	—	—	—	T	T	1	2.2
Chloride Na	3	4.8	2	4.2	1	2.2	—	—
Chloride K	—	—	—	—	1	2.2	2	4.4
Sulphate Na	4	6.3	T	T	—	—	—	—
Nitrate Na	1	1.6	—	—	—	—	—	—
Nitrate K	—	—	—	—	—	—	1	2.2
Salinity	63	100	48	100	45	100	45	100
No. analyses	6		3		10		11	

(30 mg. L^{-1}). On the basis of European experience such low alkalinity would be characteristic of lakes with a low productivity. On a relative scale, this is probably also true in tropical waters.

Both sulphate and chloride are present in small concentrations reflecting the general paucity in these two ions of truly continental African waters. The silica content (10–15 mg. L^{-1}) is relatively high, much higher than the one reported from L. Mweru.

The concentrations in inorganic nitrate-nitrogen and phosphate-phosphorus are most variable but always very low, rarely exceeding 20–25 µg. L^{-1}. While the former is thought to be mainly of atmospheric origin, the latter would essentially originate from the leaching of the soils. Both these important elements therefore appear to have their replenishment limited to the rainy season only.

Generally, concentrations in anions progressively increase along the length of the reservoir towards the dam, the greatest average change occuring between Basins II and III. This is particularly the case for total alkalinity, nitrate, and phosphate. The greatest variations in concentrations are observed in the upper two basins, directly under the influence of the Zambezi River.

13.3.10. MAIN SALTS IN SURFACE WATERS

Although of restricted applicability it has been thought of general interest to present

Table 78. Annual chemical cycle in Basin III (Sta. 3332) in 1964–65 (mg.L^{-1}, acc. MITCHELL 1970 and pers. data).

Month	Calcium		Magnesium		Sodium		Potassium		Nitrate-N μg. L^{-1}		Phosphate-P μg. L^{-1}	
	S	B	S	B	S	B	S	B	S	B	S	B
August '64	8.9	9.2	1.8	2.2	3.7	2.7	1.4	0.5	45	123	Nil	11
September	11.3	9.4	1.1	1.1	3.4	2.7	1.3	0.5	34	106	12	21
October	11.1	8.9	0.9	2.6	3.4	3.4	1.2	1.1	30	34	10	15
November	9.9	10.4	2.1	1.0	3.2	2.6	1.2	0.5	36	134	6	19
December	13.6	8.7	1.0	2.4	3.6	3.0	1.1	0.7	12	107	14	26
January '65	8.4	8.4	3.1	2.6	3.3	2.5	1.0	0.6	12	121	27	36
February	9.5	—	2.5	—	3.5	—	1.0	—	18	—	27	—
March	8.8	—	2.4	—	3.8	—	1.4	—	20	—	—	—
May (Overturn)	9.6	9.6	2.3	1.5	3.9	(9.1)	1.0	0.7	14	15	72	57
June	9.2	8.4	2.3	2.3	3.8	3.6	0.9	0.7	87	26	—	42
July	11.5	—	1.4	—	3.8	—	1.0	—	39	—	15	—
August	13.0	8.9	0.5	2.5	4.1	3.6	1.1	0.7	11	42	23	30
September	10.5	9.7	2.0	2.5	4.0	3.6	1.1	0.7	9	40	23	35

S = surface water, B = bottom water (about 50-m depth).

Table 79. Annual chemical cycle in Basin IV (Sta. 5043, surface water) in 1964–1965 (mg.L^{-1}, acc. MITCHELL, 1970 and pers. data).

Month 1964–1965	Calcium	Magnesium	Sodium	Potassium	Nitrate-N μg.L^{-1}	Phosphate-P μg.L^{-1}
July 1964	9.7	1.8	3.2	1.0	28	15
August	8.9	2.3	3.4	1.0	38	20
September	8.9	2.7	3.5	1.3	13	8
October	10.8	1.6	3.4	1.3	27	32
November	10.0	2.3	3.1	1.1	32	6
December	13.4	0.2	3.3	1.1	17	8
January 1965	10.2	3.0	3.3	1.3	23	15
February	10.1	3.1	3.6	1.1	18	11
March	9.6	2.4	3.8	1.4	21	—
April	11.2	1.0	4.1	1.1	30	7
June (Overturn)	8.4	2.3	3.6	0.7	18	70
July	10.6	1.9	3.9	1.1	24	19
August	9.2	2.3	4.1	1.3	22	44
September	10.8	2.3	4.0	1.1	9	33
October	11.3	2.0	4.0	1.1	15	16

here (Table 77) the average results of analyses showing the main salines present in the surface water at station 6348B during the filling period of the lake (HARDING, 1962, 1964a, 1965).

On a weight basis calcium bicarbonate is by far the predominant salt, the rest of the fixed carbonate being present mostly in equal weight of magnesium and sodium bicarbonate. Chloride is combined either with sodium or with potassium, sulphate being present as a sodium salt only. Most probably this salt remained undetected in later analyses.

13.3.11. SEASONAL VARIATIONS OF THE MINERAL CONTENT

It has already been stressed earlier how great were the variations of the ionic composition in the surface waters of the upper two basins, where the direct influence of the Zambezi River is preponderant. As its flow decreases the mineral content of the lake water progressively increases and conversely, the difference between high-flood (around March-April) and low-flood season being well marked. Differences are high for total solids (k_{20}), calcium, sodium, fixed CO$_3$ (bicarbonate), nitrate, and phosphate, the inflowing river water being particularly deficient in these ions under high flow conditions (Table 13).

The annual chemical cycle in surface waters has been more closely followed in the lower two basins where lacustrine conditions prevail. Results are presented in Tables

Table 80. Annual chemical cycle in Basin IV (Sta. 6047) in 1964–1965 (mg. L^{-1}, acc. MITCHELL, 1970 and pers. data).

Date 1964–1965	Temperature °C		pH		k_{20} µmhos		TA meq.L^{-1}		Ca mg.L^{-1}		Mg mg.L^{-1}		Na mg.L^{-1}		K mg.L^{-1}		NO$_3$–N µg.L^{-1}		PO$_4$–P µg.L^{-1}	
	S	B	S	B	S	B	S	B	S	B	S	B	S	B	S	B	S	B	S	B
14 Sep. '64	26.3	26.0	7.4	7.2	72	73	0.82	0.86	8.9	11.5	2.7	3.2	3.7	3.8	1.4	1.4	13	91	6	18
5 Oct.	28.0	28.0	7.4	7.3	76	75	0.82	0.88	11.9	11.1	1.5	1.0	3.5	3.5	1.6	1.5	21	86	61	66
16 Nov.	—	—	7.4	—	70	—	0.82	—	11.9	8.3	0.5	3.1	3.8	3.6	1.4	1.2	13	92	14	16
22 Dec.	—	—	—	—	—	—	—	—	9.4	11.4	2.3	1.6	3.2	3.2	1.3	1.2	15	112	6	19
28 Jan. '65	—	—	—	—	—	—	—	—	9.6	8.8	1.5	2.9	3.2	3.4	1.3	1.3	21	180	36	62
4 Feb.	—	—	—	—	—	—	—	—	8.7	8.7	2.9	2.9	3.4	3.0	1.1	1.2	22	133	18	18
3 Mar.	29.1	22.0	8.6	8.2	72	70	0.92	0.82	9.6	8.8	2.0	2.4	4.0	3.9	1.9	1.5	45	111	29	29
6 Apr.	27.3	22.0	7.8	6.8	71	72	0.76	0.84	9.6	12.7	1.9	—N	4.2	4.1	1.5	1.5	15	74	29	25
4 May	25.7	21.2	7.2	6.8	—	—	0.82	0.78	8.8	9.6	2.3	1.5	4.0	3.8	1.2	1.0	13	117	7	35
25 May	24.1	21.8	7.2	6.8	71	70	0.88	0.84	8.8	11.9	2.9	1.0	3.9	3.9	1.0	1.0	16	19	24	37
3 June	23.6	21.3	7.2	6.8	72	68	0.90	0.78	8.8	9.6	2.9	1.9	3.9	3.8	1.0	1.0	20	71	—	—
2 July	22.5	19.1	7.1	7.1	71	72	0.84	0.84	12.2	9.6	1.0	1.9	4.1	3.8	1.4	1.1	14	21	30	8
30 Aug.	24.2	21.5	7.6	7.0	71	76	0.86	0.84	10.8	10.0	1.9	0.5	4.1	3.0	1.2	0.5	10	49	N	25
7 Oct.	23.9	21.7	7.9	7.2	83	79	0.84	0.76	10.5	10.5	2.5	2.0	4.2	4.0	1.1	1.0	11	45	17	34

S = surface water, B = deep water (ab. 50-m depth).

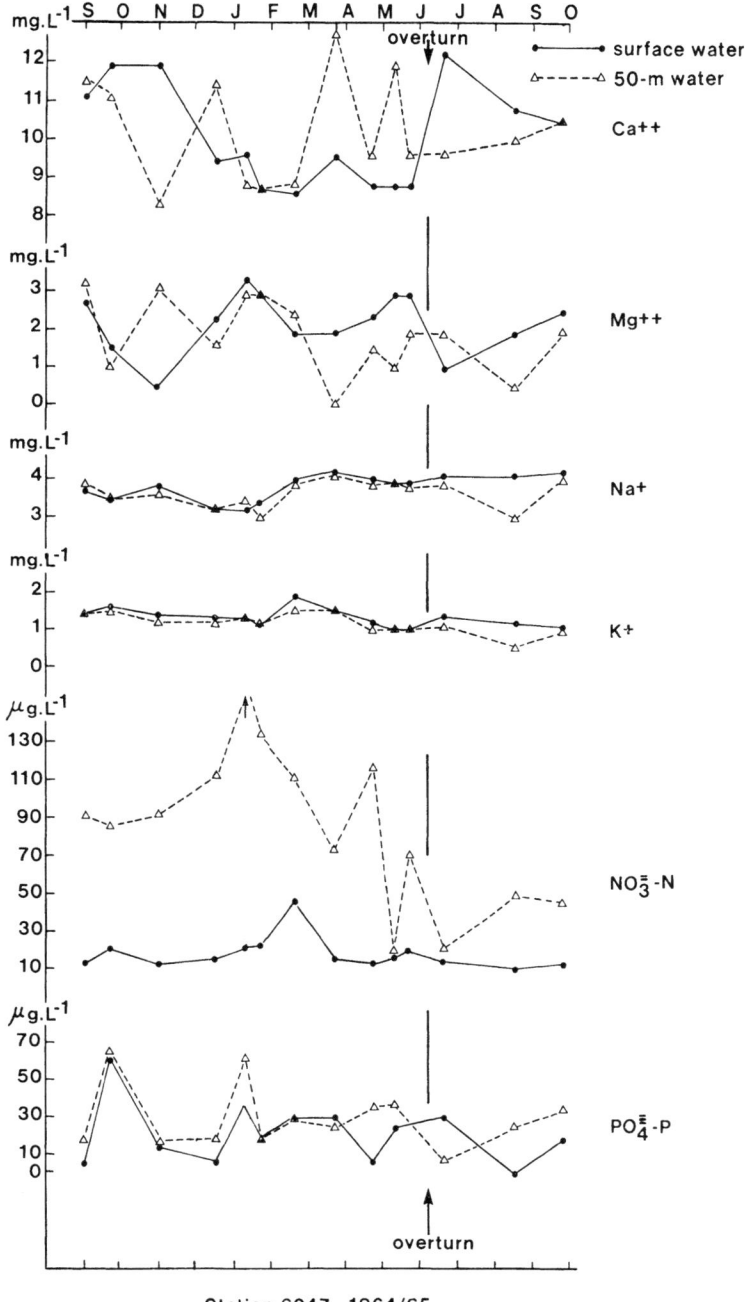

Fig. 53. Variation of the chemical constituents in Basin III (Station 3332) in 1964–65.

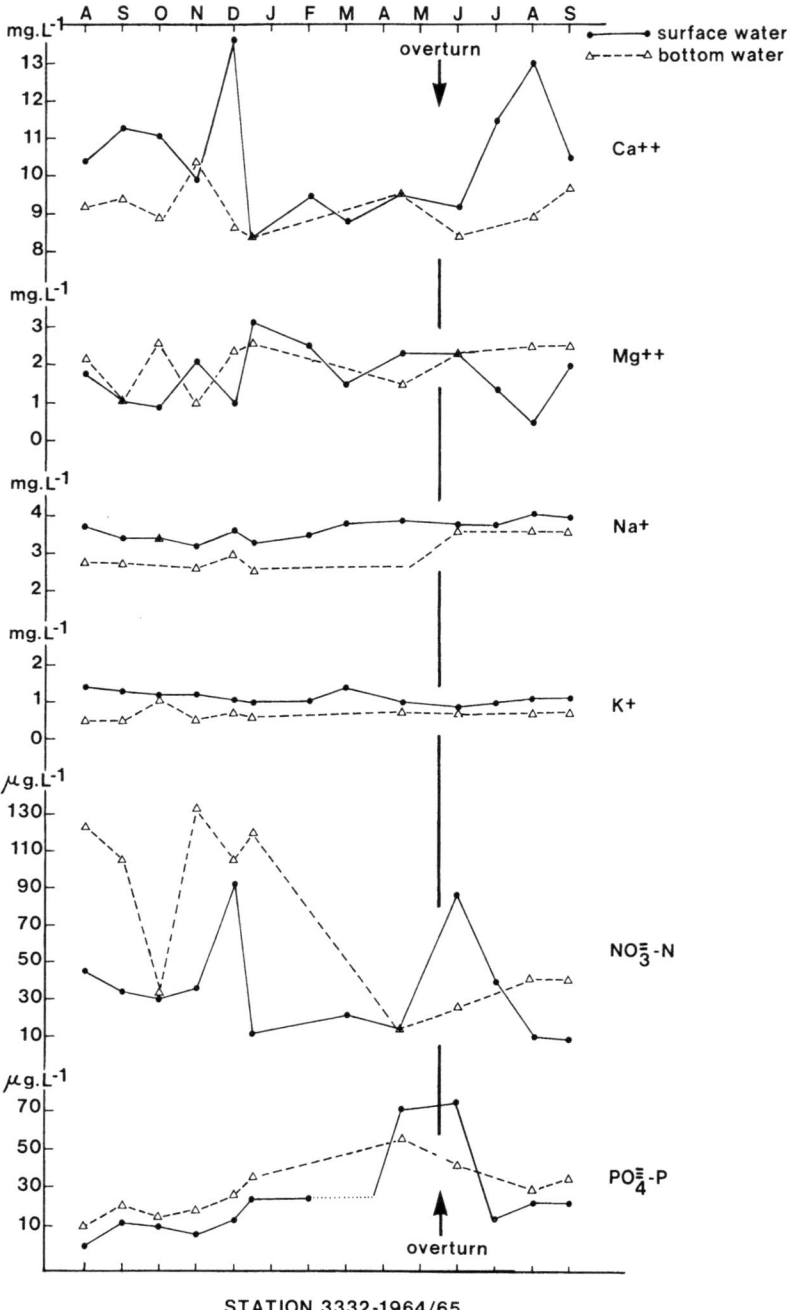

Fig. 54. Variation of the chemical constituents in Basin IV (Station 6047) in 1964–65.

78, 79 and 80. For stations 3332 and 6047 in particular, they are illustrated in Figures 53 and 54.

In Basin III (Sta. 3332) two factors appear to play a major role in governing the ionic composition of pelagic surface waters: the onset of the rainy season and the increased vertical circulation. The rains started in November 1964. There was an important simultaneous increase in calcium, nitrate, and phosphate contents. Similarly a second important increase for the same ions was recorded as vertical circulation proceeded. These sharp rises in concentrations were generally followed by sharp decreases the next month, most probably due to intensified biological uses. Variations in potassium and sodium were much less pronounced. Magnesium varied more wildly.

In Basin IV (Sta. 6047) the variations in nutrient concentrations appear to be much less related to the onset of the rains but the relationship to increased vertical mixing subsists: all cationic concentrations increase (in particular Ca and Mg). But the rise in NO_3-N and PO_4-P contents is not as marked as in Basin III. A sharp rise in the nitrite content of surface waters coincides with total circulation (COCHE, 1968). It takes a few more weeks to observe the corresponding rise in the nitrate content, a phenomenon which greatly increases biological activity: phytoplankton blooms (Cyanophyceae) and sudden appearance of mature medusae *(Limnocnida rhodesiae)*. Simultaneously the nitrite progressively disappears from the surface waters (op. cit.).

13.4. Mineral content of deep waters

The mineral content of deep waters is most closely correlated with the thermal behaviour of the lake. As stratification establishes itself a chemocline simultaneously develops at the depth of maximum density change, coinciding then with the thermocline and the oxycline. Later in the year cycle as the metalimnetic barrier opens and as vertical circulation progresses deeper and deeper, the chemocline gradually disappears. The chemical quality of the water mass becomes more uniform with complete circulation. In Basins I and II intermediate density currents occur which change the water quality according to the Zambezi River inflow. Along the floor of Basins III and IV old Zambezi water can be traced below the 55m-depth in the river channel as far as the dam wall.

Generally conductivity and total alkalinity decrease with increased depth. They may rise again either close to the bottom sediments or in presence of hydrogen sulphide (BEGG, 1970).[19] In 1965 the hypolimnetic conducticity (k_{20}) varied mostly between 58 and 68 micromhos, total alkalinity ranging from 0.60 to 0.66 meq.L^{-1} in the lacustrine part of the lake.

Detailed data on deep water chemical composition were given earlier (COCHE, 1968 – Tables 4 to 10). Further data are provided in Tables 78 (Basin III) and 80 (Basin IV), being graphically compared to surface waters in Figures 53 and 54.

---

[19] Since late 1969 the tendency for the bottom waters to generally exhibit lower conductivities and total alkalinities than surface waters seems to be reversed, suggesting a build-up of nutrients within the profundal zone of the lake, in particular in the Sanyati and Kariba Sub-Basins (G. BEGG, pers. comm.).

Among the major cations, sodium and potassium concentrations are practically always lower in deep water than in surface water. The relationship is much more variable for calcium and magnesium. Nitrate nitrogen may reach relatively high concentrations in deep waters (up to 180 μg. L^{-1}). The difference in phosphate-phosphorus content is much smaller between surface and deep layers. It also remains more constant throughout the year cycle except when vertical circulation develops.

In L. Kariba there is not any significant difference in the major ionic composition of surface and deep water. The latter also belongs to the calcico-carbonate group of waters. But pH being always lower than 8.2, all fixed carbonate are here always present under the form of bicarbonate.

13.5. Relation between the chemical composition of lake and inflow waters

13.5.1. THE ZAMBEZI RIVER AND THE LAKE

The major inflow in L. Kariba is the Zambezi River (Chapt. 5). In average it provides about 77 percent of the total annual gains in water. Major losses occur by outflow (discharge through turbines or spillage), accounting for about 86 percent of the total annual losses in water (Table 23). Furthermore the theoretical replacement time is relatively small (3.1 years – Sec. 7.2). As a result it is no surprise to find that the chemical nature of the Lake and Zambezi waters are broadly similar. The average ionic proportions in 1965 were practically identical in both environments (Table 72). As far as abundance of the major ions was concerned, the lake water was on average slightly richer than the river water in calcium, sodium, potassium, and fixed carbonate but poorer in magnesium. These differences are better expressed when comparing the cationic ratios (Table 74). The quantitative contribution of the Zambezi River will be discussed later (Sec. 13.6.2).

As mentioned earlier great differences however exist from basin to basin along the longitudinal axis of the lake (Tables 72 and 74). On an annual basis Basin I and the upper two-thirds of Basin II present in fact smaller total ionic concentrations and lower total alkalinities than the Zambezi River, the average enrichment (exc. for Mg) being only observed from about the Masumo/Zambezi confluence (when passing from Sta. 2015 to Sta. 2519). These observations made from annual average values are probably correlated with the overwhelming influence of the Zambezi River flood (made of particularly dilute water) in this upper area of the Lake. From the chemical point of view the latter might be considered as the true transitional zone between the riverine and the lacustrine regions of L. Kariba.

13.5.2. THE SECONDARY RIVERS AND THE LAKE

Although the Zambezi River must be considered as the most important factor in determining the lake's chemical character in general and the upper two basins' status in particular, the local importance of the other inflowing rivers cannot be ignored (Chap. 6). While the Zambezi influence is due to the tremendous volume of its inflow, secondary rivers find their importance in the chemical nature of their waters. Because of their generally higher ionic contents (Sec. 6.3), it has even been suggested that they bring into Lake Kariba the majority of its supply in dissolved salts (BEGG, 1969). In fact they mostly affect the lake's chemical budget in the immediate vicinity of their entry. This effect is proportional to the river water salinity, the latter itself being directly related to the characteristics of the river's watershed (runoff, soils, geology, human uses, etc.). The chemical contribution of the Rhodesian affluents is particularly important due to their greater drainage areas (Sec. 6.2) and their higher ionic content comparatively to the Zambian affluents (COCHE, 1968; BEGG, 1969). The Sanyati River (Sec. 6.3.1) has been previously mentioned for its great influence on the limnological characteristics of the eastern end of L. Kariba[20].

13.5.3. THE SUPPLY OF IONS BY RAIN

Although rain is theoretically distilled water and might as such be expected to be particularly pure, it is nevertheless contaminated in a variety of ways during its descent. Apart from dust, all sorts of dissolved substances have been identified in rain. This subject has been reviewed by HUTCHINSON (1957) and GORHAM (1961). Both authors show that the atmospheric supply of ions to natural waters is far from negligible.

In average about 3.5 cu. km (35.10^{11} litres) of rain fall on Lake Kariba's surface each year (Table 23). Although this only represents seven percent of the total inflows, it probably makes a relatively important contribution to the chemical budget because of the general low salinity of the major inflow. In November 1964 one sample of Mwenda's rainfall (Sta. 3628) was analysed (MITCHELL, 1970) giving the results shown in Table 81. The conductivity at 20° C was 15 micromhos.

When the major ions concentrations are compared to those available from the literature, L. Kariba rainwater stands well above normal ranges. In particular calcium, magnesium, potassium, and fixed carbonate are present in relatively high concentrations. This might be correlated to the sampling date which coincides for 1964 with the beginning of the rainy season (Table 1 in COCHE, 1968). Most probably the general average salinity is lower. Considering the latter as only half the reported value (possibly maximum) or about 12 mg of dissolved mineral substances per litre of rain the total ionic supply would be equivalent to 42 000 metric tons per year.

CAULTON (1970) provides average contents in phosphate-phosphorus and nitrate-nitrogen from the Kariba rainfall (Table 81). Compared to GORHAM's data (1955 cited in HUTCHINSON, 1957), the PO_4-P concentration appears to be above average. The NO_3-N content on the other hand seems to be much lower than average, especially for tropical regions. For such regions DRISCHEL (1940 cited in HUTCHINSON, 1957) has calculated a mean value of 0.267 mg NO_3-N per litre. According to KUENEN (1955), as a result from particularly effective electrical discharges acting on the atmospheric nitrogen in the tropics, about 10 kg of nitrogen compounds would be added by the rain each year to every hectare of land.

---

[20] BEGG (1972) has studied the particular influence of the Sanyati River in this area of the lake, and its relationship with the vertical distribution of zooplankton.

Table 81. Chemical composition of rainwater, mg. L^{-1}.

Ion	L. Kariba Nov. 1964	Ann. avg. misc.	Mean comp. world	Kariba Jan. 1970	Zambezi R. flood water April 1965	L. Kariba Mean comp. 1965
Ca	12.05	0.6–1.2	0.1–10	—	4.73	9.32
Mg	1.50	0.1–0.3	≥0.1	—	1.47	1.95
Na	3.46	0.3–5.2	≥0.4	—	1.70	3.52
K	1.18	0.2–0.3	≥0.03	—	T	0.97
CO$_3$	3.00	(0.4–1.2)	—	—	12.0	22.7
SO$_4$	—	1.3–2.9	2.0	—	(3.8)	(3.0)
Cl	T	0.2–(2.9)	0.5	—	(0.09)	(1.0)
NO$_3$–N	—	—	0.2	0.017	0.005	<0.025
PO$_4$–P	—	(≥0.001)	—	0.007	0.030	<0.020
Reference	Mitchell 1970	Gorham 1961 Table 5	Hutchinson 1957 Table 67	Caulton 1970	Coche 1968 Table 20	Table 71

Rain versus Zambezi water composition. Only floodwater is considered because of its relatively large volume and its direct relationship to the rainfall (Table 81). The contribution of the latter to the lake budget is relatively important (volumes not being considered) for calcium, sodium, potassium, chloride and nitrate-nitrogen. All these ions are not well represented in the Zambezi River. But the latter greatly contribute in bringing bicarbonate and phosphate into the lake.

Rain versus lake surface water composition (Table 81). The ionic contents are closely similar except for fixed carbonate and phosphate. From this comparison it also appears that these last two inorganic ions are mainly supplied by the river inflows.

13.6. The chemical budget of Lake Kariba

13.6.1. QUALITATIVE ASSESSMENT OF THE CHEMICAL BUDGET

Following GORHAM (1961) who recently reviewed the supply of ions to natural waters, the principal environmental factors which interact to determine the chemical water composition are: physiography, geology, climate, and biology. Each of these factors act through several components which may result in either positive (credit) or negative (debit) influence on the chemical budget. On the basis of the data presented previously, it is now intended to briefly discuss these possible influences in the L. Kariba ecosystem with reference to the tabulated synthesis (Table 82).

a. Physiography (Chapt. 1): the presence in the Upper Zambezi catchment of swampy areas greatly affects the chemical composition of the base flood and decreases its load of potential sediments. Local floods in the Zambian rivers occur under the form of reduced flash floods originating from relatively small catchments. On the Rhodesian side of the lake on the contrary, local floods are much more important, draining larger watersheds. Marginal runoff depends on the local topography of the immediate lake area, much more tormented on the northern side. The importance of the lake's bottom topography has been stressed by BEGG (1970) who has shown how it channels the penetration of affluent waters into the lake, reducing their possible influence to the immediate vicinity of the drowned river valleys.

b. Geology (Chapt. 2): the chemical composition of L. Kariba water reflects the mineral poverty and the relative insolubility of the rocks and soils present in the Upper Zambezi catchment. The chemical nature of the secondary rivers is also directly related to the geological character of their watershed, the southern rivers being much richer in minerals than those originating from the northern plateau. The majority of the lake sediments (S. M. MCLACHLAN, 1965) are most likely to yield low ion-exchange clays, only a high organic content (possibly in the littoral zone of the lake) being able to compensate for such deficiency. Marginal runoff and shore erosion also contribute relatively little to the lake's mineral content. Chemically rich rocks such as gneiss (potassium) and basalt (calcium, magnesium, iron) although present in the littoral areas are weathering very slowly. Most abundant and less resistant Karroo sandstones and shales are leached out and very poor in minerals. But phosphate is thought to be well replenished through marginal runoff (CAULTON, 1970). Mineral springs are well represented over the lake floor, particularly in the upper two basins

Table 82. Factors and main components determining the chemical budget in Lake Kariba ecosystem.

	Physiography	Geology	Climate	Biology
Credit	Bottom topography	Zambezi river Secondary rivers	Rain – direct import of minerals – marginal runoff – local floods (Sec. rivers) – base flood (Zambezi R.)	Aquatic $\begin{cases}\text{– bacteria}\\\text{– plants}\\\text{– animals}\end{cases}$ Submerged trees
		Lake sediments Marginal runoff Shore erosion	Thermal cycle: circulation	
	Base flood Local floods Marginal runoff Sedimentation	Mineral springs	Evaporation	Terrestrial $\begin{cases}\text{– plants}\\\text{– animals}\end{cases}$ Water level fluctuations
Debit			Thermal stratification Density currents	Human interference – outflow

(MAUFFE, 1933) and they might supply salts to the lake's economy in interesting quantities.

c. Climate (Chapt. 3): the rain falling on the lake's surface contributes major ions to the chemical budget in particular calcium, sodium, potassium, chloride, and nitrate-nitrogen, all minerals not present in great quantities in the Zambezi River (Sec. 13.5.3). In the immediate vicinity of the littoral zone, rain originates marginal runoff. Further away, it provides their physical characteristics to the local floods and to the Zambezi base flood. The latter in turn directly influences the thermal conditions in Basins I and II, indirectly facilitating vertical circulation. The thermal cycle in Basins III and IV depends upon climatic conditions which bring about total circulation and recycling of the nutrients into the photic zone. Evaporation, relatively important in Lake Kariba, has a concentration action on the chemical composition of its waters.

On the debit side, thermal stratification and density currents contribute to delay the recycling process of minerals brought into the lake with river floods. In many instances the river's entry occurs in the form of a drowned density current into the lacustrine hypolimnetic layer. This phenomenon combines with bottom topography to lock the nutrients in deep waters out of the trophogenic zone until the breakdown of thermal stratification initiates their vertical transport into the illuminated suface waters.

d. Biology: aquatic living organisms exert a profound influence upon the circulation of elements in the lake ecosystem through such activities as organic production, metabolitic processes, and decomposition (GORHAM, 1961). In Lake Kariba the floating *Salvinia auriculata* forming massive mats especially in the protected reaches of the river valleys, has several effects on the chemical budget: it traps nutrients delaying their recycling (MITCHELL, 1970); it constitutes the main source of hydrogen sulphide (BEGG, 1970). Submerged trees provide living communities with a tremendous surface area for their development. The interaction between the terrestrial and the aquatic environment particularly occurs along the lake shore on mud flats (S. M. MCLACHLAN, 1971). During the water drawdown game animals graze on growing plants. As the area becomes flooded again, drowned grass and dung release ions in relatively important quantities locally (in particular potassium and phosphate but also calcium, magnesium and nitrate). Water level fluctuations not only make it possible for the terrestrial production to contribute to the lake's chemical budget but they also enable the oxidation of organic matter in exposed sediments.

On the debit side of the budget, the major loss in minerals occurs in the lake's outflow. The chemical composition of the latter varies according to the depth level from which water is drawn (turbine or spillage water). This level in turn depends on the hydroelectrical management of the reservoir.

13.6.2. QUANTITATIVE ASSESSMENT OF THE CHEMICAL BUDGET

The quantitative assessment of the main components of Lake Kariba chemical budget as just defined is far from possible in the present state of knowledge. My purpose here is to present estimates for the most important mineral inflow (Zambezi River) and for the major mineral outflow (lake discharge). Unfortunately sufficient data are

Table 83. The annual import of major dissolved minerals by the Zambezi River.

Ion	Low water: Aug-Jan. tons	High water: Feb-Jul. tons	Annual import in minerals	
			tons. year^{-1}	kg. year^{-1}. km^{-2}
Ca	132 454	177 862	310 316	559
Mg	32 167	56 765	88 932	160
Na	75 688	64 333	140 021	252
K	6 623	30 274	36 897	66
HCO_3	334 919	454 116	789 035	1 422
SO_4	28 383	143 803	172 186	310
Cl	28 383	3 406	31 789	57
Total	638 617	930 559	1 569 176	2 827
SiO_2	94 610	378 430	473 040	852
Total general			2 042 216	3 679

not available to determine the total contributions of secondary rivers and marginal runoff. These estimates would have provided additional and most valuable information towards the quantitative assessment of the chemical budget. Tentatively, they have however been roughly estimated to obtain a general idea of their relative contribution.

a. Mineral imports by the Zambezi River. The computations of the quantity of mineral substances brought into L. Kariba by the Zambezi River each year is based on flow data (Sec. 5.1) and water chemical composition (Sec. 5.2).

In recent years the average discharge has shown a tendency to increase over the long-term average for 1925–1966 (ab. 1200 cum. sec^{-1}). The higher recent average for 1959 (Table 17) has therefore been preferred and a mean discharge of 1500 cum. sec^{-1} has been used in the computations.

Lower-water conditions normally prevail from August until January, the flood season extending from February to July. From the monthly data available (Table 18) the volume of water entering the lake during this flood season represents about 80 percent of the total annual contribution of the Zambezi River which amounts to 47304.10^6 cum. Therefore the latter's seasonal discharges have been estimated as follows: – Low-water season: 9461. 10^9 litres; – High-water season: 37843. 10^9 litres.

The chemical composition of the water according to the season is given in Table 13. The silicate content has been taken as averaging 10 mg. L^{-1} over the year. The size of the Zambezi River catchment above the lake (Northern Highlands, Central Plains and Gwaai R. essentially) has been taken as 555 000 sq. km to calculate its average contribution.

On these bases the quantities of the major dissolved mineral substances brought annually into L. Kariba by the Zambezi River have been estimated (Table 83).

Table 84. Mineral exports in 1965 from Lake Kariba into the Zambezi River in metric tons.

1965 Month	Outflow 10^9. L	Ca	Mg	Na	K	HCO_3
Jan.	1 264	11 123	3 084	5 031	1 871	31 853
Feb.	2 968	26 118	7 242	11 813	4 393	74 794
Mar.	1 390	13 900	3 336	5 560	1 807	33 360
Apr.	1 223	15 581	2 862	4 990	1 480	29 352
May	2 402	22 963	4 660	9 320	2 690	59 089
June	1 345	12 320	3 134	5 353	1 506	26 631
July	1 390	16 694	1 307	5 532	1 557	33 360
Aug.	1 516	19 723	713	5 882	1 561	41 842
Sep.	1 345	13 073	3 322	5 353	1 385	35 508
Oct.	1 390	13 511	3 433	5 532	1 432	35 028
Nov.	1 223	11 619	2 996	4 868	1 345	30 820
Dec.	8 723	78 507	21 284	34 718	10 468	219 820
Year	26 179	255 132	57 373	103 952	31 495	651 457

In average more than two million tons of dissolved inorganic matter are brought into the lake each year by the Zambezi River alone. Of this total import, bicarbonate contributes 39 percent, silicate 23 percent, and calcium 15 percent. All the other ions contribute less than ten percent each.

Although high water conditions represent as much as 80 percent of the total inflow, their import contribution amounts to 64 percent only. It is particularly important for sulphate. Low waters contribute most of the chloride.

Per square kilometre of catchment, the Zambezi River imports on average 3679 kg per year in similar ionic proportions as mentioned above. The total export estimated by SYMOENS (1968) for the Bangweulu-Luapula-Mweru catchment (4530 kg yr^{-1}km^{-2}) is closely similar. But great differences exist when comparing specifically the various ions. Exports from the catchment per unit of area are smaller in the Zambezi than in the Luapula catchment for chloride (five times), magnesium (twice) and potassium (twice). On the other hand the Zambezi River exports twice as much sulphate.

b. Mineral exports from Lake Kariba into the Zambezi River. On the basis of data on L. Kariba outflow (Table 16) and on its water chemical composition (Table 15 in COCHE, 1968), the monthly mineral exports downstream into the Zambezi River have been calculated for 1965 (Table 84). The average annual contents in sulphate (3 mg. L^{-1}), chloride (1 mg. L^{-1}), and silicate (8 mg. L^{-1}) have been estimated from the lake water composition.

For 1965 the lake exports in mineral substances can be approximated in metric tons as follows:

Calcium:	255 132	Bicarbonate:	651 457
Magnesium:	57 373	Sulphate:	78 537
Sodium:	103 952	Chloride:	26 179
Potassium:	31 495	Silicate:	209 432
Cations:	447 952	Anions:	965 605

Total export: 1 413 557 tons

To this total export close to 1.4 million tons, bicarbonate contribute 46 percent, calcium 18 percent, silicate about 15 percent, and sodium 7 percent. All other ions contribute less than six percent of the total export. Most of it took place during spilling periods in February and in particular December when the water level of the reservoir had to be lowered (flood season).

But during 1965 the water losses due to spillage were relatively small because of deficient rainfall in the Upper Zambezi catchment. This resulted in a total outflow (26.2 cu km) representing 60 percent only of the average discharge (Table 23). To take this into account the above data should be multiplied by a correction factor (1.66). It results that during a normal year L. Kariba exports at least 2.3 million tons of minerals downstream. For a catchment area of 663 820 sq km (Sec. 1.1), it represents an average export of 3534 kg. yr^{-1} km^{-2}. This specific export from the lake is very similar to the specific import of the Zambezi River into its upper end (Table 83).

c. Tentative chemical budget for L. Kariba. The total contribution to the chemical budget of the secondary rivers (incl. marginal runoff) can be roughly calculated. Their average annual volume of water has been estimated at 8 cu km (Table 23). By considering their mean chemical composition as similar to the mean composition of lake water (Table 71) their total contribution would most probably be underestimated. But such an approach provides a good basis for comparison. Thus the imports of the secondary rivers amount to about 419 000 tons of minerals per year, silicate included. For a total catchment area of 108 820 sq. km this represents an average of 3850 kg.yr^{-1} km^{-2}.

According to the water budget presented earlier (Table 23) the Zambezi River contributes on a long-term average 39 000 cu km per year. Its estimated imports of minerals calculated above (see a) must therefore be proportionally reduced by a factor of 0.83. Thus it amounts to about 1695 thousand tons or 3053 kg. yr^{-1} per sq. km of catchment.

Tentatively, the annual general chemical budget for Lake Kariba can be summarized as follows (in thousand tons):

IMPORTS	– Zambezi River:	1695
	– Secondary Rivers and local runoff:	419
	– Rain (Sec. 13.5.3):	42
		2156
EXPORTS	– Outflow	2346

There is an annual deficit balance of 190 thousand tons of minerals which might possibly be accounted for by raising the contribution of the secondary rivers and of the marginal runoff, probably underestimated here. At most this would balance the budget. But still it would not account for the minerals either invested into the increase of the lake populations (plants and animals) or contributing to the enrichment of the bottom sediments.

Therefore it appears on the basis of the above rough estimations that the average annual chemical budget of L. Kariba showed to be in deficit, more minerals being exported than imported. If this was true and persistent it could only lead to a progressive impoverishment of the lake's mineral content towards a lower level of mineraliation, more closely similar to the major inflow's level.

13.7. The chemical evolution since the closure of the dam

13.7.1. EVOLUTION OF THE TOTAL MINERAL CONTENT

The Kariba dam was definitely closed in December 1958. On that date age-old lotic conditions were abruptly interrupted above the artifical barrier, being replaced by a lentic environment. The hydrological station 6348B (Boom) was established by D. HARDING from the beginning of the formation of L. Kariba and it is on the basis of his records that the chemical evolution can best be followed (HARDING, 1962, 1964a, 1965, 1966). Further data were obtained later until 1969 by others from the same station for this purpose (COCHE, 1968; BEGG, 1970). Since 1970 average data obtained by BEGG (pers. comm.) for the Sanyati and Kariba Sub-Basins at two-months intervals have been used as estimates of the water quality encountered at Sta. 6348B. The results of the measurements made through the years (January-December) have been averaged to provide annual means (Table 85). Mean salinity in particular (either directly measured or estimated from k_{20} – Fig. 51) coupled to conductivity has been used as the evolution index (Fig. 55).

Right after the closure of the dam (1959) there was a major increase in dissolved salts. The relatively low riverine salinity nearly doubled within a few months reaching its overall maximum value (2.52 meq.L^{-1} or 63 mg. L^{-1}). All related characteristics (TS,k_{20} and TA) also attained record peak values during this first year of water filling. Concurrently the explosive developments of plankton, *Salvinia*, *Pistia* and fish populations started to be observed (HARDING, 1966).

During a second phase of evolution, between the first year of filling and its completion (1959-1963), the previous chemical richness decreased. At first the rate of decrease was fairly rapid to become slower and most regular (SL ab. 0.11 meq. L^{-1} year^{-1}) during the last three years of the filling period. By the end of this phase the mean salinity had dropped to two-thirds of the peak value (1.67 meq. L^{-1}) but it was still 30 percent higher than the original salinity of the river.

The third phase of the chemical evolution started as the average operating water level of the reservoir had been reached (1963). The content of dissolved matters again increased, very little at first but faster later most probably until about 1966.

A fourth phase then started. The rate of increase in salinity apparently slowed

Table 85. Chemical evolution at Sta. 6348B.

Year	Max. water level m. asl	Total Solids, mg. L^{-1} Range	Total Solids, mg. L^{-1} Mean	Mean Salinity meq.L^{-1}	Mean Salinity mg.L^{-1}	Mean k_{20} μmhos	Mean TA meq. L^{-1}	Reference
Zambezi R.	—	26–90	55	1.28	33.2	56	0.557	Table 14
1959	450	71–91	81	2.52	62.9	111	1.133	
1960	462	69–72	70	2.01	50.5	91	0.962	Harding
1961	472	(64–66)	(65)	1.90	47.3	85	0.890	1962
1962	480	60–69	66	1.78	44.6	80	0.855	1964a
1963	488	(60–66)	(63)	(1.67)	(41.7)	(75)	(0.819)	1965
								1966
1964	483	—	(60)	(1.69)	(42.3)	76	0.812	Coche 1968
1965	486	—	(62)	1.78	43	78	0.845	
1966	485	—	(65)	(1.84)	(47)	(82)	—	
1967	484	—	(68)	1.90	(49)	85	0.840	Begg 1970
1968	485	—	(67)	1.89	(48)	84	0.828	
1969	488	—	(64)	1.80	(45)	80	0.806	Pers. data
1970	484	—	(65)	1.84	(47)	82	0.79	According to
1971	484	—	(68)	1.91	(49)	85	0.80	G. Begg
1972	485	—	(68)	1.91	(49)	85	0.74	Pers. Comm.[1]
1973	482	—	(69)	1.95	(50)	87	0.79	

[1] Mean values of average data for the Sanyati and Kariba Sub-Basins at variable intervals throughout the year.

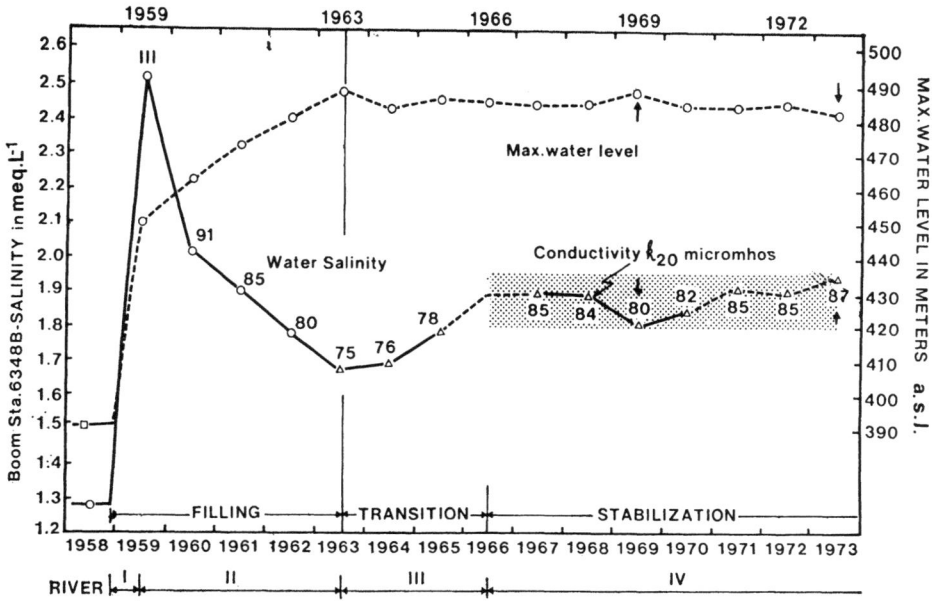

Fig. 55. The phases of chemical evolution at Station 6348B.

down from the time when the 1961 value had been approximately reached again. The latter (1.9 meq. L^{-1}) corresponds to 75 percent of the peak value and to about 150 percent of the original starting value. During this last phase of chemical evolution the mean salinity varied between 1.8 and 1.95 meq. L^{-1} (45-50 mg. L^{-1}), the mean conductivity ranging from 80 to 87 micromhos. The observed variations within these ranges appear to be inversely related to the water level (Fig. 55), values being minimum when the lake is extremely full (1969) and maximum when it is kept at low storage level (1973). A certain chemical equilibrium seems to have been reached since 1966–1967 in eastern Basin IV, eight to nine years after the closure of the dam or three to four years after the filling of the lake.

Since the closure of the Kariba dam four phases of chemical evolution might therefore be distinguished in the most lacustrine part of the reservoir on the basis of the pattern of variation of the water content in dissolved substances. Phase I coincides with the first year of rapid flooding of the valley soils (alluvia esp.) and vegetation, putting into solution vast quantities of organic and inorganic matters. During Phase II not only did the rate of flooding decrease but also the newly flooded soils were much less rich than those flooded earlier. Although still remaining above its original value the salinity dropped. Phase III started as the lake filling reached completion. Truly lacustrine conditions developed and salinity increased gradually after a particularly high level of water storage had been experienced. During Phase IV, the general chemical characteristics varied according to the storage level of the reservoir and within a relatively narrow range axed on a mean salinity about one-and-half times the original one.

221

Table 86. Chemical evolution in Lake Kariba surface waters between 1965 and 1968.

Basin	Mean total alkalinity meq. L^{-1}		Mean conductivity, k_{20} micromhos	
	1965	1968	1965	1968
I	0.584	0.524	59	60
II	0.630	0.636	62	68
III	0.817	0.824	79	86
IV	0.848	0.816	80	84
Lake	0.756	0.740	73	78
Reference	Table 72	acc. G. BEGG pers. comm.	Table 70	acc. G. BEGG pers. comm.

With most of the observations presented previously being based on data recorded during a transitional phase (1965), it is of interest to compare these results with some later ones. BEGG (pers. comm.) has kindly provided average values for conductivity and total alkalinity obtained during 1968 lake cruises. Mean annual values have been derived which may be compared with my 1965 similar data (Table 86).

The mean total alkalinity remains practically unchanged, the observed variations being too small to be considered significant. The greatest changes take place in Basin I (under the Zambezi R. influence) and in Basin IV (under the Sanyati R. and the hydroelectric management influences). All mean conductivities show a tendency to be slightly greater in 1968, in particular within Basin III, which confirms the chemical evolutionary pattern described earlier for the eastern part of the Basin IV.

13.7.2. EVOLUTION OF THE WATER CHEMICAL COMPOSITION

The evolution of the chemical composition of surface water sampled regularly at Sta. 6348B since the closure of the dam can be followed from HARDING's data (1962, 1964a, 1965, 1966) and later personal records. Annual mean compositions have been calculated to provide the basis for the estimate of the ionic proportions observed between 1959 and 1965 in the immediate vicinity of the dam (Table 87).

Results show that the cationic proportions (by equivalents) have changed when passing from purely riverine conditions to more or less pronounced lacustrine conditions: the average proportions of calcium and potassium have increased, the average proportions of magnesium and sodium decreasing. During the evolution relative proportions have not greatly varied from one year to another, except in 1962–1964 when calcium reached maximum predominance, either sodium or magnesium contents falling to a minimum. In 1965 the situation was 'normal' again. It can only be observed that these anomalies happened near the dam during the transitional period between Phases II and III of the chemical evolution, as the reservoir was reaching maximum water level in 1963.

Table 87. Evolution of the water chemical composition at Station 6348B (surface waters).

Year	Zambezi River		1959 7		1960 4		1961 10		1962 11		1964 10		1965 11	
No. samples														
Ion	meq.L^{-1}	%	meq.L^{-1}	%	meq.L^{-1}	%	meq.L^{-1}	%	meq.L^{-1}	%	meq.L^{-1}	%	meq.L^{-1}	%
Ca	0.327	52	0.720	57	0.599	59	0.569	59	0.571	64	0.527	62	0.495	55
Mg	0.155	25	0.266	21	0.185	18	0.189	20	0.179	20	0.137	16	0.192	22
Na	0.129	21	0.286	22	0.174	17	0.161	17	0.095	11	0.151	18	0.172	18
K	(0.020)	2	T	(0)	0.054	6	0.049	5	0.050	5	0.034	4	0.033	5
Cations	0.631	100	1.272	100	1.012	100	0.968	100	0.895	100	0.849	100	0.892	100
CO$_3$	0.557	85	1.133	91	0.962	96	0.890	96	0.855	97	0.812	97	0.845	—
Cl	0.019	12	0.056	5	0.038	4	0.037	4	0.028	3	(0.028)	3	?	—
SO$_4$	0.076	(4)	0.055	4	T	(0)	T	(0)	T	(0)	T	(0)	?	—
Salinity	1.28	—	2.52	—	2.01	—	1.90	—	1.78	—	1.69	—	1.78	—

Among the anions fixed carbonate gradually increased its predominance. Chloride did not vary a great deal. Sulphate was present only during the first year, disappearing later. Although it is believed that the failure to trace sulphate during all these years might be attributed to the analytical procedure involved, it remains a fact that very little sulphate has been present during the early life of the reservoir near the dam.

As far as nitrate-nitrogen and phosphate-phosphorus are concerned, a decrease in their concentrations has been reported by CAULTON (1970) for the period 1965–1970. By December 1969 they ranged from 4.5 to 13 μg. L^{-1} and from 1.6 to 4.3 μg. L^{-1} respectively, in the vicinity of the Kariba township.

The relative proportions of the main salines present in the surface waters in 1959–1962 (Sta. 6348B) are given in Table 77.

13.8. The chemical status of L. Kariba among African lakes and reservoirs

13.8.1. TOTAL MINERAL CONTENT OF SURFACE WATERS

On the basis of their total mineral content African lakes and reservoirs have been classified by TALLING & TALLING (1965) into three major classes. L. Kariba belongs to Class I grouping waters of low total ionic concentrations (k_{20} less than 600 micromhos). In fact, among East and Central African waters L. Kariba ranks very low within Class I, between L. Bangweulu and L. Victoria. It is found at a level of mineralization similar to L. Mweru which is also a typical river-lake (Table 88). The similarity with the latter has repeatedly been stressed earlier from other points of view.

In comparison with other African reservoirs the general level of minerals in L. Kariba very closely approximates L. Volta conditions. L. Kainji's level stands a little lower while L. Nasser's ionic content appears somewhat higher than the other three reservoirs considered, and closer to the average value for Africa.

13.8.2. CHEMICAL COMPOSITION OF SURFACE WATERS

The chemical composition of the surface waters in the above-mentioned lakes is compared in Table 89. For the natural lakes it is graphically represented according to KUFFERATH's (1951) method in Figure 52.[21]

In all the waters considered fixed carbonate (mostly as bicarbonate) is by far predominant among the anions, constituting at least 85 percent (by equivalents) of the total anionic content. Differences in composition are mostly to be found among the cations. L. Victoria and L. Bangweulu stand out as sodico-carbonate waters against L. Mweru and all four reservoirs where calcium is predominant (calcico-carbonate waters). This is well reflected in the cationic ratios (Table 89).

The silicate content is relatively high in all lakes considered as for example 2.8 to 10.5 mg. L^{-1} in L. Mweru and 12 to 16 mg. L^{-1} in L. Bangweulu (SYMOENS,

---

[21] The method is described in Sec. 13.3.1 (p. 192).

Table 88. Chemical composition of the surface water in some African lakes and reservoirs.

Lake	k_{20} μmhos	TS mg.L^{-1}	TIC meq.L^{-1}	TA meq.L^{-1}	Ca	Cations, mg. L^{-1} Mg	Na	K	mg. L^{-1} SO$_4$	Cl	References
Bangweulu	19– 35	39– 69[1]	0.25-0.34	0.21-0.47	1.1– 3.1	0.1– 2.5	2.5– 5.1	0.6-2.2	T-1.0	0.3– 1.0	Talling & Talling, 1965 Original data
Mweru	49–125	41– 68[1]	0.9 –1.1	0.5 –1.1	7.1– 8.2	2.1– 4.5	4.0– 5.8	0.8–2.2	1.5– 3.7	2.2– 5.0	De Kimpe, 1964 Original data
Victoria	91– 97	76–118	1.02–1.21	0.9 –1.1	5 – 7	2.3– 3.5	10 –13.5	3.7-4.2	1.8– 2.3	3.9	Talling & Talling, 1965
Kariba Ann. means	55– 81	60	0.66-0.91	0.55-0.86	7.8–10.6	1.4– 2.4	2.5– 4.1	0.4–1.3	(3.0)	(1.0)	Original data 1965
Kainji	40– 54	(35– 40)	0.6–0.8	0.5 –0.7	(4.0)	(2.4)	(0.23)	(1.6)	≤2.5	≤1.7	Henderson, 1973
Volta	65–180	60– 80	—	0.4 –1.1	3.4–10.2	2.4– 7.1	1.2– 6.8	1.5–6.0	T	1 – 4	Entz, 1969
Nasser	140–240	—	—	2.0 –3.0	23 –26	10 –11	8 –20	?	10 –25	5 –15	Entz & Ramsey, 1973
World avg.	—	—	2.35	1.71	29.86	5.11	8.28	3.13	17.77	8.16	Gorham, 1957
Africa, Avg.	182	—	ab. 1.5	—	12.5	3.8	11.0	—	13.5	12.1	Livingstone, 1963
E & W. Afr. Avg.	—	—	1.14	—	6.5	2.9	9.0	7.1	11.0	6.5	Visser in Imevbore, 1970

[1] Symoens, 1968.

Table 89. Average chemical composition of surface water in some African lakes and reservoirs (cf. Fig. 52).

Characteristic	L. Kariba		L. Volta		L. Bangweulu		L. Mweru		L. Victoria	
Nature	river–lake		river–lake		—		river–lake		closed basin	
Origin	man-made		man-made		shallow, tectonic		shallow, tectonic and swamps		shallow, tectonic	
k_{20} micromhos	73		(80)		30		80		95	
Ionic composition	meq.L^{-1}	percent	meq.L^{-1}	percent	meq.L^{-1}	percent	meq.L^{-1}	percent	meq.L^{-1}	percent
Ca	0.465	57	0.34	36	0.079	29	0.368	38	0.299	26
Mg	0.160	20	0.34	36	0.055	20	0.335	35	0.239	21
Na	0.153	18	0.19	20	0.108	40	0.211	22	0.500	44
K	0.042	5	0.08	8	0.028	10	0.048	5	0.100	9
TIC	0.820	100	0.95	100	0.270	99	0.962	100	1.138	100
CO_3	0.756	90	(0.80)	(88)	0.243	89	0.757	85	1.000	86
SO_4	0.062	7	(0.02)	(2)	0.011	4	0.040	4	0.048	4
Cl	0.028	3	0.09	(10)	0.020	7	0.096	11	0.110	10
Salinity, meq.L^{-1}	1.666		1.86		0.544		1.155		2.296	
Cationic ratios										
K:Na	0.27		0.42		0.26		0.23		0.20	
Ca:Mg	2.9		1.00		1.4		1.1		1.3	
(Ca + Mg):(Na + K)	3.2		2.52		1.0		2.7		0.9	
Reference	Pers. data 1965 Annual means		Bretschko pers. comm. weight. means		Symoens, 1968 Samfya, June 1960		De Kimpe, 1964 Means for Kilwa and Pweto stations in 1959		Talling & Talling, 1965 Mean of recent values	

1968), 13 to 27 mg. L^{-1} in L. Nasser (ENTZ & RAMSEY, 1973), and 12 to 27 mg. L^{-1} in L. Volta (G. BRETSCHKO, pers. comm.).

Phosphate-phosphorus is comparatively high in L. Nasser, varying between 0.02 and 2 mg. L^{-1}. In L. Volta its concentrations are very low (less than 0.02 mg. L^{-1}), the surface layers being very often completely depleted of this mineral even under unstratified conditions (G. BRETSCHKO, pers. comm.).

13.8.3. CONCLUSIONS

It may be concluded that, except for L. Victoria characterized by a closed basin, the water of the lakes considered as broadly similar to L. Kariba pertains to the zonal type (SYMOENS, 1968). Its quality is closely related to the climate, the soils, and the vegetation of the catchment. It is typically soft, little mineralized but rich in silica, and it mostly originates from humid tropical zones. Either calcium or sodium predominates (by equivalents) among the cations, bicarbonate and carbonate being largely dominant among the anions.

13.9. *General conclusions about the mineral content in Lake Kariba*

Lake Kariba has slightly alkaline surface waters. During thermal stratification conditions are slightly acid in the hypolimnion.

The total mineral content of pelagic waters is very small. In 1965 the lake annual average was for total solids 58 mg. L^{-1}, for salinity 42.4 mg. L^{-1}, and for conductivity (20° C) 73 micromhos. The general range of the latter is from 40 to 100 micromhos.

The average ionic composition of surface as well as deep waters is characteristic of calcico-carbonate waters. The order of dominance for the cations and for the anions is similar to the one most currently encountered in the freshwaters of the world. The spatial variation along the longitudinal axis of the lake is small. Sulphate and chloride are present in particularly small concentrations. Silica content is relatively high. The water is very soft, total alkalinity varying between 0.42 and 0.88 meq. L^{-1} fixed carbonate and averaging 0.756 meq. L^{-1}. Concentrations in nitrate-nitrogen and phosphate-phosphorus are very low and most variable rarely exceeding 20–25 μg. L^{-1}.

Seasonal variations are particularly great in riverine Basins I and II under the direct influence of the Zambezi River's discharge. In the lower lacustrine basins some relationship exists with the onset of the rainy season and vertical circulation when surface waters are enriched with minerals.

In deep water particularly the chemocycle is closely related to the thermal cycle. Generally the mineral content decreases with increased depth. But nitrate-nitrogen makes a notable exception, reaching relatively high concentrations in deep waters.

Mineral replenishment mainly takes place through river inflows. The Zambezi River which contributes in average 77 percent of the total water gains is particularly important for its imports of bicarbonate and phosphate. The importance of the secondary rivers is essentially local although their waters being more mineralized, their contribution is qualitatively important. The limited amount of annual rainfall nevertheless appears to efficiently contribute to the chemical budget in particular for calcium, sodium, potassium, chloride and nitrate, all ions not well represented in the Zambezi River.

The principal environmental factors which interact to determine the chemical composition of L. Kariba have been summarized under the headings of physiography, geology, climate, and biology. On the debit side it has been stressed how thermal stratification and density currents, combined with bottom topography, contribute to delay the recycling process of minerals brought into the lake's hypolimnion with river floods.

Basin individuality is well marked along the longitudinal axis of the lake. In particular the transitional zone between riverine and lacustrine regions roughly coincides with the limit defined between Basins II and III. More precisely it centers around the Masumo and Zambezi Rivers confluence, between stations 2015 and 2519. Great chemical changes occur in this area. The general trend is a progressive enrichment from the upper (inflow) towards the lower (dam) end of the ecosystem. It results from the progressive replacement of riverine by lacustrine conditions and from the increasing fertility of the river tributaries from the Zambezi River northeastward up to the Sibilobilo Narrows. Within Basin IV there are trenchant Sub-Basin differences from the chemical point of view, mainly due to affluent rivers such as the Lufua R., the Bumi R., and the Sanyati River.

A quantitative evaluation of the annual chemical budget has been roughly made on the basis of the average water budget and of the water chemical composition of its components. The Zambezi River imports on average from 3053 to 3679 kg. yr^{-1} km^{-2} of catchment which is equivalent to more than 2 million tons per year. Bicarbonate constitutes 39 percent, silicate 23 percent and calcium 15 percent. It has been estimated that the secondary rivers import at least 3850 kg. yr^{-1} km^{-2} of catchment or about one fifth of the total annual contribution of the Zambezi River. The rain falling directly on the lake imports at least 42 000 tons of minerals annually. Exports average 3534 kg. yr^{-1} km^{-2} of catchment above the Kariba dam, mainly constituted by 46 percent fixed carbonate, 18 percent calcium, 15 percent silicate and 7 percent sodium. The chemical budget showed a tendency in 1965 to be in deficit which, if true, would point out to a progressive future impoverishment of the lake towards a lower level of mineralization.

Since the closure of the dam the chemical evolution has been followed at a station close to the outflow. At least four phases have been distinguished. Phase I lasted the first year and corresponded to a rapid and important increase in the water salinity towards a peak value. Phase II, lasting from the second to the fifth year, was characterized by a decrease in the water salinity down to two thirds of the peak value but still 130 percent of the original river's salinity. Phase III (transition) started when the filling of the reservoir was completed and ended about three years later. A new increase in salinity was observed until about 75 percent of the peak value or 150 percent of the original river's salinity. This average level characterized Phase IV during which variations occurred above and below the mean value according to the storage level of the reservoir.

Among other African water bodies, the low mineral content classifies L. Kariba next to L. Mweru and L. Volta, slightly above L. Bangweulu and L. Kainji but below L. Victoria. Except for the latter, all these lakes have zonal waters: originating from humid tropical zones they are soft, very little mineralized, and rich in silica. Bicarbonate are predominant. On the contrary to most East and Central African lakes, the ionic composition of Lake Kariba water is not very different from most temperate lakes where calcium dominates among the cations.

On the basis of the classes of productivity proposed by NISBET & VERNEAUX (1970) for French freshwaters, Lake Kariba in 1965 exhibited chemical characteristics of low productivity potential as follows:

Class	Productivity index Range	Lake Kariba 1965 mean	Units
2	$10 \leqslant$ hardness < 20	11.3	$mg.L^{-1}$ (Ca + Mg)
2	$25 \leqslant TA < 50$	45.4	$mg.L^{-1}$ HCO_3
2	$10 \leqslant PO_4 < 50$	gen. < 25	$\mu g.L^{-1}$
1	$NO_3 < 1$	< 0.1	$mg.L^{-1}$

It is also customary to consider in freshwaters the total quantity of dissolved salts (estimated as total solids content, salinity, or conductivity) as a general indication of potential productivity. But in presence of relatively considerable quantities of carbonate and bicarbonate, the estimation of the true status of waters regarding their biologically valuable nutrients might profitably take these into account (ANON., 1955). To this effect the values for total alkalinity expressed as units of conductivity (k'_{20}) are substracted from the observed values for total conductivity (k_{20}). The former can be easily determined from the data provided by RUTTNER (1963, p. 59) and corrected for the chosen standard reference temperature, 20°C in the present case.

Results (Table 90) show how the ranking of some African lakes (for ex. L. Malawi and L. Sake) differs according to the type of index which is chosen, either total conductivity or the difference between k_{20} and k'_{20}. But Lake Kariba does not change its relative position and remains towards the bottom of the ranking list, between L. Bangweulu and L.Mweru.

The differences between the two conductivities also point out how small is the content of the electrolytes remaining after accounting for the alkaline fixed carbonate. The true trophic status of most African waters appears much lower than it might be thought on the basis of total conductivity alone. Thus in Lake Kariba water only 4 micromhos out of a total 73 micromhos represent the average quantity of non-carbonated salts available for biological production, which confirms its very low potential on chemical bases.

Table 90. Alkalinity-conductivity (k_{20}') compared with total conductivity (k_{20}) in some African lakes.

Lake	Reference	k_{20} μmhos	Total alkalinity meq.L^{-1}	Total alkalinity k_{20}' μmhos	$k_{20} - k_{20}'$
Lake Mweru Wantipa, Zambia	Pers. data	600	4.3	355	245
Lake Edward	TALLING & TALLING 1965	900	9.8	790	110
Lake Albert	TALLING & TALLING 1965	720	8.0	620	100
Lake Luhondo, Ruanda	TALLING & TALLING 1965	212	1.5	130	82
Lake Kyoga	ANON. 1955	320	2.8	241	79
Lake Tanganyika	TALLING & TALLING 1965	610	6.7	535	75
Lake Chisi, Zambia	Pers. data	136	1.0	92	44
Lake Sake, Ruanda	TALLING & TALLING 1965	232	2.4	208	24
Lake McIlwaine, Rhodesia	VAN DER LINGEN 1960	88	0.8	70	18
Lake George	TALLING & TALLING 1965	190	2.0	174	16
Lake Victoria	TALLING & TALLING 1965	95	0.9	80	15
Mazoe Dam, Rhodesia	VAN DER LINGEN 1960	240	2.8	230	10
Savory Dam, Rhodesia	VAN DER LINGEN 1960	270	3.2	260	10
Lake Malawi	TALLING & TALLING 1965	210	2.4	203	7
Lake Mweru	Pers. data	60	0.62	55	5
Lake Bangweulu	Pers. data	25	0.26	22	3
Lake Kariba	Pers. data 1965	73	0.76	69	4

SECTION IV

CONCLUSIONS

14. THE GENERAL TROPHIC STATUS OF LAKE KARIBA WITH PARTICULAR REFERENCE TO FISH PRODUCTION

14.1. Relationship of trophic status to environmental factors

On the bases of detailed studies made on Canadian lakes RAWSON (e.g. 1960) has shown how biological productions (harvest) of plankton, benthos, and fish depend on morphometric, edaphic, and climatic factors. In fact during the last twenty years the relationships existing in lacustrine environments between potential fish harvest and fish catch in particular and physico-chemical factors have received considerable interest from biologists. Among the most favoured indices of productivity were mean depth (e.g. RAWSON, 1952, 1955), area (e.g. ROUNSEFELL, 1946), and dissolved nutrients estimated either as total solids (e.g. RAWSON, 1951; NORTHCOTE & LARKIN, 1956) or as total alkalinity (e.g. MOYLE, 1956). The dependence on mean depth of fish harvest in African lakes has recently been pointed out by FRYER & ILES (1972).

Those factors have also been combined into productivity indices in an attempt to reach better correlation with the potential fish harvest. HAYES & ANTHONY (1964) defined such an index based on a multiple regression involving lake area, mean depth, and total alkalinity which accounts for 67 percent of the variability of the index. They estimated that 20 percent was due to the area function, 29 percent to mean depth, and 18 percent to total alkalinity. Taking another approach RYDER (1965) defined the morphoedaphic index (MEI = ratio of total solids to mean depth) as a means of estimating the potential fish harvest of north-temperate lakes. It provides an approximately linear fit for long-term average fish catch data from a set of relatively homogeneous lakes on a log-log plot. Later JENKINS (1967) found the morphoedaphic index as the most useful of the abiotic variables tested to relate to the total fish standing crop in a large set of typical north-American reservoirs. He obtained a good curvilinear fit against fish crop data. Recently attempts have been made to similarly apply the MEI concept to a series of limnologically similar lakes in East and Central Africa (REGIER et al., 1971). The original concept has sometimes been slightly changed, conductivity being used as an estimate of total solids content (HENDERSON & WELCOMME, 1974).

Therefore to help situate Lake Kariba trophic status from the fishery point of view in particular, RYDER's original index has been applied to some African lakes previously considered. In addition the concept of 'true chemical trophic status' (k-k') has been introduced into the morphoedaphic index in replacement of total solids content for comparison purposes (Table 91). From both conceptual points of view Lake Kariba is characterized by minimum productivity indices which points out to a very low potential fish harvest under tropical conditions.

Such conclusion based on morphometric and edaphic conditions is reinforced by the following limnological characteristics, previously discussed:

Table 91. Productivity indices for some African lakes.

Lake	Mean depth m	Morphoedaphic index		Trophic index	
		Tot. solids mg.L^{-1}	Index	$k_{20}-k_{20}'$[1] μmhos	Index
Mweru Wantipa	3[4]	429[4]	143.00	245	81.67
Kyoga	6[4]	258[2]	43.00	79	13.17
George	2[2]	264[2]	132.00	16	8.00
Albert	25[2]	565[2]	22.60	100	4.00
Edward	34[2]	521[2]	15.32	110	3.24
Bangweulu	4[4]	50	12.50	3	0.75
Mweru	7	76[5]	10.86	5	0.71
Victoria	47.4[2]	97[2]	2.05	15	0.32
Kariba	29.2	58	1.99	4	0.14

[1] See Table 90 for references.
[2] REGIER, et al. 1971.
[3] HUTCHINSON, 1957.
[4] WELCOME, 1972.
[5] TALLING & TALLING, 1965.

1. Water level fluctuations (Sec. 7.3.2): the spawning peaks of *Tilapia* and *Sarotherodon* ssp. (October – December) coincide with the falling water level (May-Dec.) which most probably has a profound influence on the breeding cycle of these fish. Specifically all *Tilapia* nurseries being mostly situated over very gently sloping beaches are adversely affected, the marginal aquatic vegetation is destroyed, and young fish are forced out (DONELLY, 1969). Water drawdown also eliminates an important part of the food supply (MCLACHLAN, 1970b). A drop in the water level from 488 m down to 482 m for example exposes as much as 630 sq. km of lake bottom (Table 28) or more than 10 percent of the total area underwater.
2. Depth of the trophogenic zone (Sec. 10.3.5): this zone where primary productivity takes place is relatively shallow. It averages only about 10 m in the upper two basins and about 16 m in Basins III and IV.
3. Primary productivity is probably low as attested by relatively high depths of visibility (Sec. 10.2) and by a small hypolimnetic areal deficit in dissolved oxygen (Sec. 12. 1.3).
4. Dissolved oxygen (Sec. 12.1.6): only in the two upper riverine basins is the trophogenic zone well oxygenated throughout the year. In the eastern part of the lake, the dissolved oxygen content limits the fish distribution within part of the illuminated zone for at least five months of the year.
5. Chemical quality (Sec. 13.9) of the pelagic lake water exhibits several characteristics known to be correlated with low biological productivity.

14.2. Past, actual, and future trophic status of Lake Kariba

Under temperate conditions limnological research has shown that in new reservoirs biological development passes through three stages during the initial years (ZHADIN & GERD, 1963). First, organic matter accumulates on the bottom. In a second stage it is mineralized which results in a great development of biological populations. Finally the nutrient content in the water increases and algae begin to form in great quantities in the plankton. Most probably these three initial stages in reservoir development happened in Lake Kariba during the filling phase as attested by several records (BALINSKY & JAMES, 1960, BOUGHEY, 1963, HARDING, 1962 & 1964a).

On a longer-term basis, the general biological development of Lake Kariba also seems to have followed a sequence similar to what has been experienced in temperate climates according to W. C. BECKMAN citing I. I. LAPITSKY (p. 154 in RZOSKA, 1966). During the first phase of the reservoir's maturation, there is a sudden increase in standing crops. This is followed by a depressed productivity which lasts in Russia from 6 to 10 years in latitudes below 50°. The next phase is characterized by a gradual increase in biological productivity towards a stabilized equilibrium.

In the tropics it has been observed that exceptional fish harvests in newly formed reservoirs may be expected within one to two years after the operating water level has been reached (F. HENDERSON, pers. comm.). This would correspond in L. Kariba to the first phase of reservoir maturation and MITCHELL (1970) has defined its status as mesotrophic for 1962-64.

Therefore the second biological phase most probably started around 1964. CAULTON (1970) has reported a decrease in phosphate and nitrate contents between 1964 and 1969. MITCHELL (cited in VAN DER LINGEN, 1973) from his own studies in relation with the autecology of *Salvinia* concluded in 1971 that the trend to oligotrophy was clear. The same conclusion has been reached in the present study from different points of view (Sec. 14.1). If reservoir maturation evolves as observed in southern Russia, this second phase should end around 1974 at the latest when a gradual increase in productivity might possibly take place. A more optimistic view that the third phase of maturation had been already, or nearly, reached was expressed by CAULTON (1970) following his studies in early 1970.

What will be the trophic status of Lake Kariba during the future third biological phase is difficult to predict. Ultimate productivity as pointed out by MITCHELL (cited in VAN DER LINGEN, 1973) will depend on the nutrient balance and on the degree of chemical recycling. It will be a function not only of the minerals entering and leaving the Kariba ecosystem but also of the effectiveness with which they will be used by the biological communities. Unfortunately general limnological conditions, such as Zambezi water inflow, lake morphometry and thermal cycle, will remain mostly unchanged and will probably keep productivity at a low level. Except for some particular areas (Sengwa, Sibilobilo, Sanyati for example) and part of the littoral zone (marginal runoff, organic matter mineralization of bottom muds, water mixing), most of Lake Kariba risks to retain its general oligotrophic status in the future, relative to other tropical water bodies.

LITERATURE CITED

ALLISON, G. F. 1965. Functions of the Central African Power Corporation relating to Lake Kariba, pp. 21–23 In: LKFRI, Research Symposium (1965), Kariba (cyclostyled).
ALLISON, G. F. 1969. The hydrological management of Lake Kariba, pp. 1–8. In: Some problems of aquatic ecosystems in Southern Africa. *Limnol. Soc. Sthrn. Africa Suppl. Newslett.* 13.
ANONYMOUS, 1955. East African Fisheries Research Organization. Annual Report 1954–1955. EAFFRO, Jinja, Uganda. 40 p.
ANONYMOUS, 1962. Programme and guide to Livingstone-Kariba excursion. Meet. Ass. Afr. Geol. Surveys (1962). North Rhod. Geol. Surv., Lusaka (cyclostyled).
ANONYMOUS, 1969. Hydrological year book 1963–64. Rhodesia, Hydrol. Branch, Minister of Irrigation and Lands, Govt. Printer, Salisbury, Rhodesia. 299 p.
ATTWELL, R. I. G. 1970. Some effects of Lake Kariba on the ecology of a flood plain of the mid-Zambezi Valley of Rhodesia. *Biol. Conserv.* 2: 189–196.
BALINSKY, B. I. & G. V. JAMES. 1960. Explosive reproduction of organisms in the Kariba Lake. *S. Afr. J. Sci.* 56: 101–104.
BALON, E. K. 1971a. Age and growth of *Hydrocynus vittatus* Castelnau, 1861, in Lake Kariba, Sinazongwe area. pp. 89–118. In: A. G. COCHE (Ed) *Fish. Res. Bull.*, Zambia 5. Centr. Fish. Res. Inst., Chilanga, Zambia.
BALON, E. K.1971b. Replacement of *Alestes imberi* Peters, 1852 by *A. lateralis* Blgr. 1900 in Lake Kariba, with ecological notes. pp. 119–162. In: A. G. COCHE (Ed). *Fish Res. Bull.*, Zambia 5. CFRI, Chilanga, Zambia.
BEETON, A. M. 1958. Relationship between Secchi disc readings and light penetration in L. Huron. *Trans. Amer. Fish Soc.* 87: 73–79.
BEETON, A. M. 1965. Eutrophication of the St. Lawrence Great Lakes. *Limnol. Ocean.* 10: 240–254.
BEGG, G. W. 1969. Observations on the water quality and nature of the affluent rivers of Lake Kariba, with reference to their biological significance. *Limnol. Soc. Sthrn. Africa, Suppl. Newslett.* 13: 26–33.
BEGG, G. W. 1970. Limnological observations on L. Kariba during 1967 with emphasis on some special features. *Limnol. Ocean.* 15: 776–788.
BEGG, G. W. 1972. The hydrology at the mouth of the Sanyati Gorge, Lake Kariba. With particular emphasis on the influence of thermal stratification on the vertical distribution of zooplankton. Lake Kariba Fish. Res. Inst., Proj. Rep. 15, Kariba, Rhodesia. 54 p. (mimeo).
BIRGE, E. A. 1915. The heat budgets of American and European lakes. *Wisc. Acad. Sci. Arts Lett.* 18: 166–213.
BIRGE, E. A. 1916. The work of the wind in warming a lake. *Wisconsin Acad. Sci. Arts Lett.* 18: 341–391.
BISWAS, S. 1968. Hydrobiology of the Volta River and some of its tributaries before the formation of the Volta Lake. *J. Sci. Ghana* 8: 152–166.
BLANC, H. & J. DAGET, 1957. Les eaux et les poissons de Haute-Volta. *Mém. Inst. fr. Afr. Noire* 50: 101–168.
BLAXTER, J. H. S. 1965. Effect of change of light intensity on fish. pp. 647–661 In: ICNAF Spec. Publ. 6
BOND, G. 1953. The origin of thermal and mineral waters in the Middle Zambezi Valley and adjoining territory. *Geol. Soc. S. Afr., Trans.* 56: 131–148.
BOND, G. 1965. Preliminary observations on the development of shore line features on Lake Kariba. pp. 33–42. In: LKFRI, Research Symposium (1965). Kariba (cyclostyled).
BOND, G. & J. D. CLARK, 1954. The Quaternary sequence in the Middle Zambezi Valley. *S. Afr. Archaeol. Bull.* 9 (6): 115–130.

BOUGHEY, A. S. 1963. The explosive development of a floating weed vegetation on Lake Kariba. *Andansonia* 3: 49–61.
BOUGHEY, A. S. 1965. Early shore line vegetation on Lake Kariba. pp. 47–50. In: LKFRI, Research Symposium (1965). Kariba (cyclostyled).
BOWMAKER, A. P. 1960. A report on the Kariba Lake area and Zambezi River prior to inundation and the initial effects of inundation with particular reference to the fisheries. p. 100–127. In: FAO/ETAP Report 1299 (2). FAO, Rome.
BOWMAKER, A. P. 1968. Preliminary observations on some aspects of the biology of the Sinamwenda estuary, Lake Kariba. *Rhod. Sci. Ass., Proc. Trans.* 53: 3–8.
BOWMAKER, A. P. 1970. A prospect of Lake Kariba. *Optima* 20: 68–74.
CAULTON, M. S. 1970. A quantitative analysis of the mode of replacement of inorganic phosphates and nitrates in the waters of Lake Kariba during the period December 1969–January 1970. *Limnol. Soc. Sthn. Afr. Newslett.* 15: 52–61.
CENTRE TECHNIQUE FORESTIER TROPICAL. 1970. Etude de l'aménagement piscicole du Lac de Kossou. Défrichements dans l'assiette du lac pour l'exercice de la pêche. Rapp. Gouvern. Côte d'Ivoire. 37 p. (cyclostyled).
CENTRE TECHNIQUE FORESTIER TROPICAL. 1972. Etude de l'aménagement piscicole du Lac de Kossou. Le peuplement de poissons du Bandama Blanc en pays Baoulé. 126 p. (cyclostyled).
CHILD, G. 1968. Behaviour of large mammals during the formation of Lake Kariba. Kariba Studies, Trustees Nat. Mus. Rhodesia. 123 p.
CLEMENTS, F. 1959. Kariba. The struggle with the River God. Methuen, London. 223 p.
COCHE, A. G. 1965. Limnological research programme for Lake Kariba. pp. 63–65. In: Lake Kariba Fish. Res. Inst., Research Symposium (June 1965), Kariba (cyclostyled).
COCHE, A. G. 1968. Description of physico-chemical aspects of Lake Kariba, an impoundment, in Zambia-Rhodesia. 68 p. Central Fish. Res. Inst., Chilanga, Zambia.
COCHE, A. G. 1969. Aspects of physical and chemical limnology of Lake Kariba, Africa. A general outline. pp. 166–122. In: L. E. OBENG (Ed.). Man-made lakes. The Accra Symposium. Ghana Univ. Press, Accra. 398 p.
COCHE, A. G. 1971. Lake Kariba Basin: a multi-disciplinary bibliography, annotated and indexed. 1954–1968. pp. 11–87. In: A. G. COCHE (Ed). *Fish. Res. Bull. Zambia* 5. Centr. Fish. Res. Inst., Chilanga, Zambia.
COLSON, E. 1960. The social organisation of the Gwembe Tonga. Kariba Studies I. Manchester Univ. Press, Manchester. 234 p.
DE KIMPE, P. 1964. Contribution à l'étude hydrobiologique du Luapula-Moero. *Annales, Sci. Zool., Musée Roy. Afr. Centr., Tervuren, Belgium.* 128 p.
DE SWARDT A. M. J. & A. R. DRYSDALL, 1964. Precambrian geology and structure in central Northern Rhodesia. *Mem. Geol. Surv. N. Rhod.* 2. 82 p. Govt. Printer, Lusaka.
DONELLY, B. G. 1969. A preliminary survey of *Tilapia* nurseries on Lake Kariba during 1967/68. *Hydrobiologia* 34: 195–206.
DOUDOROFF, P. & D. L. SHUMWAY, 1970. Dissolved oxygen requirements of freshwater fishes. FAO Fish. Techn. Pap. 86. 291 p. FAO, Rome.
DUNCAN, C. P. 1964. Seasonal occurrence of thermocline off the South-West Cape, 1955–61. Div. Sea Fish., Invest. Rep. 50. 15 p. Cape Town, Rep. Sth. Afr.
DUSSART, B. 1966. Limnologie. L'étude des eaux continentales. 677 p. Gauthier-Villars, Paris.
ECKEL, O. 1950. Über die numerische und graphische Ermittlung der Stabilität von Gewässern nach W. Schmidt. *Schweiz. Z. Hydrol.* 12: 38–46.
EIFAC, 1973. Water quality criteria for European freshwater fish. Report on dissolved oxygen and inland fisheries. Techn. Pap. 19. 10p. FAO, Rome.
ENTZ, B. 1969a. Caractéristiques limnologiques du lac de Volta le plus grand lac artificiel d'Afrique. *Newslett.* 5 (4): 10–17. Intern. Hydrol. Decade, UNESCO, Paris.
ENTZ, B. 1969b. Observations on the limnochemical conditions of the Volta Lake. pp. 110–115 In: OBENG L. E. (Ed.). Man-made lakes The Accra Symposium. Ghana Univ. Press, Accra. 398 p.
ENTZ, B.1973. Morphometry of Lake Nasser. Lake Nasser Development Centre, Working Paper 2. Aswan. 81 p. (mimeo).

ENTZ, B. & B. RAMSEY, 1973. Some physical, chemical and general limnological characteristics of Lake Nasser reservoir (UAR). In ACKERMANN, W. C., G. F. WHITE & E. B. WORTHINGTON (Eds.). Man-made lakes: their problems and environmental effects. Geophys. Monogr. Ser. 17. Amer. Geophys. Union, Washington.

ELEY, R. L., N. E. CARTER & T. C. DORRIS, 1967. Physicochemical limnology and related fish distribution of Keystone reservoir. pp. 333–357. In: Reservoir Fishery Resources Symposium, Reserv. Comm. Sthn. Div. Amer. Fish. Soc., Georgia, USA.

FAO/UNDP, 1971. Volta Lake Research, Ghana. Physico-chemical conditions of Lake Volta, Ghana. FI: SF/GHA 10. Techn. Rep. 1. FAO, Rome. 77 p.

FRYER, G. & T. D. ILES, 1972. The cichlid fishes of the great lakes of Africa. Their biology and evolution. Oliver and Boyd, Edinburgh. 641 p.

GAIR, H. S. 1959. The Karroo System and resources of the Gwembe District, north-west section. *Zambia, Geol. Surv. Dept., Bull.* I. Govt. Printer, Lusaka. 88 p., 2 maps.

GOLTERMAN, H. L. (Ed.) 1969. Methods for chemical analysis of fresh waters, IBP Handbook 8. Blackwell Sci. Publ., Oxford. 166 p.

GORHAM, E. 1957. The ionic composition of some lowland lake waters from Cheshire, England. *Limnol. Ocean.* 2: 22–27.

GORHAM, E. 1961. Factors influencing supply of major ions to inland waters, with special reference to the atmosphere. *Geol. Soc. Amer. Bull.* 72: 795–840.

HAMMERTON, D. 1972. The Nile River. A case history. pp. 171–214: In: R. T. OGLESBY et al. (Ed). River ecology and man. Academic Press, London.

HARDING, D. 1962. Research on Kariba. pp. 32–40. In: Joint Fish. Res. Organ. Ann. Rep. 10 (1960). Govt. Printer, Lusaka.

HARDING, D. 1964a. Research on Lake Kariba. pp.25–50. In: Joint Fish. Res. Organ., Ann. Rep. 11 (1961). Govt. Printer, Lusaka.

HARDING, D. 1964b. Hydrology and fisheries in Lake Kariba. *Verh. Intern. Verein. Limnol.* 15: 139–149.

HARDING, D. 1965. Research on Lake Kariba, 1962–1963. pp. 38–54 In: Fish Res. Bull. Zambia. Vol. 1 – Dept. Fish., Res. Div., Chilanga, Zambia (cyclostyled).

HARDING, D. 1966. Lake Kariba. The hydrology and development of fisheries. pp. 7–20. In: R. H. LOWE – McConnell (Ed.). Man-made lakes. Symp. Inst. Biol. 15. Acad. Press, London.

HAYES, F. R. & E. H. ANTHONY, 1964. Productive capacity of North American lakes as related to the quantity and the trophic level of fish, the lakes dimensions, and the water chemistry. *Trans. Amer. Fish. Soc.*, 93: 53–57.

HENDERSON, F. 1973. A limnological description of Kainji Lake 1969–1971. Nigeria, Kainji Lake Project, FI: DP/NIR 66/524/1 – FAO, Rome. 47 p.

HENDERSON, F. & R. L. WELCOME, 1974. The relationship of yield to morpho-edaphinic index and number of fishermen in African fisheries. CIFA Occas. Pap. 1. 19 p. FAO, Rome.

HEM, J. D. 1959. Study and interpretation of the chemical characteristics of natural waters. Geol. Surv. Water – Supply, Paper 1473. US Gov. Printing Off., Washington. 269 p.

HITCHON, B. 1958. The geology of the Kariba area. Zambia, Geol. Surv. Dept. Rep. 3. Govt. Printer, Lusaka. 41 p., 4 maps.

HOWE, G. M. 1953. Climates of the Rhodesias and Nyasaland according to the Thorthwaite classification. *Geogr. Rev.* 43: 525–539.

HUTCHINSON, G. E. 1957. A treatise on limnology. Vol. I – Geography, physics and chemistry. John Wiley, New York. 1029 p.

IMEVBORE, A. M. A. 1970. The chemistry of the River Niger in the Kainji Reservoir area. *Arch. f. Hydrobiol.* 67: 412–431.

JACKSON, P. B. N. 1960a. Ecological effects of flooding by the Kariba Dam upon Middle Zambezi fishes. pp. 277–284. In: Proc. First Fed. Sci. Congress. Salisbury, Rhodesia.

JACKSON, P. B. N. 1960b. Hydrobiological research at Kariba. *New Scientist* 7 (117): 877–880.

JACKSON, P. B. N. 1961. Ichthyology. The fish of the Middle Zambezi. Kariba Studies. Manchester Univ. Press, Manchester. 36 p.

JARMAN, P. J. 1965. Wildlife ecology. pp. 60–62. In: LKFRI, Research Symposium (1965). Kariba (cyclostyled).
JENKINS, R. M. 1967. The influence of some environmental factors on standing crop and harvest of fishes in U.S. reservoirs. pp. 298–321 In: Reservoir Fishery Resources Symposium, Sthn. Div. Amer. Fish. Soc. University of Georgia, Athens. 569 p.
JUBB, R. A. 1961. An illustrated guide to the freshwater fishes of the Zambezi River, Lake Kariba, Pungwe, Sabi, Lundi and Limpopo Rivers. S. Manning, Bulawayo. 171 p.
JUBB, R. A. 1967. Freshwater fishes of southern Africa. A. A. Balkema, Cape Town. 248 p.
KÖPPEN, W. 1931. Grundriss der Klimakunde. 2nd Ed., Walter de Gruyter Co., Berlin.
KÖPPEN, W. 1936. Das geographische System der Klimate. In: Handbuch der Klimatologie, Bd. 1, Teil C., Berlin.
KUENEN, P. H. 1955. Realms of water. Some aspects of its cycle in nature. Cleaver – Hume Press Ltd., London. 327 p.
KUFFERATH, J. 1951. Représentation graphique et classification chimique rationnelle en types des eaux naturelles. *Bull. Inst. Roy. Sci. Nat. Belg.* 27. Nos. 43, 44 and 45. Brussels
LAGUS, C. 1959. Operation Noah. W. Kimber, London. 176 p.
LAW, A. B. 1965. A report on research into winds and waves on Lake Kariba. pp. 11–14. In: LKFRI, Research Symposium (1965). Kariba (cyclostyled).
LEVRING, T. & G. R. FISH, 1956. The penetration of light in some tropical East African waters. *Oikos* 7: 98–109.
LIVINGSTONE, D. A. 1963. Chemical composition of rivers and lakes. In: FLEICHER, M. (Ed). Data of geochemistry. 6th Ed. U.S. Geol. Surv. Prof. Paper 440-G. 64 p.
LONEY, P. E. 1966. Preliminary report on the geology of the Kariba District of Rhodesia. pp. 14–15. In: Univ. Leeds, Res. Inst. Afr. Geol., 10th Ann. Rep. (1964–65).
MACRAE, F. B. 1938. Some notes on part of the Gwembe Valley in Northern Rhodesia. *Geogr. J.* 446–449.
MATHESON, G. 1969. The pre-Katangan Basement geology of the Masuku-Kabanga area, Southern Province, Zambia. Ph.D. thesis, Univ. Leeds, U.K. 123 p.
MAUFFE, H. B. 1933. A preliminary report on the mineral springs of SouthernR hodesia. Rhod. Geol. Surv. Bull. 23. Salisbury, Rhodesia. 78 p., 1 map.
MCLACHLAN, A. J. 1969. The effects of aquatic macrophytes on the variety and abundance of benthic fauna in a newly created lake in the tropics (Lake Kariba). *Arch. f. Hydrobiol.* 66: 212–216.
MCLACHLAN, A. J. 1970a. Submerged trees as a substrate for benthic fauna in the recently created Lake Kariba (Central Africa). *J. Appl. Ecology* 7: 253–266.
MCLACHLAN, A. J. 1970b. Some effects of annual fluctuations in water level on the larval chironomid communities of Lake Kariba. *J. Anim. Ecol.* 39: 79–90.
MCLACHLAN, S. M. 1965. Physical and chemical characteristics of bottom deposits. pp. 43–46. In: LKFRI, Research Symposium (1965). Kariba (cyclostyled).
MCLACHLAN, S. M. 1971. The rate of nutrient release from grass and dung following inmersion in lake water. *Hydrobiologia* 37: 521–530.
MITCHELL, D. S. 1969. The ecology of vascular hydrophytes on Lake Kariba. *Hydrobiologia* 34: 448–464.
MITCHELL, D. S. 1970. Autecological studies of *Salvinia auriculata* Aubl. Ph.D. Thesis, Univ. London. 669 p.
MONTGOMERY, H. A. C., N. S. THOM & A. COCKBURN, 1964. Determination of dissolved oxygen by the Winkler method and the solubility of oxygen in pure water and sea water. *J. Appl. Chem.* 14: 280–296.
MORTIMER, C. H. 1969. Physical factors with bearing on eutrophication in lakes in general and in large lakes in particular. pp. 340–368 In: Eutrophication: causes, consequences, correctives. Nat. Acad. Sci., Wash., D.C.
MOYLE, J. B. 1956. Relationships between the chemistry of Minnesota surface waters and wildlife management. *J. Wildl. Mgt.* 20: 303–320.
MUNCASTER, P. A. 1965. Hydrological research activities on Lake Kariba. pp. 15–20. In: LKFRI, Research Symposium (1965), Kariba (cyclostyled).

NEUMANN, J. 1959. Maximum depth and average depth of lakes. *J. Fish. Res. Bd. Canada* 16: 923–927.
NEWTON, A. R. 1963. The geology of the country between Choma and Gwembe. Zambia Geol. Surv., Report 8. Govt. Printer, Lusaka. 33 p. 3 maps.
NISBET, M. & J. VERNEAUX, 1970. Composantes chimiques des eaux courantes. Discussion et proposition de classes en tant que bases d'interprétation des analyses chimiques. *Ann. Limnol.* 6 (2): 161–190.
NORTHCOTE, T. G. & P. A. LARKIN, 1956. Indices of productivity in British Columbia lakes *J. Fish. Res. Bd. Canada* 13: 515–540.
ORTH, R. 1939. Zur Kenntnis des Lichtklimas der Tropen und Subtropen sowie des tropischen Urwaldes. *Gerlands Beitr. Z. Geophysik* 55: 52–102.
RAWSON, D. S. 1950. The physical limnology of Great Slave Lake. *J. Fish Res. Bd. Canada.* 8: 1–66.
RAWSON, D. S. 1951. The total mineral content of lake waters. *Ecology* 32: 669–672.
RAWSON, D. S. 1952. Mean depth and the fish production of large lakes. *Ecology* 33: 513–521.
RAWSON, D. S. 1955. Morphometry as a dominant factor in the productivity of large lakes. *Verh. Intern. Verein. Limnol.* 12: 164–175.
RAWSON, D. S. 1960. A limnological comparison of twelve large lakes in northern Saskatchewan. *Limnol. Ocean.* 5: 195–211.
REEVE, W. H. 1963. The geology and mineral resources of Northern Rhodesia. North. Rhod. Geol. Surv., Bull. 3. 213 p. + maps. Govt. Printer, Lusaka.
REGIER, H. A., A. J. CORDONE & R. A. RYDER, 1971. Total fish landings from fresh waters as a function of limnological variables, with special reference to lakes of East-Central Africa. Fish Stock assessm. Afr. inl. waters, Wkg. Pap. 3. FAO, Rome. 13 p.
REID, G. K. 1961. Ecology of inland waters and estuaries. Reinhold Pub. Corp., New York. 375 p.
REYNOLDS, B. 1968. The material culture of the peoples of the Gwemble Valley. Kariba Studies III. Manchester Univ. Press, Manchester. 262 p.
RILEY, G. A. 1941. Plankton studies. IV. Georges Bank. *Bull. Bingham Oceanogr. Coll.* 7 (4): 1–73.
ROBERTS, R. H., D. C. MULLINS & E. A. BARNETT, 1960. Report on the islands of Lake Kariba for the Government of Northern and Southern Rhodesia. Kariba Lake Coord. Comm., Kariba. 132 p. (cyclostyled).
RODHE, W. 1949. The ionic composition of lake waters. *Verh. Intern. Verein. Limnol.* 10: 377–386.
ROUNSEFELL, G. A. 1946. Fish production in lakes as a guide for estimating production in proposed reservoirs. *Copeia* 1946: 29–40.
RUTTNER, F. 1963. Fundamentals of limnology. 3rd Ed., Univer. Toronto Press, Toronto. 295 p. Transl. D. G. FREY and F. E. J. FRY.
RYDER, R. A. 1965. A method for estimating the potential fish production of North-temperate lakes. *Trans. Amer. Fish. Soc.*, 94: 214–218.
RZOSKA, J. 1966. The biology of reservoirs in the USSR. pp. 149–157. In: R. H. LOWE–MCCONNELL (Ed). Man-made lakes. Symposia of the Institute of Biology No. 15. Acad. Press, London. 218 p.
SAUBERER, F. 1962. Empfehlungen für die Durchführung von Strahlungsmessungen an und in Gewässern. Mitt. Intern. Verein. Limnol. 11. 77 p. E. Schweizerbart'sche, Stuttgart.
SCUDDER, T. 1962. The ecology of the Gwembe Tonga. Kariba Studies II. Manchester Univ. Press, Manchester. 274 p.
SCUDDER, T. 1968. Social anthropology, man-made lakes and population relocation in Africa. *Anthropological Quart.* 41: 168–175.
SCUDDER, T. 1970. Ecology and development: the Kariba Lake Basin. In: M. T. FARVAR & J. MILTON (Eds.). The careless technology: ecology and international development. Natural Hist. Press.
SHAND, I. H. R. 1960. The potential of the Zambezi River. pp. 76–80. In: Proc. First Fed. Sci. Congress (May 1960). Salisbury, Rhodesia.

SMITH, S. H. 1962. Temperature correction in conductivity measurements. *Limnol. Ocean* 7 (3).
SMITH, S. R. 1968. Outline Programme for hydro-electric development in West Africa to 1980. pp. 158–188. In: NEVILLE R. & W. M. WARREN (Ed). Dams in Africa. An interdisciplinary study of man-made lakes in Africa. Frank Cass & Co. Ltd., London.
STRICKLAND, J. D. H. 1958. Solar radiation penetrating the ocean; a review of requirements data and methods of measurement, with particular reference to photosynthetic productivity. *J. Fish. Res. Bd. Canada* 15: 453–493.
SWEERS, H. E. 1968. Two methods of describing the 'average' vertical temperature distribution of a lake. *J. Fish. Res. Bd. Canada* 25: 1911–1922.
SYMOENS, J. J. 1968. La minéralisation des eaux naturelles. Explor. Hydrob. Bass. Lac Bangweolo Luapula, Vol. 2, Fasc. 1. 199 p. Cercle Hydrobiol. Brux. Brussels.
TAIT, C. C. 1967. Kafue River and flood plain. Hydrological data. *Fish. Res. Bull. Zambia* 3: 26–28. Govt. Printer, Lusaka, Zambia.
TALLING, J. F. 1957. The longitudinal succession of water characteristics in the White Nile. *Hydrobiologia* 11: 73–89.
TALLING, J. F. 1965. The photosynthetic activity of phytoplankton in East African lakes. *Int. Revue ges. Hydrobiol.* 50: 1–32.
TALLING, J. F. 1966. The annual cycle of stratification and phytoplankton growth in Lake Victoria (East Africa). *Int. Revue ges. Hydrobiol.* 51: 545–621.
TALLING, J. F. 1969. The incidence of vertical mixing, and some biological and chemical consequences, in tropical African lakes. *Verh. Intern. Verein. Limnol.* 17: 998–1012.
TALLING, J. F. & I. B. TALLING, 1965. The chemical composition of African lake waters. *Int. Rev. Ges. Hydrobiol.* 50: 421–463.
TAVENER-SMITH, R. 1960. The Karroo System and coal resources of the Gwembe District, south-west section. Zambia, Geol. Surv. Dept., Bull. 4. Govt. Printer, Lusaka. 84 p., 2 maps.
THOMASSON, K. 1965. Notes on algal vegetation of Lake Kariba. Nov. Acta. Reg. Soc. Scient. Upsaliensis, Ser.4, 19: 1–34.
TONOLLI, V. 1969. Introduzione allo studio della limnologia. Instituto Italiano Idrobiologia, Verbania, Pallanza. 385 p.
TRAPNELL, C. G. & J. N. CLOTHIER, 1957. The soils, vegetation, and agricultural systems of North-Western Rhodesia. Report of the ecological survey. 69 p. Govt. Printer, Lusaka.
TYLER, J. E. 1968. The Secchi disc. *Limnol. Ocean.* 13: 1–6.
VALLENTYNE, J. R. 1957. Principles of modern limnology. *Amer. Scient.* 45: 218–244.
VAN DER LINGEN, M. I. 1960. Some observations on the limnology of water storage reservoirs and natural lakes in Central Africa. Proc. First Fed. Sci. Congr., Salisbury, Rhodesia.
VAN DER LINGEN, M. I. 1973. Lake Kariba: early history and south shore. In: ACKERMANN, W. G., G. F. WHITE & E. B. WORTHINGTON (Eds.). Man-made lakes: their problems and environmental effects. Geophys. Monogr. Ser. 17. Amer. Geophys. Union, Washington.
VERDUIN, J. 1956. Primary production in lakes. *Limnol. Ocean* 1: 85–91.
VINER, A. B. 1970a. Hydrobiology of Lake Volta, Ghana. I. Stratification and circulation of water. *Hydrobiologia* 35: 209–229.
VINER, A. B. 1970b. Hydrobiology of Lake Volta, Ghana. II. Some observations on biological features associated with the morphology and water stratification. *Hydrobiologia* 35: 230–248.
VOLLENWEIDER, R. A. 1961. Photometric studies in inland waters. I. Relations existing in the spectral extinction of light in water. *Mem. Ist. Ital. Idrobiol.* 13: 87–113.
VOLLENWEIDER, R. A. 1969. A manual on methods for measuring primary production in aquatic environments. IBP Handbook No. 12, 1st Ed. Blackwell's Scient. Publ., Oxford, England. 213 p.
WELCH, P. S. 1948. Limnological methods. McGraw-Hill, New York. 381 p.
WELCOMME, R. L. 1972. The inland waters of Africa. *CIFA Techn. Pap.* 1. FAO, Rome 117 p.
WELLINGTON, J. H. 1946. A physiographic regional classification of South Africa. *S. Afr. Geogr. J.* 28: 64–86.
WESTLAKE, D. F. 1965. Some problems in the measurement of radiation under water: a review. *Photochemistry and Photobiology* 4: 849–868.

Whitney, R. R. 1969. Schooling of fishes relative to available light. *Trans. Amer. Fish. Soc.*, 98: 497–504.
Woodhead, P. M. J. 1966. The behaviour of fish in relation to light in the sea. pp. 337–403. In: H. Barnes (Ed.). Oceanography and Marine biology, Vol. 4. Hafner Publ. Co., New York.
Zhadin, V. I. & S. V. Gerd, 1963. Fauna and flora of the rivers, lakes, and reservoirs of the USSR. 626 p. Israel Prog. Scient. Transl., Jerusalem (Transl. from Russian, 1961).

RESERVOIRS

Characteristics		L. Kariba	L. Nasser/Nubia	L. Kainji	L. Volta	L. Kossou[2]
REFERENCES		Personal data	ENTZ 1973; ENTZ & RAMSEY 1973	HENDERSON 1973 IMEVBORE 1970	ENTZ 1969b; FAO/UNDP 1971; BRETSCHKO, pers. comm.	LESSENT (pers. comm.), KRZELJ (pers. comm.)
GEOGRAPHY	Country	Rhodesia–Zambia	Egypt/Sudan	Nigeria	Ghana	Ivory Coast
	Latitude	16°28′–18°04′S	23°58′–20°27′N	9°30′–10°35′N	6°15′–9°10′N	7°–8°N
	Longitude	26°42′–29°03′E	30°35′–33°15′E	4°25′–4°45′E	1°40′W –0°20′E	5°25′–5°45′W
	Altitude, m a.s.l.	485	180	142	85	204
	Vegetation type	forest + wooden savannah	desert	forest + Guinean savannah	forest + Guinean savannah	forest + Guinean savannah
METEOROLOGY	Climate (Köppen)	BSwh′	BWhs	Aw	Amw–Aw	Amw′–Aw
	Mean air temp. °C	24.3–24.7	(24)	(26.5)	26.4–27.8	26.3
	Abs. min temp. °C	2.8	5	(mean) 14.1 (Dec. + Feb.)	20 (mean 25.7)	13 (mean 21.1)
	Abs. max temp °C	40.6	48	(mean) 39.0 (Mar.–May)	(mean) 28.0	40 (mean 31.3)
	Rainfall year, mm	610	—	660–1352	890–1712	1150–1400
HYDROLOGY	Gross evapor., mm	2500–3600	3824	1500–2000	1398	1500
	River catchment	Zambezi	Nile	Niger	Volta (Black–White)	White Bandama
	Area catchment above dam, sq. km	823 200	2 400 000	1 600 000	394 000	32 400
	Annual inflow, 10^6 cum	39 000 (recent 47 300)	84 000	80 000	37 000	5 680
	Avg. ann. inflow, cu. m. $\sec^{-1}$	566–2516 (1200)	450–9 500	(800–2000)	1 167	57–320 (179)
	River flow ratio (min:max)	1:13	1:17	1:16	—	1:5.6
	River flood season	Mar.–Apr.	Aug.–Oct.	(White floods (Jul.–Sep.)) (Black flood (Dec.–Feb.))	Jul.–Sep.	Sep.–Oct.
DAM	Site	Kariba	Aswan	Kainji	Akosombo	Kossou
	Closed in:	December 1958	May 1964	August 1968	May 1964	February 1971
	Normal water level in:	March 1963	(? 1980)	1968	1968	(? 1978)
	Hydro-electric max. capacity, Kw	1 500	2 100	960	882	175.5

MORPHOMETRY[1]	Length, km	277	482	137	180
	Mean breadth, km	19.4	13.0	9.3	8.9
	Area, sq. km	5364	6276	1280	1600
	Volume, cu. km	156	158	15.8	29.5
	Max. depth, m	93	130	50	54
	Mean depth, m	29.2	25.2	12.3	18.4
	Length shoreline, km	2164	8804	720	3500
HYDROLOGY	Ann. gross evaporation, cu.km	4.2	24	1.9–2.56	2.4
	Water level max. m	489	180	142	206
	Water level normal, m	485	170–175(?)	—	204
	Water level fluctuations, m	3–4	7–10	9–10	3
	Water exchange ratio	1:4.0 (1:3.1)	1:2	4:1	1:5.2
WATER PHYSICS	Temp. range, °C	17–32	16–32	23–31	23.8–34.1
	Thermal cycle	warm monomictic	warm monomictic	(±monomictic)	warm dimictic
	Temp. homothermy, °C	20–22 (Jun.–Jul.)	18.0	23–27.5	≤27 (Dec.–Feb.)
					(Jul.–Sep.)
	Thermal strat. period	Oct.–June	May–Aug.	Feb.–mid May	Mar.–Jun. and Oct.–Nov.
	Depth of mixed layer, m	15 – 25	13 – 20	15 – 20	5 – 10
	Secchi D.V., cm	50 –1060 (405)	20 –400	10 –300	10 –300
	Depth euphotic zone, m	2 – 24 (10–16)	1 – 10	ab. 1–8	—
WATER CHEMISTRY	pH	6.8– 8.9	7.1– 9.4	6.0– 8.0	6.8– 9.0
	Conductivity k_{20} μmhos	50 – 115 (72)	210 –250	40 – 54	36 –128
	Total solids, mg/L	40 – 70	260	(35 –40)	—
	Diss. oxygen surf. mg/L	6 – 10	2 – 7		
	Oxygen stratification	Oct.–June	June–(Dec.)	Mar.–mid May; Oct.–Nov.	(1 – 10) Oct.–Nov. & Mar.–Jun.
FISHERY MANAGEMENT	Bush clearing, type	large areas totally cleared and dispersed around the lake, within the littoral zone	none	Foge Island and access strips	– Front of the dam – Strips 100 m-wide for navigation and for fishing (ab. 950 km)
	Total area, sq. km	954	—	512	95
	Area, percent surf. area	17.8	—	40	6

[1] For additional data on morphometry, see Table 38.
[2] Physical and chemical data for 1973, filling phase.

ANNEX II
LIST OF ABBREVIATIONS AND SYMBOLS

a	conductivity coefficient for total solids (μmhos per meq.L^{-1} of TS)
asl	elevation above mean sea level (m, ft)
A_i	insulosity (percent or sq.km)
A_o	area (surface-) at contour of depth 0 or 485 m asl (sq.km)
A_z	area at contour of depth z (sq.km)
AHI	annual heat index (cal.cm^{-2}m^{-1}.yr^{-1})
AT_p	average percentile transmission
AVAC	average vertical attenuation coefficient per metre over depth z_P, natural log basis (ln units.m^{-1})
b	conductivity coefficient for salinity (μmhos per meq.L^{-1} of SL)
$\bar{b}$	mean breadth (km)
B	work of the wind (gcm.cm^{-2})
B III	Basin III, Lake Kariba
BAHB	Birgean annual heat budget (cal.cm^{-2})
c	equivalent conductivity (μmhos per meq.L^{-1} of TIC)
cal	gram-calorie
C	centigrade, degree Celsius
d	conductivity coefficient for total alkalinity (μmhos per meq.L^{-1} of TA).
d_{zs}	water density corresponding to summer mean temperature of the stratum at depth z (g. cm^{-3})
d_{zw}	water density corresponding to winter mean temperature of the stratum at depth z (g.cm^{-3})
D_L	development of shoreline at water level 485 m asl
D_V	development of volume at water level 485 m asl
DO	dissolved oxygen (mg.L^{-1})
DOE	dissolved oxygen content, epilimnion (mg.L^{-1})
DOH	dissolved oxygen content, hypolimnion (mg.L^{-1})
DV	depth of visibility, Secchi-disc transparency (cm, m)
ft	foot = 0.305 m
G	total work needed to maintain hypothetical homothermy (gcm.cm^{-2})
ha	hectare, 10^4 sq.m.
H	homothermy, isothermy
HC	heat content above 4° C (cal.cm^{-2})
HOD	hypolimnetic oxygen depletion (mg.cm^{-2})
ID	illuminated depth (m)
IP	index of light penetration
k_{20}	conductivity, specific electrical – at 20° C (micromhos.cm^{-1})
k_{20}'	conductivity (20° C) due to total alkalinity (micromhos)
K2/69	Lake Kariba, cruise in February 1969
K_t	conductivity coefficient for temperature correction
Kw	megawatt, 10^6 watts
l	length (km)
ln	natural logarithm
ly	langley (cal.cm^{-2})
L	litre
L_o	length of shore line at the 485-m water level (km)
L_z	length of the contour at depth z (km)
λ	wavelength of light (nm)

meq	milligram equivalent
MHC	maximum heat content (cal.cm^{-2})
MHI	maximum heat index (cal.cm^{-2}.m^{-1})
MVAC	mean vertical attenuation coefficient (ln units.m^{-1})
μg	microgram, 10^{-6} gram
μmhos	micromhos.cm^{-1}
nm	nannometer = millimicron = 10^{-9} metre
OD	optical depth (ln units.m^{-1})
pH	hydrogen-ion concentration
ppb	part per billion = μg.L^{-1}
ppm	part per million = mg.L^{-1}
PVR	partial volume ratio (Table 32)
q	coefficient for temperature correction of conductivity
r	correlation coefficient
RH	residual heat above 4° C (cal.cm^{-2})
RHOD	rate (average) of hypolimnetic oxygen depletion (mg.cm^{-2}.day^{-1})
RT	reduced thickness (cm, Table 34)
S	stability (gcm.cm^{-2})
SAST	South-African Standard Time
SB	Sub-Basin of Lake Kariba
SD	Stirred depth, mixed depth (m)
SL	salinity (mg.L^{-1}, meq.L^{-1})
T	thermocline, plane of maximum thermal gradient
$\bar{T}(z)$	mean temperature curve of water temperatures
T_p	percentile vertical transmission per metre
T_z^λ	relative light intensity for λ at depth z, in percent of the light present at the water surface (T_o^λ).
T_z^T	relative total light intensity (photometer response) at depth z in percent of the total light present at the water surface, T_o^T
$\bar{T}_{zs}$	mean temperature of the stratum at depth z at the time of maximum heat content (C)
$\bar{T}_{zw}$	mean temperature of the stratum at depth z at the time of minimum heat content or homothermy (C)
TA	total alkalinity (meq.L^{-1})
TG	thermal gradient (C.m^{-1})
TH	total hardness (mg.L^{-1}, meq.L^{-1})
TI	tropicality index (cal.cm^{-2}.m^{-1})
TIC	total ionic concentration (meq.L^{-1})
TS	total solids, dissolved (mg.L^{-1}, meq.L^{-1})
V	volume of water
V_z	volume of the water present between the horizontal plane of depth z and the bottom floor of the lake
VAC	vertical attenuation coefficient per metre, natural log basis (ln units.m^{-1})
WC	work constant (cm^2, Table 34)
z	depth below the water surface (cm, m)
$\bar{z}$	mean depth (m)
z_E	depth of the euphotic zone (cm,m)
z_g	depth of the geometrical centre of gravity of the lake (m)
z_m	maximum depth (m)
z_P	depth where the relative intensity of the light of the wavelengths least strongly attenuated equals one percent (cm, m)
z_r	relative depth
$\bar{z}_T$	mean depth of thermocline (m)
$\bar{Z}(t)$	mean depth curve of water temperatures

PART II

FISH PRODUCTION OF A TROPICAL ECOSYSTEM

by

Eugene K. Balon

This I dedicate
to Holly,
who like Henry James believes that to be damn critical
is the only thing to be for all else is humbug.

CONTENTS OF PART II

	A parable	255
1.	Introduction and acknowledgements	257
2.	Methods	265
2.1.	*Glossary and usage of terms*	278
3.	Age and growth studies	280
3.1.	*The green bream, Sargochromis codringtoni* (Boulenger, 1908), *by* J. HOLČIK	280
3.2.	*The plain squeaker, Synodontis zambezensis* Peters, 1852, *by* K. CHITRAVADIVELU	298
3.3.	*The red-breasted bream, Tilapia rendalli* (BOULENGER, 1896), *by* J. BASTL	311
3.4.	*The Zambezi barbel, Clarias gariepinus* (Burchell, 1822), *by* K. PIVNIČKA	318
3.5.	*The spotted squeaker, Synodontis nebulosus, the butter catfish, Schilbe mystus, the vundu, Heterobranchus longifilis, and the electric catfish, Malapterurus electricus, by* S. FRANK	325
3.6.	*The silver catfish, Eutropius depressirostris* (Peters, 1852), *by* K. ČERNY	333
3.7.	*The kurper bream, Sarotherodon mossambicus mortimeri* (Trewavas, 1966), *by* I. KRUPKA	343
3.8.	*The cornish-jack, Mormyrops deliciosus* (Leach, 1818) *and the bottlenose, Mormyrus longirostris* Peters, 1852	349
3.9.	*The parrotfish, Hippopotamyrus discorhynchus* (Peters, 1852) *and the bull-dogfish, Marcusenius macrolepidotus* (Peters, 1852), *by* A. KIRKA	380
3.10.	*The rednose mudsucker, Labeo altivelis* Peters, 1852	394
3.11.	*The silver robber, Micralestes acutidens* (Peters, 1852), *by* J. HOLČIK	406
3.12.	*The Darling's dwarf bream, Haplochromis darlingi* (Boulenger, 1911), *by* E. K. BALON & E. D. MUYANGA	419
4.	Total production, available production and yield of major fish taxa from Lake Kariba	428
4.1.	*Species size and growth intensity*	428
4.2.	*Stock size assessment values*	437
a.	*Density and mortality rates*	437
b.	*Biomass*	439
c.	*Production*	440
d.	*Yield*	441
5.	The eels	446
5.1.	*Taxonomic status*	448

5.2.	*Limits of distribution*	449
5.3.	*Age, growth and size of resource*	450
6.	Fish production of the drainage area and the influence of ecosystem changes on fish distribution	459
6.1.	*The Kalomo River*	459
6.2.	*The Elephant Stream*	472
6.3.	*The relationship of Lake Kariba to the Victoria Falls*	478
6.4.	*The Kafue River Floodplain: an example of pre-impoundment potential for fish production,* by J. M. KAPETSKY	497
7.	The success and failure of the clupeid introduction	524
7.1.	*The air lift of clupeid juveniles from Lake Tanganyika*	524
7.2.	*Successful reproduction and distribution in Lake Kariba,* by J. G. WOODWARD	526
7.3.	*The compromising density*	536
	Concluding discussion	542
	Epilogue	554
	Literature cited	558
Appendix A.	List of symbols used	574
Appendix B.	Efficiency of cove-rotenone samples, *by* G. P. BAZIGOS	575
Appendix C.	Tables of mean sizes of individual species, life intervals and growth intensity	595
Appendix D.	Tables of individual species and single sample production computations	603
Appendix E.	Morphometry of sampling sites and standing crop tables	625
Appendix F.	Time of annulus inception: a pond experiment, *by* E. K. BALON & E. M. CHADWICK	643
Appendix G.	Lepidological study: key scales of Lake Kariba fishes	647

A Parable

'... the emphasis on the whole should not blind us to the fact that recognition of the components is also essential. (...) the approach can be likened to understanding a cow by running the entire live animal through an enormous meat grinder, and scientifically analyzing the resultant hamburger. A monkey might slip in, and never be noticed. The resultant science might express a perfect bovine placidity, but it would take a sharp mental eye to be aware of the monkey business'.

Frank E. Egler (1964)

We are approaching the 16th anniversary of the creation of Lake Kariba. Interest in the lake has dwindled in direct proportion to the dramatic events and problems which surrounded its formation. There has not been anything as spectacular as 'Operation Noah' since the dam was closed in 1958. Although many problems continue to plague the area, none equal the 'glamour' of the animal rescue operation – which has since been viewed with considerable misgivings, as have other endeavors undertaken prior to impoundment as well as after. Of course, it is always easy to be critical with the benefit of hindsight; however, there do exist analogous enterprises whose lessons could and should have been heeded somewhat more strenuously than has perhaps been done.

There was the case of the Danube River. The largest European international waterway is already dammed wherever possible, and mainly in the upper zone. Ecologically, the most vital part of the river, which has a large inundation area but cataracts as well, is at the end of the foothill zone. For centuries this was the most fertile farmland of Austria, Slovakia and Hungary. It was also the best fishing area. In the 17th century the capturing by cannon fire of 5 m long sturgeons called 'hausen' was attended by royalty, although soon afterwards overfishing restricted these giants to the river delta. Flood dams were raised and strengthened, mainly beginning with the 19th century, and these limited the inundation area and substantially decreased the production of fish. The inundation area, however that remained within the limits of the flood dams which were built at some distance from the river bed, still succeeded in maintaining an important fish production and aquatic life. This area, with its regularly fluctuating water level managed to maintain a fairly effective self-purifying process (the river's capacity to absorb pollution). The groundwater in the adjacent farmlands remained at the same time at a useful level.

But technocrats still found the river cataracts too wild and too dangerous for boat traffic and, more important, completely unutilized at this convenient location. What a pity to waste so much potential electric energy! With the help of a ruler and calculator the most economically advantageous plans for water traffic and electric energy generation were drawn up. At the time (1964–1965), I had been studying the Danube area ecosystem for some 10 years and was able to present some convincing arguments on behalf of environmental conservation...

The river was to be diverted into a deep channel to ensure year round safe water traffic. It was considered a disturbing side effect that the underground water level would drop substantially and the rich farming area turn into a desert. But it was believed that the electric energy would pay for losses and enable the purchase of food elsewhere. This was stated authoritatively. We objected asking: 'For whose benefit will the street lights be lit in 1980 if we have nothing to eat then', and we pointed out the certain losses in self purification ability and fish production which would result. Notwithstanding all the evidence pointing against the implementation of their plans, the technicians persisted. With politicians interested in giant investments and in view of the nature of the regime I was threatened with punishment. The powers that be even confiscated a published issue of a cultural weekly paper which contained an article on this subject. Events that followed, however, postponed the realization of this project, but for how long, nobody knows. From then on I became especially sensitive to a minority's ability to force its decisions on the majority, regardless of adverse social and environmental results.

I have become aware that we must not seek justification in some general hope that any large scale environmental changes will be of benefit to all mankind, but must be assured of such benefits. International competition among people for short term objectives, and the profit won through isolated enterprises do not in themselves assure any long term benefits for mankind; and in spite of their glib predictions, the projects in many cases, fail to provide long term benefits even for the promoters. I became a believer in formulations such as McKINLEY's (1964): 'A study of the true wealth of a piece of land dedicated quite simply to being itself might easily be worth a man's lifetime.' I found, however, it was impossible to avoid man-made environmental changes even in a very remote area of the world. Can we prevent when we understand

> 'how those present broke the ice with tense haste; how, flouting the lofty interest of ichthyology and elbowing each other to be first, they tore off pieces of the prehistoric flesh and dragged it over to the bonfire to thaw it and bolt it down'?
>
> A. I. SOLZHENITSYN in
>
> 'The Gulag Archipelago, 1918–1956',
> The New York Times, 30 December, 1973

1. INTRODUCTION AND ACKNOWLEDGEMENTS

The original idea was to study in an identical way and within reasonable confidence limits all fish taxa present in Lake Kariba. But time was running short. It may take another two years to conclude the work on more numerous samples and on all 40 species, of which many are rare and without commercial value. The 21 species selected represent more than 97% of the total standing stock of Lake Kariba endemic fishes and thus fully justify the preparation of this work as it is, although the limited samples will determine the general validity of the values presented. We should view them as an introduction and example for similar future studies. From the practical point of view nothing is lost because any values presented will always be better than no data at all. However, more satisfactory statistical evaluation and the examination of the predator-prey production relationship had to be omitted. Should there be available time in the future it is hoped that a possibility will present itself to treat these aspects as well.

The first papers, written some time ago to serve as samples, are published in the Fisheries Research Bulletin of Zambia and concern *Hydrocynus vittatus*, *Alestes lateralis* and *Brachyalestes imberi imberi* (BALON, 1971a, 1971b). Only part of their data is, therefore, included in chapter 4 of the present study. Studies evaluating all the other species are included in chapter 3 and are presented with the differences of approach of individual authors. I decided to arrange the papers according to their methodological importance, as they are discussed by the two groups of authors working with scales and vertebrae respectively. In the individual papers the authors treat the Lake Kariba fish species arranged according to the percentage of total number as stated in the first general article (BALON, 1973a). The order in respect to standing stock is a little different (Table 1). Our subsequent statements are based upon the latter, as in my opinion, it reflects better the nutritive importance of the respective species.

There were two reasons for the cooperation of so many authors: (1) the samples were too large for one person to deal with, (2) determination of age and validity of the annuli for this region were unknown and controversial. To prove the validity of the annuli with sufficient confidence would have been an extremely arduous project for one person; by obtaining the cooperation of many skilled workers I hoped to achieve this objective in a shorter time and with greater confidence in its objectivity.

With regard to the terms 'production' and 'yield' and the meaning attached to them, I must repeat here the points already stated in the methodological outline (BALON, 1972). Parallel with the development of the production concept new terms were invented and defined. Many adjectives were attached to the term production, such as: actual, whole, gross, net, global, primary, potential, real, total, secondary, effective, etc. It might well lead to misunderstanding and misinterpretation of the contents if these terms are used without explanation.

The most accurate interpretation of fish production is that 'Production is defined

Table 1. Recent list of Lake Kariba fish species in order of species' average size, local preferrence, and biomass per surface area.

Mean percentage of total standing stock	Economically preferred species
16.00	*Sarotherodon mossambicus mortimeri
15.10	*Mormyrops deliciosus
9.30	*Tilapia rendalli
8.60	*Clarias gariepinus
5.10	*Hydrocynus vittatus
3.00	*Mormyrus longirostris
2.30	*Heterobranchus longifilis
2.20	*Sargochromis codringtoni
1.00	*Labeo altivelis
0.70	Sargochromis giardi
0.40	Distichodus schenga
0.30	Labeo congoro
0.30	Sarotherodon andersoni
0.30	Haplochromis carlottae
0.04	Distichodus mossambicus
0.03	Serranochromis robustus jallae
0.03	Serranochromis macrocephalus
64.70	
	Secondary species
15.90	*Hippopotamyrus discorhynchus
8.00	*Malapterurus electricus
3.20	*Synodontis zambezensis
1.50	*Eutropius depressirostris
0.30	*Marcusenius macrolepidotus
0.09	*Schilbe mystus
—	*Anguilla nebulosa labiata
—	*Limnothrissa miodon
29.00	
	Accompanying species
4.80	*Alestes lateralis
0.90	*Haplochromis darlingi
0.30	Barbus fasciolatus
0.30	*Synodontis nebulosus
0.10	Barbus unitaeniatus
0.10	Pseudocrenilabrus philander
0.10	*Brachyalestes imberi imberi
0.01	*Micralestes acutidens
0.01	Labeo cylindricus
0.01	Barbus lineomaculatus
0.007	Barbus poechii
0.003	Labeo lunatus
0.0005	Aplocheilichthys johnstoni
—	Barbus paludinosus
—	Barilius zambezensis
6.60	

* Fish studied in this volume (see p. 432–436 and the complete pictorial list).

as the total elaboration of fish tissue during any time interval Δt, including what is formed by individuals that do not survive to the end of Δt (IVLEV, 1945, 1966)' (or 'regardless of the ultimate fate of that tissue') (CHAPMAN, 1967, 1968). It is formulated by RICKER (1946) as

$$A = G\bar{B}.$$

Based on a similar formula the numerical estimation of production proposed by RICKER (l.c., 1958, 1969) was later developed into a simple graphical method by ALLEN (1951). Although many authors have used this method of estimating fish production, it has also been subject to some criticism (CHAPMAN, 1968: NIKOLSKY, 1965). Nevertheless, for the estimation of production according to IVLEV's definition, it is the only operational method so far available.

More meaningful, from the fishery management point of view, is that part of production which is represented by the total elaboration of fish tissue during any time interval Δt, of fish surviving at the end of Δt. This production value can be estimated 'by multiplying individual weight increase by the mean number of fish present in each interval of interest' (CHAPMAN, 1967). It was used by BURMAKIN & ZHAKOV (1961), HOLČIK (1970a, 1970b) and by myself. It can be interpreted as the available production unaffected by mortality, whether caused by predation or exploitation. It is the surviving part of IVLEV's production, which, in turn, can be interpreted as the total production. The available production also represents the fish mass value which is available within socio-economic limits for fish catch expansion, regardless of the actual situation in exloitation. Thus the related 'yield' can be interpreted as a harvestable part of available production or available yield. No other meaning should be attached to the terms production and yield throughout all the single species sections unless stated otherwise. The production *sensu lato* is evaluated in the chapter entitled 'Total production, available production and yield of major fish taxa from Lake Kariba' (p. 428). MATHEWS (1970), whose article was published and came to my attention after this study had already been written, gives yet other possible estimates of production.

Some changes occurred in the estimates of the area of Lake Kariba inhabited by fish. The original calculations were based on those given by COCHE (1968) and were 342 980 ha from 5,250 km^2 of the total lake area. Later COCHE (see Part I, chapter 8) recalculated the given areas in greater detail and came to a value for the total lake surface of 5,364 km^2. Thus the main average area of Lake Kariba inhabited by endemic riverine fish in 1968 and 1971 was 202 200 ha, corresponding to the 0–20 m depth area at normal water level of 485 m. This value represents 37.7% of the total lake area, a much lower percentage than in the previous calculation. The majority of the authors in this volume base their final calculations on the erroneous first published value as stated in my first general article (BALON, 1973a). Since the relative values per hectare and studied locality are more meaningful I decided to leave these wrong absolute estimates in the single sections of chapter 3 untouched and recalculated them only in the final summaries of chapter 4 according to the last area estimate of 202 200 ha. As a matter of fact the yearly water level fluctuations also leave this area estimate very

Table 2. Brief description of localities sampled quantitatively with fish eradicants on Lake Kariba in 1968–1971.

Sample	Lake basin	Locality	Date	Area, ha	Max. depth, meters	Topography	Sampling methods
A	4	cove near Siavonga shore	March 20 to 23, 1968	0.818	16	deep rocky cove with sand beach end; submerged trees, small groups of drifting *Salvinia*, and a few bunches of *Ceratophyllum demersum* and *Chara*; clean rocks and sand on bottom	rotenone in 0.114 ppm concentration; area blocked by small mesh net on surface only
B	4	inshore waters of island N 102 near Siavonga	March 21 to 22, 1968	0.476	21	high steep rocky shore and a few submerged trees and bushes; rock, gravel, and sandy bottom with little mud and detritus; little *Potamogeton pusillus*	electric discharge in a 20 m² area with four 226 g plastic explosives and one 150 m cord
C	3	Chikanka Island N16 cove	June 21 to 25, 1958	1.210	5	long narrow cove with rocky and muddy shores; *Salvinia* mats cover a quarter of the area at the inland end; bottom of soft mud and gravel; groups of *Phragmites*, *Lagarosiphon*, *Nymphaea*, and *Cyperus*	toxaphene in 7 ppm concentration; area completely blocked with double small mesh nets
D	2	cove of island N 20 southwest of Sinazongwe	Dec. 11 to 14, 1968	1.147	16	valley cove of *Euphorbia* hills; rock and sand on shores, very little *Salvinia*, lots of submerged vegetation in shallow part (*Ceratophyllum*, *Lagarosiphon*, *Potamogeton sweinfurthii*), submerged trees and bushes; soft muddy bottom	toxaphene in 7 ppm concentration; area completely blocked with double small mesh nets
E	3	Chete Island N 21 cove	March 20 to 25, 1969	5.273	4	shallow cove with swampy and grassy shores; quarter of the area covered by *Salvinia*; lots of submerged *Ceratophyllum demersum* and *Potamogeton pusillus*; a few submerged bushes; bottom of	toxaphene in 7 ppm concentration; area completely blocked with double small mesh nets

Sample	Lake basin	Locality	Date	Area, ha	Max. depth, meters	Topography	Sampling methods
E₃	3	intermittent stream estuary at end of Chete Island cove	May 25, 1969	0.021	0.5	shallow swampy part of cove with numerous submerged bushes; bottom of deep mud and sand; submerged *Panicum*, *Ludwigia*, and *Typha*	toxaphene in completely blocked area
F	3	cove I in Chipepo Bay	May 26 to 31, 1969	0.191	4.5	small cove at Kota Kota foothills with steep rocky and wooded shores; completely covered by *Salvinia* mat and replete with *Ceratophyllum*, *Phragmites*, and *Vossia*; bottom of soft mud	toxaphene; area completely blocked by double net; *Salvinia* and some other plants were removed before treatment
G	3	cove II in Chipepo Bay	May 28 to 31, 1969	0.083	5.4	similar to the description given above for sample F	toxaphene; area completely blocked by double net; *Salvinia* and other plants were not removed before treatment
H	4	shore-line in Loteri Bay	Jan. 28 to 30, 1971	1.000	5.6	inshore section of a large shallow bay with steep grassy and rocky shore; muddy bottom crossed along the shore by submerged river bed; a few bunches of *Ceratophyllum*, along the shore *Phragmites* and *Vossia* with drifting *Salvinia* groups	rotenone distributed by pump inside of completely blocked square of 100 m sides
I	4	Lutele River estuary	January 30, 1971	4.390	1.5	intermittent river estuary in the lower half with muddy bottom and high grassy ravine shores, lowering into a swamp; in upper half shallow sandy bottom with frequent sand banks and flooded grass edges	rotenone distributed by pump at the upstream limit; lower limit blocked with small mesh net

much in doubt (see Part I, Table 30). Consequently the area differences are of little importance.

To enable the reader to gain a clearer understanding of the quantitative samples used, their size, topography and sampling procedure, Table 2 and Fig. 1 are enclosed. In some sections the localities are referred to only by the symbols used in this table. More about the topography of sampling sites and standing crop is then given in Appendix E.

The primary purpose of chapter 4 is to evaluate the age and growth data of 20 major fishes of Lake Kariba given in chapter 3 of this volume. The species selected represent over 97 % of the average standing stock of indigenous fishes in the years 1968 and 1971. However the substantial stock of catadromous eels *Anguilla nebulosa labiata*, discovered later (BALON, 1973b) as inhabitants of the deeper lake regions, was studied and evaluated long after the main parts of this monograph were written. The eel study will be published separately although a concise version is given here as chapter 5. Following the assessment of this report, the introduced Tanganyikan anchoveta *Limnothrissa miodon* began successfully to occupy (BALON, 1971c) the open lake habitats. Again, some results of this introduction were studied later and are given in a preliminary form in chapter 7. This is a combination of parts of JOHN WOODWARD'S Terminal Report and my own survey. Consequently, the validity of the species composition recorded originally (Table 1) is subject to modification.

Most of the samples proved to be insufficient for obtaining data within good statistical confidence limits. Thus, their validity for the whole lake in the years 1968 to 1971 is limited. These limitations in relation to lake topography, size of samples and sampling biases are evaluated in a separate paper (Appendix B). Some of the data on fish succession in Lake Kariba (Table 3), historical relationships to adjacent areas and pre-impoundment production potential are evaluated in other studies (BALON, 1947b, c), but are abridged here in different form as chapter 6.

At closure of the dam on December 2, 1958 the water started to rise from 391 m above sea-level and the average operating water level of 485 m was reached four-and-a-half years later, in April 1963. In the new lacustrine conditions density of some riverine fish has decreased drastically, while the rising water and high nutrient load have brought an explosive increase in density and biomass of other species (BALINSKY & JAMES, 1960; JACKSON, 1960; JUBB, 1960; HARDING, 1964, 1966). Later, species from above Victoria Falls and adjacent streams invaded Lake Kariba. At the same time (1965) a rapid decrease in fish catches occurred. Though at the time of writing single species density was still highly variable [*A. lateralis* started to expand into the free pelagic niche (BALON, 1971b) and then gave up in competition with the introduced anchoveta; spawning grounds for *Micralestes acutidens* are being recovered by wave action along some shores (see section 3.11), etc.] from 1966 onward, upon stabilization of the nutrient load, the fish population as a whole became relatively stabilized (COCHE, 1968, 1971; BEGG, 1970). On the basis of this study it may be possible to prove this definitively in the future. If so, these may become the first known values of ecological fish production in a tropical lake.

The papers of HOLČIK have furnished data for the section on *Sargochromis codringtoni* and *Micralestes acutidens*, those of CHITRAVADIVELU on *Synodontis zam-*

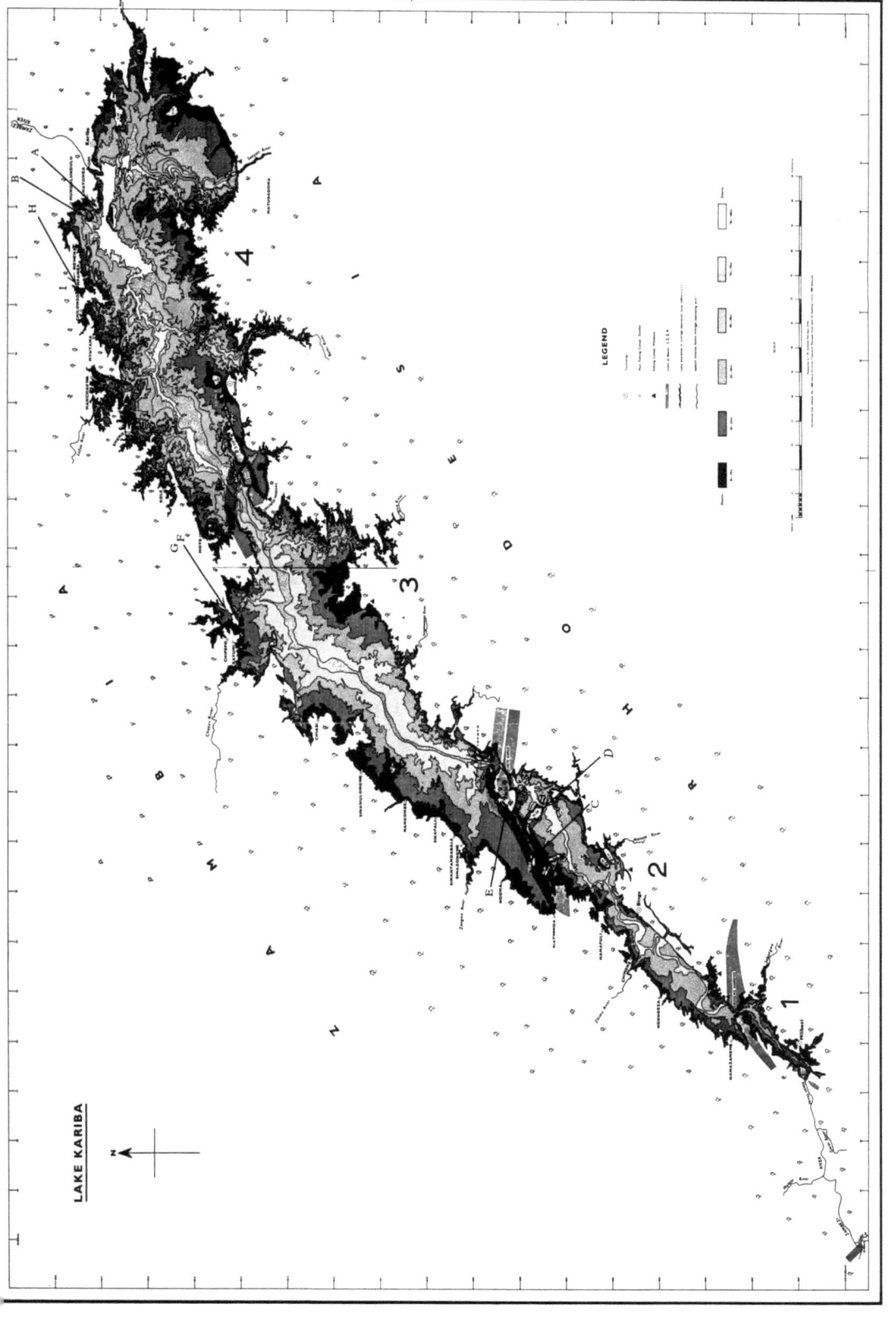

Fig. 1. Bathymetric map of Lake Kariba. Quantitative sampling localities are marked with arrows and corresponding (Table 2) letters. (Comb. G. COCHE).

Table 3. List of fish taxa of Lake Kariba[1] presented in a phylogenetic order with proposed English and vernacular names (*marked studied species)

1. *Anguilla nebulosa labiata* Peters, 1852	*African mottled eel	Mukunka
2. *Limnothrissa miodon* (Boulenger, 1906)	*Tanganyikan anchoveta	Kapenta
3. *Mormyrops deliciosus* (Leach, 1818)	*Cornish-jack	Muyanda
4. *Hippopotamyrus discorhynchus* (Peters, 1852)	*Parrotfish	Manquela
5. *Marcusenius macrolepidotus* (Peters, 1852)	*Bull-dogfish	Nembele
6. *Mormyrus longirostris* Peters, 1852	*Bottlenose	Amalompi
7. *Hydrocynus vittatus* Castelnau, 1861	*Tigerfish	Mcheni
8. *Alestes lateralis* (Boulenger, 1900)	*Alestes	Mbale
9. *Brachyalestes imberi imberi* (Peters, 1852)	*Spot-tail	Mumbele
10. *Micralestes acutidens* (Peters, 1852)	*Silver robber	Chitaka
11. *Distichodus mossambicus* Peters, 1852	Sikapapa	Jenga
12. *Distichodus schenga* Peters, 1852	Schenga	Kafulofulo
13. *Barbus poechii* Steindachner, 1911	Spot-tail barb	Mpifu
14. *Barbus paludinosus* Peters, 1852	Ubiquitous barb	Ncenga
15. *Barbus unitaeniatus* Günther, 1866	Striped barb	Msindi
16. *Barbus lineomaculatus* Boulenger, 1903	Spotted barb	Gwinya
17. *Barbus fasciolatus* Günther, 1868	Red-banded barb	Chitakala
18. *Labeo cylindricus* Peters, 1852	Golden mudsucker	Nguridzi
19. *Labeo lunatus* Jubb, 1963	Sailfin mudsucker	Linyonga
20. *Labeo rubropunctatus* Gilchrist & Thompson, 1913	Red-spotted mudsucker	Mumbu
21. *Labeo congoro* Peters, 1852	Purple mudsucker	Mucise
22. *Labeo altivelis* Peters, 1852	*Rednose mudsucker	Mutuba
23. *Barilius zambezensis* (Peters, 1852)	Barred minnow	Myamkaula
24. *Schilbe mystus* (Linneaus, 1762)	*Butter catfish	Lubangu
25. *Eutropius depressirostris* (Peters, 1852)	*Silver catfish	Mvenga
26. *Heterobranchus longifilis* Valenciennes, 1840	*Vundu	Muzunda
27. *Clarias gariepinus* (Burchell, 1822)	*Zambezi barbel	Mubondo
28. *Malapterurus electricus* (Gmelin, 1789)	*Electric catfish	Chikwinya
29. *Synodontis zambezensis* Peters, 1852	*Plain squeaker	Chikolo
30. *Synodontis nebulosus* Peters, 1852	*Spotted squeaker	Korokoro
31. *Aplocheilichthys johnstoni* (Günther, 1893)	Johnston's topminnow	Pipinka
32. *Sarotherodon andersoni* (Castelnau, 1861)	Three-spot bream	Njinji
33. *Sarotherodon mossambicus mortimeri* (Trewavas, 1966)	*Kurper bream	Muchele
34. *Tilapia rendalli* (Boulenger, 1896)	*Red-breasted bream	Mudile
35. *Sargochromis giardi* (Pellegrin, 1904)	Pink bream	Sieo
36. *Sargochromis codringtoni* (Boulenger, 1908)	*Green bream	Seao
37. *Serranochromis macrocephalus* (Boulenger, 1899)	Purple-faced bream	Mulumbo
38. *Serranochromis robustus jallae* (Boulenger, 1896)	Olive bream	Nenbwe
39. *Haplochromis carlottae* (Boulenger, 1906)	Rainbow bream	Likumbwa
40. *Haplochromis darlingi* (Boulenger, 1911)	*Darling's dwarf bream	Imbalata
41. *Pseudocrenilabrus philander* (Weber, 1897)	Moffat's dwarf bream	Hangalole

[1] For details see BALON (1974b).

bezensis, of BASTL on *Tilapia rendalli*, of PIVNIČKA on *Clarias gariepinus*, of FRANK on *Synodontis nebulosus, Schilbe mystus, Heterobranchus longifilis* and *Malapterurus electricus*, of ČERNY on *Eutropius depressirostris*, of KRUPKA on *Sarotherodon*[1] *mossambicus mortimeri*, of KIRKA on *Hippopotamyrus discorhynchus* and *Marcusenius macrolepidotus. Anguilla nebulosa labiata, Hydrocynus vittatus, Alestes lateralis, Brachyalestes imberi imberi, Mormyrops deliciosus, Mormyrus longirostris* and *Labeo altivelis* were studied by me and *Haplochromis darlingi* by me and MUYANGA. J. M. KAPETSKY wrote the section on the Kafue River and G. P. BAZIGOS Appendix B.

Thus several individual sections were written by different authors who also carried on the age and growth studies of given species from quantitative data, and fish or scale samples provided by me. The sections not written by me bear the names of the authors after each title.

Unfortunately the hiatus between the preparation and publication of the manuscript renders the original wording somewhat superfluous. It was to make an impact at the time of writing. However, some of my general concerns as expressed in the Parable, Concluding Discussion and Epilogue were later shared by others (FARVAR & MILTON, 1972; DUSSART *et al.*, 1972). Nevertheless for sentimental reasons I have decided to leave intact the original version, amending only sections finished later and adding some new ones.

I would first of all like to thank the FAO staff, particularly A. G. COCHE, J. ZNAMENSKY, K. WATANABE, Project Manager L. S. JOERIS and Officers (T. EDELMAN, J. MUBANGA, P. MUSHINGE, W. GAY) from the Department of Wildlife, Fisheries and National Parks (Chilanga) who were responsible for giving me a carte blanche for my research concept and for their generous assitance with material, equipment and staff requirements. The accomplishment of this study, in a very real sense, was made possible by the technical assistance of E. D. MUYANGA, L. C. MAKAYI, W. PENZYA, J. M. CHINYAMA, L. KAWAYO and many project fishermen, as well as University of Zambia students W. MULENGA and P. NKHUNIKA. Many ideas and formulations were suggested while consulting several FAO working papers of H. A. REGIER and upon his personal encouragement during my stay at the University of Toronto. The text has received a rigorous going over by W. C. BECKMAN (FAO Rome), H. F. HENDERSON (FAO Rome), R. M. JENKINS (Fayetteville), J. M. KAPETSKY (Ann Arbor), R. A. RYDER (Thunder Bay), and part of it by H. A. REGIER (Toronto) and E. TREWAVAS (London), and incorporates many of their comments. To all of them many thanks.

Finally, I would like to express my sincere gratitude to J. HOLČIK (Laboratory of Fishery Research and Hydrobiology, Slovak Agricultural Academy, Bratislava) and S. FRANK (Laboratory of Ichthyology, Department of Zoology, Charles University, Prague) who organized most of the work of chapter 3 at their laboratories and brought it to an effective end, and to J. TAYLOR who helped with the Index.

---

[1] The generic name *Sarotherodon* for mouth-brooding *Tilapia* was accepted after TREWAVAS (1973) along with the replacement of *Hemihaplochromis* by *Pseudocrenilabrus*.

2. METHODS

> 'Density, growth, age structure, and harvest pressure are all interrelated, and the interactions and relationships have yet to be delineated in terms that have predictive value.'[2]

Most of the methods used have been presented in a separate methodological paper (BALON, 1972) and in the preliminary evaluation of Lake Kariba standing crops (BALON, 1973a). At the time of sampling, the indigenous fish stocks, except the eel, were distributed in 202 200 ha of inshore waters less than 20 m deep. This depth limit of the distribution was obtained by evaluating electroacoustic records and gill-net sets. The distribution of species within this area is largely unknown. The density of single species was obtained from quantitative samples in known blocked areas treated with fish eradicants (toxaphene, rotenone). Age determination was done by assessment of annuli from key scales and vertebrae; the growth was back-calculated for the previous growth seasons on the basis of empirical body-scale or body-vertebra relationships. The mean weights were interpolated from the weight-length plots on a double logarithmic scale, of which a more detailed explanation can be found in a special study (BALON, 1974b).

The production concept used and its terminology is explained in the introduction to this Part. We neither estimated density nor the mortality and growth parameters for short periods of the same year class of fish stock, as is done in most fish production studies (RICKER & FOERSTER, 1948; ALLEN, 1951; COCHE, 1967; CHAPMAN, 1965, and many others), but the annual increments and mortality rates were calculated from the density of single age groups in a sample on an annual basis. The results are biased by the differences in growth, density and mortality in different years, and also by the size-selective mortality. The 'X-type rates' were rarely applicable (RICKER, 1969), because of the nature of the material e.g. the need to average growth values from unsuccessive years etc. In general we follow the statistics of RICKER (l.c.). However, his total production computations are completed by estimation of production of survivors only, equal to HOLČÍK's (1970a) 'final net production'. As these give estimates of that part of production which is untouched by natural and actual fishing mortality, we propose to designate it as available production, hence, 'net production is the better estimate of what is available to other trophic levels' (MANN, 1969). The harvestable part of this available production gives the amount by which the actual catch can be extended (or depreciated).

The following relative indices were computed to obtain the mean single species size, life intervals and growth intensity rates (BALON, 1964c, 1968a):

---

[2] Please note that all maxims in this study, if not stated otherwise, are excerpts from HANDLER's (Editor), 1970 'Biology and the Future of Man', Oxford University Press, New York, xxiv + 936 pp.

Absolute increments
$$h_i = l_i - l_{i-1} \qquad (1),$$
and the index of the species average size
$$\phi H = \sum_{i=1}^{i_j+a} h_i \qquad (2).$$
Further, from the specific weight rate of growth
$$Cw = \frac{\bar{w}_i - \bar{w}_{i-1}}{\bar{w}_{i-1}} 100 \qquad (3),$$
the index of stock weight growth intensity is
$$\phi Cw = \sum_{i=1}^{i_j+a} Cw_i \qquad (4).$$
Where i = single age groups, l = standard lengths, j = juvenile period, a = adult period, $\bar{w}$ = interpolated mean weights.

Although in temperate regions the index of species average size was sufficient to arrange the fish in order of decreasing commercial importance (BALON, 1967a) it failed in the case of Lake Kariba fish because of different local preferences and greater willingness to utilize the smallest species, if sufficiently abundant. Thus I decided to introduce a new index of the species commercial importance
$$\phi E = \frac{\phi H}{N'/A} \qquad (5),$$
which also incorporates the ratio of species density (N') to the total production (A).

The age and growth sections in this volume use the values of density found in single samples. The back calculated growth values, however, give data to the end or beginning of an annual growth season, reflected by annulus formation. HOLČIK (see section 3.1) thus finds that the density values have to be adjusted considering the time (number of days) which elapsed from the time of annulus formation to the time of sampling (t) and the rate of mortality (Z). It can be computed as
$$N' = N(1 + \frac{Zt}{345}) \qquad (6),$$
where N = abundance at the time of the sample. The densities found in single samples, however, are heavily biased in two principal ways: (i) by a size-selective collection in the sampled area, e.g. the juveniles in order of decreasing size are less completely sampled, while the predation of birds, piscivorous fish, etc., have the same effect on the bias (LOUBENS, 1969); (ii) by immigration or emigration of different age groups into the different habitats sampled, as a result of preference for sheltered habitat, spawning or feeding grounds. These biases have probably influenced the quantitative samples more strongly than any variability in year class strength. We did not find any evidence of the latter. In order to eliminate this bias to some degree and to obtain some kind of mean density values I interpolated the data from the single logarithmic stock composition diagram (BALON, 1963b) in all cases evaluated here (Table 4, Fig. 2). This is well demonstrated in the stocks of *A. lateralis* (Fig. 3, 4) where the very small

Table 4. Number of specimens per 1 ha (N) and initial biomass for *Hydrocynus vittatus* stock from cove of Chikanka Island (locality C), using data of BALON (1971a) and adjusting the number of fish at time of annulus formation ($\bar{Z} = 0.37$, $t = 200$).

Age groups	Number of specimens per 1 ha at time of			Standard length in cm of 1 specimen at the time of		Weights in g of 1 specimen at the time of		Biomass of age groups in kg at the time of	
	sampling	annulus formation		sampling	annulus formation	sampling	annulus formation	sampling	annulus formation
		real	interpolated						
0	183	(188)	(50)	10.0		17		3.11	
I	28	29	28	24.5	18	298	160	8.34	4.48
II	20	21	16	28.3	28	471	420	9.42	6.72
III	10	10	9	35.5	35	950	800	9.50	7.20
IV	1	1	5	34.3	39	830	1150	0.83	5.75
V			3	48.9	48	2720	2110	—	6.33
VI	2	2	2	51.2	49	3212	2300	6.42	6.40
Total	244	63	63					37.62	36.88

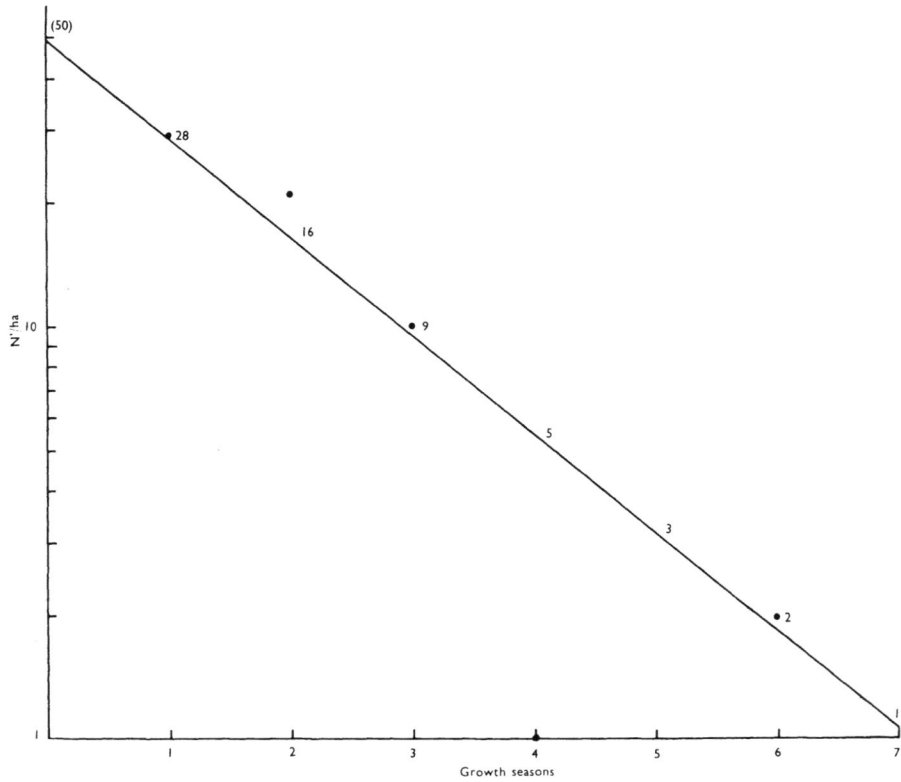

Fig. 2. Single logarithmic stock composition diagram for *Hydrocynus vittatus* from Chikanka Island cove (C). Real values are represented as dots, interpolated values as numbers.

juveniles were less completely sampled and in the stocks of *T. rendalli* (Fig. 5, 6, 7). It is of no importance whether the graphical interpolation of stock density is done before or after the correction to the density at the time of annulus formation (see Tables 63 and 64).

Most probably it is also due to the migratory pattern of the same age groups towards certain habitats that the density fluctuation in different localities are so large. The absence of some age groups in certain localities forced us to omit these areas from further calculations and explain the necessity for estimations from the data of one locality only.

In the back-calculated growth of most fish studied a 'paradoxical phenomenon of R. Lee' was noted, i.e. the computed length from older fish for the early ages were bigger than those computed from younger fish. Probably this was due to the faster growth of all fish in the early years of Lake Kariba because of the high nutrient contents (BALINSKY & JAMES, 1960; HARDING, 1964; MCLACHLAN, 1971). The paradoxical phenomenon of fast growth values was noted mainly where the back calculations enter into the initial years of the lake. This may complicate the usual pattern of the size-

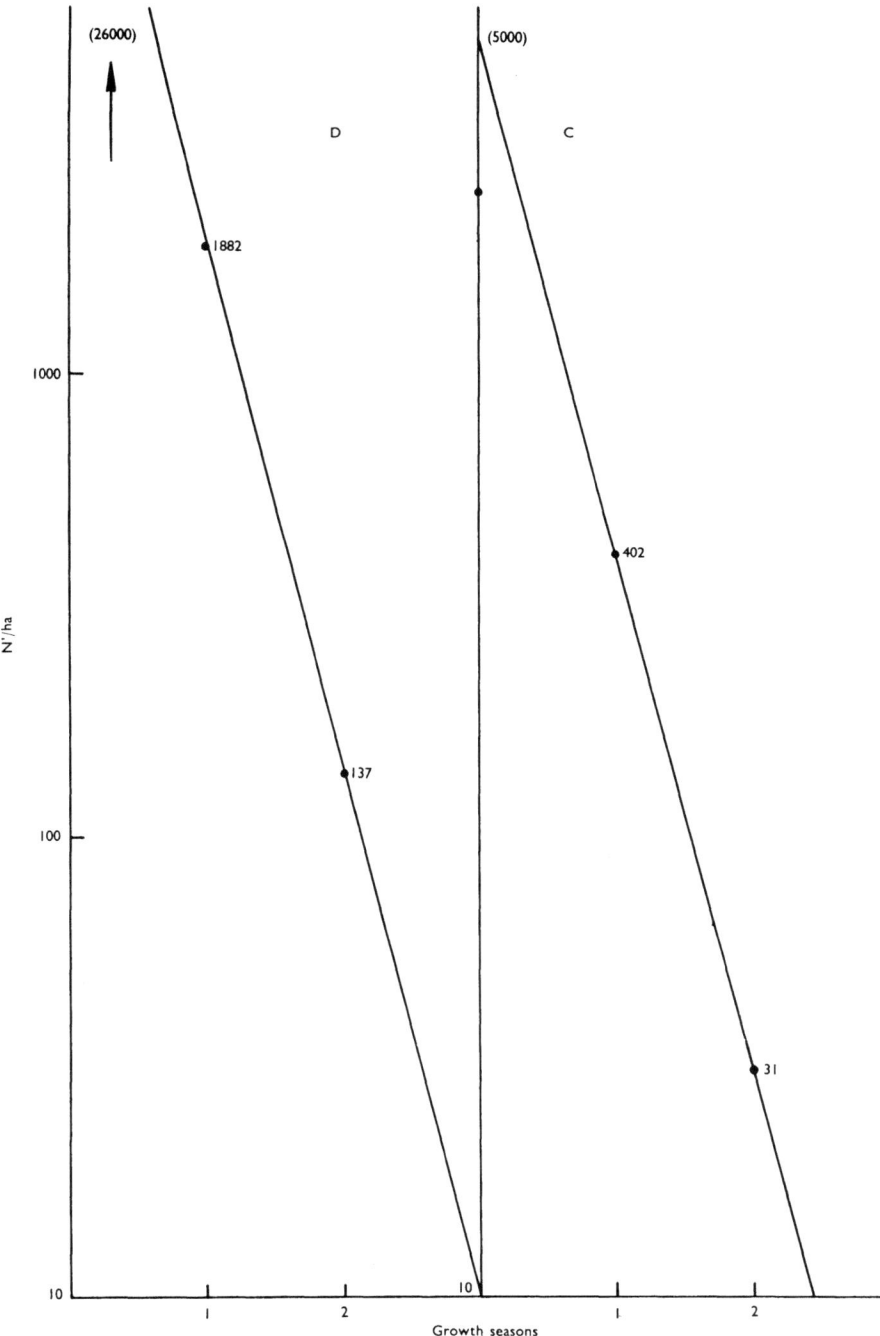

Fig. 3. Single logarithmic stock composition diagrams for *Alestes lateralis* from Buffalo Island cove (D) and Chikanka Island cove (C).

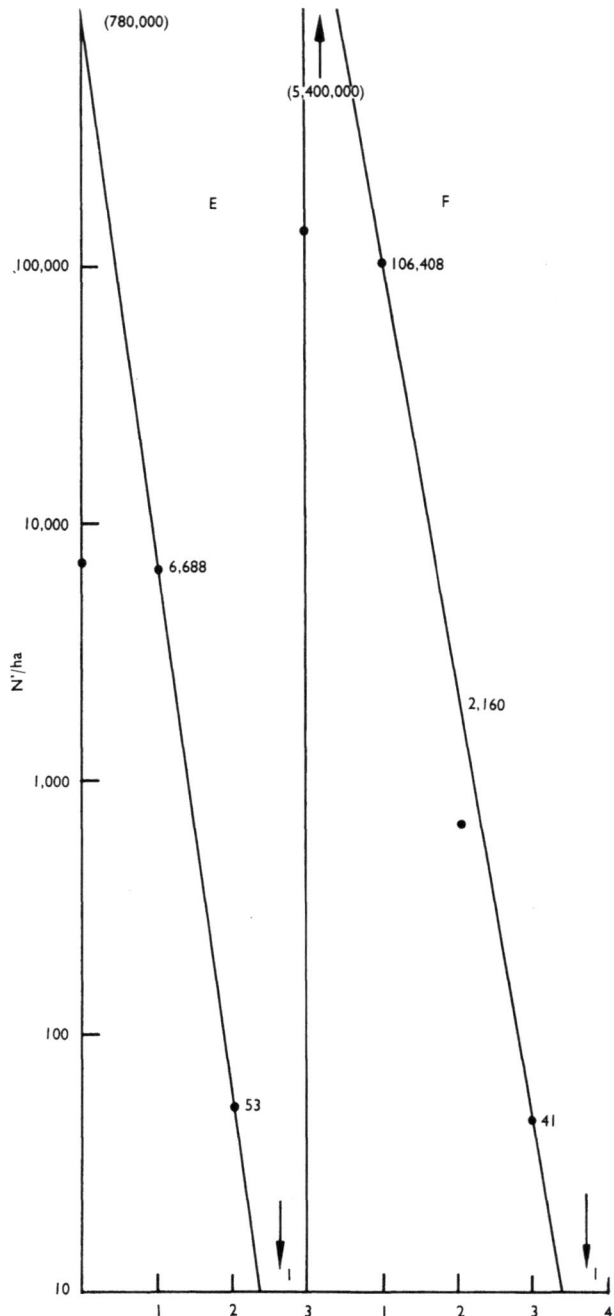

Fig. 4. Single logarithmic stock composition diagrams for *Alestes lateralis* from Chete Island cove (E) and Chipepo Bay cove (F).

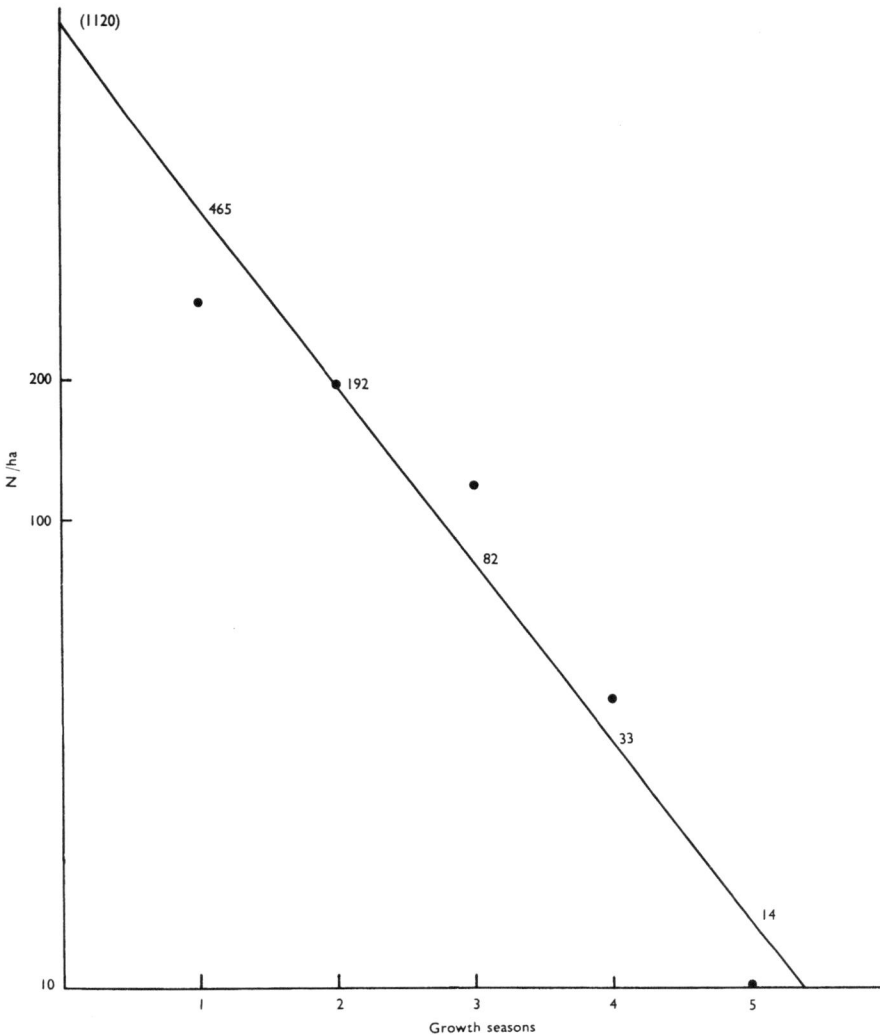

Fig. 5. Single logarithmic stock composition diagram of *Tilapia rendalli* from Chikanka Island cove for the interpolation of mean density.

selective mortality. Thus in most cases we were unable to use the rates of the same growth season (X-type rates). Where, however, the rates for the same growth season were applied they produced lower values as compared to the rates for various growth seasons (Table 5). They were also the same as RICKER's (1969) data. The following of his equations were used:

For the computation of the average instantaneous growth rate (for various growth seasons)

$$G = \log_e \bar{w}_i - \log_e \bar{w}_{i-1} \tag{7}.$$

271

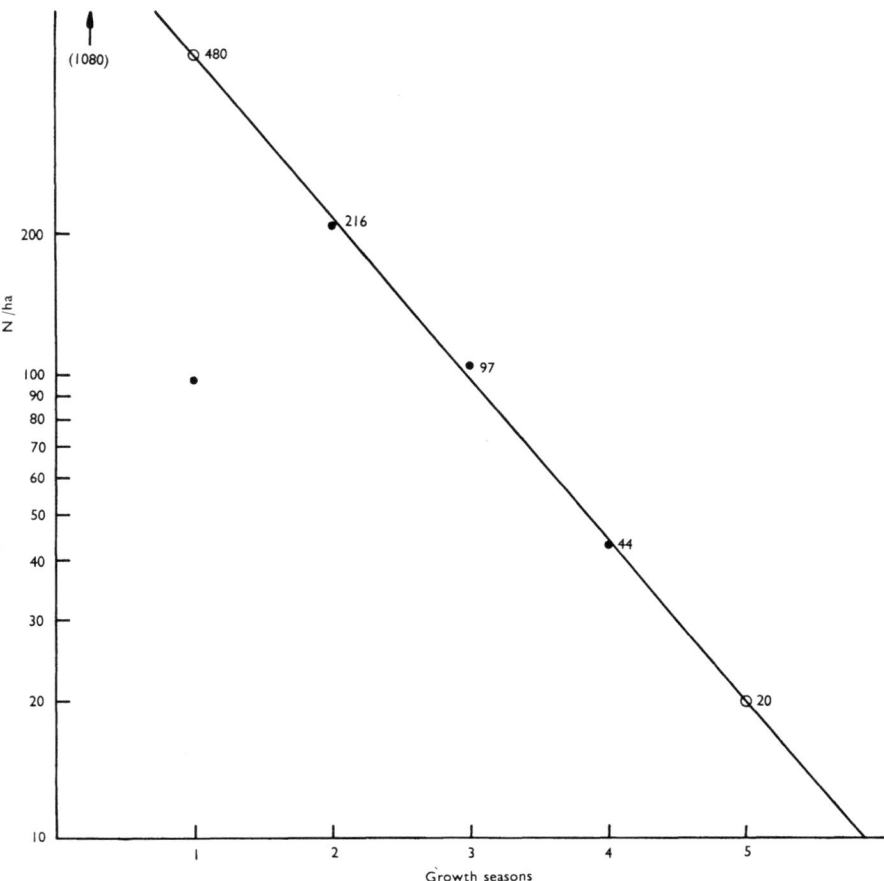

Fig. 6. Stock composition diagram of *Tilapia rendalli* from Chete Island cove.

From this the instantaneous growth rate of the same growth season can be expressed as

$$G_x = \log_e (\bar{w}_x)_i - \log_e (\bar{w}_x)_{i-1} \tag{8},$$

where $\bar{w}$ = interpolated mean weight of one specimen for various growth seasons, and $\bar{w}_x$ = interpolated mean weight of one specimen of the same growth season. The value for $\bar{w}_0$ is in most cases unknown and must be obtained by extrapolation. If we assume that the mean weight over the interval $0 - 1$ is $1/2$ of the final weight, where $\bar{w}_1 = \bar{w}_0 \, e^G$ and

$$\bar{w} = \frac{\bar{w}_0}{G} (e^G - 1) \text{ or } \frac{\bar{w}_1 \, e^{-G} (e^G - 1)}{G} \ldots = G = 2 - 2e^{-G},$$

then G solved by trial and error will be 1.59 for all species. Although using this value would be more consistent I decided to use $\bar{w}_0 = 1$ as concrete proof is lacking for the

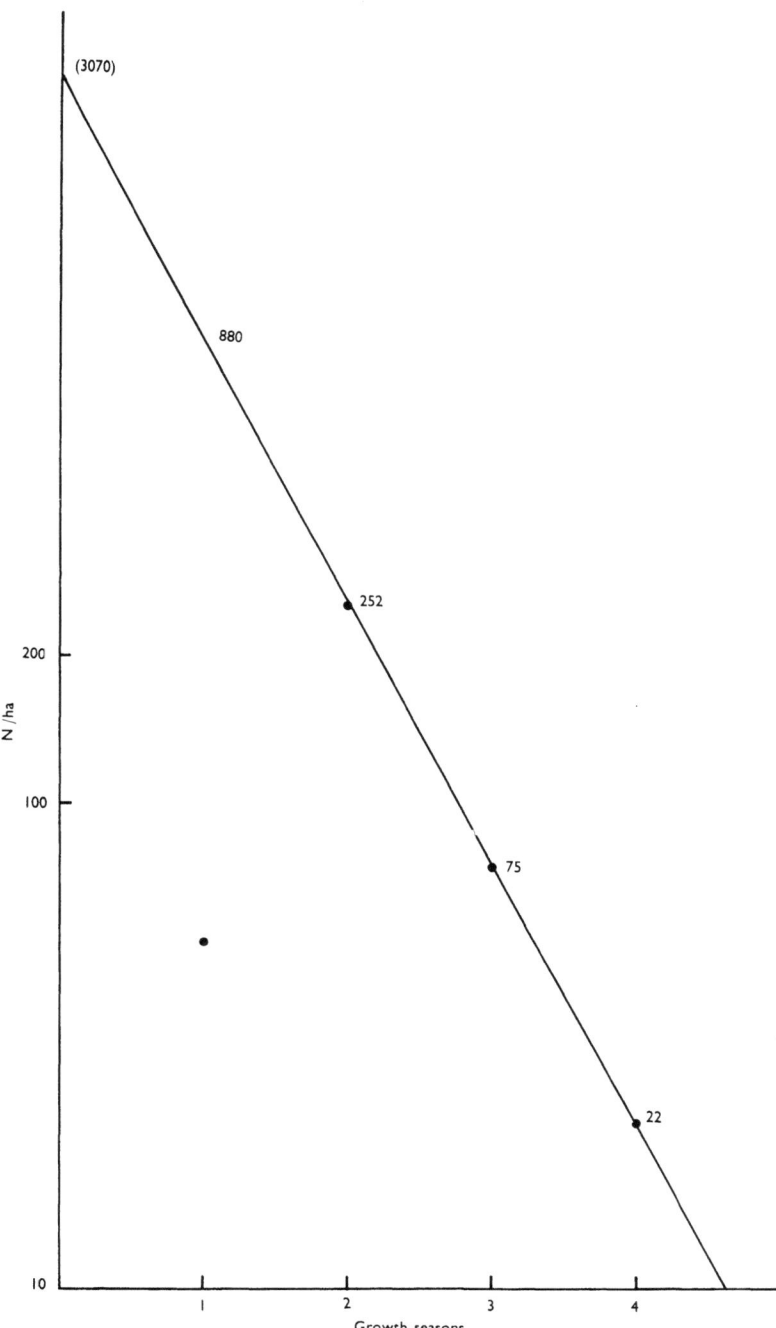

Fig. 7. Stock composition diagram of *Tilapia rendalli* from Chipepo Bay cove (F) (the relative density N/ha given before correction to the annulus formation).

Table 5. Computation of production for *Synodontis zambezensis* stock from Lake Kariba (after data from locality F – Chipepo Bay cove, 1969).

Growth seasons		0	1	2	3	4	5	6	7	8	Total
Interpolated $\bar{N}$/ha at time of annulus formation	N'	(1212)	593	301	152	77	40	20	10	5	1198
Weights in g of 1 specimen $\begin{cases} \text{average} \\ \text{same growth season} \end{cases}$	$\bar{w}$		10	24	47	80	140	210	300	440	
	$\bar{w}_x$		8	18	43	68	121	198	300	440	
Initial biomass in kg/ha $\begin{cases} \text{after } \bar{w} \\ \text{after } \bar{w}_x \end{cases}$	B'		5.93	7.22	7.14	6.16	5.60	4.20	3.00	2.20	41.45
	B'_x		4.74	5.42	6.54	5.24	4.84	1.19	3.00	2.20	33.17
Rate of increase in biomass	H	+1.5879	+0.1974	−0.0113	−0.1482	−0.0952	−0.2879	−0.3364	−0.3101		
'X' rate of increase in biomass	H_x	+1.3647	+0.1329	+0.1874	−0.2217	−0.0786	−0.2008	−0.2776	−0.3101		
Instantaneous growth rate (average)	G	2.3026	0.8755	0.6721	0.5318	0.5597	0.4054	0.3567	0.3830		
Instantaneous growth rate (for the same growth season)	G_x	2.0794	0.8110	0.8708	0.4583	0.5763	0.4925	0.4155	0.3830		0.7608
Instantaneous rate of total mortality	Z	0.7147	0.6781	0.6834	0.6800	0.6549	0.6933	0.6931	0.6931		0.6863
Mean biomass	$\bar{B}$	14.58	8.10	6.32	5.79	5.06	3.67	2.57	1.89		47.98
'X' mean biomass	$\bar{B}_x$	10.06	5.66	7.30	4.67	4.73	1.07	2.64	1.89		38.02
Production ($= G\bar{B}$) in kg/ha/yr	A	33.57	7.09	4.25	3.08	2.83	1.49	0.92	0.72		53.95
Production ($= G_x\bar{B}_x$) in kg/ha/yr	A_x	20.92	4.59	6.36	2.14	2.73	0.53	1.10	0.72		39.09
Available production in kg/ha/yr	P'	5.93	4.21	3.50	2.54	2.40	1.40	0.90	0.70		21.58
P' in A ratio (in %)	P'/A	18	59	82	82	85	94	98	97		40
Available yield in kg/ha/yr	$Y_{r'}$				2.54	2.40	1.40	0.90	0.70		7.94

former assumption (see Appendix D38). Consequently some 0 – 1 values of biomass and production are highly inflated, the expected error varies proportionally to the difference from $\bar{w}_1 = 5$ among species (HENDERSON, personal communication).

The instantaneous rate of total mortality is

$$Z = - (\log N'_i - \log N'_{i-1}) \, 2.3026 \tag{9}$$

and the mean daily rate of mortality

$$\dot{Z} = \frac{Z}{365} \cdot 100 \tag{10}.$$

The following expressions will differ according to whether the rates for various growth seasons (average rates) or rates of the same growth season (X-type rates) are used. They are determined by G or G_x and we can limit ourselves here to one expression (see Table 5 for both expressions).

The initial biomass can be written as

$$B' = N_i \, \bar{w}_i \tag{11}.$$

The rate of increase in biomass is

$$H = G - Z \tag{12},$$

the mean biomass

$$\bar{B} = B'(e^H - 1)/H \tag{13}$$

and the total production

$$A = G\bar{B} \tag{14}.$$

We can estimate the latter also by ALLEN's (1951) graphical method, using again a composite stock throughout various growth seasons instead of the same stock followed throughout the same growth season, i.e. the interpolated densities plotted against the interpolated mean individual weights in single age groups of the same sample (Fig. 8). But here also we do not know the weight which corresponds to the density of 0 age group as well as the final age group and we have to work with an assumption. Consequently this method may produce slightly different values (Table 6).

The available production computation can be written in the form

$$P' = N_i' \, (\bar{w}_i - \bar{w}_{i-1}) \tag{15}.$$

To emphasize once more, by the term 'yield' we understand the harvestable part of production, above the minimum harvestable size determined graphically for each species from corresponding growth values. Thus the harvestable part of the total production will be total yield and the harvestable part of the available production the available yield. Yield will therefore be a measure of the amount of fish tissue grown per unit area and unit time above the minimum harvestable size. It indicates clearly a value given forth by a natural process in accordance with the etymological meaning of this word. It would be better to replace the present usage of yield as a designation for the harvested portion of production by such terms as harvest, catch or landing, since production used in the latter meaning is also being replaced.

The theory of intervals in the life history of fish was applied to 'guesstimate' from the specific weight rate of growth (Cw) the limits of juvenile, adult and senective

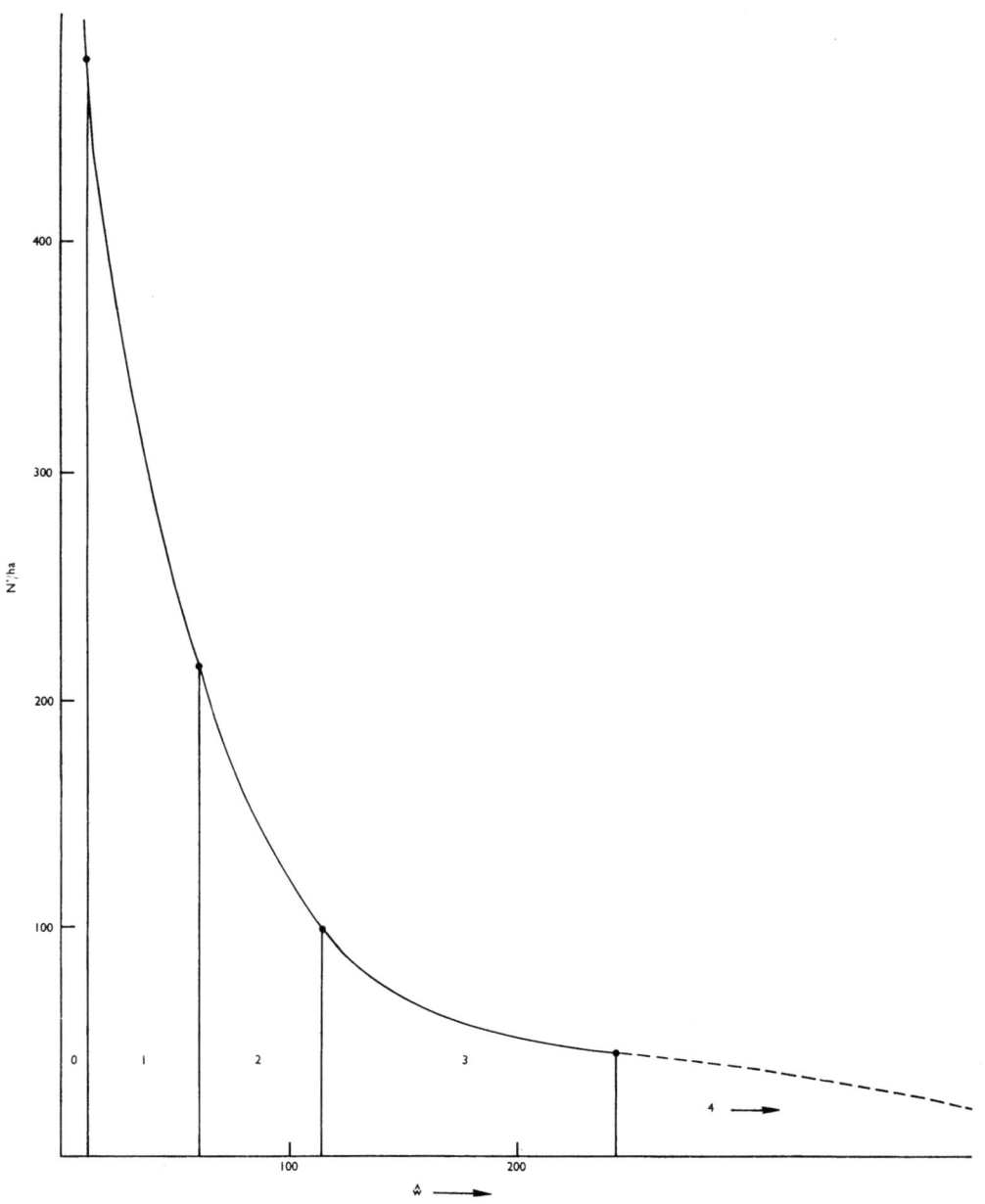

Fig. 8. Allen's curve for graphical production estimation of *Tilapia rendalli* from Chete Island cove.

Table 6. Computation of production for Tilapia rendalli stock from Lake Kariba (after data from locality E – Chete Island cove, 1969).

Growth seasons		0	1	2	3	4	5	Total
Interpolated $\bar{N}$/ha at time of annulus formation	N'	(1080)	480	216	97	44	20	857
Weights in g of 1 specimen	$\bar{w}$		12	60	115	243	(300)	
Initial biomass in kg/ha	B'		5.76	12.96	11.15	10.69	6.00	46.56
Rate of increase in biomass	H		-3.6888	$+0.8112$	-0.1500	-0.0424	-0.5779	
Instantaneous 'X' growth rate	G_x		2.4849	1.6095	0.6506	0.7481	0.2107	1.1408
Instantaneous rate of total mortality	Z		6.1737	0.7983	0.8006	0.7905	0.7886	1.8703
Mean biomass	$\bar{B}$		1.52	19.94	10.35	9.88	4.57	46.26
Production ($= G\bar{B}$) in kg/ha/yr	A		3.78	32.09	6.73	7.39	0.96	50.95
Graphical total production (kg/ha/yr)	$\bar{A}$		7.00	20.40	9.80	9.95	6.70	53.85
Available production in kg/ha/yr	P'		5.76	10.37	5.33	5.63	1.14	28.23
Available yield in kg/ha/yr	$Y_{P'}$		—	—	5.33	5.63	1.14	12.10
Total yield in kg/ha/yr	Y_A		—	—	6.73	7.39	0.96	15.08

periods (BALON, 1958, 1960, 1968a, 1971d), with the emphasis on the determination of the average age of first sexual maturity. Although proved empirically on fish of the temperate regions (BALON, 1967a), no such proof was collected in most studied species from Lake Kariba.

2.1. Glossary and usage of terms

This section was prepared in order to avoid misinterpretations of terms used. It is not a proposal for common use or unification of language, although it may contribute towards it. More specific descriptions and statistical equations of most terms presented here are given in other sections. I deviate from RICKER (1958, 1968), and WINBERG (1971), according to my discretion, only in the most pressing cases. However, some synonyms, as for example, abundance and density, have to be distinguished in order to satisfy more recent needs for more detailed terms.

ABUNDANCE (n or N): Number of individuals (n) per unit area (N) at any time (parallel to standing stock).
ANNULUS: Annual mark on fish scale, vertebra, otolith, etc. incepted at the beginning of new growth season (for details see CHUGUNOVA, 1963).
BIOMASS (B): Wet weight of single species, taxon, stock or population per unit area at a specific time from interpolated values.
  *INITIAL BIOMASS (B'): Interpolated values of mean wet weight ($\bar{w}$) multiplied by density at the beginning of annual growth season (N') per unit area.
  *MEAN BIOMASS ($\bar{B}$): Average biomass present during unit of time per unit area.
  *ICHTHYOMASS (B): Synonym of biomass to emphasize the value of a whole fish population.
CARRYING CAPACITY: Maximum level of nutrient contents which sustains definite population density and production in relation to habitat topography.
CATCH (E): Exploited (harvested) part of standing stock.
  *ACTUAL CATCH (E'): Wet weight of individuals removed from the fish population per unit area and unit time.
  *LANDED CATCH (E_{Landed}): Part of actual catch utilized by man (i.e. brought ashore).
  *SUSTAINED CATCH ($\bar{E}$): Part of yield (see below) which can be removed as a steady resource by fishing within specific socio-economic limits.
  *MAXIMUM SUSTAINED CATCH (E_{max}): Largest part of yield which can be removed by fishing due to optimal socio-economic limits.
DENSITY (N'): Number of individuals in single taxon, stock or population per unit area from interpolated values at a specified time (parallel to biomass).
ECOLOGICAL PRODUCTION: Measure of the total amount of tissue increment per unit area and unit time, synonym for TOTAL PRODUCTION in general or common expression.
  *AVAILABLE PRODUCTION (P'): Tissue elaboration of fish surviving at the end of unit time.
  *FINAL PRODUCTION ($\bar{P}$): Available production plus actual catch (P' + E'); in unexploited (virgin) populations equal to AVAILABLE PRODUCTION.
  *TOTAL PRODUCTION (A): Total elaboration of fish tissue within unit time including that which is formed by individuals that die and are removed during this unit time.
HARVEST: Synonym for CATCH.
MINIMUM HARVESTABLE SIZE: Length, weight and age of species determined biologically (first spawning, growth intensity) as the best lower limit for sustained yield.
MORTALITY (Z): Coefficient of total number or weight of individuals which do not survive

to the end of specific time unit (same as TOTAL MORTALITY or ELIMINATION: part of production removed within unit time by predators and decomposers).

*FISHING MORTALITY (F): Coefficient of mortality caused by actual catch.

*NATURAL MORTALITY (M): Coefficient of mortality due to natural causes (predation other than man, emigration, natural death).

PERIODS OF LIFE (j, a, s): Major intervals in life history limited by reproductive activity, growth intensity, and frequently, morphological changes, etc. (juvenile, adult, senective).

POPULATION: Qualitative designation of a group of individuals of single species, taxon or all fish in a large topographically interconnected ecological unit (area, drainage, lake, river).

SPECIES: Group of similar individuals each of which is potentially or actively able to reproduce, by pairing with any other individual of the opposite sex; determined taxonomically.

STOCK: Topographically unisolated part of population, sometimes isolated spatially.

STANDING CROP: A major portion of standing stock within determined limits of sampling method used.

STANDING STOCK (B_t): Empirical value of wet weight of stock per unit area at any time (e.g. of sample).

TAXOCENE: Group of related species in a specific habitat; or, part of stock associated with a particular habitat.

TAXON: Group of similar individuals of unknown reproductive quality and not sufficiently determined taxonomically (e.g. in process of adaptation).

TURNOVER RATIO ($A/\bar{B}$): Total production per unit biomass (estimated per annum for all age groups, or per individual age groups).

YIELD (Y): Harvestable part of production elaborated by natural process in general.

*AVAILABLE YIELD ($Y_{P'}$): Harvestable part of available production.

*MAXIMUM SUSTAINED YIELD (Y_{max}): Largest portion of yield which can be removed without deleterious effects on production in biological terms.

*SUSTAINED YIELD ($\bar{Y}$): Harvestable part of final production, ACTUAL CATCH plus AVAILABLE YIELD.

*TOTAL YIELD (Y_A): Harvestable part of total production.

3. AGE AND GROWTH STUDIES

3.1. The green bream *Sargochromis codringtoni* (Boulenger, 1908)

by JURAJ HOLČIK

The green bream belongs to the relatively less important group of fish species inhabiting the watershed of the Upper and Middle Zambezi River as well as the Kafue and Okavango Rivers. With the exception of some general data (e.g. JACKSON, 1961a; JUBB, 1967) little is known about the biology of this fish, its age, growth, abundance, standing stock and production. It is found in the deeper and more sheltered waters of large rivers, particularly when older; also its abundance and standing stock in Lake Kariba seems to be larger in greater depths. Our results show the abundance and standing stock of the green bream in Lake Kariba to be 117 individuals and 15.96 kg per ha respectively. The purpose of this section is to determine the age composition, growth and available production of this fish.

We utilized two sources of information: (1) the samples of scales of *S. codringtoni* obtained from a part of the fish collection from the cove samples in Lake Kariba and (2) the basic data on the abundance, standing stock and length frequencies of poisoned stocks from the mentioned coves. The samples of scales and data originate from the following localities: C – Chikanka Island N 16, 31 specimens, D – Island N 20 southwest of Sinazongwe, 45 specimens, E – Island N 21 or Chete, 78 specimens, F – Chipepo Bay Cove I, 14 specimens, G – Chipepo Bay Cove II, 45 specimens.

The scales of the green bream as well as of the other cichlids were taken from the first row of scales above the lateral line where the curve of the lateral line reaches the peak. Dry scales were interpreted in the laboratory by projector with a 17.5 magnification. The scales of some bigger fish had to be cleaned by soaking in a solution of about 50% HCl in order to remove the remains of skin. On the scales we measured the distance between focus and the edge of the longer (i.e. ventral) lobe, as well as the distance between the focus and annuli. Values obtained were used to contruct the body-scale relationship which was linear with a correction of 21.55 mm (Fig. 9).

Concerning designation of age groups I prefer to use the symbols 1+, 2+, 3+ etc. to avoid errors and misunderstanding. The Roman symbols I, II, III etc. used in temperate regions of the Northern Hemisphere are related to the current year and comprise fish with a different number of annuli in accordance with the time of annulus formation. In the case of the tropics, mainly in the Southern Hemisphere, this type of designation is less suitable. For instance the fish which would hatch at the end of December of the current year belong to age 0, but a few days later, after January 1st of next year, it should be placed into age group I.

The primary data were processed into data on age composition and growth rate following routine methods (see BALON, 1971a, 1972). The back calculations of the growth rate were performed on Lea's board by means of the Rosa Lee method, using

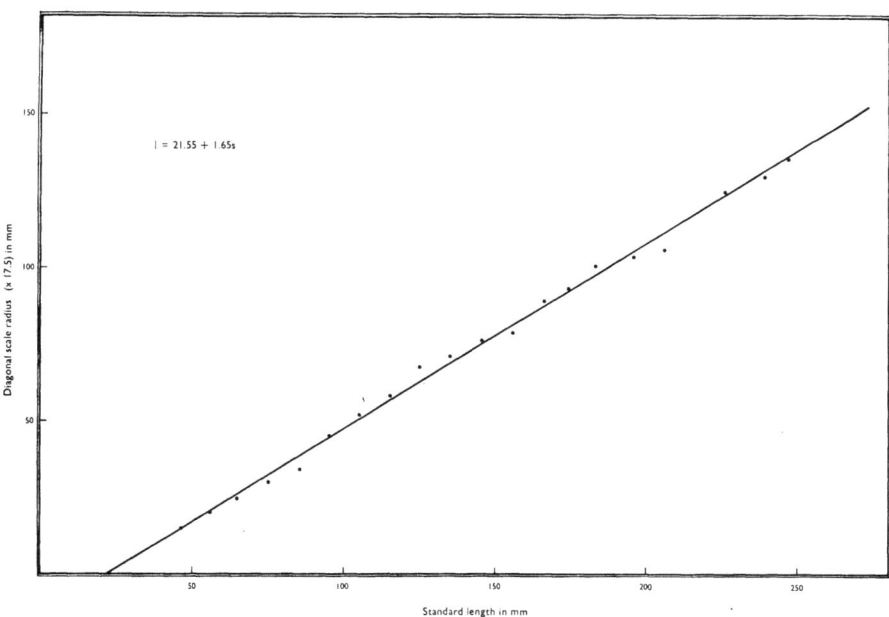

Fig. 9. Body-scale relationship in *Sargochromis codringtoni*.

22 mm as the correction value. The average length of fish was computed by the method suggested by VAN OOSTEN (1953) which saved a lot of work (the scale measurements for each year of life were averaged as well as the length of fish used and the average lengths were computed from these data). The sum total of back calculated lengths for each stock is represented by the simple arithmetical mean as suggested by PEŇÁZ (1968).

The production calculation was essentially the same as that used earlier by HOLČIK (1970a, b,) and BALON (1971a, 1972) but modified in accordance with PIVNIČKA's (1971) suggestions, in that we took into account the data of weight increments of different age groups in each locality and not the total average data of weight growth of the population. Considering this, the abundance and standing stock of age groups in samples had also to be adjusted back up to the beginning of the growth season (= time of annulus formation). This was assumed to be on November 1st because (1) the fish taken in December showed not only the new annulus but also the new growth; (2) the beginning of the warm season in the Lake Kariba area is October (COCHE, 1968). Calculations were made by taking into consideration the mean instantaneous mortality rate and the number of days that elapsed from the date of supposed time of annulus formation till the end of actual growth (= day when the fish stock was poisoned and removed).

Further adjustments were necessary because (1) of the difference between the actual standing stock found after poisoning and the standing stock obtained from length-weight relationship; (2) the actual standing stock also contained young-of-the-year fish which should be excluded from production calculations because these fish are not present in the population at the beginning of the growth season; and (3) there is a difference between the standing stock at the time of poisoning and biomass at the beginning of the growth season (though in reality this difference does not manifest itself – see further).

All differences found were expressed in percentages (weight adjustment 90.85%, adjustment for the lack of 0+ age group 92.71%, adjustment for the initial 99.12%) so the actual average standing stock of green bream (15.96 kg/ha) decreased to 13.32 kg/ha which represents the initial biomass present at the time of annulus formation.

Values of available production obtained were further calculated as the percent of available production to initial biomass (P'/B') for each locality separately (as with the other calculations mentioned above). The average sum P'/B' was used to calculate the absolute available production of Lake Kariba taking into consideration the surface area inhabited by fish (i.e. 342 980 ha = 65% of the total water surface of Lake Kariba – see BALON, 1971a, 1972) as well as the relative available production per 1 ha. The data on abundance and biomass were obtained by totaling and averaging the abundance/biomass data found in localities C, D, E, F and G. The computation of the available yield was based on the amount harvestable, as established from the intercept of the average length increments for the whole population in percentages of length of the first growth season and the length in percentages of the final growth season curve (Fig. 10). The intercepted values show the second age group with an average weight of 90 grams and an average length of 130 mm. It is in accordance with the occurrence of presumable spawning marks in fish from Chipepo Bay. Thus it was stated that the fish of the third age group inclusive, can be removed by fishing. Further calculations are to be made in three ways. First the age composition of all the stocks is considered to be based on the sum of all fish in the samples and on the sum of available production. The sum of production of the 3rd and 4th age group is expressed in percentages to the whole sum of available production. In this case the percentage calculated was 40.73 and the estimated available yield 102.9 metric tons or 3.0 kg/ha. The two additional ways are essentially the same but in the second total annual mortality rate is considered (69%). The hypothetical population is constructed on this basis. The percentage of harvestable yield calculated in this way is 33.33 and the whole available yield in Lake Kariba is 844 metric tons or 2.46 kg/ha. In the third way the actual age composition in particular localities is considered and the percentage of harvestable production calculated separately for each locality is averaged. In this case it was possible only in three localities out of five due to the lack of some size groups in the scale sample and the impossibility of establishing the age composition and growth of the given fish stock. I am taking into consideration the result of the third method because by the first the age composition of the sample investigated need not be in accordance with the actual situation in the population and the total mortality rate calculated does not seem to be correct. Besides, the available production (as well as all adjustments mentioned above) was determined only in three suitable localities so the calculation of available yield suffers from the same flaws.

Scales of the *S. codringtoni* resemble those of other members of the family Cichlidae. They are oval with a well developed convex caudal part. The oral portion of the scale is moderate or straight and its edge is serrated. On the oral part of the scale there are

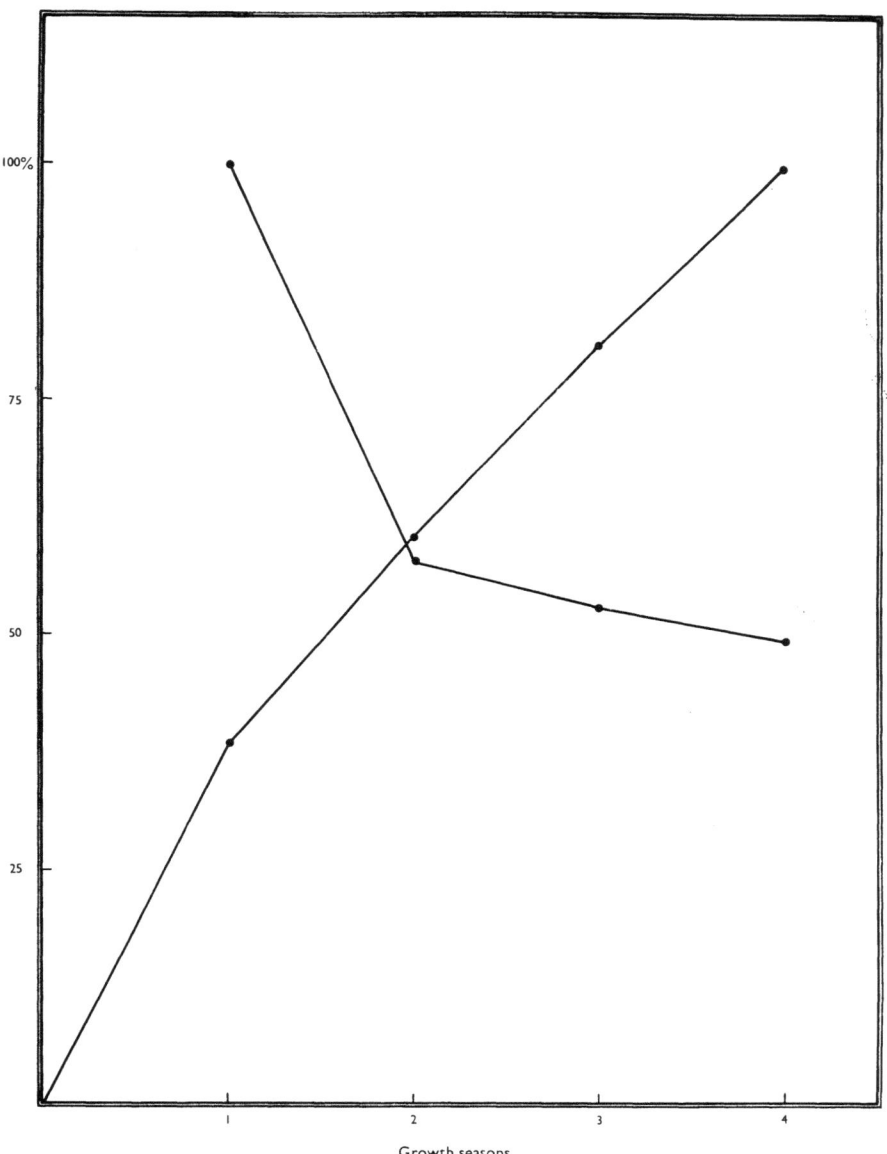

Fig. 10. Curves for the minimum harvestable size interception of the green bream in Lake Kariba. Upper curve – length increments in percent of the first growth season; lower curve – standard length in percent of the length of last growth season.

Fig. 11. Scale of 3+ old green bream. Locality G, standard length 182 mm. Three annuli from which the second and third are probably also the spawning marks.

Fig. 12. Scale of 1+ old green bream. Locality E, standard length 99 mm. Beside the normal annulus also the juvenile mark is visible.

radial canals numbering 15–20 which end on the scale edge between individual serrae. The placement of sclerites is regular and circular. Annulus is visible mainly on the diagonal and lateral parts, less on the caudal, and very rarely on the oral part. Characteristics of annulus are essentially the same as on *Hydrocynus vittatus* (BALON, 1971a) and its determination is possible as the new sclerites cut through old ones. This cutting begins at the diagonal part and is most visible in the lateral. It disappears on the caudal part of the scales. Here the first sclerite of the new increment runs obliquely to the regular sclerites of the previous growth season. In some samples the annulus beginning from the second year is very clearly visible also on the oral part. Where it cuts through there appears a relatively wide belt of eroded sclerites. This type of annulus which differs significantly from normal (Fig. 11) ones can be identical with the spawning mark of the cyprinoid and salmonoid fish of the temperate region to which it is strikingly similar (see CHUGUNOVA, 1959). It is remarkable however, that it was observed only in fish caught in locality F and G (Chipepo Bay). In 91 % of all investigated fish it was also possible to determine the juvenile mark characterized by the pronounced density of sclerites on the oral part of the scale (Fig. 12, 13). The length back calculated to this mark was 43 mm, in average ranging from 38 to 47 mm.

Fig. 13. Scale of 2+ old green bream. Locality G, standard length 132 mm. Two annuli and juvenile mark present.

Together with other investigators (e.g. CHUGUNOVA, 1959; HOLČIK, 1960, 1967b; BALON, 1971a), we believe this phenomenon to be due to a change in diet. Generally, the scales of the green bream are quite readable and only 4% of fish had to be set aside mostly because they had regenerated scales.

The time sequence of samples collected does not enable us to determine the time of annulus formation. All fish already had a formed annulus as well as a new scale increment. The green bream sample caught at locality D (sampled on 11–14 December 1968) had the smallest increment in scales. I can only state that the annulus inception at the end of a current year was formed long ago. According to the available hydrometeorological data (COCHE, 1968) I would assume that the annulus formation

Table 7. Age and length composition of *Sargochromis codringtoni* stock in cove of Chikanka Island (locality C).

Length groups	n				$\sum$
	0+	1+	2+	3+	
71– 80	1				1
81– 90	1				1
101–110	2				2
111–120	1				1
121–130		5			5
131–140		9			9
141–150		3	1		4
151–160		1	1		2
161–170		1	3		4
171–180		1	5		6
181–190			1	3	4
191–200		1	1		2
201–210			2		2
211–220				1	1
221–230				1	1
$\sum$	5	21	14	5	45

Table 8. Age and length composition of *Sargochromis codringtoni* stock in north cove of Island 21 (Chete) south from Sinanzongwe (locality E).

Length groups	n				$\sum$
	0+	1+	2+	3+	
31– 40	1				1
41– 50	1				1
51– 60	4				4
61– 70	8				8
71– 80	3	3			6
81– 90		25			25
91–100		25			25
101–110	3	28			31
111–120		25			25
121–130		11			11
131–140		14			14
141–150		14	3		17
151–160			14		14
161–170		3	11		14
171–180			3	6	9
181–190				3	3
191–200				3	3
201–210				8	8
$\sum$	20	148	31	20	219

Table 9. Age and length composition of *Sargochromis codringtoni* stock in cove of east shore of Chipepo Bay (locality F).

Length groups	n					Σ
	0+	1+	2+	3+	4+	
41– 50	1					1
51– 60	1					1
61– 70	1					1
71– 80	5					5
81– 90	9					9
91–100		5				5
101–110		3				3
111–120		2				2
121–130		3				3
131–140			1	1		2
141–150				1		1
161–170				1		1
171–180				1		1
191–200				1	2	3
231–240					1	1
241–250					1	1
251–260					1	1
Σ	17	13	1	5	5	41

Table 10. Age composition of *Sargochromis codringtoni* population from Lake Kariba (in % of total catch, localities D and G in % of sample).

Locality – Age	0+	1+	2+	3+	4+
Chikanka Island (C)	11.1	46.7	31.1	11.1	
Island 20 (D)	12.2	73.2	14.6		
Island 21 (E)	9.1	67.6	14.1	9.1	
Chipepo Bay (F)	41.5	31.7	2.4	12.2	12.2
Chipepo Bay (G)	8.8	58.8	11.8	11.8	8.8
Total (in % of sample)	13.7	59.5	15.1	9.8	1.3
	(10.8)	(61.9)	(16.0)	(9.3)	(2.1)

in this species is probably connected with the beginning of the warm season (October) when the mean air temperature ranges from 22 to 29°, and not with the rainy season (December-April). It seems, however, that the rainy season accelerates the growth rate.

The analysis of age showed that the green bream can be classified as a short life-span fish. It attains at the most 3 or 4 years. In some localities only 2 and 3 year old fish

Table 11. Age composition, seasonal total mortality rate (Z'), survival rate (S) and instantaneous rate of total mortality (Z) of *Sargochromis codringtoni* under study.

Locality	Age groups	1	2	3	4	$\sum$
Chikanka Island (C) (Total population)	n	21	14	5	—	40
	%	52.5	35.0	12.5		100.0
	Z'		0.3364	0.6430		0.4566
	S		0.6636	0.3570		0.5434
	Z		0.41	1.03		0.61
Island 20 (D) (Aged part)	n	30	6			37
	%	83.3	16.7			100.0
	Z'		0.8001			0.8001
	S		0.1999			0.1999
	Z		1.61			1.61
Island 21 (E) (Total population)	n	148	31	20		199
	%	74.4	15.6	10.0		100.0
	Z'		0.7899	0.3560		0.7163
	S		0.2101	0.6440		0.2836
	Z		1.56	0.44		1.26
Chipepo Bay (F) (Total population)	n	13	1	5	5	24
	%	54.2	4.2	20.8	20.8	100.0
	Z'		0.9227		0.000	0.4230
	S		0.0773		1.000	0.5770
	Z		2.56			0.55
Chipepo Bay (G) (Aged part)	n	20	4	4	3	31
	%	64.5	12.9	12.9	9.7	100.0
	Z'		0.8001	0.000	0.2517	0.6054
	S		0.1999	1.000	0.7384	0.3946
	Z		1.61		0.29	0.93

Table 12. Mean seasonal total mortality rate ($\bar{Z}'$), survival rate ($\bar{S}$) and mean day rate of mortality ($\bar{Z}$) of *Sargochromis codringtoni* under study.

Chikanka Island (C) (Total population)	$\bar{Z}'$	46	%
	$\bar{S}$	54	%
	$\bar{Z}$	0.17%	
Island 20 (D) (Aged part of population)	$\bar{Z}'$	80	%
	$\bar{S}$	20	%
	$\bar{Z}$	0.44%	
Island 21 (E) (Total population)	$\bar{Z}'$	72	%
	$\bar{S}$	28	%
	$\bar{Z}$	0.35%	
Chipepo Bay (F) (Total population)	$\bar{Z}'$	42	%
	$\bar{S}$	58	%
	$\bar{Z}$	0.15%	
Chipepo Bay (G) (Aged part of population)	$\bar{Z}'$	61	%
	$\bar{S}$	39	%
	$\bar{Z}$	0.25%	

Table 13. Back calculated growth values of *Sargochromis codringtoni* from Lake Kariba.

Locality	Age groups	Year of hatching	N	Mean standard length at time of capture	l_j	Back calculated standard lengths (mm) for the previous growth seasons			
						l_1	l_2	l_3	l_4
C (Chikanka Island, 21.6.1968)	0+	1967/1968	4	93	48				
	1+	1966/1967	14	143	44	111	147		
	2+	1965/1966	9	171		86	146	181	
	3+	1964/1965	3	197		87			
	Total		30		46	95	147	181	
D (Island 20, 11.–14.12.1968)	0+	1967/1968	5	63	34				
	1+	1966/1967	30	107	42	88			
	2+	1965/1966	6	112	31	74	100		
	Total		41		38	81	100		
E (Island 21, 21.–25.3.1968)	0+	1968/1969	6	68	38				
	1+	1967/1968	53	111	43	86			
	2+	1966/1967	11	161	44	90	135		
	3+	1965/1966	7	193	40	84	141	172	
	Total		77		41	87	138	172	

Table 13 (continued)

Locality	Age groups	Year of hatching	N	Mean standard length at time of capture	Back calculated standard lengths (mm) for the previous growth seasons				
					l_j	l_1	l_2	l_3	l_4
F (Chipepo Bay, 26.–31.5.1969)	0+	1968/1969	3	67	55				
	1+	1967/1968	3	119		102			
	2+	1966/1967	1	133		78	109		
	3+	1965/1966	4	193	39	75	130	178	
	4+	1964/1965	1	210		101	131	185	203
	Total		13		47	89	123	182	203
G (Chipepo Bay, 28.–31.5.1969)	0+	1968/1969	3	161	39				
	1+	1967/1968	20	96	40	82			
	2+	1966/1967	4	141	40	82	131		
	3+	1965/1966	4	180	45	85	126	166	
	4+	1964/1965	3	244	42	77	137	182	226
	Total		34		41	82	131	174	226
Grand average	1+		120	112	46	94			
	2+		31	151	40	82	124		
	3+		18	191	41	83	136	174	
	4+		4	236	42	89	130	184	216
	Total		174		42	87	130	179	216

Table 14. Lengths, weights and increments for all studied samples of *Sargochromis codringtoni* from Lake Kariba.

Locality	Parameter	Means at the end of growth seasons			
		1	2	3	4
Chikanka Island (C)	Standard length (mm)	95	147	181	
	Length increments (mm)	95	52	34	
	Length increments in % of length of first season	100	54.8	35.8	
	Weights in grams	24	130	250	
	Weights increments in grams	24	106	120	
Island 20 (D)	Standard length (mm)	81	100		
	Length increments (mm)	81	19		
	Length increments in % of length of first season	100	25.9		
	Weights in grams	20	37		
	Weights increments in grams	20	17		
Island 21 (E)	Standard length (mm)	87	138	172	
	Length increments (mm)	87	51	34	
	Length increments in % of length of first season	100	58.6	39.1	
	Weights in grams	26	108	218	
	Weights increments in grams	26	82	110	
Chipepo Bay (F)	Standard length (mm)	89	123	182	203
	Length increments (mm)	89	34	59	21
	Length increments in % of length of first season	100	38.2	66.3	23.6
	Weights in grams	28	78	260	370
	Weights increments in grams	28	50	182	110
Chipepo Bay (G)	Standard length (mm)	82	131	174	226
	Length increments (mm)	82	49	43	52
	Length increments in % of length of first season	100	59.8	52.5	63.5
	Weights in grams	21	96	220	490
	Weights increments in grams	21	75	124	270

were found. It must be noted, however, that occasionally it can reach a higher age. From the sample collected in locality F (Chipepo Bay) the standard length of one specimen was 382 mm and its age was 6+. JUBB (1967) stated the maximum weight for this species to be about 1.13 kg and JACKSON (1961a) writes about fish taken in Lake Kariba which measured 350 mm of the total length and weighed about 1.07 kg. The age composition of the population is different in separate localities because in some of them only 2 or 3 age groups were found.

Surprising is the absolutely small number of the young-of-the-year specimens which suggests that the sampling of poisoned stock could not be performed thoroughly or that for various reasons the young fish disappeared from the blocked off area. The

Table 15. Mean values of lengths, weights and increments of *Sargochromis codringtoni* population from Lake Kariba.

	Means at the end of growth seasons			
	1	2	3	4
Standard length (mm)	87	130	179	216
Increments in mm	87	43	49	37
Lengths increments in % of length of first season	100	49.4	56.0	42.5
Length in % of length of final season	40.3	60.2	82.9	100
Weights in grams	26	90	240	440
Weight increments in grams	24	76	140	240

Table 16. Absolute increments in cm (h), indexes of the species average size (φH), specific weight rates of growth (Cw) and the indexes of the population weight growth intensity (φCw) of *Sargochromis codringtoni* from Lake Kariba.

Locality	index	1	2	3	4
Chikanka Island (C)	h	9.5	5.2	3.4	
	φH			6.03	
	Cw	'2400'	441.66		92.31
	φCw			266.99	
Island 20 (D)	h	8.1	1.9		
	φH		5.0		
	Cw	'2000'	85		
	φCw		85		
Island 21 (E)	h	8.7	5.1	3.4	
	φH			5.73	
	Cw	'2600'	315.38		101.85
	φCw			208.62	
Chipepo Bay (F)	h	8.9	12.3	18.2	20.3
	φH			14.93	
	Cw	'2800'	178.57	233.33	42.31
	φCw			151.40	
Chipepo Bay (G)	h	8.2	13.1	17.4	22.6
	φH			15.33	
	Cw	'3100'	357.14	129.16	122.72
	φCw			203.00	
All material	h	8.7	4.3	4.9	3.7
	φH			5.40	
	Cw	'2600'	246.15	166.60	83.33
	φCw			165.36	

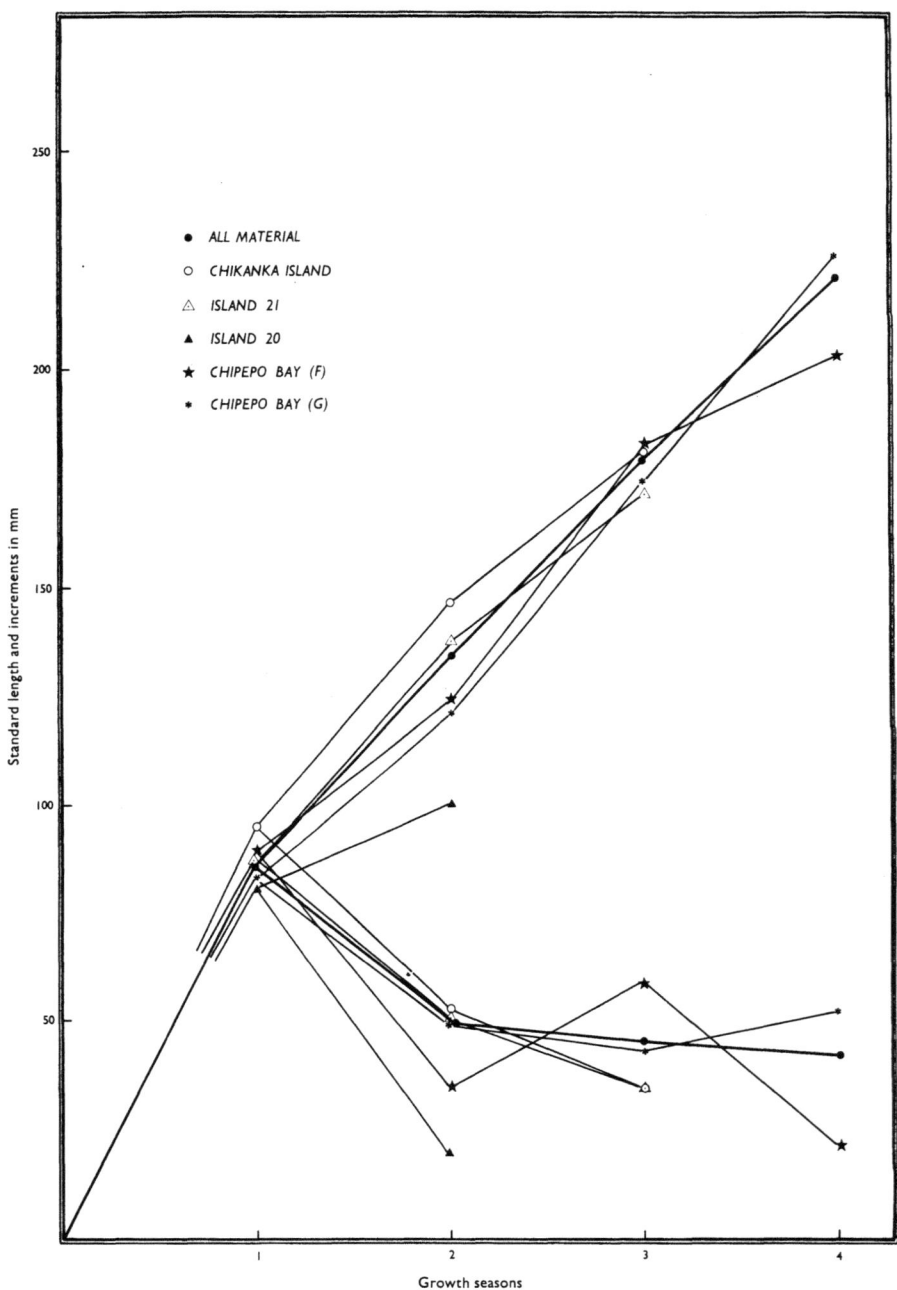

Fig. 14. Linear growth and increments of *Sargochromis codringtoni* in Lake Kariba.

Table 17. Differences between the standing stock of green bream at the time of poisoning and at the beginning of growth season.

Locality	Mean biomass in kg/ha	
	at the time of poisoning	at the beginning of growth season
Chikanka Island (C)	5.94	5.20
Island 21 (E)	3.83	3.67
Chipepo Bay (F)	25.01	24.79
Mean	11.32	11.22
Difference in percent	99.12	

Table 18. Biomass of *Sargochromis codringtoni* stock in cove of east shore of Chipepo Bay (locality F) in relationship to number of specimens per 1 ha.

Age group	Number of specimens per 1 ha	Standard length in cm of 1 specimen	Weight in g of 1 specimen	Weight of age group in kg
0+	89	67	28	2.49
1+	68	119	69	4.69
2+	5	133	104	0.52
3+	26	193	320	8.32
4+	26	210	410	10.66
	214			26.68

values of age composition and three stock samples are summarized in Tables 7–10, and values of mortality and survival in Tables 11, 12. From the last table it follows that the mortality (and survival) rate in each locality is different. Except for mortality due to natural causes the possibility of errors has to be considered as arising from inadequate sampling. This may account for the lack of fish in age group 0+. The total mean annual mortality rate seems to be very low for such a short life-span fish as the green bream. In the populations of *Cichlasoma tetracanthus* and *Lepomis macrochirus* from Cuba (HOLČIK, 1970b) we found much higher values. They range from 76 to 95% in the first species and 82 to 99% in the second. These two species also reach the age of only 5 and 3 years, respectively.

The data which characterize the growth rate both in length and weight are summarized in Tables 13–16 and in Fig. 14. As can be seen, the growth rate in each locality is somewhat different from the others, suggesting that identical conditions do not obtain here. Whether this is due to a different food composition or to differences in abundance and biomass will have to be determined in future studies. It seems, however, that the population density in this case plays only a negligible role because of a

Table 19. Biomass of *Sargochromis codringtoni* stock in cove of Chikanka Island (locality C) in relationship to number of specimens per 1 ha.

Age group	Number of specimens per 1 ha	Standard length in cm of 1 specimen	Weight in g of 1 specimen	Weight of age group in kg
0+	4	93	32	0.13
1+	17	143	127	2.16
2+	12	171	210	2.52
3+	4	197	320	1.28
	37			6.07

Table 20. Biomass of *Sargochromis condringtoni* stock in north cove of Island 21 (Chete) (locality E) in relationship to number of specimens per 1 ha.

Age group	Number of specimen per 1 ha	Standard length in cm of 1 specimen	Weight in g of 1 specimen	Weight of age group in kg
0+	4	68	12	0.05
1+	28	111	56	1.57
2+	6	161	170	1.02
3+	4	193	310	1.24
	42			3.88

higher growth rate in localities with higher values of abundance and standing stock (coefficient ϕH for stock of green bream from Chipepo Bay is higher through the standing crop of fish here reaches 555–1225 kg/ha in contrast to other localities where the standing crop of fish was found to be 276 to 482 kg/ha, but the species average size index is less).

The range of fish lengths attained in different localities is considerable. For instance, in locality E (Chete) the 0+ age group at the time of capture shows a range from 31 to 110 mm of standard length, the 1+ group ranges from 71 to 180 mm, the 2+ group from 141 to 180 and the 4+ group from 171 to 210 mm (Table 8). It points to the fact that the spawning period of this species is probably very extended, as is the case with other tropical species (HOLČIK, 1970b). The juvenile mark appeared in all age groups under study.

The single 6 year old specimen was excluded from all computations because of its abnormal growth which is as follows (in mm of standard length): 1–107, 2–282, 3–313, 4–347, 5–371, 6–375.

The similarity of the standing stocks of different cove fish stocks, before and after adjustment to the beginning of the growth season, is very interesting. The same

Table 21. Available production and yield determination in 3 stocks of Sargochromis codringtoni from Kariba Lake. Calculated per 1 ha, number of fish adjusted back to the beginning of growth season ($\bar{Z}$ = 0.17, 0.35, 0.15% and t = 233, 139, 206 days respectively).

Age group	n	Mean weight of 1 specimen in grams (weight in previous year)	Mean weight increment of 1 specimen in grams	Total increment of age group per year in kg	Production in kg	in % of biomass	Harvestable yield in kg	in % of production
Chikanka Island (C)								
1	25	56 (0)	56	1.400				
2	18	128 (25)	103	1.854	3.998	76.83	0.744	18.61
3	6	250 (126)	124	0.744				
Island 21 (E)								
1	46	26 (0)	26	1.196				
2	10	98 (28)	70	0.700	2.582	70.41	0.686	26.57
3	7	213 (115)	98	0.686				
Chipepo Bay (F)								
1	88	43 (0)	43	3.784				
2	7	51 (18)	33	0.231	12.485	50.36	8.470	67.84
3	35	235 (88)	147	5.145				
4	35	355 (260)	95	3.325				

Mean available production in kg/ha (3 localities)   6.36
Available production in percent of biomass (11.22 kg)   56.64
Available production in Lake Kariba (for 342 980 ha)   2 586   metric tons
Mean available yield in kg/ha (3 localities)   3.30
Available yield in percent of available production (6.36 kg)   51.93
Available yield in Lake Kariba   1 344   metric tons

values obtain and (Table 17) are in agreement with our previous findings that 'The abundance and mainly the standing crop of fish populations under normal circumstances are relatively very stable values both in time and space, in lakes, reservoirs and rivers' (HOLČIK, 1970c). On the other hand, the similarity of values confirms the accuracy of the mortality calculated. Considering this fact it seems to be superfluous to adjust the standing stock found at the time of sampling to that at the beginning of the growth season. However, the stability of ichthyomass concerns the total weight of the whole ichthyocenosis (which is only rarely composed of one species), while the biomass of one species changes seasonally very often as shown in the same paper. In view of this the green bream can be considered a stationary fish which does not migrate extensively. However, this statement needs to be validated by exact observations.

The values of growth increments together with the number of fish in age groups were used for computing the available production and yield in accordance with methods mentioned earlier. We took into consideration only three localities where the sample of scales included all length groups found after poisoning of the given cove, so the age composition and frequency of age groups of stock could be reconstructed (Tables 18–20). The data obtained are tabulated in Table 21 and it follows from them that the P'/B' ratio in these three localities (all belonging to the Basin III – central part of Lake Kariba) range from 50.36 to 76.83%. The average value is 56.64% (sum of available production from all three localities to the sum of biomass found in them). When this ratio is used in relation to the biomass of *S. codringtoni* equal to 13.32 kg/ha (in accordance with Balon, 1971a, 1972, 1973a, we considered only 65% of the total surface area inhabited by fish) then the whole available production of this species amounts to 2 586 metric tons or 7.54 kg/ha. The minimum harvestable size was determined to be 130 mm of standard length and 90 grams of weight (i.e., the fish from the second year). As sexual maturity is probably attained at this age (actual data are lacking and we presume it only from the values of the growth rate and presumed spawning marks), it is assumed the 3 year old and older fish can be removed. The percentage of available yield to available production found in three localities amounts to 54.93% in average, so that the available yield in Lake Kariba amounts to 1 344 metric tons.

3.2. The plain squeaker *Synodontis zambezensis* Peters, 1852

by KARTHIGESU CHITRAVADIVELU

S. zambezensis is a siluroid fish inhabiting Africa from the Webi Shebeli to the Zambezi (BOULENGER, 1911). An attempt was made to ascertain age and growth of this fish. In previous studies hard parts like the vertebrae, otoliths, dorsal and pectoral spines, cleithrum, operculum and supra occipital crests, were used. LEWIS (1949), HOOPER (1949), MENON (1950), APPELGET & SMITH (1951), FRANK (1955), HENSEL (1966), HOCHMAN (1966) and KIRKA (1969), used vertebrae to study the growth of various species. We found the use of vertebrae in *S. zambezensis* a good substitute

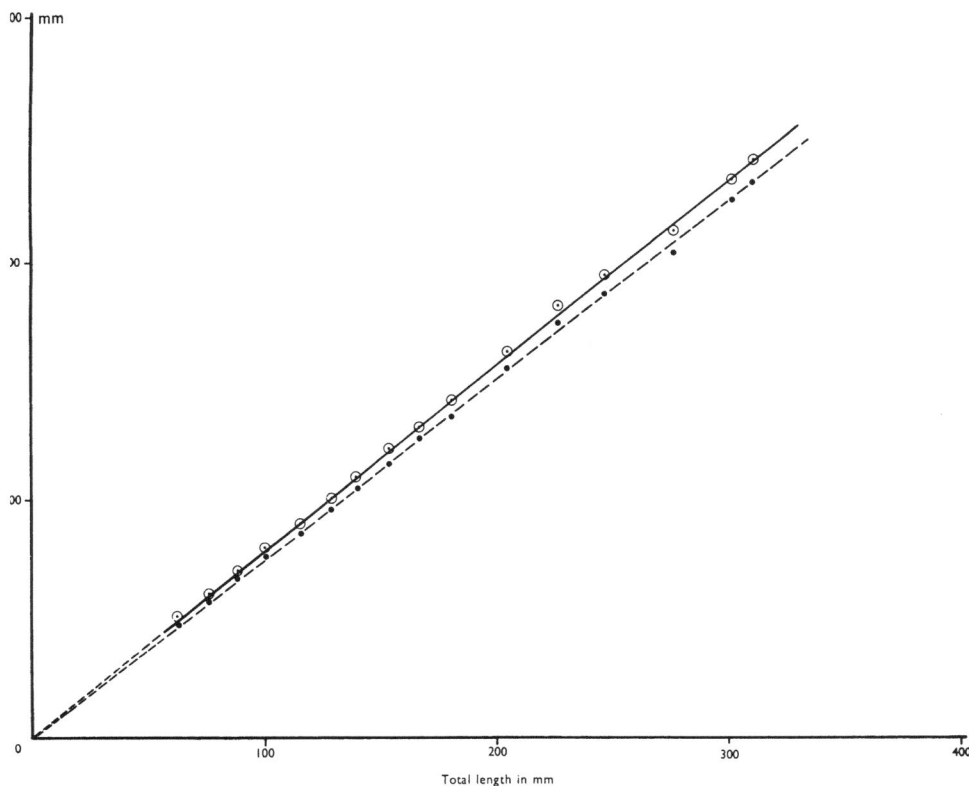

Fig. 15. Graph showing the relationship between standard lengths (1) and (2) and total length in *S. zambezensis*. Abscissa, total length in mm; ordinate, standard lengths in mm – (1) in broken lines and (2) in continuous lines.

for scales. They had well defined annual marks in spite of some doubt as to valid annuli inception in tropical fishes (DE BONT, 1967).

The material used in this investigation was collected in 1968 and 1969 from blocked-off piscicide cove samples. It was preserved in formalin and subsequently made available for study. Details concerning the methodology and localities are described by BALON (1971a, 1971b, 1972 and 1973a). In all, 279 specimens of *S. zambezensis* were studied. The age of 11 specimens could not be determined.

Two methods were employed to measure standard lengths to the nearest mm: (1) from the anterior most extremity of the fish to the anterior end of the median caudal fin rays; and (2) from the anteriormost extremity of the fish to the end of the fleshy part of the caudal peduncle. Total lengths were also measured to the nearest mm. Only standard length measured by method (2) was used in all the calculations. The conversion graphs for the other two lengths are given (Fig. 15).

A portion of the vertebral column, not more than 10 mm in length, was removed with the flesh, from a region between the dorsal and adipose fin. The anterior incision

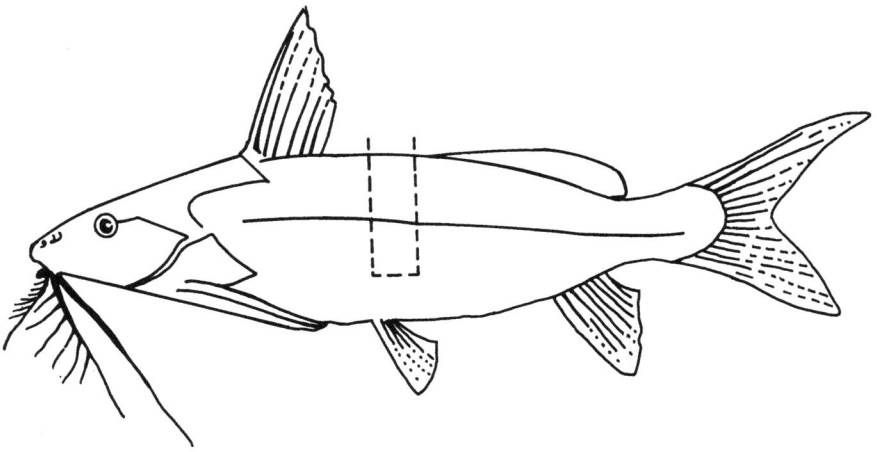

Fig. 16. S. zambezensis; the broken lines show the area from which the vertebrae were removed.

was always made opposite the origin of the pelvic fins (Fig. 16). The length of column that was removed was so chosen, depending on the length of the specimen, that the portion enclosed three or four vertebrae. The flesh was removed and the vertebrae separated from each other. The line of articulation of the vertebrae was located by holding it against the light. The separation of the vertebrae was performed with a small, thin, sharp scalpel, starting from the ventral region. If the blade of the scalpel is inserted in the correct place it passes with little effort along the line of articulation.

Benzene, ether, acetone and pepsin in hydrochloric acid were experimented with in an attempt to remove the tissue which adhered to the centrum and which makes the recognition of the annuli difficult. A freshly prepared solution of 0.7% pepsin in 0.2% hydrochloric acid (APPELGET & SMITH, 1951), was found to be most effective. After removing the intervertebral substance (remains of the notochord), the vertebrae from each specimen were introduced into numbered specimen tubes containing ten milliliters of the acidified pepsin solution and thermostated at 38°C for 24 hours. Once this treatment was over, the vertebrae were removed, enclosed in absorbent paper and stored in envelopes.

The number of annuli was determined from the face of the centrum, in reflected light, with the aid of a stereo microscope. Diameter of the centrum and the diameter of the different annuli were measured using an optical micrometer mounted on a binocular microscope. Narrow, dark bands alternating with deep, wide and pale white bands were seen in reflected light. The narrow bands were interpreted as annual marks. The accessory marks were indistinct and were not deep. The number of annual rings was considered to indicate the age of the fish in years. A plus sign was added to the age in order to indicate the growth beyond the last annual ring.

The standard lengths were grouped by 10 mm intervals and the average lengths were plotted against the corresponding average diameter of the centrum. Relationship between standard length and diameter of the centrum is linear (Fig. 17). A regression line expresses the relationship by the following equation: $l = 18.906 + 2.712\,d$, where

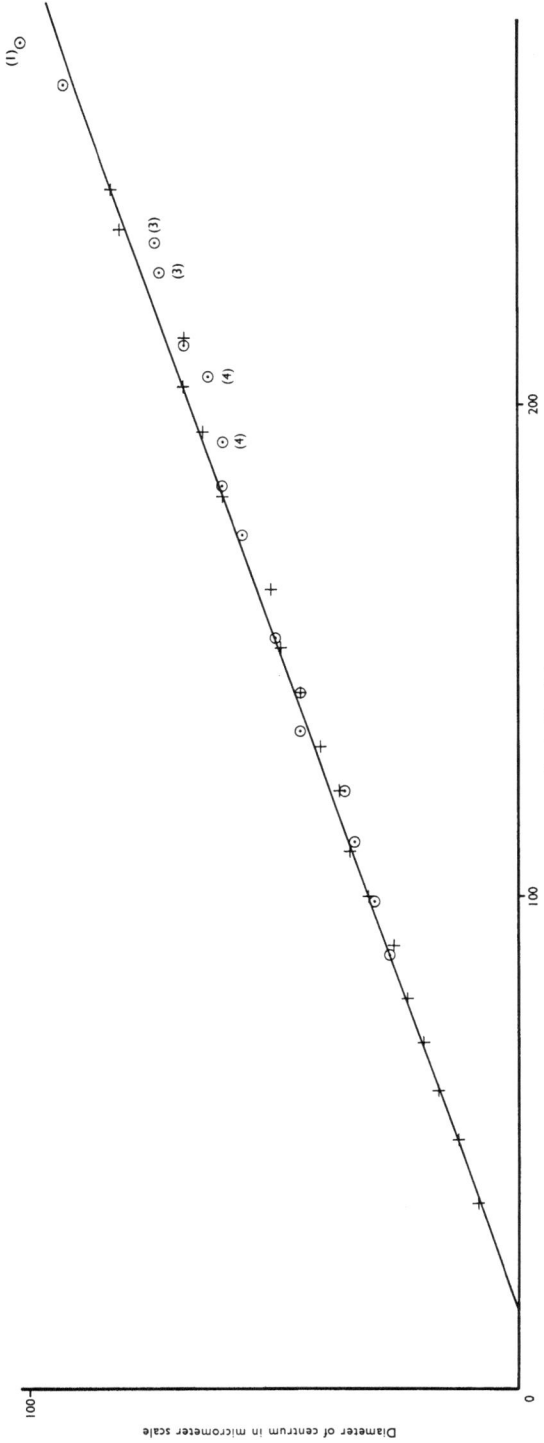

Fig. 17. Relation between standard length and diameter of centrum in *S. zambezensis*. Abscissa, standard length in mm; ordinate, centrum diameter in micrometer scales (17 micrometer scales = 1.0 mm). Circles represent the variables of the sample from Chikanka Island and the numbers in parentheses indicate the number of specimens in that length interval.

Table 22. Summary of empirical, computed standard lengths and weights of *Synodontis zambezensis* from the cove of the east shore of Chipepo Bay (locality F).

Age group	Number of specimens	Ranges and standard lengths in mm at the time of capture	Mean computed standard lengths in mm (above) and weights in grams (below)						
			l_1	l_2	l_3	l_4	l_5	l_6	l_7
0	8	(37– 66) 57							
1+	25	(67– 92) 80	68 8						
2+	30	(94–126) 107	70 9	89 18					
3+	35	(123–171) 142	76	100 26	119 43				
4+	2	(159–185) 172	73	103	119 43	138 68			
5+	4	(192–195) 193	73	95	116	146 80	168 121		
6+	4	(211–239) 224	81	103	127	149	173 134	197 198	
7+	2	(241–245) 243	73	98	130	154	187	208 230	227 300
Total	110	Average	73 10	98 24	122 47	147 80	176 140	203 210	227 300

Table 23. Summary of empirical, computed standard lengths and weights of *Synodontis zambezensis* from the cove of Lake Kariba near Siavonga (locality A) and east shore of Chipepo Bay (locality G).

Age group	Number of specimens	Ranges and average standard lengths in mm at the time of capture	l_1	l_2	l_3	l_4	l_5	l_6
0	5	(47– 95) 67						
1+	4	(73– 92) 85	69 8					
2+	31	(94–126) 108	78 12	95 22				
3+	19	(109–166) 141	86	101 27	126 52			
4+	3	(147–180) 171	109	126	141 72	157 100		
5+	3	(195–203) 199	92	122	145	166 117	184 160	
6+	1	213	100	122	133	143	154 94	171 130
Total	66	Average	89 18	113 37	136 64	155 96	169 125	171 130

Table 24. Summary of empirical, computed standard lengths and weights of Synodontis zambezensis from Chikanka Island (locality C).

Age group	Number of specimens	Ranges and standard lengths in mm at the time of capture	Mean computed standard lengths in mm (above) and weights in grams (below)							
			l_1	l_2	l_3	l_4	l_5	l_6	l_7	l_8
1+	4	(88– 98) 94	76 11							
2+	20	(101–125) 113	70 9	92 20						
3+	39	(122–173) 150	78	106 31	130 57					
4+	9	(172–192) 183	81	108	130 57	157 100				
5+	8	(183–211) 197	76	98	119	143 76	165 115			
6+	9	(208–229) 220	84	106	130	149	168 120	189 174		
7+	1	238	78	106	133	149	176	187 169	208 230	
8+	2	(264–272) 268	81	103	125	149	187	213	240 350	259 440
Total	92	Average	78 12	103 28	128 54	149 84	174 140	196 190	224 290	259 440

Table 25. Mean values of lengths, weights and increments of all samples of *Synodontis zambezensis* from Lake Kariba.

Parameters	Means at the end of each growth season							
	1	2	3	4	5	6	7	8
Standard length in mm	80	105	129	150	164	190	226	259
Length increments in mm	80	25	24	21	14	24	36	33
Length increments as % of 1st year length	100	31.3	30.0	26.3	17.5	30.0	45.0	41.3
Lengths as % of final year length	30.9	40.5	49.8	57.9	63.3	73.4	87.3	100
Weights in grams	13	30	55	87	135	177	295	440
Weight increments in grams	13	17	25	32	48	42	118	145

Table 26. Age and length composition of *Synodontis zambezensis* from the east shore of Chipepo Bay (locality F).

Range of length in mm	n								Total
	0	1+	2+	3+	4+	5+	6+	7+	
31– 40	1								1
41– 50									0
51– 60	3								3
61– 70	4	2							6
71– 80		13							13
81– 90		8							8
91–100		2	8						10
101–110			11						11
111–120			8						8
121–130			3	9					12
131–140				8					8
141–150				8					8
151–160				5	1				6
161–170				4					4
171–180				1					1
181–190					1				1
191–200						4			4
201–210									0
211–220							2		2
221–230									0
231–240							2		2
241–250								2	2
Total	8	25	30	35	2	4	4	2	110
% of total	7.3	22.7	27.3	31.9	1.8	3.6	3.6	1.8	100

l = standard length in mm and d = diameter of centrum in micrometer scales (17 micrometer scales = 1.0 mm).

Standard length-centrum diameter data revealed a correlation coefficient (r) of 0.998. Because of this high coefficient of correlation, the standard length-centrum diameter graph itself was used to compute the growth histories.

Weights at different lengths were computed from the standard length-weight double logarithmic graph.

The minimum harvestable size was determined from the crossing point of the length increments in percent of the length of the first growth season curve and the length in percent of the length of the final growth season curve (BALON, 1971a). This crossing is in the second age group and, therefore, all age groups, three and above, are harvestable. Based on this, the available yield is calculated as the sum of the available production of the third and all higher age groups.

When the study on 180 specimens was completed and the manuscript was being prepared, it was discovered that due to an oversight 99 specimens of *S. zambezensis*

Table 27. Age and length composition of *Synodontis zambezensis* from Lake Kariba near Siavonga (locality A) and the east shore of Chipepo Bay (locality G).

Range of length in mm	n							Total
	0	1+	2+	3+	4+	5+	6+	
41– 50	1							1
51– 60	1							1
61– 70	2							2
71– 80		2						2
81– 90		1						1
91–100	1	1	7					9
101–110			13	1				14
111–120			8					8
121–130			3	4				7
131–140				6				6
141–150				3	1			4
151–160				3				3
161–170				2				2
171–180					2			2
181–190								0
191–200						2		2
201–210						1		1
211–220							1	1
Total	5	4	31	19	3	3	1	66
% of total	7.6	6.1	47.0	28.8	4.5	4.5	1.5	100

from Chikanka Island (locality C) had been left behind in the depository. These specimens were also studied later. The average standard lengths in the 10 mm intervals and the corresponding average diameter of centrum when plotted in the graph for specimens from the earlier localities, did not show any significant deviation from the regression line, except beyond the 190 mm standard length (Fig. 17). This apparent deviation was presumably due to the fact that the specimens in these length intervals were not adequately represented. The same graph was also used in the computation of the growth histories of specimens from Chikanka Island.

Growth of different age groups in each locality are shown separately in Tables 22–24. The mean values of growth of all samples from Lake Kariba are given in Table 25. The smallest specimen in the sample is 37 mm and belongs to 0 age group, while the oldest and the largest is 8+ years old and 272 mm. Age groups 2+ and 3+ are predominant in all the localities. On the east shore of Chipepo Bay 1+ age group specimens are also abundant.

It has been the practice to use the corresponding vertebrae from all specimens in similar growth studies. APPELGET & SMITH (1951) used the fifth vertebra from *Ictalurus lacustris punctatus* and LEWIS (1949) appears to have used the 10th or 11th vertebra from *Ameiurus melas melas*. This was not resorted to in the present study

Table 28. Age and length composition of *Synodontis zambezensis* from Chikanka Island (locality C).

Range of length in mm	n									Total
	0	1+	2+	3+	4+	5+	6+	7+	8+	
81– 90		2								2
91–100		2								2
101–110			7							7
111–120			11							11
121–130			2	5						7
131–140				6						6
141–150				7						7
151–160				11						11
161–170				7						7
171–180				3	3					6
181–190					5	2				7
191–200					1	2				3
201–210						3	1			4
211–220						1	3			4
221–230							5			5
231–240								1		1
241–250										0
251–260										0
261–270									1	1
271–280									1	1
Total	0	4	20	39	9	8	9	1	2	92
% of total		4.3	21.7	42.2	9.9	8.7	9.9	1.1	2.2	100

because the specimens were needed for further study. However, a random check, made by measuring the centra of all the vertebrae collected from each specimen, showed that the difference in the diameter of the centra was in most cases almost nil. The maximum difference where found was only the equivalent of 3.0 mm in standard length. It appears that the error introduced here will not significantly alter the growth data.

It now appears that in *S. zambezensis*: (1) the number of annual rings in the vertebrae increases as the average length of fish increases; (2) in most cases, the length of young fish calculated from the vertebrae of older fish agrees with the average observed lengths of young fish of the same assigned age at the time of capture (Tables 22–28); and (3) in the younger age groups, the modes of length frequency distribution of all the fish in the separate samples agree satisfactorily with the modal lengths of fish assigned to each age group (Fig. 18 and 19). In the older age groups the number of specimens are few. However, the agreement up to the third age group along with (1) and (2) appears to be sufficient to give validity to the use of vertebrae in determining the age of *S. zambezensis*.

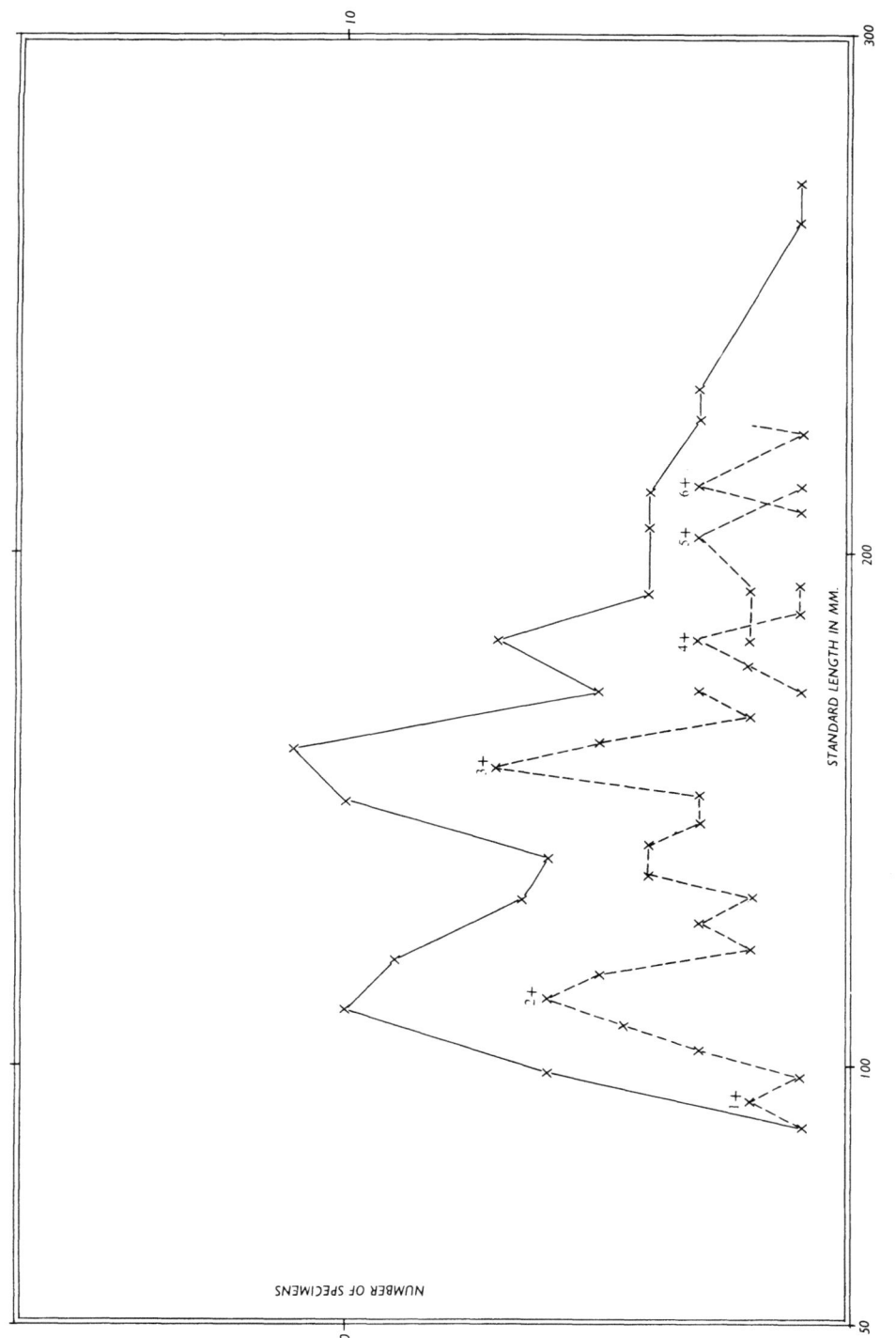

Fig. 18. Length frequency of sample of *S. zambezensis* (continuous line) collected from Chikanka Island compared with the length frequencies of fish assigned to each age group (broken lines). Abscissa, standard length in mm. Numbers above modes indicate successive age groups.

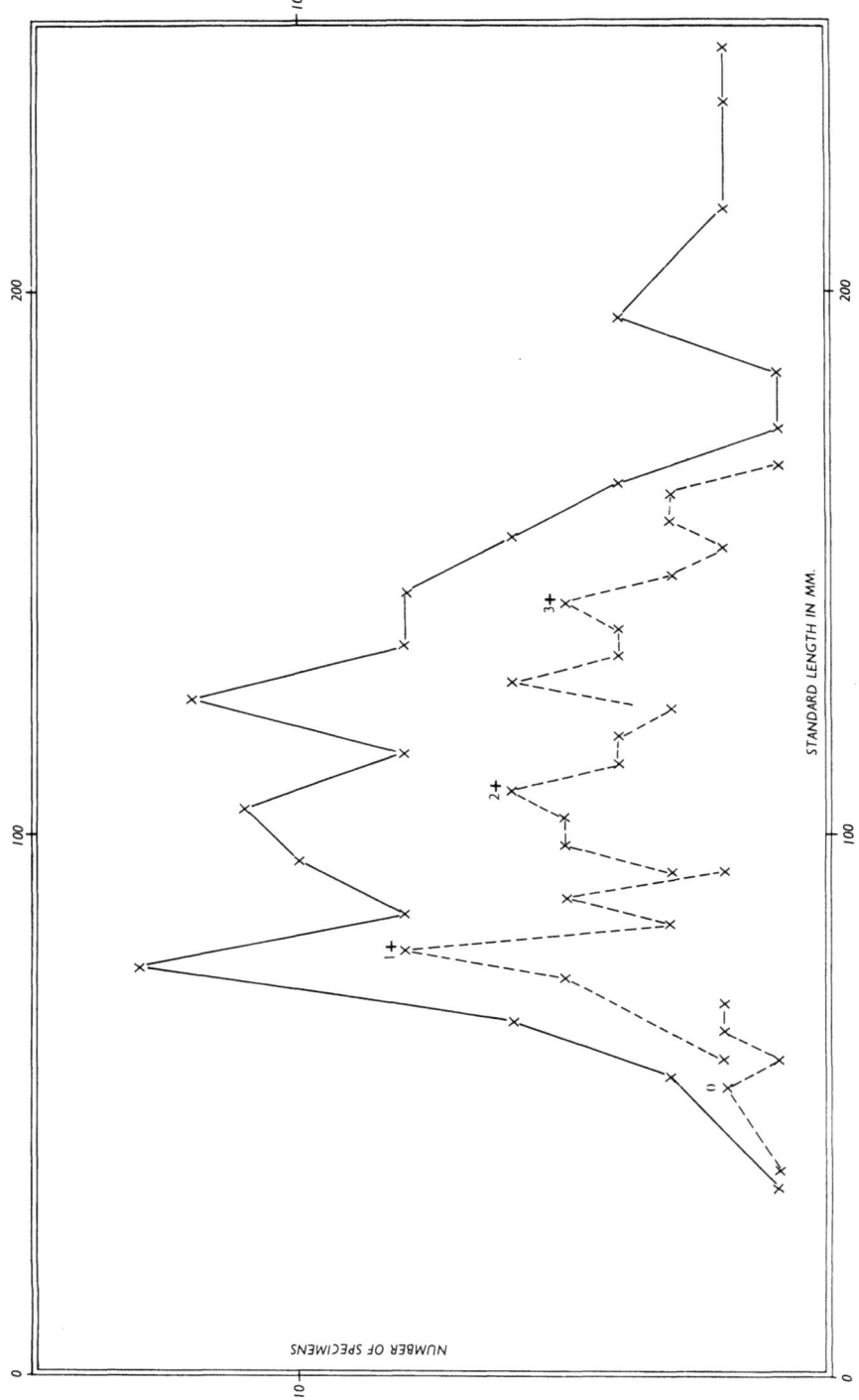

Fig. 19. Length frequency of sample of *S. zambezensis* (continuous line) collected from the east shore of Chipepo Bay compared with the length frequencies of fish assigned to each age group (broken lines). Abscissa, standard length in mm. Numbers above modes indicate successive age groups.

3.3. The red-breasted bream *Tilapia rendalli* (Boulenger, 1896)

by IVAN BASTL

The red-breasted bream forms an important component of the Lake Kariba ichthyofauna. It occurs in all localities of the lake. The densest red-breasted bream stocks were 109.5 kg/ha at Chipepo Bay (locality G) and 93.0 kg/ha in a cove of Chikanka Island (locality C). The smallest standing stocks were 27.0 kg/ha in coves at Island N20, SW from Sinazongwe (locality D), and 27.5 kg/ha in another cove of Chipepo Bay (locality F). The mean standing stock is 65.1 kg/ha, which is 12.2% of the ichthyomass (533 kg/ha) found in this lake. In this section an attempt is made to analyze the age composition and growth of this species.

In my investigation I relied, as have other authors before me, on two sources of information: (1) data on the abundance, standing stock, length frequency and length-weight relationship of this species and (2) samples of scales taken from subsamples of poisoned stocks. The exact description of method and localities was given by BALON (1971a, 1972, 1973a). Samples of scales originated from the following localities: B – Island N 102 (5 specimens), C – Chikanka Island (155), E – Island N 21 (92), F – Chipepo Bay (37). The scales were read under a trichinoscope at 45 magnification, and the longer (usually ventral) scale lobe was measured. The annuli of a usual form (Fig. 20, 21) as described in the tigerfish or the green bream, were easily identified. The body-scale relationship is linear with 16.54 intercept (Fig. 22). This value is very close to the actual size of other *Tilapia* and *Sarotherodon* species at the time of scale formation as reported by FISHELSON (1966) for *T. tholloni* (11.0–11.5 mm), *S. niloticus* (13.8–14.1 mm) and *S. macrocephalus* (10.8–11.2 mm). Back calculation of body length for the previous years was performed with the E. Lea board. The weights for the corresponding lengths were determined from the double logarithmic length-weight diagram. Computation of the age composition, mortality rates and relative growth indices was made according to methods summarized by BALON (1971a, 1972).

To estimate the production I employed the method described by HOLČIK (see Section 3.1). For this purpose the initial biomass present at the time of annulus formation was determined. As the exact time of annulus inception in this species was not ascertained, I assume that it coincides approximately with the beginning of the warm season which in the Lake Kariba area is October (COCHE, 1968). Therefore, I considered November 1st to be the time of annulus formation for all age groups of the red-breasted bream. I calculated the average biomass in localities C, E, F ($=$ 56.97 kg/ha); the standing stock based on number and weight of fish in age groups at the time of poisoning, both with the young-of-the-year fish (49.09 kg/ha) and without (48.86 kg/ha); and the biomass at the time of annulus formation (42.67 kg/ha).

The largest specimen of the red-breasted bream measured 455 mm and weighed 6,040 grams. Because scales were not taken, the age of this specimen remains unknown. The largest fish whose scales were studied by me measured 336 mm and its age was 5 years.

Age and length compositions of stocks from localities C, E and F are shown in

Fig. 20. Scale of 1+ old red-breasted bream. Locality C, standard length 68 mm.

Table 29. While in the first locality the dominating group is composed of yearlings, in the other two localities it is age group 2+, probably because sampling years were different. The low abundance of the 0+ age groups is remarkable. It indicate that this group was incompletely sampled or that the poisoned areas did not include the nursery grounds (Donnelly, 1969).

The values on mortality and survival rates are summarized in Table 30. From these it appears that the mortality rate increases with increasing age of the fish. This contradicts published data on mortality rate of other species where the highest mortality was found in younger fish, followed by a decrease in middle age and an increase during the senective period. For the time being I cannot say if this is a biological phenomenon or if it is due to some methodological shortcomings (e.g. migration or sampling bias). As can be seen, the mortality values vary in different

Fig. 21. Scale of 2+ old red-breasted bream. Locality C, standard length 95 mm.

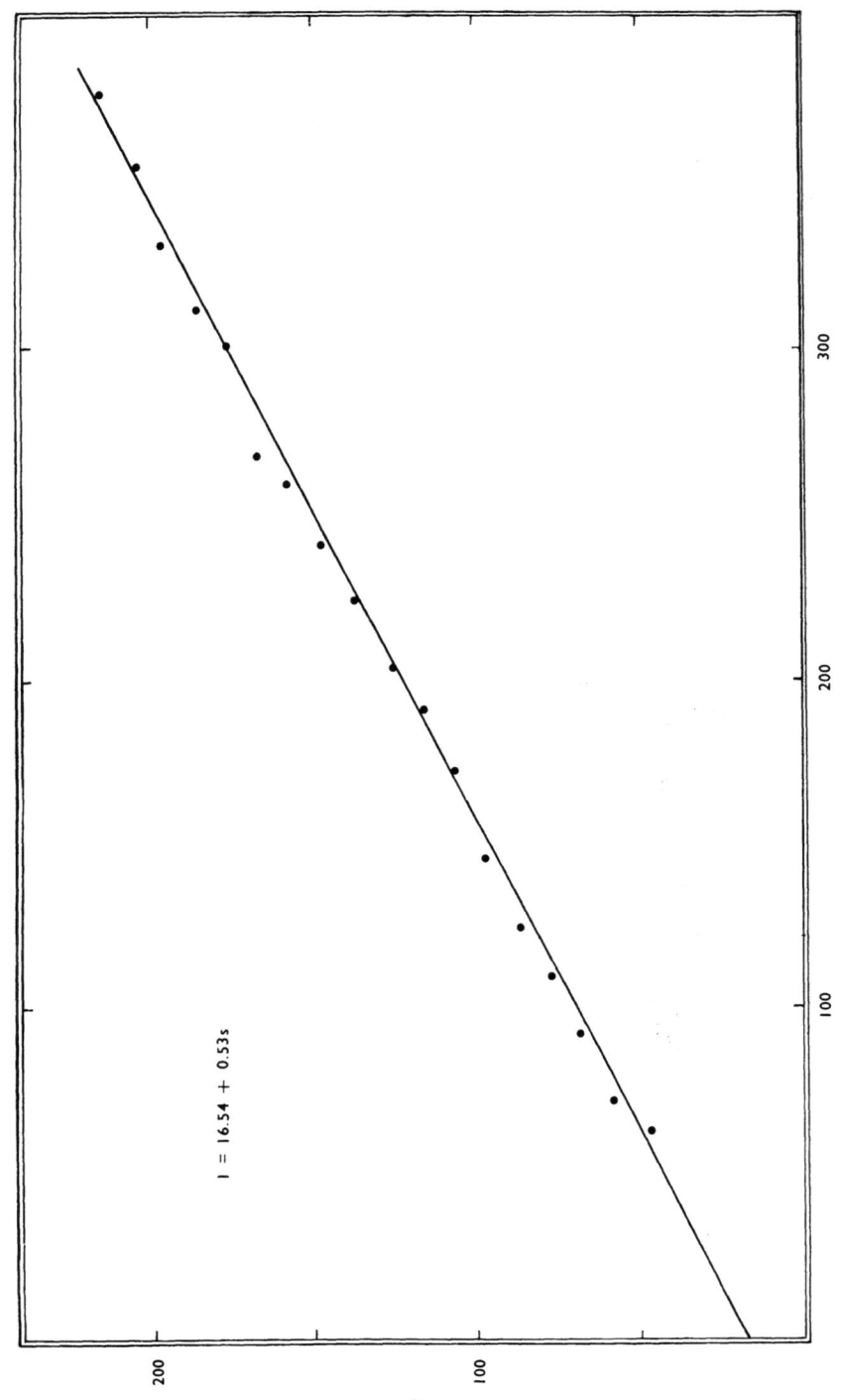

Fig. 22. Body-scale relationship of *Tilapia rendalli*.

Table 29. Age composition of *Tilapia rendalli* population from Lake Kariba (in % of total sample).

Locality \ Age	0+	1+	2+	3+	4+	5+
Chikanka Island (C)	5.0	42.2	28.4	17.0	5.9	1.5
Island 21 (E)	1.2	21.4	45.5	22.6	9.3	
Chipepo Bay (F)	11.4	13.6	54.5	15.9	4.6	
Total	2.3	26.3	41.6	21.1	8.3	0.4

Table 30. Age composition, seasonal total mortality rate (Z'), survival rate (S) and instantaneous rate of total mortality (Z) of *Tilapia rendalli* (total population).

Locality		Age groups 1	2	3	4	5	$\sum$ (years)
Chikanka Island (C)	n	245	165	99	34	9	552
	%	44.4	29.9	17.9	6.2	1.6	100.0
	Z'		0.3297	0.3995	0.6570	0.7355	0.4345 (1–5)
	S		0.6703	0.6005	0.3430	0.2645	0.5655
	Z		0.40	0.51	1.07	1.33	0.57
Island 21 (E)	n	385	817	405	166		1773
	%	21.7	46.1	22.8	9.4		100.0
	Z'	—	0.5034	0.5893			0.5323 (2–4)
	S	—	0.4966	0.4107			0.4677
	Z	—	0.70	0.89			0.76
Chipepo Bay (F)	n	6	24	7	2		39
	%	15.5	61.5	17.9	5.1		100.0
	Z'	—	0.7077	0.7135			0.7106 (2–4)
	S	—	0.2923	0.2865			0.2894
	Z	—	1.23	1.25			1.24

localities. In order to compare them more exactly, the mortality was calculated separately for age groups 2–4 in locality C. The values are: $Z' - 0.4984$; $S = 0.5016$; $Z = 0.69$. From this comparison it follows that the mean annual mortality for this locality is the lowest (50% only) in spite of having the highest abundance and biomass (481 spec./ha and 93 kg/ha), while in locality F – Chipepo Bay, the mortality was highest (71%) but the abundance and biomass were lowest (230 spec./ha and 27 kg/ha).

Back calculated lengths and weights for previous years of life and the relevant indexes are shown in Tables 31–33. The mean values do not include the oldest and

Table 31. Back calculated growth values of *Tilapia rendalli* from Lake Kariba.

Locality	Age groups	Year of hatching	N	Mean standard length at the time of capture (mm)	Back calculated standard lengths (mm) for the previous growth seasons				
					l_1	l_2	l_3	l_4	l_5
B 20.3.1968	1+	1966/1967	2	94	78				
	4+	1963/1964	1	201	75	134	163	187	
	5+	1962/1963	2	223	67	117	151	178	205
	Total		5		73	126	157	183	205
C 21.6.1968	0+	1967/1968	6	71					
	1+	1966/1967	63	89	65				
	2+	1965/1966	45	143	73	115			
	3+	1964/1965	29	182	71	116	154		
	4+	1963/1964	10	216	67	110	163	198	
	5+	1962/1963	2	231	60	101	138	186	208
	Total		155		67	111	152	192	208
E 21.–25.3.1969	0+	1968/1969	10	52					
	1+	1967/1968	27	92	63				
	2+	1966/1967	32	132	65	107			
	3+	1965/1966	16	155	63	101	130		
	4+	1964/1965	7	183	63	105	139	164	
	Total		92		64	104	135	164	
F 26.–31.5.1969	0+	1968/1969	5	59					
	1+	1967/1968	6	86	63				
	2+	1966/1967	21	116	66	99			
	3+	1965/1966	4	137	61	97	122		
	4+	1964/1965	1	175	67	104	136	162	
	Total		37		64	100	129	162	
Grand average	1+		98	90	67				
	2+		98	130	68	107			
	3+		49	158	65	105	135		
	4+		19	194	68	113	150	178	
	5+		4	227	64	109	145	182	207
	Total		268		66	109	143	180	207

Table 32. Lengths, weights and increments for all studied samples of *Tilapia rendalli* from Lake Kariba.

Locality	Parameter	Means at the end of growth seasons				
		1	2	3	4	5
Island 102 (B)	Standard length (mm)	73	126	157	183	205
	Lengths increments (mm)	73	53	31	26	22
	Lengths increments in % of length of first season	100.0	72.6	42.5	35.6	30.1
	Weight in grams	18	106	206	350	500
	Weights increments in grams	18	88	100	144	150
Chikanka Island (C)	Standard length (mm)	67	111	152	192	208
	Length increments (mm)	67	44	41	40	16
	Lengths increments in % of length of first season	100.0	65.7	61.2	59.7	23.9
	Weight in grams	14	72	197	403	520
	Weights increments in grams	14	58	125	206	117
Island 21 (E)	Standard length (mm)	64	104	135	164	
	Lengths increments (mm)	64	40	31	29	
	Lengths increments in % of length of first season	100.0	62.5	48.4	45.3	
	Weight in grams	12	59	133	245	
	Weights increments in grams	12	47	74	112	
Chipepo Bay (F)	Standard length (mm)	64	100	129	162	
	Lengths increments (mm)	64	36	29	33	
	Lengths increments in % of length of first season	100.0	56.3	45.3	51.6	
	Weight in grams	12	51	110	240	
	Weights increments in grams	12	39	59	130	

Table 33. Mean values of lengths, weights and increments of *Tilapia rendalli* population from Lake Kariba.

Parameter	Means at the end of growth season				
	1	2	3	4	5
Standard length (mm)	66	109	143	180	207
Lengths increments (mm)	66	43	34	37	27
Lengths increments in % of length of first season	100.0	65.2	51.5	56.1	40.9
Standard length in % of length of final season	31.9	52.7	69.1	87.0	100.0
Weight in grams	14	69	127	312	510
Weights increments in grams	14	55	58	185	198

biggest fish from Chikanka Island (5+ and 336 mm), the growth of which was extraordinarily good: 1–114, 2–226, 3–298, 4–315, 5–328 mm standard length.

Comparing growth values attained in different localities one can see differences among them. It is remarkable that the fastest growth was reached in the locality with the higher population density (locality C – abundance 481 spec./ha, biomass 93 kg/ha, ϕH in first 4 growth seasons 4.80) while in localities with lower population densities the intensity of growth was lower (e.g. locality F – abundance 230 spec./ha, biomass 27 kg/ha, ϕH 4.05). JUBB (1967) characterizes *Tilapia melanopleura* (old synonym for *T. rendalli*) as being herbivorous, which proved to be so at least in adult specimens (FISH, 1955; LE ROUX, 1956). Based on the data of BALON (1971a, 1972) we concluded that the higher stock density of the red-breasted bream coincides with the lower density of submerged plants. HOLČÍK (Section 3.1) found it to be so for *Sargochromis codringtoni*. It seems that the factors which limit the growth intesity of this species are not only stock density and availability of food (if this species is exceptionally herbivorous) but also some which cannot be explained at present. The paradoxical phenomenon of Rosa Lee (Table 31) which BALON (1971a) observed on *Hydrocynus vittatus* and HOLČÍK (Section 3.11) on *Micralestes acutidens* was also displayed by the red-breasted bream, though not very distinctly. This phenomenon suggests that *T. rendalli* is also a species with portional spawning and that the fish hatched from first portions gradually disappear from the population as they are consumed by predators. HOLČÍK (l.c.) believes that this is a certain adaptation of the species to enable it to escape pressure from predators. The minimum harvestable size of redbreasted bream was determined graphically and is equal to that of fish in the third year of life (standard length 143 mm and weight 127 grams).

3.4. The Zambezi barbel *Clarias gariepinus* (Burchell, 1822)
by KAREL PIVNIČKA

Clarias gariepinus is fourth among the economically preferred species from Lake Kariba (BALON, 1972, 1973a). It appears that the abundance and standing stock of the *Clarias gariepinus* in Lake Kariba was 65 specimens per ha and 65.8 kg/ha respectively, on the average. My study is directed toward the determination of age composition and growth of this species.

The first three vertebrae, just behind the median bony cover of the skull, were used in the growth study. Vertebrae were collected from specimens from the following localities: C – Chikanka Island (72 specimens); D – Island N 20, south-west of Sinazongwe (5 specimens); F – Chipepo Bay cove I (1 specimen); G – Chipepo Bay cove II (47 specimens). The method of growth determination in fishes using vertebrae is summarized by FRANK (Section 3.5). The centrolateral diameter of vertebrae was always measured with a micrometer mounted on a binocular microscope. The growth rate was back calculated according to the Rosa Lee method using Lea's board with 20 mm as the correction value. Body-vertebrae radius relationship was linear (Fig. 23).

The minimum harvestable size was established from the crossing point of the average length increments for all populations, in percentages of length of the first growth season curve and the length in percentages of length in the final growth season curve.

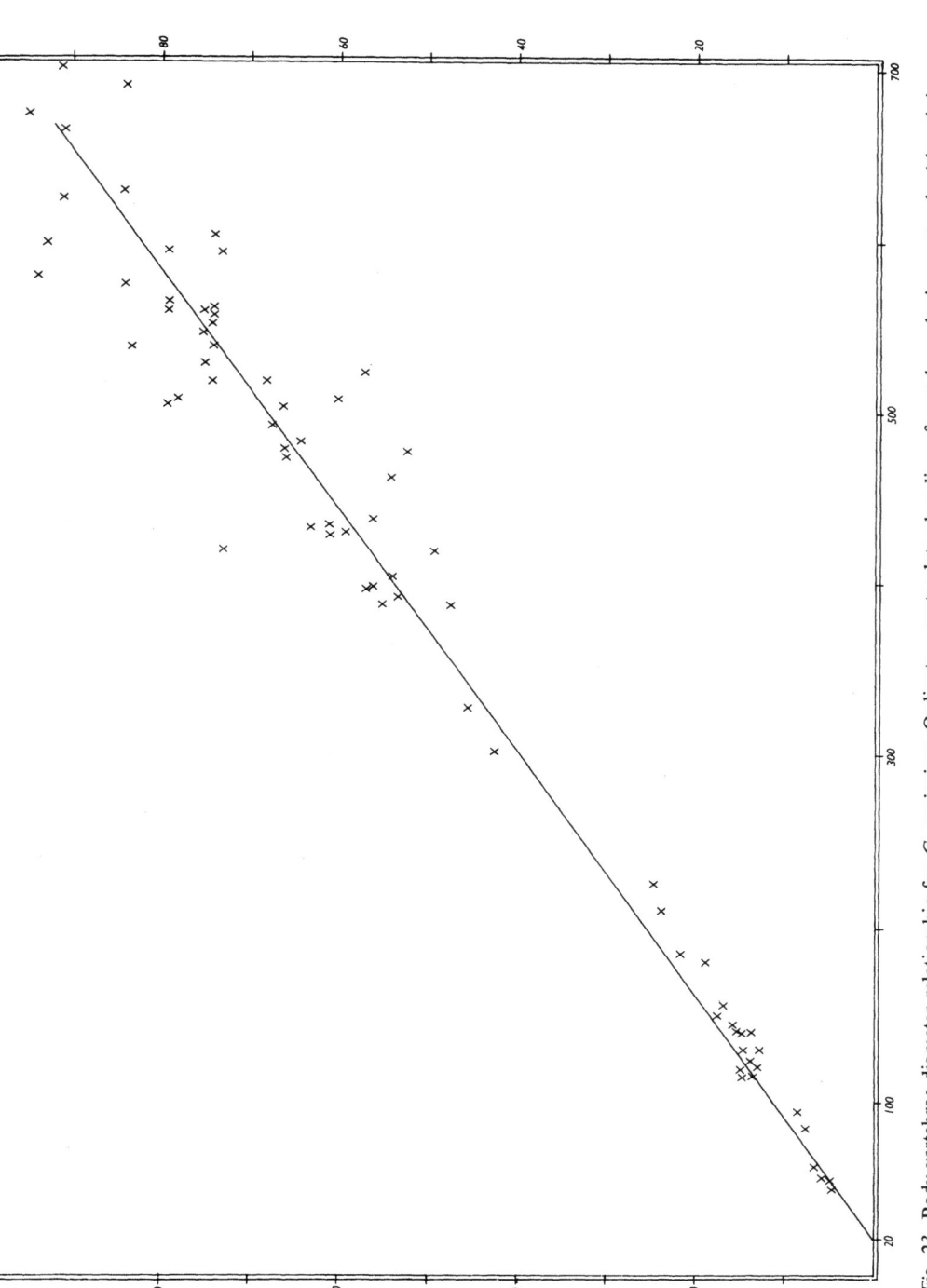

Fig. 23. Body-vertebrae diameter relationship for *C. gariepinus*. Ordinate – centro-lateral radius of vertebra; abscissa – standard length in mm.

The number in different age groups is small and, therefore, the calculation of mortality based on the decrease in number of fish in succeeding age groups (HOLČIK, 1970a, Section 3.1), may be inaccurate. This method of calculating mortality assumes equal strength of newly-born year classes, which does not exist in nature. This method was criticized several times by NIKOLSKY (1945). PIVNIČKA (1971) demonstrated that, in the case of the Kličava Valley Reservoir, the values of mortality calculated by this method were very different from those calculated from two estimations of fish abundance in two different periods.

Growth in length is shown in Table 34. Only growth data from locality C can be used for computation of production. Growth data from the other localities, D, F, and G, are given to provide an overall view. The comparisons of growth among the different localities are without value, however, the values of weight increments applied in the production calculations are of the utmost importance.

In view of the fact that I had at my disposal sufficient representation of age groups only from locality C, I have tried to estimate the available production by the following methods to compensate for the inadequacy. From individual samples I calculated the average number of fish for the localities C, D, E, G as 65 specimens per ha. Localities E_3 and F were not taken into consideration because the samples contained too many young fish. From the average weights of one fish in different localities (C – 0.86 kg; D – 1.06 kg; E – 1.58 kg; G – 0.92 kg), it appears reasonable to make the assumption that the age composition of all of them is approximately the same. I modified the age composition in locality C, where 89 fish were caught but only 72 fish were sent for growth studies. Older age groups were distinctly predominant in the sample of 72 fish. The length distribution of $0+$, $1+$ and $2+$ age groups did not overlap and so it was possible that in the $0+$ age group there were originally 13 specimens instead of 1; in the $1+$ age group 11 specimens instead of 10; and in the $2+$ age group 11 specimens instead of 7. Beginning with the $3+$ age group, I kept a number of fish (54) in these age groups as they were in the sample.

From the age composition reconstructed in this way, the annual available production between two seasons of annulus formation were calculated as the average from two different values, as mentioned previously.

The value of annual available production is 14.8 kg/ha (Table 35) or 34% of the standing stock of 43.6 kg/ha. Production calculated using grand average weight increments is 12.9 kg/ha. Quantitative samples in most localities were conducted after the time of annulus formation. The value of the standing stock at time of sampling was 65.8 kg/ha on the average. A difference in biomass is evident because in the computation we used the same number of fish and age composition in both periods. The entire production of this species available in the whole lake, assuming that it occupies 65% of the total area, i.e. 342 980 ha (BALON, 1971, 1972), is 5077 metric tons annually. The minimum harvestable size was determined to be 520 mm standard length and 1880 grams, i.e. fish from the seventh year (Table 36). I suppose that fish 8 years old and over can be removed. Available yield is 57% of available production, so the yield in Lake Kariba is 8.4 kg/ha or 2881 metric tons annually for the entire lake.

In the case of the roach *(Rutilus rutilus)* and perch *(Perca fluviatilis)* in the Kličava

Table 34. Back calculated growth values of *Clarias gariepinus* from Lake Kariba.

Locality	Age groups	N	Mean standard length at time of capture	Back calculated standard length (mm) for the previous growth season										
				l_1	l_2	l_3	l_4	l_5	l_6	l_7	l_8	l_9	l_{10}	l_{11}
C 21.6.1968	0+	1	55											
	1+	10	122 (95–140)	84										
	2+	7	177 (140–210)	85	132									
	3+	1	225	74	119	172								
	4+	7	377 (300–425)	86	157	227	313							
	5+	8	433 (385–505)	82	134	191	278	356						
	6+	10	498 (430–560)	81	130	180	270	344	409					
	7+	9	521 (490–555)	78	124	165	230	301	373	437				
	8+	10	585 (542–697)	79	130	184	259	333	393	456	523			
	9+	5	623 (575–670)	87	123	195	250	300	368	429	494	557		
	10+	2	676 (665–687)	96	140	185	214	318	393	453	509	563	620	
	11+	2	698 (660–735)	65	110	158	220	285	362	438	495	524	599	648
	Total	72		82	131	183	254	319	383	443	505	548	610	648
D 11–14.12.1968	1+	1	105	89										
	2+	1	125	86	120									
	3+	3	172 (150–210)	83	122	166								
	Total	5		86	121	166								
F 26–31.5.1969	6+	1	490	88	142	185	247	346	388					
G 28–31.5.1969	0+	25	68 (50– 85)	76										
	1+	11	106 (92–123)	94	143									
	2+	9	200 (155–255)	83	143	218								
	3+	1	285	84	129	175	294							
	5+	1	455					362						
	Total	47		84	138	197	294	362						

Table 35. Biomass (B), Production (P) and Yield (Y) of *Clarias gariepinus* from Lake Kariba (locality C), Chikanka Island, calculated up to the time of annulus formation.

Age group	Number per 10 ha	W g	Using W_n and W_{n-1} of the same age group				Using W_n of succeeding age groups			
			B kg/10 ha	w g	P kg/10 ha	Y	w g	P kg/10 ha	Y kg/10 ha	
0	95									
1	80	6.8	0.54	6.8	0.54		6.8	0.54		
2	80	26	2.0	19	1.52		19	1.52		
3	8	58	0.46	39	0.31		32	0.26		
4	51	320	16.32	192	9.79		262	13.36		
5	58	500	29.0	260	15.08		180	10.44		
6	73	750	54.75	290	21.17		250	18.25		
7	66	900	59.4	330	21.77		150	9.90		
8	73	1630	118.99	590	43.07	43.07	830	60.59	60.59	
9	36	1920	69.12	570	20.52	20.52	290	10.44	10.44	
10	15	2700	40.5	650	9.75	9.75	780	11.7	11.7	
11	15	3010	45.15	610	9.15	9.15	310	4.65	4.65	
Total	650		436.1		152.67	82.49		141.9	87.4	
per 1 ha	65		43.6		15.3	8.2		14.1	8.7	

W = absolute weights in grams; w = weight increments in grams.

Table 36. Mean values of lengths and increments of Clarias gariepinus population from Lake Kariba.

	Means at the end of growth season										
	1	2	3	4	5	6	7	8	9	10	11
Standard length (mm)	82	131	183	254	319	383	443	505	548	610	648
Increments (mm)	82	49	52	71	65	64	60	62	43	62	38
Length increments in % of length of first season	100	60	63.5	86	79	78	73	75	52	75	46
Lengths as % of final year length	12.7	20.3	28.3	39.1	49.2	59.4	68.5	78	84.5	94.5	100

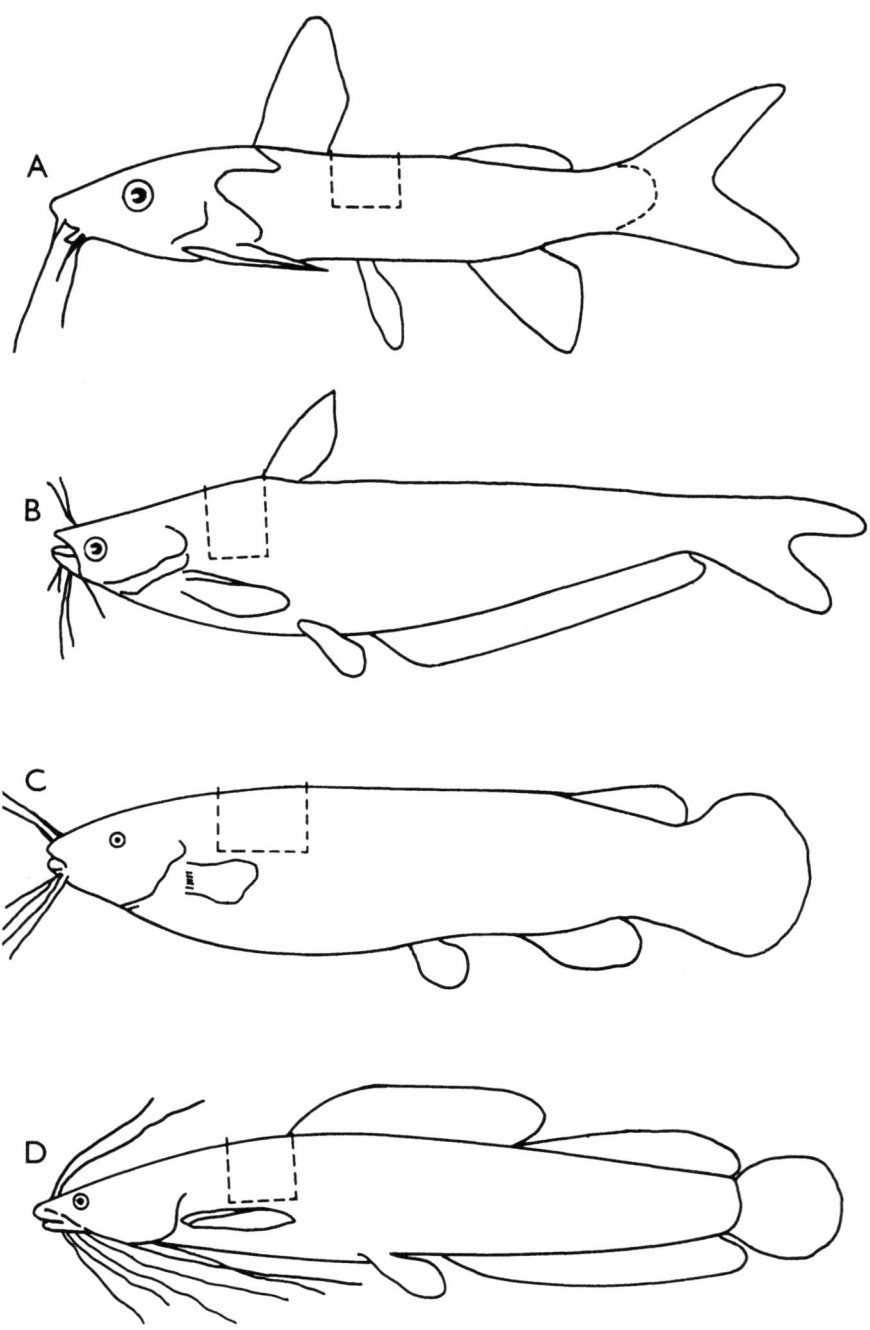

Fig. 24. Broken lines show the area from which vertebrae were removed: a – *Synodontis nebulosus*; b – *Schilbe mystus*; c – *Malapterurus electricus*; d – *Heterobranchus longifilis*.

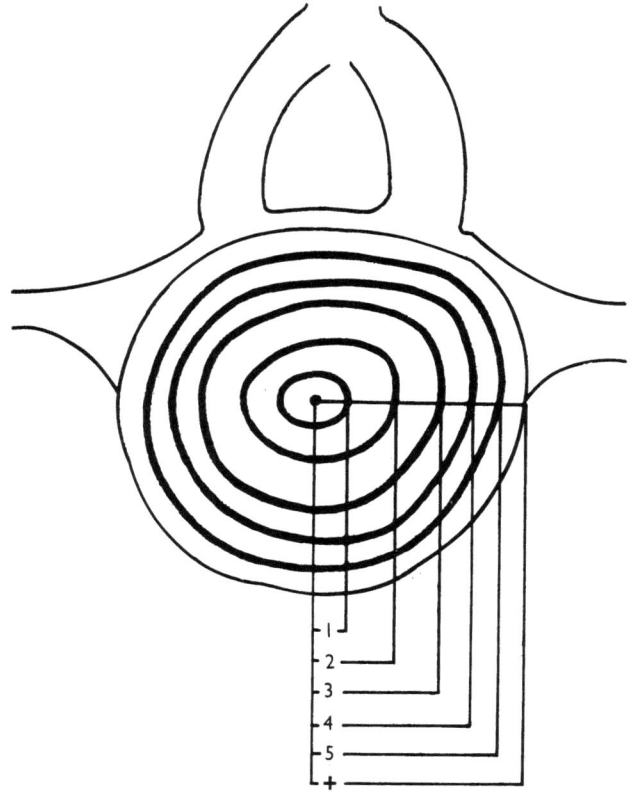

Fig. 25. Mode of measurement of yearly vertebral increments.

Reservoir (Central Bohemia), it was found that the value of the total production of 0 age group, calculated on the basis of exponential changes of biomass using the equation $A = G\bar{B}$, fluctuated in the range of 60–85% of the production of all age groups, including the 0 age group. It means that the production from 1 age group and above is 15–40% of the whole production (PIVNIČKA, 1971). Adding the production of 0 age group to our production in Lake Kariba, production might increase up to 37–99 kg/ha.

3.5. The spotted squeaker *Synodontis nebulosus*, the butter catfish *Schilbe mystus*, the vundu *Heterobranchus longifilis* and the electric catfish *Malapterurus electricus*

by STANISLAV FRANK

For growth analyses, we used the centra of vertebrae. The location of vertebrae samples from the four species, and the mode of measurement of the vertebral year marks (annuli), is shown in Fig. 24 and 25. From each specimen only the centro-

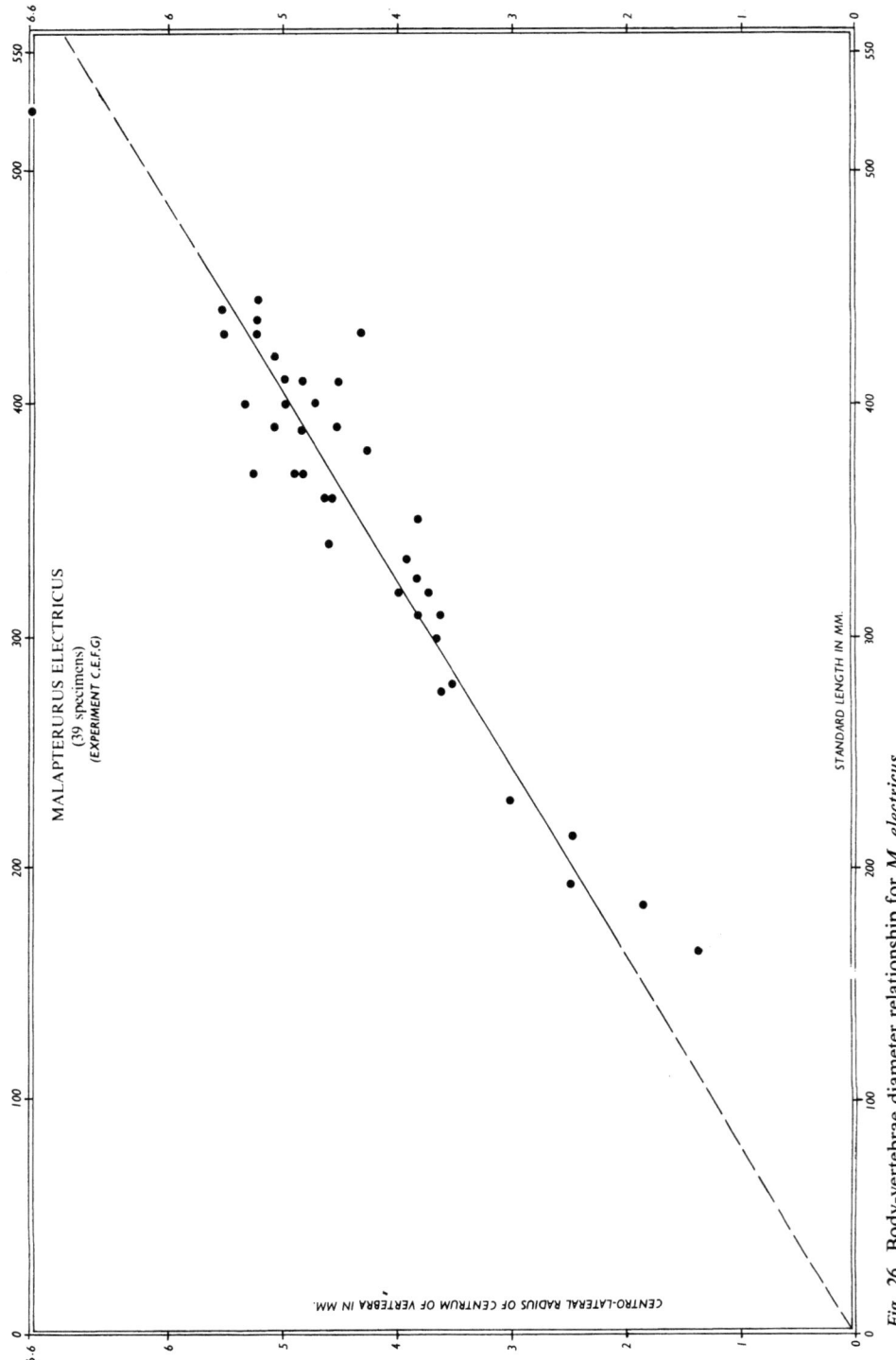

Fig. 26. Body-vertebrae diameter relationship for *M. electricus.*

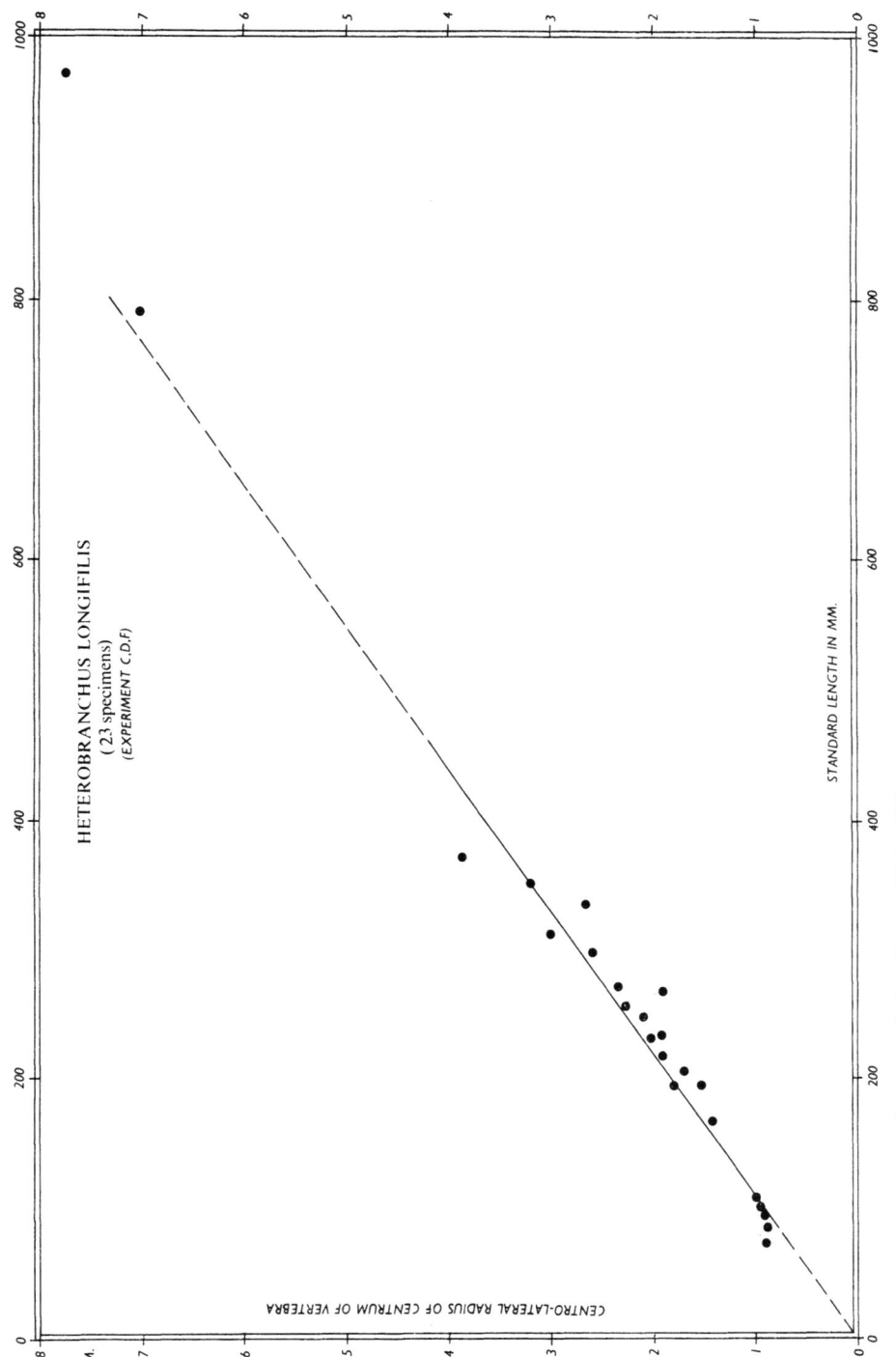

Fig. 27. Body-vertebrae diameter relationship for *H. longifilis*.

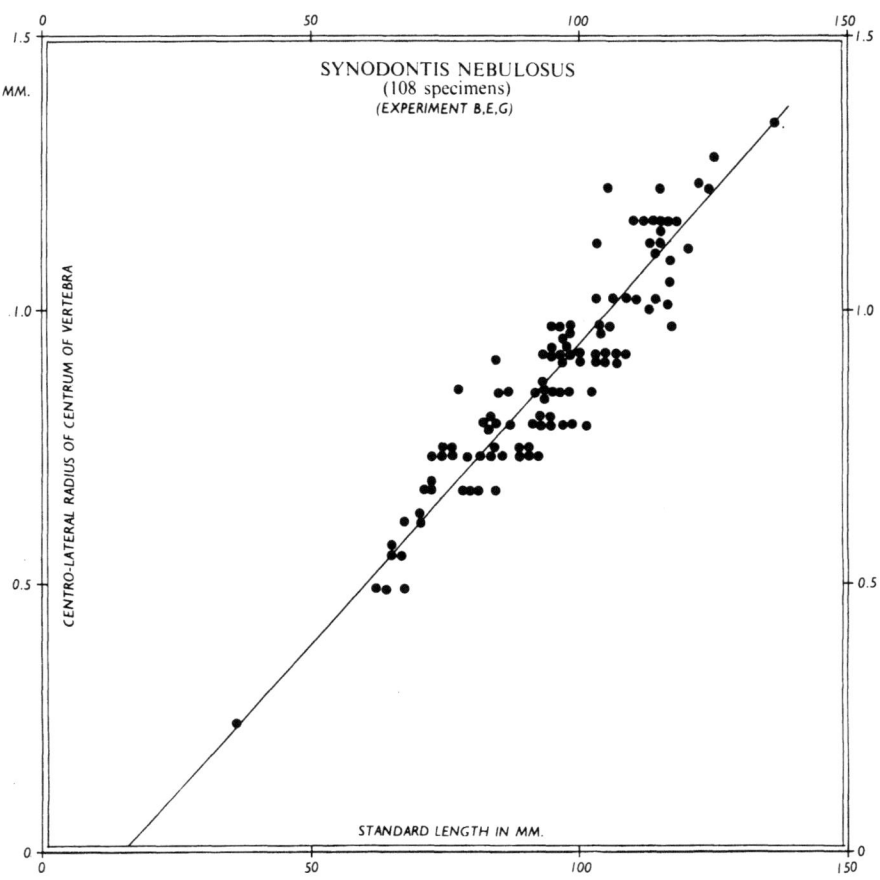

Fig. 28. Body-vertebrae diameter relationship for *S. nebulosus*.

lateral radius of one centrum of a vertebra was measured. The feasibility of employing the vertebrae for age determination in different siluroid fishes has already been confirmed by many authors, e.g. LEWIS (1949), HOOPER (1949), APPELGET & SMITH (1951), HRBAČEK et al. (1952), BIZJAJEV (1952), HRUŠKA & OLIVA (1953), FRANK (1955), HENSEL (1966), and HOCHMAN (1966). For back calculations of growth rates, the method of E. Lea was used, without correction in *M. electricus* and *H. longifilis* (Fig. 26 and 27), with a correction of 15 mm in *S. nebulosus* (Fig. 28), and with a correction of 20 mm in *S. mystus* (Fig. 29).

Published statements about the actual size of the four species examined by us are very rare. BOULENGER (1911) claims that *S. nebulosus* reaches a standard length of 150 mm, *S. mystus* of 340 mm, *H. longifilis* of 720 mm and *M. electricus* of 850 mm. In our material we have found that the largest specimen of *S. nebulosus* was smaller, measuring only 136 mm, *S. mystus* only 221 mm and *M. electricus* only 540 mm.

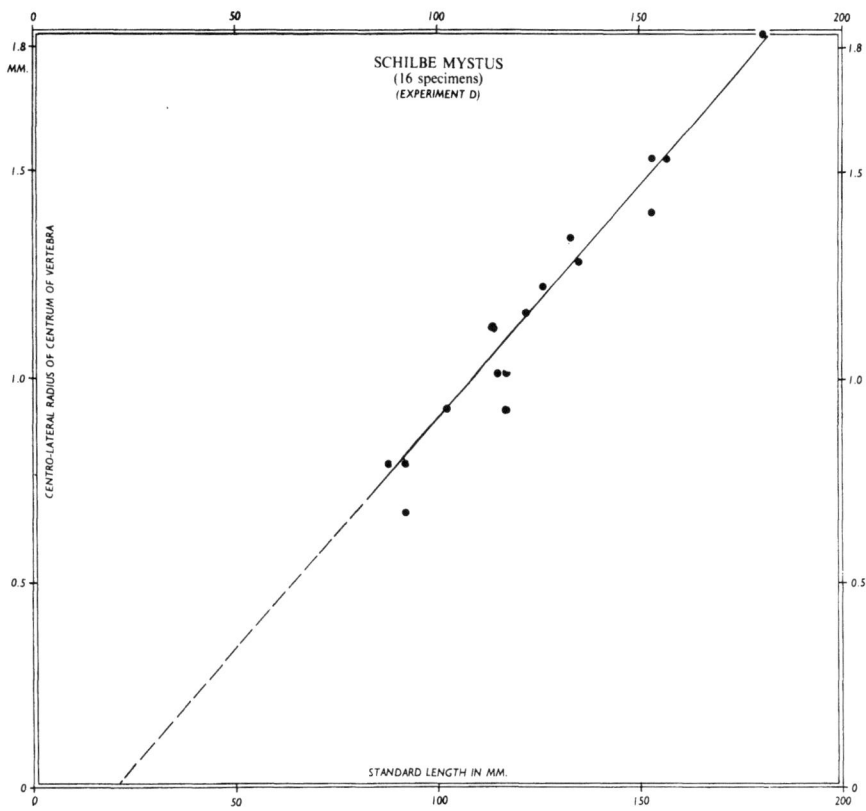

Fig. 29. Body-vertebrae diameter relationship for *S. mystus*.

Conversely the largest *H. longifilis* was much longer than the one already cited, reaching a standard length of 1170 mm. *S. nebulosus* is the slowest growing of these species, with the largest specimens attaining a maximum age of 5 years (Table 37). Only a slightly faster growth rate was confirmed for *S. mystus* whose upper age limit seems to be 6 years (Table 38). *M. electricus* is faster growing and the upper age limit of this species is 11 years (Table 39). Of the four species investigated, the fastest growth was observed in *H. longifilis* which attains an age of at least 13 years (Table 40). No vertebrae were collected from the largest *H. longifilis* because specimens measured in the field were too large for preservation and transportation. There was therefore no possibility of determining upper age limits for this species.

Annual increments of *H. longifilis* and *M. electricus* older than 7 years are very high. This fact is contrary to the experience with most other fish species. From the economical point of view, the most important species are *M. electricus* and *H. longifilis*, which reach not only a large size, but also large annual increments (Table 41).

Table 37. Back calculated growth values of *Synodontis nebulosus* (Experiment B – 10; E – 42; G – 56 specimens).

Age group	Number of specimens	Average standard lengths at capture in mm	Year of life (calculated standard lengths to the end of each vertebrae annulus)					Experiment
			1	2	3	4	5	
2+	2	93	43	70				
4+	7	117 (113–124)	42	57	79	101		B
5+	1	125	41	57	78	94	110	
0+	1	36						
2+	12	89 (72– 98)	50	74				
3+	19	101 (96–107)	46	69	89			E
4+	9	111 (103–122)	41	62	77	101		
5+	1	136	42	59	86	103	120	
1+	9	66 (62– 70)	48					
2+	31	84 (71– 95)	45	70				G
3+	8	99 (95–103)	43	67	88			
4+	8	114 (110–117)	42	60	78	99		
Total	108	Average	44	64	82	100	115	

Table 38. Back calculated growth values of *Schilbe mystus* (Experiment D).

Age group	Number of specimens	Average standard lengths at capture in mm	Year of life (calculated standard lengths to the end of each vertebrae annulus)					
			1	2	3	4	5	6
2+	3	91 (88– 92)	55	82				
3+	7	116 (102–126)	54	82	105			
4+	4	143 (133–153)	53	80	105	130		
5+	1	156	47	69	102	123	145	
6+	1	180	52	73	95	117	138	164
Total	16	Average	52	77	102	123	141	164

Table 39. Back calculated growth values of *Malapterurus electricus* (Experiment C – 7; E – 13; F – 3; G – 16 specimens).

Age group	Number of specimens	Average standard lengths at capture in mm	Year of life (calculated standard lengths to the end of each vertebrae annulus)									
			1	2	3	4	5	6	7	8	9	10
3+	2	173 (163–183)	56	101	140							
4+	2	203 (193–214)	46	85	117	172						
5+	1	230	47	74	116	160	204					
6+	11	315 (275–350)	55	100	149	200	250	286				
7+	15	390 (360–430)	56	103	148	199	259	304	357			
8+	6	426 (400–445)	53	100	154	199	234	289	336	382		
9+	1	430	55	87	126	160	224	262	304	343	394	
10+	1	525	53	98	147	185	239	292	330	371	425	495
Total	39	Average	53	93	137	182	235	287	332	365	405	495

Table 40. Back calculated growth values of *Heterobranchus longifilis* (Experiment C – 7; D – 2; F – 14 specimens).

Age group	Number of specimens	Average standard lengths at capture in mm	Year of life (calculated standard lengths to the end of each vertebrae annulus)											
			1	2	3	4	5	6	7	8	9	10	11	12
1+	5	92 (73–105)	62											
2+	1	165	72	109										
3+	5	208 (195–230)	71	119	179									
4+	6	261 (235–295)	75	129	185	234								
5+	3	332 (310–350)	64	123	180	245	300							
6+	1	370	67	99	155	214	256	321						
11+	1	790	64	132	192	272	352	430	468	538	558	646	696	
12+	1	975	82	158	234	332	426	498	556	602	670	720	774	920
Total	23	Average	69	124	187	260	333	416	512	570	614	683	735	920

Table 41. Average calculated lengths and weights at the end of each year of life of all age groups.

Fish species	Average calculated lengths and weights																							
	Growth seasons																							
	1		2		3		4		5		6		7		8		9		10		11		12	
	mm	g	mm	g	mm	g	mm	g	mm	g	mm	g	mm	g	mm	g	mm	g	mm	g	mm	g	mm	g
Synodontis nebulosus	44	2	64	6	82	13	100	23	115	35														
Schilbe mystus	52	2.1	77	6.5	102	15	123	26	141	37	164	53												
Malapterurus electricus	53	3.9	93	21	137	62	182	150	235	310	287	530	332	820	365	1,100	405	1,500	495	2,800				
Heterobranchus longifilis	69	7.9	124	22	187	82	260	220	333	490	416	990	512	1,870	570	2,600	614	3,200	683	4,200	735	5,600	920	11,500

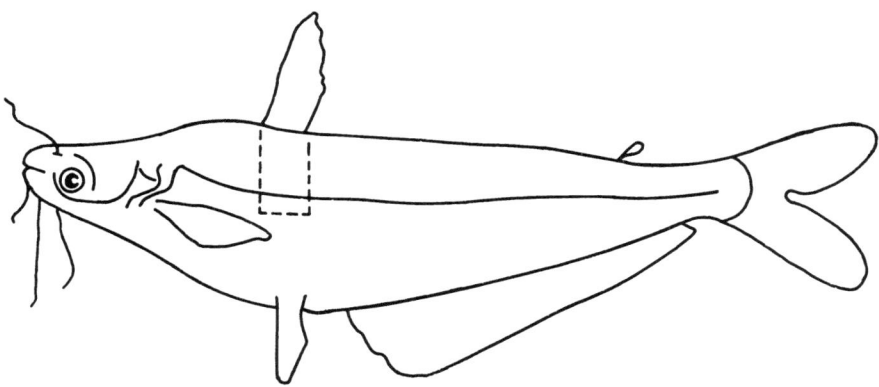

Fig. 30. E. depressirostris; the broken lines show the area from which the vertebrae were removed.

3.6. The silver catfish *Eutropius depressirostris* (Peters, 1852)

by KAREL ČERNY

BALON (1973a) classifies the silver catfish in Lake Kariba as a secondary species and according to him it forms 0.74% of the total number of 40 species found in this lake. Age and growth of *E. depressirostris* have not been previously investigated. Vertebrae exhibited comparatively well-defined annuli and they were used to determine age and growth rates.

The fishes studied were collected in the years 1968 and 1969. Only subsamples from the entire collection were studied. The fish came from the following localities: C – Chikanka Island (31 specimens), D – Island N 20, south-west of Sinazongwe (87 specimens), F – Chipepo Bay cove I (95 specimens), and G – Chipepo Bay cove II (50 specimens). Details concerning the sampling methodology of fishes in the field and detailed descriptions of localities and species composition of the fish stocks, are given by BALON (1971a, 1972, 1973a).

Two hundred and sixty-three specimens of *E. depressirostris* were utilized in this study. Standard lengths in millimeters to the anterior end of the median caudal fin rays were measured and vertebrae were used for the determination of age and growth rate. A partial bibliography of the use of this method is given by FRANK (Section 3.5). The posterior incision was made behind the last ray of the dorsal fin and only some vertebrae in the direction to the cranium were removed (Fig. 30). The rest of the musculature having been mechanically cleaned, 2–3 vertebrae were separated with a small sharp scalpel and the intervertebral substance (remains of the notochord) were then removed. The annuli in the centra of the vertebrae were easily identifiable, and it was not necessary to give treatment with acidified pepsin solution (APPELGET & SMITH, 1951). The vertebrae were inspected in reflected light with the aid of a stereo-microscope, and the radii of the different annuli were measured with an optical micrometer. The number of annuli from the centrolateral radius of vertebrae were determined as indicated by FRANK (see previous Section).

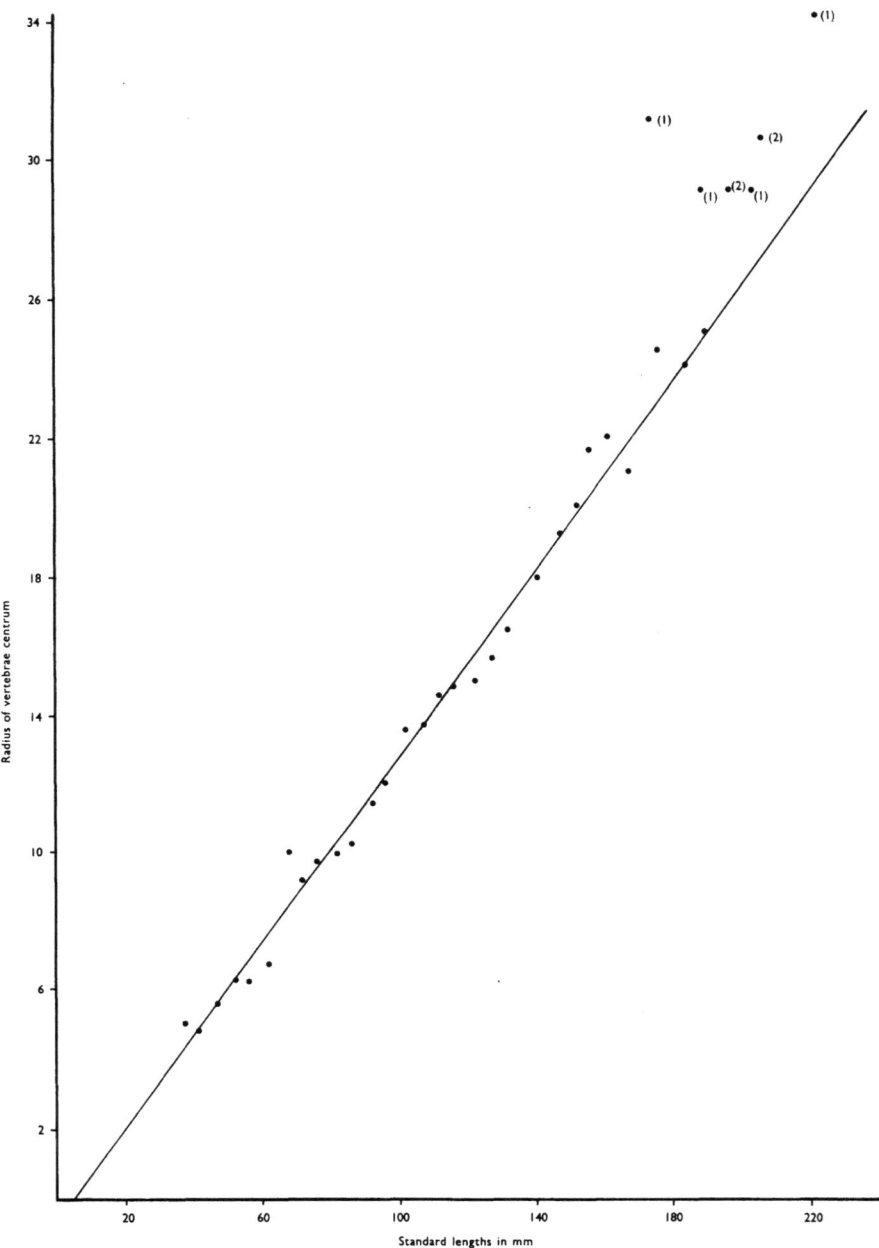

Fig. 31. Relationship between standard length and centro-lateral radius of vertebrae centrum in *E. depressirostris*. Abscissa – standard length in mm; ordinate – centro-lateral radius in micrometer scales (17 micrometer scales = 1.0 mm). Numbers in parentheses indicate the number of specimens in that length interval.

Table 42. Standard lengths and weights of *Eutropius depressirostris* from Chikanka Island (locality C).

Age group	Number of specimens	Ranges and standard lengths in mm at the time of capture	Mean computed standard lengths in mm (above) and weights in grams (below)					
			l_1	l_2	l_3	l_4	l_5	l_6
1+	1	102	65 4					
2+	18	(110–160) 133	62 3.6	101 14				
3+	8	(137–169) 154	58	95 12	126 26			
4+	3	(168–184) 176	64	95	119 23	146 40		
6+	1	223	69	96	128	147	172 66	197 95
Total	31	Average	61 3.4	99 13	124 25	146 40	172 66	197 95

Table 43. Standard lengths and weights of *Eutropius depressirostris* from Island 20, south-west of Sinazongwe (locality D).

Age group	Number of specimens	Ranges and standard lengths in mm at the time of capture	Mean computed standard lengths in mm (above) and weights in grams (below)			
			l_1	l_2	l_3	l_4
0+	1	76				
1+	41	(67–112) 85	66 4.2			
2+	20	(89–149) 111	64 3.9	95 12		
3+	22	(113–158) 137	60	90 10.5	118 22	
4+	3	(136–146) 142	63	94	116 20	139 36
Total	87	Average	64 3.8	92 11	118 22	139 36

The relationship between standard length and centro-lateral radius of the centrum was determined for fishes from all the four localities together and was found to be more or less linear (Fig. 31). The standard lengths were grouped in 5 mm intervals and the average lengths were plotted against the corresponding average radii of centra.

On the basis of the linear like relationship in empiric range between standard length and centro-lateral radius of the centrum, LEE's (1920) method was used for the back-calculation of standard lengths in preceding years of life, with a correction value of 6 mm which was extrapolated graphically.

Table 44. Standard lengths and weights of *Eutropius depressirostris* from cove I of Chipepo Bay (locality F).

Age group	Number of specimens	Ranges and standard lengths in mm at the time of capture	Mean computed standard lengths in mm (above) and weights in grams (below)					
			l_1	l_2	l_3	l_4	l_5	l_6
0+	30	(39– 72) 53						
1+	17	(80–128) 101	66					
			4.2					
2+	32	(99–148) 125	64	98				
			3.8	13				
3+	7	(130–150) 142	70	100	121			
				14	23			
4+	5	(153–174) 160	61	88	110	137		
					18	34		
5+	2	(175–190) 182	71	98	120	142	164	
						38	58	
6+	2	(211–212) 211	55	82	105	134	164	185
							58	80
Total	95	Average	65	97	115	137	164	185
			4	12.5	20	34	58	80

Body weights at different lengths were computed from the standard length-body weight relationship on a double logarithmic scale (BALON, 1974b). The symbols 1+, 2+, 3+, etc., are used for the succeeding age groups.

In the calculation of biomass, available production and available yield, growth values from this study were used in conjunction with the data provided from field samples. Available production and yield were calculated using the three different weight increments (CHITRAVADIVELU, Section 3.2). The computation of the available yield was based on the minimum harvestable size of the fishes. The latter was established for all localities as proposed by BALON (1961, 1971a, 1972). On this basis the available yield in all localities was calculated as the sum of the available production of all age groups beginning with the third.

The rate of growth of fishes from all four localities was evaluated separately. The results are shown in Tables 42 to 45, together with mean wet weights for each group. The total average values of growth for all samples from Lake Kariba are arranged in Table 46. The smallest specimen in the samples is 35 mm (from locality G) and belongs to 0 age group, while the oldest and largest specimen is 6+ years old and 223 mm (from locality C). In all localities the predominant age groups are 1+ and 2+ (with the exception of locality C, where 1+ age group is very small). In locality F, 0+ specimens were abundant, as were 3+ fish from locality D. This indicates that *E. depressirostris* is a fish with a rather short life span. The growth rates in different localities are similar. The highest rate of growth was found in locality C, where the abundance and standing stock was lowest. In *E. depressirostris* stock density appears to be responsible for the difference in growth rates, although HOLČIK (Section 3.1) attributes a negligible role to population density for the difference in rates of growth

Table 45. Standard lengths and weights of *Eutropius depressirostris* from cove II of Chipepo Bay (locality G).

Age group	Number of specimens	Ranges and standard lengths in mm at the time of capture	Mean computed standard lengths in mm (above) and weights in grams (below)				
			l_1	l_2	l_3	l_4	l_5
0+	9	(35– 64) 49					
1+	19	(82–117) 106	67				
			4.4				
2+	9	(113–126) 120	64	95			
			3.8	12			
3+	4	(128–152) 143	64	90	114		
				10.5	20		
4+	7	(156–197) 173	61	89	122	144	
					24	39	
5+	2	(195–204) 199	61	88	120	145	174
						40	69
Total	50	Average	65	92	119	144	174
			4	11	22	39	69

Table 46. Mean values of lengths, weights and annual increments of all samples of *Eutropius depressirostris* from Lake Kariba.

Parameters	Means at the end of each growth season					
	1	2	3	4	5	6
Standard length in mm	64	95	119	141	168	189
Annual length increments in mm	64	31	24	22	27	22
Annual length increments as % of 1st year length	100	48.4	37.5	34.4	42.2	34.4
Length as % of final year length	33.9	50.3	63	74.6	88.9	100
Weights in grams	3.7	11.9	22	37	62	85
Annual weight increments in grams	3.7	8.2	10.1	15	25	23

in *Sargochromis codringtoni*. There are considerable ranges of fish lengths attained in different localities. In locality D, for example, the largest specimens of age 2+ have a higher standard length than the smallest specimens of age 4+.

Biomass, available production and yield were determined for each locality separately (Tables 47 to 51). The sample from locality C had 31 specimens. The abundance in this locality was 25 specimens per ha. When the 31 fishes from this locality were divided according to age, it was not possible to compute accurately the biomass and, therefore, the number of fish per 10 ha was used in the computations.

With the exception of locality G, the values of available production calculated on the basis of the difference in weight between succeeding age groups is lower than the values of available production calculated using the mean weight at the end of the

Table 47. Biomass (B'), available production (P') and available yield ($Y_{P'}$) of *E. depressirostris* from Chikanka Island (locality C), calculated up to the time of annulus formation.

Age group	Number per ha	Using W_i and W_{i-1} of the same age group					Using W_i of succeeding age groups				Using grand average weight				
		W	B' kg/ha	w	P' kg/ha	Y' kg/ha	w	P' kg/ha	Y' kg/ha		W	B' kg/ha	w	P' kg/ha	Y' kg/ha
1	8	4	0.003	4	0.003		4	0.003			3.4	0.003	3.4	0.003	
2	145	14	0.203	10.4	0.151		10	0.145			13	0.189	9.6	0.139	
3	65	26	0.169	14	0.091	0.091	12	0.078	0.078		25	0.163	12	0.078	0.078
4	24	40	0.096	17	0.041	0.041	14	0.034	0.034		40	0.096	15	0.036	0.036
6	8	95	0.076	29	0.023	0.023	29	0.023	0.023		95	0.076	29	0.023	0.023
Total	250		0.547		0.309	0.155		0.283	0.135			0.527		0.279	0.137

Average

B' kg/ha 0.537
P' kg/ha 0.290
$Y_{P'}$ kg/ha 0.142

P'/B' in % 54.00
$Y_{P'}/P'$ in % 48.97

W = absolute weights in grams; w = weight increments in grams.

Table 48. Biomass (B′), Production (P′) and Yield (Y$_{P'}$) of *Eutropius depressirostris* from Island 20, south-west of Sinazongwe (locality D), calculated up to the time of annulus formation.

Age group	Number per ha	Using W$_n$ and W$_{n-1}$ of the same age group				Using W$_n$ of succeeding age groups				Using grand average weight				
		W	B′ kg/ha	w	P′ kg/ha	Y$_{P'}$ kg/ha	W	w	P′ kg/ha	Y$_{P'}$ kg/ha	W	B′ kg/ha	w	P′ kg/ha	Y$_{P'}$ kg/ha
0	3	6.4	0.02								6.4	0.02			
1	122	4.2	0.51	4.2	0.51		4.2		0.51		3.8	0.46	3.8	0.46	
2	60	12	0.72	8.1	0.49		7.8		0.47		11	0.66	7.2	0.43	
3	66	22	1.45	11.5	0.75	0.75	10		0.66	0.66	22	1.45	11	0.73	0.73
4	9	36	0.32	16	0.14	0.14	14		0.13	0.13	36	0.32	14	0.13	0.13
Total	260		3.02		1.89	0.89			1.77	0.79		2.91		1.75	0.86

Average

B′ kg/ha 2.97
P′ kg/ha 1.80   P′/B′ in % 60.61
Y$_{P'}$ kg/ha 0.85   Y$_{P'}$/P′ in % 47.22

W = absolute weights in grams; w = weight increments in grams.

Table 49. Biomass (B′), Production (P′) and Yield ($Y_{P'}$) of *Eutropius depressirostris* from cove I of Chipepo Bay (locality F), calculated up to the time of annulus formation.

Age group	Number per ha	Using W_n and W_{n-1} of the same age group					Using W_n of succeeding age groups			Using grand average weight				
		W	B′ kg/ha	w	P′ kg/ha	$Y_{P'}$ kg/ha	w	P′ kg/ha	$Y_{P'}$ kg/ha	W	B′ kg/ha	w	P′ kg/ha	$Y_{P'}$ kg/ha
0	154	2.2	0.34							2.2	0.34			
1	87	4.2	0.37	4.2	0.37		4.2	0.37		4	0.35	4	0.35	
2	164	13	2.13	9.2	1.51		8.8	1.44		12.5	2.05	8.5	1.39	
3	36	23	0.83	9	0.32	0.32	10	0.36	0.36	20	0.72	7.5	0.27	0.27
4	26	34	0.88	16	0.42	0.42	11	0.29	0.29	34	0.88	14	0.36	0.36
5	10	58	0.58	20	0.20	0.20	24	0.24	0.24	58	0.58	24	0.24	0.24
6	10	80	0.80	22	0.22	0.22	22	0.22	0.22	80	0.80	22	0.22	0.22
Total	487		5.93		3.04	1.16		2.92	1.11		5.72		2.83	1.09

Average B′ kg/ha 5.82
P′ kg/ha 2.93
$Y_{P'}$ kg/ha 1.12

P′/B′ in % 50.34
$Y_{P'}$/P′ in % 38.23

W = absolute weights in grams; w = weight increments in grams.

Table 50. Biomass (B'), Production (P') and Yield ($Y_{P'}$) of *Eutropius depressirostris* from cove II of Chipepo Bay (locality G), calculated up to the time of annulus formation.

Age group	Number per ha	Using W_n and W_{n-1} of the same age group					Using W_n of succeeding age groups				Using grand average weight				
		W	B' kg/ha	w	P' kg/ha	$Y_{P'}$ kg/ha	W	w	P' kg/ha	$Y_{P'}$ kg/ha	W	B' kg/ha	w	P' kg/ha	$Y_{P'}$ kg/ha
0	130	1.8	0.23								1.8	0.23			
1	275	4.4	1.21	4.4	1.21		4.4		1.21		4	1.10	4	1.10	
2	130	12	1.56	8.2	1.07		7.6		0.99		11	1.43	7	0.91	
3	58	20	1.16	9.5	0.55	0.55	8		0.46	0.46	22	1.28	11	0.64	0.64
4	101	39	3.94	15	1.52	1.52	19		1.92	1.92	39	3.94	17	1.72	1.72
5	29	69	2.00	29	0.84	0.84	30		0.87	0.87	69	2.00	30	0.87	0.87
Total	723		10.10		5.19	2.91			5.45	3.25		9.98		5.24	3.23

Average
B' kg/ha 10.04
P' kg/ha 5.29
$Y_{P'}$ kg/ha 3.13

P'/B' in % 52.69
$Y_{P'}/P'$ in % 59.17

W = absolute weights in grams; w = weight increments in grams.

Table 51. Summary of Mass (B'), Production (P') and Yield ($Y_{P'}$) of *Eutropius depressirostris* in Lake Kariba.

Locality	B' kg/ha	P' kg/ha	P'/B' in %	$Y_{P'}$ kg/ha	$Y_{P'}/P'$ in %
C – Chikanka Island	0.537	0.29	54.00	0.142	48.97
D – Island 20, south-west of Sinazongwe	2.97	1.80	60.61	0.85	47.22
F – cove I of Chipepo Bay	5.82	2.93	50.34	1.12	38.23
G – cove II of Chipepo Bay	10.04	5.29	52.69	3.13	59.17
Average for Lake Kariba	4.84	2.60	54.41	1.13	48.40

formation of the ultimate and penultimate annuli. This is in agreement with PIVNIČKA's results (Section 3.4) for *Clarias gariepinus* and differs from those of CHITRAVADIVELU (Section 3.2) for *Synodontis zambezensis*.

The highest available production was found in locality G. It is 5.29 kg/ha. The lowest available production was found in locality C. It is 0.29 kg/ha. In locality D the production is 1.80 kg/ha and in locality F 2.93 kg/ha.

The available production/biomass ratio in these four localities varies from 50.3% to 60.6%. The available yield/production ratio ranges from 38.2% to 59.2%. The mean available production of *E. depressirostris* in Lake Kariba is 2.60 kg/ha or 54.4% of the biomass and the average available yield is 1.31 kg/ha or 48.4% of the available production. The minimum harvestable size is reached in the third year of life, when the average standard length of *E. depressirostris* is 119 mm and average body weight is 22 g.

The calculated biomass in the localities D, F and G differs considerably from results of the standing stock. For instance, the calculated biomass in locality F is 5.82 kg/ha and standing stock at the time of the experiments was 11.05 kg/ha. The situation is similar in two other localities. The differences in localities F and G result from the experiments made some months after the time of annulus formation. According to BALON (1971a, 1972), annuli are formed during December to March. Because the biomass in the present work was computed for the time of annulus formation, it is necessarily lower.

Fish from locality D, which were collected from 11–14 December 1968, all had recently formed annuli. The difference, at this locality, between the biomass of BALON's total sample and that calculated from my sub-sample may be due to the sub-sample's biased age distribution. In locality C, the calculated biomass is different from that of other localities and is slightly higher than the standing stock at the time of sampling. It is caused by the fact that the number of fish in the investigated sample corresponds to the number of fish actually collected after poisoning of the locality, but in the investigated sample there were fish of longer standard lengths.

3.7. The kurper bream *Sarotherodon mossambicus mortimeri* (Trewavas, 1966)

by IVAN KRUPKA

Sarotherodon mossambicus is one of the native African species that has been artifically distributed because of its importance for warm water fish culture. Outside of its endemic occurrence it was first recorded in Java just before World War II, however, its presence there has not been explained. Gradually the bream spread over all of Indonesia, the Malay Peninsula, Philippines, Thailand, Taiwan, Southern Korea, Israel, etc., and in the Western Hemisphere over the British West Indies, Haiti and Texas (NORMAN, 1963).

In addition to these, various countries have bred the kurper bream or have conducted experiments with introductory breeding. A short review as well as an outline of the biology of *S. mossambicus*, in natural and laboratory conditions were given by MIRONOVA (1969).

According to BALON's data (1972, 1973a), the kurper bream in Lake Kariba is the most abundant species of the economically preferred group. Its share is 10% of the total number of fish present in the lake and it was found in all localities studied. The densest stock of kurper bream was located in the estuary of the intermittent brook on Chete Island (locality E_3) – 218.49 kg per one hectare and in a cove of the eastern shore of Chipepo Bay (locality G) – 224.53 kg/ha. The smallest stock density occurred in a neighboring cove of the eastern shore of Chipepo Bay (locality F) where the standing stock was only 8.41 kg/ha. All mentioned values represent the whole standing stock, including the young-of-the-year fish.

I had at my disposal samples of kurper bream scales from the following localities: Island N 102 (locality B – 3 specimens); Chikanka Island (locality C – 146 specimens); Island N 20 (locality D – 94 specimens); Island N 21 (locality E_1 – 154 specimens); locality E_2 (33 specimens); locality E_3 (90 specimens); Chipepo Bay I (locality F – 41 specimens) and Chipepo Bay II (locality G – 33 specimens).

Scale reading and measuring (diagonal – mostly ventral – scale radius) was done at $45 \times$ magnification. The body-scale relationship of *S. mossambicus mortimeri* is linear (Fig. 32). The extrapolated correction value is 11 mm. The scale formation probably sets in at this size, as is evident from comparison made with FISHELSON's (1966) data. Back calculated body lengths for previous years of life were obtained by means of R. Lee method. Corresponding weights were read off the log – log length-weight nomogram (BALON, 1974b).

Age composition of the bream population, mortality and survival rate and relative indexes of length and weight growth were estimated according to BALON's (1971a, 1972) methods. Available production and yield was determined for only five localities from which all population data were available. Available production/yield data were ascertained by methods previously employed by HOLČIK (Section 3.1). Material at my disposal did not enable me to determine the exact time of annulus formation although it probably coincides with the beginning of the warm season which, according to COCHE (1968) is October. It was estimated hypothetically for November 1st.

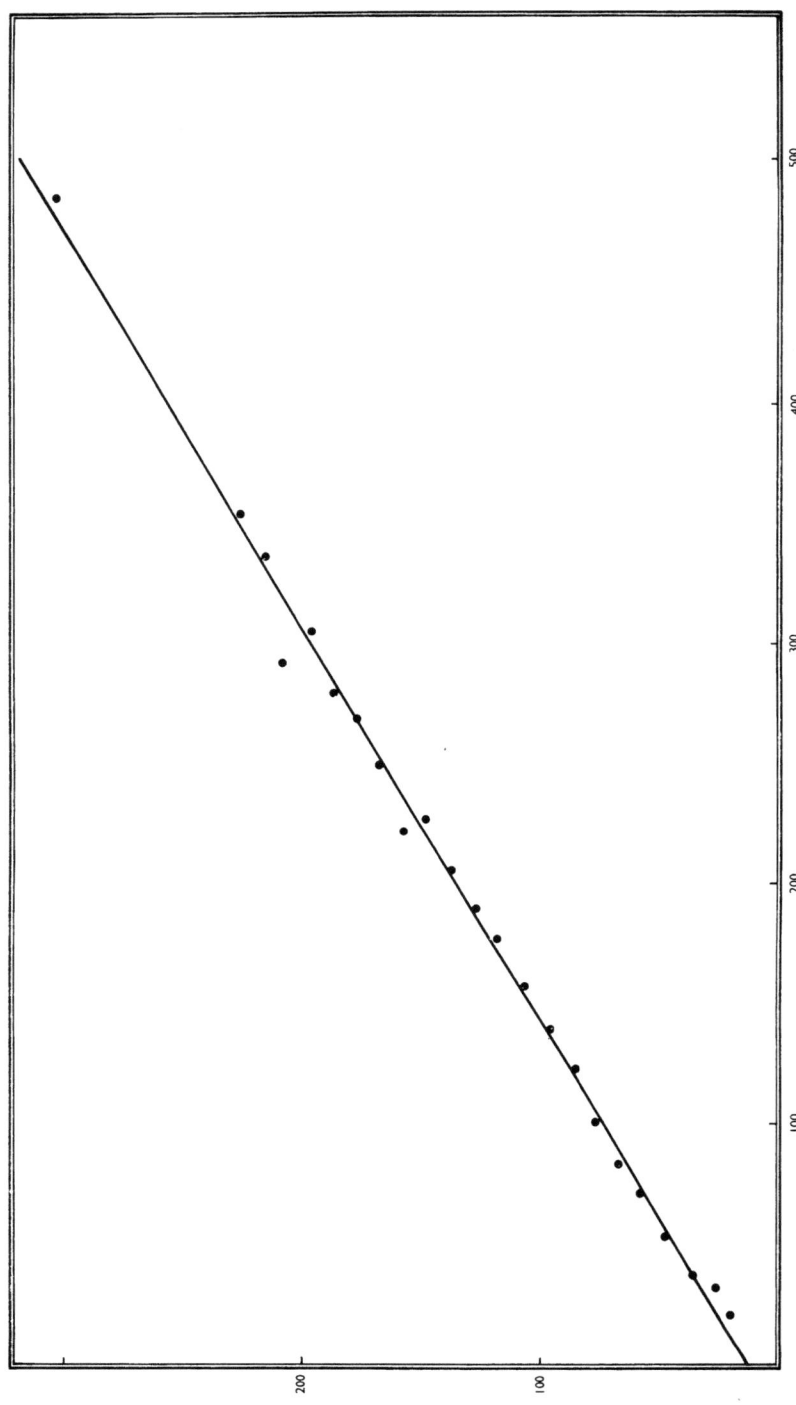

Fig. 32. Body-scale relationship of S. mossambicus mortimeri.

Fig. 33. Scale of 1+ old *S. mossambicus mortimeri*. Locality E, standard length 72 mm.

The following values of biomass were used: total standing stock of bream found after block-off sampling – 97.07 kg/ha, total standing stock without the young-of-the-year-fish – 90.79 kg/ha, total biomass calculated to the time of annulus formation – 93.96 kg/ha. Calculating available production and available yield for the whole lake I have taken into consideration 65% of the total lake area, i.e. 342 980 ha to be inhabited by fish.

The scales of kurper bream are mostly oval (Fig. 33). Annulus is visible on the diagonal or the lateral part of scales and only rarely on the caudal part (Fig. 34). The annulus is easily discernible in most fishes with the exception of the older ones where some accessory checks are found. The annulus characteristic is the same as in *Tilapia rendalli* (BASTL, section 3.3) or *Sargochromis codringtoni* (HOLČIK, section 3.1). In 69% of the fish the juvenile mark was found on the oral part of scale. The values of back-calculated lengths for this mark ranged from 25 to 39 mm of standard length (34 mm in average).

The largest specimens at my disposal measured 390 mm of standard length and weighed 2900 grams at the age of 8+. Fishes older than 4+ occur only rarely.

According to BROCK (1954), one specimen of *S. mossambicus* bred in a New York aquarium reached 7 years, but ROUX (1961) writes that this species can live up to 11 years, at which time it reaches the size of 38–39 cm. He states the weight of 700

Fig. 34. Scale of 4+ old *S. mossambicus mortimeri*. Locality E, standard length 208 mm.

Table 52. Age composition of *Sarotherodon mossambicus mortimeri* stocks from Lake Kariba (in % of total sample).

Locality – Age	0+	1+	2+	3+	4+	5+	6+	8+
Chikanka Island (C)	2.32	50.84	27.43	13.92	5.49			
Island 20 (D)	9.82	60.70	23.58	4.59	1.31			
Island 21 (E)	21.01	23.77	25.95	11.71	16.99	0.57		
Brook (E₂)	82.82	10.00	3.34	2.82	1.03			
Chipepo Bay I (F)	38.00	32.00	30.00					
Chipepo Bay II (G)	17.39	39.13	6.52	8.70	13.04	6.52	4.35	4.35
Total	23.93	30.65	23.28	9.98	11.63	0.43	0.05	0.05

grams at a length of 36 cm. In comparison with our material this weight seems to be rather low, because our specimens of similar length (344 mm) weighed 2100 grams.

Only localities C, D, E_2, F and G were evaluated for age and length composition of *S. mossambicus mortimeri* (Table 52). Other localities were excluded from calculations (B – small sample, E_1 – higher number of specimens in older age groups, unsuitable for mortality/survival calculations, E_2 – only the age groups 0 and I were present).

The highest mortality rate in localities C and D was between the second and third year of life; in locality E_3 and G between the first and second (Table 53). The total annual mortality rate varied with different stocks from 6 to 67%. Differences in total mortality rate of *T. rendalli* from Lake Kariba were recorded also by BASTL (section 3.3). He found a paradoxical occurrence of the highest mortality rate in localities with the lowest stock density. With *S. mossambicus mortimeri* the situation seems to be similar: the highest mortality (67%) was found in locality D, where the biomass was 22.51 kg/ha (i.e. second highest stock density). But in locality F, where the lowest biomass was recorded (10.05 kg/ha) the mortality also shows only small values – 6%. In the latter, however, only specimens of 1+ and 2+ were present.

No essential differences were found in the growth of the kurper bream in Lake Kariba. The exception is locality G (Chipepo Bay II) where some specimens with an abnormally fast growth rate were found (two fishes in age group 5, one specimen in age group 6 and two specimens in age group 8) and locality E (Chete Island). The shores of these two localities have a high density of wild ungulates (Table 54 to 57).

The fastest growth in bream occurs in the first year of life – 55 mm on the average. In subsequent years the growth evens out or decreases a little as can be seen from the following mean increments: in the second year 37 mm, in the third 35 mm and in the fourth 34 mm.

For this species bred in North Vietnamese ponds, LE KUANG LONG and others (1961) record a length of 32 cm and weight of 850 grams at 10 months of age. But ATZ (1954 according to MIRONOVA, 1969) mentions the length of 40 cm and weight of 2555 grams for the same age. In all these cases, however, the fishes were intensively fed. In laboratory conditions (MIRONOVA, 1969) this species attained 413

Table 53. Age composition, seasonal total mortality rate (Z') survival rate (S) and instantaneous rate of total mortality (Z) of *Sarotherodon mossambicus mortimeri* (total population).

Locality		Age groups							
		1	2	3	4	5	6	8	
Chikanka Island (C)	n	241	130	66	26				463
	%	52.05	28.08	14.25	5.62				100.0
	Z'		0.4621	0.4934	0.3750				0.4934
	S		0.5379	0.5066	0.6250				0.5066
	Z		0.62	0.68	0.47				0.68
Island 20 (D)	n	278	108	21	6				413
	%	67.31	26.15	5.08	1.45				100.0
	Z'		0.6094	0.8060	0.7135				0.6671
	S		0.3906	0.1940	0.2865				0.3329
	Z		0.94	1.64	1.25				1.10
Chete[1] stream (E$_2$)	n	39	13	11	4				67
	%	58.21	19.40	16.42	5.97				100.0
	Z'		0.6638	0.3229	0.0952				0.5551
	S		0.3362	0.6771	0.9048				0.4448
	Z		1.09	0.39	1.01				0.81
Chipepo Bay I (F)	n	16	15						31
	%	51.61	48.39						100.0
	Z'		0.0582						0.0582
	S		0.9418						0.9418
	Z		0.06						0.06
Chipepo Bay II (G)	n	18	3	4	6	3	2	2	38
	%	47.37	7.89	10.53	15.79	7.89	5.26	5.26	100.0
	Z'		0.8330			0.4984	0.3297		0.5323
	S		0.1670			0.5016	0.6703		0.4677
	Z		1.79			0.69	0.40		0.76

[1] see Elephant Stream.

grams when two years old and 750 grams or more when three years old. *S. mossambicus* is a highly adaptable species – its size strongly depends on life conditions. In big reservoirs such as the large African lakes *S. mossambicus* reached maximal sizes while in small water bodies its growth is very slow. It is obvious that the growth is strongly influenced by the quality of food. *S. mossambicus* is a species which feeds mostly on plants (LORANT, 1959; MATTHES, 1966). Larger specimens eat mosquito larvae, midges and other insects and worms but also small fishes (KRAEV, 1966). The question which is as yet unanswered is the influence of population density on the growth of this and related species. BASTL (Section 3.3) found better growth of *T. rendalli* in localities with highest density values. HOLČIK (Section 3.1) found that the same applies to the *Sargochromis codringtoni*. In *S. mossambicus mortimeri*, however, this phenomenon has not been observed.

The standing stock of kurper bream in particular localities of Lake Kariba was higher at the time of poisoning than at the time of supposed date of annulus formation (Table 58 to 62). The only exception is locality G (Chipepo Bay II) where the abnormally fast growth of some older fishes influenced subsequent results. A considerable difference between two biomasses can also be due to calculation errors of mortality data, annulus formation determination and migration of population, as discussed by HOLČIK (section 3.1). Because of the same error in the available production calculation, the P'/B' ratio, however, should be correct. The ratio in different localities varied on an average from 35.30% to 74.56% with a mean of 46.32%. The highest available production of bream was determined for locality G – 103.94 kg/ha and the lowest was in locality F – 2.85 kg/ha, however, here only the youngest fishes were present (age groups 1 and 2). The mean annual available production, as calculated from five localities, is equal to 43.52 kg/ha. Considering this figure to be valid for the whole area of Lake Kariba inhabited by fish (342 980 ha) the available production of kurper bream is 14 926 metric tons.

Minimum harvestable size of kurper bream in Lake Kariba is reached approximately in the third year of life when the fish attains about 127 mm of standard length and 90 grams in weight. The mean percentage of available yield to available production is 73.99%. It means that it may be feasible to expand the harvest to 32.20 kg/ha yearly. Considering the entire lake area inhabited by fish, the whole available yield of kurper bream in Lake Kariba could be 11 044 metric tons.

3.8. The cornish-jack *Mormyrops deliciosus* (Leach, 1818) and the bottlenose *Mormyrus longirostris* Peters, 1852

Although I decided to treat both species together, mainly because of their similar catch size and commercial value, biologically they differ considerably. The cornish-jack *(Mormyrops deliciosus)* becomes in adult life a predator and a solitary animal. The bottlenose *(Mormyrus longirostris)* feeds upon aquatic insects and can be seen in small schools, especially in the juvenile period. Furthermore the latter has a well developed caudal electric organ which produces an electric current. When a living fish is grasped by the caudal peduncle it transmits a shock in the form of rapid pulses.

In my 10 quantitative samples the cornish-jack was present only in five as 3 to 361 (mean 159) specimens per ha and as 7 to 205 (mean 55) kg/ha. The bottlenose was present in the same five samples as 1 to 60 (mean 25) specimens/ha and 0.16 to 61 (mean 18) kg/ha. They probably live in the same habitats. The cornish-jack is far more abundant in Lake Kariba than the bottlenose. Only the single sampling area of locality H, situated close to a rivulet estuary, had a greater abundance of bottlenose during the spawning season – 19 kg/ha as compared to 7 kg/ha of cornish-jack. However, in commercial catches the proportion of the bottlenose is usually higher than that of the cornish-jack. Generally speaking, both species belong to the less abundant taxa of Lake Kariba and when placed in samples in order of numerical occurrence they came seventh and eighth (cornish-jack 0.47% and bottlenose 0.05% of the total number of all fish). However, principally because of their large size, the order in respect to weight

Table 54. Back calculated growth values of *Sarotherodon mossambicus mortimeri* from Lake Kariba.

Locality	Age groups	Year of hatching	n	Mean standard length at time of capture	Back calculated standard lengths (mm) for the previous growth seasons							
					l_j	l_1	l_2	l_3	l_4	l_5	l_6	l_7	l_8
Island 102 (B) 20 March 1968	0+	1967/68	2	67	25								
	1+	1966/67	1	92	25	57							
	Total		3		25	57							
Chikanka Island (C) 21 June 1968	0+	1967/68	5	40	33								
	1+	1966/67	54	68	34	51							
	2+	1965/66	30	113	35	55	82						
	3+	1964/65	38	158	35	55	90	123					
	4+	1963/64	19	191	37	54	95	135	166				
	Total		146		35	54	89	129	166				
Island 20 (D) 11–14 December 1968	0+	1967/68	12	34	30								
	1+	1966/67	52	77	29	59							
	2+	1965/66	32	124	33	64	102						
	3+	1964/65	6	145	32	58	88	129					
	4+	1963/64	2	160	32	53	87	140	155				
	Total		94		31	58	93	134	155				
Island 21 (E) 21–25 March 1969	0+	1968/69	32	41	33								
	1+	1967/68	61	83	34	52							
	2+	1966/67	43	128	34	54	92						
	3+	1965/66	13	160	33	55	97	123					
	4+	1964/65	4	191	35	49	100	123	157				
	5+	1963/64	1	337	38	42	123	160	310	328			
	Total		154		35	50	103	135	233	328			
(E₁)	0+		31	40	34								
	1+		2	58	31	45							
	Total		33		33	45							

Table 54. (continued).

Locality	Age groups	Year of hatching	n	Mean standard length at time of capture	Back calculated standard lengths (mm) for the previous growth seasons							
					l_j	l_1	l_2	l_3	l_4	l_5	l_6	l_7	l_8
Chete Stream (E_2) 25 May 1969	0+	1968/69	34	35	33								
	1+	1967/68	28	76	34	50							
	2+	1966/67	12	118	33	56	89	111					
	3+	1965/66	12	132	32	52	85	121					
	4+	1964/65	4	175	34	52	82	116	153				
	Total		90		33	52	85	116	153				
Chipepo Bay I (F) 26–31 May 1969	0+	1968/69	11	51	34	55							
	1+	1967/68	18	72	36	55	100						
	2+	1966/67	12	122	33	75	100						
	Total		41		34	65	100						
Chipepo Bay II (G) 28–31 May 1969	0+	1968/69	4	52	34	54							
	1+	1967/68	19	70	37	42	93						
	2+	1966/67	2	121	35	61	103	139					
	3+	1965/66	2	162	33	57	88	128	176				
	4+	1964/65	1	195	39	61	153	205	235	320			
	5+	1963/64	2	328	39	59	136	288	319	330	354		
	6+	1962/63	1	371	37	78	207	260	306	324	334	353	359
	8+	1961/62	2	370	37								
	Total		33		36	59	130	204	259	325	344	353	359
Grand average (without 5–8 age group)	1+		235	74	33	53	93						
	2+		131	121	34	58	93	125					
	3+		71	151	34	56	93	129	161				
	4+		30	182	35	53	90	127	161				
	Total (without 5–8 age group)		467		34	55	92	127	161				

Table 55. Lengths and increments for all studied samples of *Sarotherodon m. mortimeri* from Lake Kariba.

Locality	Parameter	Means at the end of growth seasons							
		1	2	3	4	5	6	7	8
Chikanka Island (C)	Standard length (mm)	54	89	129	166				
	Length increments (mm)	54	35	40	37				
	Length increments in % of length of first season	100.0	64.8	74.1	68.5				
	Standard length in % of standard length of final season	32.5	53.6	74.7	100.0				
	Weight in grams	7	32	90	210				
	Weight increments in grams	7	25	58	120				
Island 20 (D)	Standard length (mm)	58	93	134	155				
	Length increments (mm)	58	35	41	21				
	Length increments in % of length of first season	100.0	60.3	70.7	36.2				
	Standard length in % of standard length of final season	37.4	60.0	86.5	100.0				
	Weight in grams	8	36	110	174				
	Weight increments in grams	8	28	74	64				
Island 21 (E)	Standard length (mm)	50	103	135	233	328			
	Length increments (mm)	50	53	32	98	95			
	Length increments in % of length of first season	100.0	106.0	64.0	196.0	190.0			
	Standard length in % of standard length of final season	15.2	31.4	41.2	71.0	100.0			
	Weight in grams	5	52	113	610	1700			
	Weight increments in grams	5	47	61	497	1090			

Table 55. (continued).

Locality	Parameter	Means at the end of growth seasons							
		1	2	3	4	5	6	7	8
(E₁)	Standard length (mm)	45							
	Length increments (mm)	45							
	Length increments in % of length of first season	100.0							
	Standard length in % of standard length of final season	100.0							
	Weight in grams	4							
	Weight increments in grams	4							
(E₂)	Standard length (mm)	52	85	116	153				
	Length increments (mm)	52	33	31	37				
	Length increments in % of length of first season	100.0	63.5	60.0	71.1				
	Standard length in % of standard length of final season	34.0	55.6	75.8	100.0				
	Weight in grams	6	28	70	170				
	Weight increments in grams	6	22	42	100				
Chipepo Bay I (F)	Standard length (mm)	65	100						
	Length increments (mm)	65	35						
	Length increments in % of length of first season	100.0	53.8						
	Standard length in % of standard length of final season	65.0	100.0						
	Weight in grams	12	46						
	Weight increments in grams	12	34						
Chipepo Bay II (G)	Standard length (mm)	59	130	204	259	325	344	353	359
	Length increments (mm)	59	71	74	55	66	19	9	6
	Length increments in % of length of first season	100.0	120.0	125.4	93.2	11.9	32.2	15.3	10.2
	Standard length in % of standard length of final season	16.4	36.2	56.8	72.1	90.5	95.8	98.3	100.0
	Weight in grams	9	100	400	820	1680	2100	2250	2300
	Weight increments in grams	9	91	300	420	860	420	150	50

Table 56. Mean values of lengths, weight and increments of *Sarotherodon mossambicus mortimeri* population from Lake Kariba.

Parameter	Means at the end of growth seasons			
	1	2	3	4
Standard length (mm)	55	92	127	161
Increments in mm	55	37	35	34
Length increments in % of length of first season	100	67.3	63.6	61.8
Length in % of length of final season	34.2	57.1	78.3	100
Weight in grams	7	36	90	200
Weight increments in grams	7	29	54	110

Table 57. Absolute increments in cm (h), indexes of the species' average size (φH), specific weight rates of growth (Cw), and the indexes of the population weight growth intensity (φCw) of *Sarotherodon mossambicus mortimeri* from Lake Kariba.

Locality	Index	1	2	3	4	5	6	7	8
Chikanka Island (C)	h	5.4	3.5	4.0	3.7				
	φH				4.15				
	Cw	700	357.14	181.25	133.33				
	φCw				223.91				
Island 20 (D)	h	5.8	3.5	4.1	2.1				
	φH				3.88				
	Cw	800	350.00	205.55	58.18				
	φCw				204.57				
Island 21 (E)	h	5.0	5.3	3.2	9.8	9.5			
	φH				6.56				
	Cw	500	940.00	117.31	439.82	178.69			
	φCw				418.96				
Chete stream (E_2)	h	5.2	3.3	3.1	3.7				
	φH				3.82				
	Cw	600	366.67	150.00	142.86				
	φCw				219.84				
Chipepo Bay I (F)	h	6.5	3.5						
	φH				5.00				
	Cw	1200	283.33						
	φCw				283.33				
Chipepo Bay II (G)	h	5.9	7.1	7.4	5.5	6.6	1.9	0.9	0.6
	φH				4.49				
	Cw	900	1011.11	300.00	105.00	104.88	25.00	7.14	2.22
	φCw				222.19				
All material without age groups 5–8	h	5.5	3.7	3.5	3.4				
	φH				4.02				
	Cw	700	414.28	150.00	122.22				
	φCw				228.83				

Table 58. Biomass of *Sarotherodon m. mortimeri* stock in cove of Chikanka Island (locality C) in relationship to number of specimens per 1 ha (number of fish at the time of annulus formation adjusted according to $\bar{Z} = 0.19\%$ and t = 233 days).

Age group	Number of specimens per 1 ha at the time of		Standard length in cm of 1 specimen at the time of		Weight in grams of 1 specimen at the time of		Mass (weight) of age group in kg at the time of	
	poisoning	annulus formation	poisoning	annulus formation	poisoning	annulus formation	poisoning	annulus formation
0+	9		4.0		3		0.03	
1+	199	311	6.8	5.1	14	6	2.79	1.87
2+	108	169	11.3	8.2	65	25	7.20	4.23
3+	55	86	15.8	12.3	180	87	9.90	7.48
4+	21	33	19.1	16.6	340	217	7.14	7.16
Total	392	599					27.06	20.74

Table 59. Biomass of *Sarotherodon m. mortimeri* stock in cove of Island 20 (locality D) in relationship to number of specimens per 1 ha (number of fish at the time of annulus formation adjusted according to $\bar{Z} = 0.30\%$ and t = 40 days).

Age group	Number of specimens per 1 ha at the time of		Standard length in cm of 1 specimen at the time of		Weight in grams of 1 specimen at the time of		Mass (weight) of age group in kg at the time of	
	poisoning	annulus formation	poisoning	annulus formation	poisoning	annulus formation	poisoning	annulus formation
0+	39		3.4		1.7		0.07	
1+	243	274	7.7	5.9	21	9	5.10	2.47
2+	94	106	12.4	10.4	85	50	7.99	5.30
3+	18	20	14.5	12.9	140	95	2.52	1.90
4+	5	6	16.0	15.5	190	170	0.95	1.02
Total	399	406					16.63	10.69

Table 60. Biomass of *Sarotherodon m. mortimeri* stock in cove of Brook (locality E_2) in relationship to number of specimens per 1 ha (number of fish at the time of annulus formation adjusted according to $\bar{Z} = 0.22$ and $t = 205$ days).

Age group	Number of specimens per 1 ha at the time of		Standard length in cm of 1 specimen at the time of		Weight in grams of 1 specimen at the time of		Mass (weight) of age group in kg at the time of	
	poisoning	annulus formation	poisoning	annulus formation	poisoning	annulus formation	poisoning	annulus formation
0+	15380		3.5		2		30.76	
1+	1857	2852	7.6	5.0	20	6	37.14	17.11
2+	620	952	11.8	8.9	75	32	46.50	30.46
3+	524	805	13.2	11.1	108	65	56.59	52.33
4+	190	292	17.5	15.3	250	168	47.50	49.06
Total	18571	4901					218.49	148.96

Table 61. Biomass of *Sarotherodon m. mortimeri* stock in cove of Chipepo Bay I (locality F) in relationship to number of specimens per 1 ha (number of fish at the time of annulus formation adjusted according to $\bar{Z} = 0.01\%$ and $t = 206$ days).

Age group	Number of specimens per 1 ha at the time of		Standard length in cm of 1 specimen at the time of		Weight in grams of 1 specimen at the time of		Mass (weight) of age group in kg at the time of	
	poisoning	annulus formation	poisoning	annulus formation	poisoning	annulus formation	poisoning	annulus formation
0+	100		5.1		6		0.60	
1+	84	88	7.2	5.5	16	7.5	1.34	0.66
2+	78	81	12.2	10.0	83	46.0	6.47	3.73
Total	262	169					8.41	4.39

Table 62. Biomass of *Sarotherodon m. mortimeri* stock in cove of Chipepo Bay II (locality G) in relationship to number of specimens per 1 ha (number of fish at the time of annulus formation adjusted according to $\bar{Z} = 0.21\%$ and t = 208 days).

Age group	Number of specimens per 1 ha at the time of		Standard length in cm of 1 specimen at the time of		Weight in grams of 1 specimen at the time of		Mass (weight) of age group in kg at the time of	
	poisoning	annulus formation	poisoning	annulus formation	poisoning	annulus formation	poisoning	annulus formation
0+	96		5.2		6		0.58	
1+	218	337	7.0	5.4	15	7.3	3.27	2.46
2+	36	56	12.1	9.3	80	37	2.88	2.07
3+	48	74	16.2	13.9	200	125	9.60	9.20
4+	72	111	19.5	17.6	350	255	25.20	28.31
5+	36	56	32.8	32.0	1750	1600	63.00	89.60
6+	24	37	37.1	35.4	2500	2150	60.00	79.55
8+	24	37	37.0	35.9	2500	2250	60.00	83.25
Total	554	708					224.53	294.44

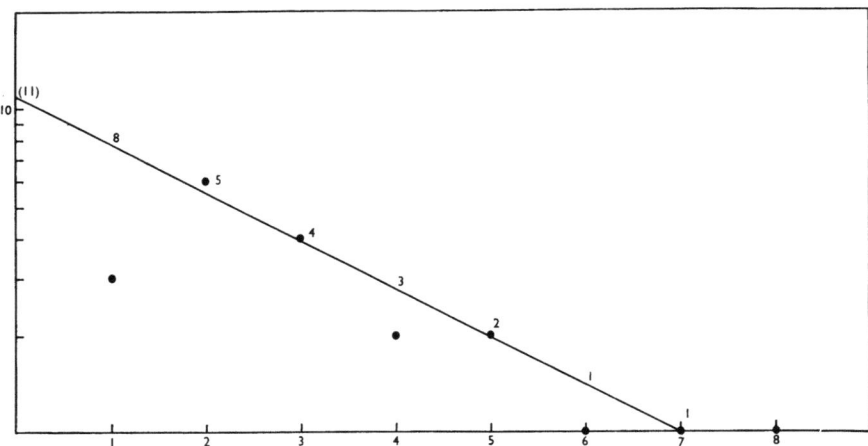

Fig. 35. Single logarithmic diagram for interpolation of the cornish-jack's age composition.

shows the cornish-jack as the second most important fish (15% in weight) and the bottlenose as the sixth (3% in weight).

This section is limited exclusively to the age and growth data needed for the production evaluation of these two species in Lake Kariba. For this reason only the larger quantitative samples can be used, and because of the low density of both these large mormyrids, only one locality proved to be suitable. We abandoned the evaluation of all other material, as it was impossible to relate the larger part of it to the area, though some better knowledge of the age span and size ranges may be obtained from the remainder.

Because of the scarcity of the whole size range distribution in the samples I decided to use the cornish-jack material only from locality D – Buffalo Island cove (sampled with toxaphene on 11–14 December 1968 in an area of 1.147 ha). This represents 272 specimens. From localities E, F, G grouped together an additional 118 specimens were studied and recalculated for the total area of 5,547 ha. The final numerical age composition was graphically interpolated on the assumption that the sampling bias is far more disastrous to the final conclusion than any possible year class strength. This was done on a semilogarithmic diagram (Fig. 35 and 36). The results were not significantly different whether this was done before or after the adjustment for the time of annulus formation (Table 63 and 64).

For the bottlenose we were able to use only the 53 specimens in the sample from locality D and these were treated in a similar way. As in other species, quite strong differences in abundance, biomass and thus in production were established depending on topography and season; and therefore this material cannot be treated as a good representative sample of the whole lake. If we later do appear to treat this material as representative, the bias involved must always be taken into account and the figures treated as preliminary values.

On the average, the scales of the cornish-jack are a little smaller in relation to the

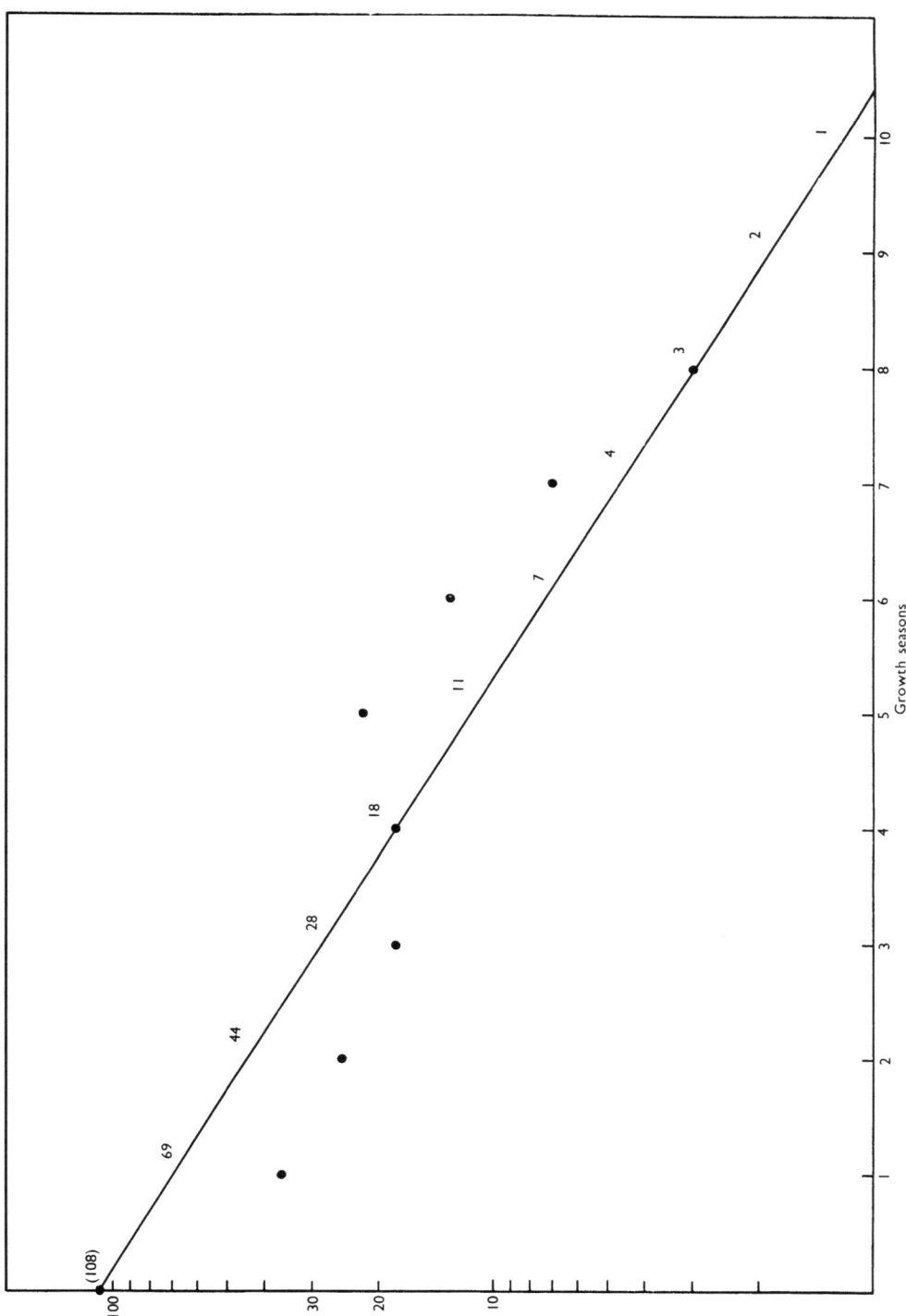

Fig. 36. Single logarithmic diagram for interpolation of the bottlenose's age composition.

Table 63. Computation of numerical age composition for *Mormyrops deliciosus* stock from Buffalo Island cove (locality D).

Age groups		0	I	II	III	IV	V	VI	VII	VIII	IX	X	Total
Number in sample	N	117	39	28	19	19	24	14	8	4			272
Number in sample per ha	N/ha	102	34	24	17	17	21	12	7	3			135
Graphically interpolated N/ha	Ñ/ha	(100)	64	41	26	17	11	7	5	3	2	1	177
Instantaneous rate of mortality	Z		0.4462	0.4453	0.4554	0.4251	0.4352	0.4520	0.3364	0.5109	0.4055	0.6931	mean 0.4605
Number per ha at time of annulus formation ($\bar{Z} = 0.13$, t = 40)	N'/ha	(105)	67	43	27	18	12	7	5	3	2	1	185

Table 64. Computation of numerical age composition for *Mormyrops deliciosus* stock from Buffalo Island cove (locality D).

Age groups		0	I	II	III	IV	V	VI	VII	VIII	IX	X	Total
Number in sample	N	117	39	28	19	19	24	14	8	4			Σ 135
Number in sample per ha	N/ha	102	34	24	17	17	21	12	7	3			
Instantaneous rate of mortality	Z	1.099	0.348	0.345	0.000	0.211	0.559	0.539	0.847			mean 0.5642	
Number per ha at times of annulus formation ($\bar{Z} = 0.15$, t = 40)	N'/ha	(108)	36	25	18	18	22	13	7	3	2	1	Σ 142
Graphically interpolated N'/ha	Ñ'/ha	(108)	69	44	28	18	11	7	4	3	2	1	Σ 187

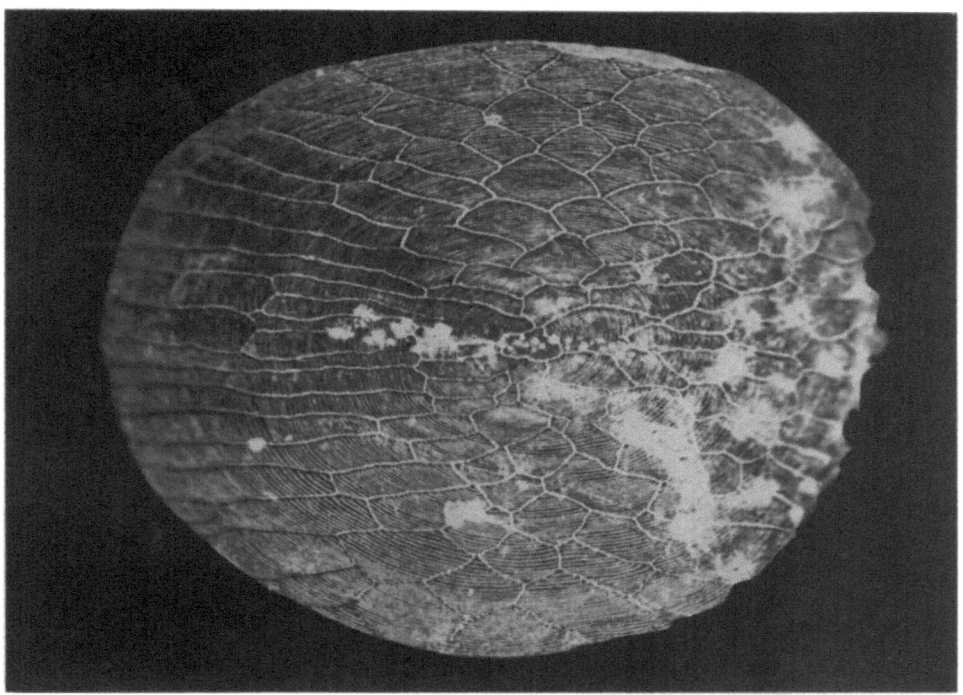

Fig. 37. Key scale of the cornish-jack from Lake Kariba, standard length 295 mm.

same body length, than the scales of the bottlenose; the extreme sizes in both species, however, overlap. This refers to the key scales collected from the rows above the lateral line and the beginning of the anal fin. In the cornish-jack these scales are fairly regular, and of an oval or rounded shape with the caudal part slightly wider. In the bottlenose this shape is more often distorted by irregular bends and lobes in the dorso-ventral parts (Fig. 37 and 38). Numerous radial canals in both species form parallel lines in the oral part of the scales; however, in the caudal and dorso-ventral parts these canals form a distinct reticulating pattern. The annuli are visible as a border between the narrow band and a wide band of circuli, sometimes with a more or less interrupted circulus in this border line. Usually no cut circuli are to be found in the diagonal parts of the scales and sometimes only an interrupted circulus without a narrow band forms an annulus (Fig. 39 and 40). Generally speaking, the annuli in both species cannot be determined as easily as in other species and a higher degree of errors must be taken into account.

In both species 9 to 11% of the key scales were regenerated or deformed and their age could not be ascertained. Some of the specimens had a juvenile mark; the cornish-jack from locality D had 50% juvenile marks, from localities E, F, G 27%. In the case of the bottlenose 41% of specimens had juvenile marks on their scales. Although the body-scale relationship of the bottlenose forms a wide correlation-field (VOVK,

Fig. 38. Key scale of the bottlenose from Lake Kariba, standard length 450 mm.

Fig. 39. Key scale of the cornish-jack from Lake Kariba, standard length 245 mm.

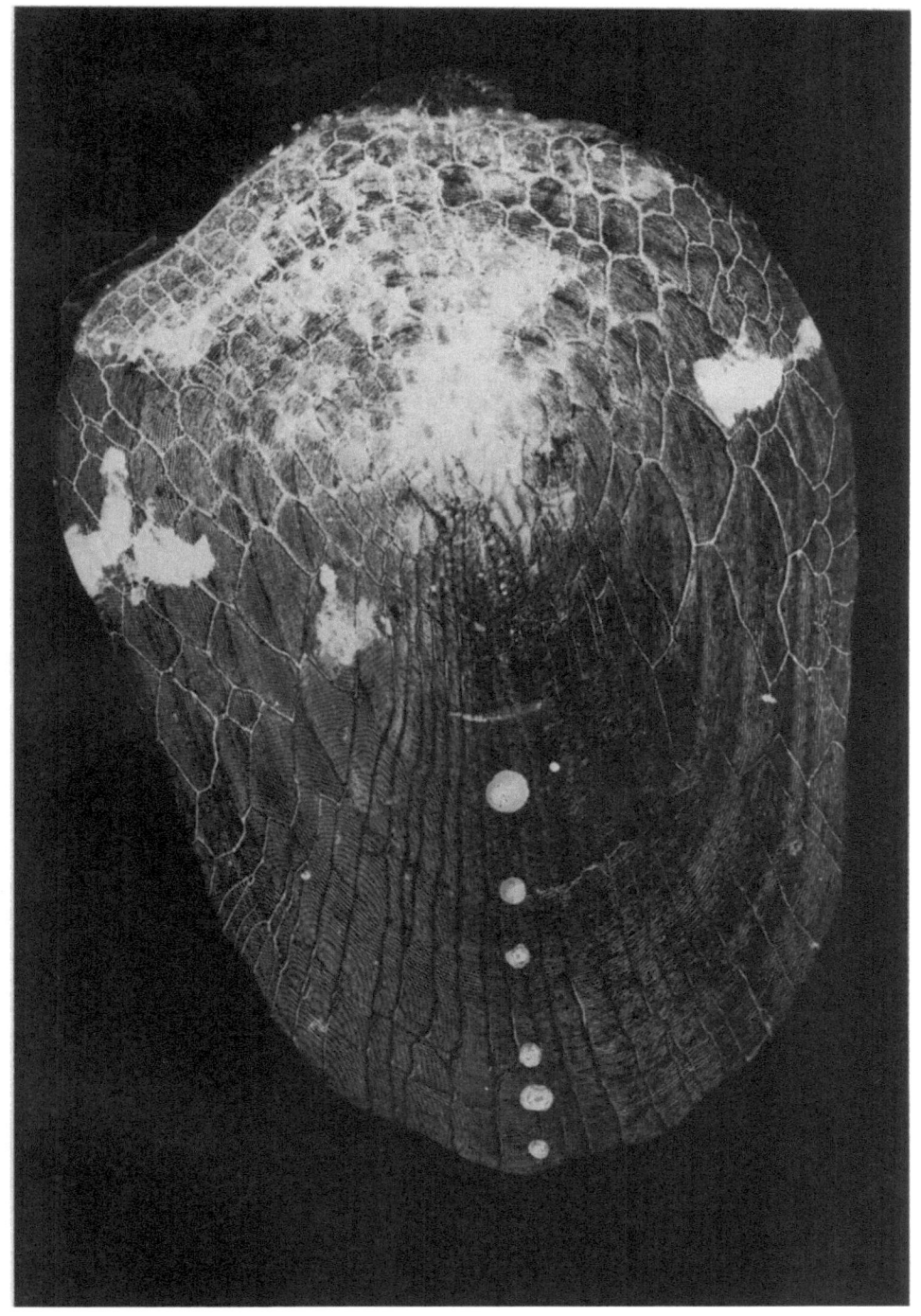

Fig. 40. Key scale of the cornish-jack from Lake Kariba, locality D, standard length 665, number of annuli 6.

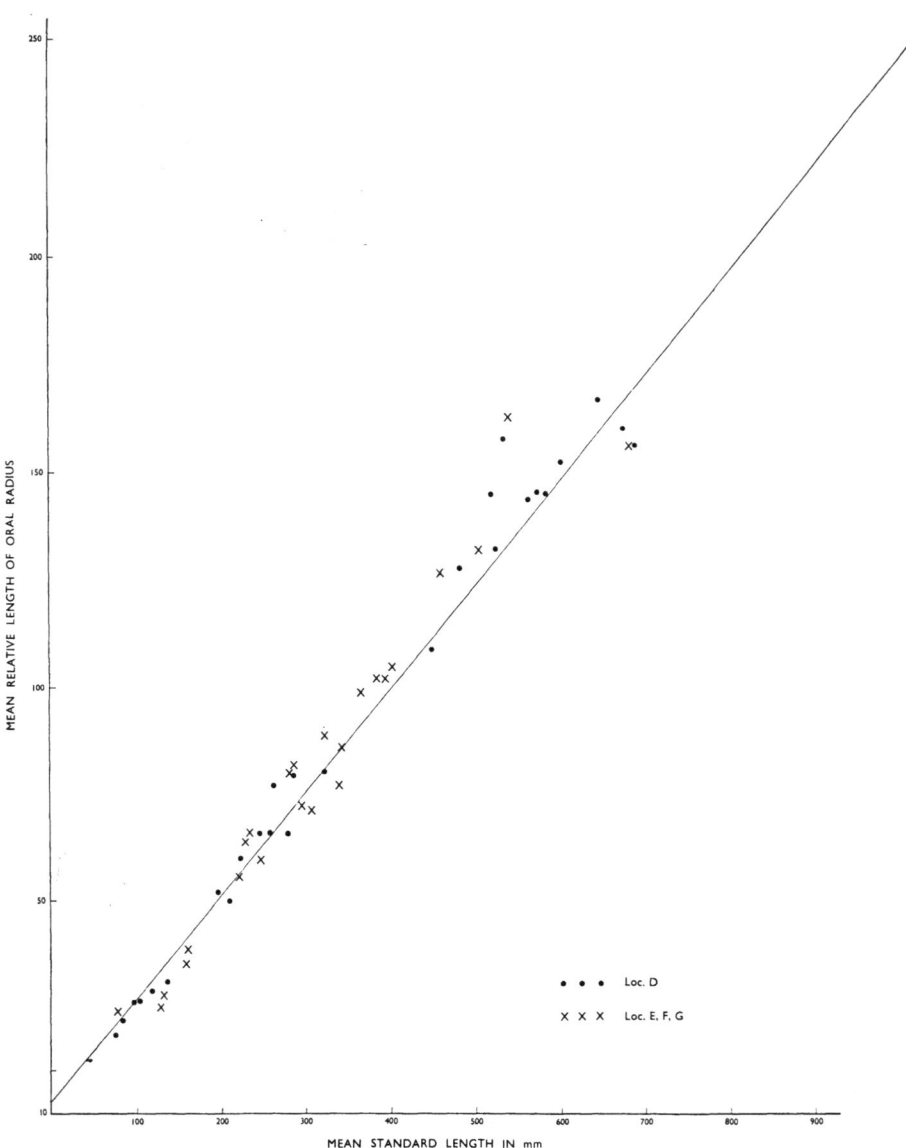

Fig. 41. Empiric curve of the scale radius and standard length relationship of the cornishjack from Lake Kariba after 10 mm group averages.

1956) and the empiric curve drawn according to the perimeter curves has a fine S-shape, for practical use (BALON, 1963b) the relationship was considered to be linear in both species (Fig. 41 and 42).

Because of the relatively long life span and large length frequency distribution, even as large a sample as nearly 300 specimens from one locality is insufficient for ascer-

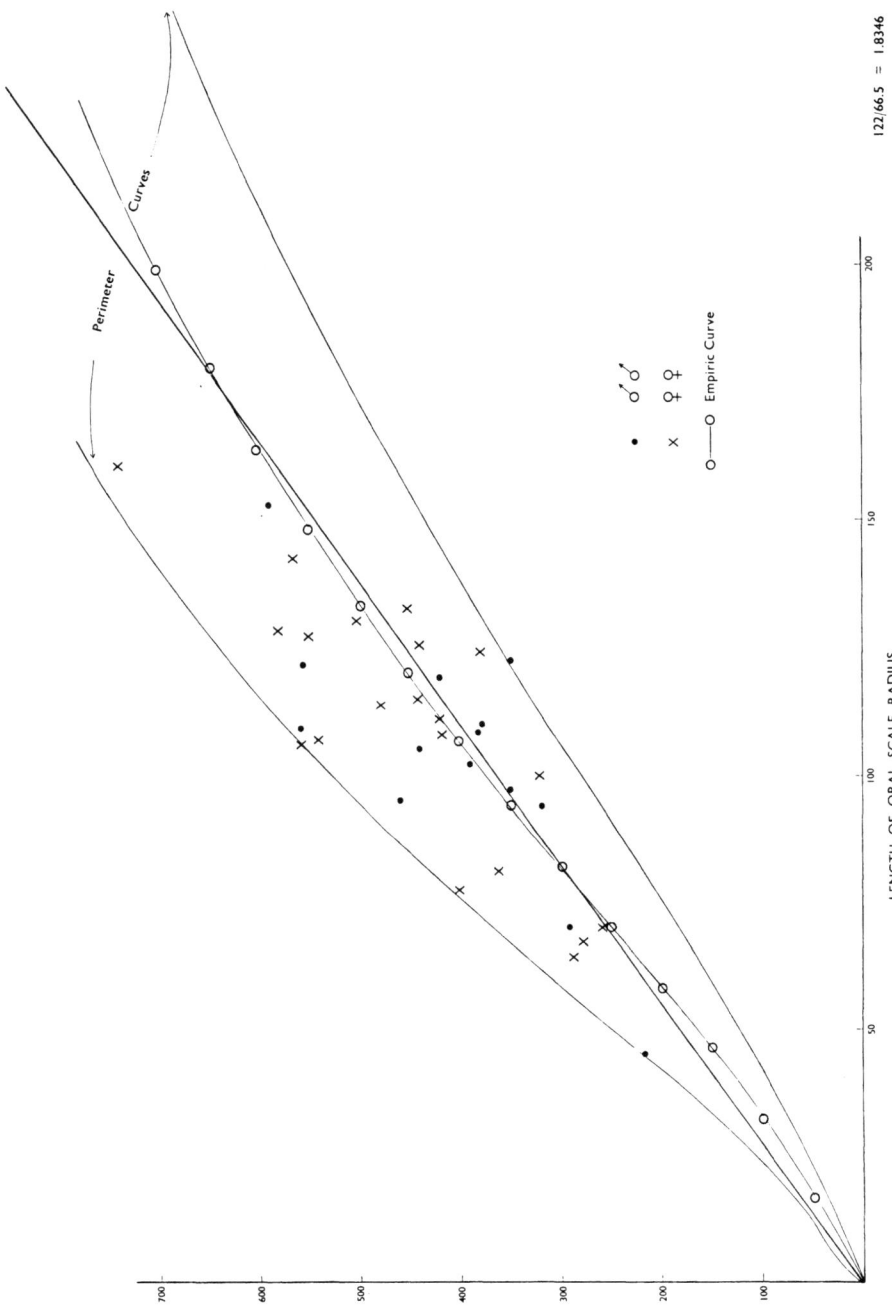

Fig. 42. Correlation field, perimeter and empiric curves of the bottlenose's scale radius and standard length relationship.

Table 65. Age and length composition of *Mormyrops deliciosus* stock from cove of Buffalo Island (locality D = 1.147 ha).

Length groups in mm	N in age groups									Σ n
	0	I	II	III	IV	V	VI	VII	VIII	
11– 20	1									1
41– 50	1									1
61– 70	3									3
71– 80	8									8
81– 90	14									14
91–100	29									29
101–110	36	5								41
111–120	16									16
121–130	6									6
131–138	1									1
141–150	1									1
181–190	1									1
201–210		5								5
211–220		2	1							3
221–230		2	1							3
231–240		5								5
241–250		2	4							6
251–260			1							1
261–270		2								2
271–280		4								4
281–290		3	4							7
291–300		4								4
301–310			3							3
311–320			4							4
321–330		1	3							4
331–340					1					1
341–350		1								1
351–360		1	2							3
361–370		1								1
371–380		1	1							2
381–390			1							1
401–410			1	3						4
411–420			1	2						3
421–430				2						2
431–440				2						2
441–450				1						1
451–460				1		1				2
461–470					1					1
471–480				1	1	1				3
481–490				2	1					3
491–500				2	3	2				7
501–510			1	2	2					5
511–520						3				3
521–530						5				5
531–540						1	2			3
541–550						3				3
551–560						2	2			4

Table 65. (continued)

Length groups in mm	\multicolumn{9}{c}{N is age groups}									Σ n
	0	I	II	III	IV	V	VI	VII	VIII	
561–570					1	1	1			3
571–580					3	1	1			5
581–590							1	4		5
591–600				1			1			2
601–610					1					1
611–620					1	1	1			3
621–630					1	1	1			3
631–640					1					1
641–650					1					1
651–660					1	1				2
661–670							1			1
671–680							1			1
681–690								1	1	2
701–710						1	1	1	1	4
711–720								1		1
741–750							1			1
761–770								1		1
811–820									1	1
821–830									1	1
Σ N_{0-8}	117	39	28	19	19	24	14	8	4	272

taining the age composition in good confidence limits. For this reason most of the samples cannot be used and we are limited with both species to one locality where either for seasonal reasons or because of the short distance from the old Zambezi bed, the concentration of studied mormyrids was fairly high. The Buffalo Island cove (locality D) was blocked and treated with fish eradicant on December 11–14, 1968. Most of the 0 age group specimens in this sample were late spawners descendants of the previous main spawning season at the beginning of the year; a few of the smallest, however, can be considered as offspring of the current spawning season. The same phenomenon can explain the wide size range in single age groups (Table 65). We also tried to group three additional localities – E, F, G – to obtain a usable sample (Table 66 and 67) of cornish-jack. Nowhere did we find an animal more than 8 years old. To the VIII age group belongs also the biggest specimen of cornish-jack, over 1 m standard length and 9 kg in weight from Chete Island cove (E). Eight years was also found to be the oldest age attained in the bottlenose, though in smaller specimens.

So far we have no evidence about the behaviour of either the cornish-jack or the bottlenose. However, the differences in diel distribution and topographical preferences will, probably, explain the differences found in stock density. Too little is known about the nocturnal and daytime movements of both mormyrids, but, whichever predominates, they are not evenly distributed along the lake shoreline and our density

Table 66. Age and length composition of *Mormyrops deliciosus* stock from coves of Chete Island and Chipepo Bay (localities E, F, G = 5.547 ha).

Length groups in mm	N in age groups									Σ n
	0	I	II	III	IV	V	VI	VII	VIII	
31– 40	3									3
41– 50	6									6
51– 60	4									4
61– 70	3									3
71– 80	2									2
91– 100	2									2
121– 130	4									4
131– 140	1									1
141– 150	1									1
151– 160	1									1
161– 170	2									2
221– 230		1								1
231– 240		1	1							2
241– 250			1							1
281– 290			1							1
291– 300		3	2							5
301– 310		1	3							4
311– 320		1	2							3
331– 340		1	1	1	1					4
341– 350		1								1
351– 360					1	1				2
361– 370		1	1							2
381– 390		1	2							3
391– 400		1	2							3
401– 410			1	2						3
411– 420			1	1						2
421– 430		1		1						2
431– 440			1	4						5
441– 450				3	1					4
451– 460					1	2				3
461– 470		1		1						2
471– 480				1						1
481– 490			1							1
491– 500			4	2	1	1				8
501– 510			1	1						2
511– 520				1	1	3				5
521– 530						2				2
531– 540							1			1
541– 550							1			1
551– 560							1	1		2
561– 570							1			1
581– 590								2		2
641– 650					1					1
651– 660					1	1				2
681– 690								1	1	2
751– 760									1	1
1001–1010									1	1
Σ N$_{0-8}$	29	14	25	18	8	10	4	4	3	118

Table 67. Computation of numerical age composition for *Mormyrops deliciosus* stock from coves of Chete Island and Chipepo Bay (localities E, F, G).

Age groups		0+	1+	2+	3+	4+	5+	6+	7+	8+	Total
Number in sample	N	29	14	25	18	8	10	4	4	3	Σ 118
Number in sample per ha	N/ha	5.25	2.58	4.61	3.26	1.54	1.85	0.82	0.82	0.54	Σ 21.27
Instantaneous rate of mortality	Z	0.711	0.580	0.346	0.750	0.183	0.405	0.000	0.042		x̄ 0.4311
Number per ha at time of annulus formation ($\bar{Z} = 0.12$, t = 170)	N'/ha	(6)	3	6	4	2	2	1	1	1	27
Graphically interpolated N'/ha	Ñ'/ha	(11)	8	5	4	3	2	1	1	1	25

Table 68. Age and length composition of *Mormyrus longirostris* stock from Buffalo Island cove (locality D = 1.147 ha).

Lengths groups in mm	N in age groups								Σ n
	I	II	III	IV	V	VI	VII	VIII	
211–220	1								1
251–260	1								1
271–280	2								2
281–290	1								1
291–300	1								1
311–320		2							2
331–340		1							1
341–350		2							2
361–370		2							2
371–380		2	1						3
381–390		2							2
391–400		1							1
401–410		1							1
411–420		1	1	1					3
421–430		1							1
431–440		1	1	2					4
441–450			2	3					5
451–460		1	1	1					3
471–480			1						1
481–490				1					1
491–500			1						1
501–510			1						1
511–520				1					1
541–550				1		1			2
551–560					3	2			5
561–570							1		1
571–580								1	1
581–590					1	1			2
741–750							1		1
Σ N_{0-8}	6	16	10	10	4	4	2	1	53

estimates for cornish-jack from sample D (187 specimens per ha) would probably produce over-estimates for the whole lake while the density from sample E, F and G (25 specimens per ha) would underestimate the true population density in Lake Kariba. With only 10 quantitative samples for such a large lake it was not possible to place statistically significant confidence limits on the values of stock density nor to estimate valid means. Nevertheless it is clear that the true value lies between these two values and its mean will represent the approximate density.

Consequently the stock density of bottlenose computed from the D sample will probably overestimate the true value only if recalculated for the whole lake. The value of 78 specimens per ha represents also the highest density (Table 68 and 69) and we have no means to establish the average values.

Table 69. Computation of numerical age composition for *Mormyrus longirostris* stock from Buffalo Island cove (locality D).

Age groups		0	I	II	III	IV	V	VI	VII	VIII	Total
Number in sample	N		6	16	10	10	4	4	2	1	53
Number in sample per ha	N/ha		5	14	9	9	3	3	2	1	46
Graphically interpolated N/ha	$\bar{N}$/ha	(47)	29	18	11	7	4	3	2	1	75
Instantaneous rate of mortality	Z	0.4828	0.4769	0.4925	0.4520	0.5595	0.2878	0.4055	0.6931		mean 0.4813
Number per ha at time of annulus formation ($\bar{Z} = 0.13$, t = 40)	N'/ha	(49)	30	19	12	7	4	3	2	1	78

Table 70. Back calculated growth values of *Mormyrops deliciosus* from Lake Kariba.

Locality	Age groups	Year of hatching	N	Mean standard length at the time of capture in mm	Back calculated standard length (l) in mm for the previous growth seasons and corresponding interpolated weights ($\bar{w}$) in g								
					l_j / $\bar{w}_j$	l_1 / $\bar{w}_1$	l_2 / $\bar{w}_2$	l_3 / $\bar{w}_3$	l_4 / $\bar{w}_4$	l_5 / $\bar{w}_5$	l_6 / $\bar{w}_6$	l_7 / $\bar{w}_7$	l_8 / $\bar{w}_8$
D (Buffalo Island, 11–14 December, 1968)	0	1968	37	102	67 / 2.9								
	I	1967	9	239	54 / 1.5	198 / 74							
	II	1966	5	251	81 / 5.1	160 / 39	230 / 115						
	III	1965	3	509		211 / 88	346 / 355	445 / 800					
	IV	1964	3	595		291 / 230	382 / 510	452 / 840	521 / 1260				
	V	1963	3	535		188 / 64	302 / 253	362 / 434	439 / 763	512 / 1170			
	VI	1962	3	589		245 / 140	311 / 273	385 / 510	450 / 820	502 / 1120	562 / 1560		
	VII	1961	1	582		326 / 320	400 / 580	438 / 730	464 / 906	478 / 978	528 / 1300	559 / 1500	
	Total		64		67 / 3.2	231 / 136	328 / 348	416 / 663	468 / 937	497 / 1089	545 / 1430	559 / 1500	

E, F, G (Chete Island, 20–25 March, 1969; coves of Chipepo Bay, 26–31 May, 1969)

Age	n	1968	1967	1966	1965	1964	1963	1962	1961	1960	1959
0	7	123	101/10								
I	9	288		212/92							
II	10	326		174/51	283/214						
III	2	402	77/4.4	162/42	295/240	361/440					
IV	1	340		136/24	180/55	250/143	300/250				
V	1	457		122/18	236/124	337/350	384/510	434/725			
VII	1	688		261/168	354/402	459/890	526/1310	570/1650	607/2000	682/2800	
VIII	3	745		171/48	279/200	354/410	460/895	532/1300	621/2140	686/2870	737/3450
Total	34		89/7.2	177/63	271/206	352/447	417/741	512/1225	614/2070	684/2835	737/3450

Table 71. Back calculated growth values of *Mormyrus longirostris* from Lake Kariba (after data from locality D – Buffalo Island cove).

	Age groups	Year of hatching	N	Mean standard length at the time of capture in mm	l_j $\bar{w}_j$	l_1 $\bar{w}_1$	l_2 $\bar{w}_2$	l_3 $\bar{w}_3$	l_4 $\bar{w}_4$	l_5 $\bar{w}_5$	l_6 $\bar{w}_6$	l_7 $\bar{w}_7$
♂♂	I	1967	2	253	139 29	235 158						
	II	1966	7	375	146 34	307 376	371 698					
	IV	1964	2	430	176 62	277 266	347 551	406 902				
	V	1963	2	557		321 430	394 839	460 1330	520 2000			
	VI	1962	1	590		248 188	400 862	486 1601	527 2200	553 2460	581 2900	
	Total		14		154 42	278 284	378 737	451 1278	523 2100	553 2460	581 2900	(590) (3080)

			♀♀								
I	1967	3	275		270 243						
II	1966	4	375	110 13	299 338						
III	1965	5	444		266 234	401 881					
IV	1964	3	477	98 8	277 270	384 770	446 1215				
VI	1962	2	552	120 18	194 86	229 140	279 270	307 370	326 450	355 600	
VII	1961	2	653	150 37	254 205	331 480	390 800	510 1900	565 2700	612 3400	597 3100
VIII	1960	1	580	151 38	230 146	310 380	370 680	422 1030	479 1520	537 2200	565 2700
Total		20		126 23	256 217	331 530	377 769	413 1100	457 1557	501 2067	581 2900
♂♂ + ♀♀ Grand total		34		140 32	267 250	354 633	414 1023	468 1600	505 2008	541 2483	581 2900

Table 72. Parameters for computation of harvestable size of *Mormyrops deliciosus* from Lake Kariba.

Locality			Means at the end of growth seasons							
			1	2	3	4	5	6	7	8
D (Buffalo Island, 11–14 December, 1968)	Standard length (mm)	l	231	328	416	468	497	545	559	
	Length increments (mm)	$l_i - l_{i-1}$	231	97	88	52	29	48	14	
	Length increments in % of length of first growth season	D	100	42	38	22	12	21	6	
	Standard lengths in % of length of final growth season	I	41	59	74	84	89	97	100	
	Wet weight (g)	$\bar{w}$	136	348	663	937	1089	1430	1500	
	Weight increments (g)	$\bar{w}_i - \bar{w}_{i-1}$	136	212	315	274	152	341	70	
E, F, G (Chete Island, 20–25 March, 1969, coves of Chipepo Bay, 26–31 May, 1969)	Standard length (mm)	l	177	271	352	417	512	614	684	737
	Length increments (mm)	$l_i - l_{i-1}$	177	94	81	65	95	102	70	53
	Length increments in % of length of first growth season	D	100	53	46	37	54	58	39	30
	Standard lengths in % of length of final growth season	I	24	37	48	57	69	83	93	100
	Wet weight (g)	$\bar{w}$	63	206	447	741	1225	2070	2835	3450
	Weight increments (g)	$\bar{w}_i - \bar{w}_{i-1}$	63	143	241	294	484	845	765	615

Table 73. Parameters for computation of harvestable size of *Mormyrus longirostris* from Lake Kariba.

		Means at the end of growth seasons						
		1	2	3	4	5	6	7
Standard length (mm)	l	267	354	414	468	505	541	581
Length increments (mm)	$l_i - l_{i-1}$	267	87	60	54	37	36	40
Length increments in % of length of first growth season	D	100	32	22	20	14	13	15
Standard lengths in % of length of final growth season	I	46	61	71	80	87	93	100
Wet weight (g)	$\bar{w}$	250	633	1023	1600	2008	2483	2900
Weight increments (g)	$\bar{w}_i - \bar{w}_{i-1}$	250	383	390	577	408	475	417

The juvenile marks in the cornish-jack from sample D are present in 38% of fish in the 0 age group, 22% in the I age group and 20% in the II age group. They are not found in the higher age groups and must presumably be related to the size-selective mortality (BALON, 1963; RICKER, 1969). They are formed by the time the fish reach 54 to 81 mm in standard length. Although the percentage of occurrence of this mark in other samples, as well as in the bottlenose, does not give such clear decreasing values this may be due to the low number of specimens. The juvenile mark in the bottlenose is formed when this species reaches nearly double the size of the cornish-jack (98 to 176 mm, mean 147), a fact which is probably related to the different food habits of each species.

The back calculated growth data for the cornish-jack are in Table 70, for the bottlenose in Table 71. The weight values in both tables were interpolated from the weight-length relationship double logarithmic diagrams in a special paper (BALON, 1974b). As the growth of males and females in the cornish-jack is the same, sexes are treated together. In the bottlenose the growth of females is slower, especially in the reproductive adult period.

The final means of size and growth parameters for both species are in Tables 72 and 73. Though the linear growth values in the cornish-jack and the bottlenose are similar, with a marked growth compensation pattern, the weight growth values are more favorable in the bottlenose; though the body depth increase may be the cause the material is too sparse to reach any further conclusions. The minimum harvestable size interpolated from these values is 300 mm standard length and 250 g of weight for the cornish-jack (Fig. 43). For the bottlenose of a similar size it is 360 g (Fig. 44). Both species attain this size during the second year of life.

The biggest angler's record noted by JUBB (1967) for cornish-jack is 13 kg and for bottlenose 4.5 kg. This is close to our 9 kg for the former and 3 kg for the latter. The stock density compared well with the catch records from the previous years, as noted by HARDING (1964) and COKE (1968), though as mentioned before it showed the bottlenose predominating. This reflects even better the average data obtained up to now:

Age groups	0	1	2	3	4	5	6	7	8	9	10
N'/ha of cornish-jack	(59)	38	24	16	10	6	4	2	2	1	1
N'/ha of bottlenose	(49)	30	19	12	7	4	3	2	1		

The density of cornish-jack is 163 specimens per 1 ha and of bottlenose 127 N'/ha. Within the limits of insufficient data and the bias spoken of before the total number of cornish-jack in Lake Kariba can be estimated as 32 million specimens and the total number of bottlenose as 25 million (if the inhabited area is 202 200 ha).

Thus in relation to the growth values obtained the cornish-jack represents 57 kg/ha of the initial biomass, 155 kg/ha of the mean biomass, 503 kg/ha/yr of total production ($A = G\bar{B}$), 24 kg/ha/yr of available production, 14 kg/ha/yr of available yield and 35 kg/ha/yr of the harvestable part of the total production or total yield. Thus a little less than half of the total yield – 14 kg – is unexploited. Detailed tables are in Appendix D.

The values for bottlenose based on the data from sample D will most likely over-

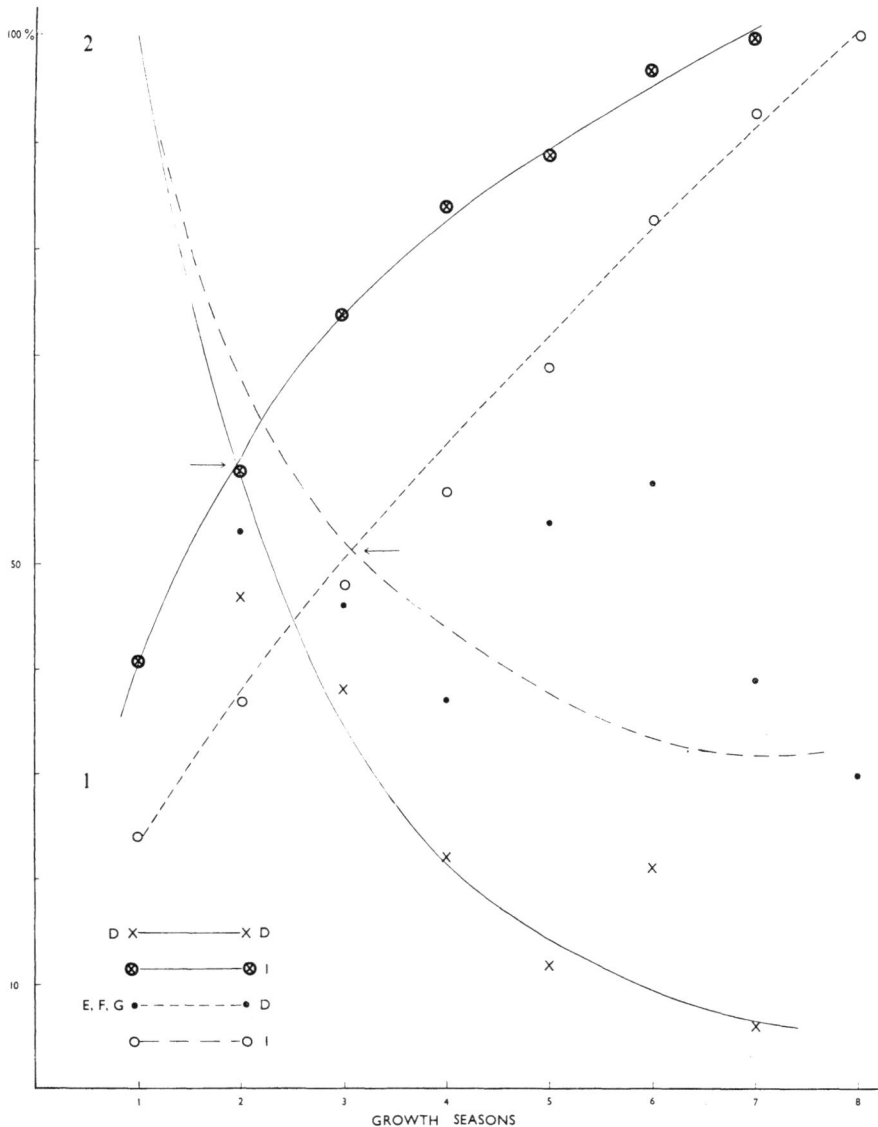

Fig. 43. Growth curves of the cornish-jack from Lake Kariba for minimum harvestable size estimation. (1) Standard length in percent of the length of the last growth season and (2) length increments in percent of the length of the first growth season.

estimate the values for the whole lake: the initial biomass is 67 kg/ha, the mean biomass 278 kg/ha, total production 1342 kg/ha/yr, the total yield 14 kg/ha/yr, the available production 31 kg/ha/yr and the available yield 15 kg/ha/yr. Here the total yield is nearly equal to the available yield. This is in full agreement with catch records,

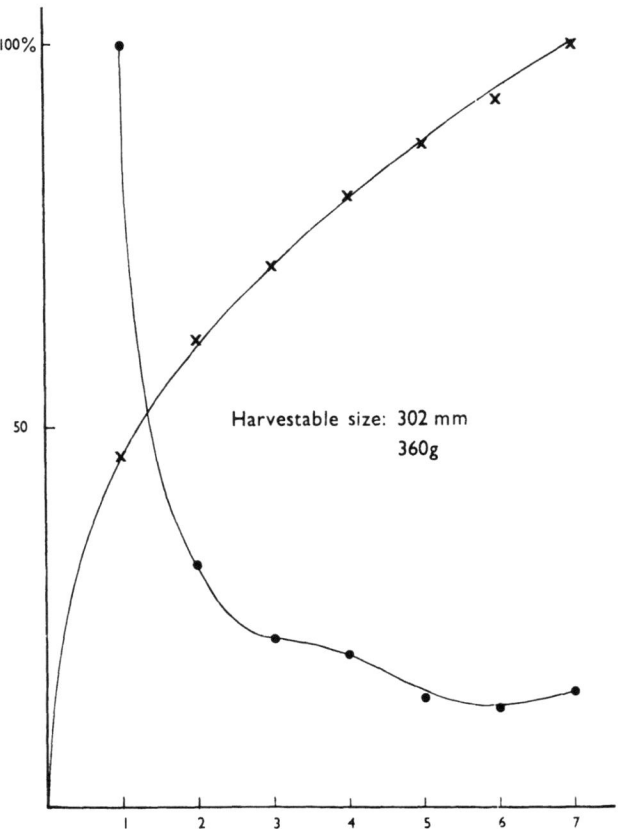

Fig. 44. Growth curves of the bottlenose from Lake Kariba for minimum harvestable size estimation.

where the bottlenose predominates, although it is the less abundant species. We do not have sufficient data available to explain this, but the greater vulnerability of the bottlenose to gillnets may well be the reason.

3.9. The parrotfish *Hippopotamyrus discorhynchus* (Peters, 1852) and the bull-dogfish *Marcusenius macrolepidotus* (Peters, 1852)

by ANTON KIRKA

The parrotfish, *Hippopotamyrus discorhynchus* (Peters, 1852) is a common species occurring in the warmer waters of the whole Zambezi River basin. It is prevalent in water-bodies up to the Congo basin and Lake Tanganyika to the north. BALON (1972, 1973a) consideres this species to be the third most frequent in Lake Kariba.

The bull-dogfish, *Marcusenius macrolepidotus* (Peters, 1852) is the most common

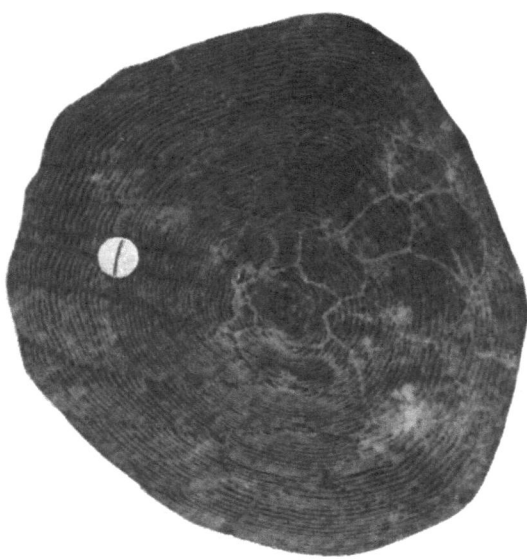

Fig. 45. The scale of *H. discorhynchus* in the age 1+.

species of the genus in Zambia, in the area of Lake Malawi and in the Pongolo river basin in Natal. The distribution of this species extends to the Congo basin and the rivers of East Africa (JACKSON, 1961a; JUBB, 1967). According to JACKSON (1961b) this species was not found in the area which was later covered by waters of Lake Kariba. Most probably this region was colonized by populations from tributaries of the lake. The bull-dogfish prefers more restricted waters, such as small rivers, swamps, lake shores, etc. The ecology of both species with regard to their age, growth, density of population and production is completely unknown.

This work is based on a sample of scales and population density data. The samples were obtained by poisoning blocked-off coves and detailed information about this can be found in BALON's (1971a, 1972, 1973a) papers. The samples of scales and other data originate from the following localities: 1. the parrotfish: locality D, Island N 20 (96 specimens); 2. the bull-dogfish: locality E, Island N 21 or Chete (58 specimens).

Dry scales were read under the trichinoscope with 45 × magnification. The oral diameter of a scale was used for reading and measuring the annuli. It was found that the body-scale relationship of both species is linear with correction values 35.0 mm for the parrotfish and 30.3 mm for the bull-dogfish. Back calculated growth values for the previous years of life were obtained by R. Lee's method; the corresponding weight values were extrapolated from the length-weight nomograms (BALON, 1974b).

Because the material at our disposal did not allow exact determination of annulus formation November 1st was assumed to be the time it occurred (HOLČIK, section 3.11; BASTL, section 3.3). I also assume that the beginning of the growth season and time of annulus formation is connected with the coming of the warm season, which is the beginning of October in the area of Lake Kariba (COCHE, 1968).

Fig. 46. The scale of *H. discorhynchus* in the age 2+.

Methods for the calculation of age composition of populations, mortality, survival, and the relative growth indexes are summarized by BALON (1971a, 1972). For the calculation of biomass and available production and yield of bull-dog fish, the complete sample was used. In parrotfish the weight of 21 fishes (9.975 kg), not represented in the subsample of scales, had to be subtracted from the data of the whole sample of standing stock. All resulting data for this species are therefore biased by this shortcoming. Subsequent computation of biomass, available production and available yield was performed using the method described by HOLČIK (Section 3.1). The standing stock found at the time of sampling was diminished by the weight of 0+ age group and then recalculated back to the time of annulus formation. The P'/B' ratio was then calculated and used to estimate the production for the whole population living in the lake, considering that only 65% of the whole area of the lake i.e. 342 980 ha, is inhabited by fish. Also the available yield was recalculated only for that area.

The scales of both species are cycloid, oval and smaller in *H. discorhynchus* than in

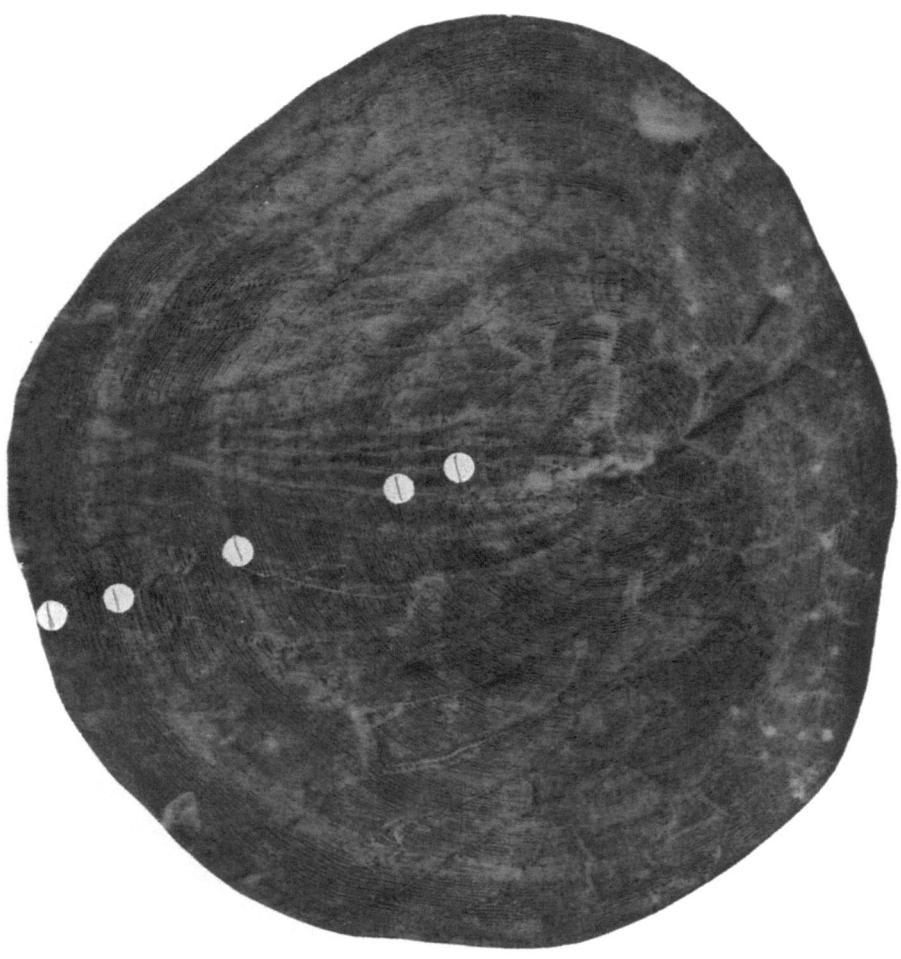

Fig. 47. The scale of *H. discorhynchus* in the age 5+.

M. macrolepidotus. JUBB (1967) described two basic types of cycloidal scales: longitudinally striated and radially striated. It seems, however, that *H. discorhynchus* and *M. macrolepidotus* forms a third group with reticulary striated cycloid scales. The caudal side of the scale shows the 6 angle reticulation which is changed by the 5 angle reticulation on the lateral side and then by a 4 angle reticulation. The central part of the scale displays 4–8 secondary radial channels. They end on the irregular area around the focus but not directly in the latter. In older *M. macrolepidotus* fish these channels are often connected transversely and form elongated 4 angle fields.

Annuli are easily visible in the oral part of the scale. After removing the epidermis the diagonal and in some cases also the caudal part displays the annuli. They are

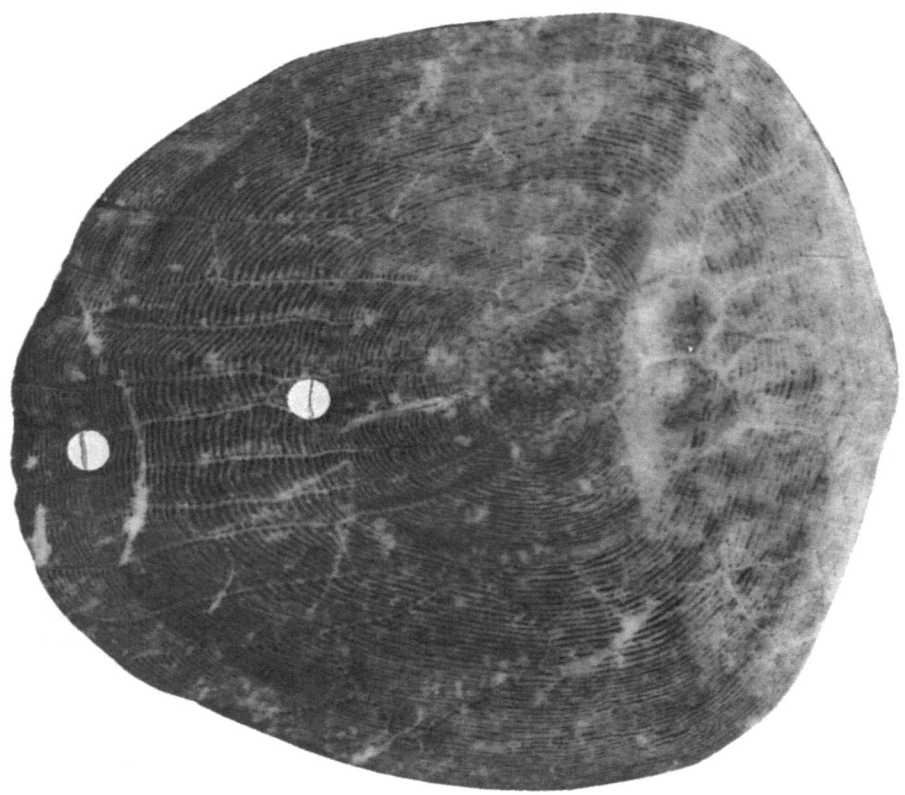

Fig. 48. The scale of *M. macrolepidotus* in the age 2+.

manifested as the transition of 2–3 dense sclerites to the band of wider ones around the whole scale (Fig. 45 to 49).

The age analysis is difficult, however, because both species obviously respond to the warm as well as to the rainy season. At the beginning of the rainy season some annuli-like checks form mainly on the lateral part of the scale. They gradually disappear on the oral part and it is necessary to consider them as accessory checks.

H. discorhynchus: Because of lack of age data on this species and because our sample of scales was not complete it is impossible to determine the potential age for this fish. The largest specimen found measured 576 mm of standard length and weighed 3.75 kg. According to our observation the parrotfish belongs among the species which reach middle age. Specimens older than five years were comparatively rare (Table 74). It seems, however, that there is a difference in age between sexes. The males reach maximally five years (they were not present in the age group 5+) but the age of females can be twice this (Table 75). The number of males in age groups 2+, 3+ and 4+ is higher than the number of females.

Fig. 49. The scale of *M. macrolepidotus* in the age 4+.

Table 74. Age and length composition of *Hippopotamyrus discorhynchus* stock in south cove of Island N20, south-west from Sinazongwe (11–14.12.1968).

		0+	1+	2+	3+	4+	5+	Σ
21– 30		2						2
31– 40		3						3
51– 60			2					2
61– 70			10					10
71– 80			12					12
81– 90			8					8
91–100			9	11				20
101–110				38				38
111–120			28	70				98
121–130				134	17			151
131–140				78	79			157
141–150					135			135
151–160					72			72
161–170					10	32		42
171–180						19		19
181–190						11		11
191–200							9	9
Total	n	5	69	331	313	62	9	789
	%	0.63	8.75	41.95	39.67	7.86	1.14	100.0

Table 76. Age composition, seasonal total mortality rate (Z′), survival rate (S), and instantaneous rate of total mortality (Z) of *Hippopotamyrus discorhynchus* in south cove at Island N20.

		Age groups					
		1	2	3	4	5	Σ
All material	n	69	331	313	62	9	784
	%	8.80	42.22	39.92	7.91	1.15	100.0
	Z′	—	0.0582	0.8021	0.8549		0.4566 (2–5)
	S	—	0.9418	0.1979	0.1451		0.5434
	Z	—	0.06	1.62	1.93		0.61
♂♂	n	30	167	198	34	—	429
	%	6.99	38.93	46.15	7.93	—	100.0
	Z′	—	—	0.8245	—		0.8245 (3–4)
	S	—	—	0.1755	—		0.1755
	Z	—	—	1.74	—		1.74
♀♀	n	39	164	115	28	9	355
	%	10.99	46.20	32.39	7.88	2.54	100.0
	Z′	—	0.2953	0.7631	0.6770		0.5034 (2–5)
	S	—	0.7047	0.2369	0.3230		0.4966
	Z	—	0.35	1.44	1.13		0.70

Table 75. Male and female age and length composition of *Hippopotamyrus discorhynchus* stock in south cove of Island N 20, south-west from Sinazongwe (11–14.12.1968).

	1+ ♂♂	1+ ♀♀	2+ ♂♂	2+ ♀♀	3+ ♂♂	3+ ♀♀	4+ ♂♂	4+ ♀♀	5+ ♂♂	5+ ♀♀	Σ ♂♂	Σ ♀♀
51– 60	1	1									1	1
61– 70	8	2									8	2
71– 80	4	8									4	8
81– 90	3	5									3	5
91–100		9										9
101–110			11	11							11	11
111–120	14	14	27	42							27	56
121–130			28	33	17						42	33
131–140			101	78	59	20					118	98
141–150					67	68					59	68
151–160					50	22					67	22
161–170					5	5	21	11			50	16
171–180							13	6			26	6
181–190								11			13	11
191–200										9	—	9
Total n	30	39	167	164	198	115	34	28	—	9	100.0	100.0
%	6.99	10.99	38.93	46.20	46.15	32.39	7.93	7.88	—	2.54	100.0	100.0

Table 77. Age and length composition of *Marcusenius macrolepidotus* stock in north cove of Island 21 (Chete), south from Sinazongwe (locality E).

	2+	3+	4+	5+	6+	Σ
101–110	4					4
111–120	6					6
121–130	7					7
131–140	9					9
141–150	29	3				32
151–160	9	21				30
161–170		30	6			36
171–180			16			16
181–190		2	4			6
191–200		3				3
201–210				3		3
231–240					1	1
241–250				1	1	2
Total n	64	59	26	4	2	155
%	41.29	38.06	16.77	2.58	1.30	

Table 79. Age composition, seasonal total mortality rate (Z'), survival rate (S), and instantaneous rate of total mortality (Z) of *Marcusenius macrolepidotus* (Island 21).

		Age groups					
		2	3	4	5	6	Σ
All material	n	64	59	26	4	2	155
	%	41.29	38.06	16.77	2.58	1.30	100.00
	Z'		0.0769	0.5596	0.8459	0.4984		0.4055
	S		0.9231	0.4404	0.1541	0.5016		0.5945
	Z		0.08	0.82	1.87	0.69		0.52
♂♂	n	27	26	14	4	1	72
	%	37.50	36.11	19.44	5.56	1.39	100.00
	Z'		0.0392	0.4621	0.7135	0.7509		0.3687
	S		0.9608	0.5379	0.2865	0.2491		0.6313
	Z		0.04	0.62	1.25	1.39		0.46
♀♀	n	37	33	12	—	1	83
	%	44.58	39.76	14.46	—	1.20	100.00
	Z'		0.1042	0.6358		0.9163		0.4401
	S		0.8958	0.3642		0.0837		0.5599
	Z		0.11	1.01		2.48		0.58

Table 78. Male and female age and length composition of *Marcusenius macrolepidotus* stock in north cove of Island 21 (Chete), south from Sinazongwe (locality E).

	2+ ♂	2+ ♀	3+ ♂	3+ ♀	4+ ♂	4+ ♀	5+ ♂	5+ ♀	6+ ♂	6+ ♀	Σ ♂	Σ ♀
101–110	—	4									—	4
111–120	2	4									2	4
121–130	1	6									1	6
131–140	2	7									2	7
141–150	19	10	—	3							19	13
151–160	3	6	9	12							12	18
161–170			12	18	6	—					18	18
171–180					5	11					5	11
181–190			2	—	3	1					5	1
191–200			3	—							3	—
201–210							3	—			3	—
231–240									—	1	—	1
241–250							1	—	1	—	2	—
Total n	27	37	26	33	14	12	4	0	1	1	72	83
%	37.50	44.58	36.11	39.76	19.44	14.46	5.56		1.39	1.20	100	100

Table 80. Back calculated growth values of *Hippopotamyrus discorhynchus* from the south cove of Island N 20.

Material	Age groups	Year of hatching	n	Mean standard length at time of capture	Back calculated standard lengths (mm) for the previous growth seasons				
					l_1	l_2	l_3	l_4	l_5
All material	1+	1966/1967	32	80	70				
	2+	1965/1966	28	117	79	103			
	3+	1964/1965	23	151	76	111	133		
	4+	1963/1964	11	173	82	112	136	156	
	5+	1962/1963	2	195	76	116	149	169	182
	Total		96		77	111	139	163	182
♂♂	1+	1966/1967	14	76	70				
	2+	1965/1966	17	113	76	101			
	3+	1964/1965	15	151	74	111	134		
	4+	1963/1964	6	170	83	118	138	155	
	5+	1962/1963	—	—	—	—	—	—	—
	Total		52		76	110	136	155	—
♀♀	1+	1966/1967	18	83	71				
	2+	1965/1966	11	124	84	108			
	3+	1964/1965	8	152	78	113	136		
	4+	1963/1964	5	177	79	106	136	160	
	5+	1962/1963	2	195	76	116	149	169	182
	Total		44		78	111	140	165	182

Table 81. Back calculated growth values of *Marcusenius macrolepidotus* from Island N 21.

	Age groups	Year of hatching	n	Mean standard length at time of capture	Back calculated standard lengths (mm) for the previous growth seasons					
					l_1	l_2	l_3	l_4	l_5	l_6
All material	2+	1966/1967	28	134	88	121				
	3+	1965/1966	17	166	100	132	153			
	4+	1964/1965	9	180	90	126	154	170		
	5+	1963/1964	2	227	102	137	169	181	217	
	6+	1962/1963	2	239	94	136	162	190	218	232
	Total		58		95	130	160	180	218	232
♂♂	2+	1966/1967	11	137	92	123				
	3+	1965/1966	9	173	105	139	159			
	4+	1964/1965	5	180	98	129	150	170		
	5+	1963/1964	2	227	102	137	169	181	217	
	6+	1962/1963	1	245	93	135	163	195	227	236
	Total		28		98	133	160	182	222	236
♀♀	2+	1966/1967	17	132	86	119				
	3+	1965/1966	8	159	94	125	145			
	4+	1964/1965	4	182	80	125	158	167		
	6+	1962/1963	1	233	94	137	161	189	209	226
	Total		30		89	127	155	178	209	226

Table 82. Biomass of *Hippopotamyrus discorhynchus* stock from the south cove of Island 20 in relation to number of specimens per 1 ha (p = values at the time of poisoning; a = values at the beginning of annulus formation).

Age groups	Number of specimens per 1 ha		Standard length in cm of 1 specimens		Weight in g of 1 specimen		Mass (weight) of age group in kg	
	p	a	p	a	p	a	p	a
0+	4	—	3.3	—	0.6	—	0.002	—
1+	60	64	8.0	7.0	9.8	6	0.590	0.384
2+	289	309	11.7	10.3	32	22	9.248	6.798
3+	273	292	15.1	13.3	72	49	19.656	14.308
4+	54	58	17.3	15.6	112	80	6.048	4.640
5+	8	9	19.5	18.2	160	132	1.280	1.188
Total	688	732					36.824	27.318

Table 83. Production and yield determination of *Hippopotamyrus discorhynchus* from the south cove of Island N 20. Calculated per 1 ha, number of fish adjusted back to the beginning of growth season ($\bar{Z}$ = 0.17% and t = 40 days).

Age group	n	Mean weight of 1 specimen in grams (weight in previous year)	Mean weight increment in g of 1 specimen	Total increment of age group per year in kg	Available production		Available yield	
					in kg	in % of mass	in kg	in % of production
1+	64	6	6	0.38				
2+	309	22 (9)	11	3.40				
3+	292	49 (27)	22	6.42	12.11	44.33	8.33	68.79
4+	58	80 (52)	28	1.62				
5+	9	132 (100)	32	0.29				

Table 84. Biomass of *Marcusenius macrolepidotus* stock in north cove of Island 21 in relation to number of specimens per total poisoned area (a = values at the time of annulus formation; p = values at the time of poisoning).

Age groups	Number of specimens per total poisoned area		Standard length in cm of 1 specimen		Weight in g of 1 specimen		Mass (weight) of age group in kg	
	p	a	p	a	p	a	p	a
2+	64	78	13.4	12.1	38	28	2.43	2.18
3+	59	72	16.6	15.3	76	60	4.48	4.32
4+	26	32	18.0	17.0	98	82	2.55	2.62
5+	4	5	22.7	21.7	200	172	0.80	0.86
6+	2	2	23.9	23.2	238	220	0.49	0.44
Total per 1 ha	155	189					10.74	10.42
	29	36					2.04	1.98

Table 85. Production and yield determination of *Marcusenius macrolepidotus* from Island 21 (locality E). Number of fish adjusted back to the beginning of growth season ($\bar{Z} = 0.14$ and $t = 139$ days).

Age group	n	Mean weight of 1 specimen in grams (weight in previous year)	Mean weight increment in g of 1 specimen	Total increment of age group per year in kg	Available production		Available yield	
					in kg	in % of mass	in kg	in % of production
2+	78	28 (7)	21	1.64				
3+	72	60 (36)	23	1.66				
4+	32	82 (60)	22	0.70	4.45	42.71	2.81	63.15
5+	5	172 (101)	71	0.36				
6+	2	220 (177)	43	0.09				

The above observations compare well with the mean seasonal total mortality rates (Table 76), which in males is 83% but in the females only 50% in respective age groups. Seasonal total mortality rates increased with increasing age.

M. macrolepidotus: This species also belongs among the species with a medium life span. Specimens older than seven years occur only rarely (Table 77). No difference between the life span of females and males was observed (Table 78). Both sexes can be found in age group 6+.

The mean seasonal total mortality rate for the sample investigated is 41%, for males 37% and for females 44%. The mortality rate increased up to the fourth year and in subsequent years decreased gradually (Table 79).

H. discorhynchus: The fastest growth is achieved in the first age group when the fish attain 70–82 mm standard length. The growth in older age groups gradually decreased (Table 80). The growth of males is slower than that of females even during their first year of life. The difference gradually increased, and in age group 4+ reached more than 10 mm.

M. macrolepidotus: The growth values for this species (Table 81) are higher than those for the parrotfish despite the smaller size which they attain. The males of this species show faster growth than females. The angler's record for this species given by JUBB (1967) is about 350 g; unfortunately length is unknown.

H. discorhynchus: The biomass of this species (adjusted for the lack of young-of-the-year fish) is 71.53 kg/ha. As the P'/B' ratio found in the sample under study is 44.33%, the annual available production of parrotfish is 31.71 kg/ha (Table 82 and 83).

Minimum harvestable size determined graphically is attained by specimens almost three years old, at a standard length of 103 mm and body weight of 22 grams. Owing to this the ratio of available yield to the production is 68.79%, which means that the available yield can reach 21.81 kg/ha annually.

M. macrolepidotus: The population density of the bull-dog-fish is much smaller in comparison with the parrotfish, the biomass reached only 1.75 kg/ha. Estimated avilable production and available yield is 1.75 and 0.47 kg/ha/yr (Table 84 and 85). These results are in agreement with JACKSON's opinion (1961) that the bull-dogfish is an important food fish in the Bangweulu swamps but is not caught in significant numbers anywhere else.

3.10. The rednose mudsucker *Labeo altivelis* Peters, 1852

The rednose mudsucker *(Labeo altivelis)* was commercially the most important fish in the Middle Zambezi prior to the creation of Lake Kariba. According to JACKSON (1961b), specimens of this species in one pool in 1957 amounted to 11% of the total number of fish sampled and it is 'the most abundant single component of the fish population of the area, (...) forming the biggest part of the catch in nearly all nets.' Before the creation of Lake Kariba it formed, together with the closely related purple mudsucker *(Labeo congoro)* the largest proportion in catches of HARDING's (1964) experimental gillnets in the Zambezi River, though some decline in the same gillnets occurred at the early stages of flooding (1960). The declining proportion of this fish in experimental catches is not reflected in the commercial catches (COKE, 1968) and

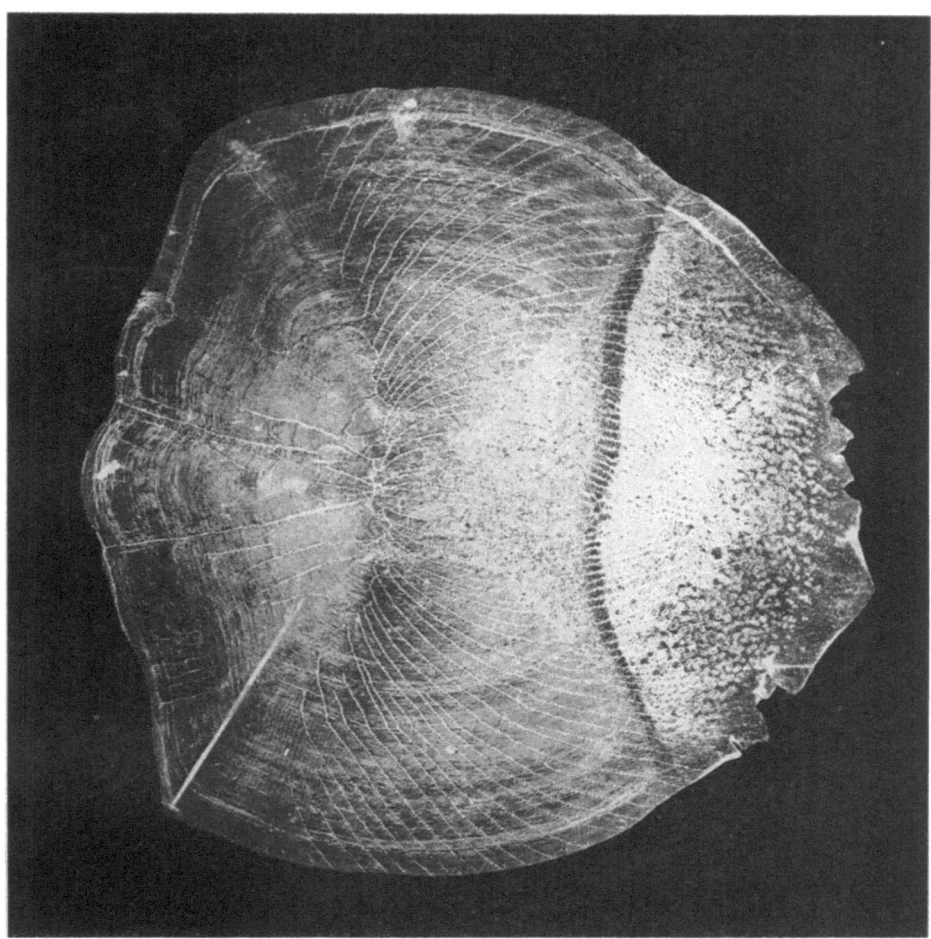

Fig. 50. Key scale of the rednose mudsucker *(Labeo altivelis)*. Lake Kariba, locality G, standard length 290 mm.

in 1962 it still represented 21%, making it the second most frequently caught fish after the kurper bream (HARDING, 1966). In spite of the inaccurate catch statistics, the rednose mudsucker seems to manifest a steady decline in catches since the formation of the lake. In the period 1968 to 1971 it occurs only in 50% of our samples, as 0.18% of the total number and 1% of the total weight. Hence it is the eighth species in abundance and ninth in standing stock. It is represented by 5–96 (mean 42) specimens per ha and 1–12 (mean 6) kg/ha. Nevertheless even at this time it is fairly well represented in the commercial catches.

Because of these facts we thought it useful to establish the age composition and growth of this species so as to be able to compute the production. Because of the low representation in samples, however, it is possible to evaluate only one locality, i.e. D, Buffalo Island cove which is situated close to the old Zambezi River bed. It is a

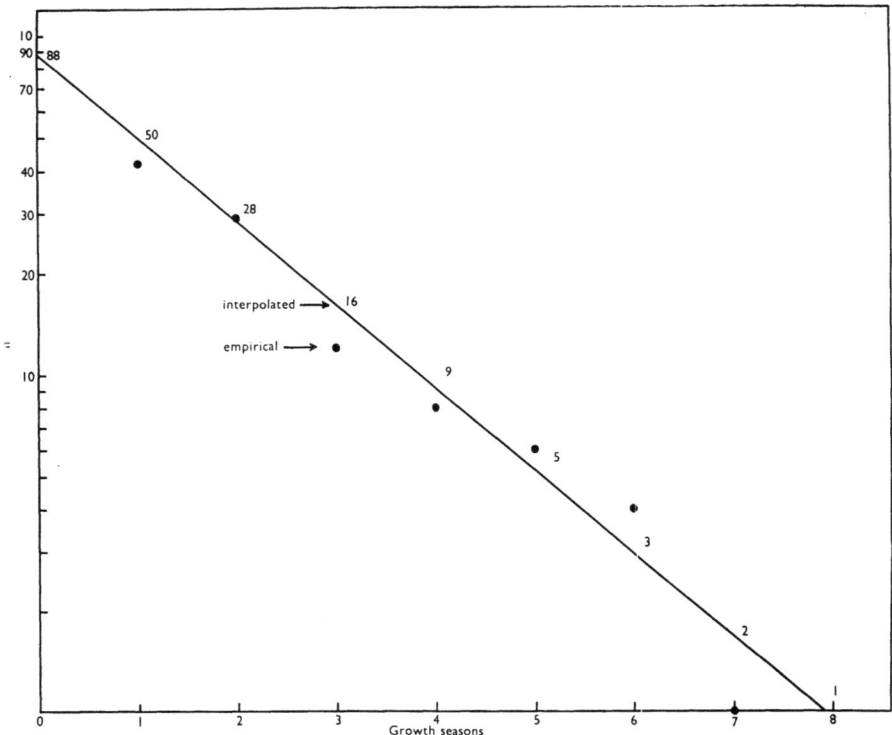

Fig. 51. Single logarithmic diagram for interpolation of the rednose mudsucker's age composition.

locality which has been used many times before because it yields the largest samples. Thus the values computed must be considered as maximum values and will probably overestimate the size of the rednose mudsucker population in the whole lake.

From 90 preserved specimens of sample D key scales were extracted from the row above the lateral line and ventral fins. These are cycloid with a parallel fan-like distribution of radial canals in the caudal and dorso-ventral parts. Only single radial canals are present in the oral part, their number increasing with age, especially the secondary ones (Fig. 50). The annuli are hardly visible. They form an interrupted circulus or a few circuli, which cut the previous circuli in the dorso-ventral parts but only in some specimens. Dense circuli only rarely occur in the annulus region. From the 90 specimens studied I found 4 juvenile specimens with regenerated key scales and 1 in which the scales were unidentifiable. Among the adults 8 males had regenerated key scales and the scales of 4 females could not be identified.

The processing follows the same system developed earlier (BALON, 1963, 1971a, 1972). The studied sample was recalculated for the whole sample and the age composition interpolated from the single logarithmic diagram to eliminate sampling bias (Fig. 51). For back calculation of linear growth a nomogram was constructed (Fig. 52) from the empiric curve (VOVK, 1956). The standard length and diagonal scale radius

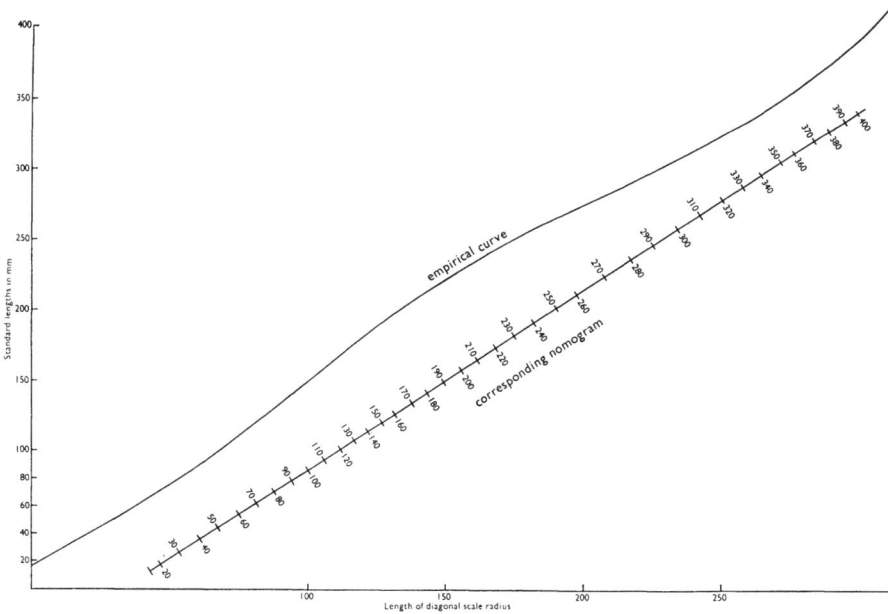

Fig. 52. Scale radius and standard length relationships of the rednose mudsucker from Lake Kariba.

relationship of juveniles, males and females overlap in the common correlation field.

The largest catch record of the rednose mudsucker is noted by JUBB (1967) in the Sanyati River, Lake Kariba in 1963. The fish weighed 2.155 kg and most likely reached 434 mm in standard length. In the Zambezi River prior to the impoundment, JACKSON's (1961b) largest specimen was 1.162 kg and 372 mm. In our samples the largest specimens did not exceed 1 kg; we have, however, collected larger specimens from gillnet catches in Sinazongwe – 40 cm long and 1.63 kg – which, according to some fishermen, are not particularly rare.

According to the estimated age composition (Table 86 and 87) the sexual maturity of some males is reached in the first age group, though most of the males mature in the second age group. In females it is in the third age group that the majority of specimens mature, though the first mature females can occasionally occur in the first age group and more frequently in the second age group. But in the fourth age group some specimens may still be immature. The mean age of sexual maturity determined according to the specific weight rate of growth (Cw) – the third growth season – is thus in full agreement with the empirical findings. In Lake Kariba the fifth age group of the rednose mudsucker begins to show signs of senility (BALON, 1960, 1964c); the species appears, on an average, to have only two years of full reproductive maturity.

In sample D where the highest density was found, it was 102 specimens per ha of which 44 were juveniles, 26 males and 31 females. In sample C there were only 8 specimens per ha, in sample E 13, in sample F 5 and in sample G 96, all of which were juveniles.

Table 86. Age and length composition of *Labeo altivelis* stock from cove of Buffalo Island (D = 1.147 ha).

Length groups in mm	N in age groups									Σ N
	0	I	II	III	IV	V	VI	VII	VIII	
71– 80		2								2
81– 90		4								4
91–100		13								13
101–110		12	3							15
111–120		7	8							15
121–130		2	10							12
131–140		2	3	2						7
141–150			3	6						9
151–160				1	1					2
161–170			1	2						3
171–180				1						1
181–190			1							1
221–230					1					1
231–240					2					2
241–250					2					2
251–260					2	1	2			5
261–270						2				2
271–280						1				1
281–290						1				1
291–300						1	1			2
301–310							1			1
331–340								1		1
n		42	29	12	8	6	4	1		102
Graphically interpolated										
Ñ	(88)	50	28	16	9	5	3	2	1	114
Ñ/ha	(77)	44	24	14	8	4	3	2	1	100
Juv. Ñ/ha	(77)	34	4	2	3					43
%		77	15	12	33					
♂♂ Ñ/ha		9	12	2		3				26
%		20	50	12		67				
♀♀ Ñ/ha		1	8	10	5	1	3	2	1	31
%		3	35	76	67	33	100	100	100	

The juvenile marks were present in only two juveniles and in two females of a standard length of 38 and 50 mm (mean 43). The back calculated lengths are slowest for juveniles. Both sexes grow at a similar rate until the third growth season, although the females always lag behind. After the third growth season, however, the growth rate of the males increases and that of the females decreases sharply so that the differences become very large (Table 88, Fig. 53). The slower growth of juveniles

Table 87. Computation of numerical age composition for *Labeo altivelis* stock from Buffalo Island cove at the time of annulus formation.

Age groups		0	I	II	III	IV	V	VI	VII	VIII	Total
Instantaneous rate of mortality	Z	0.5595	0.6063	0.5390	0.5595	0.6931	0.2878	0.4055	0.6931		0.5430
Number per ha at time of annulus formation ($\bar{Z} = 0.15$, t = 40)	N'/ha	78	45	24	14	8	4	3	2	1	102
Same for juveniles		78	34	4	2	3					44
Same for males			9	12	2	3	3				26
Same for females			1	8	10	5	1	3	2	1	31
Specific weight rate of growth	Cw	1300	346	219	109	55	13	86	13	15	
Life periods		juvenile				adult			senective		

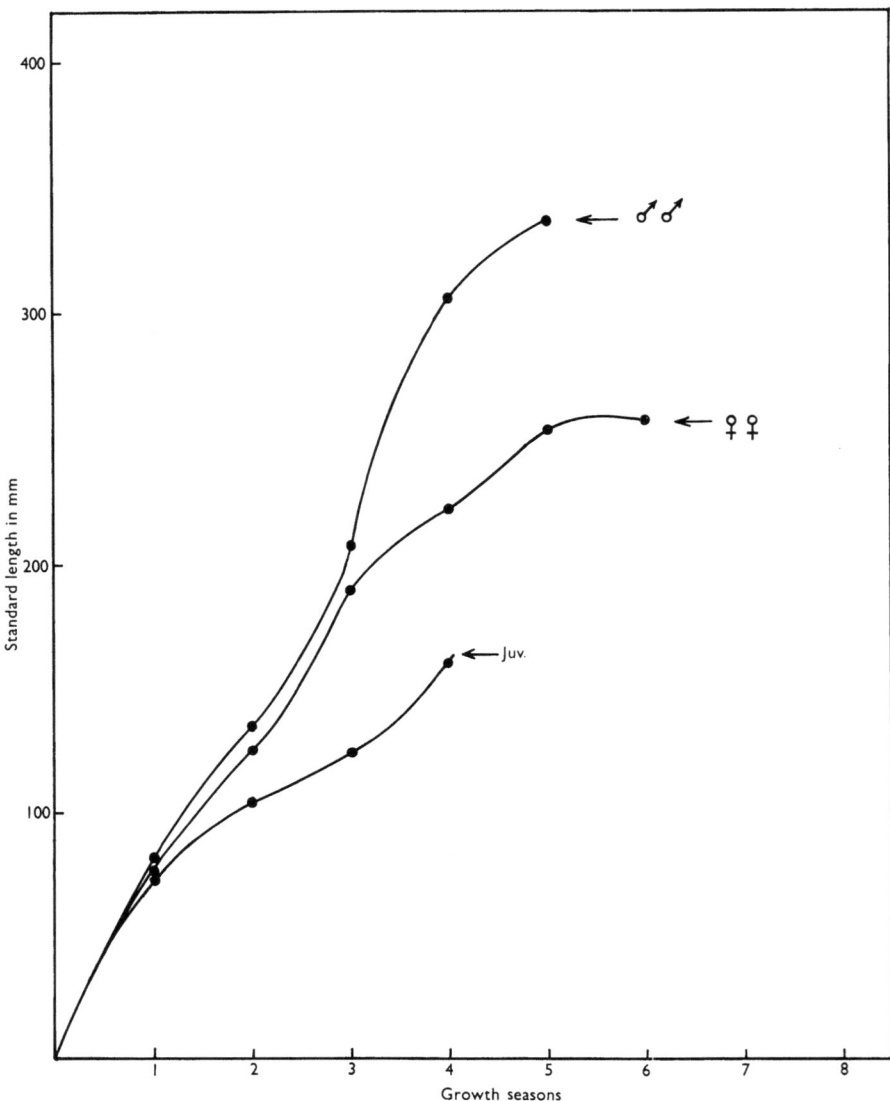

Fig. 53. Linear growth curves of juveniles, females and males of the rednose mudsucker from Lake Kariba.

probably indicates an earlier rate of maturation and a certain type of selective mortality in fast growing animals (RICKER, 1969). According to these values the best minimum harvestable size of the rednose mudsucker (Table 89) can be established at a standard length of 24 cm and a weight of 400 g (Fig. 54). The maximum age found in quantitative samples was 6 years, while the largest specimens from gillnet catches were 9 years old.

The commercial importance of the rednose mudsucker in Lake Kariba decreased,

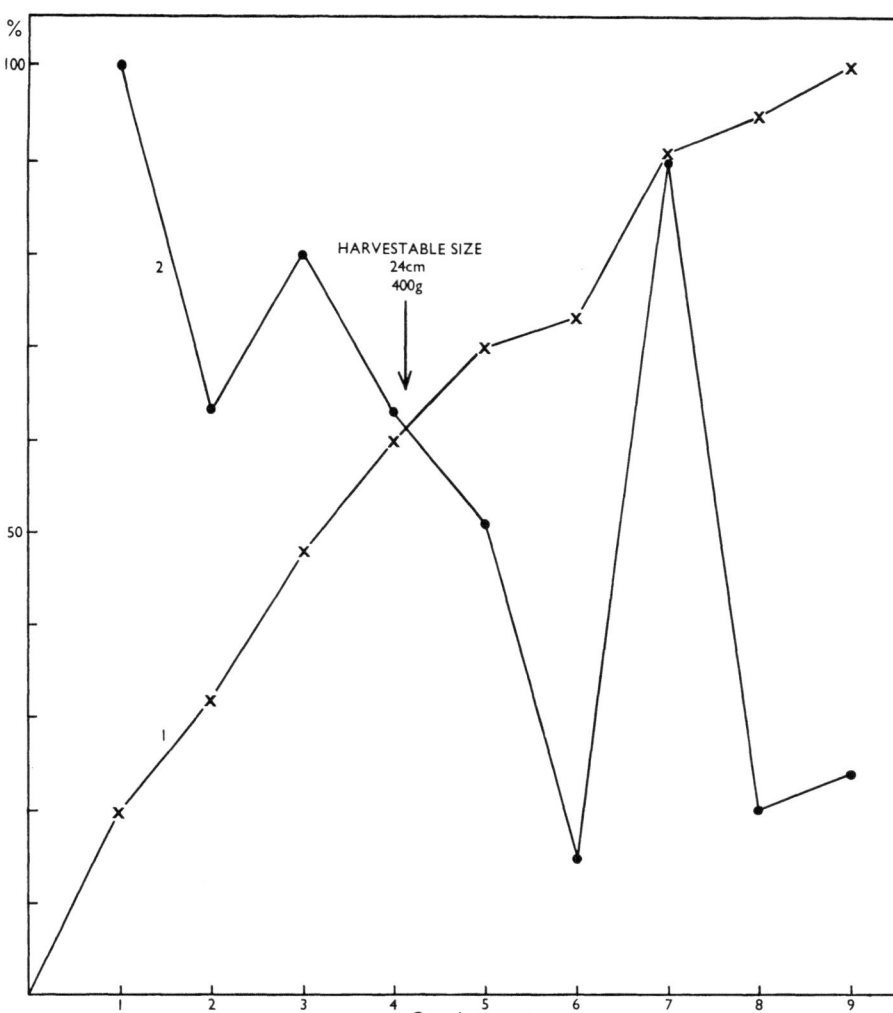

Fig. 54. Growth curves of the rednose mudsucker from Lake Kariba for minimum harvestable size estimation. (1) Standard lengths in percent of the length of the last growth season, (2) length increments in percent of the length of the first growth season.

though it is still represented in every catch by single specimens. Seasonally, particularly in river estuaries, it can amount to an important proportion in catches. The total production of juveniles is 2.19 kg/ha/yr, of males 5.24 and of females 5.17. The maximum total production of this species is estimated at 12.6 kg/ha/yr, which for the inhabited lake area gives 25×10^5 kg/yr. The available production (Tables 90 to 92) in juveniles is 0.64, in males 1.82 and in females 2.45 – in total 4.91 kg/ha/yr. The available yield according to the established harvestable size is 1.01 kg/ha/yr, which, since these values are somewhat inflated because of the selected sample, brings the rednose mudsucker in Lake Kariba to the limit of exploitation.

Table 88. Back calculated growth values of *Labeo altivelis* from Lake Kariba.

Buffalo Island cove (D)	Age groups	Year of hatching	N	Mean standard length at the time of capture in mm	Back calculated standard lengths (l) in mm for the previous growth seasons and corresponding interpolated weights (w̄) in g								
					l_j / $\bar{w}_j$	l_1 / $\bar{w}_1$	l_2 / $\bar{w}_2$	l_3 / $\bar{w}_3$	l_4 / $\bar{w}_4$	l_5 / $\bar{w}_5$	l_6 / $\bar{w}_6$	l_7 / $\bar{w}_7$	l_8 / $\bar{w}_8$	l_9 / $\bar{w}_9$
Juveniles	I	1967	27	103	50 / 3.4	88 / 18								
	II	1966	3	131		67 / 8	121 / 48							
	III	1965	1	150		66 / 8	102 / 30	132 / 64						
	IV	1964	1	160		68 / 9	90 / 20	116 / 43	160 / 110					
	Total mean				50 / 3.4	72 / 10	104 / 32	124 / 54	160 / 110					
Males	I	1967	7	104		82 / 15								
	II	1966	10	131		75 / 11	115 / 42							
	III	1965	1	147		59 / 5	103 / 31	147 / 86						
	V	1963	2	343		110 / 36	187 / 172	267 / 500	306 / 760	337 / 1000				
	Total means					81 / 14	135 / 67	207 / 234	306 / 760	337 / 1000				

Table 88. (continued)

Females	I	1967	1	80		51							
						3.6							
	II	1966	7	123	38	80	112						
					1.5	14	40						
	III	1965	6	143		79	116	149					
						13	43	88					
	IV	1964	2	249		83	170	237					
						15	130	350					
	V	1963	1	270		78	118	210	245	268			
						13	44	245	390	500			
	VI	1962	2	257		79	111	166	198	239	257		
						13	37	126	200	360	448		
	Total mean				38	75	125	190	221	253	257		
					1.5	11	54	180	290	440	448		
Gill-nets Sinazongwe 1959	IX	1959	2	400		69	135	201	266	295	331	365	381	400
1968						9	68	220	492	680	960	1260	1420	1630
All material	I	1967	35	102	48	83								
					3.0	15	113							
	II	1966	20	128	38	72	41	146	214	313				
					1.5	10	112	85	263	815				
	III	1965	8	144		72	40	194	289	239	257			
	IV	1964	3	216		10	135	200	622	360	448			
						76	67	256	198	295	331			
	V	1963	3	319		12	168	440	165	204	680	960		
						101	127							
	VI	1962	2	257		29	111							
						79	37	125			257			
	IX	1959	2	400		13	135	201	266	295	331	365	381	400
						69	67	220	492	680	448	1260	1420	1630
						9					960			
	Grand total mean				43	79	129	192	242	282	294	365	381	400
					2.1	13	58	185	386	600	678	1260	1420	1630

Table 89. Parameters for computation of harvestable size of *Labeo altivelis* from Lake Kariba.

		\multicolumn{9}{c}{Means at the end of growth season}								
		1	2	3	4	5	6	7	8	9
Standard length (mm)	l	79	129	192	242	282	294	365	381	400
Length increments (mm)	$l_i - l_{i-1}$	79	50	63	50	40	12	71	16	19
Length increments in % of length of the first growth season	D	100	63	80	63	51	15	90	20	24
Standard lengths in % of length of the final growth season	I	20	32	48	60	70	73	91	95	100
Wet weight (g)	$\bar{w}$	13	58	185	386	600	678	1260	1420	1630
Weight increments (g)	$\bar{w}_i - \bar{w}_{i-1}$	13	45	127	201	214	78	582	160	210

Table 90. Computation of production for *Labeo altivelis* stock from Lake Kariba (after data from locality D – Buffalo Island cove, 1968. Juveniles only).

Growth seasons		0	1	2	3	4	Total
Interpolated N'/ha at time of annulus formation	N'	78	34	4	2	3	44
Weights in g of 1 specimen	$\bar{w}$		10	32	54	110	
Initial biomass in kg/ha	B'		0.34	0.13	0.11	0.33	0.91
Rate of increase in biomass	H	+1.4723	−0.9768	−0.1701	+0.3060		
Instantaneous growth rate	G	2.3026	1.1632	0.5232	0.7115		
Instantaneous rate of total mortality	Z	0.8303	2.1400	0.6933	0.4055		
Mean biomass	$\bar{B}$	0.77	0.08	0.10	0.39		1.34
Production (= $G\bar{B}$) in kg/ha/yr	A	1.77	0.09	0.05	0.28		2.19
Available production in kg/ha/yr	P'	0.34	0.09	0.04	0.17		0.64

Table 91. Computation of production for *Labeo altivelis* stock from Lake Kariba (after data from locality D - Buffalo Island cove, 1968. Males only).

Growth seasons		0	1	2	3	4	5	Total
Interpolated N'/ha at time of annulus formation	N'		9	12	2		3	26
Weights in g of 1 specimen	$\bar{w}$		14	67	234	760	1000	
Initial biomass in kg/ha	B'		0.13	0.80	0.47		3.00	4.40
Rate of increase in biomass	H	+2.6391	+1.2778	−0.5412	+0.4848	−0.8241		
Instantaneous growth rate	G	2.6391	1.5656	1.2507	1.1779	0.2745		
Instantaneous rate of total mortality	Z		0.2878	1.7919	0.6931	1.0986		
Mean biomass	$\bar{B}$	0.64	1.62	0.36		2.04		4.66
Production (= $G\bar{B}$) in kg/ha/yr	A	1.69	2.54	0.45		0.56		5.24
Available production in kg/ha/yr	P'	0.13	0.64	0.33		0.72		1.82
Available yield in kg/ha/yr	$Y_{P'}$					0.72		0.72

Table 92. Computation of production for *Labeo altivelis* stock from Lake Kariba (after data from locality D - Buffalo Island cove, 1968. Females only).

Growth seasons		0	1	2	3	4	5	6	7	8	Total
Interpolated N'/ha at time of annulus formation	N'		1	8	10	5	1	3	2	1	31
Weights in g of 1 specimen	$\bar{w}$		11	54	180	290	440	448	500	520	
Initial biomass in kg/ha	B'		0.01	0.43	1.80	1.45	0.44	1.34	1.00	0.52	6.99
Rate of increase in biomass	H	+2.3979	−0.4884	+0.9809	−0.2162	−1.1926	−1.0806	−0.2957	−0.6538		
Instantaneous growth rate	G	2.3979	1.5911	1.2040	0.4769	0.4169	0.0180	0.1098	0.0393		
Instantaneous rate of total mortality	Z		2.0795	0.2231	0.6931	1.6095	1.0986	0.4055	0.6931		
Mean biomass	$\bar{B}$	0.04	0.34	3.05	1.32	0.26	0.82	0.88	0.38		7.09
Production (= $G\bar{B}$) in kg/ha/yr	A	0.10	0.54	3.67	0.63	0.11	0.01	0.10	0.01		5.17
Available production in kg/ha/yr	P'	0.01	0.34	1.26	0.55	0.15	0.02	0.10	0.02		2.45
Available yield in kg/ha/yr	$Y_{P'}$					0.15	0.02	0.10	0.02		0.29

These low yield rates, however, should not only be connected with fishing intensity, but may also demonstrate the effect of the species retreating because of lacustrine conditions. The relatively extreme low production of juveniles strongly supports this view.

3.11. The silver robber *Micralestes acutidens* (Peters, 1852)

by Juraj Holčík

The family Characidae is represented by 4 species in Lake Kariba. Their share and role in the ichthyocenosis of the lake is not equal. *A. lateralis* is the most abundant (59% in number) of the 41 species found in this lake and *H. vittatus* occupies 6th place (3%). On the other hand, *M. acutidens* forms only a negligible part of the population with 0.41% of the total number of fish (all data according to Balon, 1971a, 1972) and as it is small in size its weight share is only 0.01%. The biology of this species is unknown except that it is an insectivorous animal forming schools and that it performs spawning migrations up the flooded streams during the first heavy rains (Jackson, 1961a; Jubb, 1967).

I have studied two sub-samples of fish from population density surveys (Balon, 1973a). One sub-sample (18 specimens) was collected at the cove of Chikanka Island (June 21–25, 1968) and the second (73 specimens) was sampled in the cove at Island N 20, southwest of Sinazongwe (December 11–14, 1968). The fish were preserved in formalin, sent to Czechoslovakia, and worked on in the laboratory. Each fish was measured (standard length), weighed, and its sex ascertained. Key scales were taken from the first row above the lateral line just above the base of the pectoral fin (as proposed by Balon, 1972). The measurements were taken with precision of 0.1 mm and the weights with a precision of 0.01 grams. The ovaries of all females from Chikanka Island were removed and the number of eggs and their size was ascertained employing the gravimetric method and the egg measuring trough respectively. Dry scales were read by means of the Zeiss-Lesegerät projector with 17.5 magnification. The diagonal radius was used to read the annuli and to measure their distance from the scale center. The relationship between the body and scale size was linear, showing a correction value of 14.2 m ($l = 14.2 + 1.05046s$). The length-weight graph was constructed to extrapolate weights for back calculated lengths. All other methods concerning the determination of age composition of stock, mortality, growth rate and available production were in accordance with those used earlier (Holčík, section 3.1). I did not make a calculation of yield because of the small size of the fish and the low population density.

The scales (Fig. 55 to 57) are cycloidal, oval and thin. The area of the scale is divided by radii (grooves) into 4–6 regular parts. The oral part has a typical concave margin. Sclerites are concentric, regular and in the caudal part more or less waved. In fish of a bigger size (above 40 mm) the caudal sclerites are irregular, not concentric and mostly without connections. The annulus can be detected as a fine line visible only on the diagonal and caudal part. Normally only one diagonal part displays the annulus. The new increment usually cut off the older sclerites. In some cases the new increment

Fig. 55. Scale of 1+ old *M. acutidens*. Chikanka Island, male of 46.4 mm and 1.57 g in standard length and weight respectively.

is waved so the annulus becomes very clear. Only the normal annuli were present in all fishes. No juvenile marks or spawning checks were observed. Two of 81 fish had to be excluded because of invisible annuli or accessory checks. Four other fish were also excluded due to a complete lack of scales.

At present the time of annulus formation is unknown, but I can state that all fish caught at the beginning of December (locality D) also showed new increments of different sizes – the bigger the fish the smaller the increment. From COCHE's (1968) hydrometeorological data I assumed that the time of annulus formation falls at the end of October in young fish and the middle of November in the older. November 1st was tentatively ascertained to be the beginning of the growth season for *M. acutidens*. The retardation of the growth season in older fish is obviously due to their sexual maturation and subsequent spawning as in the case of some fish from temperate regions (BECKMAN, 1943; BALON, 1955; HOLČIK, 1967b).

Fig. 56. Scale of 2+ old female of *M. acutidens* from Island N 20 cove. Standard length 53.0 mm, body weight 2.82 g.

Fig. 57. Scale of 3+ old female of *M. acutidens* from Island N 20 cove. Standard length 64.5 mm, body weight 5.25 g. Note the accessory check between the first and second annulus.

Table 93. Age composition (in percentages) of *Micralestes acutidens* under study.

Locality		Age group			
		0+	1+	2+	3+
Chikanka	Aged sample	50.0	50.0			100.0
Island	Total population	43.2	50.0	6.8		100.0
Island N 20	Aged sample	24.5	65.3	8.2	2.0	100.0
	Total population	20.4	73.8	5.3	0.5	100.0

Table 94. Sex composition of *Micralestes acutidens* stock from Chikanka Island and from Island 20 (in percent of total catch).

Locality	Sex	Age groups			Total population
		1	2	3	
Chikanka Island	♂♂	55.6	—	—	48.0
	♀♀	44.4	100.0	—	52.0
Island 20	♂♂	62.3	—	—	57.7
	♀♀	37.7	100.0	100.0	42.3

Table 95. Age and length composition of *Micralestes acutidens* stock in cove of Chikanka Island.

Length groups	Age (N)			Sum
	0+	1+	2+	
31–40	14			14
41–50	5	21		26
51–60		1	3	4
Total	19	22	3	44

Table 96. Age and length composition of *Micralestes acutidens* stock in cove of Island N 20.

Length groups	Age (N)				Sum
	0+	1+	2+	3+	
21–30	1				1
31–40	9	6			15
41–50	28	112			140
51–60		20	10		30
61–70				1	1
Total	38	138	10	1	187

Table 97. Age composition, seasonal total mortality rate (Z'), survival rate (S) and instantaneous rate of total mortality (Z) of *Micralestes acutidens* under study.

Locality	Age groups	1	2	3	Sum
Chikanka Island	n	22	3		25
	%	88.0	22.0		100.0
	Z'		0.8577		0.8577
	S		0.1423		0.1423
	Z		1.95		1.95
Island N 20	n	138	10	1	149
	%	92.6	6.7	0.7	100.0
	Z'		0.9272	0.9017	0.9250
	S		0.0728	0.0983	0.0750
	Z		2.63	2.32	2.59

The oldest fish in our samples were three years old. The maximum size attained was 64.5 mm standard length both in aged sample and in all poisoned stocks of this fish in Lake Kariba. BOULENGER (1909) gives 65 mm for the total length and JUBB (1967) writes that this species only rarely exceeds 3 inches (= 7.6 cm) in length (probably total length). The mean age of *M. acutidens* is 1–2 years because of the rare occurrence of older fish (Table 93). I can confirm JUBB's (1967) observation that the larger fish are females; all fish older than 1 year were females (Table 94). I assume collection of small fish was insufficient or the nursery grounds were located elsewhere. This assumption is based on the relatively low number of young of the year fish both in whole samples as well as in sub-samples (Table 95 and 96).

As expected the mortality rate, in such a short life span fish, is very high (Table 97), varying from 86.4% per year in stock from Chikanka Island, up to 92.6% per year in the sample from Island N 20 for fish one year or older. By the second year of life the mortality of males is 100%, but that of females only 70–82% calculated separately for Chikanka and Island N 20 stock. It is obvious that the mortality rates of males must be higher as females attain an older age. Using the value of a mean annual surviving rate of 7.4% found in the population from Island N 20 it is possible to find that from more than 18 thousand yearlings only 100 reach the third year of life (Table 98).

The number of females and males in the whole population seems to be equal (Table 94). This ratio, however, differs with different age groups. The number of males in age group 1 predominates over the females. Because of the complete lack of males in subsequent age groups this disproportion evens out.

The sexual maturity of the silver minnow already sets in during the first year of life. The first matured males and females were found among the young-of-the-year fishes at standard lengths of 36 and 40 mm respectively. In the sample from Chikanka Island all yearlings have ripe sexual products. A similar situation was found in the sample from Island N 20, where 91% of all fish showed fully developed gonads. Also the

Table 98. Relative abundance and ichthyomass of *Micralestes acutidens* after mean survival rate $\bar{S} = 7.4$.

Growth seasons	Number of surviving specimens ($\bar{S} = 7.4$)	Standard length (mm) of 1 spec.	Weight (g) of 1 spec.	Mass weight of age group in kg
1	18260	43.3	1.14	20.82
2	1351	56.5	3.70	5.00
3	100	63.5	5.00	0.50

sharp decrease of absolute length increments after attaining the first year indirectly confirms the age of sexual maturation.

All females from the Chikanka Island sample were dissected, the gonads were removed and the number and quality of eggs were ascertained. The results summarized in Table 99 show that *M. acutidens* is a fish with portional spawning. The gonads contain three groups of eggs differing both in number and size. The first portion contained eggs measuring 0.58–0.84 mm in diameter and numbering 210–1507 pieces. The second and smaller portion contained 57–374 eggs measuring 0.46–0.62 mm. The amount of eggs in the third portion was 51–398 and their size varied from 0.32 to 0.52 mm. The number of eggs increases with increasing length, weight and age of fish. The relative number of eggs in the first portion increases but the relative number of eggs in the third portion decreases with increasing length and age of fish. A similar situation can also be observed in size of eggs in the respective groups. The number of eggs and their size in the second portion seems to be constant.

There is a remarkable difference between the growth rate of males and females (Table 100 to 102). The latter grow faster than the former, already at the beginning of their life. Very interesting is the range of size attained by different groups. In the stock from Chikanka Island the standard length of young-of-the-year varied from 32.3 to 44.0 mm, while the standard length of yearlings fluctuates between 41.4 and 50.8 mm at the time of capture and from 32.0 to 47.0 mm at the length of first growth season (l_1). The sample from Island N 20 which is composed of more fish shows the following size variability of a standard length at the time of capture: 0+ 29.5–47.2 mm, 1+ 35.5–53.7 mm, 2+ 53.0–60.4 mm. One can see the decreasing range of variation with increasing age. This fact is in accordance with portional spawning and subsequent differences in the mortality rate of fish from different egg portions, as found in such temperate zone species as *Rhodeus sericeus amarus* and *Scardinius erythrophthalmus* (HOLČIK, 1960, 1967c). The fish hatched from the earlier portions gradually disappear from the population as they are eaten by predators, so the subsequent age classes are composed mostly from fish hatched from eggs which were laid later. The disappearance of bigger fish was observed also in fish which spawn only once and this is known as the phenomenon of Rosa Lee (HOLČIK, 1969). The silver minnow, however, presents a rather complicated picture. The increasing length attained at the end of the first year in older age groups shows that the higher age is reached only by fish with better growth.

Table 99. Fecundity data from 5 females of *Micralestes acutidens* from Chikanka Island.

No.	Standard length (mm) at time of capture	Body weight (grams)	Age	Weight of ovaries (grams)	Number and size (mm) of eggs in portions											Total number of eggs
					1			2			3			
					N	% in total number	diameter	N	% in total number	diameter	N	% in total number	diameter	
1	47.8	1.80	1+	0.073	210	32.9	0.68	192	30.1	0.60	236	37.0	0.50	638
2	49.0	1.85	1+	0.228	607	52.1	0.66	207	17.8	0.46	350	30.1	0.32	1164
3	49.0	2.20	1+	0.247	808	55.9	0.74	374	25.9	0.60	264	18.2	0.52	1446
4	49.4	1.80	0+	0.093	437	80.2	0.58	57	10.5	0.50	51	9.3	0.42	545
5	50.8	2.73	1+	0.621	1507	69.5	0.84	265	12.2	0.62	398	18.3	0.52	2170
Mean	49.2	2.07		0.252	714	59.8	0.70	219	18.4	0.56	260	21.8	0.46	1193

Table 100. Back calculated linear growth of *Micralestes acutidens* from Lake Kariba.

Locality	Sex	Age group	Year of hatching	Mean standard length (mm) at time of capture	Back calculated lengths (mm)			N
					l_1	l_2	l_3	
Chikanka Island	Both sexes	1+	1966/67	46.9	42.2			9
		Total			42.2			9
	♂♂	1+	1966/67	45.1	39.8			5
	♀♀	1+	1966/67	49.2	45.3			4
Island 20	Both sexes	1+	1966/67	46.0	41.5			32
		2+	1965/66	55.9	43.4	53.9		4
		3+	1964/65	64.5	45.0	59.0	63.5	1
		Total			43.3	56.5	63.5	37
	♂♂	1+	1966/67	45.9	41.2			18
	♀♀	1+	1966/67	48.2	43.8			11
		2+	1965/66	55.9	43.4	53.9		4
		3+	1964/65	64.5	45.0	59.0	63.5	3
		Total			44.1	56.5	63.5	18

Table 101. Mean of back calculated linear growth of *Micralestes acutidens* from Lake Kariba.

Age group	Year of hatching	Mean standard length (mm) at time of capture	Back calculated lengths (mm)			N
			l_1	l_2	l_3	
Both sexes						
1+	1966/67	46.5	41.9			41
2+	1965/66	55.9	43.4	53.9		4
3+	1964/65	64.5	45.0	59.0	63.5	1
Total			43.4	56.5	63.5	46
♂♂						
1+	1966/67	45.5	40.5			23
♀♀						
1+	1966/67	48.7	44.6			15
2+	1965/66	55.9	43.5	53.9		4
3+	1964/65	64.5	45.0	59.0	63.5	1
Total			44.4	56.5	63.5	20

Table 102. Mean values of lengths, weights and increments of *Micralestes acutidens* under study; absolute increments in cm (h), indexes of the species average size (φH), specific weight rates of growth (Cw) and the indexes of the population weight growth intensity (φCw).

Locality	Parameter	Means at the end of growth season		
		1	2	3
Chikanka Island	Standard length (mm)	42.2		
	Increments in mm	42.2		
	Standard length increments in % of length of first season	100.0		
	Body weight (grams)	1.15		
	Weight increments	1.15		
Island N 20	Standard length (mm)	43.3	56.5	63.5
	Increments in mm	43.3	13.2	7.0
	Standard length increments in % of length of first season	100.0	30.5	16.2
	Body weight (grams)	1.15	3.70	5.00
	Weight increments	1.15	2.55	1.30
Mean	Standard length (mm)	43.3	56.5	63.5
	Increments in mm	43.3	13.2	7.0
	Standard length increments in % of length of first season	100.0	30.5	16.2
	Body weight (grams)	1.14	3.70	5.00
	Weight increments	1.14	2.56	1.30
Island N 20	h	4.34	1.31	0.7
	φH		1.91	
	Cw	115.00	221.73	35.14
	φCw		123.96	
All material from Lake Kariba	h	4.33	1.32	0.7
	φH		1.90	
	Cw	114.00	224.56	35.14
	φCw		124.57	

This paradoxical phenomenon of Rosa Lee which was observed by BALON (1971a) also in tigerfish from Lake Kariba is probably the adaptation of species to enable escape from predators, as suggested by NIKOLSKY (1963, 1965).

As mentioned above the abundance and biomass of *M. acutidens* is very small in comparison with other species; the density of stocks varies greatly (Table 103). The reasons can be seen in abiotic and biotic environmental conditions. BALON (1973a and Table 2 in Part II of this Monograph) introduces also a short description of the characteristics of each locality investigated by him. Comparing his observations with the density of stocks, one can see that this species forms the densest population in such localities where the shore line is covered with rocks, sand or gravel and where there

Table 103. Comparison of abundance and mass of *Micralestes acutidens* from cove of Island 20 found at the time of poisoning and calculated to the beginning of growth season after mean daily mortality rate $\bar{Z} = 0.71\%$ and $t = 40$ days.

Growth seasons	Number of specimens per 1 ha at the time of		Standard length (mm) of 1 specimen at the time of		Body weight (grams) of 1 specimen at the time of		Mass weight (kilograms) of age group at the time of	
	poisoning	annulus formation	poisoning	annulus formation	poisoning	annulus formation	poisoning	annulus formation
1	120	160	46.0	41.5	1.59	1.16	0.191	0.186
2	9	11	55.9	53.9	3.12	2.50	0.028	0.028
3	1	1	64.5	63.5	5.25	5.00	0.005	0.005
Sum	130	172					0.224	0.219

Table 104. Mass of *Micralestes acutidens* stock in cove of Island N 20 in relationship to number of specimens per 1 ha of area.

Growth seasons	Number of specimens per 1 ha	Standard length (mm) of 1 spec.	Weight (g) of 1 spec.	Mass weight of age group in kg
0+	33	39.5	0.97	0.025
1+	120	46.0	1.59	0.191
2+	9	55.9	3.12	0.028
3+	1	64.5	5.25	0.005
Sum	163			0.249

is only scant vegetation. On the other hand the more water plants are found and the muddier the shore the smaller the density of *M. acutidens*. This fish is obviously very sensitive to lack of oxygen, and the presence of a hard bottom is probably a necessary prerequisit for successful spawning. The unsuccessful breeding of *M. acutidens* in captivity (JUBB, 1967) confirms this statement. The silver minnow belongs probably to the litophilous (or psammophilous) ecological groups (KRYZHANOVSKY, 1949).

The available production has been calculated only for the stock from Island N 20 where the sample contained all age groups. It was therefore possible to reconstruct age composition as well as growth increments. The data of abundance and weight of fish in different age groups (Table 104), with the mean daily mortality rate (0.71%) and the days that elapsed from presumed time of annulus formation to the day of sampling (= 40 days) were used in order to calculate the density and biomass of fish at the beginning of the growth season (Table 105). This was done according to the method described earlier (HOLČIK, section 3.1). Only a small difference in the biomass of these two periods can be seen. This would indicate that the mortality calculated is correct, the determination of the beginning of growth season is realistic and the fish in the studied cove do not migrate. The standing stock found at the time of poisoning (0.256 kg/ha) adjusted to the lack of the young-of-the-year fish (12.5% of the total biomass) and to the beginning of growth season (97.77% of the standing stock at the time of poisoning) was used to calculate the P'/B' ratio which is 90.98% (Table 106). This means that the available production for the season 1966/1967 was 0.199 kg/ha with biomass being equal to 0.219 kg/ha. If we relate the value of this locality to the whole lake, then available production during this period was 26 metric tons and the corresponding biomass 28 metric tons at a density equal to 14×10^6 specimens of *M. acutidens*. In other words Lake Kariba contains 0.084 kg/ha of this fish, available production of which is equal to 0.076 kg/ha/yr and a density of 41 specimens/ha.

As is evident from our observations and from the sparse literature data of *M. acutidens*, this species of fish can be classified as rheophilous, insectivorous and litophilous (or psammophilous). Its original habitats are rivers and streams and the biotope is characterized by water containing sufficient amounts of oxygen, rocky, gravel or sandy bottoms and shores, and sparse vegetation. The fish avoid swampy

Table 105. Abundance and ichthyomass of *Micralestes acutidens* in different localities of Lake Kariba in relation with their topographical characteristics.

Locality	Maximal depth (m)	Surface area (ha)	Topography	Number of fish per		Weight of stock (kg) per	
				1 ha of area	1 km of shore line	1 ha of area	1 km of shore line
Chikanka Island	5	1.210	Long narrow cove with rocky and muddy shores. *Salvinia* mats cover quarter of area at the island end. Bottom soft mud and gravel. Groups of *Phragmites, Lagarosiphon, Nymphaea* and *Cyperus*.	36	6	0.06	0.01
Cove of Island 20 SW of Sinazongwe	16	1.147	Valley cove of *Euphorbia* hills. Rock and sand on shores, very little *Salvinia*, lot of submerged vegetation in shallow part (*Ceratophyllum, Lagarosiphon, Potamogeton*), submerged trees and bush. Bottom of soft mud.	163	49	0.32	0.10
Chete Island cove	4	5.273	Shallow cove with swampy and grassy shores. Quarter of area covered by *Salvinia*. Lot of submerged *Ceratophyllum* and *Potamogeton*. Few submerged bushes and trees. Bottom of gravel, rocks and soft mud.	2	1	0.004	0.002
Cove I in Chipepo Bay	4.5	0.191	Small cove at Kota Kota Hills valley with steep rocky and forest shores. Completely covered by *Salvinia* mat and full of *Ceratophyllum, Phragmites*. Bottom of soft mud.	5	1	0.007	0.001
Mean				52	14	0.098	0.028

Table 106. Determination of available production in Island 20 stock of *Micralestes acutidens* (calculated per 1 ha of area).

Age group	N	Mean weight of 1 specimen in grams (weight in previous year)	Mean weight increment of 1 specimen in grams	Total increment of age group per year in kg	Production	
					in kg	in percent of biomass
1	160	1.16 (0)	1.16	0.186		
2	11	2.50 (1.34)	1.16	0.012	0.199	90.87
3	1	5.00 (3.85)	1.15	0.001		

waters, grassy shores and muddy bottoms. After Kariba Dam was built the fish were enclosed in the lake and the survivors retreated to coves and bays where the environmental conditions approached the conditions of rivers. The reproduction of this species probably sharply decreased as did the density of its stocks. This happened mainly during the first years of Lake Kariba's existence. The destruction of the grassy bottom and the formation of sandy and gravel shores by wave action probably improved the spawning conditions. The *M. acutidens* may also adapt gradually to lentic waters and change its rheophilous nature as is the case with some species in temperate zone impoundments (BALON, 1962a, 1964b; HOLČIK, 1966, 1970a). The type and extent of acquired adaptations should be the subject of further observation. Due to its low density the role of *M. acutidens* in contemporary Lake Kariba ichthyocenosis is obviously negligible.

3.12. The Darling's dwarf bream *Haplochromis darlingi* (Boulenger, 1911)

by EUGENE K. BALON & EPHRAIM D. MUYANGA

The natural lakes of Africa abound with the species of genus *Haplochromis* – in fact they are the most numerous of all fishes in these lakes. This diversity is known to be a result of adaptive radiation which enables utilization of the numerous ecological niches in such lakes. Most of these species are very closely related and it is not easy to distinguish between them. Though small in size, they form flocks of considerable magnitude and this is the reason for recent interest in their exploitation as a virgin resource, for example in Lake Victoria. GREENWOOD (1958, 1966) found over ninety *Haplochromis* species in that lake, while more recently, investigators (CORDONE, personal communication) speak of more than 200 species. JACKSON (1971), however, only settled for some 120 species. Writings on other natural lakes mention fewer *Haplochromis* species but always list numerous species of closely related genera which demonstrate a similar evolutionary trend (POLL, 1956; FRYER, 1960; FRYER & ILES, 1972). Rivers are in general considered habitats with fewer environmental differentiations and are thus inhabited, if at all, by single *Haplochromis* species, only. In the Zambezi River, in the area flooded by the Kariba impoundment, but prior to the formation of the lake, only Darling's dwarf bream *Haplochromis darlingi* was known (JACKSON, 1961b). It is a small mouth-brooder which inhabited, in moderate numbers, shallow running waters of the main river bed and tributaries. In the newly impounded lake this species became common everywhere and it is now the fourth most numerous taxon (5.2% of the total number of fish in 10 quantitative samples in 1968 and 1971). Because of its small size it is not exploited and forms only 0.9% of the average ichthyomass. In our samples there were 136 to 6286 (mean 1611) specimens per ha and 0.33 to 21 (mean 5) kg per ha of Darling's dwarf bream.

We decided to include this species in the Lake Kariba fish age and growth studies in anticipation of some interesting evolutionary future trends caused by the great experiment – the creation of a new lake. It may prove of importance to have these data if and when these evolutionary speciations or adaptations appear. The high density also justifies its inclusion in this study, as it will probably be of some importance in the grand total estimation of production.

H. darlingi was joined in the second phase of the Lake Kariba succession by four other closely related species. *Pseudocrenilabrus philander*, which came from upland rivers, streams and backwaters of the Middle Zambezi some time around 1964, occurs in 70% of our samples as 18 to 2319 (mean 531) specimens per ha and 0.04 to 3.4 (mean 0.78) kg per ha; larger breams such as *Haplochromis carlottae*, *Sargochromis (Haplochromis) codringtoni* and *S. giardi*, which came from the Upper Zambezi even later, most likely via the edge of Victoria Falls, were present in 12%, 75% and 25% of our samples respectively, as 2 to 61 kg per ha. It will be of great interest to see how far these cichlids are able to originate a new speciation in an environment full of free ecological niches or to what extent they are already too specialized to be able to do so.

H. darlingi was present in all 10 quantitative samples in blocked areas of Lake Kariba treated with fish eradicants, and numbered in total 5,027 specimens. Because of the narrow length-frequency range, from 1 to 8 cm with some exceptions 9 to 10 cm, there was no need to evaluate large samples. Subsampling was done by 10 mm length groups. The methods were the same as those described in other sections of this chapter and in the special methodological outline (BALON, 1972). Six most typical localities with the most complete size range for this species were finally evaluated.

Only the standard length was measured, the weights being interpolated from the length-weight relationship diagrams. Key scales were collected from the place above the lateral line and above the base of the pelvic fins. The regular ctenoid scales had well developed annuli in the form of interrupted and irregular circuli. They were easily visible in the oral and dorso-ventral parts (Fig. 58).

The age composition from the subsamples was then recalculated for the whole sample (Table 107) according to the percentage of distribution. In order to correct the sampling bias the stock density values per ha and in single age groups were interpolated from a semi-logarithmical diagram. Then according to the computed daily rate of mortality ($\dot{Z}$) and the time which elapsed from the annulus formation (t) the final stock density values were estimated.

For back calculation of linear growth a nomogram of standard length and scale radius size was constructed (Table 108, Fig. 59). Then from the means of length and

Fig. 58. Key scale of *Haplochromis darlingi* from Lake Kariba. Standard length 46 mm, first annulus at the scale edge.

size measurements the average lengths for previous growth seasons for each age group were computed on Lea's board with the help of this nomogram (VAN OOSTEN, 1953).

The largest Darling's dwarf bream found was from Buffalo Island cove (D). It was 10 cm long, weighed 27 g and it was four years old. Only rarely do single specimens exceed 8 cm in standard length and thus the average maximum age of Lake Kariba *H. darlingi* population can be said to be three years.

Younger fish most likely inhabit only shallow inshore parts of the lake, while the older can be found in deeper waters, or perhaps, the younger fish occupy only the shallow parts of the lake overgrown with water plants while the older occupy in addition free inshore waters. The first habitat represents localities F and G and here only fish of the first two age groups were found. The rest of the localities represented mixed habitats and here the older fish also occurred. In the water plant habitats the mean biomass ($\bar{B}$) amounted to 3069.5 for the first growth season and 2209.– for the second (total 5278.5), while in the other localities it was only 0.1557, 0.1963 and 0.6548 in the three successive growth seasons, a total of 1.0068. The instantaneous rate of total mortality (Z) also reflects these differences, which are 4.4091 and 4.4632 for plant habitats and 3.3783, 3.8255 and 4.1800 for the others.

Juvenile marks were found on the scales of most, though not all of the fish. The back calculated lengths show that these marks appeared at 17 to 43 mm of standard

Table 107. Computation of *Haplochromis darlingi* stock density according to the total sample of locality E.

	0	I	II	III	Total
21–30	430	108			538
31–40		925			925
41–50		301	100		401
51–60			250		250
61–70		13	67	54	134
71–80				1	1
Totals	430	1347	417	55	2249
N/ha	81	255	80	10	426
N̄/ha	760	225	83	10	1078
N'	1119	331	122	15	1587

length (mean 30 mm). For the time being it is hard to tell what caused these marks, though most likely, as is the case with other fishes, the change in diet and habitat may have been the cause.

Substantial differences in growth rates were found in different localities (Table 109). The instantaneous growth rate (G) for successive growth seasons ranged from 0.6592–3.4610, 0.2921–0.9808 and 0.914–0.6418. The means for plant habitats (F, G) amounted to 2.1664 and 0.3229 respectively; for the mixed habitats (localities A, C, D and E) the mean instantaneous growth rates decrease. They are 0.8602, 0.6432 and 0.2997.

These topographical differences of the *H. darlingi* population can be demonstrated even better by the final production computations (Table 110). In successive growth seasons production rapidly decreases in the young stocks from plant habitats. In other habitats these values sometimes increase, though at a lower rate. In spite of all the great differences in production of separate stocks, on the average the production of the plant habitats is over 14 times higher than the production of the mixed habitats.

Because of an oversight, the sexes were treated together, and later we were unable to arrange corrective processing of the material which by that time had been destroyed. Hence we have to treat our values as an average of both sexes, though the unknown sex ratio will reduce their reliability. This and the fact that we do not know how representative the selected samples are for the whole lake, is the weakest part of our conclusions.

However, from the six evaluated samples three show high and three low production values. The three coves with the high production are known to be places, within the shoreline, which have an enormous density of wild animals. The shores of Chete Island cove (locality E) are permanently inhabited by elephants *(Loxodonta africana)* and several species of antelope *(Kobus ellipsiprymnus, Aepyceros melampus)* and birds *(Bubulcus ibis, Ardea cinerea, A. goliath, Haliaetus vocifer, Phalacrocorax africanus* and many others). The whole shoreline is covered with trampled grass and dung,

Table 108. Standard length and mean length of oral radius (1 mm × 26.5) of *Haplochromis darlingi* in different length groups.

Locality	Length groups in mm															
	11–20		21–30		31–40		41–50		51–60		61–70		71–80		81–90	
	Mean length in mm l	Mean length of oral radius S^e	Mean length in mm l	Mean length of oral radius S^e	Mean length in mm l	Mean length of oral radius S^e	Mean length in mm l	Mean length of oral radius S^e	Mean length in mm l	Mean length of oral radius S^e	Mean length in mm l	Mean length of oral radius S^e	Mean length in mm l	Mean length of oral radius S^e	Mean length in mm l	Mean length of oral radius S^e
A	20	10	27	9	36	15	44	19	58	26	65	27				
B	17	10	27	12	35	16	45	18	56	26	65	36				
C			29	11	36	16	46	20	57	26	62	32				
D			26	14	34	15	45	20	54	26	67	32	75	42		
E			27	13	37	18	43	20	56	27	66	31	75	33		
F					36	15	42	19	58	27	65	30	76	36	82	37
G					38	17	46	19	55	23	65	31				
Total mean values	18	10	27	12	36	16	44	19	56	26	65	31	75	37	82	37

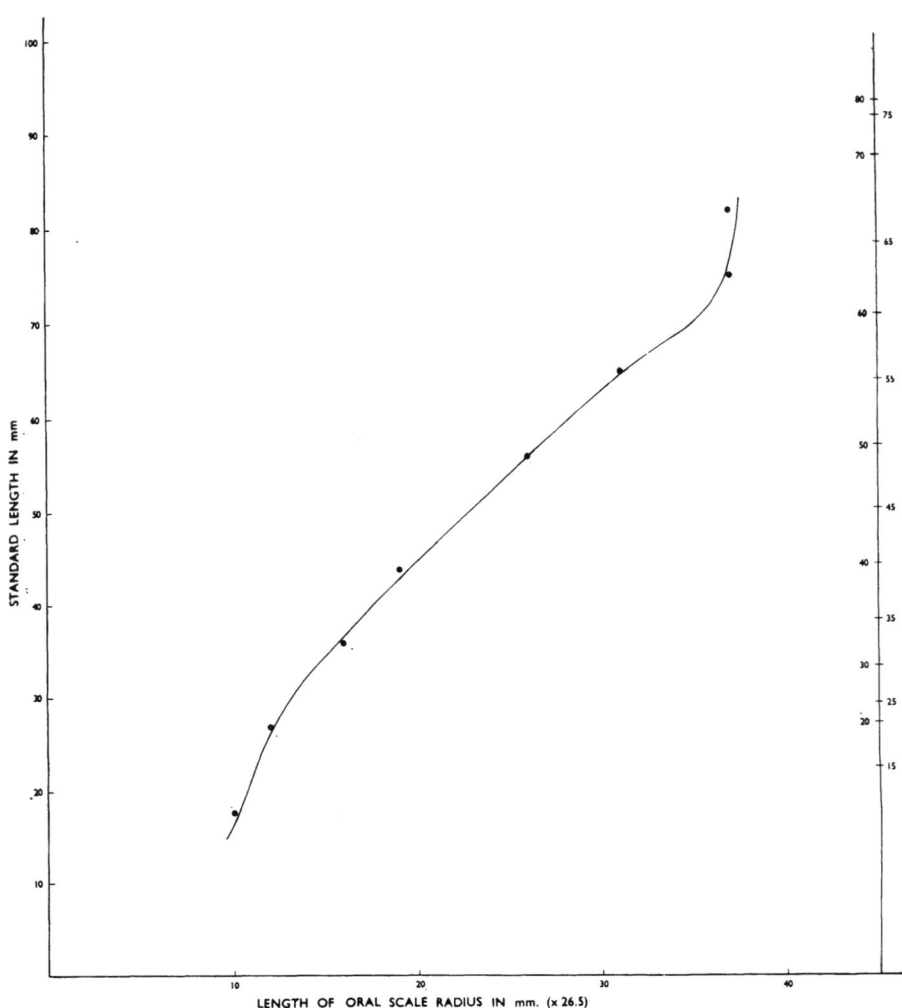

Fig. 59. Empiric curve and nomograph of the body-scale relationship of *H. darlingi* from Lake Kariba.

especially that of elephants. The same is true of the coves in Chipepo Bay at the foot of Kota Kota (F, G). It thus seems likely that here not only the plant habitats but also the nutrient release from game defecation will reflect upon the production of fish (McLachlan, 1971). Small fish with as short a life span as *H. darlingi*, with their preference for phytophilous habitats and feeding, will naturally be good indicators of the interaction between game and fish. (Plant fibres, diatoms, algae, insect larvae, copepods and detritus were found in 15 examined specimens from locality F). The three coves with a low production are all situated on islands with no game animals (C, D) or on a peninsula near highly populated areas with a few single game animals

Table 109. Back calculated growth values of *Haplochromis darlingi* from Lake Kariba.

Locality	Age groups	Year of hatching	N	Mean standard length at the time of capture in mm	Back calculated standard length (l) in mm for the previous growth seasons and corresponding interpolated weights ($\bar{w}$) in g			
					l_j $\bar{w}_j$	l_1 $\bar{w}_1$	l_2 $\bar{w}_2$	l_3 $\bar{w}_3$
A	0	1967	28	36	33 1.1			
	I	1966	8	60	35 3.4	51 3.9		
	II	1965	3	68	41 2.2	54 4.7	63 7.2	66 8.5
	Mean total		39		36 2.2	52 4.7	63 7.2	66 8.5
B	0	1967	19	33	28 0.7			
	I	1966	6	57	37 1.6	52 4.2		
	II	1965	1	68		57 5.6	66 8.2	
	Mean total		26		32 1.2	54 4.9	66 8.2	
C	0	1967	21	37	27 0.6			
	I	1966	9	53	30 0.9	45 2.8		
	II	1965	3	62	20 0.3	39 1.8	50 3.8	
	Mean total		33		26 0.6	42 2.3	50 3.8	
D	0	1968	18	40	38 1.7			
	I	1967	4	63	43 2.5	54 4.7		
	II	1966	3	70	33 1.1	44 2.6	63 8.2	
	III	1965	1	78	29 0.8	35 1.4	41 2.1	54 4.7
	Mean total		26		36 1.5	44 2.9	52 5.2	54 4.7

Table 109. (Continued)

Locality	Age groups	Year of hatching	N	Mean standard length at the time of capture in mm	Back calculated standard length (l) in mm for the previous growth seasons and corresponding interpolated weights (w̄) in g		
					l_j $\bar{w}_j$	l_1 $\bar{w}_1$	l_2 $\bar{w}_2$	l_3 $\bar{w}_3$
E	0	1968	8	26	17 0.7			
	I	1967	14	39	18 0.2	33 1.15		
	II	1966	11	59	20 0.3	37 1.6	52 4.2	
	III	1965	5	70	22 0.4	39 1.8	50 3.8	64 7.6
	Mean total		38		19 0.4	36 1.5	51 4.0	64 7.6
F	0	1968	5	48	35 1.3			
	I	1967	7	66	26 0.6	57 5.6	59 7.5	
	Mean total		12		30 0.96	57 5.6	59 7.5	
G	0	1968	8	46	36 1.5			
	I	1967	6	60	33 1.5	48 3.3		
	II	1966	1	67	36 1.5	48 3.3	54 4.7	
	Mean total		15		35 1.4	48 3.3	54 4.7	
	Mean grand total		189		30 1.2	47 3.5	56 5.5	59 6.1

(A). Nevertheless, far more evidence must be collected before this interaction can be stated definitively.

With our recently acquired knowledge it is impossible to pinpoint with any confidence the area of the lake inhabited by *H. darlingi*. Most likely its distribution will not reach the limits of general fish distribution valid for larger species as indicated on electroacoustic records. The scuba-diving tests, however, revealed *H. darlingi* nesting

Table 110. Total production (= GB̄) and available production of different *Haplochromis darlingi* stocks in Lake Kariba.

	Localities	Growth seasons			Total
		1	2	3	
Total production	A	0.1917	0.0197	0.0016	0.2130
	C	0.0227	0.2253		0.2480
	D	0.0521	0.0154	0.0177	0.0852
	E	0.2522	0.2663	1.1302	1.6487
	mean	0.1297	0.1317	0.3832	0.6446
A	F	15.9096	1.1462		17.0558
	G	1.3440	0.1746		1.5186
	mean	8.6268	0.6604		9.2872
Available production	A	0.2247	0.0551	0.0052	0.2850
	C	0.0629	0.0060		0.0689
	D	0.1372	0.0405	0.0014	0.1791
	E	0.3707	0.3050	0.0540	0.7297
	mean	0.1989	0.1016	0.0202	0.3207
P'	F	9.4795	0.3040		9.7835
	G	2.2714	0.4382		2.7096
	mean	5.8754	0.3711		6.2465

at a 15–20 m depth. It thus has a distribution in the range of other fish. Although great differences in density and size will occur in this range, the average production value – A 4.97 kg/ha/yr and P' 3.28 kg/ha/yr – may be fairly accurate.

4. TOTAL PRODUCTION, AVAILABLE PRODUCTION AND YIELD OF MAJOR FISH TAXA FROM LAKE KARIBA

4.1. Species size and growth intensity

> 'Tropical fresh waters (reservoirs and ponds) seem destined to contribute prominently to human nutrition. (...) Some, at least, are more resilient under heavy exploitation than those in other areas, and, in fact, underharvest may be our most difficult problem.'

Although in some Lake Kariba fish species a pronounced sexual dimorphism in growth was noted and the growth of some species differs in various localities, I decided to omit these differences from consideration in this chapter and to attempt a comparison of specific total means. The values for individual species are summarized in Appendix C.

We encountered species with a very long life span – over 10 years – *A. nebulosa*, *H. longifilis*, *C. gariepinus* and *M. electricus*. The majority of species form two groups of medium life span fishes: those which reach 7 to 9 years of age – *M. deliciosus*, *M. longirostris*, *H. vittatus*, *L. altivelis*, *T. rendalli* and *S. zambezensis* – and those which reach 4 to 6 years – *H. discorhynchus*, *M. macrolepidotus*, *E. depressirostris*, *S. mystus*, *S. mossambicus mortimeri*, *S. codringtoni* and *S. nebulosus*. From among the 22 species studied the short life span species, which do not exceed 3 years of age, are *L. miodon*, *A. lateralis*, *B. i. imberi*, *M. acutidens* and *H. darlingi*.

In most species the value of the specific rate of growth gradually decreases, with marked thresholds at the borders of life intervals – juvenile (j), adult (a) and senective (s). However, all of the short life span species of the family Characidae have remarkably increased values for the specific rate of growth in the adult period. The only short life span cichlid, *H. darlingi*, shows the specific rate of growth decreasing, though, as in the former species, the periods are represented by a single growth season. The majority of species studied seem on the average to reach maturity in the second growth season. Only *L. altivelis* mature in the third growth season, *C. gariepinus* in the fourth, *M. electricus* in the fifth, and *H. longifilis* in the seventh. The presence of a few years of the senective period in nearly all species studied may be another evidence of underexploitation. However, as already noted in the previous section, the time-strata of life periods were not checked empirically. The averaging, omitting sexual and topographical growth differences from consideration, probably contains errors in the estimated limits of periods. Consequently, the indices are biased to an unknown extent.

Obviously the large differences in ages, the mentioned biases and local eating customs are the reasons why 'the index of the species average size' does not provide satisfactory data for the arrangement of the species in order of commercial interest, though in temperate regions this was usually possible (BALON, 1964c, 1966a). With some excep-

Table 111. Lake Kariba major fish species arranged according to the 'index of the species average size'.

1.	*Mormyrops deliciosus*	9.08
2.	*Mormyrus longirostris*	9.02
3.	*Hydrocynus vittatus*	7.50
4.	*Heterobranchus longifilis*	6.83
5.	*Clarias gariepinus*	6.10
6.	*Brachyalestes imberi imberi*	5.80
7.	*Labeo altivelis*	5.64
8.	*Sargochromis codringtoni*	5.40
9.	*Malapterurus electricus*	4.95
10.	*Tilapia rendalli*	4.50
11.	*Marcusenius macrolepidotus*	4.36
12.	*Sarotherodon mossambicus*	4.02
13.	*Hippopotamyrus discorhynchus*	3.64
14.	*Alestes lateralis*	3.53
15.	*Eutropius depressirostris*	3.36
16.	*Synodontis zambezensis*	3.21
17.	*Schilbe mystus*	3.07
18.	*Micralestes acutidens*	2.82
19.	*Haplochromis darlingi*	2.80
20.	*Synodontis nebulosus*	2.30

Table 112. Lake Kariba major fish species arranged according to the 'index of stock weight growth intensity'.

Species		φCw
1.	*Mormyrus longirostris*	4220
2.	*Mormyrops deliciosus*	2322
3.	*Hydrocynus vittatus*	2074
4.	*Brachyalestes imberi imberi*	680
5.	*Tilapia rendalli*	418
6.	*Labeo altivelis*	406
7.	*Sarotherodon mossambicus mortimeri*	346
8.	*Marcusenius macrolepidotus*	338
9.	*Alestes lateralis*	300
10.	*Synodontis zambezensis*	222
11.	*Haplochromis darlingi*	203
12.	*Heterobranchus longifilis*	182
13.	*Clarias gariepinus*	180
14.	*Micralestes acutidens*	169
15.	*Sargochromis codringtoni*	165
16.	*Eutropius depressirostris*	164
17.	*Schilbe mystus*	159
18.	*Malapterurus electricus*	155
19.	*Synodontis nebulosus*	129
20.	*Hippopotamyrus discorhynchus*	105

tions (e.g. *B. i. imberi*), however, this index sufficiently represents (Table 111) the order of the decreasing rate of linear growth.

The specific rate of weight increase for the first growth season has an effect on the 'index of stock weight growth intensity'. It is as high as 25,000 for *M. longirostris* and only 114 for *M. acutidens*. The corresponding index thus arranges in order of decreasing growth rates the bottlenose, cornish-jack, tigerfish, with more than 2,000 (Table 112), the spot-tail, red-breasted bream, rednose mudsucker, kurper bream, bull-dogfish, with more than 300, the alestes, plus the other 11 species down to as low as 105 for the last, the parrotfish. This index ranges from 285 for huchen *(Hucho hucho)* to 60 for the roach *(Rutilus rutilus)* in the man-made Lake Orava (Temperate Region). This is remarkably lower than in Lake Kariba fish. In Lake Kariba the growth intensity is 2–15 times higher than in the Lake Orava fish species, the highest being in the larger species. At the same time, however, the index of the species average size is in favour of Lake Orava fish, which is at least for the three largest fish – huchen *(Hucho hucho)*, pike *(Esox lucius)* and trout *(Salmo trutta)*, above that of the cornish-jack and bottlenose and these are the largest species in Lake Kariba.

According to the 'index of the species commercial importance' nine of the studied taxa belong to the economically preferred species with index (ϕE) over 0.3, the six belong to the secondary species with index 0.3 down to 0.02 and five to the accompanying species with the index below 0.02 (Table 113, Fig. 60, 61a and 61b).

Table 113. Lake Kariba major fish taxa arranged according to the 'index of the species commercial importance'.

Economically preferred species		Secondary species		Accompanying species	
Mormyrus longirostris	150.3333	Marcusenius macrolepidotus	0.2506	Brachyalestes imberi imberi	0.0197
Hydrocynus vittatus	57.6923	Malapterurus electricus	0.1345	Alestes lateralis	0.0113
Mormyrops deliciosus	43.2381	Synodontis zambezensis	0.1100	Haplochromis darlingi	0.0110
Sargochromis codringtoni	2.9834	Hippopotamyrus discorhynchus	0.0345	Synodontis nebulosus	0.0076
Heterobranchus longifilis	2.2767	Eutropius depressirostris	0.0338	Micralestes acutidens	0.0002
Clarias gariepinus	1.9489	Schilbe mystus	0.0213		
Sarotherodon mossambicus	1.0388				
Labeo altivelis	0.9947				
Tilapia rendalli	0.3810				
$\Phi E > 0.3$		$0.3 > \Phi E > 0.02$		$\Phi E < 0.02$	

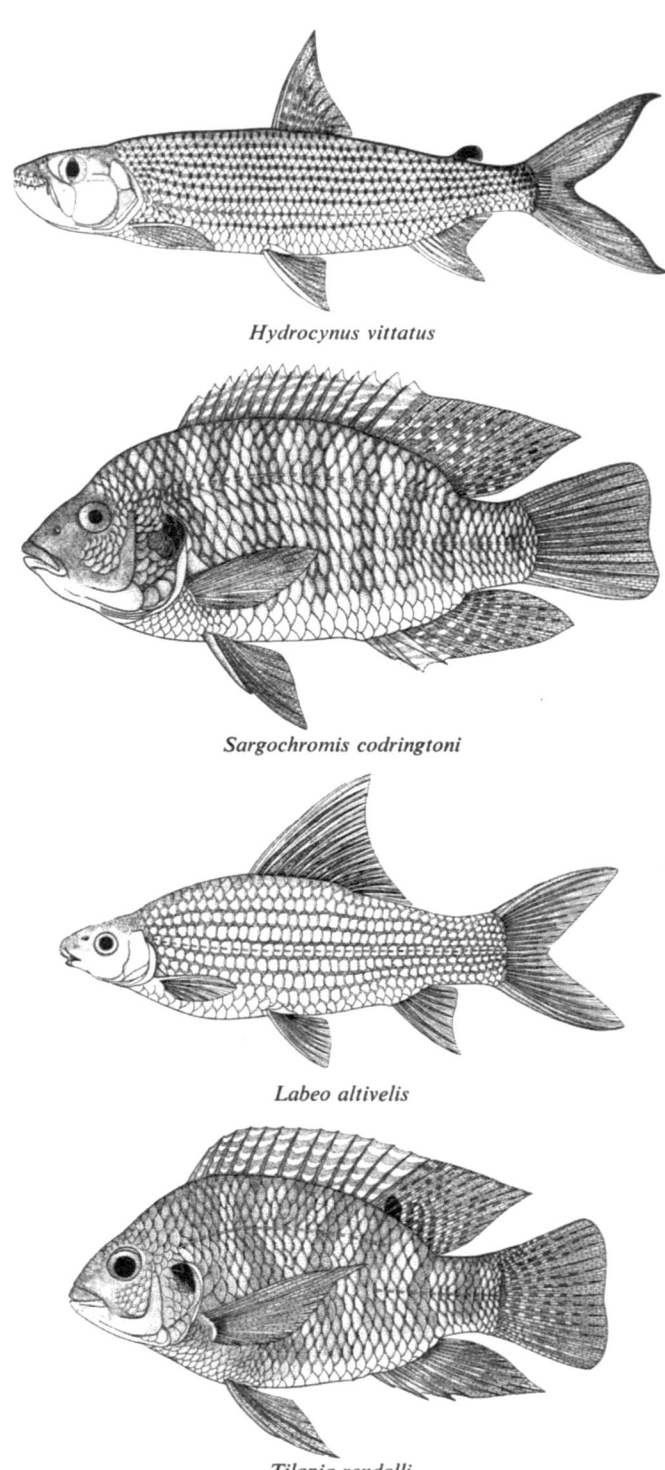

Fig. 60. Economically preferred species (original drawings by MIRIAM BARADLAI).

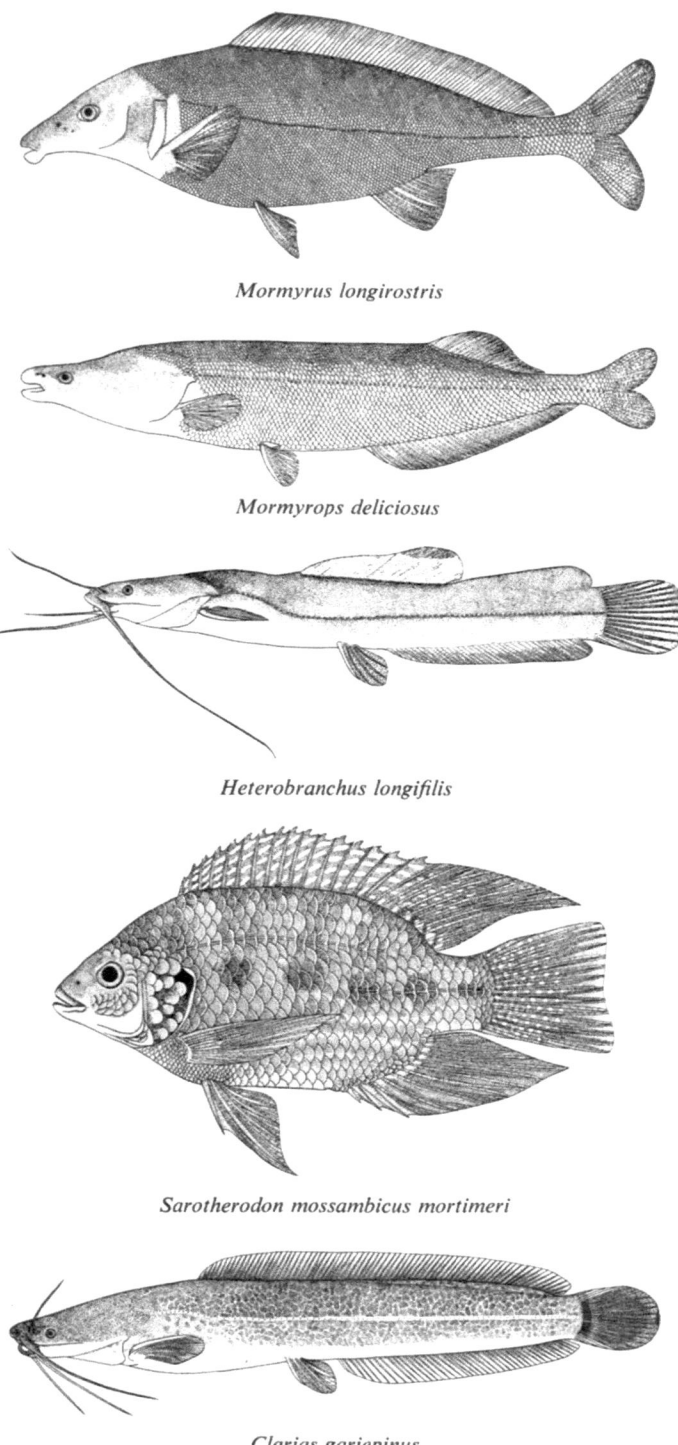

Fig. 60. Economically preferred species (concluded).

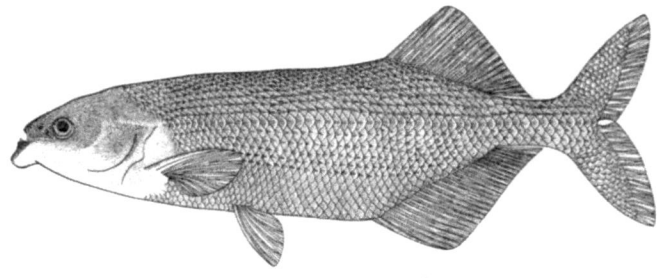

Marcusenius macrolepidotus

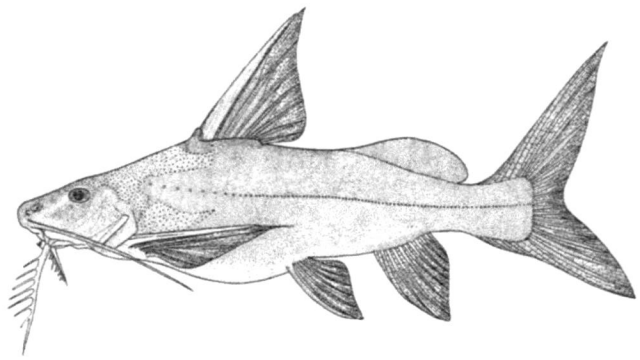

Synodontis zambezensis

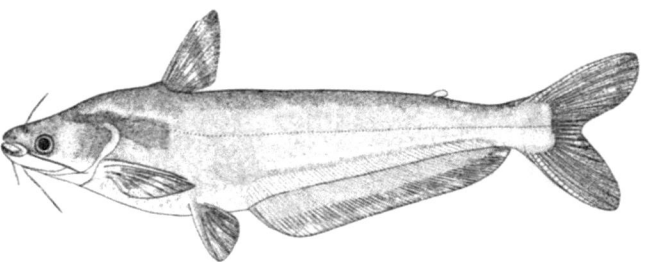

Eutropius depressirostris

Fig. 61a. Secondary species.

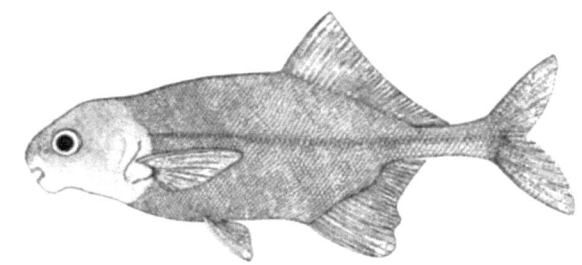

Hippopotamyrus discorhynchus

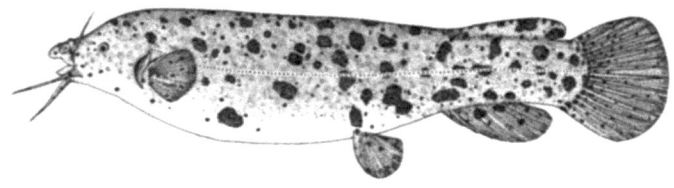

Malapterurus electricus

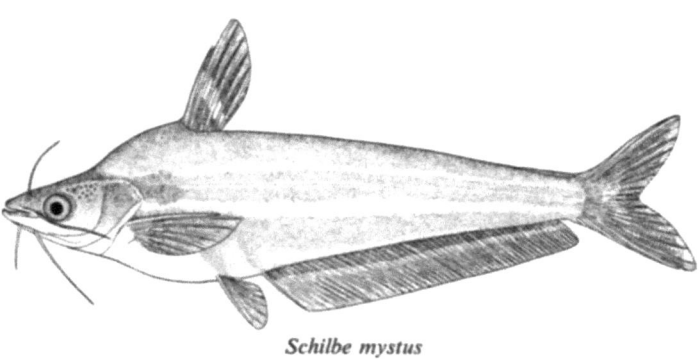

Schilbe mystus

Fig. 61a. Secondary species (concluded).

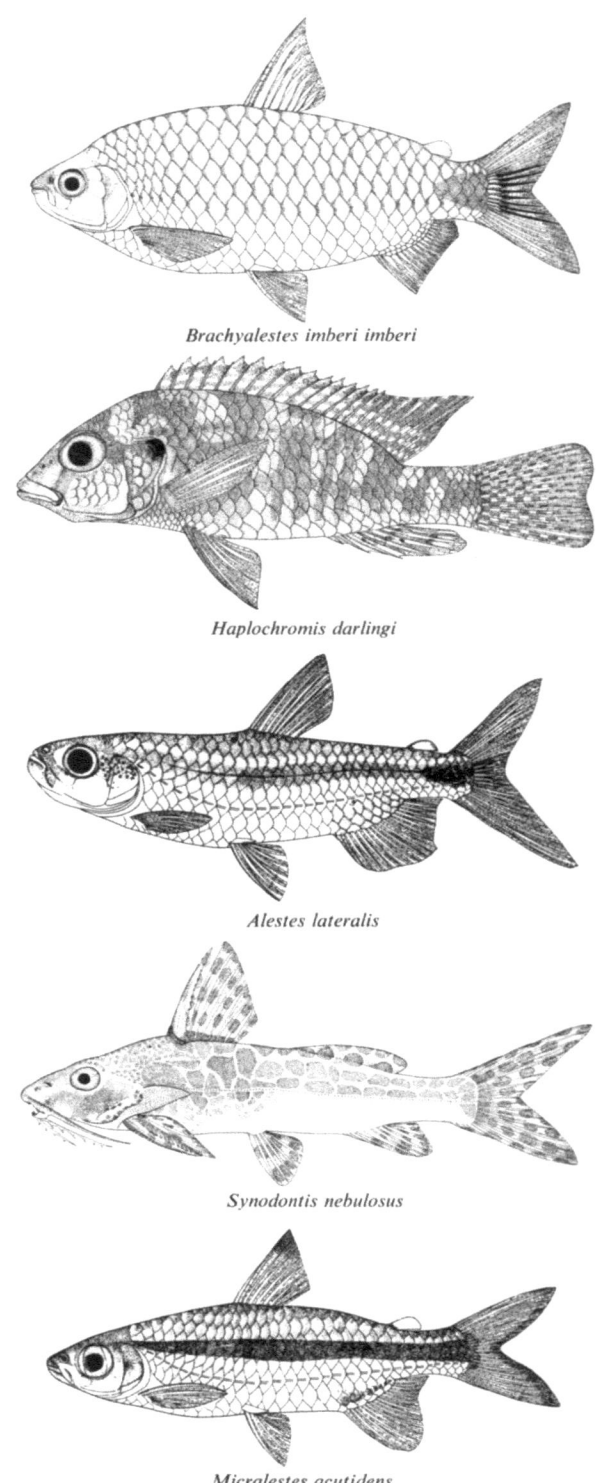

Fig. 61b. Accompanying species.

Table 114. Total average density, biomass and production values per 1 ha area of Lake Kariba (in number of specimens and kilograms).

	N'/ha		B/ha		B̄/ha		A/ha/yr		Y_A/ha/yr		P'/ha/yr		$Y_{P'}$/ha/yr	
	ranges	mean	ranges	mean	ranges	mean	ranges	mean	ranges	mean	ranges	mean	ranges	mean
Mormyrus longirostris		78		67		278		1242		14		31		15
Hydrocynus vittatus		63		37		126		472		12		17		7
Mormyrops deliciosus	25– 187	106	16 – 98	57	24 – 285	155	33 – 972	503	5 – 35	20	6 – 43	24	4 – 24	14
Sargochromis codringtoni	57– 216	116	5 – 26	13	9 – 61	28	18 – 144	64	1.24– 33	12	4 – 15	8	1 – 11	4
Heterobranchus longirostris		18		7		9		6		1		4		1
Clarias gariepinus		97		46		60		31		9		15		7
Sarotherodon m. mortimeri	386– 5337	2122	13 – 607	233	17 – 896	340	31 –1455	548	15 –1256	468	10 –339	131	4 –267	101
Labeo altivelis		102		16		19		18		3		7		2
Tilapia rendalli	786– 1229	957	36 – 59	47	45 – 75	55	51 – 112	81	8 – 26	16	27 – 38	31	5 – 19	12
Marcusenius macrolepidotus		261		6		8		15		0.64		4		0.68
Malapterurus electricus		6994		154		166		190		10		101		9
Synodontis zambezensis	174– 1198	613	6 – 33	17	7 – 38	19	8 – 39	21		4	3 – 22	10	1 – 8	4
Hippopotamyrus discorhynchus		18060		167		113		171		8		141		12
Eutropius depressirostris	232– 659	497	3 – 7	5	3 – 7	6	2 – 7	5		0.79	2 – 4	3	0.42– 0.92	0.66
Schilbe mystus		144		1.5		1		1		0.16		1		0.16
Brachyalestes imberi imberi	19– 93	56	0.12– 0.39	0.26	0.13– 0.22	0.18	0.11– 0.26	0.19	0.04– 0.05	0.05	0.10– 0.35	0.23		—
Alestes lateralis	433–108610	29454	3 –557	151	2 – 215	59	3 – 342	94	1.28– 71	20	3 –545	148		—
Haplochromis darlingi	41– 2203	760	0.1 –5375	1287	0.30–8521	1760	0.08– 17.06	3	0.03– 1.40	0.5	0.07– 9.78	2.29		—
Synodontis nebulosus		904		5		5		3		0.09		3		—
Micralestes acutidens		132		0.18		0.04		0.01		0.001		0.16		—
		61,534		2,317		3,208		3,468		599		682		189

4.2. Stock size assessment values

> 'Animal numbers in a given region are relatively constant. While they may fluctuate from year to year, particularly with changes in the environment, the fluctuations are often transitory and, over longer periods, trivial.'

One important difference between our relative values (per ha) and absolute values (per total area of the lake inhabited by fish) must be stated: the confidence in the former is limited by the number and size of samples, but is valid for every species; the confidence in the latter is much lower, since mainly intermediate instruments (echosounder) were used and the results obtained were valid for the fish population as a whole, without differentiation of species. Even if this was supported by direct observation with gill-nets and divers, it was only sufficient to adjust the echo-traces. Though both values are used simultaneously, the relative values must be recommended as more realistic.

The data for separate samples yielded mean values of single species only after correction according to the time of annulus formation and graphical interpolation of individual age group density (Appendix D). The means of different samples varied by as much as several hundred per cent above or below the total mean and so it was useless to place any confidence limits around the mean. Tables 114 and 115 give the summarized data for each species.

A. DENSITY AND MORTALITY RATES

> 'Something regulates densities within remarkably narrow limits.'

The average density of all 20 species studied is over 61 thousand individuals per inhabited hectare. The number of individuals in the whole lake may thus be over 12×10^9.

The lowest density is represented by *H. longifilis* with 18 specimens/ha, or over 3.6×10^6 for the whole lake. In order of increasing density are the following: *B. i. imberi, H. vittatus, M. longirostris, C. gariepinus, L. altivelis, M. deliciosus, S. codringtoni, M. acutidens* and *S. mystus*. These 10 species can be classified as the low density group, with less than 150 individuals per hectare each.

The medium density group of Lake Kariba fishes consists of *M. macrolepidotus, E. depressirostris, S. zambezensis, H. darlingi, S. nebulosus* and *T. rendalli*. The first species in this group has an estimated 261 and the last 957 specimens per hectare.

The last four species studied form the high density group with the widest range between and the largest variability within each species. *S. m. mortimeri* is the member of this group with the lowest population density, with a mean of 2,122 specimens per hectare. It is followed by *M. electricus* with 6,994 specimens/ha, *H. discorhynchus* with 18,060 and *A. lateralis* with 29,454 specimens/ha. The fact that this last species shows the highest density is of particular interest, as in 1963 it has become a new member of the Lake Kariba fish fauna. A detailed explanation can be found in a special study devoted to it (BALON, 1971b), abridged parts of which are presented in chapter 6.

Table 115. Average values of number, biomass and production per total inhabited area of Lake Kariba.

Per 202 200 ha	N' in n	B' in kg	B̄ in kg	A in kg	Y_A in kg	P' in kg	$Y_{P'}$ in kg
Mormyrus longirostris	15 771 600	13 547 400	56 211 600	251 132 400	2 830 800	6 268 200	3 033 000
Hydrocynus vittatus	12 738 600	7 481 400	25 477 200	95 438 400	2 426 400	3 437 400	1 415 400
Mormyrops deliciosus	21 433 200	11 525 400	31 341 000	101 706 600	4 044 000	4 852 800	2 830 800
Sargochromis codringtoni	23 455 200	2 628 600	5 863 800	12 940 800	2 426 400	1 617 600	808 800
Heterobranchus longifilis	3 639 600	1 415 400	1 819 800	1 213 200	202 200	808 800	202 200
Clarias gariepinus	19 613 400	9 301 200	12 132 000	6 268 200	1 819 800	3 033 000	1 415 400
Sarotherodon mossambicus mortimeri	429 068 400	47 112 600	68 748 000	110 805 600	94 629 600	26 488 200	20 422 200
Labeo altivelis	20 624 400	3 235 200	3 841 800	3 639 600	606 600	1 415 400	404 400
Tilapia rendalli	193 505 400	9 503 400	11 121 000	16 378 200	3 235 200	6 268 200	2 426 400
Marcusenius macrolepidotus	52 774 200	1 213 200	1 617 600	3 033 000	129 408	808 800	137 496
Malapterurus electricus	1 414 186 800	31 138 800	33 565 200	38 418 000	2 022 000	20 422 200	1 819 800
Synodontis zambezensis	123 948 600	3 437 400	3 841 800	4 246 200	808 800	2 224 200	808 800
Hippopotamyrus discorhynchus	3 651 732 000	33 767 400	22 848 600	34 576 200	1 617 600	28 510 200	2 426 400
Eutropius depressirostris	100 493 400	1 011 000	1 213 200	1 011 000	159 738	606 600	133 452
Schilbe mystus	29 116 800	303 300	202 200	202 200	32 352	202 200	32 352
Brachyalestes imberi imberi	11 323 200	52 572	36 396	38 418	10 110	46 506	—
Alestes lateralis	5 955 598 800	30 532 200	11 929 800	19 006 800	4 044 000	29 925 600	—
Haplochromis darlingi	153 672 000	260 231 400	355 872 000	606 600	101 100	463 038	—
Synodontis nebulosus	182 788 800	1 011 000	1 011 000	606 600	18 198	606 600	—
Micralestes acutidens	26 690 400	36 396	8 088	2 022	202	32 352	—
Total (in millions)	12 442	468	647	701	120	138	38

Preliminary results of a current study of *A. nebulosa labiata* seem to place this fish in the medium density group (chapter 5). It is entirely possible that (1970-71) the population of the clupeid *L. miodon*, as a descendant of the third and probably only successful stocking program in 1968 will show a higher density than *A. lateralis* (BALON, 1971c). This statement is supported by a comparison of electroacoustic traces of the commercial stock in Lake Tanganyika and the diel movements and density of the stock in Lake Kariba. Furthermore, this Tanganyikan anchoveta today occupies the whole of Lake Kariba, even such unusual habitats as shallow bays and coves and estuaries of rivers or intermittent rivulets (see chapter 7).

The remaining 18 species, as yet not studied, composed 9.3% of the mean total density of evaluated samples. It is, therefore, probable that the Lake Kariba fish population density was in the vicinity of 867 thousand individuals/ha, taking into account 6,292 individuals/ha for the 18 unstudied species, 40 individuals/ha for the eel and 800 thousand individuals/ha for the anchoveta. For the whole lake this comes to over 1.7×10^8 individuals.

With the short life span species the instantaneous rate of total mortality is clearly the highest, being for *B. i. imberi* 1.53, for *M. acutidens* 2.69, for *A. lateralis* 3.37 and for *H. darlingi* 4.05. *H. discorhynchus* with 1.92 formed an exception and this can probably be explained by the high density of this species. Due to the interpolation of density parameters in single age groups, the corresponding mortality rates varied very little, but significant differences were noted between the samples. Mortality rates between 0.7 and 1.3 exist for *S. zambezensis*, *S. mossambicus mortimeri*, *S. codringtoni*, *E. depressirostris*, *S. nebulosus*, *T. rendalli*, *M. electricus* and *M. macrolepidotus*. The lowest instantaneous rates of total mortality were computed for *S. mystus* 0.60, *L. altivelis* 0.58, *H. vittatus* 0.54, *M. longirostris* 0.49, *H. longifilis* 0.39, *M. deliciosus* 0.37 and *C. gariepinus* 0.24.

B. BIOMASS

> 'Ecology became a quantitative science when ecologists began to count organisms, but it entered its modern phase only when they also began to weigh.'

In various stages of data evaluation we encounter three various types of biomass: (i) the empirical standing stock of single species in any particular sample, (ii) the initial biomass (B') obtained from interpolated values for density and weight in a single age group, corrected to the time of the beginning of the growth season (annulus formation) and (iii) the mean biomass ($\bar{B}$) from equation 13 as a value of average biomass present during the year. We will deal here only with the last two as corrected for sampling bias, though in the first attempt at a Lake Kariba fish stock size assessment (BALON, 1973a) only the standing stock was used. Its values, however, are much lower than estimated biomass, which incorporates corrections on sampling bias and mortality.

When considering the initial biomass (B') three clearly marked groups can be seen in the Lake Kariba fish studied. To the first group, with a low biomass, below 10 kg/ha, belong: *M. acutidens* 0.18 kg/ha, *B. i. imberi* 0.26, *S. mystus* 1.5, *S. nebulosus* 5, *E.*

depressirostris 5, *M. macrolepidotus* 6 and *H. longifilis* 7 kg/ha. The second group, with 10 to 70 kg/ha, contains *S. codringtoni* 13, *L. altivelis* 16, *S. zambezensis* 17, *H. vittatus* 37, *C. gariepinus* 46, *T. rendalli* 47, *M. deliciosus* 57 and *M. longirostris* 67 kg/ha. The third group, with high initial biomass, over 150 kg/ha, consists of *A. lateralis* 151, *M. electricus* 154, *H. discorhynchus* 167, *S. m. mortimeri* 233 and *H. darlingi* with 1287 kg/ha.

Though the mean biomass ($\bar{B}$) in the majority of species studied is higher than the initial biomass, in *M. acutidens*, *B. imberi*, *S. mystus*, *A. lateralis* and *H. discorhynchus* the former is lower than the latter. The arrangement of species in order of increasing values for the first group of fish with low biomass is similar, but, for the latter species is completely different, beginning with *L. altivelis* with 19 kg/ha of mean biomass, and followed by *S. zambezensis* with 19, *S. codringtoni* 29, *T. rendalli* 55, *A. lateralis* 59, *C. gariepinus* 60, *H. discorhynchus* 113, *H. vittatus* 126, *M. deliciosus* 155, *M. electricus* 166, *M. longirostris* 278, *S. mossambicus mortimeri* 340 and *H. darlingi* with 1760 kg/ha.

In both single species and samples the initial biomass usually shows values increasing with increasing age up to a particular growth season (2 to 5), after which the values decrease again. There are, however, a few exceptions, with steadily increasing values *(H. longifilis, C. gariepinus)* and steadily decreasing values *(S. codringtoni, M. macrolepidotus, B. imberi, H. discorhynchus, A. lateralis, H. darlingi, M. acutidens)*. The mean biomass values form even more different types, though predominant here is the steadily decreasing value (see Appendix D).

The initial biomass for the 20 species studied amounted to 2,317 kg/ha and the mean biomass to 3,208 kg/ha; recalculated for the whole lake area inhabited by fish it gives 468 and 648 thousand metric tons respectively. The remaining 18 unstudied species composed 2.63% of the mean total standing stock, and may give 63 kg/ha of initial biomass and 87 kg/ha of mean biomass; the eel and anchoveta biomass can be 'guesstimated' according to density assessment as 100 and 150 kg/ha for the eel and 800 and 200 kg/ha for anchoveta respectively. The initial biomass of the Lake Kariba fish population will thus be 3280 kg/ha and the mean biomass 3645 kg/ha. For the whole lake it gives for the former a value of 6.6×10^8, and for the latter 7.4×10^8 kg. The standing stock gives a mean from all samples of 533 kg/ha or 18 thousand metric tons for the whole lake.

C. Production

> 'A population that is either large or small, relative to the carrying capacity of its environment, is less productive than a population of intermediate size. For populations valuable to man there must always be some optimum size that provides the maximum sustained yield.'

Let us first consider the total production values. The mean production of all species studied is only slightly above the corresponding mean biomass. In single species, however, these two values differ remarkably in most cases. The highest total production was estimated for *M. longirostris*, amounting to 1242 kg/ha/yr. Then follow *S. mossambicus mortimeri* with 548, *M. deliciosus* 503, *H. vittatus* 472, *M. electricus* 190 and

H. discorhynchus 171 kg/ha/yr. These six species may be considered as a group with a high total production.

A medium total production is achieved by *A. lateralis* with 94 kg/ha/yr, *T. rendalli* 81, *S. codringtoni* 64, *C. gariepinus* 31, *S. zambezensis* 21, *L. altivelis* 18 and *M. macrolepidotus* with 15 kg/ha/yr. The rest of the species studied can be designated as fish with a low total production, amounting for *H. longifilis* to 6 kg/ha/yr, for *E. depressirostris* to 5, for *H. darlingi* and *S. nebulosus* to 3 kg each, for *S. mystus* to 1, for *B. imberi* to 0.19 and for *M. acutidens* to 0.01 kg/ha/yr.

The sum of the total production thus gives 3,468 kg/ha/yr in the 20 studied species. In reality this figure will be higher, including the production of eel, anchoveta and the 18 species not studied. The total production for the whole area inhabited by fish, amounts to over 701 thousands metric tons or 1 306 kg per ha of the whole lake.

The values for available production are very different and in most cases much lower. Only in species with the lowest available production the latter is sometimes higher than the total production. The *A. lateralis* with 148 kg/ha/yr of available production, the *H. discorhynchus* with 141, *S. mossambicus mortimeri* with 131 and *M. electricus* with 101 form a group with a high production of survivors. The following five and eleven species form respectively the medium and low available production groups: *M. longirostris* 31, *T. rendalli* 31, *M. deliciosus* 24, *H. vittatus* 17, and *C. gariepinus* 15 kg/ha/yr; the *S. zambezensis* 10, *S. codringtoni* 8, *L. altivelis* 7, *H. longifilis* and *M. macrolepidotus* 4 each, *E. depressirostris* and *S. nebulosus* 3 each, *H. darlingi* 2.29, *S. mystus* 1, *B. imberi* 0.23 and *M. acutidens* 0.16.

The sum for the available production of all 20 species studied amounted to 682 kg/ha/yr and may be nearly one ton for the whole fish population. For the inhabited area of the lake the figure is at least 137 thousand metric tons per year and possibly over 200 thousand metric tons.

The largest differences between these two production values, in the first (in some instances affected by computation error – see page 272) and following few growth seasons, correspond to the highest growth increments and density parameters. Hence, from the total production of 3,468 kg/ha/yr, only 20% survived to the end of the year to form the 682 kg of available production. About 80% or 2,786 kg became victims of predators and decomposers (natural and fishing mortality).

D. YIELD

> 'Harvesting of fish is essentially a hunting economy. Rates of harvest and sustainable yields are matters of continued concern about which much remains to be learned.'

The harvestable part of production is designated here as yield. To emphasize once more this is not the definition of 'yield' used by most fish stock statisticians (to cite a few at random – BEVERTON & HOLT, 1957; RICKER, 1958; GULLAND, 1969), though my term 'available yield' is similar in meaning to the terms 'surplus production' or 'equilibrium catch'. The most important parameter for my yield estimation, therefore, became the minimum harvestable size of every species. In an attempt to establish these values objectively several ways were developed in the past (BARANOV, 1925; BEVERTON & HOLT, l.c.; RICKER, l.c.; ZAWISZA, 1961; DEMENTYEVA et al., 1961; TJURIN, 1962; HOLČIK, 1970). The most practical approach in my opinion, was that of using the first spawning age and the age of highest biomass (TJURIN, 1963; BALON, 1966a); the simplest, the approach of graphical growth increments (BALON, 1955; 1971a). In the individual papers of this volume the latter is used exclusively. The minimum harvestable size of studied species was determined as follows:

	Standard length in cm	Live weight in g	Age group
Mormyrus longirostris	30	360	II
Hydrocynus vittatus	38	1000	II
Mormyrops deliciosus	30	250	II
Sargochromis codringtoni	13	90	II
Heterobranchus longifilis	54	2000	VII
Clarias gariepinus	52	1880	VII
Sarotherodon mossambicus mortimeri	13	90	III
Labeo altivelis	24	400	IV
Tilapia rendalli	14	127	III
Marcusenius macrolepidotus	15	60	III
Malapterurus electricus	28	500	V
Synodontis zambezensis	11	40	III
Hippopotamyrus discorhynchus	10	22	III
Eutropius depressirostris	12	22	III
Schilbe mystus	13	30	IV

The sum of production values above this minimum harvestable size gives the yield values. The total yield of accompanying species is estimated as a matter of interest, though no minimum harvestable size was satisfactorily computed for these. For this reason no available yield of these species was evaluated.

No doubt the previous parameters – densities, biomasses, productions – are of great theoretical interest. The yield parameters on the other hand are of direct practical interest to the fisheries. Let me explain my concept in the following way: The total yield comprises of (i) the production of fish of harvestable size which do not survive to the end of the unit time because of death from natural causes (elimination), (ii) the production of fish of harvestable size which are caught during the time of interest and (iii) the production of fish of harvestable size which survive to the end of the time of interest. In other words, the total yield is composed of production above the minimum harvestable size divided between natural mortality, actual catch and whatever remains over – the available yield.

For example, the total yield of the 20 species studied amounted to 599 kg/ha/yr. The assessed total catch in Lake Kariba in the years 1968 and 1969 was 3 thousand metric tons, which gives for the 202 200 ha of area inhabited by fish a value of 14.8 kg/ha/yr. Because the available yield was estimated to be 189 kg/ha/yr, 395 kg/ha/yr from the total yield is lost through natural mortality. In other words, from the total yield of the whole evaluated Lake Kariba stock 66% is lost through natural mortality, 2.5% forms the actual catch and 32% is the unexploited available yield from which fishery can increase its exploitation. Expressed in absolute terms for the whole lake, while the actual catch is 3 thousand tons, it can be increased within 35 thousand tons of available yield. Since the eel and Tanganyikan anchoveta form completely unexploited populations, the available yield will in reality be much higher. The maximum sustained yield will most likely be in the vicinity of 40 thousand metric tons per year, whereas present fishing exploits only 3 thousand tons. The yield must be considerably less than 25% of the total production, as suggested by MANN (1969), because the total yield in Lake Kariba is only about 16% of the total production. The estimates of the maximum sustained yield may also, therefore, be in the vicinity of 8% of the total production for an average of all harvestable fish. ALLEN (1951) finds 7% to be the case for trout of the Horokiwi Stream.

While in a simpler system it may sometimes be possible to harvest this ecological maximum yield, in Lake Kariba a more realistic view must be taken. Because of the dendritic shoreline (see Part I, Section 8.2) and vast areas of submerged trees and bushes major parts of the yield are physically inaccesible. Furthermore, socio-economic factors limit the catches so that a realistic value for a potential catch will be substantially lower than the available yield. It should, however, not be much lower than a quarter of the available yield or 1/5 of the maximum sustained yield – about 10 thousand metric tons ($= E_{max}$).

I doubt that a solution to objective identification of a sustained catch can ever be found by speculation. For example, the model suggested by GULLAND (according to REGIER, 1970a) $C_{max} = 0.4\ MB_0$, to estimate maximum potential yield (C_{max}) from the natural mortality coefficient (M) and virgin ichthyomass (B_0), yielded an enormously inflated value if applied to Lake Kariba (biomass value corrected for virgin ichthyomass = 3,400 + actual catch = $\sim$ 4,000 kg/ha). It has to be emphasized, however, that ichthyomass in Gulland's equation is an entirely different variable, an incomplete biomass derived from selective samples (over harvested size) and recalculated for a whole area with no regard to fish distribution.

The actual catch statistics were very unsatisfactory and I have little confidence in their assessed value. But whatever this value may be, it has little to do with the available yield – the unexploited amount of fish of harvestable size. A higher value for the actual catch will only decrease the percentage which falls to natural mortality from the total yield. Because of the low confidence in the actual catch values final production was not computed.

The highest total yield is estimated for the *S. mossambicus mortimeri* – 468 kg/ha/yr, which is 23 times higher than the next highest value of total yield. This is also very clearly shown in the catch statistics, however imprecise they may be. In order of decreasing total yield follow: *M. deliciosus* 20, *T. rendalli* 16, *M. longirostris* 14, *H. vittatus* 12, *S. codringtoni* 12, *M. electricus* 10, *C. gariepinus* 9, *H. discorhynchus* 8, *S. zambezensis* 4, *L. altivelis* 3, *H. longifilis* 1, *E. depressirostris* 0.79, *M. macrolepidotus* 0.64 and *S. mystus* 0.16 kg/ha per annum.

The highest available yield is also that of *S. mossambicus mortimeri* with 101 kg/ha/yr, though the order of other species is different and may be a reflection of unequal

Table 116. Some ratios of computed stock size and production values per 1 ha area of Lake Kariba.

	G/Z	N'/A	A/B̄	B'/P'	P'/A	$Y_{P'}/Y_A$	$Y_{P'}/P'$
Mormyrus longirostris	2.0493	0.06	4.47	2.16	0.025	1.07	0.484
Hydrocynus vittatus	2.4047	0.13	3.75	2.18	0.036	0.58	0.412
Mormyrops deliciosus	2.7551	0.21	3.24	2.37	0.048	0.70	0.583
Sargochromis condringtoni	2.0229	1.81	2.21	1.62	0.125	0.33	0.500
Heterobranchus longifilis	2.5254	3.00	0.67	1.75	0.667	1.00	0.250
Clarias gariepinus	3.0350	3.13	0.52	3.07	0.484	0.78	0.467
Sarotherodon mossambicus mortimeri	1.2279	3.87	1.61	1.78	0.240	0.22	0.771
Labeo altivelis	1.4193	5.67	0.95	2.29	0.389	0.67	0.286
Tilapia rendalli	0.9439	11.81	1.47	1.52	0.383	0.75	0.387
Marcusenius macrolepidotus	0.6846	17.40	1.87	1.50	0.267	1.06	0.170
Malapterurus electricus	0.7530	36.81	1.14	1.52	0.532	0.90	0.089
Synodontis zambezensis	1.0278	29.19	1.10	1.70	0.476	1.00	0.400
Hippopotamyrus discorhynchus	0.4461	105.61	1.51	1.18	0.824	1.50	0.085
Eutropius depressirostris	0.9102	99.40	0.83	1.67	0.600	0.83	0.220
Schilbe mystus	0.9577	144.00	1.00	1.50	1.000	1.00	0.160
Brachyalestes imberi imberi	0.5944	294.74	1.06	1.13	1.210	—	—
Alestes lateralis	0.3814	313.34	1.59	1.02	1.574	—	—
Haplochromis darlingi	0.2051	253.33	0.002	562.01	0.763	—	—
Synodontis nebulosus	0.6233	301.33	0.60	1.67	1.000	—	—
Micralestes acutidens	0.1829	13200.00	0.25	1.12	16.000	—	—
		17.74	1.08	3.40	0.197	0.31	0.277

exploitation pressure on single species. *M. longirostris* shows an available yield of 15 kg/ha/yr, *M. deliciosus* 14, *T. rendalli* 12, *H. discorhynchus* 12, *M. electricus* 9, *H. vittatus* and *C. gariepinus* 7 each, *S. codringtoni* and *S. zambezensis* 4 each, *L. altivelis* 2, *H. longifilis* 1, *M. macrolepidotus* 0.68, *E. depressirostris* 0.66 and *S. mystus* 0.16 kg/ha/yr.

The unequal catch pressure upon single species can be expressed better in percentages of available yield to the total yield. Most heavily exploited seems to be *S. m. mortimeri* with only 20% left as available yield from the total yield. It is followed by *S. codringtoni* with 33% left, *H. vittatus* 58%, *L. altivelis* 67%, *M. deliciosus* 70%, *T. rendalli* 75%, *C. gariepinus* 78%, *E. depressirostris* 83% and *M. electricus* with 90% (Table 116, ratio $Y_{P'}/Y_A$) left. *S. mystus*, *S. zambezensis* and *H. longifilis* are completely unexploited with 100% left as available yield, and *M. macrolepidotus* with 106%, *M. longirostris* with 107% and *H. discorhynchus* with 150% left. This phenomenon of an available yield higher than the total yield can be explained by an unstabilized, increasing population density, by increasing production due to improving conditions or to an as yet unattained carrying capacity for the species in question. There may also be other explanations, or it may reflect a simple computational error or a peculiar sampling bias. In such cases the suggested ratio may have no real usefulness.

5. THE EELS

> 'I think it physically impossible for any fish to get up the falls and rapids which would point to the fact that the species is not katadromous.'
> P. I. R. MacLaren to Frost (1957a)

> 'The "atlantic eel problem" was reopened by Tucker 1959, 1960), who believes that the mature European eels are physically incapable of swimming 7000 km to the Sargasso Sea to breed.'
> E. M. Pantelouris et al. (1970)

On the basis of extensive studies of the eels of southern Africa, Jubb (1961b, 1964a, 1967) concluded that eels will disappear from Lake Kariba within 10 to 15 years following the dam closure, since 'mature eels will leave the lake for the sea, but small eels returning from the breeding grounds at sea will be too large to negotiate the mighty 420 ft. wall which is 700 miles from the Indian Ocean. Physical conditions in the area are not suitable for an overland route, and entry via the tail-race and turbines quite impossible' (Jubb, 1960b, p. 119). Prior to the construction of Kariba Dam, numerous eels of the species *Anguilla nebulosa labiata* were recorded from the section of the Zambezi River which Lake Kariba now occupies (Frost, 1957a; Jubb, 1961a; Jackson, 1961b). General acceptance of Jubb's (1964b) erroneous conclusions and pecularities in the distribution of eels in the lake observed later, had the unfortunate consequences that the existence of eels in Lake Kariba was not considered at all. The occasional eels caught by sport fishermen were explained away as survivors from the pre-dam period and because eels were rarely marketed, as local fishermen feared the snake-like creature, they remained unrecorded.

During the ichthyological survey of the Kalomo River eels were unexpectedly collected below Siengwazi Fall, 335 km above Kariba Dam and 81 km above the upper limits of Lake Kariba. From the 7 eels of species *Anguilla nebulosa labiata* caught there on the 2nd October 1969, four specimens were juveniles ranging from 25 to 51 cm in length. This discovery was described in detail in a special article (Balon, 1973b).

The electroacoustic survey of the lake conducted in 1968 and 1969 revealed that fishes in Lake Kariba are limited to an inshore area of 0–20 m depth. Accordingly, during the period 1968–1971 quantitative sampling with piscicides was performed in blocked-off coves and bays in a maximum depth of 16 m. At no time did eels constitute a part of these samples. Gill-net catches from a depth of up to 12 m also did not yield any eels (Coke, 1968). The initial invasion of *Alestes lateralis* into uninhabited open lake waters which was observed in 1969 was gradually substituted by the introduced *Limnothrissa miodon* and the 38 % of the area inhabited by fish ceased to be valid or was valid only for the riverine indigenous species. Sampling conducted in 1970 with an electrofishing apparatus and toxicants along the rocky shores and stream entries still yielded no eels.

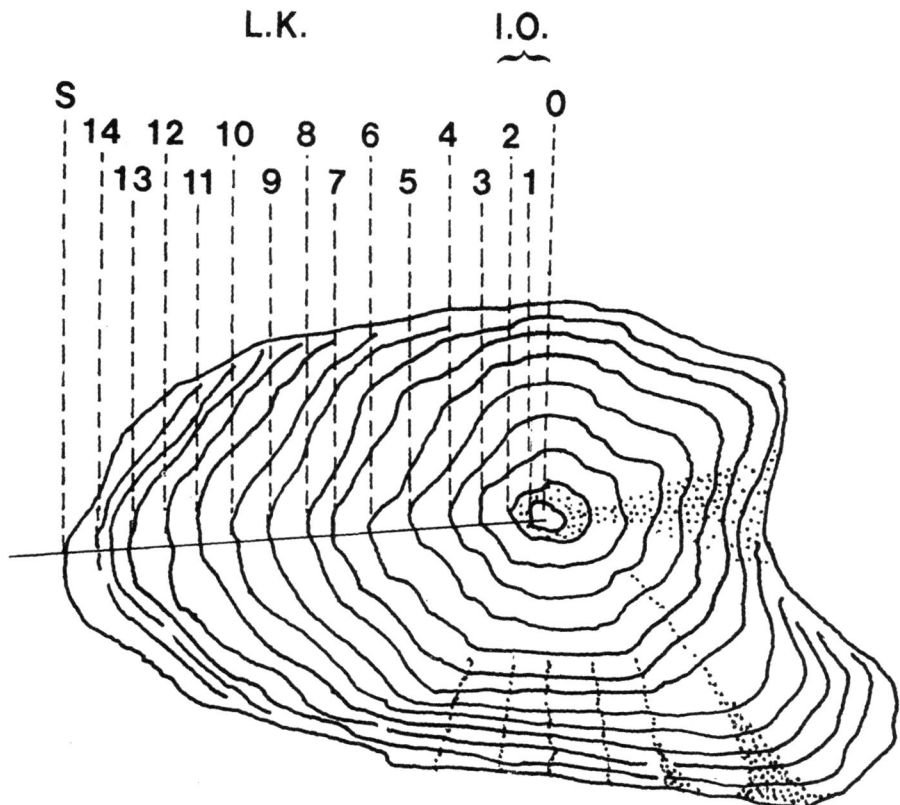

Fig. 62. Otolith of a Lake Kariba eel: 0 is the centre or beginning, S the edge or end of measured radius, 1–14 are individual annuli, I.O. the darkened opaque zone including life in the Indian Ocean and during the river ascent, and L.K. represents years spent in the feeding grounds of Lake Kariba drainage.

Long-lines of 88, 175 and 263 hooks, and traps, with both side entries (hoopnets) constructed from chicken wire (mesh size 2.5 cm) with nylon net funnels and central leading wings were set in a row of six. Unbaited or baited with chunks of fish or beef this gear was placed at different depths from the shore toward the center of the lake. The actual depths were always recorded with an echo-sounder. At certain depths abundant eels were captured. After the depth of eel occurrence was established, two sites for more extensive sampling were selected. The Namazambwe Bay in the upper part of the lake with a relatively distinct 136 ha area of a 25 to 40 m depth inhabited by eels served for eel density assessment. Near the dam at Siavonga (Banana Island N 102) an area of the same depth was selected to compare the values for catch per unit of effort and eel sizes.

Details on the history of the existence of eels in Lake Kariba, on methods of taxa identification and, sex and age determination are given in a study published separately (BALON, 1974d). It should be remembered that this concise extract from the latter study is given here only to complete the information on Lake Kariba and is supplemented by the original study.

Interpretation of Otoliths and Back-calculation of Growth. – In both European (FROST, 1945; SINHA & JONES, 1967) and American eels (GRAY & ANDREWS, 1971) the alternate opaque and transparent zones on otoliths proved to be annual. Although available evidence points plainly to the fact that the opaque rings represent the faster summer growth and the transparent dark rings the winter growth, some authors count the former, others the latter to determine the age of eels (PANTULU & SINGH, 1962). As to Kenya eels, FROST (1955) even voiced some doubts about their zones being annual.

I have previously discussed the problem of the validity of the annulus in tropical fishes (BALON, 1971a, 1972) in connection with the annuli on scales. I concluded that not climatic factors alone but the 'planetary yearly system' or 'circaannual rhythm' is responsible for intervals in seasonal development (see also Appendix F). These intervals are recorded as annual rings on the structure of hard body tissue. The concept of 'the endogenous biological rhythm' (SCHWASSMANN, 1971) or 'the physiological clock' (BÜNNING, 1967) is another expression of the same problem. The endogenous annual rhythm (circaannual) in fish growth is probably reflected in alternate and different layers of body tissues irrespective of geographical zones. This rhythm may be inherited as a sort of 'endogenous calendar' and is independent of, but regulated or timed by environmental factors. The regular formation of annuli in wild carp under constant laboratory conditions can be considered as evidence of such an endogenous calendar in action (BALON, 1974a).

Similar to the definition of the annulus on scales (CHUGUNOVA, 1963) the completed annual cycle on otoliths would be the area where the narrow (translucent) band of the slow 'winter' growth contacts the broad (opaque) band of fast 'summer' growth. As on scales, an annulus can be located after each ring of fast and slow growth (Fig. 62). The otolith center of Lake Kariba eels shows a darkened opaque zone; however, even within this zone, alternate fast-slow growth bands were located. It can therefore be inferred that the first two incomplete years represent growth in the Indian Ocean or, more probably, combined life of ocean leptocephalus', transistory elver's and early juveniles' phases – the eel's life on the move, before the ultimate freshwater feeding grounds are reached. The back-calculated lengths for these two years support such a concept.

5.1. Taxonomic status

The most distinct key characters given by EGE (1939), FROST (1955) and JUBB (1967) for species separation of African eels were the ano-dorsal value, the vertebral count and the dentition patterns of the upper jaw. In the Kariba eels the first character was within a range of values known for *Anguilla nebulosa labiata*. The eels of the Kalomo River and Lake Kariba showed a marked tendency to higher values in the ano-dorsal distance of the type noted in *A. mossambica*; but FROST (1957a) had previously observed high D values for eels from this area. The vertebral count for Kariba eels, however, was below the range of 108 to 115 given for *A.n. labiata*. The values of 97 to 107 vertebrae (x̄ 103.6) resembled those observed for *A. mossambica*. The dentition patterns on the other hand were similar to those of *A.n. labiata;* only a small number are reminiscent of the *A. mossambica* type. No separation of eel taxa was possible when based on the above characters only. Both mottled and plain eels had these characters intermingled.

There is no adequate evidence available to make any confident statement about the intermingled characters of Lake Kariba eels. Either the ranges of key characters have to be revised or the new ecosystem created by the impoundment favours a development of both morphs, plain and mottled African eels, from the same juveniles. To what extent the validity of separate species *A.n. labiata* and *A. mossambica* can be questioned from this evidence would be too soon to judge. Prior to the existence of Lake Kariba the Zambezi River was inhabited by *A.n. labiata*; only single and doubtful *A. mossambica* were recorded (FROST, 1957a). The intermingled characters of Lake Kariba eels may be attributed to the effect of deep waters, selected here as a preferred habitat. Far more, however, has to be known about the spawning grounds and larval route of African eels to enable some insight into a possible phenotypic character of presumed specific differences.

Lake Kariba could not have affected the ocean period of the eel's life or have any meaning in the selection by larvae of the one or the other species to enter the Zambezi River. The changes observed can be attributed to the effect of the new environment itself on the same migrating eel juveniles as were those prior to the creation of Lake Kariba. Should it be found later that *A.n. labiata* and *A. mossambica* are conspecific, the nomenclature will have to be revised. But failing further evidence, I am provisionally retaining the name *A.n. labiata* for both mottled and plain Lake Kariba eels.

5.2. Limits of distribution

It was established that in Lake Kariba no eels were present within the distribution limits of the indigenous riverine species – in 25 m deep inshore waters. The first eels were recorded from a depth of 25 and 33 m in Namazambwe Bay (Basin II). Eels were also taken in a 31 m depth near Banana Island (N 102) of the lowermost Basin IV; in the same basin no eels were caught in 77–79 m. Although collecting in Basins II and IV at a depth from 2 to 79 m was limited to four days of fishing with long-lines and hoopnets sets, to 6 days of fishing in 25 to 40 m at Siavonga and 23 days in 25 to 40 m depth at Namazambwe and included density estimates, it yielded enough evidence to limit the depth distribution of Lake Kariba eels from 25 to 40 m. Some individuals were, however, taken in about 20 m, some a little below 40 m. The limits of 25 to 40 m are thus not absolutely precise but convenient because of ready estimates of isobaths and area. To learn the precise depth distribution of eels in all the basins of Lake Kariba will require more extensive sampling; some seasonal changes may then be expected.

We were unsuccessful in entering Devils Gorge due to dense salvinia mats and therefore, the only eels known from above the lake were those sampled at Siengwazi Fall. Their total lengths ranged from 25 to 84 cm. The total lengths of eels at Namazambwe ranged from 66 to 134 cm, and at Siavonga from 84 to 124 cm. The males of Kariba Lake eels tended to be larger than the females. In spite of limited evidence, there seems to be a trend in distribution of smaller eels farther upstream with a greater number of larger eels in Basin IV, the last basin near the dam.

5.3. Age, growth and size of resource

The sizes of the African juvenile mottled eels recorded so far were from 54 to 176 mm in East African rivers and at a distance from the ocean of 29 to 563 km (FROST, 1957b; WHITEHEAD & VAN SOMEREN, 1959; JUBB, 1961b). The eel juveniles mounting Kariba Dam, 1018 km from the Ocean, could be 6 to 10 cm long, judging from the evidence available. After surmounting the Dam, the eel juveniles probably enter the tributaries of Lake Kariba (SMITH, 1971) and drift back into the lake only after spending about 7 years in the streams. The size of eels found in the Kalomo (see section 6.1) and Lusito (unpublished data) rivers and the back-calculated lengths for the previous years (Tab. 117) of eels in the lake should serve as evidence for this statement.

The eel juveniles enter Lake Kariba for the first time in their second year of life. At the beginning of sexual maturation, at an age of approximately 9 years, they re-enter the lake and spend another 1 to 9 years at a depth of about 25 to 40 m. The oldest fish in my samples was 18 years old. The time when the lake is reentered is probably the beginning of the catadromic migration.

Judging from the 67 specimens taken in March and April 1971 and used for age analysis, 1.5% surmounted Kariba Dam in 1953, 4.5% each in 1954 and 1955, 10.4% in 1956, 19.4% in 1957, 16.4% in 1958, 11.9% in 1959, 13.4% in 1960, 11.9% in 1961 and 6% in 1962. No change was demonstrated in the number of eels entering the Lake Kariba area before and after dam closure.

The largest eel in my Lake Kariba samples was a female. It measured 132 cm and was 18 years old. Several eels, however, over 120 cm and 3 kg were taken. The back-calculated lengths increase steadily throughout the entire life (Tab. 117) in a similar manner as the lengths of Kenya eels (FROST, 1955). When the lengths of Kenya eels were shifted plus one year, the relative faster growth becomes similar to the growth of Lake Kariba eels. From the data given by FROST (1955, p. 8, Fig. 4) this is the way it has to be done in order to bring age identifications under the same criteria. FROST (1955) counted the second broad transparent zone as the first year in fresh water, leaving the first transparent and opaque zones uncounted and considered as a record of oceanic life of unknown length. In spite of her doubts the values obtained by her were in agreement with mine.

The length-weight relationship of immature European and Lake Kariba eels is very similar (Fig. 3). The tendency of males to be heavier than females begins earlier in the European eels (DEELDER, 1970), but the extent is the same as with Lake Kariba eels, although in general the latter reached a larger size and probably began to mature later. The length-weight relationship also reflects the fact that individuals which mature leave the lake earlier for the ocean and only the lighter, slower maturing individuals remain as a residual stock of the lake. This phenomenon of a split within the population

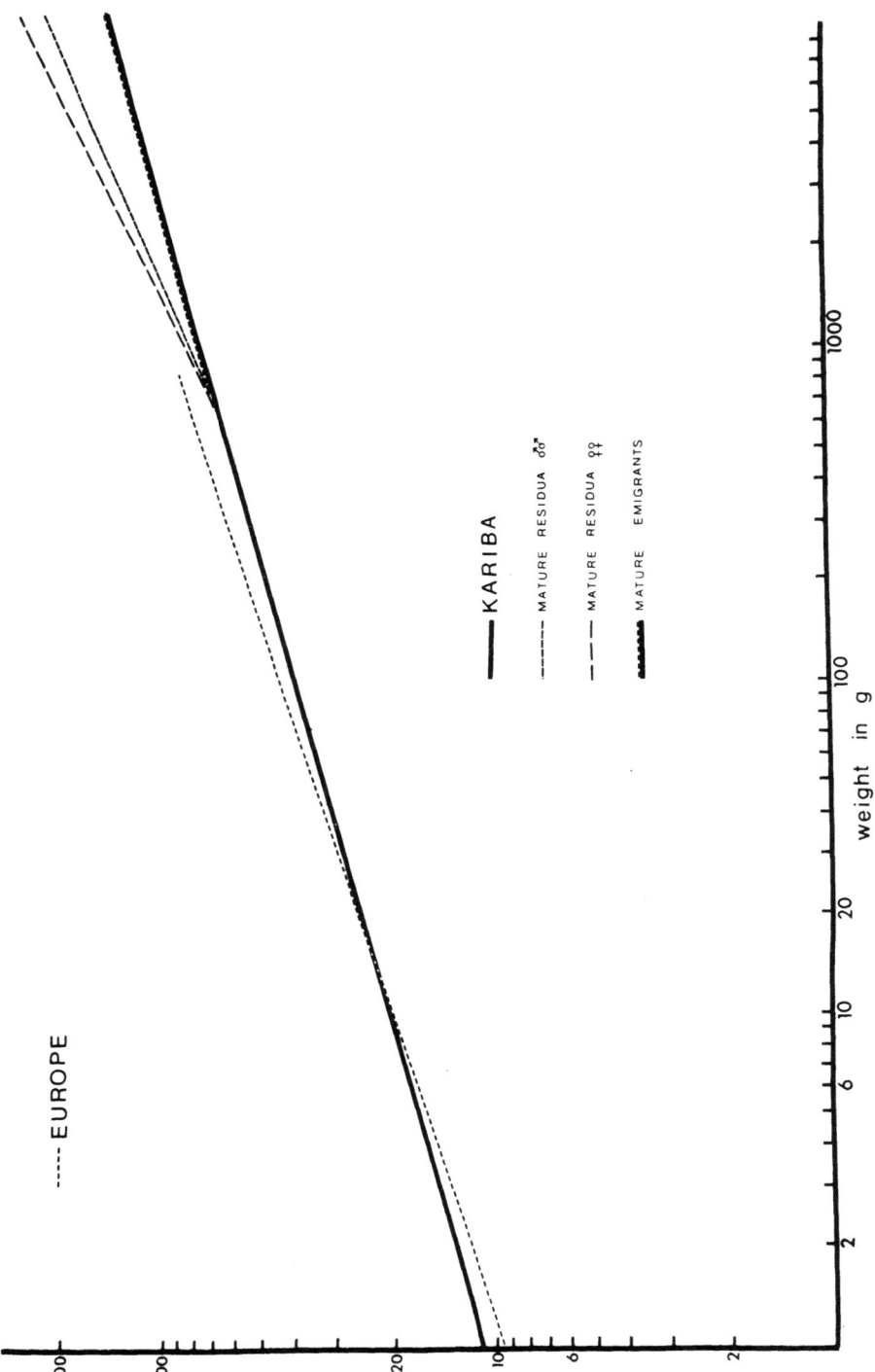

Fig. 63. Length-weight relationship of Lake Kariba eels as compared to the European eel.

Table 117. Analysis of age and growth of the eel stock in Lake Kariba.

Locality, sex	Year of hatching	N	TL, cm at the time of capture	Back-calculated total lengths in cm for the previous growth seasons																
				1	2	3	4	5	6	7	8	9	10	11	12	13	14	15	16	17	18
Namazambwe Bay ♀♀	1961	4	76.6	5	15	23	31	38	45	53	61	70									
	1960	6	86.7	5	13	22	31	38	45	53	62	71	80								
	1959	5	84.1	4	11	22	31	38	45	53	61	70	78	86							
	1958	3	100.5	5	15	23	31	39	48	57	64	71	80	87	94						
	1957	6	108.3	5	13	22	29	35	44	50	57	65	74	82	90	97					
	1956	5	108.8	5	11	20	28	35	44	51	60	66	74	82	91	98	106				
	1953	2	126.0	4	11	22	31	38	45	53	61	70	78	86	92	100	107	117	125	133	
	1952	1	132.0	5	15	20	26	34	38	45	53	61	68	75	81	88	96	102	111	120	129
	Σ x̄	32	97.8	5	13	22	29	37	45	53	61	68	77	84	91	97	106	112	121	129	129
	Min.		66.4	4	10	17	23	31	38	45	53	58	66	74	81	88	96	102	111	120	
	Max.		132.0	7	19	26	35	43	50	57	66	74	82	90	97	103	112	120	129	135	
♂♂	1960	2	93.7	4	11	22	29	37	45	54	60	70	77								
	1959	3	93.8	5	13	20	28	37	45	54	61	68	77	84							
	1958	4	114.6	5	13	22	29	38	45	53	60	68	77	84	92	98					
	1957	2	101.5	5	13	22	31	39	47	54	61	66	75	82	91	102	110				
	1956	6	112.5	5	13	22	29	37	45	54	61	70	77	86	94	102	107	116			
	1955	5	121.6	5	11	19	26	35	44	51	60	66	75	82	92	102	107	111	117		
	1954	3	122.0	5	13	22	28	35	43	50	58	65	73	80	88	94	102		117		
	Σ x̄	25	111.2	5	13	22	29	37	45	53	60	68	75	84	92	100	107	114	117	129	
	Min.		81.5	4	10	16	23	31	38	44	51	58	66	74	84	91	97	106	111		
	Max.		134.0	7	16	25	32	39	51	60	68	77	84	91	101	110	120	124	125		

Table 117. (continued)

Locality, sex	Year of hatching	N	TL, cm at the time of capture	Back-calculated total lengths in cm for the previous growth seasons																	
				1	2	3	4	5	6	7	8	9	10	11	12	13	14	15	16	17	18
Near Banana Island ♀♀	1959	1	84.0	7	15	24	32	38	45	54	61	70	77	86						
	1958	1	92.0	5	13	22	31	38	45	54	61	70	78	87						
	1957	2	110.0	4	11	20	28	37	45	53	60	68	75	84	96	101				
	1956	2	101.7	5	15	23	31	38	39	47	54	62	71	78	86	94	101			
	Σ x̄	6	99.9	5	13	22	31	38	44	51	58	66	75	82	90	98	101			
	Min.		84.0	4	11	20	28	37	31	38	45	53	61	68	77	84	91			
	Max.		110.0	7	16	25	34	39	48	55	64	73	81	88	96	106	112			
♂♂	1957	1	115.0	4	13	22	29	37	45	53	60	68	75	86	94	102				
	1955	2	117.0	4	13	19	26	35	45	53	60	68	77	87	94	104	111	118	120	
	1953	1	122.0	5	11	20	29	38	45	53	60	68	77	84	90	97	106	112	120	129
	Σ x̄	4	117.7	4	13	20	28	37	45	53	60	68	77	86	92	101	110	117	120	129
	Min.		110.0	4	11	17	25	35	45	53	60	68	75	84	90	97	106	112		
	Max.		124.0	5	15	22	29	38	47	54	61	70	78	87	96	107	114	121		
Total	x̄	67		5	13	22	29	37	45	53	60	68	76	84	91	99	106	114	119	129
	TL(cm) ♀♀													940	1100	1300	1480	1760	2000	2400	2620
	w̄(g) ♂♂			0.02	1.9	13	35	90	180	325	510	740		900	1200	1500	1800	2100	2550	2620

Table 118. Catch per set of 6 baited traps checked once per 24 hours and planted on the eel grounds (25–40 m).

Place	Date	Number of traps	Number of eels taken (per trap)	Catch per unit of effort		Bait used
				n	w̄	
Siavonga	18–23 March 1971	8	2 (0– 1)	0.25	0.57	chunks of beef
		26	4 (0– 1)	0.15	0.34	chunks of fish[1]
				x̄ 0.18	0.41	
Namazambwe	4– 8 April 1971	28	57 (0–11)	2.04	4.69	chunks of fish[1]
	7–24 May 1971	108	82 (0–11)	0.76	1.75	chunks of fish and beef[1]
				x̄ 1.02	2.35	

[1] *Synodontis zambezensis, Hydrocynus vittatus.*

Table 119. Estimation of eel density at Namazambwe Bay of Lake Kariba.

n_i	t_i	$n_i t_i$	m_i	d_i	s_i	$t_i s_i$	$n_i t_i^2$	s_i^2/n_i
3	—	—	—	—	—	—	—	—
7	3	21	—	7	—	—	63	—
4	10	40	—	4	—	—	400	—
3	14	42	—	3	—	—	588	—
7	17	119	—	7	—	—	2023	—
1	24	24	—	1	—	—	576	—
8	25	200	1	7	1	25	5000	0.125
2	32	64	—	2	—	—	2048	—
5	34	170	—	5	—	—	5780	—
11	44	484	1	9	2	88	21296	0.3636
5	55	275	—	4	1	55	15125	0.2000
14	60	840	—	13	1	60	50400	0.0714
1	71	71	—	1	—	—	5041	—
1	72	72	—	1	—	—	5184	—
2	72	144	2	2	—	—	10368	—
1	72	72	1	1	—	—	5184	—
2	72	144	2	1	1	72	10368	0.5000
4	72	288	4	4	—	—	20736	—
Σ 81	—	3070	11	72	6	300	160180	1.2600

$$\bar{N} = \frac{\Sigma n_i t_i}{\Sigma s_i} = \frac{3070}{6} = 511.6, \text{ conf. limits } 250 < \bar{N} < 1{,}101.$$

was observed in eels from Lake Ontario (HURLEY, 1972), and at approximately the same length and weight.

A consistent survey in the direction of E/f statistics was not carried out. However, data obtained for other purposes may yield some values of E/f which may be interesting. At Namazambwe the baited traps (of the type called hoopnets) were on an average able to take 2.35 kg of eels per one net lift. Some seasonal differences may occur since in April E/f amounted to 4.69 kg and in May to 1.75 kg only. There is not enough data available, however, to assert this with any certainty (Tab. 118). At Siavonga the same set of baited traps yielded an E/f average of 0.41 kg only. This could indicate a substantial decrease in eel density extending from the upper to the near dam areas of the lake.

For the density assessment, SCHNABEL's (1938) method as used by HOLČIK (1970a), was applied at the Namazambwe eel grounds. The eel density was estimated from $\bar{N} = \sum n_i t_i / \sum s_i$, Table 119, where n_i was the total number of eels in the i catch, d_i was the number of unmarked eels in the i catch, m_i was the number of eels removed (e.g. preserved for later taxonomic examination), t_i was the total number of marked eels present in the area during the catch $i + 1$, and s_i was the number of marked recaptured eels in the catch i. HOLČIK's (1970a) procedure for the computation of 95% confidence limits was accepted.

Since only one 17 day period (7–24 May 1971) at one location was used for the density estimates and there was no certainty that the estimated limits of the area in question define the true area of this particular group of eels, the significance of the values obtained should not be overestimated. Nevertheless, in connection with the data on catch per unit of effort some indication of the eel stock size can be inferred.

An average of 512 eels, within confidence limits of 250 to 1,101 were estimated. This comes to less than 4 eels per ha if 136 ha of eel grounds are accepted. The catches, however, probably indicate a higher density (Tab. 118) and even the upper limits of 1,101 eels are from this point of view an underestimation. It is apparent that the range of the habitat of the Namazambwe eels was judged incorrectly. If the habitat range is considered to be only one lobe of the previous area, 46 eels could be present per 1 ha. Whereas the catch per unit of effort was 5.7 times lower in the lake basin near Kariba Dam than the catch at Namazambwe, the density probably decreased toward the lower parts of the lake in similar proportions to the catch.

The density estimate, when based on the 136 ha area, yielded values too low to sustain the observed catches. Within the presumed eel grounds the area of only one cove will now be considered as the eel territory appearing in the capture-mark-recapture estimates. In this case the average number of 512 eels would have inhabited 11 ha. The estimated territory range of the eels made these values highly speculative, though some insight into the Lake Kariba eel potential may emerge.

Already the length-weight relationship indicated (Fig. 63) a presence of two groups of eels within the mature portion of the population: those which mature faster and leave the lake – the mature emigrants, and those which mature slower and remain in the lake for a prolonged period of time – the mature residua. If the same structure is transferred into the population diagram illustrating the density in relation to age, leaving a similar ratio between emigrants and residua after age 10, the theoretical density values can be intercepted from the regression lines (Fig. 64). Since the position of these lines has been drawn by eye the significance of values obtained should not be overestimated. I am trying rather to indicate the possibilities of this approach when more data is collected; only then will the general trend of the possible structure of eel density in Lake Kariba be established.

About 300 eel juveniles per 1 km^2 of eel grounds would have to enter Lake Kariba to sustain the density observed at Namazambwe. Approximately 2000, from the age of 2 to 9, (per the same area) should then live in the streams of the lake drainage. About 100 from these will have to retreat into the lake at the beginning of maturation each year to sustain the possible 484 emigrants and 270 residua of age 11 to 18 per 1 km^2 of eel ground area (Tab. 120).

A multiplication of the weight increments and density values should give a value of the available production. The residua could then be considered to be the same as the harvestable part of the available production or the available yield. Since they were based on the high density at Namazambwe, the values obtained – 55.8 kg/ha of available ecological production and 7.5 kg/ha of yield – would be the maximum values. If the density of eels is reflected in the catch per unit of effort then the decrease in density from Namazambwe to the dam area could also be 5.7 times lower. Estimated yield for the Namazambwe area is higher or similar to the catch of the best European eel grounds (RASMUSSEN, 1952; MÜLLER, 1962; MANN, 1963; DEELDER, 1970).

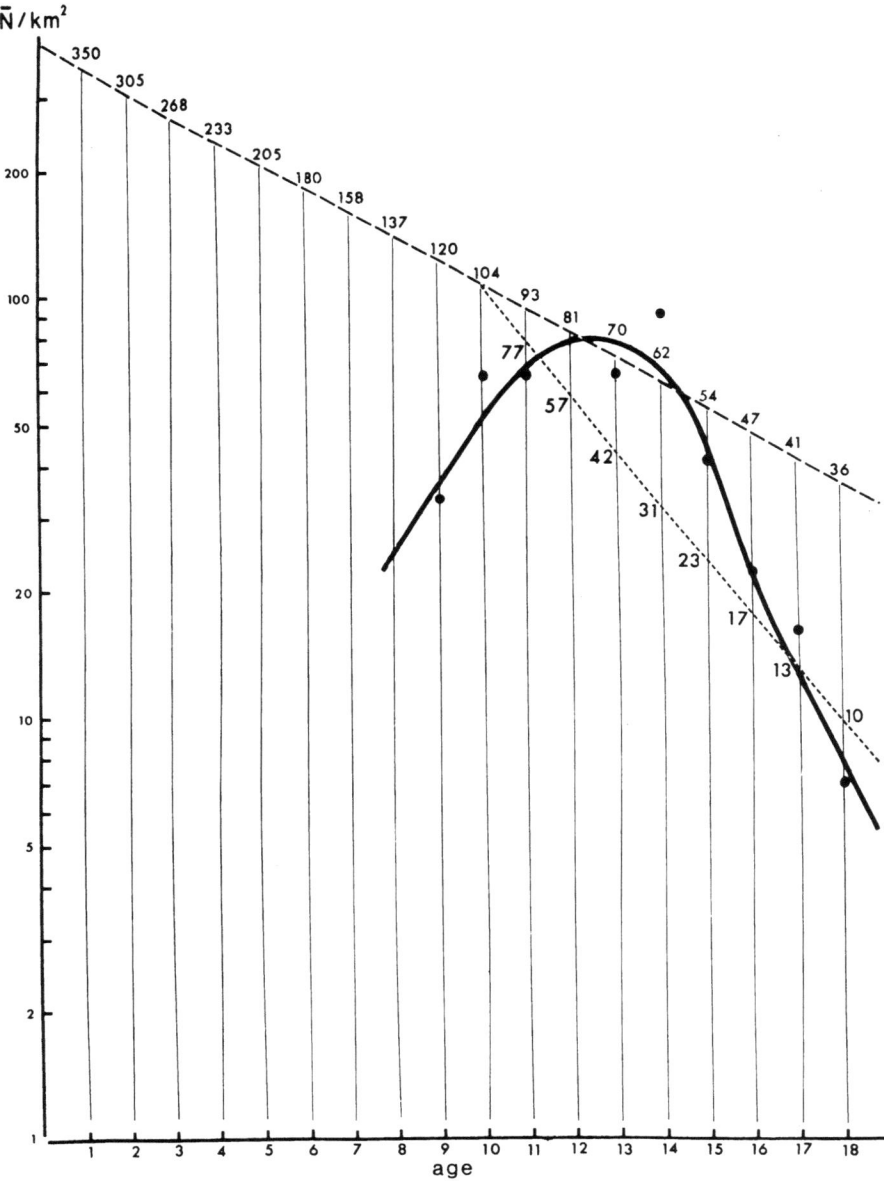

Fig. 64. Semilogarithmic diagram of the age to density structure of Lake Kariba eels (bold curve and points) with an attempt to reconstruct the total density of the drainage stream juveniles (age 1–10), mature emigrants and residua per given area (as interpolated from the length-weight regression, see Table 120).

Table 120. Speculative assessment of relative density and available production of eels in Lake Kariba.

Age	1	2	3	4	5	6	7	8	9	10	11	12	13	14	15	16	17	18	Σ
N, %	—	—	—	—	—	—	—	—	7.0	14.0	14.0	12.3	14.0	19.3	8.8	5.3	3.5	1.8	
$\bar{N}$/km²	—	—	—	—	—	—	—	—	33	65	65	57	65	90	41	24	16	7	
$\bar{w}$(g)	0.02	1.9	13	35	90	mature emigrants 180	325	510	740	1260	1800	2400	3200	4180	5500	6000	8000	8000	
						mature residua 180	158	137											
N'/km²	350	305	268	233	205	mature emigrants 180	145	185	230										484
						mature residua 180	158	137	120	104	93	81	70	62	54	47	41	36	2060
																			270
$\bar{w}_i - \bar{w}_{i-1}$	1.88	11.1	22	55	mature emigrants 90	145	185	230	520	540	600	800	980	1320	500	2000	0	0	
					mature residua				160	200	200	200	300	300	100	400	0	0	
P'(kg)/km²	0.6	3.0	5.1	11.3	emigrants 16.2	22.9	25.3	27.6	54.1	50.2	48.6	56.0	60.8	71.3	23.5	82.0	0	0	446
					residua				16.6	15.4	11.4	8.4	9.3	6.9	1.7	5.2	0	0	112
																			75

$P' = N'_i(\bar{w}_i - \bar{w}_{i-1})$

P' total   55.8 kg/ha
P' emigrants   37.1 kg/ha
P' residua = $Y_{P'}$ 7.5 kg/ha

6. FISH PRODUCTION OF THE DRAINAGE AREA AND THE INFLUENCE OF ECOSYSTEM CHANGES ON FISH DISTRIBUTION

> Contrary to popular opinion, man the hunter and gatherer did not live on the edge of catastrophe, nor did he spend the greater part of his time seeking food. Studies of the few remaining hunting cultures indicate that he worked less than we do, that his quests for food took up only a small portion of his time, and that leisure was abundant. He was relatively happy and lacked any real concern for tomorrow. His wants were restricted, and, because of his need to move, wealth was a burden...
> ROBERT LEO SMITH (1972)

At one point during my Lake Kariba studies it became clear that the local subsistence fishery, even on the smallest intermittent streams, could have some importance in the nutrition of the local population. The drainage area of a river and/or a lake *in toto* might yield a significant amount of protein and therefore should not be neglected in research programs.

Time was running out unfortunately and a survey of most Lake Kariba drainage rivers proved impossible. I sampled only four streams, two of which have here been selected for discussion in detail: the Kalomo River and the Elephant Stream. The first could illustrate the fish production potential of larger drainage streams as well as the relationship of the local people toward fish resources in pre-Kariba times. The second demonstrates the importance of even the smallest streams to indigenous fish populations.

Since no data on fish production potential are available for the Gwembe aluvium in pre-Kariba Lake times (JACKSON, 1961b) the last section of this chapter presents data on the Kafue River. These were obtained by using the same methods as were used to collect data on Lake Kariba. Although the studied Kafue River floodplain presents to some extent a habitat different from the original Gwembe Valley alluvium and a fish fauna of a different origin (see section 6.1 and 6.3), the virgin fish production potential of both areas is possibly comparable.

6.1. The Kalomo River

Stream gradients have been recognized for some time as important factors in fish distribution and stock size (SHELFORD, 1911; TRAUTMAN, 1942; LEGER, 1945; BURTON & ODUM, 1945; HUET, 1946). In Europe, HUET (1954, 1959), STARMACH (1956), and BALON (1959) attempted to test the validity of HUET's initial generalized concept, i.e., that species occurrence, diversity and stock size related to stream gradient are similar in all rivers of certain geographic area. Further data, however, disproved this simplistic concept (SOMMANI, 1953; MÜLLER, 1954; BACKIEL, 1964; BALON, 1964a, 1964b, 1967b) and its general validity within a given biogeographical area. In spite of the similarity in recent topography of different streams, HUET's hypothesis did not apply because of differences in historical origin or geomorphological record (BURTON & ODUM, 1945) of the water course. The classification of streams into orders (HORTON, 1945) would be only another variation of the gradient rule since 'stream order and gradient are closely related' (KUEHNE, 1962). Stream gradient nevertheless remained a useful tool in describing and comparing fish populations of individual streams (HYNES, 1970). The concept, as far as I know, was applied in Africa only by LELEK (1968), although it was hinted at by MARLIER (1953), OLIFF (1960, 1963) and MALAISSE (1968, 1969a) when changes of fish composition were noted along certain African streams. The Kalomo River, represents a stream with a relatively young (late-Tertiary) lower part and a much older upper part. The Kalomo has been populated by man since the Stone Age in the Pleistocene. Recently, there have occurred extensive run-offs from farm land and stresses from urban center.

The Kalomo River is the first large left tributary of the Middle Zambezi River. Its source is 216.5 km north of its entry into Batoka Gorge below Victoria Falls, approximately mid-way between the beginning of the Middle Zambezi and the upper end of Lake Kariba. The sources of the Kalomo are near the watershed of the Kafue River and the main stream runs parallel to northern lowermost rivulets of the Upper Zambezi drainage. The upper part of the Kalomo River is of low gradient owing to the low resistance of the underlying sediments; within this Miocene plateau the river eroded a valley 20 km wide and 60 m deep. The gradient at about mid-stream increases toward the late-Tertiary Zambezi floor and forms an escarpment and gorges (DIXEY, 1944).

At the time of the Miocene-Pliocene transition the Kalomo River was probably part of the Okavango (Ngami) basin, as were the Upper Zambezi, Chobe and the upper Kafue river systems. The present connection with the Middle Zambezi River (Pliocene) and the Indian Ocean was established later. Kalomo water began flowing to the Ocean first along the Matetsi River (BOND, 1964) and the Ngwezi-Matezi River and later through the area of Victoria Falls. At some time in the upper-Pliocene the Okavangian Kafue-Kalomo water system of the highveld plateau (the upper part of the present Kalomo River) was captured by the new river constellation and diverted from the Ngami Lake direction into the Indian Ocean. The diversion occurred at the spillway stream connections of the valley flank (late-Tertiary in origin). Approximately 500,000 years ago several gorges were cut but these alternate streams gradually dried up when most of the available water was concentrated in the central gorge. Numerous falls were created along the Kalomo bed. On the Zambezi River the proto-Victoria Falls eroded to the present position 99 km upstream at a rate of one kilometer every 5,500 years (CLARK, 1950) and at some stage passed the recent Kalomo River entry 86 km below.

While the waters of the Upper Zambezi, Middle Zambezi, and Kalomo rivers were joined, the majority of the Okavangian and Indian Ocean-Zambezi River fish faunas remained isolated one from another because of physical barriers such as Victoria and

Siengwazi Falls and other spectacular formations of the spillway connections such as Basokwe, Sikampatwe, Ichidi, Lombelombe and Moemba falls. It is generally accepted that the present topography had already been formed in the Pleistocene and that erosion since that time according to BOND (1963), has 'merely put the finishing touches to our present landscape'.

The existence of numerous rapids, falls and a gradually increasing gradient is evidence of Kalomo's long history. The average gradient of the lower part of the river is double that of the upper part, with variations from 1 to 19 meters per kilometer in short sections. No doubt the upper highveld plateau of the Kalomo River is older than the steep middle and lowveld part.

The sources of the Kalomo River are grassland streams located near Nakabanga at an altitude of 1368.5 m. The streams flow into a flat area of scattered woodlands and tree savanna (Munga and Miombo types of vegetation) at the southeastern edge of the Kalomo Hills. This is an area of protected forest with a fair population of wild animals and no human settlements. Due to the low gradient the streams form numerous small herbaceous swamps.

On the uppermost limits of the 6 km long irrigation reservoir, whose dam is built some 50 km from the source of the Kalomo River, a tsetse control fence marks the beginning of cultivated farm land. This area of the river valley is the most heavily populated part around Kalomo township. There are villages and large farms with about 50% of the virgin land adjacent to the river cultivated. Farther downstream the cultivated land is reduced rapidly to occasional individual farms and small local plantations. Downstream, where Nekoya Stream enters the Kalomo there are no farms; only a few villages of from 3 to 10 huts are scattered along the river. Where the Matezi River joins the Kalomo River (Fig. 65), there are no settlements whatsoever within close proximity of the river. The vegetation in that area is of pure Miombo type.

There are 22 well defined fish taxa present in the Kalomo River and one hybrid or introduced species of *Sarotherodon*. According to the classification of GREENWOOD *et al.* (1966) the species given belong to 8 families:

ANGUILLIDAE
    Anguilla nebulosa labiata Peters, 1852
MORMYRIDAE
    Mormyrus longirostris Peters, 1852
CHARACIDAE
    Hydrocynus vittatus Castelnau, 1861
    Petersius rhodesiensis Ricardo, 1943
CYPRINIDAE
    Barbus marequensis Smith, 1841
    Barbus poechii Steindachner, 1911
    Barbus paludinosus Peters, 1897
    Barbus viviparus Weber, 1897
    Barbus annectens Gilchrist & Thompson, 1917
    Barbus barotseensis Pellegrin, 1920
    Barbus radiatus radiatus Peters, 1853
    Labeo cylindricus Peters, 1852
    Barilius zambezensis (Peters, 1852)
MOCHOKIDAE
    Chiloglanis neumanni Boulenger, 1911

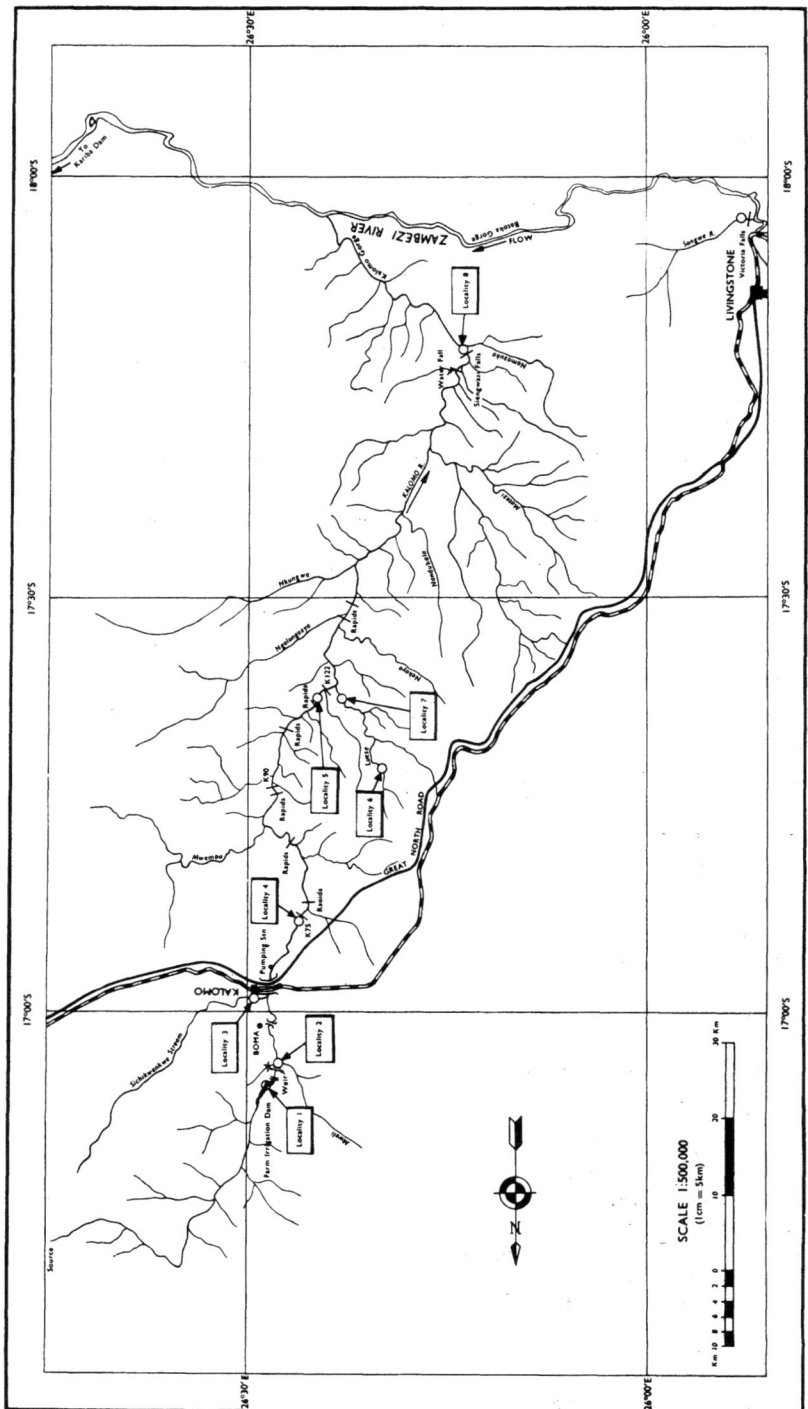

Fig. 65. Map of the Kalomo River drainage and localities sampled.

AMPHILIIDAE
Amphilius platychir Günther, 1864
CLARIIDAE
Clarias ngamensis Castelnau, 1961
Clarias gariepinus (Burchell, 1822)
Clarias theodorae Weber, 1897
CICHLIDAE
Sarotherodon andersoni (Castelnau, 1861)
Sarotherodon macrochir Boulenger, 1912 (? × *S. andersoni*)
Tilapia sparrmani Smith, 1840
Serranochromis robustus jallae (Boulenger, 1896)
Pseudocrenilabrus philander (Weber, 1897)

The sample of juvenile eels from Siengwazi Falls was the first evidence that Kariba Dam does not, as was previously assumed, form an insurmountable obstacle for elvers migrating from the Indian Ocean. This observation led me to look for and to discover an abundant eel population in Lake Kariba. From these discoveries I infer (Fig. 66) that when eel juveniles reach Siengwazi and Victoria Falls they are over the size limit at which it is possible for them to surmount such barriers.

On the basis of present topography, gradients, and to some extent fish distribution, the Kalomo River was divided into three distinct zones:

1. the highveld plateau
   – from the source downstream to the mouth of the Luese Stream. Average gradient 2.21‰. Upper Kalomo River.
2. the middleveld (escarpment)
   – from the mouth of the Luese Stream downstream to the edge of Siengwazi Fall. Average gradient 5.73‰. Middle Kalomo River.
3. the lowveld (gorge, and valley floor)
   – from base of Siengwazi Fall to the Kalomo River junction with the Zambezi River. Average gradient 6.63‰. Lower Kalomo River.

With respect to distribution along the river (Fig. 67) the fishes can be classified into four distinct groups:

a. ubiquitous
   – *B. paludinosus, L. cylindricus, B. radiatus, P. philander, A. platychir, C. gariepinus*. These are considered to inhabit the whole river. More detailed limits of distribution cannot be established, in this and subsequent groups, since too few localities were sampled. The same limitation applies to knowledge of the distribution of habitats such as rapids, lotic and lentic streams, and pools which occur frequently along the river.
b. plateau and escarpment
   – *T. sparrmani, B. poechii, P. rhodesiensis, B. barotseensis, S. andersoni* and *S. robustus* – which inhabit zones 1 & 2.
c. plateau
   – *B. annectens, B. viviparus, C. theodorae, C. ngamensis*, and *S. macrochir* (× *andersoni*). This group is restricted to the highveld plateau.
d. gorge
   – *B. marequensis, B. zambezensis, C. neumanni, A. nebulosa, M. longirostris* and *H. vittatus*. This group is composed of fishes unable to surmount Siengwazi Fall. Included in this group are the yellowfish and eel, well-known migratory animals. Three other species are known to migrate upstream for spawning. Siengwazi Fall definitely isolates this gorge group (limited to the lowveld) from the two upstream river sections.

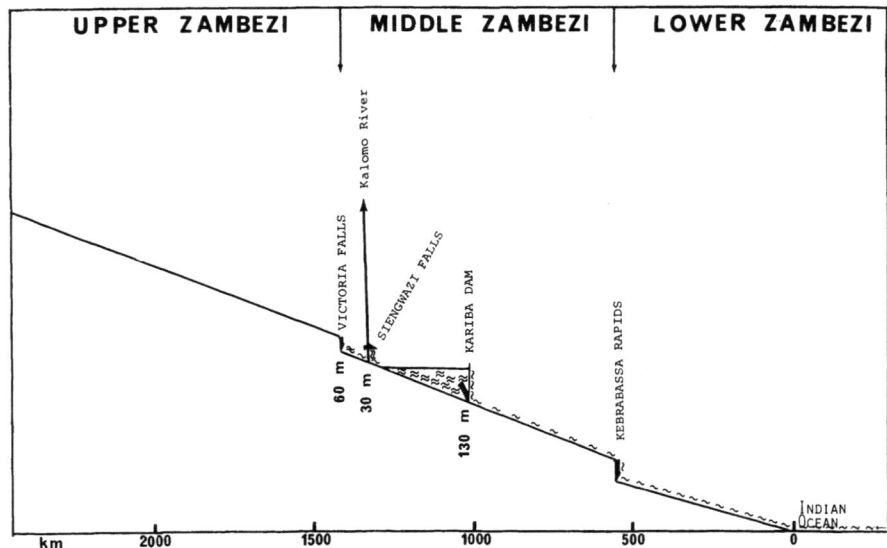

Fig. 66. Eel *Anguilla nebulosa labiata* juveniles on their way up the Zambezi River to surmount the Kebrabassa Rapids and the 130 m high concrete wall of Kariba Dam. However, when they reach Siengwazi and Victoria Falls their increased size, probably over 17 cm, diminishes their climbing ability, and the 30 and 60 m high barriers become insurmountable for them.

Most of the ubiquitous species (group a) are small. Only *L. cylindricus* and *C. gariepinus* reach a length over 20 cm, the latter species, although the largest, is rare. The plateau and escarpment group (b) consists entirely of small species rarely over 15 cm in length. Included in that group are three species, *T. sparrmani*, *S. andersoni* and *S. robustus*, which are able to grow to a much larger size in other habitats. The plateau group (c) also is represented by small animals only. However, the gorge group (d) consists overwhelmingly of large fishes.

Only a few sections of the river dry up completely, but the intermittent character of the river over much of its length and the enormous reduction of water volume is probably responsible for the preponderance of small species. The scarcity of large specimens of species known to grow to a large size could also be attributed to fishing; the shallow discontinuous habitats would be extremely vulnerable to overfishing in the dry season. Evidence of fishing was observed even in such a remote place as Siengwazi Fall. This will be discussed in another part of this section.

The use of geographical terms for the designation of river zones is consistent with my previous concept (BALON, 1959, 1964a, 1964b, 1967b; MIŠIK, 1959; HOLČIK and MIŠIK, 1962), since HUET's designation based on the dominant fish proved to be irrelevant (SOMMANI, 1953; MÜLLER, 1954). Furthermore, the Kalomo River can be considered as another example that general application of the gradient rule is not feasible for an extended area. HUET's (1959) statement that the gradient rule

'... has been found to be correct in most Western European streams' conflicts with reality and his attempt to incorporate increasingly wider areas does not seem reasonable (HUET et al., 1969). It may be possible to apply this rule in a small area of a single drainage and most importantly between two rivers of the same evolutionary history.

The Kalomo River has reverse longitudinal and cross section profiles as opposed to most streams modeled up to now (LELEK, 1968; MALAISSE, 1969a). The flat plateau stream which meanders in a wide circular groove-like dambo has the lowest gradient (0.9 to 10‰) of the upper part of the river. Farther downstream, the stream gradually carves a steep-banked bed between hillocks, causing the gradient to change frequently over the range from 3 to 19‰. It also forms an increasing number of rapids and rock outcrops. Finally the river flows into steep, narrow gorges where the bottom of the river is 300 m from the top of the surrounding hilltops. This paradoxical stream gradient, however, has not an exclusive effect on species distribution, since none of the group, characteristic of the gorge (d, except *C. neumanni*) is a noted inhabitant of steep slopes and high water velocity. They inhabit the Middle and Lower Zambezi, and half of the species in the group inhabit Lake Kariba as well.

The reverse stream gradient of the Kalomo River, however, has some relation to the distribution of cichlids since most of the species known in the river (except *P. philander*) were missing from the lowest part of the lowveld (Zone 3), which has the steepest gradient. There is a similarity in fish distribution in Ebo Stream (Ghana), which has a normal gradient, where cichlids are limited to the lowest river section with the lowest gradient (LELEK, 1968). Regardless of the location of the source, the gradient of the stream probably acts as a factor limiting fish distribution. Stream history, however, seems to be of equal importance, especially ancient drainage connections, since the origin of taxa seems to determine their recent existence in a given section of the river. In a regular gradient situation, for example, the distribution would most probably have been similar to that of the Luanza River (MALAISSE, 1968), although only in the higher taxa (at the generic and family level).

With the exceptions of the reservoir (locality 1) and the Songwe River, all fish samples represent sections of the Kalomo River 100 m in length. No estimates of area were made since the width of the river was highly variable within these short sections. The values for abundance and standing crop are therefore calculated for 100 m of shore line. The total values represent multiples of a given distance within determined zones. Due to the small number of samples the confidence limits must be very wide. Nonetheless, there are no other data available at present and it seems that their utilization for quantitative evaluation might be of some interest, provided their weaknesses and biases are borne in mind. I am certainly aware that the conclusions drawn in the following sections have very little supporting data, but if they should inspire those who remain unconvinced to present additional evidence to disprove my views my purpose will have been served in that more will be learned about a totally neglected area.

The relative size of samples (number of fish) in sampled sections reflects, however, also the water volume in a given locality at the time of sampling (Fig. 67). The abundance of any species (Table 121) is irregularly variable along the river with almost no

Table 121. Abundance, in radius index values (πr^2), of individual species from localities of the Kalomo River.

Species	Locality							
	1	2	3	4	5	6	7	8
Tilapia sparrmani	2	15	23	16	16	4	26	—
Barbus paludinosus	7	8	24	13	12	15	14	1
Barbus poechii	1	8	10	10	27	4	13	—
Petersius rhodesiensis	—	3	13	15	8	—	4	—
Labeo cylindricus	3	6	8	7	13	—	5	8
Barbus radiatus	5	4	2	10	5	3	13	1
Pseudocrenilabrus philander	—	5	14	10	6	1	3	1
Sarotherodon macrochir $\lessgtr$ andersoni	—	—	—	—	—	—	15	—
Barbus barotseensis	—	2	9	3	3	3	7	—
Sarotherodon andersoni	—	6	3	5	10	2	—	—
Serranochromis robustus	1	3	7	5	3	—	4	—
Barbus annectens	—	7	3	—	—	—	8	—
Barbus viviparus	—	3	1	2	—	5	—	—
Clarias theodorae	—	1	5	2	1	2	—	—
Amphilius platychir	—	—	1	1	—	—	—	3
Clarias ngamensis	—	1	1	1	—	—	—	—
Clarias gariepinus	1	—	—	—	—	—	1	1
Barbus marequensis	—	—	—	—	—	—	—	7
Barilius zambezensis	—	—	—	—	—	—	—	6
Chiloglanis neumanni	—	—	—	—	—	—	—	4
Anguilla nebulosa	—	—	—	—	—	—	—	3
Mormyrus longirostris	—	—	—	—	—	—	—	2
Hydrocynus vittatus	—	—	—	—	—	—	—	1
Total πr^2	11	28	49	38	44	20	46	16

general pattern, though cichlids tend to be more abundant at tributary mouths and cyprinids and catfishes in the river itself.

The eel, *A. nebulosa labiata*, had the highest standing crop (Table 122) though it was present in one locality only. It was closely followed by *L. cylindricus* and *T. sparrmani*. In the reservoir (locality 1) *C. gariepinus* formed the highest percentage (36%) of the standing crop, followed by *B. paludinosus* (20%), *L. cylindricus* (17%), *S. robustus* (10%), *T. sparrmani* (8%), *B. radiatus* (8%), and *B. poechii* (0.6%). At the second locality *T. sparrmani* formed 51%, *S. andersoni* 14%, *S. robustus* 9% and *L. cylindricus* 7% of the standing crop. At locality 3, in the Sichikwenkwe Stream, the highest standing crop (22%) was represented by *B. paludinosus*, and *T. sparrmani* contributed 19%. At locality 4 *T. sparrmani* was highest at 32% and at locality 5 *B. paludinosus* with 52%. As with abundance estimates, there was not a clear pattern of increase or decrease in standing crop along the river. This may be caused by an equal volume of water in different localities within the river bed in the dry seasons,

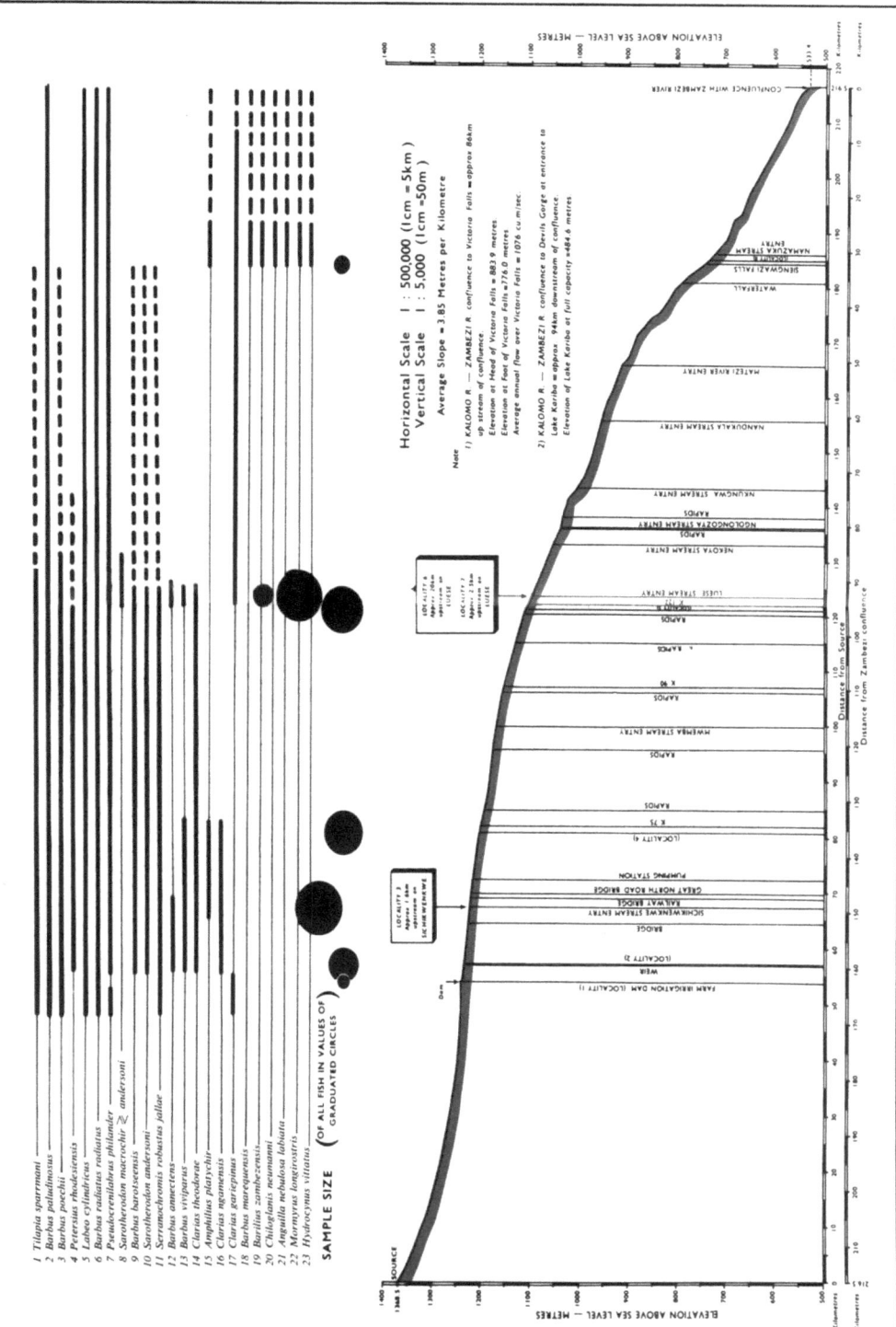

Fig. 67. Elevation diagram of the Kalomo River with approximate distribution of individual fish species (solid and broken bars) and size of samples.

and by a random distribution of rapids, shallows and pools within sampled localities or in adjacent sections of the river.

The total abundance and standing crop for the whole river were calculated from the average relative abundance and standing crop of those 100 m sections of stream (Table 123) within approximately established limits of the distribution of individual species. The greatest abundance at a single locality was reached by *Barbus poechii* – 313 specimens, by *T. sparrmani* 303 and *B. paludinosus* 264. However, *T. sparrmani* is more abundant over all localities (154 specimens per 100 m) than *B. paludinosus* (98), *B. poechii* (94), *P. rhodesiensis* (53), or *L. cylindricus* (39), etc. The highest standing crop per 100 m of the most widely distributed species was that of *T. sparrmani* (498 g); also high were *L. cylindricus* (480 g), *B. poechii* (332 g), followed by *S. andersoni* (130 g), *B. paludinosus* (125 g), *S. robustus* (105 g) and others.

The most abundant fish in the whole river were *T. sparrmani* (283,000 individuals), *B. paludinosus* (213,000), *B. poechii* (174,000) and the suspected *Sarotherodon* hybrid (137,000). For the whole river the total number was over 1,300,000 individuals. In weight (calculated from samples) the eel surpasses other species with a standing crop of 1,074 kg, followed by *L. cylindricus* (1,040 kg) and *T. sparrmani* (916 kg). Six metric tons is the estimate for all fishes present in all zones, with 2,644 kg of fish in zone 1 (highveld plateau), 440 kg in zone 2, and 2,928 kg in zone 3 (lowveld).

Zone 1 of the Kalomo River thus contained, within stated confidence limits, about 21 kg/km of fish, zone 2 approximately 7 kg/km and zone 3 as high as 91 kg/km. Compared with an increasing standing crop of 1.3, 26.1 and 31.7 kg/km in the successive downstream zones of the Luanza River (MALAISSE, 1968), the standing crop of fish in the Kalomo River was relatively high, though the recorded harvest of 103 to 447 kg/km in Mwekera Stream of the Kafue River drainage indicates much higher standing crops in rivers of this region (MORTIMER, 1965). On the other hand, standing crop values for the Kalomo River were similar to some English streams (MILLS, 1970).

On the whole, piscivorous species were not abundant in the Kalomo River. Six species can be designated as predators when adult: *S. robustus*, *Clarias* spp., *A. nebulosa*, and *H. vittatus*. Two of these were limited exclusively to the lowveld zone, the others are rare or small in size. With the two exceptions of zone 3, the piscivorous fishes in the Kalomo River formed less than 25% of the standing crop and a few percent of the abundance (Table 123).

According to my estimation of the variables in fish production in Lake Kariba, the standing crop is much lower than the mean biomass which incorporates corrections for sampling bias and mortality. I have no means, however, to follow the proper procedure for the Kalomo River and can only approximate the mean biomass and turnover ratio from comparison of the rough standing crop values with Lake Kariba. A minimum figure for the turnover ratio could be 0.5 to 1.0 and for the mean biomass at least 10 metric tons. Consequently the Kalomo River may produce about 10,000 kg of fish annually along its entire length.

Local people and the fish resource. – From earliest times population density was dependent on climatic conditions. Changes of climate in the area of the Kalomo River, from wet (pluvial) to arid (non-pluvial) phases, determined river topography and human culture. The first near-men (Australopithecines) inhabited the area in the

Table 122. Estimated standing crop in wet weight (g, above) and percentage of the sample's total weight (below) of individual species in sampled localities of the Kalomo River.

Species	\multicolumn{8}{c	}{Locality}	Total wet weight
	1	2	3	4	5	6	7	8	
Tilapia sparrmani	205	1217	325	641	286	28	489	—	3191
	8	51	19	32	12	13	34		
Barbus paludinosus	484	48	386	142	140	96	56	8	1360
	20	2	22	7	6	47	4	0.1	
Barbus poechii	15	81	192	210	1293	24	193	—	2008
	0.6	3	11	10	52	11	13		
Petersius rhodesiensis	—	7	9.3	98	44	—	12	—	254
		0.3	5	5	2		1		
Labeo cylindricus	419	169	175	382	325	—	72	1755	3297
	17	7	10	19	13		5	19	
Barbus radiatus	200	23	7	86	25	6	45	4	396
	8	1	0.4	4	1	3	3	0.05	
Pseudocrenilabrus philander	—	81	194	124	35	2	8	1	445
		3	11	6	1	1	0.6	0.005	
Sarotherodon macrochir $\lessgtr$ andersoni	—	—	—	—	—	—	341	—	431
							30		
Barbus barotseensis	—	4	40	4	9	3	9	—	69
		0.2	2	0.2	0.4	1	0.6		
Sarotherodon andersoni	—	336	22	56	226	11	—	—	651
		14	1	3	9	5			
Serranochromis robustus	252	226	74	72	68	—	87	—	779
	10	9	4	4	3		6		
Barbus annectens	—	33	4	—	—	—	13	—	50
		1	0.2				1		
Barbus viviparus	—	5	2	2	—	9	—	—	18
		0.2	0.1	0.1		4			
Clarias theodorae	—	59	124	84	11	27	—	—	305
		2	7	4	0.4	13			
Amphilius platychir	—	—	17	20	—	—	—	44	81
			1	1				0.5	
Clarias ngamensis	—	80	64	93	—	—	—	—	237
		3	4	5					
Clarias gariepinus	885	—	—	—	—	—	32	4	921
	36						2	0.04	
Barbus marequensis	—	—	—	—	—	—	—	2799	2799
								31	
Barilius zambezensis	—	—	—	—	—	—	—	11	11
								0.1	
Chiloglanis neumanni	—	—	—	—	—	—	—	10	10
								0.1	
Anquilla nebulosa	—	—	—	—	—	—	—	3334	3334
								37	
Mormyrus longirostris	—	—	—	—	—	—	—	756	756
								8	
Hydrocynus vittatus	—	—	—	—	—	—	—	368	368
								4	
Total wet weight for localities	2460	2368	1720	2015	2461	205	1446	9095	21771

Table 123. Abundance and standing crop of Kalomo River fishes by 100 m sections (A) and for whole portion of the stream inhabited by that species (B).

Species	Number of localities in which species occurred (of 7)	Abundance			Standing Crop				
		A (in N)		B (in N)	A (in g)		B (in kg)		
		range	x̄		range	x̄			
Barbus paludinosus	7	1–264	98	213,393	8– 386	125	272	ubiquitous fish group	216.8 km
Labeo cylindricus	6	20– 87	39	84,552	72–1755	480	1,040		
Barbus radiatus	7	1– 92	27	58,536	4– 86	28	61		
Pseudocrenilabrus philander	7	1– 98	29	63,491	1– 194	63	138		
Amphilius platychir	3	2– 7	4	7,949	17– 44	27	58		
Clarias gariepinus	2	1	1	4,336	4– 32	18	38		
Tilapia sparrmani	6	11–303	154	283,822	28–1217	498	916	plateau and escarp-	
Barbus poechii	6	11–313	94	174,163	24–1293	332	612	ment fish group	184.3 km
Petersius rhodensiensis	5	9–115	53	97,679	7– 98	51	93		
Barbus barotseensis	6	5– 46	18	33,174	3– 40	11	21		
Sarotherodon andersoni	5	3– 58	21	39,440	11– 336	130	240		
Serranochromis robustus	5	7– 31	15	28,382	68– 226	105	194		
Barbus annectens	3	8– 36	26	32,110	4– 33	17	20	plateau fish group	123.5 km
Barbus viviparus	4	2– 17	7	9,262	2– 9	4	5		
Clarias theodorae	5	1– 20	6	7,410	11– 124	61	75		
Clarias ngamensis	3	2	2	2,470	64– 93	79	98		
S. macrochir ≤ andersoni	1	111	111	137,085	431	431	532		
Barbus marequensis	1	30	30	9,660	2799	2799	901	gorge fish group	32.2 km
Barilius zambezensis	1	27	27	8,694	11	11	4		
Chiloglanis neumanni	1	12	12	3,864	10	10	3		
Anguilla nebulosa	1	7	7	2,254	3334	3334	1,074		
Mormyrus longirostris	1	4	4	1,288	756	756	243		
Hydrocynus vittatus	1	1	1	322	368	368	380		

Middle Pleistocene, some 500,000 years ago (CLARK, 1950, 1960, 1964). That fish were caught and consumed in these early stages of human history is indicated by rare remains of *Barbus* and *Protopterus* as well as by remnants of stone weirs (POSNANSKY, 1959 after CLARK, 1960). The climate of the early and middle Stone Age was warm and wet. As it changed and the area gradually became more and more arid the human population moved, probably northwards. By that time Victoria Falls had retreated through the Karoo basalt lavas nearly to the position of the present Songwe Gorge. This happened some 410,000 years after the main connection of the Upper and Middle Zambezi River was formed (see p. 486).

About 9,000 years ago, and after another dry and wet climatic change, the middle Stone Age hominids were replaced by the later Stone Age people. In the Kalomo River region, the bushmen-like hunters and fishermen were joined by the early nomad-farmers from the north and these two formed the various Iron Age tribes (CLARK & FAGAN, 1964). Yet another dry period decreased the size of the population, but a subsequent wet phase brought new migrants into this area (VOGEL, 1970).

From Stone and Iron Age excavations, it is evident that the highveld plateau was always more populated than were the escarpment and lowveld areas below. The population was certainly very small and consisted even in historical times of a few groups of Iron Age (Northern Bantu) farmers and hunter-gatherers, who were possibly ancestors of the Leya, Toka and Tonga tribes, and of scattered families of Stone Age Bushman hunters and gatherers. The latter were gradually absorbed or pushed southward. The people of the Kalomo Culture probably increased at some time in relation to favourable climatic conditions, but their density fluctuated as a consequence of contact with neighboring warrior tribes. Most of these changes occurred at about 1000, 1300 and 1650 A.D. (FAGAN, 1963; DAVIDSON, 1967; VOGEL, 1970; MUNTEMBA, 1970). During the best climatic conditions no more than several thousand people lived along the Kalomo River *(sensu stricto)*. The best estimate of the size of the present population along the river would be about two thousand, while in the whole Kalomo drainage area it may reach ten thousand.

The present system of the Kalomo River had formed by the time the first Australopithecines appeared (evolved?) in the area. The gorges and falls, however were still being modified by erosion when modern man arrived. Owing to the low density of the population at all times, probably no significant changes in fish species composition resulted from fishing.

The subsistence fishery consisted mainly of weirs, basket and reed traps, and to a lesser extent of spears and scoop net gear. It has probably changed very little with the ages. It is likely that occasionally an outlet was dammed and a suitable stream branch dried out. The dammed area, usually close to a village would become basket fishing grounds for gathering women, whereas specialized weir-trap and spear fishing was performed by hunting men at more distant locations. If the interpretation of archeological artifacts found in the area is correct, present day hunting and fishing remain very much the same as in the distant past. Some accommodation to modern times was made, however, and thorn spear heads were replaced by barbed iron heads. I found net sinkers of burned clay with remains of nets of recent origin at the foot of Siengwazi Fall, a point on the Kalomo River quite remote from any present day permanent settlement. I believe that I have sufficiently documented the fishing activities of that area as indicated by fishing spears, different types of scoop nets found in villages along the river, reed traps on the river banks (in my private collection), and remnants of stone weirs. In general it is safe to say that the present day Kalomo people use fishing gear similar to that of the Tonga peoples, as described by REYNOLDS (1968).

Of recent origin, however, is the rod and barbed hook frequently used by local women and children at locality 2. It was probably introduced by European farmers. Although home-made hooks were used before the advent of European colonization (LIVINGSTONE & LIVINGSTONE, 1865), those hooks were of copper or iron wire and usually without barbs. There is no evidence of wooden or bone spears and hooks, but they were certainly in use.

The local people are not selective about the size of fish, though some species are favoured more than others. For example, among the few Leya-Toka people who go on fishing expeditions as far as Siengwazi Fall and gorge region, eels are feared and considered inedible and dangerous. Fish are most accessible in the dry season at which time they are an important source of food. Children fish from an early age and grill their catch as a type of special meal. Since preference at meals is accorded to adult men, fish protein is of fundamental importance for the children. Fishes certainly play a most important role in the diet of the people along the Kalomo River though overhunting of other sources of protein indicates that the available fish cannot satisfy total protein needs.

Approximately ten metric tons of fish may be available annually throughout the Kalomo River, which means around 5 kg per capita. This equals the average fish consumption in the world (BORGSTROM, 1962). The lack of large individuals in the river could be a reflection of constant fishing although those sections which are uninhabited or inaccessible should not be subject to overfishing. Furthermore, the fish population is adapted to the intermittent nature of some parts of the river and to surviving wide ranges of climatic stresses. No really preferred fish species exist in the river with the exception of the lowveld zone. Consequently extinction because of fishing pressures is at the moment improbable.

I did not encounter the use of fish poisons by indigenous tribes, however, many who watched my sampling procedure were clearly familiar with the effects of poison on fish. A government ban on the use of fish poisons may have been imposed in the early years and enforced by European farmers. However, even in remote villages I found no plants, such as *Tephrosia vogelii*, which contained ichthyotoxine (MALAISSE, 1969b), under cultivation. Wild shrubs were, however, common to the area. Many of them can be used and of course the possibility of limited use of poisons from *Acacia nigrescens, Ademium obesum, Euphorbia candelabrum, Tephrosia vogelii* and *Lonchocarpus capassa* should not be excluded.

Judging from species diversity, abundance, and standing crop as related to the size, gradients and historical factors of the river, the endemic fish fauna seemed to be in reasonably good condition at the time I sampled it.

A fish population adapted to variable climatic conditions is more than likely vulnerable to any chemical pollutants which concentrate heavily during the limited flow in the dry season. The rapidly growing use of fertilizers and insecticides may prove to have a disastrous effect on the fishes. This effect will be increased by the diminished or interrupted flow of water between irrigation reservoirs (NEEL, 1953). Facts have rarely substantiated MAAR's (1958) statement that 'the existing fish stock should only benefit from an improved flow in the river' when reservoirs are created by dams. The 'beneficial' effect of reservoirs in increasing fish production in those rivers which are considered of 'very small fish potential in comparison with dams' (MAAR, 1958) should be carefully evaluated, taking into consideration a complete energy budget of the aquatic, and adjacent terrestrial, environments (ODUM, 1971). The example of the overrated fish potential of Lake Kariba (MAAR, 1958) as it relates to the sad facts should warn all future reservoir-happy decision makers.

*

This is undoubtedly only a hasty glance at one very limited component of the environment of the Kalomo River. Nevertheless, however limited my study, there are several factors that seem to emerge quite clearly. Some insights may have a general implication.

The species composition of the river is threatened by the stocking of farm reservoirs and by the presence of an introduced strain. Of equal and equally immediate danger could be the sewage released by the expanding Kalomo township. As fish are excellent indicators of water purity, which is so important to man and animals, a decrease in their number indicates that more is at stake than a decrease in the availability of fish protein.

It is evident that the fish fauna of the Kalomo River has been developing for one million years and for a half a million of those years has fed migrating tribes which settled along the riverbanks. Now, however, the age-long harmony seems in jeopardy. As a consequence of recently introduced technology, the number of people in this area will undoubtedly increase, as it has everywhere else in developing countries. What has, for thousands of years, helped very effectively to sustain a hunting-gathering population is suddenly on the verge of destruction.

Here it may be useful to mention the fate of another people '... The shift from a mobile hunter-gatherer way of life to a sedentary farming way of life made irrelevant the Ik's entire repertoire of beliefs, habits and traditions. Their guidelines for life were inappropriate to farming. They seemed to adapt, but at heart they remained hunters and gatherers. Their cultural templates fitted them for that one way of life. {...} They were suddenly crowded together at a density, intimacy and frequency of contact far greater than they had ever before been required to experience. {...} The Ik failed to remain human. I have put mice to the same test and they failed to remain mice' (CALHOUN, 1972).

As of the moment it is difficult to predict whether a similar fate awaits the people of the Kalomo region but the example of the Iks should serve as a warning.

Will man and fish be permitted to exist in a modicum of harmony?

6.2. *The Elephant Stream*

Chete Island, south-east of Sinazongwe, is the largest island created by the rising waters of the Zambezi River. It is 10 km long and 4.5 km wide; from the Rhodesian

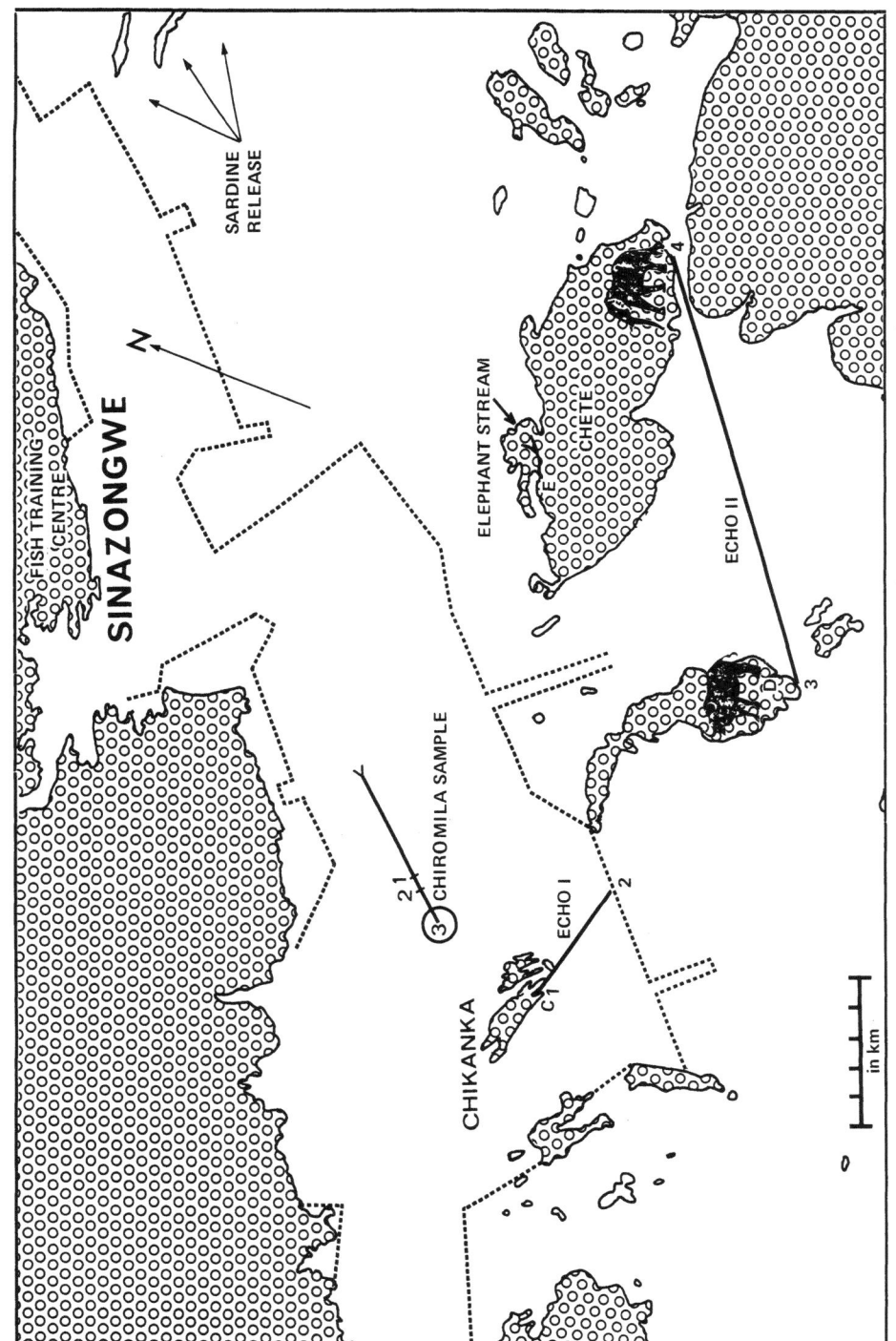

Fig. 68. A schematic plan illustrating the sites of action in this chapter and following chapter 7. For explanations see text.

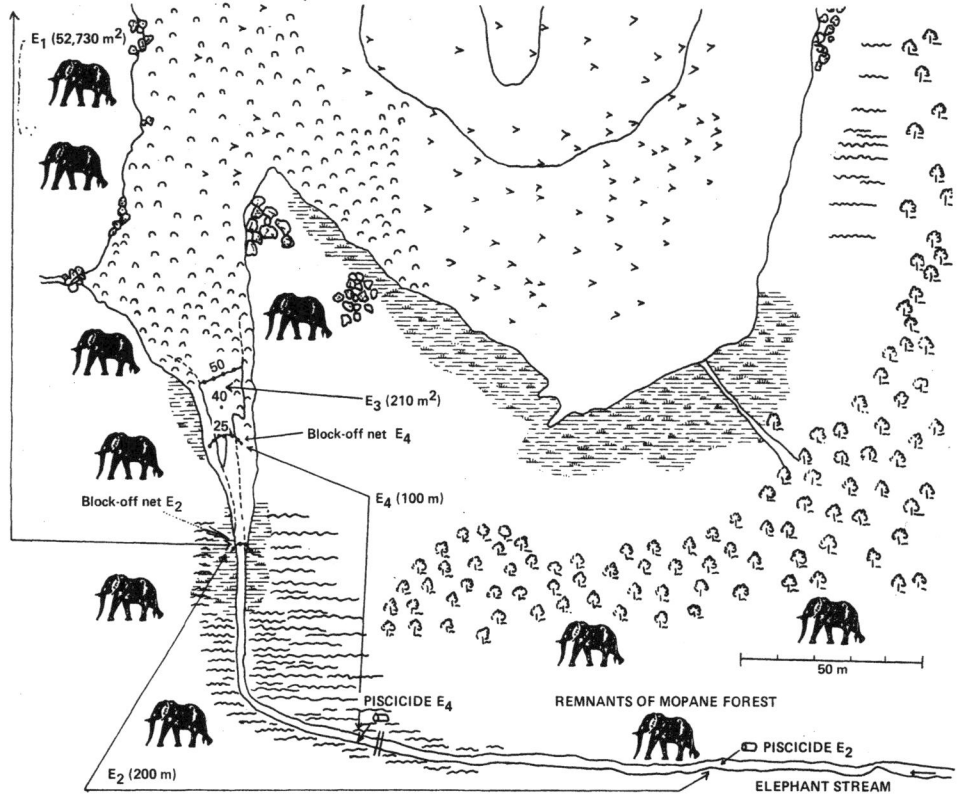

Fig. 69. End of the north cove at Chete Island and lower part of the Elephant Stream with sections treated in samples E_2, E_3 and E_4.

shore Chete is separated by the old Zambezi River bed – a narrow gorge 300 m wide and about 2 km long; from the Zambian shore at Sinazongwe it is separated by 11 km of water.

I have always been fascinated by this island with its baobab forest and varied wild life. It has the largest elephant population ever seen in such a confined area. Although the elephants play havoc with the bark of the trees, stripping it to the core for their daily repasts, the trees bloom every year and positively defy destruction. The coves offer shelter to crocodiles. Impalas and warthogs roam the island in relative tranquility and the ubiquitous cries of the fish eagle give the island a somewhat eery atmosphere.

Chete's southern cove was where I cast anchor on my frequent overnight trips to the island; the northern cove was quantitatively sampled and it yielded a most interesting fish composition. Into this cove enters an intermittent stream of surface run-off water, maximally 1300 m long after the rains (Fig. 68). After several weeks of dry weather all that remains wet of the shallow valley floor through which the stream runs in the rainy season is an occasional swampy grass patch and a coved stream entry into the lake. This stream entry has a large number of submerged bushes and trees. Remnants of mopane trees nibbled off by elephants to their broom-like trunks border both sides of the stream. This nameless stream I shall henceforth call the Elephant Stream.

Three quantitative fish samples were made in the Elephant Stream, in addition to the initial large cove sample E_1. At termination of sampling E_1 (25 March 1969) 200 m of the blocked-off Elephant Stream were sampled (E_2); this represents conditions of high water (peak of the rainy season). Two months later (25 May 1969) 210 m^2 of blocked-off stream entry were sampled (E_3); this represents conditions in the dry season. The following year, after several rainless days (7 February 1970) a blocked-off area of 100 m of average water level in the Elephant Stream yielded the last quantitative sample (E_4). The width of the stream in the section sampled was 0.5 to 2 m (Fig. 69). The stream was very shallow (maximum depth 0.5 m) with small rapids 130 m above its entry into the cove. For density estimation rounded area values of E_1 52,730, E_2 200, E_3 210 and E_4 100 m^2 were used.

The stream was inhabited by 11 fish species at most, whereas 27 species were found in the adjacent lake cove (Table 124). Only two species, *Barbus poechii* and *B. paludinosus* were present exclusively in the stream. The latter of the two was probably a survivor of the pre-impoundment situation. After the initial samples *B. paludinosus* was not found again in the Elephant Stream; it may thus be inferred that this species cannot survive in lacustrine conditions, while *B. poechii* can, as it appeared again in the third stream sample.

Only juveniles of preferred or secondary species, or adults of small accompanying species were present in the stream, most of them not exceeding 50 mm in length, although some individuals of *Clarias* were 250 mm long, *Tilapia* and *Sarotherodon* over 100 mm (Table 124). With the exception of *B. paludinosus* no species was exterminated and two months later there were already 9 species present. The density and biomass was 46 thousand and 324 kg per hectare respectively. It might be considered as evidence of a speedy recovery after chemical treatment and of a quick reoccupation of an artificially freed litoral niche. The initial stream sample yielded 12 thousand specimens per hectare and 26 kg per hectare. The last stream sample yielded 21 thousand specimens per hectare and 65 kg per hectare (Table 125). The most abundant fish were *Sarotherodon mossambicus* juveniles. This makes the Elephant Stream either a nursery ground for this species or a density or predator dependent escape habitat. *Tilapia rendalli*, although present, was for its much lower density in the stream only a moderate user of this habitat, in spite of its similar density and standing crop to the *S. mossambicus* in the adjacent cove. *Alestes lateralis* and *Haplochromis darlingi* were abundant only in the lacustrine habitat of the stream entry (E_3). *Clarias gariepinus* might be considered as the second important user of the stream itself.

In general the standing crop of 26 to 65 kg/ha found in the stream is much lower than the cove's standing crop of 320 kg/ha. But the opposite values in fish abundance indicate the importance of such a small and intermittent stream for the survival of juveniles of endemic species.

Table 124. Chete Island cove and the Elephant Stream species composition and sizes.

	Average length of 1 fish					Standard length ranges			
	E_1	E_2	E_3	E_4	E_1	E_2	E_3	E_4	
Heterobranchus longifilis	788	—	—	—	105–1170	—	—	—	
Clarias gariepinus	499	177	75	135	85– 780	141–245	49–128	34–255	
Mormyrops deliciosus	478	—	—	—	299–1005	—	—	—	
Malapterurus electricus	417	—	—	—	246– 540	—	—	—	
Mormyrus longirostris	397	—	—	—	397	—	—	—	
Labeo altivelis	189	—	—	58	70– 325	—	—	52– 64	
Schilbe mystus	180	—	—	—	155– 221	—	—	—	
Hydrocynus vittatus	158	—	59	39	30– 610	—	40– 81	39	
Synodontis zambezensis	158	—	—	—	65– 310	—	—	—	
Marcusenius macrolepidotus	158	—	—	—	109– 250	—	—	—	
Distichodus schenga	155	—	—	—	112– 263	—	—	—	
Eutropius depressirostris	152	—	—	—	100– 260	—	—	—	
Hippopotamyrus discorhynchus	147	—	—	—	50– 355	—	—	—	
Sargochromis codringtoni	140	—	31	—	35– 245	—	25– 36	—	
Tilapia rendalli	138	44	41	31	27– 455	28– 60	18–130	26– 47	
Sarotherodon mossambicus mortimeri	120	37	43	42	20– 565	22– 65	17–195	11–105	
Synodontis nebulosus	116	—	—	—	70– 175	—	—	—	
Brachyalestes imberi	113	—	—	28	60– 155	—	—	27– 30	
Labeo cylindricus	100	—	—	—	100	—	—	—	
Barbus unitaeniatus	45	52	49	51	25– 87	50– 56	40– 57	46– 58	
Micralestes acutidens	43	—	—	30	25– 52	—	—	30	
Haplochromis darlingi	40	33	48	—	18– 73	25– 42	35– 60	—	
Alestes lateralis	39	—	47	—	15– 87	—	33– 55	—	
Pseudocrenilabrus philander	35	37	41	34	20– 65	30– 42	36– 46	30– 40	
Barbus lineomaculatus	34	35	—	41	34	35	—	41	
Barbus fasciolatus	29	—	—	—	18– 54	—	—	—	
Aplocheilichthys johnstoni	25	—	—	77	25	—	—	—	
Barbus poechii	—	82	—	—	—	75– 94	—	74– 79	
Barbus paludinosus	—	50	—	—	—	50	—	—	
Number of species	27	9	9	11					

Table 125. Chete Island cove and the Elephant Stream assessments of fish abundance and standing crop.

	Number of fish per 1 ha				Average weight of 1 spec. in g				Weights in kg per 1 ha			
	E_1	E_2	E_3	E_4	E_1	E_2	E_3	E_4	E_1	E_2	E_3	E_4
Heterobranchus longifilis	2	—	—	—	9155	—	—	—	19.10	—	—	—
Clarias gariepinus	37	550	4000	700	1593	61	5.4	28	58.62	3.35	21.71	1.96
Mormyrops deliciosus	11	—	—	—	1166	—	—	—	12.60	—	—	—
Malapterurus electricus	8	—	—	—	1717	—	—	—	13.67	—	—	—
Mormyrus longirostris	0.2	—	—	—	850	—	—	—	0.16	—	—	—
Labeo altivelis	13	—	—	500	242	—	—	5.2	3.25	—	—	2.60
Schilbe mystus	0.6	—	—	—	64	—	—	—	0.04	—	—	—
Hydrocynus vittatus	153	—	714	100	226	—	3.8	0.1	33.18	—	2.71	0.01
Synodontis zambezensis	50	—	—	—	122	—	—	—	6.11	—	—	—
Marcusenius macrolepidotus	29	—	—	—	70	—	—	—	2.05	—	—	—
Distichodus schenga	4	—	—	—	161	—	—	—	0.67	—	—	—
Eutropius depressirostris	10	—	—	—	55	—	—	—	0.57	—	—	—
Hippopotamyrus discorhynchus	642	—	—	—	71	—	—	—	45.72	—	—	—
Sargochromis codringtoni	41	—	1905	—	146	—	0.9	—	6.01	—	1.67	—
Tilapia rendalli	349	100	1095	600	196	3.7	11.4	1.2	68.36	0.37	12.48	0.72
Sarotherodon mossambicus mortimeri	434	9250	18571	16900	124	2.1	12.5	3.2	56.24	19.42	232.14	54.08
Synodontis nebulosus	5	—	—	—	37	—	—	—	0.19	—	—	—
Brachyalestes imberi	4	—	—	300	45	—	—	0.4	0.20	—	—	0.12
Labeo cylindricus	0.2	—	—	—	25	—	—	—	0.01	—	—	—
Barbus unitaeniatus	270	150	1857	300	2.04	3.0	2.5	2.8	0.55	0.04	4.62	0.84
Micralestes acutidens	2	—	—	100	137	—	—	0.4	0.004	—	—	0.04
Haplochromis darlingi	428	1150	6286	—	2.4	1.1	3.3	—	1.03	1.26	21.00	—
Alestes lateralis	5426	—	11238	—	1.3	—	2.4	—	7.09	—	27.09	—
Pseudocrenilabrus philander	87	450	381	1300	1.5	1.5	2.1	1.2	0.13	0.67	0.81	1.56
Barbus lineomaculatus	0.4	50	—	100	7.5	0.8	—	1.4	0.003	0.04	—	0.14
Barbus fasciolatus	71	—	—	—	0.7	—	—	—	0.05	—	—	—
Aplocheilichthys johnstoni	0.2	—	—	—	0.3	—	—	—	0.0005	—	—	—
Barbus poechii	—	700	—	300	—	13	—	10.5	—	0.91	—	3.15
Barbus paludinosus	—	50	—	—	—	2.6	—	—	—	0.13	—	—
Total	8678	12450	46047	21200	41.53	9.9	7.0	4.94	335.60	26.19	324.23	65.22

6.3. The relationship of Lake Kariba to the Victoria Falls

The entire problem of the sudden replacement of *Brachyalestes imberi* by *Alestes lateralis* in Lake Kariba and the fish invasion via Victoria Falls was covered in detail in separate studies (BALON, 1971b, 1974c) and is here added in abridged form.

The alestes case. – *Alestes lateralis* was considered until recently a fish appearing exclusively in the Upper Zambezi system. JACKSON (1961a) summarized its distribution: 'Occurrence of this species is limited to the Upper Zambezi and neighbouring waters on either side (e.g., Okavango and the type comes from Lake Dililo, Katanga), and the Kafue.' CRASS (1964), notes: 'It has not been found in the middle or lower Zambezi, except at one locality close to the sea, nor is it recorded from any part of Southern Rhodesia, the eastern Transvaal or Mozambique. One specimen is recorded from Hluhluwe Game Reserve, two from Elcheleselwane to the west of Lake St. Lucia and two from Umpangazi Lake on the eastern shores of Lake St. Lucia. (...) All three localities are within 30 miles of each other and are far removed from the lower Zambezi. A single specimen was collected in 'Natal' ...'. On the contrary, the waters of the Middle Zambezi were exclusively inhabited by *Brachyalestes imberi* as proved by JUBB's (1958) records who found 1 specimen in the Middle Zambezi at the confluence of Umsengadzi in 1950 and 12 specimens in the Lower Sabi River at the confluence with the Lundi River. JUBB (1961a) summarizes its occurrence as 'widely distributed in the Middle Zambezi, its tributaries and rivers south of it to the Pongola River system in Natal', and later (JUBB, 1967) adds that 'it is also found in the Congo basin'. JACKSON (1961a) wrote about the distribution of *Brachyalestes imberi* as follows: 'A widespread little fish of the Zambezi and Upper Congo system, also in Nyasaland and Southern Rhodesia, but absent from Upper Zambezi and Kafue, where its place is taken by *A. lateralis*. Some find it difficult to separate the two, but the differences are really quite well marked and use of the key should obviate any future difficulty in this respect. They are most abundant in our area in the Middle and Lower Zambezi and its tributaries such as the Luangwa, but are not uncommon, though not numerous, in the Rhodesian Congo. In the Kariba area they have always been one of the most numerous species present, but have taken advantage of the changed ecological conditions brought about by the closing of the dam to multiply to an enormous extent. Because of fear of predation by the tiger-fish (JACKSON, 1961c) only the very largest are found at any distance from shore, but large shoals of average-sized individuals are found near vegetation and in shallow water.'

In the part of the Middle Zambezi which was to disappear under the surface of the future lake, by using dynamite and seines, JACKSON found 644 specimens of *B. imberi*. This was in November of 1957 near Sinazongwe and it represented 75% of the total catches. Shortly before the water level rose in Lake Kariba, HARDING (1964) using experimental gillnets caught this species (less than 1%) in the Zambezi when the river was low and in backwaters; shortly after the closure of the dam in December 1958 he was catching in the same nets in the marginal waters around the lake as well as in the mouth of the Sanyati River 3% of *B. imberi* out of the total catch of all species, 3% in the Kariba Gorge, 1% in the Zambezi River below the dam and 3% in the river at Chirundu.

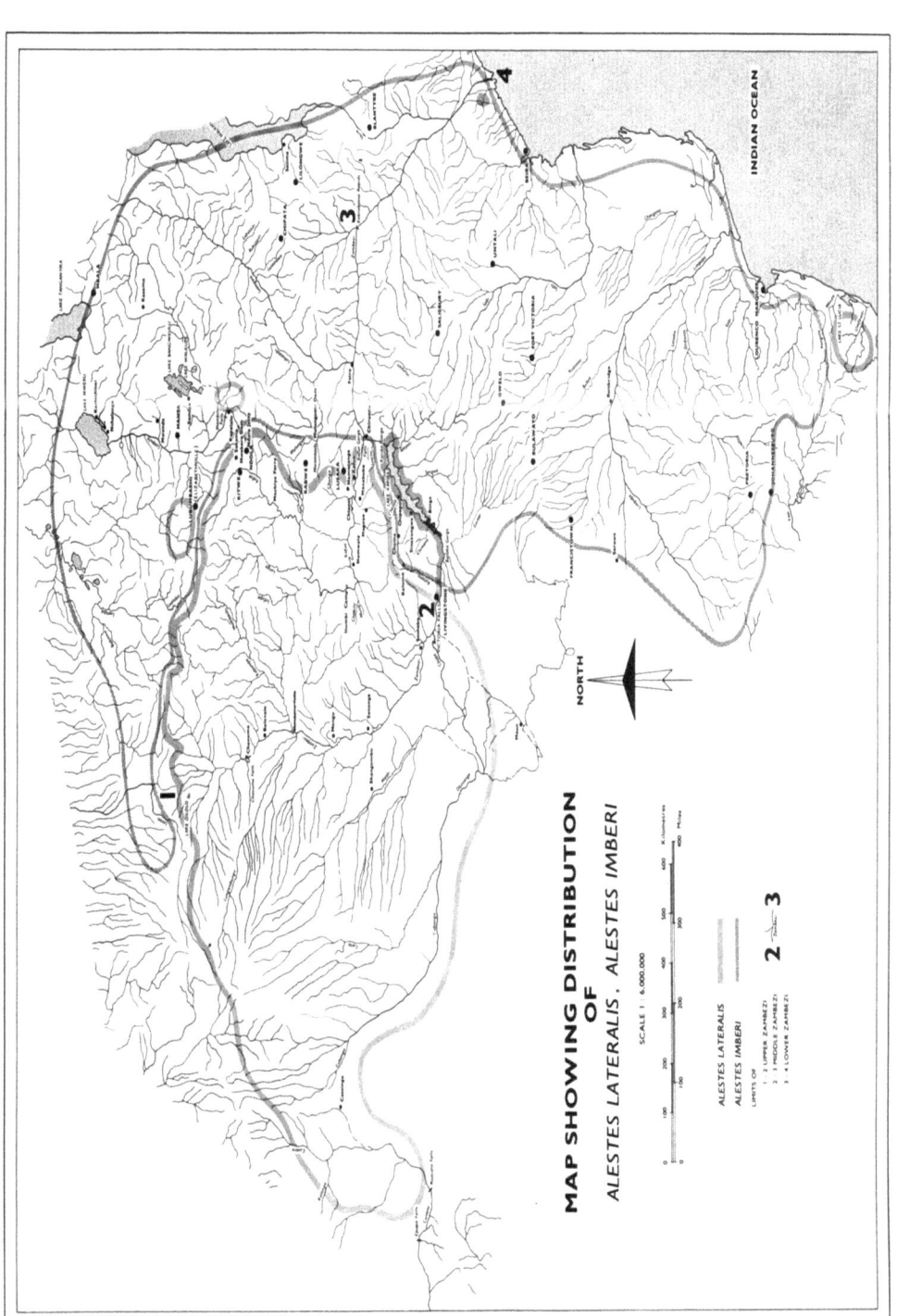

Fig. 70. Geographical distribution of *A. lateralis* and *B. imberi.*

In September 1960, after Kariba Dam had been closed for 20 months, HARDING (1964) recorded 32% of *B. imberi* out of the total number of fish caught, in experimental gillnets in the area of the Sanyati River, 18% in the Sibilobilo cleared area, 5% in the open lake, 22% in the Sinazongwe estuary, and 8% in the Chezia estuary. In January 1961 *B. imberi* represented 2% of the catches of the experimental gillnets placed in the area of the open lake, 1.2% in the cleared area of Sanyati West, 33.8% in the Sibilobilo River, 32.2% in the Chezia River, 27.7% in the Luzilikulu River, 17.2% in the Sinazongwe estuary, 53.2% in the Chimene estuary, 12% in the cleared area of Binga and 62.7% in the Sanyati Gorge. Even the landing statistics recorded in 1961 by professional fishermen contained 25 specimens of this fish in the Simamba landing area, 12 in Chipepo, 11 in Sinazongwe and 24 in Mwemba (SOULSBY, 1963). Similar statistics in 1962 mentioned 11 *B. imberi* at Simamba, 36 at Chipepo, 69 at Sinazongwe and 4 at Mwemba.

This would seem to indicate very strongly that until 1962 the waters of the Middle Zambezi were inhabited exclusively by *B. imberi* and that it was one of the most numerous fishes in the Zambezi in the area of the Gwenbe Valley and later in the newly formed reservoir. Prior to this *A. lateralis* had never been found there.

It is probable that the first specimen of *A. lateralis* was caught by HARDING in Lake Kariba in 1963 near the island north east of the Sibilobilo Narrows (HARDING, personal communication). This was accomplished with the assistance of a rifle. Unfortunately no evidence was preserved. In September 1964 BELL-CROSS' field notes reveal the following details: '7 specimens were caught at minute islands near Sibilobilo River in the south west corner of Island S117 of Lake Kariba. The length of measured specimens ranged from 55 to 70 mm; number of scales in lateral line 28–30; body depth 4.0–4.3; scale count from origin of dorsal fin to lateral line 5–5.5'. At that time the finding was a surprise as can be seen from a remark made by BELL-CROSS: '*Alestes* sp. Considered to be *A. lateralis* because of higher l.l. count, higher scale count between dorsal and lateral line, narrower depth. Lack of spot post to operculum'. During the same expedition BELL-CROSS caught only 4 *B. imberi* 0.5 mile from the Training Center at Sinazongwe and 32 *B. imberi* inside of Chimene River mouth, by surface gillnet sets (BELL-CROSS, 1968b). In studying the feeding habits of *Hydrocynus vittatus* in 1964 and 1965, MATTHES (1968a) observed large shoals of *A. lateralis* in the lake. In the stomach content of the tigerfish under study he found 5.3% of *A. lateralis*. He makes no mention at all of the presence of *B. imberi* and this is the reason for the assumption that at that time it had already been replaced by *A. lateralis*. In the course of my research at Lake Kariba (1968–1971) 10 quantitative samples (A, B, C, D, E, E_3, F, G, H, I) yielded 76,978 specimens of *A. lateralis*. These ranged in standard lengths from 10–122 mm. This species was absent only in one locality (I). The abundance ranged from 43 to 58,115 specimens per ha ($\bar{x}$ 13,400). *B. imberi* was present only in 7 of the same samples, in a size range of 30–195 mm, 161 specimens in total. The abundance range was 2–361 specimens per ha ($\bar{x}$ 58).

With regard to the specific conditions of geographical distribution (Fig. 70) and sudden replacement of the original species *(B. imberi)* by a new species *(A. lateralis)* in Lake Kariba, I have considered it necessary to verify at least some of the taxonomic characters of both species. Those which were selected should be sufficient to provide evidence of the taxonomic differences. But in no case did I intend to perform a taxonomic revision of the species.

In the original article, of which this is an abridged version, comparison was made of the observed values for meristic and morphometric characters with those mentioned by previous authors (eg. BOULENGER, 1909; GILCHRIST & THOMPSON, 1913; JACKSON, 1961a; JUBB, 1961a; 1967; CRASS, 1964) and it was found that they were generally in agreement. But in the majority of the cases I have found a greater range than had been previously mentioned, with the exception of BOULENGER's 3.5 scales above the lateral line and IV spine rays on dorsal fin. I have also confirmed the difference in the morphology of the scales of both species as observed by BOULENGER (1909); it can be observed in the anastomosing canals between the radial canals of the caudal part covered by epidermis of the scales of *B. imberi*.

Thus *A. lateralis* differs significantly in many characters and coloration from *B. imberi*. It is unlikely that it might be the ecotype or morpha (according to BERG's 1948 concept) of the same species. Consequently it was necessary to search for another explanation of the sudden replacement of the original species by another in Lake Kariba.

As has already been stated there are such significant differences in body structure between *A. lateralis* and *B. imberi* that the degree of differentiation of both species cannot be conspecific. As a matter of fact they both belong to different genera. The Kafue River, however, harbors mostly large specimens of *A. lateralis*. Their bodies measure deeper (Fig. 71) and have some related morphometric and meristic characters that are so distinctly different from Lake Kariba *A. lateralis* as to constitute a taxonomical relevance (Table 126).

The fecundity of *A. lateralis* is extremely low. Lake Kariba individuals have on an average 300 eggs per female as compared to *B. imberi* which had 14,000 eggs per female (Table 127). The lower egg production is also noticeable in *A. lateralis* of the Kafue River whose individuals attain the same size as *B. imberi* in Lake Kariba. The Kafue *Alestes* population, however, had a much higher egg count per female ($\bar{x}$ 5,000) than the Lake Kariba population. Speedy replacement of *B. imberi* by *A. lateralis* in Lake Kariba took place in spite of the much lower *Alestes* fecundity. In general it can be stated that the males of both species grow more slowly than the females. In *A. lateralis* the slower growth of males is even more pronounced. The annual increment in standard length and weight is substantially greater for *B. imberi*. In Lake Kariba both species occur together with an additional 40 species of fishes.

A. lateralis with 42% of the total number of fish is the most abundant. But its share in the ichthyomass is small as it constitutes only a little over 0.7%. In spite of this fact it has been calculated that its annual production for the whole of Lake Kariba is 19 thousand metric tons and perhaps more as compared to 38 metric tons of *B. imberi*. From an analysis of length frequency it was inferred that both species, including the largest specimens, are small enough to be consumed by the known stock of *Hydrocynus vittatus*. The highest abundance of tigerfish, however, is in the length range of *B. imberi*. The young tigerfish are, therefore, unable to feed on the *B. imberi* and so the entire stock of *A. lateralis* remains the only suitably sized prey for the extensive stock of tigerfish.

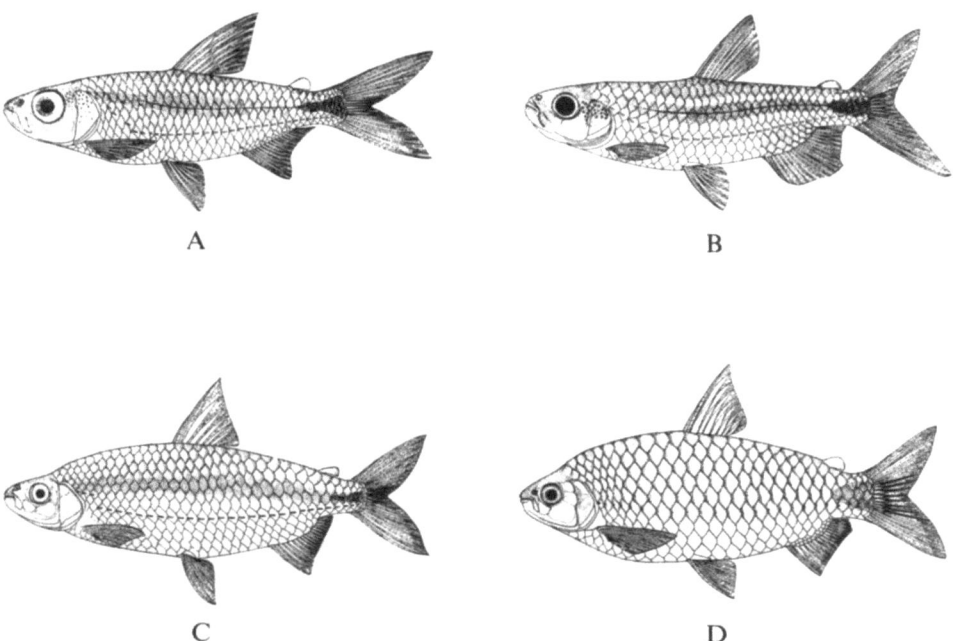

Fig. 71. A – *Alestes lateralis* female 4.7 cm standard length; B – male 5.4 cm long from Lake Kariba; C – female 13.9 cm long from Kafue River. D – *Brachyalestes imberi imberi* female 13.9 cm long from Lake Kariba.

Alestes and *Brachyalestes* will obviously also be devoured by other predators, but at the present time there is no information available on this issue. It is very probable that both species are preyed upon preponderantly by the tigerfish.

The special distribution of the closely related species, *A. lateralis* and *B. imberi*, is explained by BELL-CROSS (1965b, 1968a, b) by the time difference of distribution. He mentions that *A. lateralis* 'is found in several tributaries of the Luapula River and are separated from the main river by physical barriers (BELL-CROSS, 1965b). It appears that a later invasion of the closely related *B. imberi* displaced *A. lateralis* in all waters that *B. imberi* was to colonize. A similar situation exists in the lower Zambezi system where *A. lateralis* occurs in an isolated tributary. In its early migration, *A. lateralis* managed to move as far south as Zululand whereas the more recent invader *B. imberi* was unable to penetrate so far south.' A similar suggestion is made by CRASS (1964) when he says that *A. lateralis* 'has perhaps been exterminated over much of its former range by more successful competitors'.

The higher values of many meristic characters of *A. lateralis* would according to DOGEL's (1954) oligomerisation law support this explanation. According to this explanation *A. lateralis* should be evolutionarily older (HOLČIK & DUYVENÉ DE WIT, 1964) than *B. imberi* which has a smaller number of meristic characters. This fact, however, does not necessarily have anything in common with the distribution and it

Table 126. Some meristic and morphometric characters of *A. lateralis* from the Kafue River (n 67) and Lake Kariba (n 100).

	Kafue River		Lake Kariba	
	range	x̄	range	x̄
Standard length, mm	116 –140	124.4	34 –59	48.3
Count of pore-bearing l.l. scales	27 – 34	30.8	23 –33	31.0
Scales above l.l.	6 – 7	6.1	4 – 5½	5½
Scales below l.l.	2 – 3	2	1½–2	2
Rays of the dorsal fin	II–III 6 – 8	7	II 5 – 8	8
Rays of the anal fin	II–III 14 – 17	15.4	II–III 14 –17	III 15
Count of gill rakers	28 – 37	32.1	27 –34	30.8
Length of head in % of SL	22.0– 25.2	23.2	22.7–33.3	37.2
Depth of body in % of SL	29.9– 35.8	33.0	19.8–30.2	24.8

Table 127. Fecundity of *A. lateralis* from Lake Kariba and Kafue River and *B. imberi* from Lake Kariba.

Length groups mm	SL mm x̄	w̄ g x̄	Egg counts x̄ left/right ovary[1]	Total egg counts gravimetric (n)/counted (n)
	A. lateralis		Lake Kariba	
21– 30	29.6	0.5	—/—	58(1)/—
31– 40	38.9	1.4	—/—	314(53)/214(15)
41– 50	45.1	2.0	—/—	381(111)/ 276(20)
51– 60	54.6	3.4	—/—	620(14)/ 362(4)
x̄				378(179)/ 261(39)
	A. lateralis		Kafue River	
71– 80	80.0	20.7	2010/ 1983	—/ 3993(1)
91–100	97.0	21.2	725/ 739	—/ 1464(7)
101–110	105.0	28.7	1877/ 1822	—/ 3699(23)
111–120	116.0	35.4	2636/ 2482	—/ 5118(21)
121–130	127.0	44.7	3853/ 3618	—/ 7471(9)
131–140	135.0	52.5	3676/ 3756	—/ 7432(1)
x̄			2463/ 2400	—/ 4863(62)
	B. imberi		Lake Kariba	
91–100	95.0	45.1	—/—	8931(1)/—
111–120	115.0	65.9	4490/ 4513	7237(1)/ 9003(2)
121–130	125.0	76.3	—/—	12359(2)/—
131–140	135.2	92.1	4564/ 3557	12126(1)/ 8121(1)
141–150	145.0	97.0	—/—	12272(1)/—
151–160	152.3	104.6	13684/13147	—/26832(1)
x̄			7579/ 7072	8817(6)/14652(4)

[1] Counted only.

determines only the order of differentiation of both species. There appears to be a certain similarity with the evolution of trout and salmon, where as a consequence of an excessive population density of the stationary fresh water predecessors a separation of new stocks occurred from which probably originated the migratory, anadromous stock of future sea trout and salmon (BALON, 1968b). As a result of the much older evolutionary age of Characidae a similar differentiation could produce a specific taxa. I cannot however imagine how the later invasion of *B. imberi* could cause the extermination of the original population of *A. lateralis*.

The recent case in Lake Kariba complicates this question even more because here took place the displacement of 'the successful competitor' *B. imberi* by *A. lateralis*. This happened in a strikingly short time if the researchers' accuracy of findings and subsequent conclusions are accepted and if their observations were significant. Even if it is a species with a short reproduction cycle it would appear that the time is too short if the new species did not originate directly from the population of the original species. But it was proved that this cannot be the case.

BELL-CROSS' (1968b) explanation of the first catch in 1964 of *A. lateralis* from Lake Kariba is that 'it appears possible that it gained access to the Middle Zambezi via the Victoria Falls, and that the changing ecological conditions in the establishment of Lake Kariba favoured its survival.' But at that time he did not yet know that *A. lateralis* would shortly be the most numerous fish in Lake Kariba and that *B. imberi* would almost disappear. What caused the previously fully operating isolation barrier of the Victoria Falls between the different ichthys of the Upper and Middle Zambezi to stop working? The change of food sources for *Alestes* is almost out of the question as stated by Peter (1967): 'In *Alestes* we have a typical example of a fish with no strict feeding regime.' The only predator-prey relationship appears to be in favor of *B. imberi* which because of its size is out of reach for the greater part of the *H. vittatus* population. But the highly abundant *H. vittatus* existed in the Middle Zambezi together with the abundant population of *B. imberi* before the formation of Lake Kariba. This abundance of *B. imberi* and tigerfish increased even during the unstable period of the first years of the formation of the lake. How did it come about that *A. lateralis* appeared here?

B. imberi is more adapted to river conditions and its spawning takes place in the rainy season on the newly flooded shallow areas inaccessible to predators. Its development in Lake Kariba could have continued successfully until 1963 when the water of the lake was continuously rising. That year, at the beginning of the rainy season the water level was lowered for reasons of safety for the first time. In April of 1964 after the floodgates had been closed the level stabilized and even increased somewhat until the beginning of the next rainy reason. Since 1964 the opening of the gates to decrease the water level prior to the rainy season has been a yearly practice. The decrease in the first year could have prevented the spawning of *B. imberi*.

A. lateralis is able to spawn on submerged flora such as the roots of *Salvinia* without any regard to the fluctuations of the water level and does not require freshly flooded plants for spawning. Those individual specimens which were carried over the Victoria Falls and survived the fall and escaped predators could have without further recruitment created a population suitable for an explosive development in a short time by means of successful spawning and penetration into the area liberated by remnants of *B. imberi*. Moreover fishes accustomed to taking shelter in submerged plants were better able to escape from predators in the newly created conditions than fishes accustomed to living on uncovered river beds.

Many of these things were surmised by JACKSON (1960) when he wrote: 'Physical conditions in Middle Zambezi have of course been recently very drastically altered by man, and it is of great interest that a permanent vegetation is already beginning to form in the river below the Kariba Dam. This will probably increase when the dam reaches retention level in 1963/64, and it is quite possible that within the next decade some small species from upland streams within the system will colonise the Middle Zambezi below the dam.' He could not know that after the filling of the lake, at the beginning of the rainy season each year the floodgates of the dam would be opened, creating more normal conditions below the dam but that above the dam the conditions which he predicted would actually occur.

Other invaders from the Upper Zambezi. – During my work on Lake Kariba the original list of 28 fish species was increased to 41. Prior to impoundment these 28 species existed in that portion of the Zambezi River which flowed through the Gwembe Valley (JACKSON, 1961b), the area later to become Lake Kariba. Most of the new species are clearly invaders from the Upper Zambezi, via the Victoria Falls. Probably the newly created man-made environment (Lake Kariba) 137 km below the great Falls, has changed the initial age old isolating mechanism. Consequently this mechanism cannot consist solely of a physical barrier.

Nevertheless, only 13 species (20%) of the 64 species restricted to the Upper Zambezi River (BELL-CROSS & KAOMA, 1971; BELL-CROSS, 1972 adjusted according to my records) have so far invaded Lake Kariba. Why not the others, or, which species will come next? A good collection of Upper Zambezi species inhabiting the edge of Victoria Falls may at least provide some clue. The fishes from the edge of the Falls are most susceptible to being washed over the edge when the rain swollen Zambezi hits the Victoria Falls. They should constitute the most probable invaders of Lake Kariba.

The Victoria Falls, known locally as 'Mosi-o-Tunya' – 'the smoke that thunders' are a unique barrier for fish distribution. DAVID LIVINGSTONE (1857), the first European to see them, was so anxious not to overestimate the size of his discovery, that he underestimated them. His error was subsequently rectified by WILLIAM BALDWIN (1863), the second European visitor to the Falls. Nevertheless LIVINGSTONE's impressions recorded in his diary (CLAY, 1964) are still accurate enough to be quoted:

'The falls are singularly formed. They are simply the whole mass of the Zambesi waters rushing into a fissure or rent made right across the bed of the river. In other falls we have usually a great change of level both in the bed of the river and adjacent country and after the leap the river is not much different from what it was above the falls. But here the river flowing rapidly among numerous islands and from 800 to 1,000 yards wide [exact width 1692 m] meets a rent in the bed at least 100 feet deep [exact depth 70 to 100 m, see Fig. 72] and at right angles with its course, or nearly

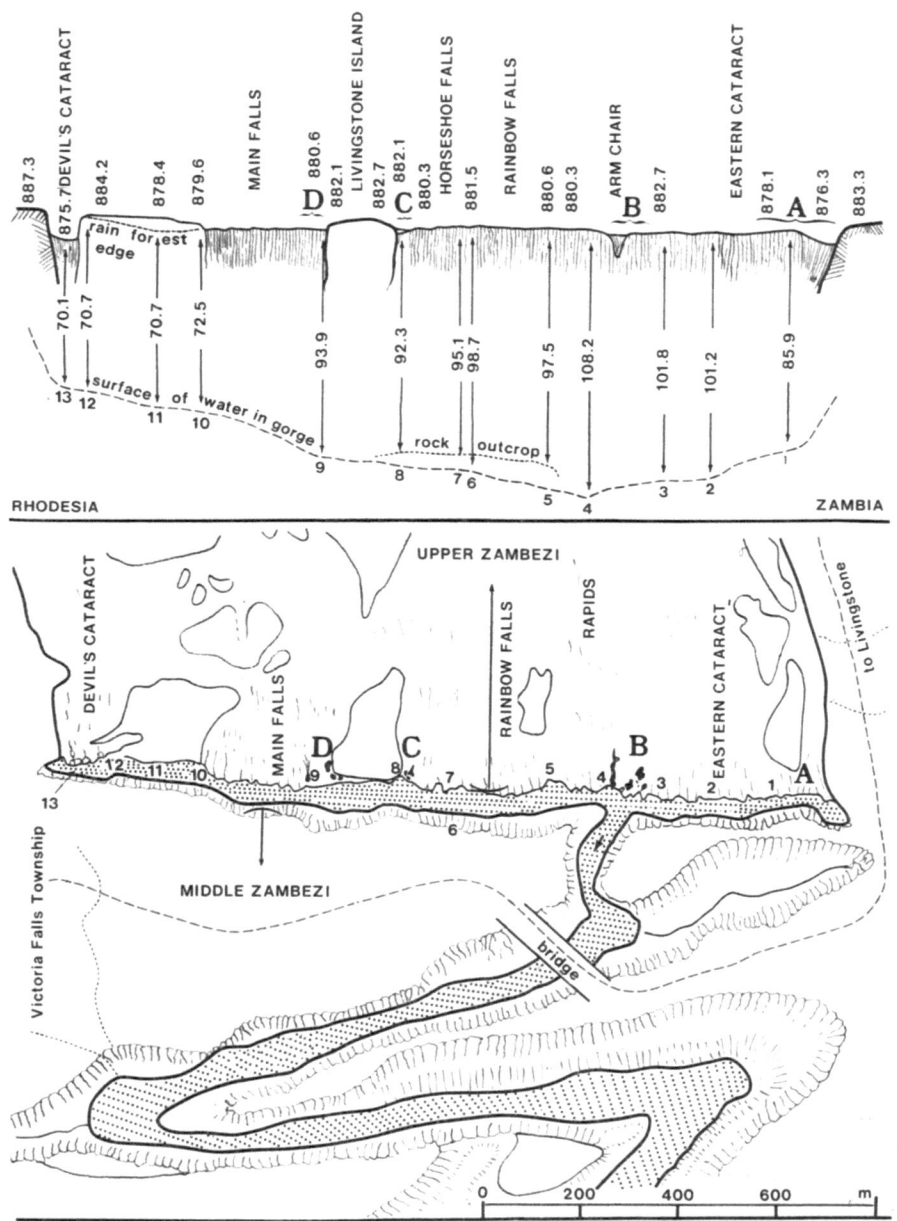

Fig. 72. Sketch of the front and top views of the Victoria Falls. All measurements in meters.

due east and west, leaps into it and becomes a boiling white mass at the bottom ten or twelve yards broad. Its course is changed also. It runs or rather rolls and wriggles from east to west until it reaches what above was its left bank and then turns a corner and follows or rather is guided by the fissure away in its usual route of S.E. and by E. The lips of the rent are in some parts not more than fifty or sixty feet apart. The southern lip is straight, and except at the west corner which seems inclined from a split in it to fall into the gulph is straight and level with the general bed of the river above, its wall is quite perpendicular. The northern lip is jagged, several pieces having fallen off, and five or six parts have the edge worn down a foot or two. In these when the water is low, as it now is, the falls divide themselves and from each ascends a column of vapour which rises from 200 to 250 ft.'

No endemic fish, with the possible exception of eels when small enough (BALON, 1973b), can climb the vertical basalt rocks or swim up the falling mass of water, not even during the dry season when the water level is at its lowest and the entire edge of the falls is frequently exposed revealing rock formations interrupted here and there by small islands of clustered trees and an occasional stream of water rushing over the jagged edge of the chalcedony and basalt lip.

The Victoria Falls were formed 99 km downstream from its present position, approximately 500,000 years ago (CLARK, 1950). There the separate Miocene-Pliocene rivers, the pre-Upper Zambezi (drained south into Okavango-Ngami basin) and the pre-Middle-Lower Zambezi (entering the eastern Indian Ocean) joined for the first time through a newly diverted spillway system to form a united Zambezi River as it is known today. Although waters of the two separate drainage areas were united, fish faunas of separate western Kasai-Congo and eastern Nile-Congo origins remained to a large extent isolated. After union of the two previously separate rivers the Victoria Falls were believed to become the isolating barrier.

The initial theory (LIVINGSTONE, 1857) that the Zambezi River bed had been split by a geological cataclysm and the cleft consequently formed 'the largest known curtain of falling water in the world' (BOND, 1964) was later denied by geologists (LAMPLUGH, 1907). It was speculated that the downcutting of the Zambezi backed-up the river's 'nick point' from the ocean into rocks favourable for the formation of a waterfall. A more complete explanation of this speculation, however, which also accounts for the fish distribution (JUBB & FARQUHARSON, 1965), was presented later.

In the Pliocene the Upper Zambezi River and its probable tributaries, the Upper Kafue-Kalomo systems (see section 6.1), flowed south together with the present Chobe-Okavango into the Ngami-Makarikari Lake and Limpopo Valley (Fig. 73). The Middle Zambezi was then a separate river with headwaters somewhere in the region of the present Matetsi River. A tectonic upwarp in the Caprivi area broke the connection of the Upper Zambezi with the Ngami-Makarikari drainage and the waters of the former escaped eastward along a depression and spilled over the sheer drop at the old Middle Zambezi headwaters. Here, at point of origin of the Batoka Gorge, began the existence of the united Upper-Middle Zambezi River (LAMPLUGH, 1907) and Victoria Falls (Fig. 74). The recession mechanism of the Victoria Falls and the gorges is well documented, for example, in CLARK (1950) and BOND (1963, 1964).

The monolithic surface of the basalt rocks along the entire length of the gorges

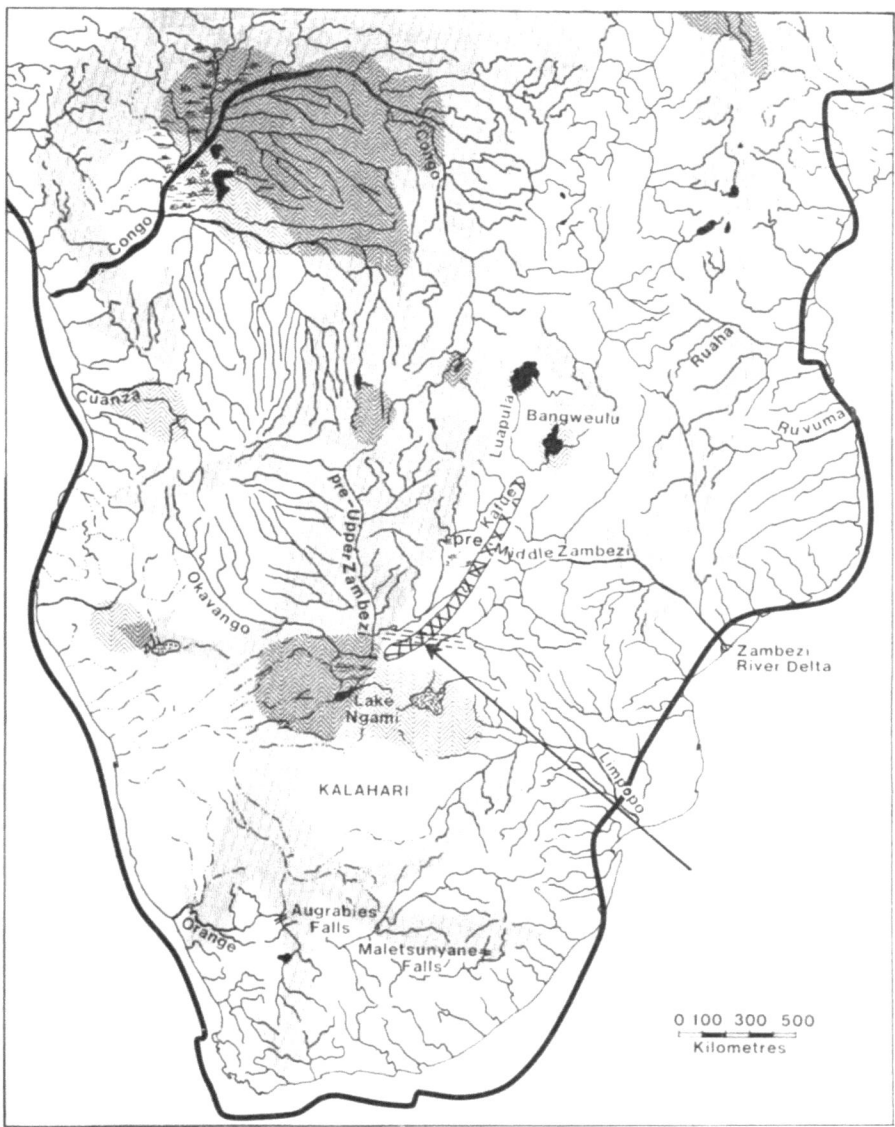

Fig. 73. An idea of the hydrographic situation in the Miocene-Pliocene, and major routes of fish invasions into the area of the separate pre-Upper and pre-Middle-Lower Zambezi rivers (interior basins based on the 'Pre-Rift map' in HOWELL & BOURLIÈRE, 1963 and on some routes on the maps in JUBB & FARQUHARSON, 1965, as well as on FRYER & ILES' 1972 illustrations attempting to recapture the past of the Great Lakes of Africa). The area of the future union is pointed out by an arrow and marked by interrupted lines. Ancient basins – broken-line screen, invasion routes – vertical-line screen.

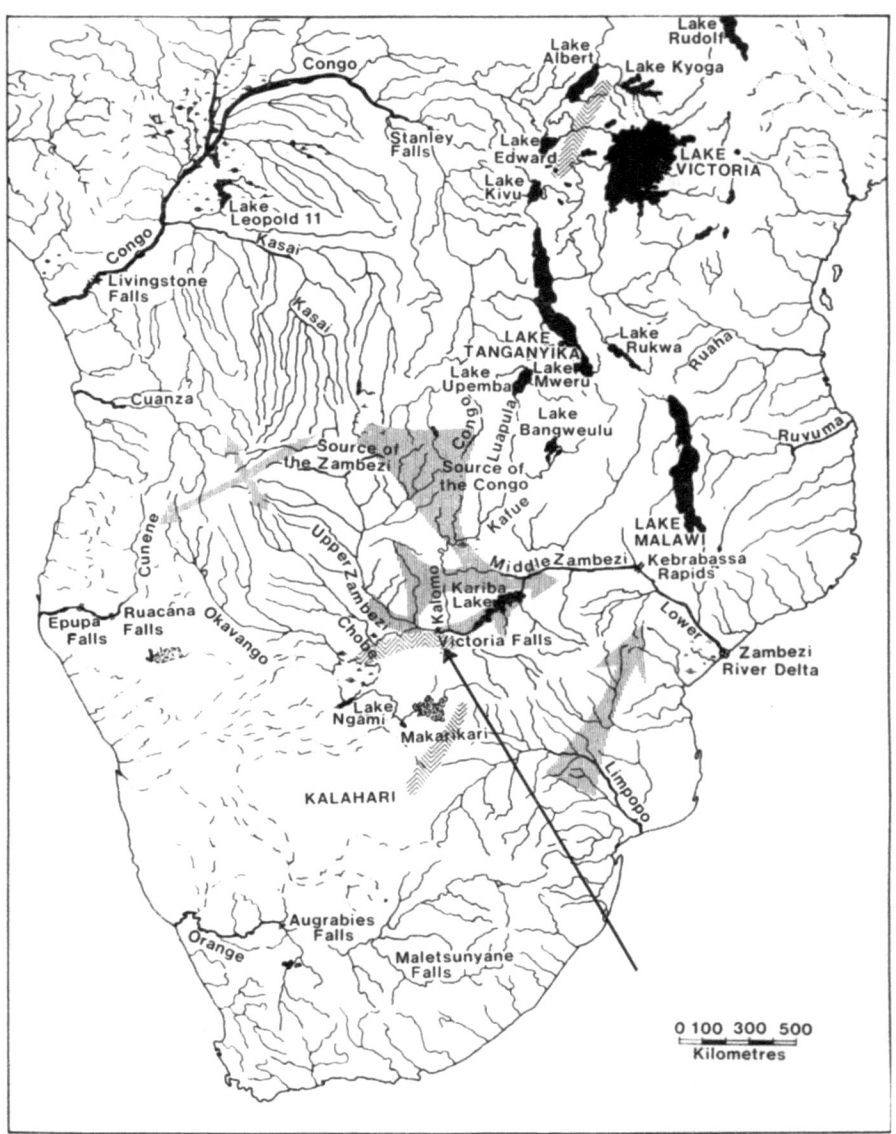

Fig. 74. The tectonic upwarp in early Pleistocene (broken-line screens) diverted the pre-Upper Zambezi from the southern Ngami into the eastern Middle Zambezi. The initially prevalent north-south longitudinal invasion of fishes established new latitudinal routes (vertical-line screens). The Victoria Falls is pointed out by an arrow.

has a system of fissures probably formed when the lava cooled. These were later enlarged by upwarp tectonic movements. The fissures form distinct fault lines. The gorges are eroded along such fault lines where a protective barrier of solid lava had been dissipated. When the river was spilling over an edge of solid basalt the erosion was slowed down until the water found the weakest point in the edge. (This is actually the most recent status of the Falls with the weakest point appearing in the growing fissure of Devil's Cataract, the second weakest point being the gradually deepening cleft of Cataract Island). When the river succeeded in eroding its way into another fault line, the process of upstream cutting was greatly speeded up.

The lower gorges between Chimamba Rapids and the entry of the Matetsi River (64 km) have taken 248,000 years to erode, the next 35 km 162,000 years and the last 2 km approximately 3,000 years. The whole recession of the Victoria Falls started some time before the beginning of the Middle Pleistocene.

In the second gorge the water rises to about 30 m in the rainy season. During the dry season the lowest water level was found to be 15 m. The Zambezi River has a slope of 325 m between the bottom of the Falls and the point of origin of the gorges at the Matetsi River. The walls of the gorge are 109 m high at Victoria Falls, 197 m at Chimamba Rapids 40 km below and 253 m at Namuruba 8 miles upstream from the entry of the Matetsi River. The walls are vertical in the uppermost gorges or slope at a very steep angle farther downstream. This character is maintained throughout the entire length of the gorges. Size and form of the contemporary Victoria Falls are illustrated on Figure 72.

During the lowest water level, usually just before the first rains in November, a substantial part of the edge of Victoria Falls is accessible on foot from the Zambian side. Some streams about 1 m deep had to be negotiated in the areas of the Eastern Cataract and Rainbow Falls, but no major obstacles were encountered on the way to Livingstone Island. In order to avoid the hazards of deep potholes, fast flowing streams and slippery rocks, some sections had to be by-passed upstream from the edge.

First sampling with the Smith-Root Type Electrofisher took place in inshore streams areas between protruding tree roots and pools 10 m above the edge of the Falls (Fig. 72). Quantitative sampling was not possible because the lowest water level was missed by postponing the trip till after the first rainfall. By November 25 even the northernmost stream of the Eastern Cataract was impassable.

The following year, however, the goal was achieved and selected potholes (abrasion holes), pools and streams (Fig. 72B, C, D) along the entire accessible edge of the falls were quantitatively sampled (November 3–4, 1971). An overdose of rotenone solution, Pro-Noxfish, was added and mixed with the residual water of the stagnant potholes. It was also added to the uppermost limit of pools with flowing water to be sampled. On site B sampling took place in three potholes of stagnant water and in one fissure with three deep pothole-pools with a substantial stream of water cascading down 1–2 m high steps extending from the upper (c) to the lower (a) hole. On the northern side of Livingstone Island three shallow pools were sampled. Of these only the largest had a steady in and out stream of water. Between the southern shore of the same island and the lip of the Falls another three shallow pools of stagnant water were sampled as well as one with a stream flowing through it.

Table 128. Number of specimens sampled at different edge sites of the Victoria Falls.

Sample site	A	B				C			D	
Pothole or pool		1	2	3	4	5a	5b	5c	6	7
Petrocephalus catostoma					1					
Marcusenius ansorgi			1			2	1			
Marcusenius macrolepidotus			3	12	2	1				
Hippopotamyrus discorhynchus			31	41	6	14		2		5
Micralestes acutidens	12									
Barbus m. codringtoni	22	24	136	173	154	23	3	7	85	45
Labeo c. annectens	4		19		53				13	8
Barilius zambezensis	11	2								1
Leptoglanis rotundiceps						1				
Schilbe mystus			14			7				
Amphilius platychir	1	2	19	7	16	2		1		22
Clarias submarginatus				1	1	3			2	4
Clarias theodorae			2		1					
Synodontis leopardinus					1					
Synodontis nebulosus			1							
Synodontis nigromaculatus			4		2					
Tilapia rendalli			2				7		1	1
Sarotherodon andersoni				8		15				
Serranochromis macrocephalus	1									
Serranochromis thumbergi			2							
Hemichromis fasciatus										1
Haplochromis darlingi		1	16	1	7	22	5	1	8	2
Pseudocrenilabrus philander						8			2	
Mastacembelus mellandi					1				1	32
Number of taxa	6	6	13	7	14	7	5	3	8	9
Total number of specimens	51	63	269	198	260	75	18	9	117	116
Number of specimens per 1 m^2 of water surface		27	20	25	11	10	45	22	7	34

The taxonomic status of the large yellowfish and golden mudsucker was evaluated in separate studies. It was concluded that individuals of these species from the edge of the Falls differ sufficiently from those below the Falls to guarantee a probable genotypic isolation and separate taxonomic status.

Twenty-four species of Upper Zambezi fishes were found on the extreme edge of the Falls (Table 128). However, the actual number of species occasionally occurring at the edge may be higher. Two species *(Micralestes acutidens* and *Serranochromis macrocephalus)* sampled at higher water and in strong current in the Eastern Cataract (A) serve as evidence that this is so. These two species were not present in the more extensive and quantitative samples from stagnant or slow flowing waters of potholes and pools.

Table 129. The total number of specimens sampled at edge sites B, C and D in 11 potholes and pools of the Victoria Falls (total surface area 75 m²).

Species	N	N/1 m²
Petrocephalus catostoma	1	0.01
Marcusenius ansorgi	4	0.05
Marcusenius macrolepidotus	18	0.24
Hippopotamyrus discorhynchus	99	1.32
Barbus m. codringtoni	648	8.64
Labeo c. annectens	95	1.27
Barilius zambezensis	3	0.04
Leptoglanis rotundiceps	1	0.01
Schilbe mystus	21	0.28
Amphilius platychir	69	0.92
Clarias submarginatus	11	0.15
Clarias theodorae	3	0.04
Synodontis leopardinus	1	0.01
Synodontis nebulosus	1	0.01
Synodontis nigromaculatus	6	0.08
Tilapia rendalli	11	0.15
Sarotherodon andersoni	23	0.31
Serranochromis thumbergi	2	0.03
Hemichromis fasciatus	1	0.01
Haplochromis darlingi	63	0.84
Pseudocrenilabrus philander	10	0.13
Mastacembelus mellandi	34	0.45
Total	1,125	15.00

In single potholes or pools 3 to 14 species were present, at a density of 7 to 45 specimens per 1 m² of water surface. Only *Barbus marequensis codringtoni*, the most abundant species, was represented in all holes. Some species not represented in most holes appear in high density in a single hole; for example, *Schilbe mystus* was confined to two of the stagnant holes, whereas *Mastacembelus mellandi*, the second most abundant fish at the edge of the Main Falls (D7), was in one pool which was fed by a strong stream of water.

In all samples only small individuals were represented the largest being a specimen of *Labeo cylindricus annectens*, 184 mm in standard length. Cichlids and the abundant *B. marequensis codringtoni* were represented by the smallest individuals. Some measured only 11 mm (SL). In the total average *B. m. codringtoni* was represented by 8.6 specimens per 1 m² of water surface sampled, followed by *Hippopotamyrus discorhynchus* (1.32), *L.c. annectens*, (1.27), *Amphilius platychir* (0.92) (see Table 129 for mean contribution of each species as number captured per m² of surface sampled).

No doubt the species composition varies not only within habitats, but seasonally as well. The recorded density may reflect species concentration following the last phase of low water level. It is possible that fishes gradually drifted into potholes and remained there in the estimated density until the first rains depending on the rate of survival in the unusually stagnant conditions. However, a similar composition and density found in non-stagnant holes undermines the assumption of a mortality dependant density in stagnant potholes. I have no evidence that the fish are not able to return to a safe distance above the edge of the Falls. However, because of the well known inability of fish of a similar size to overcome strong currents (GRAY, 1968), it is assumed that most of them were flushed over the edge and carried into the gorges when the first gush of rain swells the river above the Falls.

In the construction of Table 130 only better documented data are incorporated. Nevertheless there is a very distinct probability that some known distributions were overlooked or that new invasions occurred between the sampling of the data and writing of this section. 1. The Upper Zambezi fishes (marked +) are listed for the general distribution (BELL-CROSS, 1971, 1972; BELL-CROSS & KAOMA, 1971). Those for which presence above the Victoria Falls has been verified are marked ++ (JUBB, 1964b). 2. The fishes from the edge of the Victoria Falls are listed according to my samples, as described in detail in the previous section. 3. Fishes present in the Batoka Gorge and Kalomo River lowveld are listed according to JUBB (1964b) and BALON (1974c). The Kalomo River below Siengwazi Fall is considered to be a new Middle Zambezi tributary. 4. Fishes of that portion of the Middle Zambezi River which flowed through the Gwembe Valley before the area was inundated are listed according to JACKSON (1961b). 5. Lake Kariba fishes are listed from my samples (BALON, 1974b). 6. Finally, the fishes of the Lusito River are given as a probable example of a species composition in streams of the Lake Kariba drainage and as fishes of a Middle Zambezi tributary with no barrier to fish movements. This list is based on my survey.

Eleven fish taxa, *Petrocephalus catostoma*, *Marcusenius ansorgi*, *Barbus marequensis codringtoni*, *Labeo cylindricus annectens*, *Clarias submarginatus*, *Clarias theodorae*, *Synodontis leopardinus*, *S. nigromaculatus*, *Serranochromis thumbergi*, *Hemichromis fasciatus* and *Mastacembelus mellandi*, were present at the edge of Victoria Falls but nowhere below. Both highly abundant and less abundant species were represented (Table 129). At different times all of these taxa may have been flushed over the edge into the Middle Zambezi River, so probably none have survived to populate the river below the Falls.

Three other species from the edge of the Victoria Falls, *Leptoglanis rotundiceps*, *Amphilius platychir* and *Sarotherodon andersoni*, were collected in the Kalomo River lowveld and in the Lusito River. They may represent relicts of an ancient invasion through the spillway systems.

Marcusenius macrolepidotus, *Alestes lateralis*, *Labeo lunatus*, *Schilbe mystus*, *Serranochromis macrocephalus*, *S. robustus jallae*, *Sargochromis giardi* and *Haplochromis carlottae*, present at the edge of Victoria Falls, are known to be recent invaders of Lake Kariba, their only place of occurrence in the Middle Zambezi area. Most likely the new man-made environment was essential to the survival of these 8 species.

Another 10 species, *Hippopotamyrus discorhynchus*, *Hydrocynus vittatus*, *Micralestes acutidens*, *Barbus barotseensis* (?), *B. fasciolatus*, *Syndontis nebulosus*, *Aplocheilichthys johnstoni*, *Tilapia rendalli*, *Sargochromis codringtoni* and *Haplochromis darlingi*, may have invaded the Middle Zambezi via Victoria Falls prior to the impoundment of Lake Kariba. Whereas 10 other species, *Kneria auriculata*, *Barbus annectens*, *B. lineomaculatus*, *B. paludinosus*, *B. poechii*, *B. radiatus*, *B. unitaeniatus*, *Barilius zam-*

Table 130. Verified records of fish distribution in the Zambezi River area adjacent to the Victoria Falls.

	1. Upper Zambezi River	2. Victoria Falls edge	3. Gorges & Kalomo lowveld	4. Gwembe Valley Zambezi before Lake Kariba	5. Lake Kariba	6. Lusito River
Protopterus annectens brieni				+		+
Anguilla nebulosa labiata			+	+	+	+
Limnothrissa miodon[1]					+	
Mormyrops deliciosus				+	+	
Petrocephalus catostoma	+	+				
Marcusenius ansorgi	+	+				
Marcusenius macrolepidotus	++	+			+	
Marcusenius castelnaui	+					
Hippopotamyrus discorhynchus	++	+		+	+	+
Mormyrus lacerda	++					
Mormyrus longirostris			+	+	+	
Mormyrus ellenbergeri	+					
Kneria auriculata	+					+
Kneria polli	+					
Hydrocynus vittatus	++		+	+	+	+
Alestes lateralis	++				+	
Brachyalestes imberi			+	+	+	
Micralestes acutidens	+	+		+	+	
Rhabdalestes rhodesiensis	+					
Hepsetus odoe	++					
Distichodus mossambicus				+	+	
Distichodus schenga				+	+	
Hemigrammocharax machadoi	+					
Hemigrammocharax multifasciatus	+					
Nannocharax macropterus	+					
Barbus afrohamiltoni	+					
Barbus afrovernayi	+					
Barbus annectens	+					+
Barbus barnardi	+					
Barbus barotseensis	+			+		
Barbus bellcrossi	+					
Barbus eutaenia	+					
Barbus fasciolatus	++		+	+	+	
Barbus haasianus	+					
Barbus lineomaculatus	+				+	+
Barbus manicensis	+					
Barbus marequensis marequensis			+	+		+
Barbus m. codringtoni	++	+				
Barbus multilineatus	++					
Barbus neefi	+					

Table 130. (continued)

	1. Upper Zambezi River	2. Victoria Falls edge	3. Gorges & Kalomo lowveld	4. Gwembe Valley Zambezi before Lake Kariba	5. Lake Kariba	6. Lusito River
Barbus paludinosus	++		+		+	+
Barbus poechii	+				+	+
Barbus puellus	+					
Barbus radiatus	+		+			+
Barbus tangandensis	+					
Barbus thamalakensis	+					
Barbus trimaculatus			+			+
Barbus unitaeniatus	+				+	+
Barbus viviparus	+					
Coptostomabarbus wittei	+					
Labeo altivelis			+	+	+	+
Labeo congoro			+	+	+	
Labeo rubropunctatus					+	
Labeo cylindricus subsp.			+		+	+
Labeo c. annectens	++	+				
Labeo lunatus	+				+	
Barilius zambezensis	++	+	+	+	+	+
Engraulicypris brevianalis	+					
Auchenoglanis ngamensis	++					
Leptoglanis rotundiceps	+	+				+
Eutropius depressirostris				+	+	+
Eutropius yangambianus	+					
Schilbe mystus	++	+			+	
Amphilius platychir	+	+	+			
Heterobranchus longifilis				+	+	
Clarias gariepinus	++		+	+	+	+
Clarias ngamensis	++					
Clarias submarginatus	+	+				
Clarias theodorae	+	+				
Clariallabes platyprospos	+					
Malapterurus electricus			+	+	+	
Chiloglanis neumanni	++		+	+		+
Chiloglanis swierstrai						+
Synodontis leopardinus	+	+				
Synodontis nebulosus		++	+		+	+
Synodontis nigromaculatus	++	+				
Synodontis woosnami	++					
Synodontis zambezensis			+	+	+	+
Aplocheilichthys hutereani	+					
Aplocheilichthys johnstoni	++			+	+	
Aplocheilichthys katangae	+					

Table 130. (concluded)

	1. Upper Zambezi River	2. Victoria Falls edge	3. Gorges & Kalomo lowveld	4. Gwembe Valley Zambezi before Lake Kariba	5. Lake Kariba	6. Lusito River
Hypopanchax jubbi	+					
Sarotherodon macrochir	+ +					
Sarotherodon mossambicus mortimeri			+	+	+	+
Sarotherodon andersoni	+ +	+			+	+
Tilapia rendalli	+ +	+	+	+	+	+
Tilapia sparrmani	+ +					
Serranochromis r. jallae	+ +				+	
Serranochromis angusticeps	+					
Serranochromis macrocephalus	+ +	+			+	
Serranochromis thumbergi	+ +	+				
Serranochromis longimanus	+					
Hemichromis fasciatus	+ +	+				
Sargochromis giardi	+				+	
Sargochromis codringtoni	+ +		+	+	+	+
Haplochromis carlottae	+ +				+	
Haplochromis darlingi	+ +	+	+	+	+	
Haplochromis frederici	+ +					
Pseudocrenilabrus philander	+	+	+		+	+
Pelmatochromis ruweti	+					
Ctenopoma ctenotis	+					
Ctenopoma multispinis	+ +					
Mastacembelus mellandi	+ +	+				
Total number of species	82	24	21	27	41	28

[1] Introduced from Lake Tanganyika.

bezensis, *Clarias gariepinus* and *Chiloglanis neumanni*, may have done so by a route other than Victoria Falls since their present distribution is limited to headwater tributaries of the Zambezi River.

Before the recent invasion of the Upper Zambezi fishes into Lake Kariba, JUBB (1964b, p. 130) described the influence of Victoria Falls upon fish distribution in the following words:

'When considering the Victoria Falls as a physical barrier it is important to remember that the geological history of the Batoka Gorge must be taken into account as well. We understand that the Batoka Gorge is the result of erosion by the Zambesi along lines of geological weakness. This retrogression has taken a very long time so, as we see things today, it must be realised that the fish fauna of the Upper Zambesi River system has been geographically isolated from that of the Middle Zambesi for a considerable period of geological time. The Victoria Falls, and whatever physical barrier existed in the Batoka Gorge during its erosion, have been impassable to all fish trying to move upstream. It appears quite evident that no fish attempting to go down the Falls now would survive the shock.'

Some recent invaders into Lake Kariba (see p. 492) disprove the assumption that no fish are able to survive the drop over the edge of the Falls. Speculation that the supersaturation of water with gases, especially with nitrogen and the pressurized entry into the narrow and deep gorges creates a hostile section in which every fish carried over the edge of Victoria Falls may be destroyed, is also obviously incorrect. Furthermore, in the beginning there were no inhospitable gorges and the water of the Upper Zambezi probably spilled relatively gently into headwater streams of the Matetsi-Middle Zambezi River.

Before the union of the originally separate pre-Upper and pre-Middle Zambezi rivers, their fishes were dissimilar and of different origins. But so were the rivers. They matured in pre-Miocene times, over a long spell of tectonic calm. Their headwaters eroded back through the central plateau and moved towards each other. Stream capture or discharge into alternate catchments was plentiful. Some points of contact with the Kasai-Congo system still persist in the Upper Zambezi headwaters (FRADE & PINTO, 1961; JUBB, 1964c; BELL-CROSS, 1965a). In the Pleistocene the Upper Zambezi was already a widely diversified stream with numerous lateral-flood lakes and swamps. This was in the area above the Middle Zambezi headwaters, somewhere in the vicinity of contemporary Sesheke. It was in such an environment that adaptive radiation took place filling all the available niches left unoccupied by the invaders from the headwaters.

In the Pliocene-Pleistocene transition the Middle Zambezi sources were on the edge of a central plateau. Their headwaters probably formed fast running, and in some places intermittent streams which cut into valley floors, sometimes forming gorges. Floods during the rainy season were extensive but inundations occurred only temporarily and in limited areas. Probably none, or very few, lateral flood waters were to be found. This river was invaded mainly by fish species from the drainage of pre-Rift Valley headwater catchments of the eastern Nile-Congo invasion route. Paleographic evidence of the united Rukuru-Ruvuma rivers (FRYER, 1959) implies this and BELL-CROSS (1972, 1973) assumed this to have occurred. Only species capable of transfer via some Upper Zambezi and Middle Zambezi tributaries entered both systems (JUBB, 1964b), but probably survived only if there was a suitable niche available.

Here I should like to introduce the story of the Victoria Falls and speculate on possible causes of a functional barrier between these two separate fish faunas. These two faunas had been developing, no doubt, in relative isolation for a considerable time. It is most likely that the available niches in both rivers were well filled, when the tectonic warping diverted the flow of the Upper Zambezi from the Makarikari-Ngami direction into the headwaters of the Middle Zambezi. At that time both rivers achieved a maximum diversity of fish fauna in relation to available habitats. The two rivers did not have similar habitats. Saturated niches and differences in habitats and not the physical barrier of the Victoria Falls may have been the effective barrier against intermingling of species.

Successive pluvial episodes (COETZEE, 1964; LIVINGSTONE, 1967, 1971; LIVINGSTONE & KENDALL, 1969) and progressive erosion may later have created some habitat changes and made possible successful invasions by single species. Since the Upper Kalomo and Kafue rivers later became tributaries of the Middle Zambezi, the invasion did not necessarily always take place via Victoria Falls.

The invasion of Lake Kariba can serve as evidence in support of the above theory. During the initial stages of filling, conditions in the impoundment actually simulated for a prolonged period of time the floods or rainy seasons typical of this region. Middle Zambezi fishes, well adapted to such conditions, were able to keep pace with the expanding water volume. Endemic species increased enormously their abundance and production (JACKSON, 1960). This was the result of the increase in nutrient load from the flooded terestrial habitats, and the fact that freshly flooded areas provide year round spawning grounds. At that time not one Upper Zambezi fish had as yet invaded Lake Kariba.

In 1963 the filling phase of the new lake was completed. Since then, for reasons of safety, the spillgates of the dam have been opened before the start of the first rains in order to lower the water level in the lake during most of the rainy season. The 'natural' conditions were suddenly reversed and the endemic species were no longer able to utilize the new habitats even though the nutrient load began to decrease. Lacustrine conditions to which the Middle Zambezi fishes were not adapted (JACKSON, 1960), prevailed more and more. In 1963 and 1964, before this condition was firmly established, an invasion of new species from the Upper Zambezi had begun via Victoria Falls.

This invasion of new species, however, is limited for the same reasons as was the invasion before the advent of Lake Kariba: the available niches will sooner or later be filled and probably no new invaders will be able to enter the competition. Some adaptive radiation of older colonizers may also enter the picture, speeding up the filling of available niches.

6.4. The Kafue River Floodplain: an example of pre-impoundment potential for fish production

by JAMES M. KAPETSKY[3]

Background. – In 1970, a dam was completed in the Kafue Gorge, 35 km downstream from the Kafue floodplain and at present another dam is under construction at Itezhi-Tezhi at the upstream end. The author, along with other biologists, participated

---

[3] The field work was carried out while the author was employed by the University of Michigan and the University was under contract to the UNDP/FAO. However, much of the information presented is based on additional data collected over and above the requirements of the contract. The material presented represents preliminary results selected from dissertation work still in progress, in partial fulfillment of the requirements for the degree of Doctor of Philosophy in the Horace H. Rackham School of Graduate Studies at the University of Michigan, Ann Arbor.

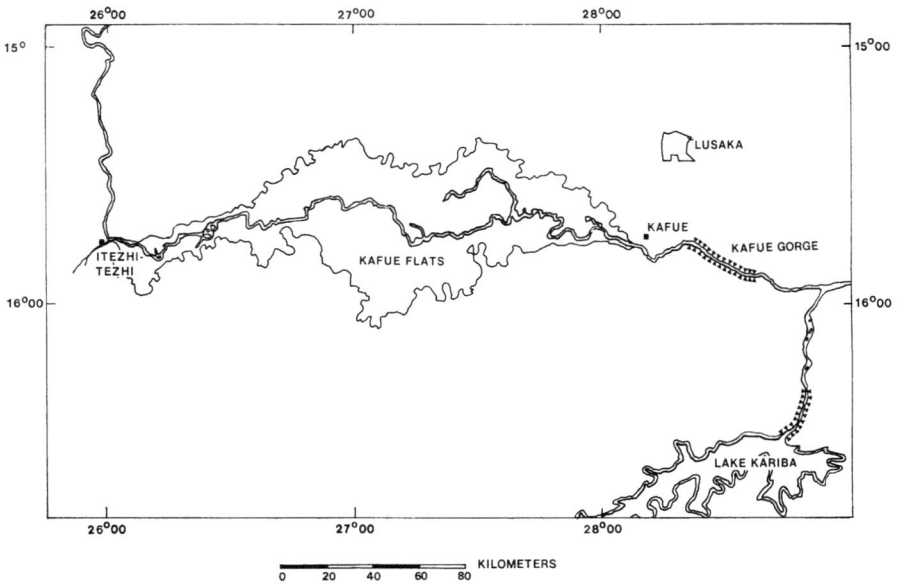

Fig. 75. The Kafue River and floodplain in southern Zambia.

in the last large-scale ichthyobiological studies to be conducted under the natural flood regime on the Kafue floodplain system (LAGLER et al., 1971; CHAPMAN et al., 1971). The results of these investigations and those of earlier studies (CAREY, 1971; WILLIAMS, 1971) provide a sound basis from which to extend the existing information on standing stocks, fish distribution, and physical-chemical measurements to include the concept of fish production. Thus the objective here is to present some first approximations of fish production potential in a floodplain system under natural hydrological conditions. In so far as the author knows the production estimates presented here are the only values available from a sub-tropical floodplain system.

The Kafue floodplain is the low lying part of the vast (14,000 km^2) Kafue Flats. Situated in the lower reaches of the 1480 km long Kafue River about 140 km northwest of Lake Kariba at the closest point, the floodplain extends eastward from Itezhi-Tezhi for 240 km to Kafue township, about 70 km above the Kafue-Zambezi confluence (Fig. 75). The Kafue River meanders over the floodplain for a distance of about 400 km with a gradient of approximately 3 cm/km (FAO Survey, 1968).

Grasses are the dominant vegetation types on the Kafue floodplain with rice grass, *Oryza barthi*, extending over more than 2200 km^2. Other grasses such as *Vossia cuspidata* and *Echinochloa* spp. are also abundant (VAN RENSBURG, 1968). In former times the grasses and other kinds of vegetation provided grazing for vast herds of wild ungulates, prominently the red lechwe, *Kobus leche kafuensis*, an antelope (J. BUTTS, A. JONES, personal communication). In the terrestrial part of the floodplain system these wild animals have largely been replaced by cattle of which about 250,000 may seasonally graze there as the flood waters recede (VAN RENSBURG, 1968). In the aquatic system the hippopotamus has been affected somewhat by the presence of man but the crocodile was intensively hunted and its numbers have been considerably reduced (J. BUTTS, personal communication). Birds, numbering 337 species (DOWSETT, 1966), are abundant. Some species such as the white pelican, *Pelecanus onocrotalus*, are seasonal inhabitants and arrive in multitudes during the recession of the flood.

In addition to seasonal cattle herding, the floodplain also provides a year around fishery, perhaps involving 1,000 people including licensed fishermen, their helpers, and fish traders. Through the early years of this century fishing was a seasonal occupation during the period of low water levels, practiced with traditional gears by local tribesmen (PIKE & CAREY, 1965; BROWNRIDGE, 1968). During the 1950's the Kafue fishery underwent a marked change brought about by the availability of inexpensive nylon gill nets and a shortfall in the usual supply of fish from other areas of Zambia. At first, gill nets were used seasonally with seine nets, but by 1958 fishing became a year around occupation (WILLIAMS, 1960). From the beginning of full-time fishing through 1970, the annual catch has averaged 5,788 metric tons and ranged from 2,450 metric tons to 10,250 metric tons (Fishery Statistics, Central Statistical Office, Lusaka).

The Kafue fish fauna consists of 67 species (BELL-CROSS, 1972), of which about 19 species are exploited by gill nets and seines of a minimum stretched mesh size of 7.6 centimeters. Although the Kafue fishes are believed to be similar to those of the Upper Zambezi system (BELL-CROSS, 1968a and Section 6.1–6.3), 21 species occur both in the Kafue and in Lake Kariba. Of the species in common between the two systems, 5 have been studied in Lake Kariba, and the production and yield parameters of one species, *Tilapia rendalli* have been estimated for both systems (Section 3.3 and herein).

Flooding of the plain is caused by inputs from three sources – direct rainfall, flows from tributaries, and overspill from the Kafue River. Usually, local rainfall and tributary flows cause flooding beginning in December. Later, flood waters from catchments above the Kafue Flats arrive at Itezhi-Tezhi with increasing volume and the river overflows its banks causing additional flooding. Maximum flooding is attained in February or March in the western end of the plain, but not until April or May at the eastern end (FAO Survey, 1968). The contribution of each of the inputs to flooding varies and therefore the timing, extent and duration of flooding also vary from year to year.

Over an 11-year period the area of maximum inundation ranged from about 3200 km^2 to 5000 km^2 with an average maximum inundation at peak flood of about 4300 km^2. The most extensive flooding is of short duration, perhaps for only a few days. The mean area annually flooded for a four month period has been estimated as 2800 km^2 (SWECO, 1967).

In May, when the rainy season ends, water levels begin to recede, and the plain usually becomes dry three or four months later. In 1970, the floodplain became dry by early September; however, in 1969, much of the plain remained inundated throughout the dry season.

In years when the floodplain becomes entirely dry about 1,460 km^2 remain inundated. The area remaining inundated consists of former river channels, oxbows, and small 'pond-like' and large 'lake-like' bodies of water collectively referred to as lagoons. Some lagoons may become isolated from the river channel as the dry season progresses, and other shallow lagoons may dry completely. With continued evaporation and transpiration through the course of the dry season the area in river channel and lagoons may become reduced to about 1,210 km^2 (FAO Survey, 1968).

At 970 to 980 meters above sea level, 500 meters above Lake Kariba, the Kafue floodplain is cooler than the Lake Kariba area with mean monthly air temperatures about 4 °C lower. From July, 1969 through June, 1970, mean monthly air temperatures on the eastern floodplain ranged from a low of 16.8 °C in July to 25.1 °C in October, and the mean monthly extremes were from 8.2 °C in July to 32.2 in October (Department of Meteorology, Lusaka). Over nearly the same period, from May, 1969 to May, 1970, water temperatures, averaged through the water column, ranged from 17.7 °C to 29.1 °C with extremes of 17 °C to 35 °C (CHAPMAN et al., 1971) (Fig. 76).

Annual variations in the water chemistry of the floodplain system are a function of the flood regime and the periods of high and low primary production closely associated with its phases. Lowest nutrient concentrations occur during the period of rising water levels (TAIT, 1967a). At this time primary production in the aquatic system is at its highest annual levels. For example, some floodplain grasses that were from 15 to 30 cm in height at the beginning of December were from 90 to 130 cm in height by mid-month (VAN RENSBURG, 1968).

As the water levels recede the leaves and stems of floodplain grasses decompose and nutrient concentrations increase due in part to concentration and in part to decreasing rates of primary production, a pattern which is evident in the annual variation in water level, conductivity, and pH (Fig. 76).

In general, nutrient concentrations in the Kafue are higher than in the richest areas of Lake Kariba (Basins III and IV). Total dissolved solids in the Kafue River range from 110 to 252 ppm and total hardness is from 90 to 203 ppm (SWECO, 1967). Although most attempts to measure phosphates indicated no measurable concentrations, when values could be obtained they were as high as 0.34 ppm (TAIT, 1967a).

Dissolved oxygen concentration may be the most important chemical variable influencing fish distribution in the Kafue floodplain system (CHAPMAN et al., 1971). As the heavy rains begin in November and December, allochtonous organic material is washed into the aquatic system by rainfall runoff, and as the floodwaters rise over the plain additional organic substances such as animal manure and vegetation are inundated. The result is a high biochemical oxygen demand which may completely deoxygenate large areas of the floodplain except in the vicinity of its margin. Over a three-week period in January, 1970, the inflow of organically laden floodwaters to a large lagoon caused a drop from 7.5 ppm to 0.5 ppm in dissolved oxygen concentration (Fig. 76). As the flood progresses, the oxygen demand is gradually satisfied, and by the time maximum water levels are attained most of the plain is reoxygenated. However, beginning about the time of recession of the flood the rate of decomposition of organic material again increases as vegetation dies. Oxygen concentrations in some areas may again decrease to levels critical for some fishes (CHAPMAN et al., 1971;

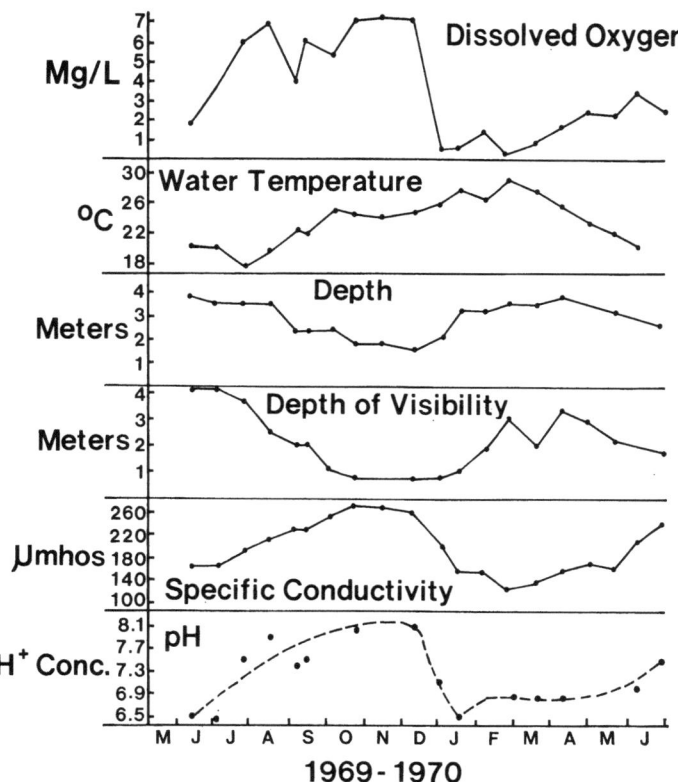

Fig. 76. Annual variation in the physical and chemical parameters of the Kafue floodplain system (adapted from SCULLY, 1972).

LAGLER et al., 1971) and some fish mortalities may take place (TAIT, 1967b; SCULLY, 1972).

In the following sections results of the investigation of growth, abundance and standing stock, mortality, and production and yield of *Sarotherodon andersoni*, *Sarotherodon macrochir*, and *Tilapia rendalli* are presented. These species are in the first group selected for study because of their biological and economic importance in the Kafue floodplain ecosystem. Collectively, the three species comprised from 47 to 59 percent of the total standing stock in three series of population estimates made in 1970 (LAGLER et al., 1971, and herein). As a group, the three species are the most ubiquitous fishes in the commercial catch and are preferred by fishermen, traders, and consumers over the group second in abundance consisting of two species of *Clarias*.

The spawning seasons of the Kafue *Tilapia* and *Sarotherodon* extend from October through January (CHAPMAN et al., 1971), coinciding with the beginning of the warm, rainy season. Spawning may occur for the mouthbrooders, *Sarotherodon*, before the flood (WILLIAMS, 1971). Filamentous algae, vascular plant fragments and various crustaceans have been found in the stomachs of each of the species (CAREY, 1968);

Table 131. Total length-scale radius relationships, standard length- total length relationships, and weight-length relationships for Sarotherodon andersoni, Sarotherodon macrochir, and Tilapia rendalli.

Species	Geometric mean regression	Sample size
Sarotherodon andersoni	Total length = 12.5422 + 1.2197 (Sample radius × 40)	617
	Standard length = −1.6700 + 0.8038 (Total length)	220
	Log_{10} weight = −4.91516 + 3.07986 (Log_{10} total length)	220
Sarotherodon macrochir	Total length = 5.8148 + 1.1191 (Scale radius × 40)	733
	Standard length = −5.0242 + 0.8214 (Total length)	226
	Log_{10} weight = −5.07125 + 3.17372 (Log_{10} Total length)	226
Tilapia rendalli	Total length = 0.4083 + 1.1205 (Scale radius × 40)	395
	Standard length = −3.0755 + 0.8183 (Total length)	173
	Log_{10} weight = −5.04972 + 3.15923 (Log_{10} total length)	173

however, *S. andersoni* may be primarily a bottom feeder (MORTIMER, 1960), and *T. rendalli* may feed mostly on vascular plants (CHAPMAN et al., 1971).

Acknowledgements. – In particular the author wishes to express his appreciation to Dr. KARL F. LAGLER for counsel in planning the investigation, and to DONALD J. STEWART for his assistance with the field work.

Growth of Sarotherodon andersoni, Sarotherodon macrochir, and Tilapia rendalli. – Materials for the study of growth were collected from June to mid-November, 1970, in up to 62 quantitative samples of fish stocks, made in the principal habitat types of the central and eastern floodplain areas. In order to reflect the temporal and spatial distribution of the collections, they were grouped into 20 data sets. Each data set included key scales and random samples of lengths and weights obtained in a similar manner to that already described (Chapter 2).

Scales were read on an Eberbach projector at 40X magnification either mounted between glass slides or as celluose acetate impressions. Scale measurements were made along the ventral (longest) radius of the scale, on a perpendicular to the anterior axis. The body length-scale radius, weight-length, and standard length-total length relationships were described by using geometric mean regressions (Table 131).

The annuli on the scales of the three species are remarkably well defined and their characteristics are similar to those already described for *Sargochromis codringtoni* by HOLČIK (Section 3.1). In contrast to the scales of some north temperate species, such as *Lepomis macrochirus*, there is no pronounced density of sclerites proximal to the annulus nor is there a marked wider spacing between the sclerites distal to the annulus. False annuli may be identified according to the criteria given by CHUGUNOVA (1959).

Criteria used in the recognition of new annuli at the scale margin were the appearance of the mark on all of the scales of the sample, and subsequent growth distal to the new annulus. According to these criteria annulus formation began in October, 1970; however, some of the scales which were not counted as having an annulus at the margin gave the appearance of iminent mark formation. On these scales the outermost sclerites delimited a 'disturbed' region. Often there were wide spaces between the extreme marginal sclerites, and other marginal sclerites were discontinuous in a fashion not characteristic of the margins of scales observed from collections taken earlier in the year.

In mid-October, 1970, newly formed annuli appeared on a small percentage of the scales of *S. andersoni* and *S. macrochir* taken from the Ceres section of the river (Table 132). Collections made in the Chikunka section of the river about three weeks later showed a much higher proportion of new annuli, 40 per cent for *S. andersoni* and 53 per cent for *S. macrochir*. In the Luwato lagoon, at the upstream end of the Chikunka section, the scales of the same two species indicated that new annulus formation was well advanced. No new annuli were found on the scales of *T. rendalli* from the Ceres and Chikunka river sections; however, 8 per cent of the samples from the Luwato lagoon exhibited recent annulus formation.

Juvenile marks were easily seen on the scales of *T. rendalli* and occurred at a mean back-calculated total length of 42 mm (31 mm in standard length). These marks were also observed on the scales of *S. andersoni*, but at a lower frequency of occurrence than for *T. rendalli*. The mean total length at which these marks were formed was 45 mm (35 mm in standard length). Juvenile marks were not apparent on the scales of *S. macrochir*.

Significant differences in growth between the sexes of *S. andersoni* and *S. macrochir* can first be demonstrated after completion of the third season of growth, based on the results of one-way analyses of variance of the scale measurements of males and females in the same age group (Fig. 77). Insufficient sample sizes of older aged *T. rendalli* precluded extensive analysis; however, there was no significant difference between male and female growth at the completion of the first and second seasons of growth.

The mean back-calculated lengths attained by *S. andersoni* and *S. macrochir* at the end of the first and second seasons of growth exhibit a trend toward decreasing average lengths among the fish of older age groups (Table 133). The results of one-way analysis of variance confirmed that significant differences existed among the scale measurements to the first and second annuli for each species.

Therefore, the occurrence of Lee's phenomenon (RICKER, 1969) was suspect and further tests were made. These tests, regressions of the mean scale measurements to the first annulus and to the second annulus on the ages attained by the various age groups, indicated that there was no mathematical basis for Lee's phenomenon when significance of regression was used as the test criterion. Moreover, it has been shown that annual variation in the flood regime can account for much of the year to year variation in the first year growth of *S. andersoni* (DUDLEY, 1972).

Despite the apparent differences in growth among age groups, and although there is dimorphic growth between sexes, the scale measurement data from various age groups have been pooled to form the composite representation of growth used by other authors (Table 134) (Sections 3 and 4). This is to enable comparisons with the results of growth studies for Lake Kariba *Sarotherodon* and *Tilapia* and is not meant to suggest a belief of stability in the system which is implicit in such a manipulation.

Applying the relative indices of growth (see Chapter 2), species average size (Φ h)

Table 132. New annulus formation on the scales of *Sarotherodon andersoni*, *Sarotherodon macrochir*, and *Tilapia rendalli*.

Species	Sample areas and dates of samples								
	Ceres section 15–17 October, 1970			Chikunka section 7–11 November, 1970			Luwato lagoon 13–14 November, 1970		
	New annuli	Percentage of new annuli	Total sample	New annuli	Percentage of new annuli	Total sample	New annuli	Percentage of new annuli	Total sample
Sarotherodon andersoni	2	2.8	72	51	39.5	129	19	31.1	61
Sarotherodon macrochir	5	7.6	66	39	52.8	74	23	82.1	28
Tilapia rendalli	0	0.0	88	0	0.0	77	9	7.6	118

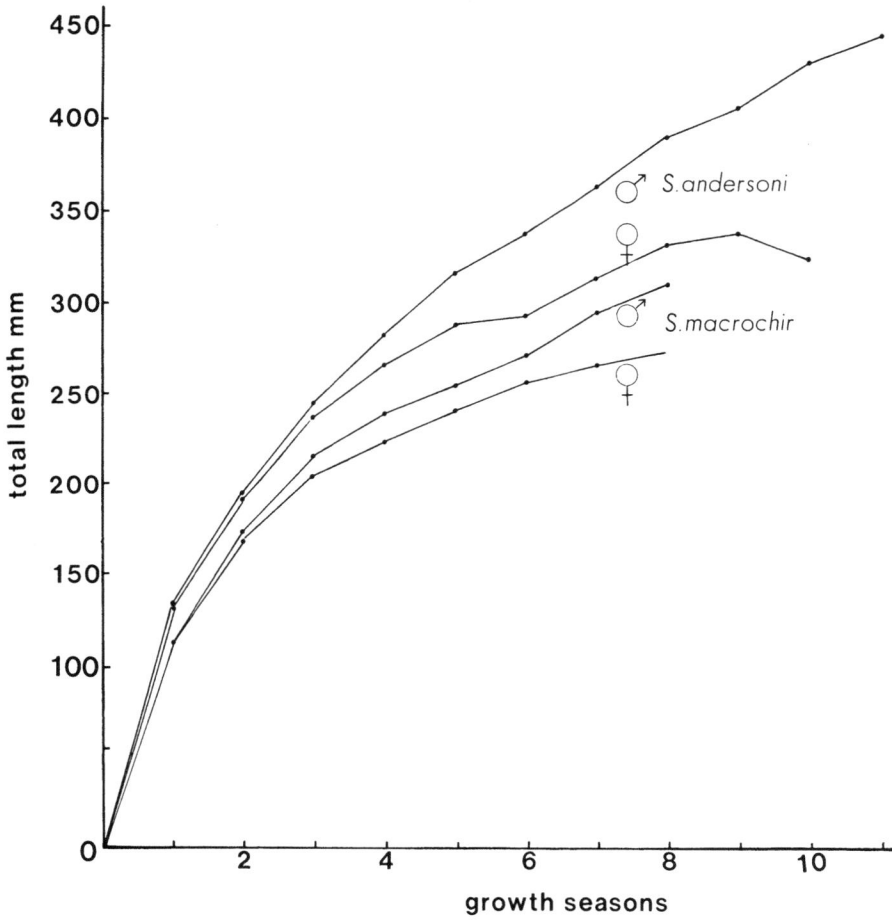

Fig. 77. Average total lengths at the end of growth seasons of male and female *Sarotherodon andersoni* and *Sarotherodon macrochir*.

and stock weight growth intensity (Φ Cw), *T. rendalli* is foremost among the three species with indices Φ h = 3.5 and Φ Cw = 743, *S. andersoni* is intermediate (Φ h = 3.2, Φ Cw = 439), and *S. macrochir* is last (Φ h = 3.0, Φ Cw = 400).

The growth of *Tilapia rendalli* in Lake Kariba has been studied by BASTL (Section 3.3). In the first year of growth the Lake Kariba *T. rendalli* attain only 59 percent of the standard length of the Kafue *T. rendalli*; however, in subsequent years of growth the Lake Kariba *T. rendalli* converges in length on its Kafue conspecific. By the end of the fifth season of growth, the last season for which growth data were available for Kariba, the Lake Kariba population has attained 92 percent of the standard length of the Kafue population (Fig. 78).

In terms of weight, the Kafue *T. rendalli* is surpassed by the *T. rendalli* of Lake

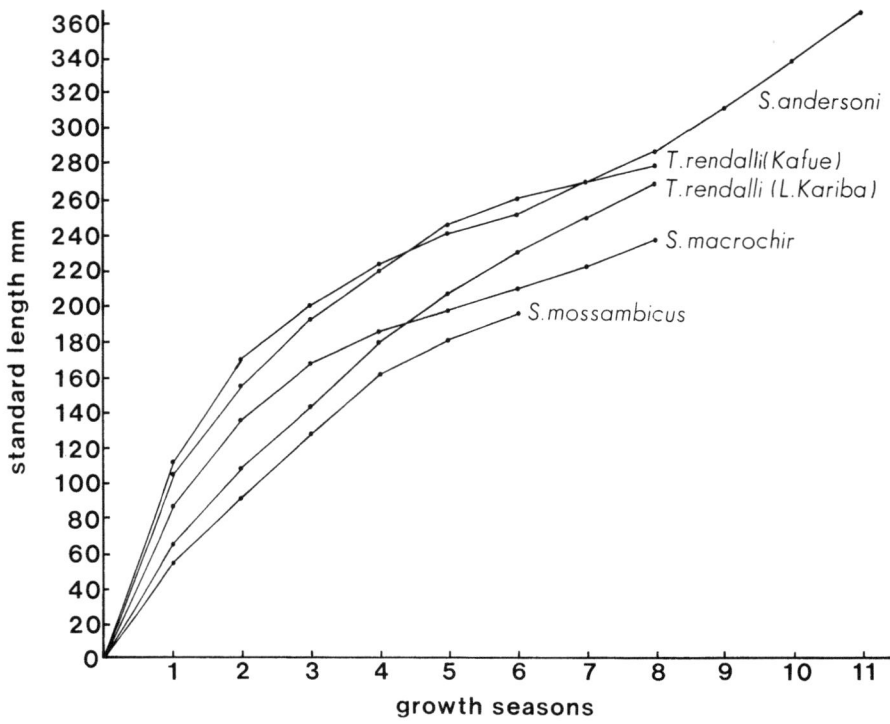

Fig. 78. Average standard lengths at the end of growth seasons of *Sarotherodon* and *Tilapia* of the Kafue floodplain and of Lake Kariba.

Table 133. Mean total lengths attained at the end of the first and second seasons of growth by age groups of *Sarotherodon andersoni* and *Sarotherodon macrochir*.

	Sarotherodon andersoni			*Sarotherodon macrochir*		
	First growth season	Second growth season		First growth season	Second growth season	
Age group	Total length mm	Total length mm	Sample size	Total length mm	Total length mm	Sample size
1	134	—	261	129	—	205
2	127	204	112	105	182	217
3	123	189	106	105	164	166
4	122	194	46	96	166	46
5	121	188	31	98	156	28
6	127	183	12	113	173	14
7	127	176	7	123	174	12
8	131	199	10	104	164	7
9	121	184	6			

Table 134. Composite representation of the growth of *Sarotherodon andersoni*, *Sarotherodon macrochir*, and *Tilapia rendalli*.

	Number of growth seasons										
	1	2	3	4	5	6	7	8	9	10	11
Sarotherodon andersoni											
Mean total length, mm	134	194	242	276	307	324	337	358	389	377	444
Length increment, mm	134	50	48	34	31	17	13	31	21	−11[1]	57
Mean weight, g	43	135	267	400	556	656	741	892	1152	1046	1731
Weight increment, g	43	92	132	133	156	100	85	151	260	−106[1]	685
Sample size	593	332	220	114	68	37	25	18	8	2	1
Sarotherodon macrochir											
Mean total length, mm	112	172	211	231	247	262	276	295			
Length increment, mm	112	60	39	20	16	15	14	19			
Mean weight, g	27	106	202	269	333	402	474	585			
Weight increment, g	27	79	96	67	64	69	72	111			
Sample size	695	490	273	107	61	33	19	7			
Tilapia rendalli											
Mean total length, mm	141	212	248	277	298	310	334	344			
Length increment, mm	141	71	36	29	21	12	24	10			
Mean weight, g	55	199	327	464	585	662	838	920			
Weight increment, g	55	144	128	137	121	77	176	82			
Sample size	327	79	21	6	3	3	2	2			

[1] The interpolated length is 444 mm and the corresponding weight is 1438 g.

Kariba after the fourth season of growth. This apparent anomaly can be explained by differences in the weight-length relationships for the two populations. At 207 mm of standard length, the length attained at the end of the fifth season of growth the corresponding weight of the Kariba *T. rendalli* is 520 g (see Section 3.3), however, for a Kafue *T. rendalli* of equal standard length the corresponding weight is 366 g, equivalent to only 70 percent of that attained by the Lake Kariba population. The apparent substantial difference in weights for the same length may indicate real weight differences between the two populations.

A plot of standard lengths at the end of successive growth seasons for *Sarotherodon mossambicus* (Fig. 77) exhibits essentially the same pattern as that for *T. rendalli* of Lake Kariba. When compared to the plots of standard length for Kafue species, it is apparent that a different pattern of growth of *Tilapia* and *Sarotherodon* exists between the two systems. For the *Tilapia* and *Sarotherodon* of Lake Kariba growth is slow for the first two growth seasons relative to the Kafue species; however, as growth decreases substantially in later years for the Kafue *Sarotherodon* and *Tilapia* their Lake Kariba congeners maintain nearly the same rate of growth for several more growth seasons. This pattern is most evident from a comparison of the length plots of the Lake Kariba and Kafue *Tilapia rendalli* (Fig. 78).

Abundance and standing stock. – After the floodplain has dried the lagoons and river channel are the only habitats remaining which can support fish life, and as the dry season progresses the extent of these habitats is successively reduced. Based on a total of 25 population estimates made when these habitats were approaching their minimum expanses in 1970, the mean abundance of the 31 species encountered in lagoons was 11,053/ha and the mean abundance of 24 species found in the river channel was 3,894/ha. The corresponding average standing stock was 680 kg/ha for lagoons and 333 kg/ha for the river channel (Table 135). In lagoons, *S. andersoni*, *S. macrochir*, and *T. rendalli* collectively comprised 53 percent of the standing stock; in the river channel the three species accounted for 73 percent of the mass. Weighted for the relative extent of lagoon and riverine habitats in the eastern floodplain sample area, the mean abundance of fishes was 8,191/ha and their mean weight was 541 kg/ha. The average standing stock of the three species group was 317 kg/ha, 59 percent of the total for all fishes.

The total sample area included the eastern floodplain from Luwato lagoon east to its terminus at Kafue (Fig. 79). Within this area the river channel was sampled in three sections, Kafue, Ceres, and Chikunka, in consideration of apparent ecological differences among the sections and also on the basis of varying commercial fishing pressure. Lagoons were sampled according to broad ecological categories. Luwato lagoon was the only large 'river-like' type in the sample area, Chanyanya lagoon exemplified the large 'lake-like' kind of lagoon of which there were four, and an unnamed lagoon confluent with the Chikunka river section served to approximate fish faunal abundance and standing stock in the small narrow lagoons.

Methods employed to make population estimates and evaluations of their efficiencies have been presented in detail elsewhere (LAGLER *et al.*, 1971). Therefore, only a summary is given here. All of the riverine population estimates were made with a

Species	Lagoons				River channel			
	Mean standing stock		Mean abundance		Mean standing stock		Mean abundance	
	kg/ha	Range	N/ha	Range	kg/ha	Range	N/ha	Range
*Sarotherodon andersoni	180.34	1.49–709.03	839	11– 2627	101.25	9.57–354.72	837	90–3571
*Tilapia rendalli	100.04	14.26–251.70	706	236– 2084	20.87	0.19–104.11	187	3–1003
*Sarotherodon macrochir	84.03	27.07–150.18	737	127– 1295	123.90	4.73–541.56	1327	54–5314
*Tilapia sparrmani	63.50	0.69–177.70	1091	26– 2842	23.24	0.0 –245.97	414	0–4348
*Synodontis macrostigma	42.54	0.0 –247.69	2215	0–13125	1.63	0.0 – 10.33	77	0– 403
*Clarias ngamensis	40.65	0.79–213.49	134	5– 635	2.66	0.0 – 16.47	11	0– 73
Alestes lateralis	36.81	3.65– 63.60	2347	302– 4190	9.40	0.41– 86.40	492	27–4025
*Clarias gariepinus	30.58	3.47– 96.34	28	9– 60	10.54	0.0 – 70.57	17	0– 167
*Schilbe mystus	18.48	0.0 – 71.79	355	0– 1034	12.48	0.03–158.10	178	3–2350
*Hepsetus odoe	16.12	0.54– 39.95	85	7– 226	18.04	0.0 – 45.77	87	0– 264
*Marcusenius macrolepidotus	14.65	0.0 – 86.76	1008	0– 6000	2.07	0.0 – 17.85	94	0– 941
Serranochromis angusticeps	12.88	1.40– 33.96	62	20– 124	0.97	0.0 – 5.98	10	0– 59
*Sargochromis giardi	10.93	0.20– 42.32	145	4– 611	0.20	0.0 – 1.37	3	0– 24
Barbus poechii	7.41	0.0 – 21.60	628	0– 2124	0.90	0.0 – 4.97	62	0– 216
*Haplochromis carlottae	5.45	0.39– 16.84	62	8– 176	1.04	0.0 – 7.03	15	0– 107
*Serranochromis macrocephalus	4.24	0.29– 9.93	27	7– 58	0.20	0.0 – 1.54	2	0– 19
Barbus unitaeniatus	2.14	0.0 – 6.44	199	0– 585	0.23	0.0 – 1.80	20	0– 145
*Sargochromis codringtoni	1.93	0.0 – 7.57	24	0– 112	0.11	0.0 – 1.01	1	0– 7
*Labeo molybdinus	1.91	0.0 – 6.33	19	0– 64	2.89	0.0 – 18.34	27	0– 199
Clarias stappersi	1.31	0.0 – 7.16	7	0– 36	0.02	0.0 – 0.39	.01	0– 3
Clarias theodorae	1.07	0.0 – 5.16	29	0– 140	—	—	—	—
*Serranochromis thumbergi	0.64	0.0 – 1.78	6	0– 18	0.39	0.0 – 4.01	4	0– 29
Ctenopoma multispinis	0.59	0.0 – 4.08	36	0– 248	—	—	—	—
*Serranochromis robustus	0.56	0.0 – 2.56	3	0– 10	0.05	0.0 – 0.55	.70	0– 9
Pseudocrenilabrus philander	0.39	0.0 – 2.32	188	0– 1128	—	—	—	—
*Mormyrus lacerda	0.37	0.0 – 2.22	3	0– 20	0.03	0.0 – 0.50	.20	0– 3
Marcusenius castelnaui	0.22	0.0 – 1.32	46	0– 276	—	—	—	—
Tilapia ruweti	0.07	0.0 – 0.40	11	0– 64	—	—	—	—
Petrocephalus catostoma	0.02	0.0 – 0.64	4	0– 13	0.24	0.0 – 1.91	28	0– 225
Barbus paludinosus	0.01	0.0 – 0.80	9	0– 56	—	—	—	—
Totals	679.88		11053		333.35		3894	

* species of commercial importance.

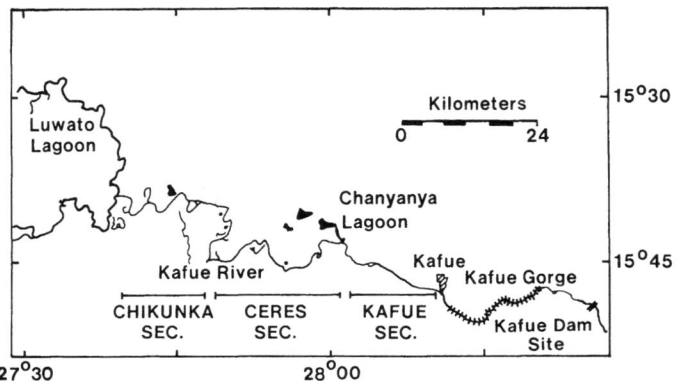

Fig. 79. Lagoons and river channel sections in the eastern Kafue floodplain area.

100 m × 6 m acentric, variable mesh bag seine with meshes (stretched) from 1.3 cm in the bag to 7.6 cm in the wings. This seine was sometimes used alone, and also in conjunction with a 200 m × 5 m × 2.5 cm blocking net. Seine hauls were made within the blocked area and the blocking net was fished for the residual. In the Luwato lagoon the 200 m × 5 m blocking net was used as a seine, and in Chanyanya lagoon it was set to enclose a 0.25 ha area in which 75 percent emulsifiable toxaphene was pumped as a fish eradicant. In the unnamed lagoon no blocking device was used; however, the shape of the sample area was such that it became narrowly constricted at both ends and a dense 'curtain' of rotenone was maintained at these narrow points white the remainder of the sample area was treated. In total, 19 population estimates were made in the river channel and 6 were made in lagoons.

Using the age determinations, length frequencies, and estimates of abundance per hectare which were available for each of the river sections and for each of the lagoons sampled, the numbers per hectare of each age group of each of three species were tabulated and averaged for the riverine habitat and for the lagoon habitat. Because the total extent of lagoon and riverine habitats in the sample area was different, 60 percent and 40 percent, respectively, it was necessary to weigh the mean abundances of each age group of each species in each of the two habitat types in order to obtain a representative overall mean abundance per hectare. The abundances (Table 136), so weighted, provide the basis for estimates of density and biomass.

Density, biomass, and mortality. – The transformations from empirical abundance to stock density were made on the assumption that the two parameters could be directly equated. This assumption can be supported on the basis of the following evidence. The population estimates, from 7 October to 14 November, 1970, closely coincided with the end of the current growth season and the beginning of a new season of growth. This is evident from the occurence of newly formed annuli which began for two of the species in mid-October and for the other species in mid-November (Table 132). Further, the new 'hydrological year' commenced in the eastern floodplain area when

Table 136. Mean abundance per hectare of age groups of *Sarotherodon andersoni*, *Sarotherodon macrochir*, and *Tilapia rendalli* in riverine and lagoon habitats, and weighted mean abundance per hectare in the eastern floodplain area.

Habitat	N/ha of age group											
	0	1	2	3	4	5	6	7	8	9	10	11
Sarotherodon andersoni												
River Channel	0	814	13	8	1	0.5	0.7	—	—	—	—	—
Lagoons	0	509	111	116	58	35	5	2	2	1	0.3	0.3
Weighted mean abundance	0	631	72	73	35	21	3	1	1	0.6	0.2	0.2
Sarotherodon macrochir												
River channel	9	1243	40	23	8	2	1	0.8	—			
Lagoons	39	522	46	51	32	33	14	—				
Weighted mean abundance	27	810	87	32	22	21	9	0.3				
Tilapia rendalli												
River channel	18	166	2	1	—	—	—	—	—			
Lagoons	234	403	54	7	4	—	—	—	4			
Weighted mean abundance	148	308	33	9	2	—	—	—	2			

water levels began to rise on 9 November. Finally, the beginning of the spawning period of the three species, in October (CHAPMAN et al., 1971; WILLIAMS, 1971), coincides with the timing of the population estimates.

The youngest age groups of some of the species probably were not quantitatively represented in the stock samples because of the selectivity of the fishing gears. Older age groups might not have been sampled adequately due to their non-random distribution relative to the habitats which were sampled and probable immigration and emigration of some age groups. Therefore, in order to minimize these biases, the density of each age group of each species was estimated by interpolation. The method used for interpolation differed from the graphical technique presented previously (Section 2) in that regression of the common logarithm of empirical density on age was used to estimate the parameters for calculation of interpolated density (Table 137).

Because the density of each age group of each species has been mathematically interpolated, the respective instantaneous mortality rates are constant from age group to age group, and are estimated as 2.30259 X (slope of the regression). Initial biomass is calculated as the product of the density of an age group and the mean weight of an individual of the age group at the beginning of the growth season.

Production and yield. – For the approximation of production and yield the estimates derived above for growth and density have been incorporated into the formulae discussed previously (see Section 2). With these procedures the total production of *Sarotherodon* and *Tilapia* in the eastern floodplain area is 461 kg/ha/yr. When adjusted for the much larger extent of lagoons in the floodplain system relative to the eastern sample area, the total production of these three species is increased to 580 kg/ha/yr, equivalent to about 70,000 metric tons for the entire floodplain area remaining inundated at lowest water levels in 1970.

In order to consider simultaneously the parameters of production and yield, the minimum harvestable size of each of the three species is first determined, and then production and yield are considered for the eastern floodplain and for the floodplain system in its entirety.

Minimum harvestable size has been considered from the viewpoint of the graphical method of determination (Section 2) and also from the standpoint of the results obtained by considering the age of first reproduction. For two of the species, *T. rendalli* and *S. macrochir*, the sizes attained at first spawning and the minimum harvestable sizes obtained from the graphical plots were nearly the same. For these two species the minimum harvestable size has been set at the sizes attained at the end of the second season of growth (Table 134). It is for *S. andersoni* that there is a large disparity between the results of the graphical plot and the age of first reproduction.

The graphical intercept for the minimum harvestable size of *S. andersoni* occurs near the end of the second season of growth at a total length of 189 mm; however, the available evidence on size at first reproduction suggests that most spawners are more than 300 mm in total length, and that in the Kafue *S. andersoni* less than 240 mm in total length have not as yet been observed to reproduce (MORTIMER, 1960; DUDLEY, 1972). The ages corresponding to these lengths are three and younger for the non-spawners and five and older for the group in which most spawners occur. However,

Table 137. Empirical density, interpolated density, initial biomass, and density parameters for *Sarotherodon andersonii*, *Sarotherodon macrochir*, and *Tilapia rendalli*.

Sarotherodon andersoni	Age group											Total
	0	1	2	3	4	5	6	7	8	9	10	11	
Empirical mean density N/ha	—	—	631	72	73	35	21	3	1	1	0.6	0.2	837.8
Interpolated density N/ha	1557	697	312	140	63	28	13	6	3	1	0.5	0.2	2820.7
Initial biomass kg/ha	1.56	29.97	42.12	37.38	25.20	15.57	8.53	4.45	2.68	1.15	0.72	0.35	169.68
Density regression parameters	$\text{Log}_{10} N'/\text{ha} = 3.19220 - 0.34896 (\text{Age})$												

Sarotherodon macrochir	Age group									Total
	0	1	2	3	4	5	6	7	8	
Empirical mean density N/ha	—	27	810	87	32	22	21	9	0.3	1008.3
Interpolated density N/ha	3739	1343	483	173	62	22	8	3	1	5834
Initial biomass kg/ha	3.74	36.26	60.86	34.95	16.68	7.33	3.22	1.42	0.58	165.04
Density regression parameters	$\text{Log}_{10} N'/\text{ha} = 3.57270 - 0.44456 (\text{Age})$									

Tilapia rendalli	Age group									Total
	0	1	2	3	4	5	6	7	8	
Empirical mean density N/ha	—	148	308	33	9	2	—	—	—	8008.38
Interpolated density N/ha	6457	1251	242	47	9	2	0.3	0.07	0.01	
Initial biomass kg/ha	6.46	68.81	48.16	15.37	4.18	1.17	0.20	0.06	0.01	144.42
Density regression parameters	$\text{Log}_{10} N'/\text{ha} = 3.81000 - 0.71268 (\text{Age})$									

it is known from collections in dams and ponds in Zambia that *S. andersoni* may reproduce at ages as young as 11 months and at sizes as small as 180 mm (MORTIMER, 1960). The dimorphic growth data for the males and females of *S. andersoni* exhibit significant differences after completion of the third season of growth (Fig. 77), suggesting that *S. andersoni* might spawn in the Kafue system at sizes smaller than those observed.

The minimum harvestable size of *S. andersoni* has been conservatively set at the length attained at the end of the third season of growth, 242 mm (Table 134). It is of interest to note that the modal length of *S. andersoni* caught in commercial seine nets (7.6 cm mesh) in the latter part of 1970 was 190 mm, and that this species has been harvested at this size and at smaller sizes for several years.

Based on the densities characteristic of the lagoons and river channel in the eastern floodplain area, *Tilapia rendalli* has the highest value of total production, 198 kg/ha/yr, *Sarotherodon macrochir* is second with 145 kg/ha/yr, and *S. andersoni* is last with 119 kg/ha/yr (Table 138). The total production: mean biomass ($A/\bar{B}$) ratios are in the same order, 1.59, 1.0, and 0.75, respectively. However, the total yield and available yield estimates do not exhibit the same order. *S. macrochir* is foremost in these parameters with 39 kg/ha/yr and 23 kg/ha/yr, followed by *S. andersoni* with 23 kg/ha/yr and 15 kg/ha/yr; *T. rendalli* is last with 18 kg/ha/yr in total yield and 8 kg/ha/yr in available yield.

For the approximately 3 880 ha of the eastern floodplain inundated at lowest water levels during 1970 (derived from planimeter measurements on 1: 50 000 maps), total production of the three species might amount to 1700 mt and total yield about 300 mt. The available yield estimate of 170 mt, suggests that additional catches of *Sarotherodon* and *Tilapia* could have been removed from the system without harm to stocks.

The densities and hence the production and yield estimates made for each species in the eastern floodplain area were based on the abundance of populations in lagoons and in the river channel, with the grand mean abundances weighted for the relative extent of these two habitat types in the sample area. To extend these production and yield estimates over the entire floodplain system, it is necessary to recalculate the grand mean abundance for each species and to reinterpolate the densities to reflect the much larger extent of lagoon habitats on the floodplain. In the total floodplain system, lagoons and adjacent marsh areas occupy 96 percent of the total surface area (115,700 ha) and the river channel only about 4 percent (5,300 ha) (derived from SWECO, 1967; FAO Survey, 1968). In effect, with a weighting factor of only 4 percent, the riverine population estimates become insignificant in the calculation of the overall mean abundances for each species, and the result is a shift in the mean abundances almost entirely toward the characteristics of lagoon populations. Relative to the river channel population estimates, these characteristics are a greater abundance of individuals of older ages, and a lesser abundance of fishes of younger ages (Table 136). Therefore, there are increases in the mathematically interpolated density and initial biomass estimates and decreases in the instantaneous mortality rates. In turn, the production and yield estimates for each species are higher for the entire floodplain system than for the eastern floodplain area *per se*.

Table 138. Production and yield estimates for Sarotherodon andersoni, Sarotherodon macrochir and Tilapia rendalli in the eastern floodplain area.

Sarotherodon andersoni

Growth seasons		0	1	2	3	4	5	6	7	8	9	10	11	Total/Mean
Interpolated density N/ha	N'	1557	697	312	140	63	28	13	6	3	1	0.5	0.2	2820.7
	Z								741	892	1152	1438	1731	
Mean weight g		1	43	135	267	400	556	656						
Initial biomass kg/ha	B'	1.56	29.97	42.12	37.38	25.20	15.57	8.53	4.45	2.68	1.15	0.72	0.35	169.68
Rate of increase in biomass	H		2.9577	0.3406	−0.2979	−0.3993	−0.4742	−0.6381	−0.6817	−0.6150	−0.5477	−0.5817	−0.6179	
Instantaneous growth rate	G_x		3.7612	1.1441	0.5056	0.4042	0.3293	0.1654	0.1218	0.1885	0.2558	0.2218	0.1855	0.6621
Instantaneous mortality rate	Z							−0.8035 −						0.8035
Mean biomass kg/ha	$\bar{B}$		9.61	35.71	36.43	30.82	20.01	11.51	6.18	3.32	2.06	0.87	0.57	157.09
Total production kg/ha/yr	A		36.14	40.85	18.42	12.46	6.59	1.90	0.75	0.62	0.53	0.19	0.11	118.56
Available production kg/ha/yr	P'		29.27	28.70	18.48	8.38	4.37	1.30	0.51	0.45	0.26	0.14	0.06	91.92
Total yield kg/ha/yr	Y_A		—	—	—	12.46	6.59	1.90	0.75	0.62	0.53	0.19	0.11	23.15
Available yield kg/ha/yr	$Y_{P'}$		—	—	—	8.38	4.37	1.30	0.51	0.45	0.26	0.14	0.06	15.47

Table 138. (Continued)

Sarotherodon macrochir

Growth seasons		0	1	2	3	4	5	6	7	8	Total/Mean
Interpolated density N/ha	N'	3739	1343	483	173	62	22	8	3	1	5834
Mean weight g	Z	1	27	106	202	269	333	402	474	585	
Initial biomass kg/ha	B'	3.74	36.26	51.20	34.95	16.68	7.33	3.22	1.42	0.59	155.39
Rate of increase in biomass	H		2.2722	0.3440	−0.3788	−0.7372	−0.8102	−0.8353	−0.8589	−0.8132	
Instantaneous growth rate G_x			3.2958	1.3676	0.6488	0.2864	0.2134	0.1883	0.1647	0.2104	0.7969
Instantaneous mortality rate	Z					−1.2036 →					
Mean biomass kg/ha	B̄		14.32	43.28	42.73	24.72	11.43	4.97	2.16	0.97	144.58
Total production kg/ha/yr	A		47.19	59.19	27.55	7.08	2.44	0.93	0.35	0.21	144.94
Available production kg/ha/yr	P'		34.92	38.16	16.61	4.15	1.41	0.55	0.22	0.11	96.13
Total yield kg/ha/yr	Y_A		—	27.55	27.55	7.08	2.44	0.93	0.35	0.21	38.56
Available yield kg/ha/yr	$Y_{P'}$		—	16.61	16.61	4.15	1.41	0.55	0.22	0.11	23.05

Table 138. (Concluded)

Tilapia rendalli

Growth seasons		0	1	2	3	4	5	6	7	8	Total/Mean
Interpolated density N/ha	N'	6457	1251	242	47	9	2	0.3	0.07	0.01	8008.38
Mean weight g	Z	1	55	199	327	464	585	662	838	920	
Initial biomass kg/ha	B'	6.46	68.81	48.16	15.37	4.18	1.17	0.20	0.06	0.01	144.42
Rate of increase in biomass	H		2.3663	−0.3550	−1.1443	−1.2911	−1.4093	−1.5173	−1.4053	−1.5476	
Instantaneous growth rate G_x			4.0073	1.2860	0.4967	0.3499	0.2317	0.1237	0.2357	0.0934	0.8531
Instantaneous mortality rate	Z					−1.6410 −					1.6410
Mean biomass kg/ha	B̄		26.35	57.92	28.68	8.63	2.24	0.60	0.11	0.03	124.56
Total production kg/ha/yr	A		105.61	74.49	14.25	3.02	0.52	0.07	0.03	0.01	198.00
Available production kg/ha/yr	P'		57.55	34.85	6.02	1.23	0.24	0.02	0.01	0.00	109.92
Total yield kg/ha/yr	Y_A		—	—	14.25	3.02	0.52	0.07	0.03	0.01	17.90
Available yield kg/ha/yr	$Y_{P'}$		—	—	6.02	1.23	0.24	0.02	0.01	0.00	7.52

Individually, the recalculated total production estimates amount to 232 kg/ha/yr for *T. rendalli*, 193 kg/ha/yr for *S. macrochir*, and 155 kg/ha/yr for *S. andersoni*. The corresponding total and available yields are, respectively, 24 kg/ha/yr and 10 kg/ha/yr for *T. rendalli*, 41 kg/ha/yr and 23 kg/ha/yr for *S. macrochir*, and 32 kg/ha/yr and 22 kg/ha/yr for *S. andersoni*. Extended over the entire area of the floodplain at low water, 121,000 ha, the approximation of total production for the three species is 70,000 mt, total yield is 12,000 mt, and available yield is 7,000 mt.

Obviously, the confidence to be placed in the production and yield estimates extended over the entire floodplain is less than for the same parameter estimates made for the eastern floodplain area in which the population estimates were made.

Discussion and conclusions. – Having approximated production and yield for three of the most important species of the Kafue floodplain system, of importance is the reliability of these estimates in relation to the sampling error of the field methods and within the context of the simplifying assumptions which are necessary for the analysis and interpretation of the data. The information presented herein regarding the time of annulus formation and the values obtained for back-calculated growth are confirmed by results obtained for the same species in the Kafue system by DUDLEY (1972). In 1970, annulus formation on the eastern floodplain began in mid-October and its progress was followed through mid-November. In 1969, on the central floodplain, annulus formation also began in October, and for two species extended into January, 1970 (DUDLEY, 1972). In comparing the results from successive years it can be inferred that annulus formation may vary in time and by area in different years.

Despite differences in times, locations and methods of collection, measurement along different scale radii, and the use of different kinds of regressions (predictive, geometric mean), for describing the body length-scale radius relationship, the results obtained for back-calculated growth by DUDLEY (1972) are remarkably similar to those presented herein. For those age groups in which sample sizes were adequate for confidence in the results, the differences were insignificant. For example, in the case of *Sarotherodon andersoni* the mean absolute difference in the back-calculated lengths for the first seven age groups was 6 mm, for the same age groups of *S. macrochir*, 8 mm, and for the first six age groups of *Tilapia rendalli*, 4 mm. With the time span of annulus formation known, and with the close agreement between the results of two workers in back-calculated growth, the ageing method for these three species in the Kafue system can be considered as validated. Therefore, inasmuch as the growth measurements contribute to the estimates of production and yield, it can be assumed that acceptable levels of accuracy have been attained.

The mathematically interpolated estimates of density may also be checked according to the type of reproductive behavior of each species. *Sarotherodon andersoni* and *S. macrochir* are mouthbrooders and *Tilapia rendalli* is a bottom spawner in which no oral incubation occurs. Therefore, the number of eggs and young produced by the former two species would be expected to be less than the number produced by the latter species (FRYER & ILES, 1972). The interpolated values of density show this trend, with lesser numbers interpolated for *S. andersoni* and *S. macrochir* than for *T. rendalli* for the beginning of the first year of life.

Although the trends for density of the youngest fishes are in the expected direction for each species, it is likely that their absolute densities have been underestimated as can be demonstrated using information on the fecundity of two of the species. That the estimates of density of individuals at the beginning of the first and second years of life are critical to the values obtained for the production estimates is evident because most of the species' production occurs during these years of life (Table 138).

The average number of ripe eggs found in the ovaries of *S. andersoni* is 429 with a range from 349 to 567 (MORTIMER, 1960). Assuming for the moment that about one-half of the males and females of *S. andersoni* from age five through age nine are spawners, as may be the case (DUDLEY, 1972), and further assuming a 1:1 sex ratio, then from the interpolated densities, 28 pairs could spawn (Table 137), and from the empirical abundances about 23 pairs could be spawners (Table 136). Based on the mean fecundity given above, total egg production from the mathematically interpolated spawners might amount to about 12,000 eggs/ha/yr. Based on empirical abundances about 10,000 eggs/ha/yr might be produced. In contrast, the mathematically interpolated density at the beginning of the first year of life is 1557/ha for *S. andersoni*. The same disparity between the density at the beginning of the first year of life and the density expected from information on fecundity and abundance of spawners can also be demonstrated for *S. macrochir*; however, the apparent underestimate only amounts to about 50 percent as compared to 85 percent for *S. andersoni*. Thus, it is possible that the production of individuals through the first year of life might have been underestimated in the cases of *S. andersoni* and *S. macrochir*. No reliable data are available from which to evaluate the density estimates for *T. rendalli* in this respect.

About the population estimates which underlie the calculations of mean abundance and interpolated densities, little can be said except that to improve the statistical properties of the means could require years of research on the temporal and spatial distribution of floodplain fishes with the methods presently available, and in the end it is doubtful whether this effort could be justified from the standpoint of the pressing need of lesser developed countries for information upon which to base the exploitation of their fishery resources.

Realizing the limitations of the estimates of production and yield made for *Sarotherodon* and *Tilapia* it is still of interest to speculate on the magnitudes of these parameters when all of the fishes of the system are considered. In so doing, production values can be considered in several phases related to the recent history of the floodplain and to its exploitation. The first phase, of indefinite time span, occurred when the floodplain was most nearly in its natural state, before the human population was as large as it is presently and when the wild ungulates, and not cattle, were the dominant animals of the terrestrial system. If, the lechwe (antelopes) were as abundant over the entire floodplain as they are now on one of the game parks set aside for them (about 800/km^2 as derived from GALLAGHER et al., 1972) and if other animals such as the hippopotamus were also more abundant then aquatic production may have been higher than in post-cattle times. The lechwe feed on vegetation at the shoreline of the floodplain as the waters recede, and, adapted to the floodplain existence, they also wade out to depths of shoulder height and graze on emergent and submerged vegetation. Thus, some of their manure is recycled directly into the aquatic system, the remainder arrives later with the rainy season run-off and inundation.

The second phase which may have influenced the fish production potential of the floodplain system began with the increasing human populations and the subsequent replacement of wild ungulates on the plain by cattle. This phase probably began during the first decades of the present century. VAN RENSBURG (1968), discussing utilization of the floodplain vegetation by cattle states: 'The herbage on the floodplain grasslands are largely wasted. Vast areas of tall grasses get burnt repeatedly and are not grazed and utilized. Apart from cattle concentrations in a few areas (...) most of these vast floodplain regions are void and empty.' Thus, if it can be assumed that the lechwe and other wild ungulates were not only more abundant in pre-cattle days, but also more efficient in utilizing the floodplain resources, then it is reasonable to believe that nutrients were more rapidly and efficiently cycled between the terrestrial and aquatic systems, and consequently fish production may have been higher (see Section 3.12 and Concluding Discussion).

The third phase, during which the fish production potential of the system may have changed, began with year around exploitation of the fishery resource and a shift from traditional methods of capture to the more efficient nylon gill nets and seines. This phase, beginning in the mid-1950's and continuing through 1970, may have resulted either in a decrease in fish production over that existing during the previous phase or it may have engendered an increase, according to how one interprets the available evidence. This much is known: numbers of fishermen, numbers of nets, and fishing effort increased from the early 1950's through 1958 when the maximum historical catch of 10,250 mt was attained. At this time overfishing was believed to be iminent (WILLIAMS, 1960), but not until the catch fell to 2,450 mt in 1960 was a restriction on fishing instituted. This restriction, a ban on seining, was largely ineffective because it was not enforced, and by 1963 the gill net catches made during the low water periods were doubled in the official statistics to account for the catch made by illegal seining (Annual Report, Department of Game and Fisheries for 1964 (1965)). In 1966 the ban on seining was lifted in consideration of increasing annual catches. If one takes the position that since the mid-1950's the fish exploitation has had a significant effect on the fish stocks to the point of over-exploitation, then production of the exploited species would probably have been temporarily reduced. However, if one chooses to believe that the fishery has had a significant effect on stocks, but not to the point of over-exploitation, then it is possible that fish production has actually increased through changes in the age structures of the populations of exploited fishes, from the dominance of the older, slower growing individuals to a greater abundance of the younger, faster growing fishes (BALON, 1963b).

At this point, during the production phase corresponding to the time when the field work was completed, it is possible to speculate in quantitative terms about the fish production potential of the floodplain system. In approximating the fish production of the three species, *Sarotherodon* and *Tilapia*, about 60 percent of the standing stock was accounted for, and because the time of the population estimates coincided with the beginning of a new growth season, the standing stock estimates made for other species can be reasonably equated with their initial biomass, less the mass of the younger age groups of the larger species which were not sampled adequately, and less the mass of the smaller species which were not estimated in relation to their actual abundance. Further, the production estimates for the three Kafue cichlids are within the range of the production values for one conspecific and another closely related species from Lake Kariba perhaps indicating a similarity of production potential for the areas inhabited by fishes. Finally, about one-half of the species found in Lake Kariba also occur in the Kafue, and many of the other Kafue species have congeners in Lake Kariba. Thus, it would seem logical as a first crude approximation, to estimate the fish production potential of the Kafue floodplain system on the basis of the production values already calculated for *Sarotherodon* and *Tilapia* from the Kafue system (580 kg/ha/yr), and to this add the product of the initial biomass of species other than the *Sarotherodon* and *Tilapia* studied (306 kg) and the total production: initial biomass ratio ($A/\bar{B} = 1.5$) pertaining to Lake Kariba which gives 459 kg/ha/yr for the total production of species in the Kafue system other than the three species of *Sarotherodon* and *Tilapia*. Thus the total production potential of Kafue floodplain fish stocks might be approximated by 580 kg/ha/yr + 459 kg/ha/yr = about 1000 kg/ha/yr, for all species.

The ecological production approximated for the Kafue floodplain system is quite low in comparison with the same parameter estimate for Lake Kariba (3468 + kg/ha/yr; see section 4), although the production values presented for three of the Kafue species fall within the range of corresponding values for the Kariba congeners (Section 3.3, 3.7). Setting aside for the moment the obvious reasons for these differences such as biases in populations estimates (Appendix B), we may speculatively consider some possible sources for this discrepancy from a broad view of the Kafue system.

The Kafue floodplain fish populations exist in a system which undergoes high amplitude ecological fluctuations seasonally and from year to year. Variation in most environmental parameters can ultimately be traced to variation in the flood regime and it is probably these flood-related fluctuations which determine the magnitude of variations in fish production of the floodplain system. That the amplitudes of the fluctuations in the Kafue system are higher than in Lake Kariba (in its present condition) is evidenced by the apparent higher mortality rates, greater growth increments, lower weight: length relationship, and larger annual variations in growth of Kafue species relative to the same or closely related species in the Lake Kariba ecosystem.

The relatively high amplitude fluctuations characteristic of the system are beneficial from the standpoint of fish production because these environmental variations may not permit the establishment of fish populations of high biomass and low productive potential. However, fluctuations in the extreme may occur, and fish production for periods of several years may be substantially reduced, but due to the adaptation of these populations to the fluctuations, recovery is probably rapid.

The highest amplitude of fluctuation may be attained with the combination of a delayed onset and early termination of the rains, below average rainfall over the flats, low volume of flood, rapid recession of the flood waters, and subsequent contraction and ultimate drying of many floodplain lagoons during the dry season. It is evident that such a combination of conditions would result in decreased growth rates, increased mortality rates and ultimately in decreased fish production in one year. If the same conditions were to prevail in successive years the effect on fish production could be severe.

Although it is apparent that successive low water years could reduce fish populations, it does not follow that the opposite condition, in its extremes, would necessarily promote high fish production. Although a greater expanse of water on the plain would provide a greater area in which aquatic primary production could occur, the anaerobic conditions attendant with the flooding of the plain might also be proportionally more extensive with no net gain in habitat at the floodplain margin. Although some of the Kafue fishes *(Clarias)* are air breathing, it is likely that their food sources are aerobically produced which would confine them for most of the time to oxygenated habitats. Further, if most of the increased inundated area supported grassland vegetation, then a corresponding gain in secondary production might still not be realized. Epiphytic producers might not be able to utilize the expanded grassland areas efficiently because the thick emergent vegetation above the surface and dense mats of dead leaves and stems on the surface appear to limit the penetration of light in the water column thereby reducing the potential production of epiphytes for grazing by fishes and other organisms, as suggested by SCULLY (1972). Furthermore, even assuming that the greater grassland area could be utilized by fishes as the water levels rise and the anaerobic conditions moderate, the areas which might again become oxygen defficient due to the rotting of grasses as the water levels recede would probably be proportionally as large as in other years. Finally, in the case where one high water level year is followed by another, the conditions produced by decomposing vegetation left from the previous year appear to be more severe during the recession of the flood waters in the subsequent year. While the reduction in dissolved oxygen might not cause mass fish mortalities, the resultant stress might be such that much of the energy input would be required to maintain rather than to increase body weight.

It is apparent then that high amplitude ecological fluctuations do occur in the Kafue floodplain system and even though the sequence of events and magnitudes of effects may not be exactly as postulated above, it must also be apparent that the Kafue fish populations are especially adapted to these fluctuations and that such mechanisms of adaptation are incorporated in the population dynamics. In such a system with high amplitudes of ecological fluctuations and corresponding adaptive mechanisms to cope with the fluctuations, it would seem that 'overfishing' and the recovery of the stocks from 'overharvest' are primarily engendered by the environment, that is, by the natural system, in comparison with which the oscillations set in motion by man are probably inconsequential.

From the hypothesis given above, a parallel can be drawn between the Kafue floodplain system and Lake Kariba. Although there is a vast surface area potentially available for fish production in both Lake Kariba and on the Kafue flood plain (for the Kafue for only part of the year), much of these areas cannot be efficiently utilized by the respective fish populations because of adverse physical and chemical factors.

Further, although Lake Kariba probably does not experience the magnitude of fluctuations characteristic of the Kafue system, fluctuations of similar magnitude to those in the Kafue floodplain probably occurred in the alluvium of the Gwembe valley in pre-Kariba times.

7. THE SUCCESS AND FAILURE OF THE CLUPEID INTRODUCTION

In order to present a complete account of the introduction of anchoveta some published data especially from the article of BELL-CROSS & BELL-CROSS (1971) have to be repeated.

Two species of freshwater clupeids, *Stolothrissa tanganicae* and *Limnothrissa miodon*, are endemic to Lake Tanganyika. They form dense schools which are exploited commercially by Greek fishermen and for food by local inhabitants. Sun dried clupeids are an important food item which is marketed as far away as the Zambian and Zaire Copperbelts.

Both Tanganyikan clupeids are small and have a short life span. They usually attain 12 cm in length (maximum 15–17 cm) and live up to 1 year (some 2–3 years). At about 6 months they spawn for the first time (MATTHES, 1968b). They are harvested at some distance from the shore. These clupeids are known to perform diel vertical movements and seasonal in and off shore migrations.

The attempt by Belgian biologists (COLLART, 1960) to transfer *S. tanganicae* to Lake Kivu, though unsuccessful, formed a precedent and presented a temptation for repetition of this experiment for Lake Kariba where the open water niche was un-inhabited by fishes.

7.1. The air lift of clupeid juveniles from Lake Tanganyika

'Investigations into the practical aspects of transferring clupeids into Lake Kariba was initiated by C. C. TAIT early in 1963. Various methods of capture and transportation were tried out resulting in a first airlift of 350 fry of *Limnothrissa miodon* from Mpulungu, Lake Tanganyika, to Sinazongwe, Lake Kariba, on 25 February 1963. Some 45 per cent of the fry survived the flight. Half of them were immediately placed in a large lake-side storage dam but all died within a few minutes. By the following morning only 14 fish were still alive in the transport containers. They were moved into a keep net placed in the lake where they grew successfully for over 3 months. But a storm then destroyed the net and the fish disappeared.

The partial success of this pilot project called for further research. In early 1966 Dr. H. MATTHES, FAO Fishery Biologist with the Lake Kariba Fisheries Research Institute, moved to Mpulungu to start investigating the biology of clupeid fishes. He concluded that *Limnothrissa miodon* (Fig. 1) would be the most suitable species to introduce into Lake Kariba (MATTHES, 1968a, 1968b).' (BELL-CROSS & BELL-CROSS, 1971). Gradually a suitable method of catching anchoveta larvae was developed. The air lifts from Lake Tanganyika to Kariba roughly matched the description given by BELL-CROSS & BELL-CROSS (1971): 'At dawn a catching team, consisting of seven men, proceeded by launch to known inshore haunts of clupeid fry shoals. The most productive areas were often adjacent to sandy shores but the availability of fry of the required size (1.2 to 2.5 cm) was unpredictable and to some extent dependent on weather conditions. For example, wind or clouds made spotting difficult. Shoals would sometimes disappear from their normal inshore haunts for up to a week at a time for no known reason. On spotting a shoal in water of 1 metre depth or less, part of the catching team armed with lusenga scoop nets (90 cm diameter in mosquito gauze) would disembark behind the fry while two men would position themselves some 20 m in front of the shoal with the transport containers. The success of the operation depended on the ability of the former part of the crew to gently chase the fry towards the partly submerged containers. These containers were made of heavy white plastic, cylindrical with a wide mouth, and 20 gallons in capacity. To facilitate the entry of the fry a large funnel was placed in the mouth. The fry normally showed no hesitation in swimming in. Experimental work also indicated that approximately 300–400 fry per gallon was the optimum density for transportation, but this target depended on the experience and judgement of the catching crew. Once the required number of fry had entered the container it was hoisted into the launch where excess water and dead or injured fry were siphoned out. Aircraft load limitations restricted the quantity of water per container to 15 gallons.

Early operations during 1967 centred at Mpulungu. Later, the catching venue was changed to Kassaba Bay (Fig. 2) where the airstrip was situated on the lake shore, whereas Mbala airport was 20 miles from Mpulungu.

The *Limnothrissa* fry were then flown to Sinazongwe, Lake Kariba, where a vehicle ferried the containers 2 miles to a launch waiting in the fisheries harbour. The fry were planted in the lake at pre-determined stations at first in water between 7 and 15 m in depth and up to 20 miles from the harbour. But because of heavy mortalities these stations were moved to deep water areas closer to the harbour.' (see Fig. 68). 'Total flying time from Lake Tanganyika to Sinazongwe averaged 5.2 hours including a 20-minute stop at Lusaka City airport for refuelling the aircraft. The flights were made at an altitude of 8,000–10,000 ft between Lake Tanganyika and Lusaka (mode 8,500 ft) and at 6,500 ft from Lusaka to Sinazongwe. Frequent inspections of the containers at Lusaka airport suggested that mortality reached a peak between Lake Tanganyika and Lusaka. The pilot of the aircraft, A. SPENCE, who had previously worked with the Fisheries Research Section, reported a possible correlation between turbulent flying conditions and fry mortality. The containers were only three-quarters full and the resulting water movement during flight turbulence might have caused injury to the young fish. Further observations at Lusaka airport indicated a correlation between mortality and density of fry in the containers. Survival rate was higher when the container was stocked to maximum capacity and under these conditions the fry moved together around the can in a single direction. If the can contained less than the optimum number of fry either through mortality or poor stocking the movement of the fry became sporadic and non-directional which might have resulted in physiological stress'.

In 1967 about 440,000 larvae and early juveniles of *Limnothrissa miodon* were flown from Lake Tanganyika to Lake Kariba; an additional 280,000 were flown in 1968. Estimating a 46–54% mortality, approximately 360,000 live *L. miodon* were planted into Lake Kariba during the 1967 and 1968 airlifts.

7.2. Successful reproduction and distribution in Lake Kariba

by JOHN G. WOODWARD

First documented evidence of the presence of *L. miodon* in Lake Kariba became available when numerous juveniles and some adult clupeids were found in the stomach of a tigerfish captured on 25 July 1969 near Sinazongwe. The sighting of clupeid schools reported on earlier occasions was consequently viewed more favourable and a search for further evidence began.

Experimental fishing for clupeids commenced in the Sinazongwe area in September (BALON, 1971c) and the first catch was made on 11 September 1969. Three sampling methods were used, all based on the propensity of *L. miodon* to be attracted to light.

The 'lusenga net' most commonly employed in the Lake Tanganyika fishery consists of a wooden frame and a bag of mosquito netting or of 8 mm (stretched) knotless nylon. The bag measures about 3.0 to 4.0 m and has an aperture of about 1.5 to 2.0 m in diameter. The bag is mounted on a pole about 2 m long (measurements vary according to user's preference). The netsman fishes from a canoe equipped with pressure kerosene lights and when a school has been accumulated the lusenga is dipped into the water and moved through the mass of fish.

The lift nets used on Lake Kariba were merely the bags of lusenga nets mounted on metal frames of about 1.5 to 2.0 m^2. When fishing the lift net is suspended over the side of a canoe on two ropes and when a school of fish has been accumulated the lift net is drawn up rapidly from beneath them.

The chiromila is an open water seine net (COULTER & ZNAMENSKY, 1971). It is drawn under and around a school of fish and then closed (pursed) by means of a pursing rope running through rings attached to the footrope. It is a semi circle of netting with floats closely spaced around the rim and weights spaced along the bottom edge. The net is not simply a flat sheet but is constructed of numerous individual panels so as to hang in the water as a concave scoop. The chiromila is used on Lake Tanganyika by a number of commercial fishing units. Two chiromila nets were used on Kariba. The first was built throughout of 8 mm knotless nylon and measured 26 meters along the headrope. It was 11 meters deep at the center and was used in this form from June 1970 to December 1970, when it was slightly enlarged to measure 34 meters along the headrope by 13 meters at the center. The second chiromila net was borrowed from Lake Tanganyika and was constructed of 12 mm nylon with a central bag of 8 mm nylon. This net measured 60 meters along the headrope by 18 meters deep at the center. On Lake Tanganyika the common chiromila size is 90 meters along the headrope by 28 meters deep at the center. The center depths quoted refer to the actual depths which the nets hang in the water when set. The chiromila nets were fished from a 30 foot skiff of the type employed on Lake Tanganyika. The net was shot around a lightboat that had previously been towed out to the chosen fishing site.

Each fishing method was based on light attraction. For lusenga and lift netting two paraffin pressure lamps each of 500 candlepower (cp) were used, mounted on the prow of a canoe for lusenga netting and over the side for lift netting. On Lake Tanganyika the lusenga netsmen commonly use 2 × 380 candlepower lamps. For fishing with the small chiromila net two 500 cp lamps were used but with the larger net three systems were tried: 3 × 500 cp pressure lamps, 1 × 300 Watt electric bulb mounted above water and 1 × 250 Watt bulb operated underwater. The latter two systems were powered by a 300 Watt generator carried in the lightboat. The underwater light was the most powerful of the lights employed as it was of the 'photoflood' type and although nominally rated at 250 Watt it gave a light output equivalent to a conventional bulb of 500–600 Watt. This light was run naked underwater without mishap, the twinflex cables having been soldered directly to the base terminals and then sealed with plastic resin. This proved a simple and reliable method of obtaining an intense light at modest depths. The only precaution necessary before switching it on was to place the light in the water as a hot bulb submerged in cold water would shatter immediately.

The Lake Kariba anchovetas tended to be at a depth of at least one meter and were difficult to catch. This tendency was commented on by a fishguard who had fished in Lake Tanganyika where the fish apparently come more readily to the surface. The reason for the different behaviour of clupeids in Lake Tanganyika is not known. Possibly the presence of the Nile perch (*Lates* sp.) beneath the schools tends to force them upwards. We found that by gradually dimming the lights the anchoveta could be induced to rise. From this we concluded that although the fish are attracted by bright lights they prefer to remain within a certain optimum level of illumination. This suggests that when powerful electric lights are employed to attract a school they should be fitted with a dimming device to concentrate the school prior to setting the net.

When fishing with the lusenga net we found that the fish evaded the net with ease. As the net was moved through the water towards them, they would scatter briefly and then return to their original position when the net was withdrawn. The avoidance was based mainly on visual stimulus for if the lights were suddenly switched off and the net immediately moved a catch would be made. Based on these observations we eventually evolved the following lift net procedure: 1) the lift net was hung in the water under the lights; 2) when a school had gathered the lights were dimmed to concentrate the fish; 3) the lights were suddenly removed and the net immediately hauled.

The accompanying map (Fig. 80) shows the areas of Lake Kariba that were successfully fished for clupeids. From May 1970 eight tours were made, four in each direction up and down the lake. *Limnothrissa miodon* was found to be present in all parts of the lake. The rate of expansion must have been quite phenomenal for the juveniles were all introduced within 15 miles of Sinazongwe (Fig. 68), the majority within 5 miles, yet by June 1970 specimens had been taken from the first (Sebungwe) basin and from Siavonga near the dam wall, respectively 72 and 176 km from Sinazongwe. (In retrospect it can be seen that sampling at various points along the lake should have commenced as soon as the introduction was completed. In this way the rate of spread could have been accurately measured.)

Extensive sampling revealed that anchoveta was not only occupying the open waters of the lake but had spread into those bays and river estuaries that form a prominent feature of the Northern shoreline. Catches were made in the almost

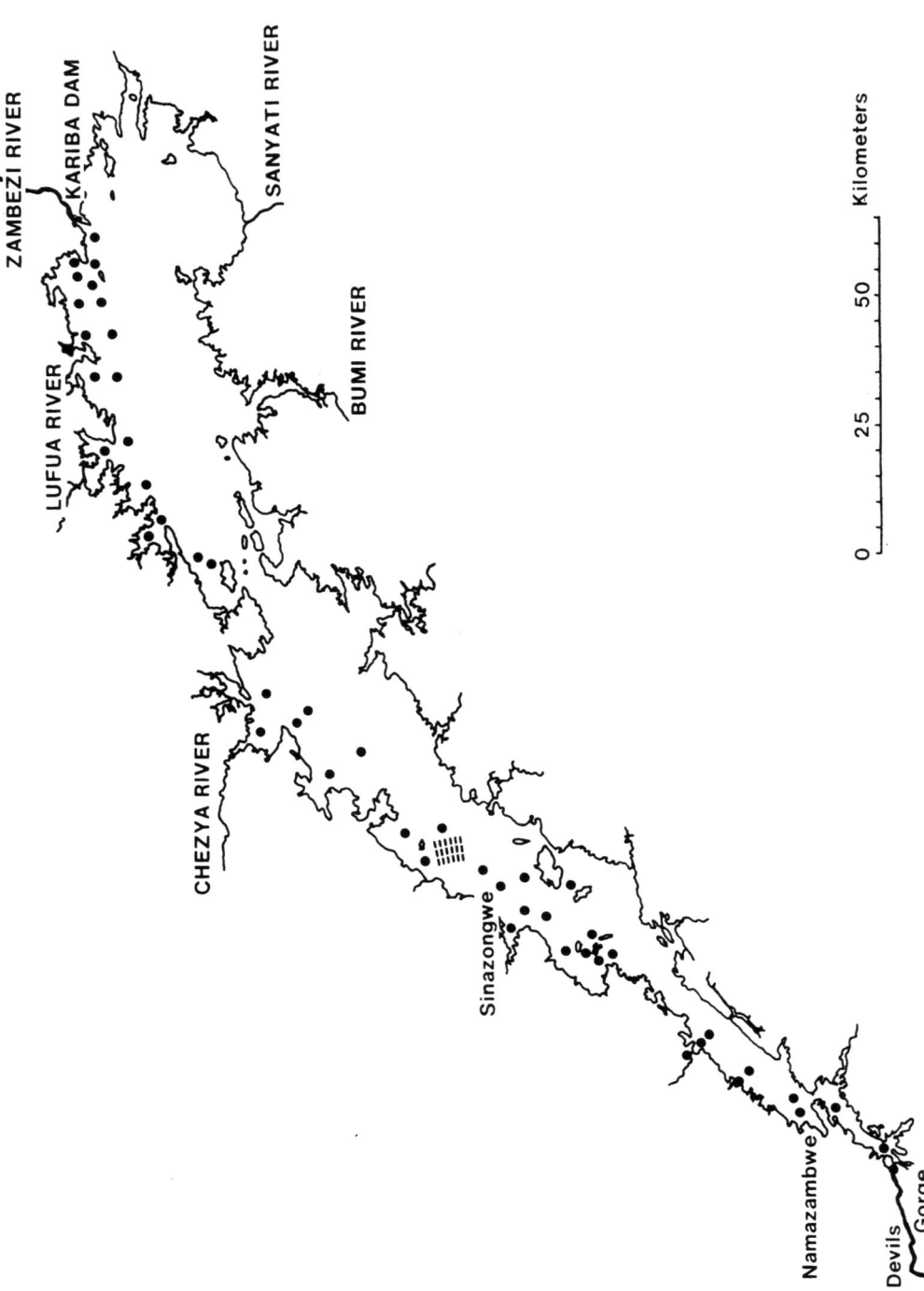

Fig. 80. Sites of Lake Kariba successfully fished for clupeids (circles). The square east of Sinazongwe marks the place of the most frequent clupeid releases.

landlocked Chipepo Harbour and Loteri Bay which were quite as large as catches made in the open waters offshore. A tour to the mouth of the Zambezi in November 1970 revealed the presence of anchoveta in Devil's Gorge. Specimens were also taken in the deep gorges of the Chimini and Lufua rivers.

The question which arises is what has caused this rapid spread of *L. miodon* to all parts of the lake. At the time of introduction the open waters of Lake Kariba were virtually unoccupied although there was evidence that *Alestes lateralis* was beginning to occupy the open waters; competition from this species would surely have been much less than the competition that *L. miodon* encounters in Lake Tanganyika from the closely related species *Stolothrissa tanganicae*. The principal predator of Lake Kariba, *Hydrocynus vittatus*, was confined to more inshore waters up to 25 meters in depth in contrast to the situation on Lake Tanganyika where clupeids are preyed on by the pelagic *Lates* sp. and *Luciolates* sp.

Lusenga netting, the first method used, averaged 11 individuals of anchoveta per boat-night catch as compared to Lake Tanganyika where a lusenga fisherman can take up to 100 kg per night. As the low catch was partly attributable to the fishermen's lack of expertise in operating such gear and their shock at the prospect of fishing in rough waters, total darkness and behind blinding lamps, the lusenga net was abandoned as a sampling instrument after only a few weeks. It was, however, resurrected in January 1970 when an accomplished Tanganyika fisherman joined the unit. The net proved no more successful in experienced hands and was again abandoned.

With the benefit of hindsight it is obvious that lusenga fishing could not have succeeded on Kariba given the low density of the clupeid population. The lusenga net is a very inefficient instrument, its success demanding that a large mass of fish be accumulated under the lights. From this mass relatively few fish are dipped by the netsman. The situation on Lake Tanganyika illustrates this well. A lusenga fisherman using a canoe with two 380 cp lights may catch 50–100 kg of clupeids; a chiromila net shot around the same canoe might take as much as 500 kg or more. The lusenga net barely scratches the surface of the accumulated school.

Observations of the behaviour of anchoveta when fishing with lusenga nets led to the construction of lift nets. It had been noted that the anchoveta did not come right to the surface but stayed down at a depth of at least 1 meter or more, which placed them on the edge of the lusenga net's fishing range. The lift net fished deeper than the lusenga and could be hauled through the water at a higher speed. Average catch per boat-night with lift nets was 41 individual anchovetas.

The first chiromila net was about 1/4 the size of those in use on Lake Tanganyika. It was later enlarged slightly as mentioned earlier. Average catch per haul in this net was 3.5 kg. From March 1971 a larger chiromila was used, about 2/3 the size of the Lake Tanganyika models. Used in conjunction with more powerful lighting systems this net gave an average catch per haul of 23.1 kg. On Lake Tanganyika the purse-seine catch was between 400 and 1,000 kg per haul, depending on the season. It can be seen that catches on Lake Kariba were much lower than on Tanganyika. This implies much lower population densities for the clupeid.

The factor of catchability must be taken into account, however, when comparing these catches. The methods employed depend on light attraction and light attraction

Table 139. Comparison of fishing performance of two types of light.

Type of light	paraffin 3 × 500 cp.	electric 1 × 600 Watt u.w.
Best haul	45.0 kg	40.0 kg
Least haul	10.0 kg	5.0 kg
Mean haul	21.3 kg	19.9 kg
Deviation	11.5	14.4
No. of hauls	6	6

is determined not only by the power of the lights but also by the clarity of the water. Lake Tanganyika is clearer than Kariba and COULTER (1968) refers to its transparency as being 'unusually high for a freshwater lake'. Secchi disc readings taken by COULTER at an offshore station in the South Basin in 1964–1965 varied from 50–175 cm. In contrast Kariba is a turbid lake; 50% of the light present at the surface is generally absorbed within the first two meters of water. Even in Basins III and IV the greatest observed depth for the 50% relative light intensity was only 3.5 meters (see Part I, section III/10). Thus light penetration is much lower on Lake Kariba and the degree of attraction exerted by a light of a given candlepower is less. The most powerful light employed in our fishing was the underwater unit of 250 Watt. Placing the light underwater enhances its efficiency for with a conventional above water light there is a considerable loss of light by reflection from the water's surface, particularly in rough weather. However, a comparison between the performance of this light and a conventional lightboat with three 500 cp pressure lamps revealed no significant difference between them (Table 139).

Another important factor in determining the catch seemed to be the competence with which the net was shot and hauled and the skill of the lightboatman in placing his craft at the center of the arc of the net. These factors lessen the utility of the chiromila as a gear for assessing clupeid density but they do not invalidate the comparisons with Lake Tanganyika catches, as the same biases apply equally there.

The entire lusenga and lift net catch was measured but the sample was generally too small to permit any valid conclusions to be drawn. For chiromila catches a subsample was taken from each catch which meant, in practice, that between 600 and 3000 specimens were measured from each catch. Measurement was in millimeters from the tip of the snout to the end of the median caudal ray (fork length). It was hoped that analysis of mean lengths of samples over a period of time would enable the growth rate to be deduced from the increasing values of the mean lengths. The grouping of the means from August to December 1970 suggests an approximate 10 mm/month growth rate. COULTER (1970) used this method in studying the clupeids of Lake Tanganyika, and presented data for the five years 1962–1966 relating to inshore populations of *Limnothrissa miodon*. For two of these years he concludes that the growth rate was approximately 10 mm/month. MATTHES (1968b) presents data from the northern end of Lake Tanganyika suggesting a rate of 8 mm/month up to a length of 100 mm.

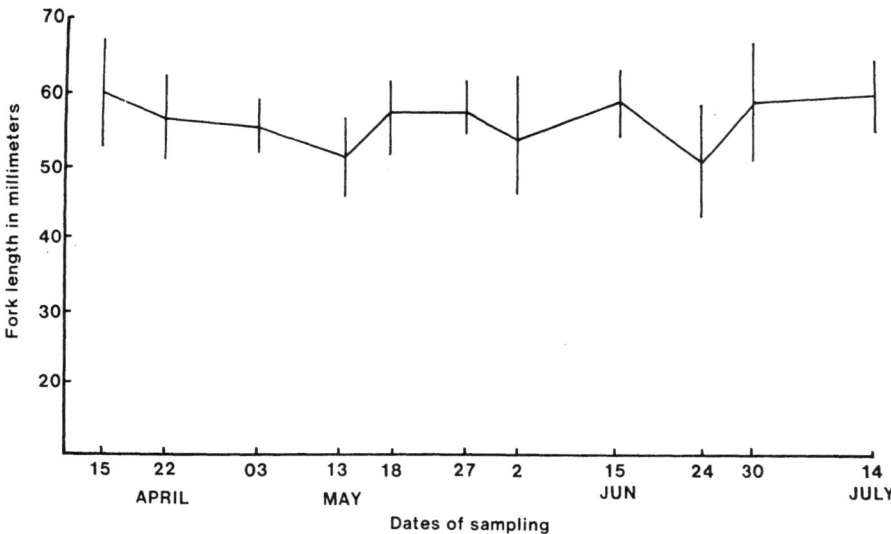

Fig. 81. Mean lengths and deviations for anchoveta during the three months 15 April to 14 July, 1971. Anchoveta were from a series of identical chiromila hauls (20–24 m depth) near Sinazongwe.

In April 1971 it was decided to make a series of hauls in one spot near Sinazongwe. Each haul was made in the same place (to within a half kilometer), approximately two kilometers off Sinazongwe Harbour in a depth of 20–24 meters and using the same lightboat and chiromila net. Little variation was found in mean length (Fig. 81) over a period of three months. Possibly the fish tend to move further out into the lake as they grow older.

The majority of the samples obtained with the chiromila net showed a normal or unimodal population curve, characteristically peaking at between 50 and 75 mm (Fig. 82). Occasionally a bimodal curve was encountered (Fig. 83). The smallest specimens taken in the chiromila net were 20–22 mm and the largest were 90–100 mm. The latter is in contrast to the situation on Lake Tanganyika where specimens of 150–170 mm are occasionally encountered (MATTHES, 1968b) and specimens of over 100 mm are common.

Gonad examinations were performed on a subsample of each night's catch. Five main phases were recognized in the development of the fish: immature, mature inactive, mature active, ripe and spent. Great difficulty was experienced in distinguishing between immature males and females of less than 50 mm and between immature and mature inactive males of less than 60 mm. The immature phase was recognized by the presence of gonads that were transparent and threadlike, the mature inactive phase by a thickening of the gonad and in the females, the assumption of a certain amount of color (generally orange); the mature active phase by the presence of a certain amount of content in the gonads and the ripe phase by the swollen gonad being obviously filled with eggs or sperm. The ripe female gonad was usually bright orange in color

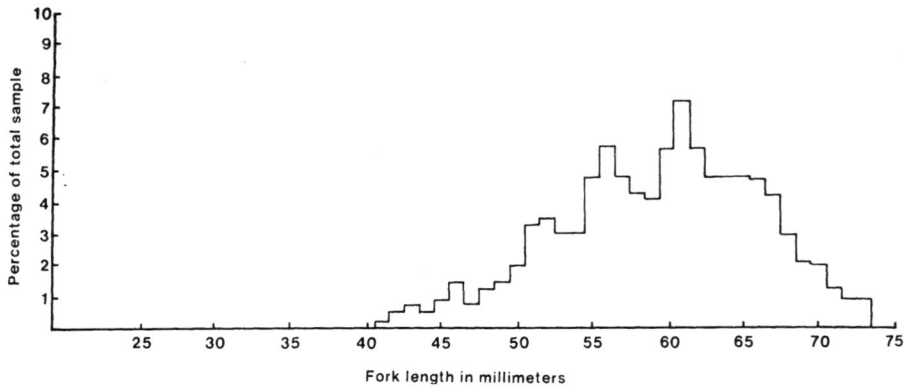

Fig. 82. Length-frequency diagram of 592 *Limnothrissa miodon* from a chiromila haul at Sinazongwe (2 July, 1970).

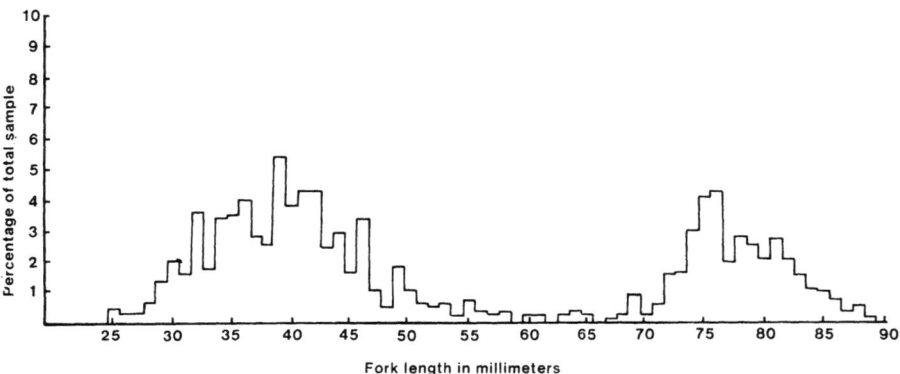

Fig. 83. Length-frequency diagram of 1,105 *Limnothrissa miodon* from a chiromila haul at Sinazongwe 25 June, 1970).

and had a bubbly appearance on the outside. The ripe male gonad was a rich cream color, shading into grey, sometimes having an orange streak down its length. The spent phase was a dubious identification. Occasionally specimens were encountered with flaccid, empty gonads and were placed in this category.

The size of first maturity was taken as the length at which 50% of the specimens were in the active – ripe phase. Data for twelve months is presented in Fig. 84 and 85 combined in periods of three lunar months. The 50% level for both sexes was only obtained in the months April-July, which represents the peak breeding season. During this period the 50% level for females was 52–56 mm. This is lower than any published figure for Lake Tanganyika. The figure for males in the same period is 71–73 mm which is more in keeping with Lake Tanganyika values.

The peak breeding season in 1970–1971 was April to mid August. This is slightly

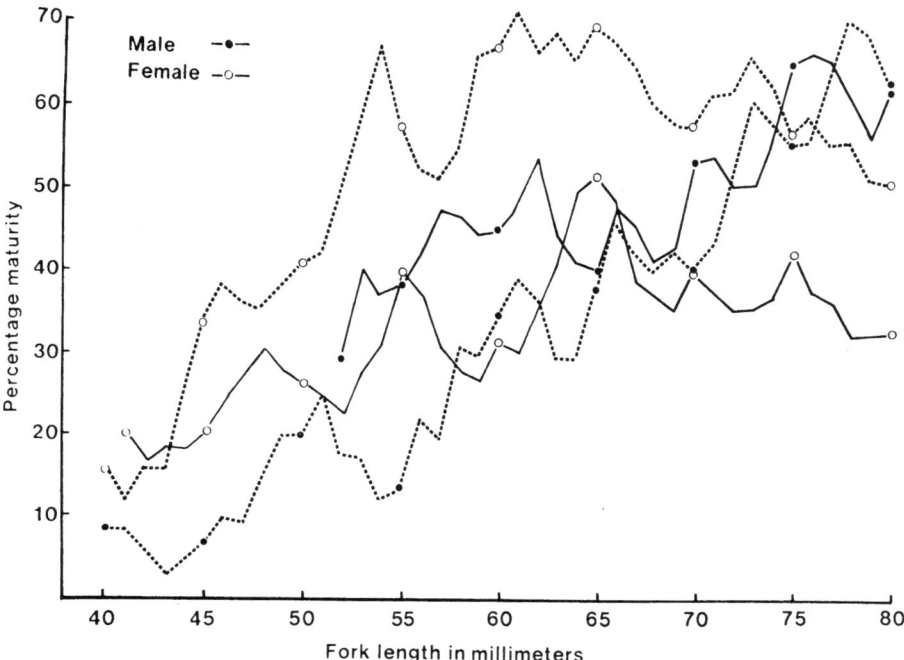

Fig. 84. Percentage maturity of *Limnothrissa miodon* in the period from 6 April to 2 July (interrupted lines) and 30 July to 29 September, 1970 (full lines).

longer than the season on Lake Tanganyika which MATTHES (1968b) reported as May–June, with secondary peaks from September to December. On Lake Kariba, however, we found the November–December period to have the lowest level of sexual activity.

In mature specimens the sex ratio (Fig. 86) varied from 1:4 (25% males) to 1:1 (55% males). No reliable data can be presented for immature specimens owing to the difficulty of distinguishing between the sexes in the smallest sizes. MATTHES (1968b) reports sex ratios on Lake Tanganyika as normally varying from 1:5 to 1:2.

The growth rate and fecundity of *L. miodon* are probably variable according to prevailing circumstances of population pressure and predation, food availability and competition, i.e. the clupeid would exhibit a high fecundity and high growth rate when population pressure was low or predation high and a low fecundity and low growth rate when population pressure was high or predation low (BALON, 1963, 1972). The ability to vary fecundity and growth rate would be a defence mechanism against predation and competition, perhaps the only real defence possessed by such small species. In terms of Lake Kariba this would mean a high fecundity and high growth rate in the early stages of colonization when the population is expanding to occupy the open waters, followed by a lowered fecundity and growth rate when the population has stabilised. *Limnothrissa* certainly matures earlier in Lake Kariba than in Lake

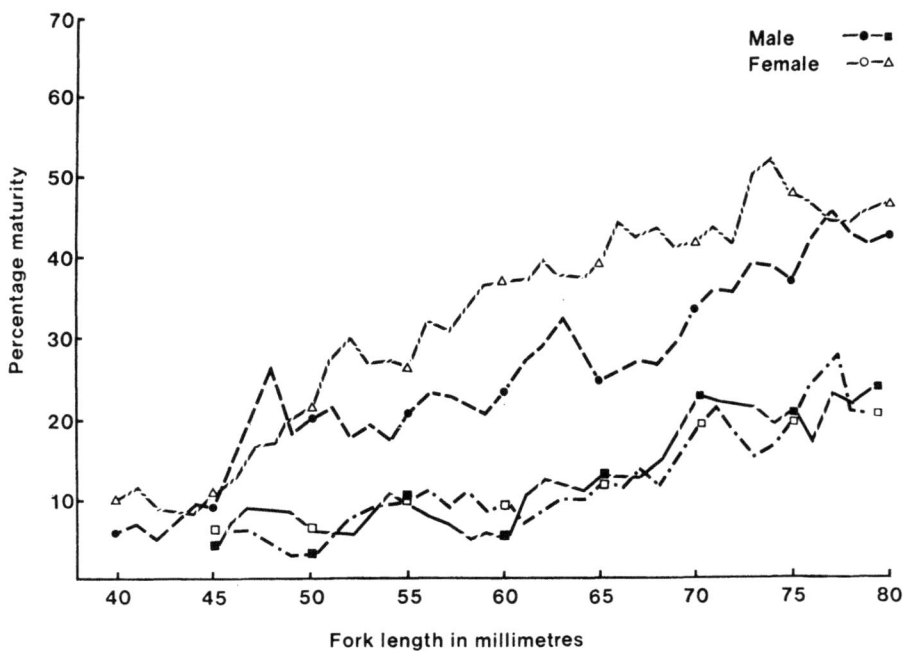

Fig. 85. Percentage maturity of *Limnothrissa miodon* in Lake Kariba sampled from 30 September to 27 December, 1970 (two lower curves) and from 28 December, 1970, to 25 March, 1971 (upper two lines).

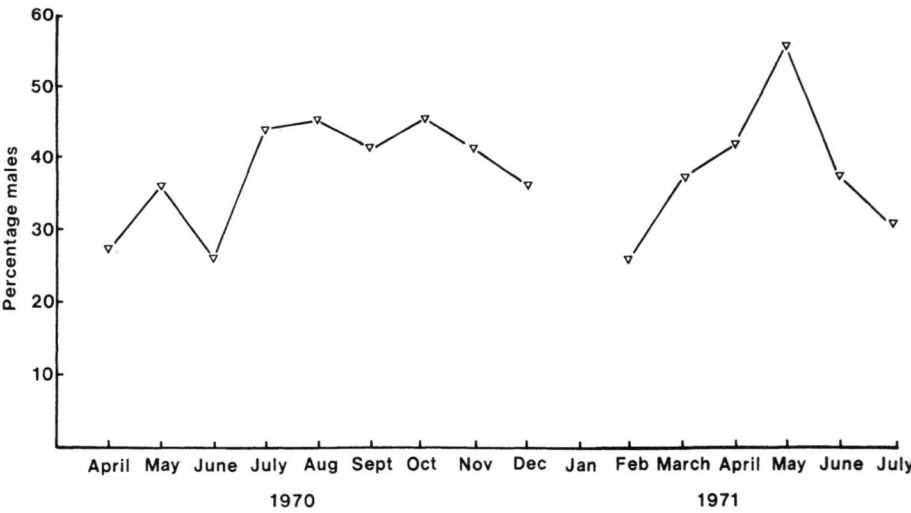

Fig. 86. Percentage of males for mature *Limnothrissa miodon* taken from Lake Kariba in samples from April 1970 to July 1971.

Table 140. Percentage *Alestes*, by number, in single Chiromila hauls.

Locality	Date	Distance from shore, meters	Depth, meters	*Alestes*	Sample size
Loteri Bay	2510	landlocked 300 m	15 m	19.2	581
Chipepo Harbour	2810	landlocked 500 m	16 m	1.6	788
Lufua R. Gorge	2710	landlocked 100 m	30 m	58.9	151
Chimini R. Gorge	2411	landlocked 100 m	29 m	68.7	14
Zambezi R. Gorge	2611	landlocked 100 m	29 m	72.6	1000

Tanganyika; on Lake Kariba ripe specimens of both sexes are commonly found at 50–55 mm, a situation unparalleled on Lake Tanganyika.

As stated earlier *Alestes lateralis* began to colonize the open waters of Lake Kariba before the introduction of clupeids. When fishing for *Limnothrissa miodon* commenced at Sinazongwe *Alestes lateralis* formed a large portion of the catch. In the first eight months, using lusenga and lift nets, alestes formed 20.5% of the catch (total catch 3,061 specimens). The great majority of these catches were made at least one mile offshore. It should not be assumed that 20.5 represents the actual percentage of alestes present in the open waters, as alestes was easier to catch than anchoveta. There are two reasons for this: a) alestes came sooner to the lights than anchoveta, perhaps because it does not descend as deep as *L. miodon* during daytime, b) alestes came closer to the surface than anchoveta, a matter of significance when fishing with a lusenga net. The two species could be easily distinguished by their different modes of swimming – alestes near the surface darting with rapid upward movements to investigate stranded insects, anchoveta remaining deeper, swimming slowly in a circular path with a steady, purposeful movement.

With the introduction of the chiromila net the percentage of alestes present in the catch fell and from June to December 1970 it was 8.7% of the total number of both species captured. The chiromila is a less selective sampling method and this may be taken as a more accurate figure. It is probable that the percentage of alestes was declining throughout this period as chiromila samples taken in the Sinazongwe area from February to August 1971 showed less than 0.05% *Alestes*.

Although *Alestes* declined in the offshore catches it still formed a large proportion of the inshore catches in the latter half of 1970 (Table 140). *Alestes* appears to have retreated from the open waters to the inshore areas. Additional evidence for this retreat comes from stomach analyses of tiger fish caught in an open water station off Sinazongwe. In 51 stomachs containing identifiable contents only *Limnothrissa miodon* was found. It may seem surprising that anchoveta, usually thought of as an open water species, should have moved into the very inshore waters of Lake Kariba. MATTHES (1968b), however, notes that in Lake Tanganyika *Limnothrissa miodon* is found 'closer to the shore than *Stolothrissa tanganicae* and also occurs at the rivermouths where the water is not too muddy'.

In Lake Volta, Ghana, one clupeid species, *Pellonula afzelius*, that was endemic to the Volta River, has successfully established itself in the open waters of the lake and became a major food source for the Nile perch (PETR, 1971).

The tigerfish is the main predator of Lake Kariba. Largely an inshore predator, it feeds on a variety of small species, principally *Alestes* and *Barbus* spp. When fishing for anchoveta in 1969 tigerfish were occasionally seen swimming rapidly beneath the lights. When the chiromila net was introduced in 1970 tigerfish began to be caught along with the anchoveta particularly in inshore areas such as the estuary of the Zongwe River, where the weight of tigerfish in a haul occasionally exceeded that of anchoveta.

In February 1971 we carried out echo soundings in various areas of Lake Kariba to determine the distribution and abundance of anchoveta. The soundings made in some deep water areas, e.g. Sebungwe Basin, Sebungwe Narrows and Chete Gorge, showed the presence of individual large fish at depths down to 20 meters. The fish were assumed to be tigerfish. Commencing on March 1971 a fleet of gill nets was set in a deep water area off Sinazongwe. The site was over the submerged valley of the Zongwe River in 35–45 meters of water about 8 km off Sinazongwe and 5 km off the eastern tip of Chete Island, outside the area considered to be occupied by the original riverine fishes of the lake. Kariba fishermen never set their nets in such areas, preferring to fish the shallow waters closer to the shore. Five gillnets were used, ranging from 50 mm stretched mesh, to 150 mm stretched mesh, each net was 92 m stretched (i.e. 46 m mounted) and was of double depth, 44 meshes deep. The nets were set at three depths over a period of 34 days: 1) to fish 0–2 meters deep (surface set) 8 days, 2) to fish 3–5 meters deep 11 days, and 3) to fish 6–8 meters deep, 15 days. *Hydrocynus vittatus* was the only species taken. A total of 262 specimens were captured and 51 of them were found to have retained identifiable stomach contents of *Limnothrissa miodon*. The majority of the tigerfish were taken in the 76 mm net but even in this net the catch was only 1.4 kg per 100 yards which is too low to be of commercial interest. Fish taken in the 76 mm net ranged in fork length from 280 mm to 330 mm, which places them in their second and third growth seasons.

There can be no doubt that tigerfish is moving out into the open waters to exploit the anchoveta as a food source. Whether it will become as important a predator of the open waters of Kariba as are the *Lates* sp. of Lake Tanganyika remains to be seen. *Hydrocynus vittatus* is essentially an inshore predator and its anadromous breeding habit will in any event remove it from the open waters for part of the year.

7.3. The compromising density

Judging from the catch per unit of effort of anchoveta in Lake Kariba as compared to that of Lake Tanganyika the abundance of *Limnothrissa miodon* in the former lake has to be considerably lower than the abundance of *Limnothrissa* and *Stolothrissa* in the latter. Although allowances were made for the difference in the size of chiromila nets as well as for the different optical properties and skills of fishermen, it is probably unwise to attach any significant confidence in the values obtained. Nonetheless they give some indication on the existing disparity in density.

On Lake Kariba the chiromila technique was improved to perfection when the haul

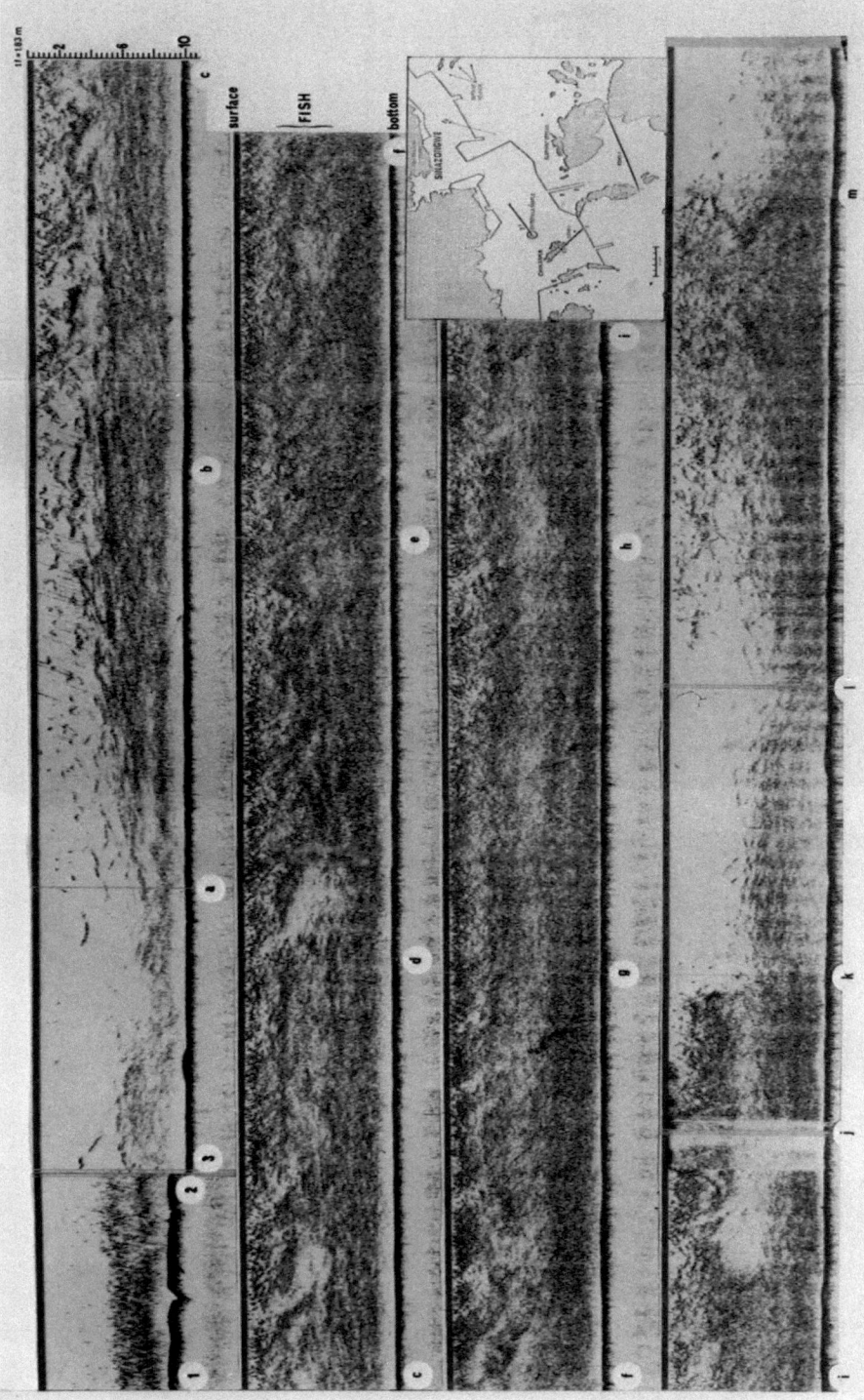

Fig 88. Echo-traces on Lake Kariba during a chiromila-set haul near Chikanka Island (2 February, 1971): 1–2 trace from a moving boat while approaching fishing site (2114–2119 hr) with anchoveta schools close to the bottom; 3 – traces from anchored boat (2125 hr) recorded after dark; a – the light boat with two kerosene lamps (500 sp each) joining the recording boat (2132), b – 10 minutes later (2142) fish attracted to the light coming up to the surface, c – after 20 minutes (2152), d – 30 minutes later (2202), e – 40 minutes later (2212) the clupeoid school spreading nearly regularly from the bottom to the surface, f – 50 minutes later (2222) lamps are dimmed, g – 10 minutes later lamps are turned on again (2232) and the fishing boat closing up (2240), h – another 10 minute period (2242), i – the chiromila net is set around the recording and light boats (2247) and j – the chiromila is closed (2254), k – lightboat moving away from the recording boat (2257) and the anchoveta schools immediately reacting by leaving the upper half of the water column; l – the lightboat returns to the recording boat (2304), followed, with some delay, by the anchoveta moving up to the surface, m – the lightboat turns off lamps, the anchoveta react by immediately leaving the upper half of the water column. (Depths are in fathoms).

Fig. 87. The boat with ELAC Echograph and floating transducer used for survey of anchoveta distribution.

of 2 February 1971 was made near Chikanka Island. It resulted in a catch of 5,322 *Limnothrissa miodon* ranging in standard lengths from 31–40 mm (40 specimens), 41–50 (3,218 specimens), 51–60 (1,547), 61–70 (428) and 71–80 (89 fish) and in a total weight of 9 kg. Again, however, the haul amounted to strikingly less than the usual catches in Lake Tanganyika (COULTER & ZNAMENSKY, 1971).

To gain more insight into the abundance and distribution of clupeids a special electroacoustic survey was performed. An ELAC Echograph Castor (LAZ 17) was fixed to a small boat, with the transducer pulled in front on a pole and mounted under a reinforced styrofoam float. The speed of the boat was maintained constant by predetermined engine revolutions and by employment of the same operators so as not to vary the weight in the boat (Fig. 87). The echograph controls and stylus were also maintained constant. The boat, together with echosounder, was transferred by truck from Lake Kariba to Lake Tanganyika.

The chiromila haul near Chikanka (Fig. 88) was performed in a shallow area of the lake (16 m depth) simultaneously with the electroacoustic operation. The recorder detected anchoveta schools in a depth of 7 to 12 m; after 5 minutes the lightboat attracted some of the fish toward the surface and after 20 minutes the clupeids were spread throughout the entire vertical water column 16 m deep, except for 1 m above the bottom and 1 m below the surface. When the lightboat or the lights were removed the fish immediately descended below 7 m; with the lights on again they re-appeared near the surface in one to two minutes.

The chiromila catch itself could have served as sufficient evidence for the type of fish that appeared on the electroacoustic charts. In most cases this was checked by

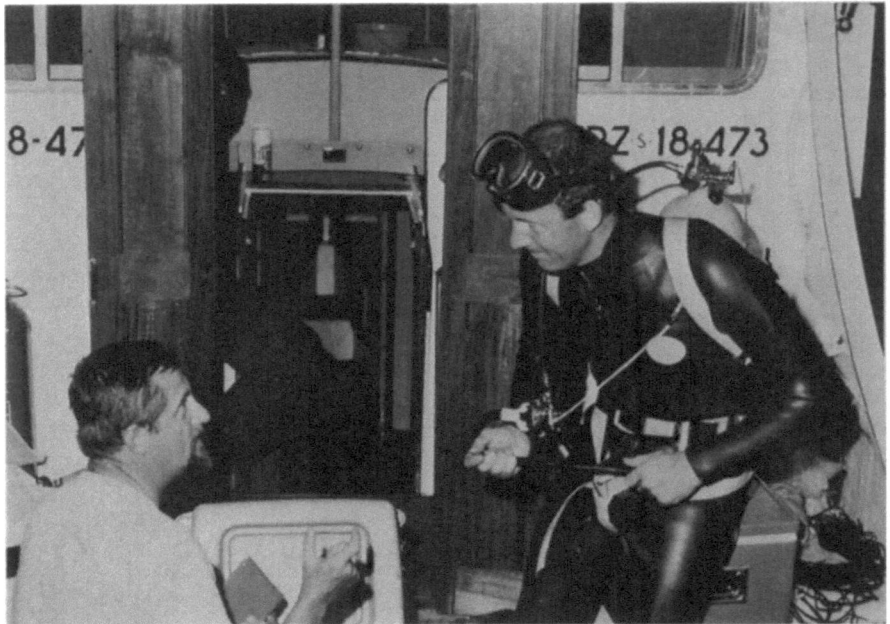

Fig. 89. Last instructions before SCUBA diver submerges for simultaneous electroacoustic and photographic registration of clupeid schools.

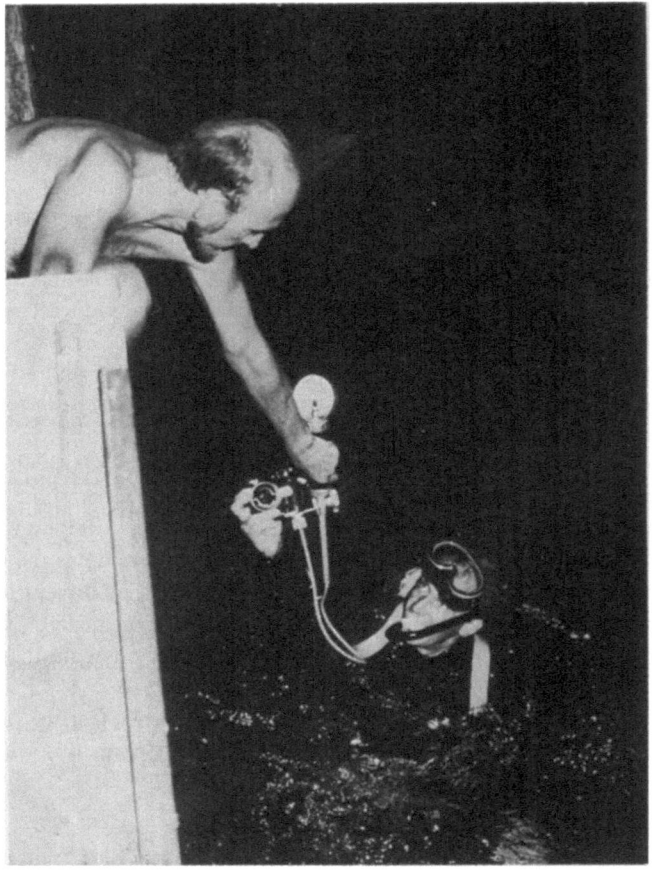

Fig. 90. Nikonos II camera with 35 mm lens was calibrated for sardine density estimates.

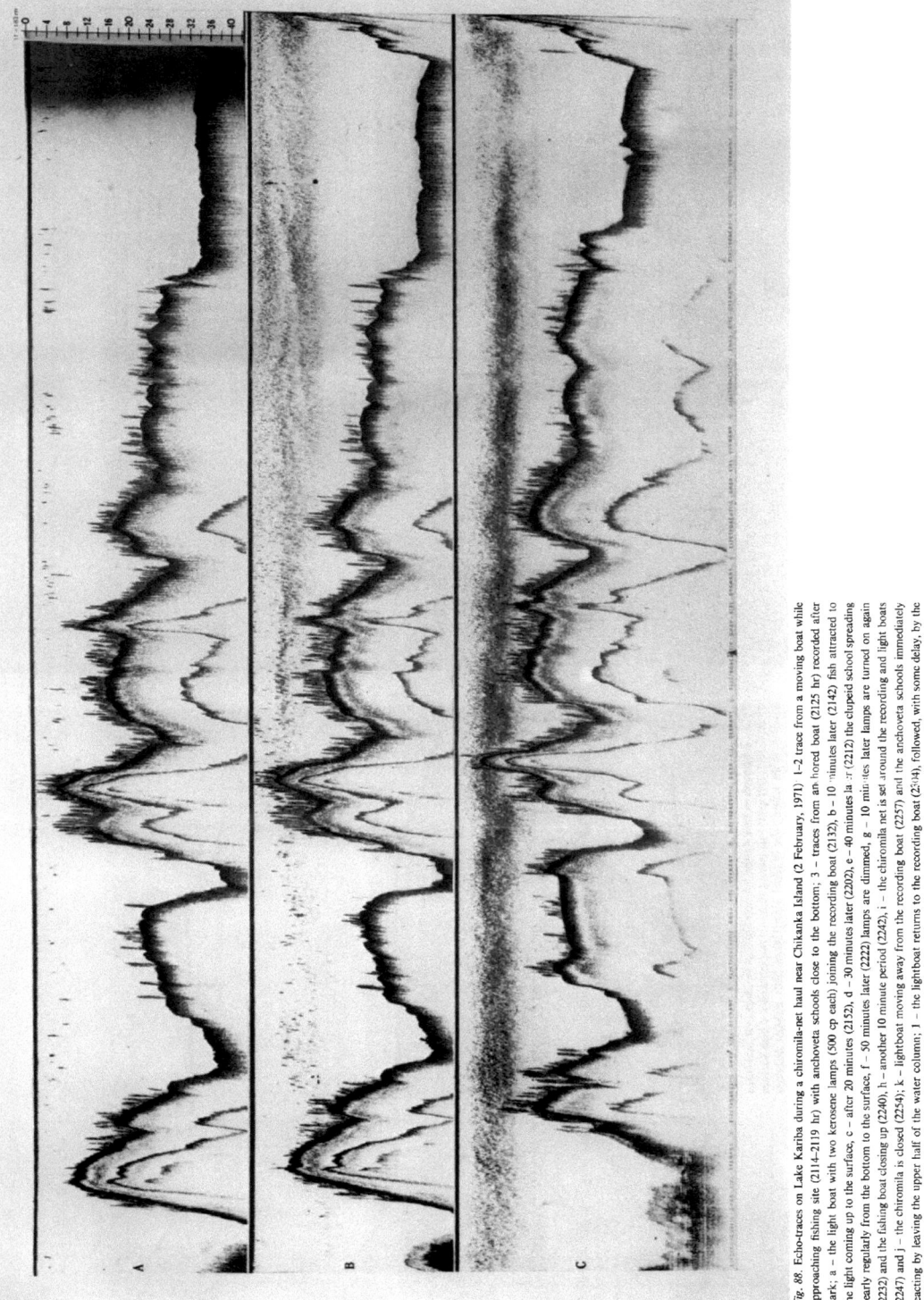

Fig. 88. Echo-traces on Lake Kariba during a chiromila-net haul near Chikanka Island (2 February, 1971) 1–2 trace from a moving boat while approaching fishing site (2114–2119 hr) with anchoveta schools close to the bottom; 3 – traces from anchored boat (2125 hr) recorded after dark; a – the light boat with two kerosene lamps (500 cp each) joining the recording boat (2132), b – 10 minutes later (2142) fish attracted to the light coming up to the surface, c – after 20 minutes (2152), d – 30 minutes later (2202), e – 40 minutes later (2212) the clupeid school spreading nearly regularly from the bottom to the surface, f – 50 minutes later (2222) lamps are dimmed, g – 10 minutes later lamps are turned on again (2232) and the fishing boat closing up (2240), h – another 10 minute period (2242), i – the chiromila net is set around the recording and light boats (2247) and j – the chiromila is closed (2254); k – lightboat moving away from the recording boat (2257) and the anchoveta schools immediately reacting by leaving the upper half of the water column; l – the lightboat returns to the recording boat (2304), followed, with some delay, by the anchoveta moving up to the surface, m – the lightboat turns off lamps, the anchoveta react by immediately leaving the upper half of the water column. (Depths are in fathoms).

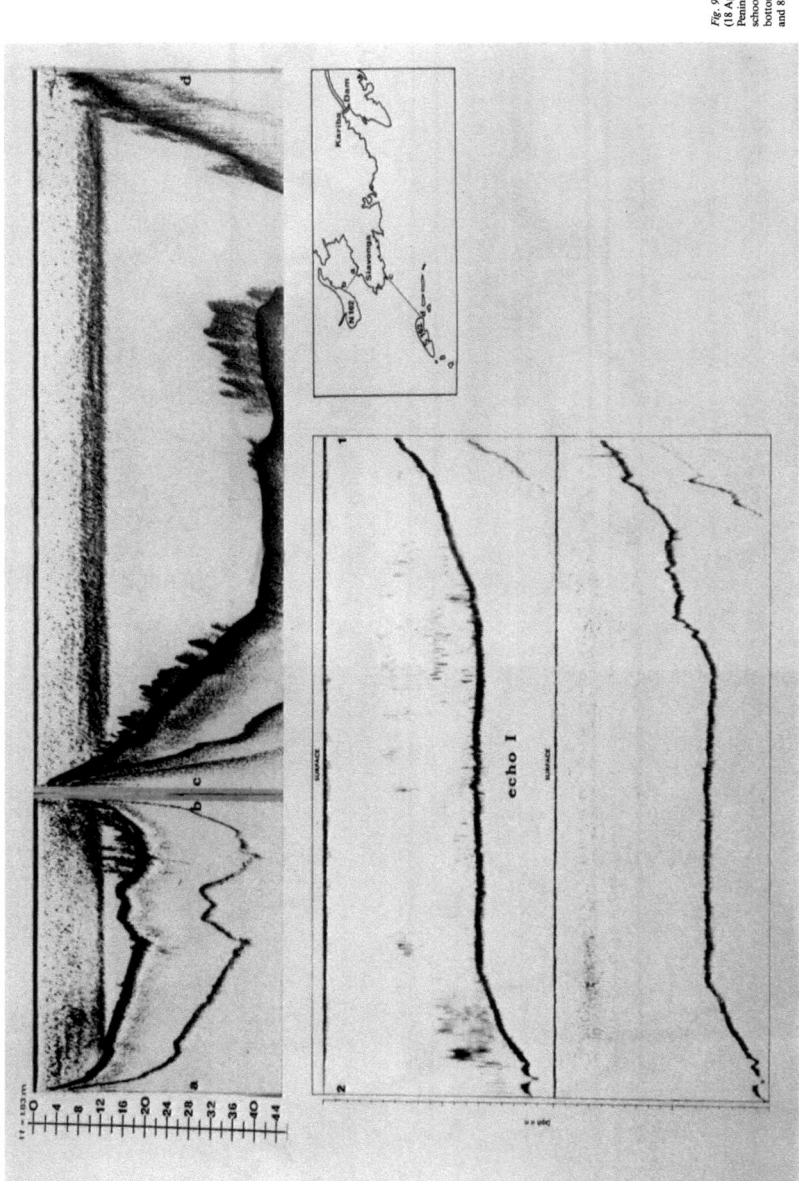

Fig. 93. Echo-traces on Lake Kariba: a–b from Siavonga Peninsula to Banana Island (18 April, 1971; 1845 to 1855 hr), evening anchoveta concentration; c–d from Siavonga Peninsula to the eastern point of island S162 (2240 to 2305), midnight anchoveta schools; echo 1 represents traces made in April 1970, with clupeids podding on the bottom at midday (above) and schooling under the surface at midnight (see Fig. 68 and 88 for location).

Fig. 91. Underwater photographs of clupeid schools were suitable for density estimates but proved useless for species separation.

SCUBA divers and underwater photography (Fig. 89 and 90). The known angle of the lens and simulated wooden fish dummies of different sizes photographed in advance enabled the calibration of the fish to water volume ratio on photographs and on simultaneously run echorecordings (Fig. 91).

In daylight the anchoveta of Lake Kariba grouped in dense 'pods' (BREDER, 1959) near the bottom 10 to 40 m deep. In twilight the anchovetas began to form schools spread from 2 to 20 m. At night, in total darkness, the schools concentrated between 7–20 m in shallow waters and 14–24 m in deeper waters (Fig. 92 and 93). Solitary fish or small groups aggregated above this schooling depth up to 1 m below the surface and down to 30 m, smaller fish remaining in general deeper. In daylight the clupeids were replaced by solitary tigerfish in a depth of 5 to 10 m, both tigerfish and anchoveta appearing together in twilight for a short period of time in a 10 m depth (Fig. 92). The density of anchoveta in Lake Kariba during nighttime schooling was assessed on the average as 8 fish per m^3 (cubic meter) of the densest traces.

In the shallow areas of Lake Tanganyika inhabited overwhelmingly by *Limnothrissa miodon* and fished mainly by local lusenga-net fishermen, the clupeids were podding in daylight near the bottom at a depth of 20 m, or above the bottom between 10 and 20 m. After dark they formed schools spread 25 m through the vertical water column from the surface to bottom, with larger individuals aggregating in the center (Fig. 94). It was assessed that sixty to eighty fish were present per 1 m^3 occupied by the densest schools in shallow water.

Fig. 96. The density of clupeids in Lake Tanganyika is so high that local inhabitants can catch them profitably even with such inefficient gear as a lusenga net.

In the deep water areas of Lake Tanganyika daylight podding occurred at 30–40 and sometimes 60 or more meters. At nighttime the fish spread in dense schools 3 to 4 km from the shore (Fig. 95). The commercial catches in this area consisted of *Stolothrissa tanganicae* and *Limnothrissa miodon*, both species possibly occurring in mixed schools. Species separation on photographs was, however, impossible. The inshore side of the schooling area was bordered by echos of larger predators (*Lates* sp. judging from purse seine catches), the clupeid schools were spread from 25 m to the surface, with larger fish below. The commercial catches were the best in this area. The clupeid abundance at densest points was estimated to be from 200 to 300 fish per m^3, approximately 30 times more than in Lake Kariba.

Commercial catches were also conducted in areas closer to shore at a distance of 1.5 to 3 km, where the clupeids were aggregated in the top 7 m from the surface, with smaller fish mixing with larger predators (?) down to 16 m (Fig. 95). The light-boat concentrations seemed to be effective mainly in the 7 to 16 m depth. The density of the clupeids here averaged 180 fish per m^3.

The vertical distribution of clupeids in Lake Tanganyika differs significantly from their distribution in Lake Kariba. In the former lake the schools were spread from the surface in a continuous layer 16 to 25 m deep; in Lake Kariba nighttime schools were rarely near the surface, most frequently the schools were detected forming a layer between 7 and 20 m. The depth of schooling clupeids was consequently 10 to 18 m in Lake Kariba and 16 to 25 m in Lake Tanganyika. In addition to the

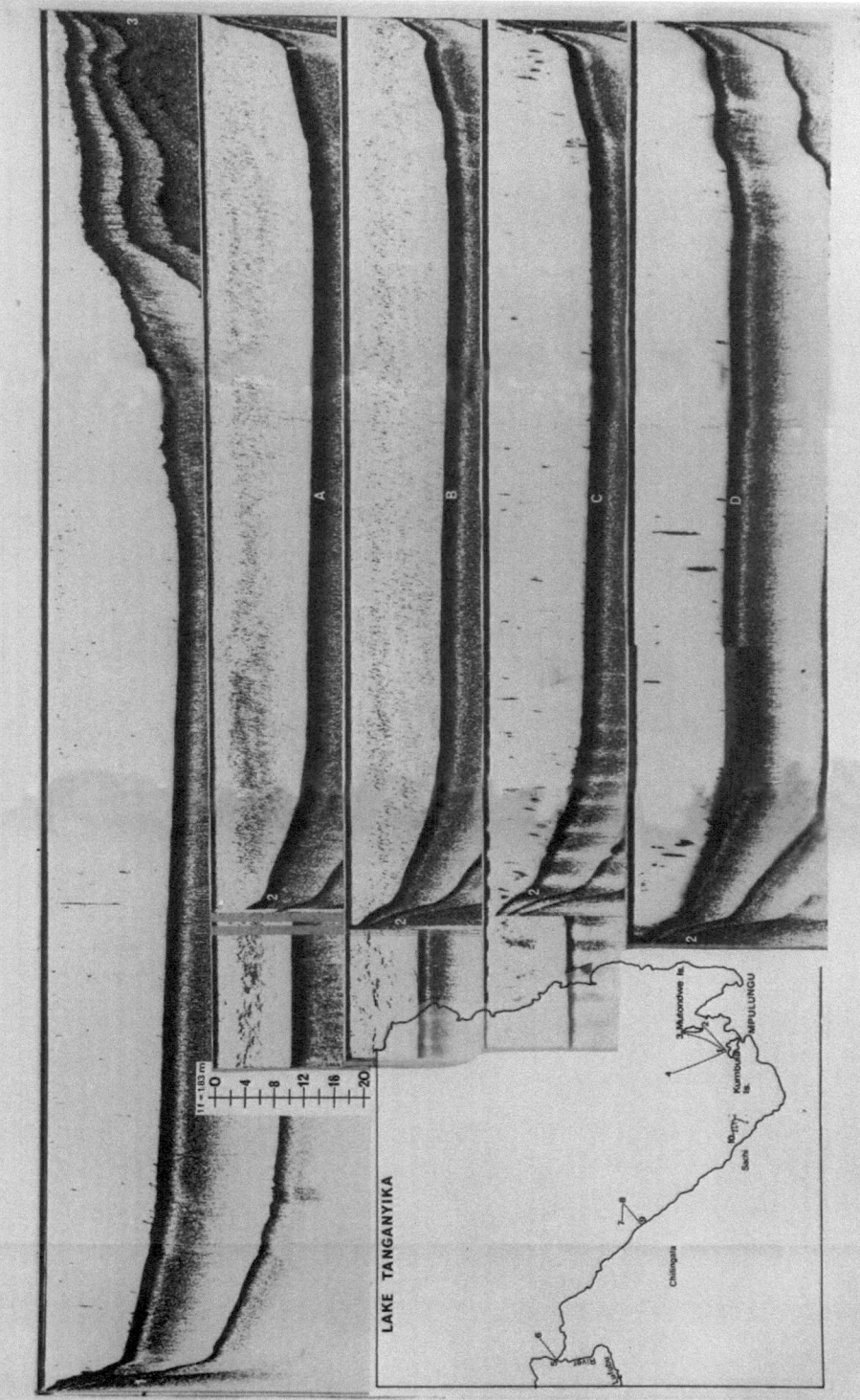

Fig. 94. Echo-traces on Lake Tanganyika, shallow part: 1–3 from Kumbula Island (28 February, 1971: 1808 to 1859 hr), sunset, no clupeids were found towards the center of the lake, vertical line in the middle is a gill-net; A – small section at the beginning is an echo-trace while drifting 500 m south of the sandbank of Mwina Point (1920 to 1929), spot of frequent lusenga-net fishing (see Fig. 96); 2–1 trace from Mwina Point to Kumbula Island (1934 to 2007) with evening schools of *Limnothrissa miodon*; B – the same, drift from 2430 to 2435 hr (1 March, 1971), 2–1 trace from 2445–0118, midnight schools of clupeids, under moonless skies; C – the same, drift from 0609 to 0614, 2–1 trace from 0619 to 1652, during sunrise with numerous clupeid poddings; D = the same 2–1 trace from 1201 to 1235 with midday clupeid pods.

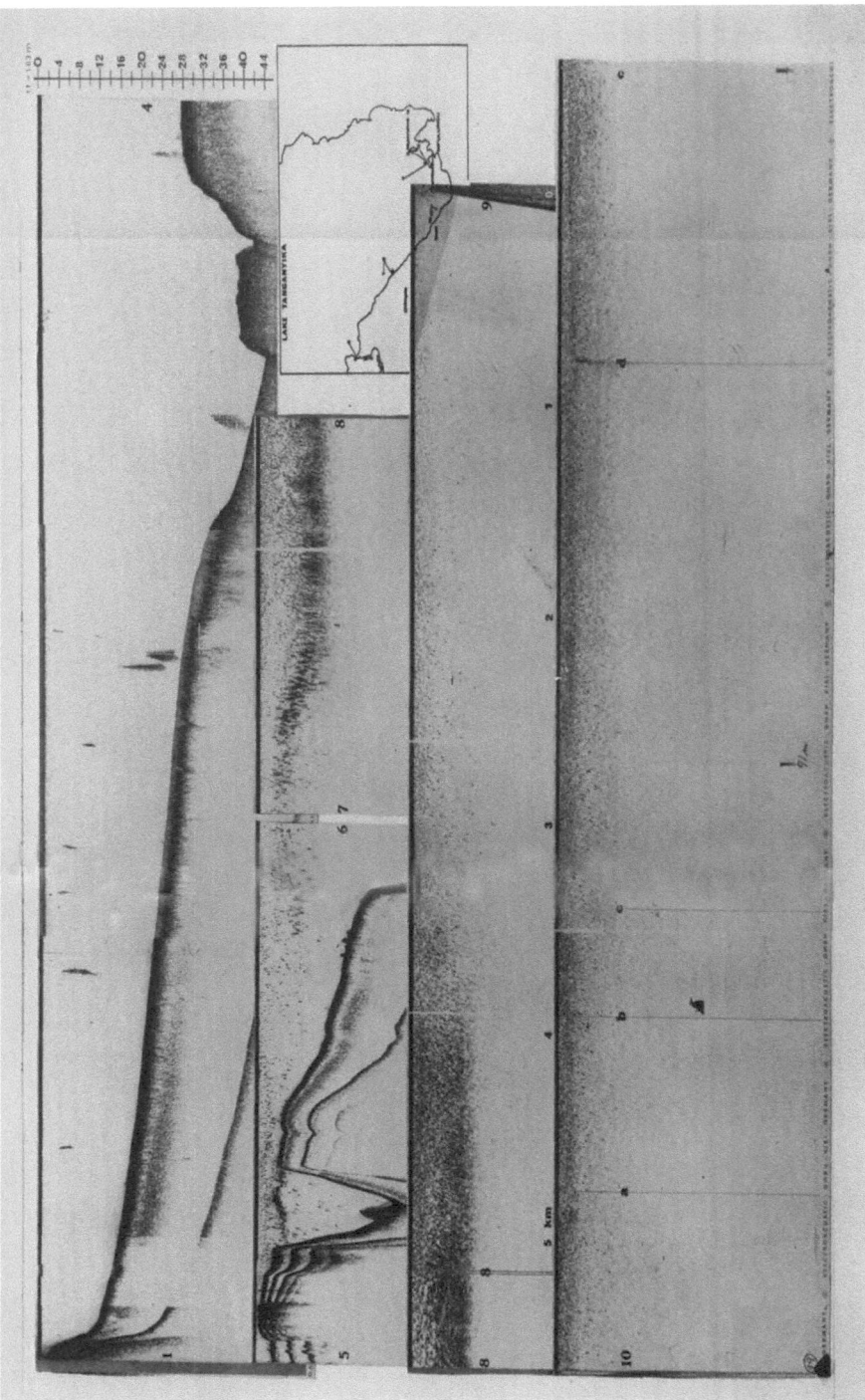

Fig. 95. Echo-traces on Lake Tanganyika: 1–4 from Kumbula Island north toward the lake center (1 March, 1971: 1045 to 1130 hr), of clupeid aggregations (pods) at daylight; 5–6 from Lufubu River entry EN toward the open lake (2 March, 1971: 1136 to 1156), a ditch 48 m in depth in front of Lufubu entry contains at its mouth larger predators below, small fish – two thin lines mark the border of brown-yellowish river water spread into the lake; 7–8 in front of sacred mountain Chilingala from 2 to 5 km off-shore (1845 to 1900), at the bordyrline of a clupeid school and an aggregation of predators (spaced vertical line marks 137 m depth to the lake bottom); 8–8 drift at point 8 above the densest clupeid school (1902 to 1905); 8–9 perpendicularly toward the shore (1906 to 1945) which is shown as a very steep gradient to the right (2.7 km depth check 135 m, 3.9 km 136 m); 10 two km off-shore at Sachi (2027 to 2120) along the light boats (lights on for 1.5 hr) of a Greek commercial purse seiner: a, b first and second lightboat passing at a distance of 100 m, c, d third and fourth lightboats passing at a distance of 10 m (91 m deep between 3rd and 4th

20–30 time lower density of clupeids in Lake Kariba, as compared to that of Lake Tanganyika (Fig. 96), the differences in vertical spread of schools must mean a further decrease in the total population density.

The reason for introducing anchoveta into Lake Kariba was to increase fishery potential. Since anchoveta failed to develop here into a fishable stock and alestes, already had a population explosion prior to the anchoveta introduction (see section 6.3) and might have had the potential of achieving similar density, little remains to view the introduction of anchoveta favourably.

CONCLUDING DISCUSSION

'... it was then obvious that managing a lake for game fish is very much like farming Africa for lions.'

Prior to the creation of Lake Kariba some predictive estimates of the future fish potential were made, and though based on only a few general parameters, one estimate at least came surprisingly close to reality. Even more interesting, it was the first estimate made by HICKLING (1956), who wrote: 'While I do not think that these favourable factors ((1) volume of inflow sufficient to change one third of the lake contents per year; (2) annual drawdown of 25 to 36 feet exposing the bottom for aeration; (3) seasonal temperature change and wind will probably break down the discontinuity layer) may bring the fish potential of Kariba Lake up to that of Lake Mweru, I think it reasonable that, as against Mweru's 19 tons per square mile, the shallower parts of Kariba Lake, taking for this purpose water less than 100 feet in depth, should produce about 13 tons of fish per square mile, or about 8,800 tons annually. Taking into account those two-thirds of the lake which are deeper, we might round off to 9,000 tons annually...' (after COCHE, 1971). Though not all quoted parameters are correct, the value of 43 kg/ha/yr (for inhabited area 202 200 ha) is close to our potential catch estimation and the 9 thousand tons of the total fish potential not significantly less than our assessment of potential harvest. His later estimates – 45 kg/ha (HICKLING, 1961), as well as the estimates of MAAR (1959, 1960) – 56 kg/ha/yr for the total lake area of 30 or 31 thousand tons/yr – are blind shots far from reality.

To cite a few examples where a similar methodology was employed (in most cases only fish larvae were not sampled), higher fish population abundance than that of Lake Kariba was observed by myself in one Danube River arm (BALON, 1966b; 258.6 fish/m^2) and some Danube River oxbows (BALON, 1963c; 4.7 to 186 sp/m^2); also by HOLČIK (1970b) in some Cuban lagunas (maximum 178 sp/m^2), and in the River Thames (MATHEWS, 1971; 97 sp/m^2). The estimated Lake Kariba density (61 sp/m^2) was close to some Elbe River backwaters (OLIVA, 1955, 1960; 12 to 69 sp/m^2). In Lake Husie of the River Danube LAC & ERTL (1961) estimated 22 specimens/m^2, in the Meuse HUET & TIMMERMANS (1966) found 17.4 sp/m^2, in Lake Zhemtchuzhnoe BURMAKIN (1962) found 6.8 sp/m^2, in the Klíčava Reservoir HOLČIK (1970a) found 1.21 sp/m^2 and in Lake Tchad drainage waters LOUBENS (1969) found 0.6 to 8.6 specimens per m^2. Thus high density is by no means a tropical region characteristic.

In some waters of the Lake Tchad region LOUBENS (1969) noted a standing crop of 562 g/m^2; the lowest found in the same region was only 4 g/m^2. HOLDEN's (1963) values for the sandy bottom of the Sokoto River dry season pools are 62.6 to 101.7 g/m^2, for muddy bottom pools 19.6 to 27.0 g/m^2 and for mixed bottom pools 58.5 to 144 g/m^2. The highest figure for the temperate region was found in the Thames River – 65.9 g/m^2 (MANN, 1965) and more than double after correction for 0+ fish (MATHEWS,

1971). Similar to the Lake Kariba standing stock value of 53.3 g/m^2 (BALON, 1973a) are values of 57 g/m^2 given by CARLANDER (1955) for North American backwaters and oxbows, by MCFADDEN & COOPER (1962) – 47 g/m^2 – for some North American streams, by OLIVA (1955, 1960) – 55.8 g/m^2 – for Elbe River backwaters, by BALON (1966c) – 49.7 g/m^2 – for a Danube River arm near Bratislava, by TURNER (1960) – 43.1 g/m^2 – for 22 Kentucky ponds, by JENKINS (1958) – 38.2 g/m^2 – for 42 Oklahoma ponds, by ALLEN (1951) – 31.1 g/m^2 – for the Horokiwi Stream in New Zealand and by HUET & TIMMERMANS (1963) – 13–30 g/m^2 – in some Belgian streams. In Danube River oxbows 25.9 g/m^2 was recorded (BALON, 1963c, 1966b) and in the Kličava Reservoir 7.6 g/m^2 (HOLČIK, 1970a) with an increase in successive years up to 17.1 g/m^2 as a result of a decline in semipredatory perch abundance and a two fold density increase in the forage roach (HOLČIK, 1970c). Again the standing stock values are very variable with no significant differences between temperate or tropical regions.

So far there do not exist many total production statistics obtained from comparable methodology. The few cited here at random show Lake Kariba with the highest total production so far recorded – 346.8 g/m^2/annum (or kcal/m^2/annum if applicable). The next highest is the River Thames' total production of 197 g/m^2/annum (MATHEWS, 1971), followed by ALLEN's (1951) Horokiwi Stream with a 54.7 trout production, HORTON's (1961) Dartmoor stream trout 10–18 g/m^2/annum production, CHAPMAN's (1965) Deer Creek with a 16 g production, Wyland Lake with a 13.6 g/m^2/yr production (GERKING, 1962) and Whistler's Bend Reservoir with a 5.7 g production (COCHE, 1967). The production which is closely related to the metabolic rates, is significantly higher in warm regions.

Available production in the Kličava Reservoir amounted to 2.75 g/m^2/yr in 1964 and rose to 3.99 g/m^2/yr in 1968 (HOLČIK, 1970c). It was over one-third and later declined to one-quarter of the standing stock. Similar values were found in the North Russian perch lakes by BURMAKIN & ZHAKOV (1961) and RUDENKO (1962). GERKING (1962) estimated available production of the bluegill sunfish in Wyland Lake at 9.1 g/m^2/yr. The available production of 22 to 27.7 g in Cuban lagunas (HOLČIK, 1970b) was, however, a much higher percentage of the standing stock; that of Lake Kariba (68.2 g/m^2/yr) more than a hundred per cent. This high production is supported by DUNN's (1970) data of 185 g/m^2/yr of the actual catch in the equatorial Lake George. The high catch is a sign of an enormous production, although because of the low depth the whole area of Lake George contributes to the production. Consequently the total production may be about 400 g/m^2/yr.

The turnover ratio (WATERS, 1969) of total production per unit biomass for the fish studied from Lake Kariba was on an average 1.08 but within individual species this varied from 4.47 to 0.002 (estimated per annum and all age groups). As this ratio illustrates how many units of biomass the stock produced within each unit of biomass existing in an evaluated time, it became an important variable. It is the highest with the first economically preferred species: *M. longirostris* enlarged 4.47 times its mean biomass in a period of a year, *H. vittatus* 3.75 times, *M. deliciosus* 3.24 times, *S. codringtoni* 2.21, *M. macrolepidotus* 1.87, *S. m. mortimeri* 1.61, *A. lateralis* 1.59, *H. discorhynchus* 1.51, *T. rendalli* 1.47, *M. electricus* 1.14, *S. zambezensis* 1.10, *B. i. imberi* 1.06, *S. mystus* 1.00, *L. altivelis* 0.95, *E. depressirostris* 0.83, *H. longifilis* 0.67,

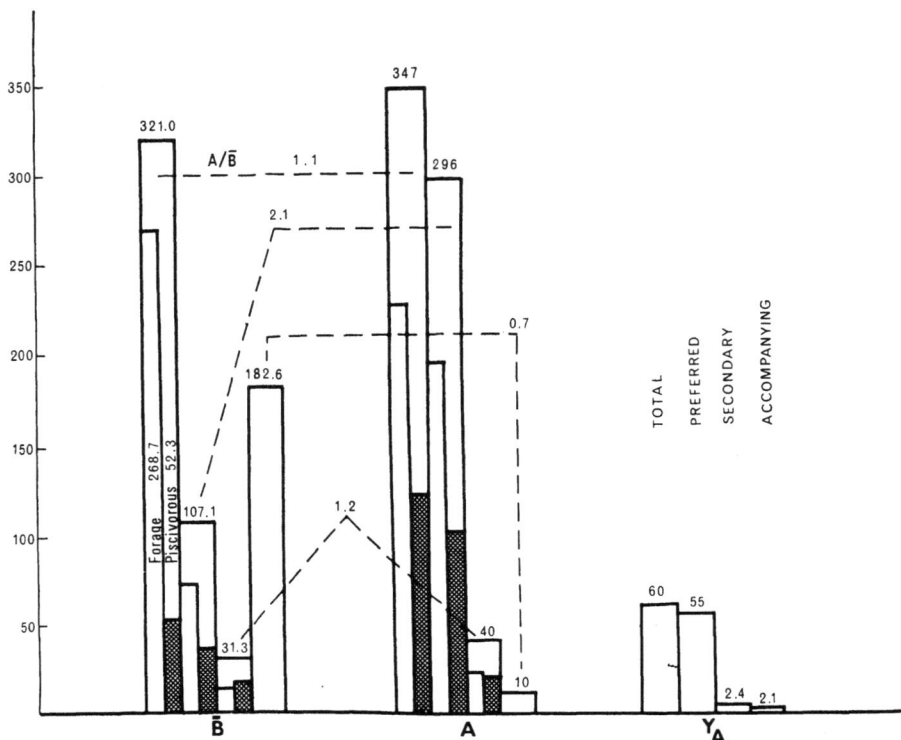

Fig. 97. The size of mean biomass ($\bar{B}$), total production (A) and total yield (Y_A) of the whole Lake Kariba fish populations, as well as forage and piscivorous fish, separately, with turnover ratio (A/$\bar{B}$) for preferred, secondary and accompanying fish groups. Ordinate and numbers in columns are in g/m²/annum, numbers at broken lines are the turnover ratio.

S. nebulosus 0.60, *C. gariepinus* 0.52, *M. acutidens* 0.25 and *H. darlingi* only 0.002 times (Table 116). This ratio for the River Thames gudgeon was 2.00 and 1.94, for bleak 1.92 and 1.32, for dace 1.75 and for roach 1.12 and 0.68 (MATHEWS, 1971), also definitely lower than for most of the Lake Kariba fish.

The mean biomass and total production of Lake Kariba fish fauna, when viewed with respect to the individual groups (Table 113) – economically preferred, secondary and accompanying species – reveal different relationships than when viewed as a whole (Fig. 97). The production of preferred species is 2.1 times the mean annual biomass. There are twice as many forage as piscivorous fish in this group; their biomass doubled, however, proportionally. The secondary species production is only 1.2 times the mean annual biomass with nearly an equal portion of forage and piscivorous species; the latter increased the biomass a little less effectively than the forage species. The accompanying species production is the lowest – 0.7 A/$\bar{B}$ – in spite of the highest mean biomass. It consists of forage fish only. The major contribution to production is made by fishes of the first group – 296 kcal/m²/annum – with only 50 kcal/m²/annum produced by the remaining two groups. It is even more pronounced in total yield when 55 kcal/m^2/annum is formed by preferred species and only 4.5 kcal by secondary and accompanying species.

The ratio of initial biomass to available production shows the initial biomass on an average 3.4 times the available production. It is similar to standing stock – available production ratio in the temperate region. It was found to be exceptionally high with *H. darlingi* – 562.01 – which when compared with a range of 3.07 to 1.02 for this ratio for the rest of the species studied is hard to explain and seems to be a result of a peculiar bias. However, in the ratio of available production to total production, which shows a gradual increase in the proportion of available production in the direction of the accompanying species, the exceptionally high value for *M. acutidens* can be explained by its increasing density with a recovery of suitable habitats in the lake (HOLČIK, see Section 3.11).

Not enough data are so far available for evaluation of the ratio of fish production to primary production though STRAŠKRABA *et al.* (1968) had illustrated some discrepancy between this ratio in temperate and tropical regions. While primary production seems to be on the average four times higher in tropical regions than in lakes of temperate regions, the fish production in Cuban lakes was found to be only two times higher. If, however, the primary production of Lake Kariba is similar to other tropical or subtropical lakes (25 g/dm^2/yr of O_2), the available production of the fish population found (68.2 g/m^2/yr) is more than four fold higher than the available fish production in similar temperate lakes, therefore, the suspected 'lower efficiency of the utilization of gross primary production throughout the trophic chain in tropical conditions' cannot be true. However, the fishing pressure in temperate waters must be higher and thus the comparison of available production with tropical waters invalid. More meaningful would be to compare the final production (available production + actual catch). Attempts to compare standing crop ($\simeq$ biomass) to primary production, however, failed completely (RUPP & DEROCHE, 1965) and the same factors suggest the lack of a realistic correlation between primary and fish production (MCCONNELL, 1963). The production of plants can be utilized at different trophic levels and various parts of it, however transformed, may never reach the fish.

The available yield of the Lake Kariba fish population suggests a possible exploitation three times higher than at present but not all species are equally exploited. Most of the underexploited species are less attractive to an industrial fisherman or fishing-crafts than to a single commercial fisherman. Only the individual fisherman can effectively exploit the given species diversity and all the different habitats along the entire shore line. Such are also the facts found by SCUDDER (1967, personal communication). Local fishermen do exist and they will increase if market facilities improve. Moreover, the traditional home-made fishing gear (e.g. traps, baskets) should not be omitted from consideration; not all species can be effectively caught in gillnets, although it was the only gear officially recommended and supplied. No doubt this has been in conflict with the local human experience (SCUDDER, 1960) and has been limiting the number of species in catches and habitats to be fished.

In order to exploit the Lake Kariba fish resources to the limits of potential catch, it is most likely that no industrial fisheries are needed, but more commercial fishermen, well distributed on islands and along the lake shore. The only limiting factors are

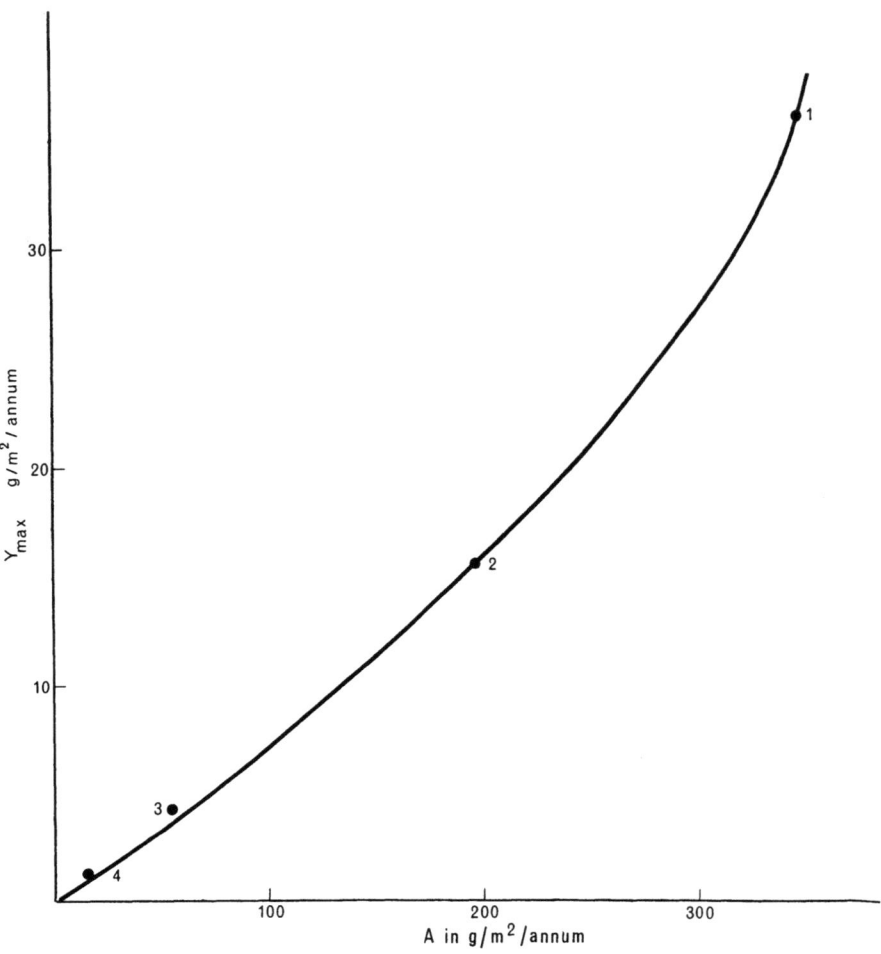

Fig. 98. Maximum sustainable yield (Y_{max}) in g/m²/annum (ordinate) expressed as 8 percent of total production (A) in g/m²/annum (abscissa) for: 1 – Lake Kariba, 2 – River Thames, 3 – Horokiwi Stream, 4 – Deer Creek values.

market, transport system and the availability of an increased range of consumer goods. The maximum sustained yield (Y_{max}) is an extrapolated value of available yield, introduced here as a measure of yield and estimated from empiric data as 8% of the total production. The given relationship (Fig. 98), positively verified on more ecosystems, can become a helpful means in assessing yield value from known ecological production or vice versa.

BRIAN CURTIS' (1949) statement that 'the "maximum sustained yield" concept is complicated in detail and in practice, but in theory it is simply that for each kind of fish there is a surplus produced over and above that needed to maintain the population at an optimum level, and that that surplus, no more and no less, is the number which should be taken by fishing' has recently been considered as more complicated than that (RADOVICH, 1973). Simply stated, however, the produced surplus is an adaptation of the population to cope in its succession with various stresses (predator, climate, etc.). In a less mature system the surplus produced will be greater and will be manifested mainly by short life span opportunists (MARGALEF, 1968). A mature balanced system will have little surplus and is by the same token more vulnerable to pressures from predators (overfishing), as has repeatedly been illustrated in the examples of the Laurentian Great Lakes (REGIER et al., 1969; REGIER & COWELL, 1972). Various stresses can slow down or reverse the process of securing stability. The Lake Kariba fish population yield, or the maximum sustained catch (E_{max}, see p. 443), should be examined also from the stability point of view, since there are doubts about its state. Thus REGIER's (1970) scepticism – 'What is the maximum sustained yield?' where the system is unstable – is justified.

Only after the four-and-a-half initial years of rising water, that is, after the establishment of the lake's final size and lacustrine abiotic and biotic characters (MCLACHLAN, 1969a, b, 1970) did the invasion of the new taxa into Lake Kariba begin. In the initial stage of filling only stock density and growth differences occurred as a consequence of the expanding aquatic environment and its large temporary load of nutrients from flooded terrestrial habitats (JACKSON, 1960; BOWMAKER, 1970). However, immediately after the lake was stabilized in its main topographical and limnological characters (COCHE, 1968 and Part I of this book), a rapid shift in taxa composition and taxa invasion occurred, an occurence to be expected in a new habitat with unoccupied niches, although the rapidity of the invasion was probably due to the tropical conditions. The indigenous *B. i. imberi*, for example, is replaced by *A. lateralis* in enormous density and in a very short time (BALON, 1971b, and section 6.3); the same route, via the edge of Victoria Falls and/or streams of the Lake Kariba drainage area was taken by a number of other species – *Labeo lunatus, Sargochromis giardi, Haplochromis carlottae, M. macrolepidotus, Barbus poechii, B. paludinosus, B. unitaeniatus, B. lineomaculatus, Labeo cylindricus, Schilbe mystus, Sarotherodon andersoni, Serranochromis* sp. and *Pseudocrenilabrus philander*.

Meanwhile, however, the early stocking program in the initial phase of the lake (1959–1962) – of pond reared *Sarotherodon macrochir* (66%) and *Tilapia rendalli* (34%) from the Kafue River failed completely (26 tons of juveniles were introduced). *T. rendalli* was an indigenous species of the Middle Zambezi and present in the lake before stocking, while *S. macrochir* was a new taxon, which, with the exception of some individuals shortly after stocking, has never been found in the lake again. However, the more risky introduction (TAIT, 1965; BELL-CROSS, 1971) of the Tanganyikan anchoveta *(L. miodon)* undertaken later was successful (see section 7). It seems that the lake in its first filling phase actively rejected any new taxon or, more precisely, the exploding density of a few dominant species (due to flooding of new areas suitable as spawning grounds) was able to control the fish taxocenes and thwart every attempt at competition. Only in the second successive phase, when the explosive development of dominant species stopped (no spawning grounds due to annual drop in water level), could the system accept new components. The anchoveta's introduction accidentally fell into this phase and it was accepted together with many other natural invaders from adjacent systems.

As I had been against the introduction of the anchoveta because of the same reasons expressed by REGIER (1968), its success came as a surprise. In spite of this, my original reasoning remains unchanged and I am still convinced that the introduction was unnecessary. The ecosystem itself was already on the way to filling the pelagic niche and though *A. lateralis* may never have reached the high density of the better adapted anchoveta, the latter will attain its density at the price of other taxa, more marketable perhaps if harvested. The common dinner table – the nutrient load (or carrying capacity) – remains unchanged with a new species introduction. But the infiltration of the dense *A. lateralis* population into the pelagic free niche in 1966 to 1969 seems later to have stopped completely, as this niche became occupied and gradually dominated by the anchoveta. Mixed catches of *A. lateralis* and *L. miodon*, common in a 1:1 ratio in 1969, gradually changed to pure catches of anchoveta (1971, see section 7.2). This happened more likely because of a limitation in available nutrients and a different level of adaptation rather than for reasons of interspecific competition (LARKIN, 1956).

To be more specific, such an explanation may simply mean that the nutrient contents and topographical factors have stabilized the production of the Lake Kariba fish population to a certain carrying capacity, though the increase of single taxa diversity and differentiation is still in progress. It is the sort of process that needs a more detailed explanation. RYDER (personal communication in lit. February 1972) believes that 'what appears to be the attainment of absolute stability in a tropical reservoir may only be an approach and gradual leveling off at the asymtote of absolute stability, the ultimate stability achieved only at some point in the distant future when a backlog of organic matter has been introduced into the system and an autocatalytic stage reached'.

There is room to speculate even further. In a new ecosystem, and a man-made lake is that, the fish population tries to fit itself into a given content of nutrients and the, at the beginning, large oscillations of its variables, a sort of adaptation trial, gradually decrease toward some ranges of stability. The 'oscillation' peaks are sometimes noticed as a cyclic phenomenon. Because of faster energy cycling in a tropical environment, one would expect the population to reach the stability ranges considerably earlier.

The alignment of production size and nutrient contents may be the first step in a succession when the system tries to achieve a balance and full functioning of the energy flow (ODUM, 1969). Due to a low taxa diversity it operates at this phase with no reserves and is probably more vulnerable than later when the balance or stability is gradually secured by an increased species diversity. MARGALEF's (1968, 1969) opinion of the stability equal to diversity is probably valid when expressed in dynamic terms: The stability of an ecosystem becomes more secure and decreases in vulnerability with increased species diversity. But the determining factor of stability will be the achievement of full utilization of given nutrients through a complete energy flow. The successive increase in taxa diversity secures only the system's stability. A changing nutrient load does not enable the system to reach stability or to become more mature as does overfishing or other cultural stresses.

MARGALEF's (l.c.) 'oligotrophy should succeed eutrophy' seems in this context to be perfectly correct. Lake Kariba's rising water volume and increasing nutrient load were in a state of unbalanced eutrophy from 1958 to 1963 (Fig. 99). From 1964 to 1967 due to an equilibrated nutrient load a kind of oligotrophic but vulnerable stability was reached. From 1968 on, this stability, without changes in quantitative variables of energy flow, has with an increase in taxa diversity been in the process of securing its state (towards an ultimate maturity).[4] The whole system reacts here like a single species, in a similar way as does adaptive radiation in evolution.

Any quantitative changes of individual elements – species – of the system do not matter; less utilized nutrients on one side are immediately more utilized somewhere else. The resultant energy budget as a whole becomes stable, as determined by the available nutrient contents.

A supporting view was presented in the form of four types of 'population dynamics' (BALON, 1963b), built on the work of NIKOLSKY (1953a, b), LAPIN & JUROVICKY (1959), and others. As a consequence of environmental and fishing intensity changes, the single species stock can overgrow into another population type, differing mainly in density, food resources, growth intensity, first age of sexual maturity, and fecundity. At the same time the total production may be unaltered. Later LECREN (1965) expressed it as follows: 'One can also postulate that there is a whole range of possible 'population-dynamics' available for exploitation in an ecosystem analogous to the range of available niches.'

Most likely the fish population production as a whole will be fairly stable, as long as the fishing intensity does not affect the minimum species diversity needed to maintain the stability ($=$ maximum sustained yield) or as long as no drastic environmental changes occur (ultimately with similar effect). At the same time single species or individual locality stocks may differ more frequently, due to variations in nutrient inflow and other local variables.

For example, the density and production of Lake Kariba *H. darlingi* is significantly higher along the shores with high game density (nutrient inflow from game defecation, see section 3.12). It is also higher in a cove situated on the old Zambezi River bed (D) with its mouth directed towards the inflowing river waters. Though only a third of the water volume in the lake is replaced each year, the entire water mass of Basins I and II is flushed out annually by the flow of the Zambezi River, which provides about 70% of the total inflow into the lake (COCHE, 1968; BEGG, 1970). As a result any loss of nutrients due to removal of fish may be negligible as long as it does not exceed the nutrient inflow and its retention by water plants. The removal of the entire maximum sustained yield is thus probably feasible without any deleterious effect upon the nutrient content.

The turnover of nutrients may also be reflected in some variables of energy flow from the different trophic levels (e.g. nitrogen (GERKING, 1962; HRBAČEK, 1969)), if such data are available, or in some physical characteristics (HAYES & ANTHONY, 1964; FISH, 1970), and it would be very useful to relate this to the production and yield of the

---

[4] The chemical phases are slightly different, see Part I, Fig. 55.

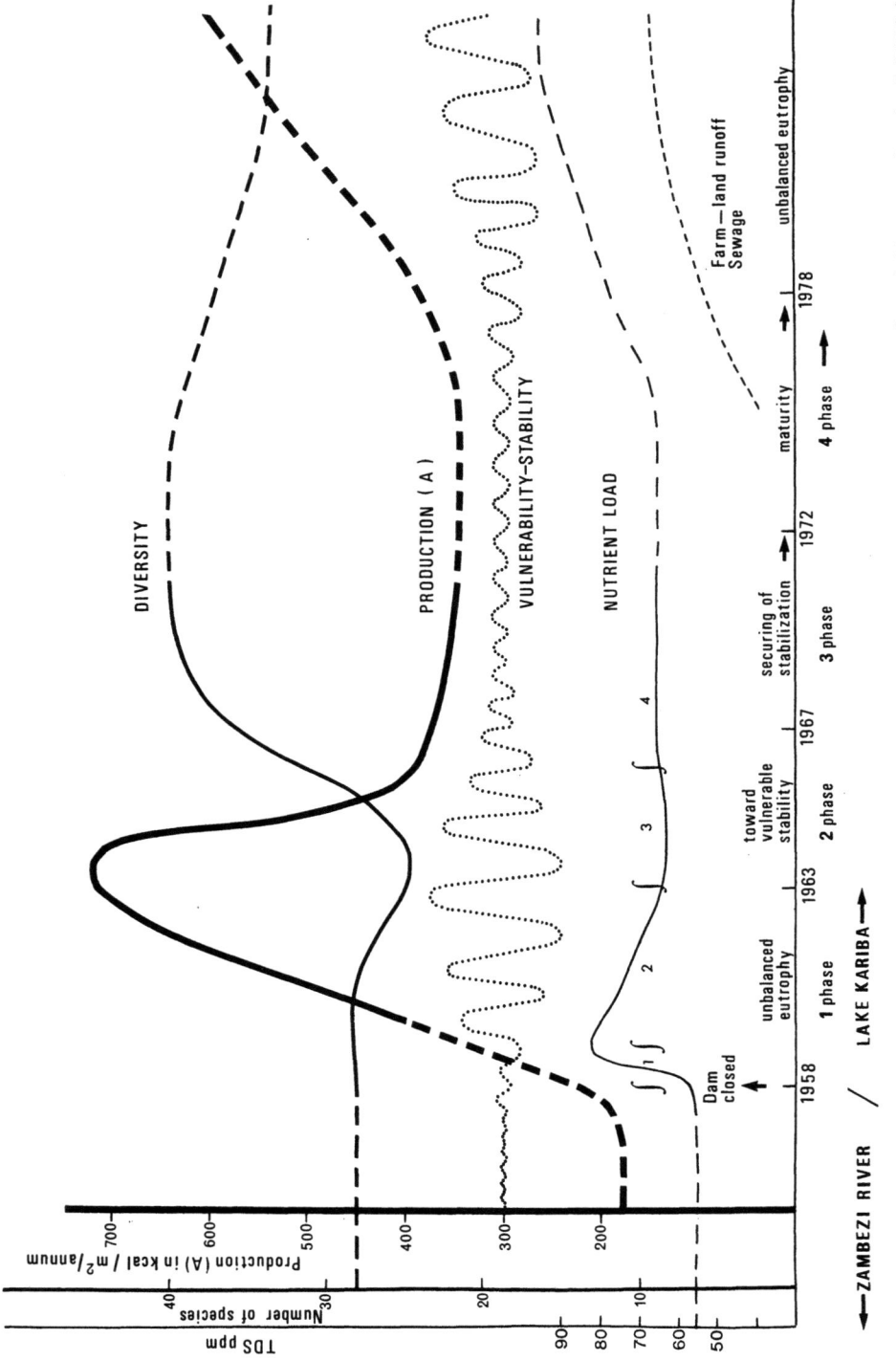

Fig. 99. Zambezi River prior to impoundment and Lake Kariba history: past, present and future in dynamic variables – nutrient load (TDS in ppm), total production (A in kcal/m²/annum), species diversity (in number of species), and theoretical image of stability (vulnerability) oscillations (ordinate). Abscissa – years and phases of succession. Chemical phases are marked along the nutrient load line.

550

fish population. A detailed review of such up-to-date trials, in relation to different environmental factors, is given by JENKINS (1968). He pointed out as one of the most useful indices, the total dissolved solids divided by mean depth (i.e. RYDER's (1965) morphoedaphic index). According to COCHE's data collected at the same time as mine the relevant Lake Kariba variables are as follows (see also Part I, Table 70):

Lake Basin	I	II	III	IV	Mean
Mean depth, m	12.6	24.0	26.5	33.2	29.2
Total dissolved solids, ppm	47	49	63	64	58

The corresponding RYDER's Index for the whole lake is 1.99. As my computation is related to the area inhabited by native fish and not to the whole lake area, significant comparisons with data of previous authors are impossible. If, however, the same procedure is applied for Lake Kariba and the actual catch recalculated for the whole lake area (5.59 kg/ha/yr) it corresponds to RYDER's Index in the bottom low part of JENKINS (l.c.) regression curve (REGIER, CORDONE & RYDER, 1971). However, I feel it is rather imprecise, as no measures of fishing intensity are included. A correlation with production and yield would be more meaningful. The correlation with standing crop already proved highly positive in U.S. reservoirs (JENKINS, 1970).

There is no doubt that there is a highly positive relationship between nutrient contents and fish production. Prior to dam closure the Zambezi River waters at Kariba contained on an average 55 ppm of total dissolved solids (see Part I, Table 85). In the rising Lake Kariba waters, during the first year of filling, TDS rose to 81 ppm. The catches doubled though with some delay as compared to the sudden decrease after 1963, when also the TDS contents declined to 63 ppm. From then on it became relatively stable at this value, with the probable help from a well-dispersed *Salvinia auriculata* and a submerged aquatic vegetation biomass which held some of the nutrients that would otherwise have been lost through tunnels of turbines and spillgates (BOWMAKER, 1970). WARREN and others (1964) indicated that the increased nutrient load will raise the production many-fold more than the corresponding nutrient increase value. This explains the enormous production increase within only moderate additions of nutrients, for example from game defecation.

If the fish production values of Lake Kariba are recalculated for the whole lake area instead for the 38% of the total area inhabited by fish it certainly will reveal a picture of low fishery potential. Actual catch (E') presented in this manner (E' 5.59 kg/ha/yr) will support the conclusions that Lake Kariba is an oligotrophic impoundment of low productivity; it may, however, have little value in terms of energy distribution and utilization. The relationship of nutrients locked in inshore waters to those of open waters is an unknown quantity. In order to draw valid conclusions on the ecological production potential of Lake Kariba these phenomena will have to be understood much better than they are now.

Realistic values for the interaction of lake and terrestrial ecosystems were noted only with reference to the differences of production of *Haplochromis darlingi* and *Sarotherodon mossambicus mortimeri*, which was substantially higher near localities with dense game concentration. It is apparent that the nutrient enrichment of the lake water from frequent dunging by game (i.e. elephants) along such shores is the most important factor affecting production here and with recent knowledge of the effects of pond fertilization it is the sort of thing that can be expected. In this particular species, however, the total production was more than 35 times higher in waters bordering game areas than in waters not bordering such areas. I am inclined to think that this interaction was underestimated until now at least in some areas suitable for the mutual survival of human and game populations. Perhaps Lake Kariba may be such a place.

At this time only a small part of the human population is engaged in fishing along the lake shore. The majority are farmers reaping a very low harvest and suffering from frequent crop failures, a consequence to be expected in an environment of a 'dry marginal climate' unsuited for conventional agriculture, except on an extensive scale (RINEY, 1963). Furthermore, promises of irrigation facilites by the dam promoters were never kept. KEAY & AUBREVILLE (1959) described the vegetation of this type as 'savannah woodland dominated by mopane' *(Colophospermum mopane)* and according to CHILD (1968) '... mopane, which has hardly been disturbed, can carry reasonably large game population, and this is significant in view of its low assessment for agricultural purposes.' Even if there is no data to enable a definite conclusion, I think that local people instead of fighting the remnants of wild ungulates to protect their fields, which are of dubious value, could gain more benefit from full utilization of the fish potential, which may in fact be increased by normal (i.e. up to the woodland's carrying capacity) concentration of wild animals. To emphasize once more, '... a mixed population of wild ungulates efficiently utilizes most of the available forms of vegetation, while domestic livestock receive sufficient nutrition from only a small part of it. Therefore, the carrying capacity is higher for wild animals than for domestic livestock ...' (TALBOT *et al.*, 1965). The wild ungulates are 'also inherently more efficient, when considered as a converter of solar energy to protein'. This was quoted by MARTIN & WRIGHT (1967) who in addition gave examples of this from evolutionary history. Thus, in the Lake Kariba area, the sorely needed protein could be provided by a balanced use of wild ungulates and fish, rather than by plant farming which, if and when water becomes available, is more suitable for the inland plateaux. Substantial differences exist in the way domestic livestock and wild ungulates obtain water. Wild ungulates are adapted to utilize water retained in and condensed on plants (TAYLOR, 1970, 1971). This is another reason why in these areas use should be made of such wild animals.

For a mopane veld of Lake Kariba's island 17 CHILD's (1968) estimate of a standing stock of large herbivores was over 13 kg/ha; however, according to DASMANN & MOSSMAN (1960, 1962), this kind of woodland under game management gives a standing stock of 44 kg/ha, and is close to the 52 kg/ha of the annual standing stock of wild ungulates at carrying capacity level of acacia-commiphora bushland estimated by TALBOT *et al.* (1965) and Serengeti savannah by GRZIMEKS (1960). In such a case at least 13 kg/ha could be harvested.

It is well documented, that 'if properly processed, most game meat is of high quality and is satisfactory for human consumption and could, therefore, contribute substantially to the production of food' (TALBOT *et al.*, 1965). Limited by the escarpment belts on both sides of Lake Kariba the area of approximately 50,000 ha could yield over 600,000 kg of meat of mixed wild ungulates. Consequently, this type of management would substantially increase the fish production of Lake Kariba. A minimum of 15 thousand tons of fish and ungulate protein could be expected from this area on a renewable sustained basis. The exploitation of both resources may be accomplished by the same people, the processing, marketing and distribution by the same units.

This can be easily achieved because the fish and wild ungulate potential is still there. Though the density of large herbivorous mammals in most of the area is already far below the carrying capacity of the ecosystem, in many isolated places the density of wild ungulates is probably in its upper limits (JARMAN, 1969) of about 150 kg/ha. This is similar to some estimates for ungulate biomass in National Parks (BOURLIÉRE, 1963). Such is the Chete Island elephant and antelope density, an ideal study area for the interaction of water and terrestrial communities, communities in the process of maturation (in the changed ecosystem), with the production potential in determined limits given by the recent nutrient load.

In the future the growth of the human population together with growth of urban centers and industries along the Zambezi River and streams of the Lake Kariba drainage area will increase the sewage release, and along with the increasing runoff from farm lands will result in a substantially larger nutrient load of the lake. In the long run this cultural eutrophication (HUTCHINSON, 1969) may raise the fish production of the lake, although it would decrease its esthetic value and stability. The man-made Lake Kariba ecosystem, on its way to maturity but still for a long time vulnerable, can easily be destroyed if certain further man-made changes take place.

EPILOGUE

> The basic right to life (to ecosystem joiners) requires revision of laws to convict anyone of willful release of waste which disorders the general life-support systems, the mineral cycles, and the complex ecological systems that keep the life-support system working efficiently and smoothly. Economic developers selfishly interested in their industry, their town, or their records of expansion must include in the costs of any new ventures the full interfacing of the activity with the life-support systems and payments in full to the public sector for any energy values heretofore regarded as free (except for processing cost) such as water, air, mineral reserves, and ecosystem area. The developer may protest that it will block development. Why should there be a God-given right to develop?
>
> HOWARD T. ODUM, 1971, p. 300

We judge the well-being of other humans by our own standards and consider this humane. But the new source of energy from fossil fuels destroyed the old mores and ethics which evolved from the knowledge of limited energy available through the natural ecosystems, and also destroyed the energy controls so imperative for the human culture and possibly for human survival. From then on people who live as a part of a balanced environment have been considered underdeveloped and are, whenever possible, doped with our 'all-economy' and/or 'all-profit' drugs to help them keep pace in an ever-accelerating race, whose end result will be exhaustion. To our surprise we have discovered primitive tribes with better developed regulatory mechanisms for survival than ours (THOMPSON, 1950). Although many may not have the ability to be objective when confronted with different life standards, more and more people recognize (I hope) that the low standard assures a better alliance between man and the earth biosphere, between man and his sources of energy. We have changed the energy source so fast that the adpatations, controls – the whole human culture – stayed far behind. We must revise our approach toward 'under developed' societies and redefine 'humanity'.

Let us limit reasoning to my field. For example, the utilization of any small fish, in order to maintain a sustained harvest at lowest energy costs, is rarely explained as a regulatory mechanism superior to our own. Rather the willingness of natural tribes to utilize the most abundant species of small fish is likely to be viewed as a temporary phenomenon connected with protein deficiency, which, with a rising standard of living, will change into the habit of cultural tribes of selectively utilizing only the larger palatable or small delicious fish species. Although utilization of small fish may be a subconscious regulatory mechanism, the fishing of the rednose mudsucker *(Labeo altivelis)* during its spawning migrations up the Luapula River was known to be customarily regulated by tribal man (through chiefs). As later events proved, enforcement of such regulation by witchcraft and ritual dances was far superior to our written laws. The weirs built in the Johnston Falls region were permitted to catch only fish which were migrating downstream back from the spawning grounds. European interference in the recent past has led to the extinction of this fishery (DE KIMPE, 1964) due to overfishing, since 'gravid fish were killed in thousands during the spawning run and caviare made from the ripe eggs in the ovaries was a popular delicacy in the Belgian Congo' (JACKSON, 1961a).

From records of all aspects of natural tribal life similar examples can be extracted. Further speculations and theorizing may be fascinating, though I fear that most generalizations which extend beyond real findings are highly subjective and can be erroneous. Broad, general conclusions should probably be formulated in spite of these objections, since most of the changes within studied systems occur faster than the systems can be scientifically monitored and analyzed. However, they should preferably be formulated not as scientific contributions, but as expressions of citizens' concern about the loss of power[5] in the world; more energy subsidies will lead to an unacceptable waste increase.

Let us, however, return to the subject. The maximum sustained yield should be viewed as an initial estimate similar to those estimates produced for purposes of planning mining activities by geological surveys. The organization of exploitation of these resources seldom immediately follows discovery or size assessments and requires experts with somewhat different skills than those of prospectors or resource estimators. Science is succeeded by technology. With respect to the requirements of the local population, the difference between resource size expressed in our case as maximum sustained yield and exploitation intensity, which we may call maximum sustained catch, remains one of the unanswered questions. Like the non-renewable resources, the extensive utilization of the fish resource needs an additional power source and energy flow from outside the system. This is an additional unanswered question about the ultimate effectiveness of such utilization.

In so far as humans are concerned the unexploited part of the resource is, generally speaking lost forever as fish protein. We do not know, however, the role such an unexploited part plays in a sustained safety margin for long term natural stresses, if the loss of protein is ultimately true where energy flow is concerned. The fish resource is, however, definitely limited; in the Lake Kariba case the given limits are determined entirely by man, who created this whole new ecosystem as his 'technological triumph' in the 'contemporary development policy'.

Let us stress: the main purpose of the Kariba Dam was to generate hydroelectric energy for the Copperbelt mines; high fish potential and efficient irrigation for farming were the major promises among numerous minor ideas for development. The 'underdeveloped' local population of Tonga (SCUDDER, 1962; COLSON, 1960; REYNOLDS, 1968) struggled with only the help of their bare hands and a naive belief ('Nyaminyami, the river God, will help us') in order that their familiar and balanced environment might survive. Some had to be forced though others volunteered to abandon the

---

[5] The rate of flow of useful energy.

alluvial valley for pre-selected resettlement areas which were in general less fertile than their native lands. At the time the fate of the local population (GADD et al., 1962) caused less concern than the animal tragedy in the same area – the hopeless 'Operation Noah' (SCUDDER, 1968; ROBINS & LEGGE, 1959; BOWMAKER, 1970). The complaints of the people were subdued by a grant of a two year government subsidy on certain essential material needs such as grain, medicines, etc. Some difficulties were alleviated by the enormous fish production of the lake; the diluted nutrients from the flooded terrestrial habitats and the rising lake waters brought an explosive reproduction, survival and growth of fish and so this well known natural phenomenon helped, in the most critical stage, to support the builders' promises. Then the lake began to stabilize and the catches decreased. By that time those responsible had long since left the scene.

LAGUS (1959), a BBC movie maker, felt the need to make a film 'on the Kariba tragedy: for it wasn't the triumph of the dam that seemed important to me but the dying of the animals'. Today I cannot be sure that it was not a tragedy for the promoters themselves, since the electricity thus gained is helping to exhaust the copper resources even faster. The benefits to the local people were dubious, the energy generated at Kariba mostly out of their reach ... and what about the gains of fish protein?

Prior to the dam closure at Kariba the total dissolved solids of the Zambezi River waters (55 ppm) indicated a low production of fish. Due to the annual flood, however, the nutrient deposit in the inundated alluvium yielded crops of plants and animals (game ungulates) which were or could have been extensively utilized. Wild animals alone, if harvested, could have yielded the same amount of protein as the lake, although there was more space for agricultural intensification and a much higher potential harvest at a lower cost of energy in the river alluvium.

After the first initial four years of rising waters and high contents of dissolved solids (81 ppm) diluted from the 500,000 ha of gradually flooded terrestrial deposits, the dissolved solids contents of the lake stabilized at 65 ppm (Fig. 99). The fish fauna of riverine origin consequently inhabited only the inshore fold of the lake (37.7% of the total lake area) and the terrestrial habitats were driven closer to the escarpment, where low farming potential is estimated. Hence the total potential of food production, in spite of later occupation of the open lake by some fish species, should not be higher than the potential of the undisturbed Gwembe Valley habitats prior to the dam. It may even be lower if the annual heat budget of about 15,000 cal/cm^2 (Part I, Sec. II), which is relatively low because of the evaporation rate and reflection of a substantial part of the solar energy by the water mass (BUTLER, 1963), means anything in this connection (VAN DER LINGEN, 1971). In addition, to utilize the lake's fish potential to the given limits requires artificial subsidies of energy at present simply not available. Also, there is some doubt whether they will be found efficient later on.

In view of this the objections I raised years ago about the Danube River dam projects (BALON, 1967a) still seems to be valid: 'For whose benefit will the street lights be lit in 1980 if we have nothing to eat then.' Who will be held responsible? Do we fear to formulate the responsibility more explicitly in spite of knowing that 'man-made lakes (...) are usually built for some *primary* purpose, but the construction or presence of reservoirs can create *secondary* problems which may involve economic loss and human suffering unless reservoir planning in its earliest stages thoroughly considers both primary and secondary aspects' (LAGLER, 1969)? 'There is no cost charged to the developer for destroying a well-balanced, stable ecological system (...) in the area developed nor for radically transforming all nearby ecosystems' (REGIER, 1970a). Here somewhere lies the moral of the parable.

> 'Unfortunately, bringing the third world into the twentieth century means, in effect, intensifying ecological malpractice.'
>
> PAUL SHEPARD in
> 'The Tender Carnivore and the Sacred Game' (1973, p.26)

LITERATURE CITED

ALLEN, K. R. 1950. The computation of production in fish populations. *N.Z. Sci. Rev.* 8: 89.
ALLEN, K. R. 1951. The Horokiwi Stream: a study of a trout population. *Fish. Bull. N.Z.* 10: 1–238.
APPLEGET, J. & L. L. SMITH, JR., 1951. Determination of age and rate of growth of channel catfish *(Ictalurus lacustris punctatus)* of the Upper Mississippi River from the vertebrae. *Trans. Amer. Fish. Soc.* 80: 119–139.
BACKIEL, T. 1964. On the fish populations in small streams. *Verh. intern. Ver. theor. angew. Limnol.* 15: 529–534.
BALDWIN, W. 1863. African hunting from Natal to the Zambezi. Bentley, London. 439 p.
BALINSKY, B. I. & G. V. JAMES, 1960. Explosive reproduction of organisms in the Kariba Lake. *S. Afr. J. Sci.* 56(4): 101–104.
BALON, E. K. 1955. Growth of the roach *(Rutilus rutilus)* and a revision of main methods of its determination. Vyd. SAV, Bratislava, 166 p. (in Czech).
BALON, E. K. 1958. About development intervals of the life of fishes. *Biuletyn Zakładu Biologii Stawów PAN* 7: 5–15 (in Polish).
BALON, E. K. 1959. Ichthyobiological characteristic of the Olza and Łucyna rivers. 50 lat Jedenastoletniej Szkoły Średniej z polskim językiem nauczania w Orłowej: 55–63 (in Polish).
BALON, E. K. 1960. Über die Entwicklungsstufen des Lebens der Fische und ihre Terminologie. *Zeitschrift für wissenschaftliche Zoologie* 164(3–4): 294–314.
BALON, E. K. 1961. Vergleich der Altersstruktur und der gesetzmässigkeit des Populationswachstums dreier Donauarten der gattung *Abramis*. *Zool Anzeiger* 167: 403–412.
BALON, E. K. 1962a. About the occurrence of rheofil fishes in branch and inundation zone of Danube near Medvedovo. *Prace Laboratoria rybarstva* 1: 55–62 (in Slovak).
BALON, E. K. 1962b. Age and growth of the chub in the Orava River and in the Orava Valley Reservoir in the years following its filling (with methodical notes). *Ibidem* 1: 79–104 (in Slovak).
BALON, E. K. 1963a. Description of key scales of Danubian breams. *Biologia (Bratislava)* 18: 273–285 (in Slovak).
BALON, E. K. 1963b. Altersstruktur der Populationen und Wachstumsgesetzmässigkeiten der Donaubrachsen *(Abramis brama, A. sapa, A. ballerus)*. *Bul. Inst. Chem. Techn. (Prague), Technology of Water* 7(2): 459–542.
BALON, E. K. 1963c. Einige Fragen über das Vorkommen und Biomasse der Fische in Inundationsseen und im Hauptstrom der Donau in der Zeit des niedrigen Wasserstandes. *Zoologischer Anzeiger* 171(11/12): 415–423.
BALON, E. K. 1964a. List and ecological characteristic of the Danubian fishes. *Hydrobiologia* 24: 441–451.
BALON, E. K. 1964b. A contribution to the knowledge of the ichthyofauna in the head water area of the Poprad River. *Sbornik prac o Tatranskom narodnom parku* 7: 140–164 (in Slovak).
BALON, E. K. 1964c. On relative indexes for comparison of the growth of fishes. *Věst. Čs. spol. zool.* 28(4): 369–379.
BALON, E. K. 1964d. A list and the ecological characterisation of Polish freshwater lampreys and fishes. *Polskie Archiwum Hydrobiologii* 12(2): 233–251 (in Polish).
BALON, E. K. 1965. Wachstum des Hechtes *(Esox lucius* L.) im Orava-Stausee. *Zeitschrift für Fischerei* 13(1–2): 113–158.
BALON, E. K. 1966a. Practical applications of the growth values of fish for rational fishery management in the Orava Dam Lake. *Buletin VUR Vodňany* 2(1): 3–12 (in Slovak).
BALON, E. K. 1966b. The ichthyomass and abundance of fish of the Danube inundation arm below Bratislava with description of toxaphene poisoning. *Biologia (Bratislava)* 21(4): 295–307 (in Slovak).

BALON, E. K. 1966c. Bemerkungen über die Fischgemeinschaften und über die Ichthyomasse eines Inundationsarmes der Donau. *Verh. Internat. Verein. Limnol.* 16: 1108–1115.
BALON, E. K. 1967a. Influence of life environment on the growth of fishes in the Orava Dam Lake. *Biologické práce* 13(1): 123–175 (in Slovak).
BALON, E. K. 1967b. Evolution of the Danube ichthyofauna, its recent state and an attempt for the prediction of further changes after building of hydro-electric power stations. *Ibidem:* 5–121 (in Slovak).
BALON, E. K. 1968a. The periodicity and relative indexes of the growth of fishes (with notes on their terminology). In: Mimeographed invited lectures of The International Conference on Ageing and Growth of Fishes, Smolenice: 115–132.
BALON, E. K. 1968b. Notes to the origin and evolution of trouts and salmons with special reference to the Danubian trouts. *Acta Soc. Zool. Bohemoslovacae* 32: 1–21.
BALON, E. K. 1971a. Age and growth of *Hydrocynus vittatus* Castelnau, 1861 in Lake Kariba, Sinazongwe area. *Fish. Res. Bull. Zambia* 5: 89–118.
BALON, E. K. 1971b. Replacement of *Alestes imberi* Peters, 1852 by *A. lateralis* Boulenger, 1900 in Lake Kariba, with ecological notes. *Ibidem* 5: 119–162.
BALON, E. K. 1971c. First catches of Lake Tanganyika clupeids (kapenta – *Limnothrissa miodon*) in Lake Kariba. *Ibidem* 5: 175–186.
BALON, E. K. 1971d. The intervals of early fish development and their terminology (A review and proposals). *Věst. Čs. spol. zool.* 35(1): 1–8.
BALON, E. K. 1972. Possible fish stock size assessment and available production survey as developed on Lake Kariba. *Afr. J. Trop. Hydrobiol. Fish.* 2: 45–73.
BALON, E. K. 1973a. Results of fish population size assessments in Lake Kariba coves (Zambia), a decade after their creation. In: Man-Made Lakes, Their Problems and Environmental Effects. *Geophys. Monogr. Ser.* 17: 149–158.
BALON, E. K. 1973b. The eels of Siengwazi Falls (Kalomo River, Zambia) and their significance. *J. Nat. Mus. Zambia* 2: 65–82.
BALON, E. K. 1974a. Domestication of the carp *Cyprinus carpio* L. Roy. Ont. Mus. Life Sci. Misc. Pub. (Toronto). 37 p.
BALON, E. K. 1974b. Fishes of Lake Kariba, Africa: length–weight relationship, a pictorial guide. T.F.H. Publications, Neptune City. 144 p.
BALON, E. K. 1974c. Fishes from the edge of Victoria Falls, Africa: demise of a physical barrier for downstream invasions. *Copeia* 3: 643–660.
BALON, E. K. 1974d. The eels of Lake Kariba: distribution, taxonomic status, age, growth and density. Ms.
BARANOV, F. I. 1925. On the dynamics of fish industry. *Biull. rybn. choz.* 8: 7–11 (in Russian).
BECKMAN, W. C. 1943. Annulus formation on the scales of certain Michigan game fishes. *Pap. Mich. Acad. Sci. Arts and Letters* 28: 281–312.
BEGG, G. W. 1970. Limnological observations on Lake Kariba during 1967 with emphasis on some special features. *Limnology and Oceanography* 15(5): 776–788.
BELL-CROSS, G. 1965a. Movement of fish across the Congo-Zambezi watershed in the Mwinilunga District of Northern Rhodesia. Proc. Central Afr. Sci. Med. Congress (Lusaka 1963), Pergamon Press, London. 415–424 p.
BELL-CROSS, G. 1965b. Addition and amendments to the check list of the fishes of Zambia. *The Puku* 3: 29–43.
BELL-CROSS, G. 1968a. The distribution of fishes in Central Africa. *Fish. Res. Bull. Zambia* 4: 3–20.
BELL-CROSS, G. 1968b. Additions and amendments to the check list of the fishes of Zambia – No. 2. *Ibidem:* 4: 99–101.
BELL-CROSS, G. 1971. Weir fishing on the Central Barotse Flood Plain in Zambia. *Ibidem* 5: 331–340.
BELL-CROSS, G. 1972. The fish fauna of the Zambezi River system. *Arnoldia (Rhodesia)* 5(29): 1–19.
BELL-CROSS, G. 1973. The fish fauna of the Buzi River system in Rhodesia and Mocambique. *Ibidem* 6(5): 1–14.

BELL-CROSS, G. & B. BELL-CROSS, 1971. Introduction of *Limnothrissa miodon* and *Limnocaridina tanganicae* from Lake Tanganyika into Lake Kariba. *Fish. Res. Bull. Zambia* 5: 207–214.
BELL-CROSS, G. & J. KAOMA, 1971. Additions and amendments to the check list of the fishes of Zambia – No. 3. *Ibidem* 5: 235–244.
BERG, A. & E. GRIMALDI, 1967. A critical interpretation of the scale structures used for the determination of annuli in fish growth studies. *Mem. Ist. Ital. Idrobiol.* 21: 225–239.
BEVERTON, R. J. H. & S. J. HOLT, 1957. On the dynamics of exploited fish populations. *Fishery Invest. (Lond. Ser. 2)* 19: 1–533.
BIZJAJEV, F. N. 1952. On the method of determining age and growth of wels (*Silurus glanis* L.). *Zool. Zhurnal* 31: 696–699 (in Russian).
BOND, G. 1963. Pleistocene environments in Southern Africa. In: African Ecology and Human Evolution, *Viking Fund Publ. in Anthropology* 36: 308–334.
BOND, G. 1964. The origin of the Victoria Falls. In: The Victoria Falls, B. M. FAGAN (Edit.), Com. Pres. Nat. Hist. Mon. Rel., Northern Rhodesia: 45–54.
BORGSTROM, G. 1962. Fish in world nutrition. In: Fish as Food, G. BORGSTROM (Edit.), Academic Press, New York, II: 267–360.
BOULENGER, G. A. 1909, 1911. Catalogue of the freshwater fishes of Africa in the British Museum (Natural History). Vol. I and II, London.
BOURLIÈRE, F. 1963. Observations on the ecology of some large African mammals. In: African Ecology and Human Evolution, *Viking Fund Publ. in Anthrop.* 36: 43–54.
BOWMAKER, A. 1970. A prospect of Lake Kariba. *Optima* 20: 68–74.
BREDER, C. M. JR., 1959. Studies in social grouping in fishes. *Bull. Am. Mus. Nat. Hist.* 117: 395–481.
BROCK, V. E. 1954. A note on the spawning of *Tilapia mossambica* in sea water. *Copeia* 1954 (1): 72.
BROWNRIDGE, H. 1968. Notes on the Kafue floodplain. Manuscript, 8 p. (seen in photocopy).
BURMAKIN, J. V. 1962. On fishery management of small lakes. Gidrobiologicheskie issl. *AN Estonskoy SSR* 3: 344–382 (in Russian).
BURMAKIN, J. V. & L. A. ZHAKOV, 1961. An attempt to determine fish production in perch lake. *Nautchno-technickeskiy biull. GosNIORCH* 13–14: 25–27 (in Russian).
BURTON, G. W. & E. P. ODUM, 1945. The distribution of stream fish in the vicinity of Mountain Lake, Virginia. *Ecology* 26: 182–194.
BUTLER, J. L. 1963. Temperature relations in shallow turbid ponds. *Proc. Okla. Acad. Sci.* 34: 90–95.
BÜNNING, E. 1967. The physiological clock. Springer-Verlag, New York Inc. 167 p.
CALHOUN, J. B. 1972. Plight of the Ik and Kaiadilt is seen as a chilling possible end for Man. *Smithsonian* 3(8): 27–33.
CAREY, T. 1968. Feeding habits of some fishes in the Kafue River. *Fish. Res. Bull. Zambia* 4: 105–109.
CAREY, T. 1971. Hydrobiological survey of the Kafue flood-plain. *Ibidem* 5: 245–295.
CARLANDER, K. D. 1955. The standing crop of fish in lakes. *J. Fish. Res. Bd. Canada* 12(4): 543–570.
CHAPMAN, D. W. 1965. Net production of juvenile coho salmon in three Oregon streams. *Trans. Amer. Fish. Soc.* 94: 40–52.
CHAPMAN, D. W. 1967. Production in fish populations. In: Symposium papers 'The Biological Basis of Freshwater Fish Production': 3–29, Blackwell, Oxford–Edinburgh.
CHAPMAN, D. W. 1968. Production. In: Methods for Assessment of Fish Production in Fresh Waters. IBP Handbook 3: 182–196.
CHAPMAN, D. W., W. H. MILLER, R. G. DUDLEY & R. J. SCULLY, 1971. Ecology of fishes in the Kafue River. *Central Fisheries Research Institute, Chilanga, Zambia, Technical Report* 2: 1–66 (mimeographed).
CHILD, G. 1968. Behaviour of large mammals during the formation of Lake Kariba. Trustees of the National Museums of Rhodesia, Salisbury, 123 p.
CHU, Y. T. 1935. Comparative studies on the scales and the pharyngeals and their teeth in

Chinese cyprinids, with particular reference to taxonomy and evolution. Biological Bulletin of St. John's University, Shanghai 2.

CHUGUNOVA, N. A. 1959. Handbook of age and growth studies in fish (in Russian). Izd.AN SSSR, Moskva, 194 p. (English translation 1963, National Science Foundation, Washington, 132 p.)

CLARK, J. D. 1950. The Stone Age cultures of Northern Rhodesia. South Africa Archaeological Society, Cape Town, 157 p.

CLARK, J. D. 1960. Human ecology during Pleistocene and later times in Africa south of the Sahara. *Current Anthropology* 1: 307–324.

CLARK, J. D. 1964. Stone Age man at the falls. In: The Victoria Falls, B. M. FAGAN (Edit.), Com. Pres. Nat. Hist. Mon. Rel., Northern Rhodesia: 55–64.

CLARK, J. D. & B. M. FAGAN, 1964. The Iron Age and the native tribes. *Ibidem:* 65–75.

CLAY, G. 1964. The discovery and historical associations. *Ibidem:* 21–44.

COETZEE, J. A. 1964. Evidence for a considerable depression of the vegetation belts during the Upper Pleistocene on the East African mountains. *Nature* 204: 564–566.

COCHE, A. G. 1967. Production of juvenile steelhead trout in a freshwater impoundment. *Ecological Monographs* 37: 201–228.

COCHE, A. G. 1968. Description of physico-chemical aspects of Lake Kariba, an impoundment, in Zambia-Rhodesia. *Fish. Res. Bull. Zambia* 5: 200–267 (special pre-issue of the regular volume 5 published in 1971).

COCHE, A. G. 1971. Lake Kariba Basin: a multidisciplinary bibliography, annotated and indexed 1954–1968. *Ibidem* 5: 11–87.

COKE, M. 1968. Depth distribution of fish on bush-cleared area of Lake Kariba, Central Africa. *Trans. Amer. Fish. Soc.* 97: 460–465.

COLLART, A. 1960. L'introduction du *Stolothrissa tanganicae* (Ndagala) au Lac Kivu. *Troisième Coll. Hydrob. et Pêche eu Eaux Douces, CCTA Publ.* 63: 7–8.

COLSON, E. 1960. The social organisation of the Gwembe Tonga. Kariba Studies I, Manchester Univ. Press. 234 p.

COULTER, G. W. 1968. Hydrological processes and primary productivity in Lake Tanganyika. Proc. 11th Conf. Great Lakes Res. 1968: 609–626.

COULTER, G. W. 1970. Population changes within a group of fish species in Lake Tanganyika following their exploitation. *J. Fish. Biol.* 2: 329–353.

COULTER, G. W. & Y. A. ZNAMENSKY, 1971. The chiromila net. A method for catching light-attracted fish in Lake Tanganyika. *Fish. Res. Bull. Zambia* 5: 215–224.

CRASS, R. S. 1964. Freshwater fishes of Natal. Shuter and Shooter, Pietermaritzburg. 167 p.

CURTIS, B. 1949. The Life Story of the Fish. Harcourt, Brace and Co.

DASMANN, R. F. & A. S. MOSSMAN, 1960. The economic value of Rhodesian game. *Rhod. Farm.* 30: 17–20.

DASMANN, R. F. & A. S. MOSSMAN, 1962. Abundance and population structure of wild ungulates in some areas of Southern Rhodesia. *J. Wild. Manag.* 26: 262–268.

DAVIDSON, B. 1967. The growth of African civilisation. East and Central Africa to the late Nineteenth Century. Longmans, Nairobi, 334 p.

DE BONT, A. F. 1967. Some aspect of age and growth of fish in temperate and tropical waters. In: S. D. GERKING (Ed.), Blackwell, Oxford: 67–88.

DEELDER, C. L. 1970. Synopsis of biological data on the eel *Anguilla anguilla* (Linnaeus) 1759. FAO Fisheries Synopsis 80. 80 p.

DEMENTYEVA, T. F., J. J. MARTI, P. A. MOISEYEV & G. V. NIKOLSKY, 1961. On the legalities of the dynamics of fish population. *Trudy soveshchanii Ichtiologicheskoy komissii AN SSSR* 13: 7–20 (in Russian).

DIXEY, F. 1944. The geomorphology of Northern Rhodesia *Trans. and Proc. Geological Society of South Africa* 47: 9–45.

DOGEL, V. A. 1954. Oligomerisation of homologic organs as one of the main ways of animal evolution. Izd. Leningradskogo univ., Leningrad.

DONNELLY, B. G. 1969. A preliminary survey of *Tilapia* nurseries on Lake Kariba during 1967/68. *Hydrobiologia* (The Hague) 34: 195–206.

DOWSETT, R. 1966. A preliminary list of the birds of the Kafue Flats. *The Puku, Occ. Pap. Dept. Game and Fish., Zambia* 4: 101–124.
DUDLEY, R. 1972. Biology of *Tilapia* of the Kafue flood-plain, Zambia: Predicted effects of the Kafue Gorge Dam. University of Idaho Graduate School. Ph.D. thesis. 50 p. (unpublished).
DUNN, I. G. 1970. Fish. In: Progress report of the African Freshwater Biological Team on Lake George, Uganda, IBP.
DUSSARD, B. H., K. F. LAGLER, P. A. LARKIN, T. SCUDDER, K. SZESZTAY & G. F. WHITE, 1972. Man-made lakes as modified ecosystems. International Council of Scientific Unions, SCOPE Report 2. 76 p.
EGE, V. 1939. A revision of the genus *Anguilla* Shaw: a systematic, phylogenetic and geographical study. *Dana Rep.* 16: 1–256.
EGLER, F. E. 1964. Pescicides – in our ecosystem. *American Scientist* 52: 110–136.
EHRLICH, P. R. & A. H. EHRLICH, 1970. Population, Resources, Environment. Issues in human ecology. W. H. Freeman and Comp., San Francisco, 383 p.
EL-ZARKA, S. 1961. Annulus formation on the scales of the cichlid fish *Tilapia zillii* (Gerv.) and its validity in age and growth studies. *Notes and Memoirs Hydrobiol. Dept. Kayed Bey, Alexandria* 62: 1–18.
FAGAN, B. M. 1963. The Iron Age sequence in the Southern Province of Northern Rhodesia. *J. Afr. Hist.* 4: 157–177.
FARVAR, M. T. & J. P. NILTON, edit. 1972. The careless technology. The Natural History Press, Garden City, N.Y. xxix + 1030 p.
FISH, G. R. 1955. The food of *Tilapia* in East Africa. *The Uganda Journal* 19: 85–89.
FISH, G. R. 1970. A limnological study of four lakes near Rotorua. *N.Z. J. mar. Freshwat. Res.* 4: 165–194.
FISHELSON, L. 1966. Untersuchungen zur vergleichenden Entwicklungsgeschichte der Gattung *Tilapia* (Cichlidae, Teleostei). *Zool. Jb. Anat.* 83: 571–656.
FRADE, F. and E. T. PINTO, 1961. Prospecções hidrobiologicas nos lagos Cameia e Dilolo (Angola). *Memórias Trabalhos do Centro de Zoologia (Lisboa)* 23: 83–114.
FRANK, S. 1955. A contribution to the biology of common bullhead, *Ameiurus nebulosus* (Le Sueur, 1819). *Věst. Čs. spol. zool.* 19: 62–81.
FROST, W. E. 1945. The age and growth of eels *(Anguilla anguilla)* from the Windermere catchment area. *J. Anim. Ecol.* 14: 26–36, 106–124.
FROST, W. E. 1955. Observations on the biology of eels (*Anguilla* spp.) of Kenya Colony, East Africa. *Colonial Office Fishery Publications* 6: 1–28.
FROST, W. E. 1957a. A note on eels (*Anguilla* spp.). *Piscator* 10(38): 104–106.
FROST, W. E. 1957b. First record of the elver of the African eel *A. nebulosa labiata* Peters. *Nature* 179: 594.
FRYER, G. 1959. The trophic interrelationships and ecology of some littoral communities of Lake Nyasa with especial reference to the fishes, and a discussion of the evolution of a group of rock-frequenting Cichlidae. *Proc. zool. Soc. Lond.* 132: 153–281.
FRYER, G. 1960. Some controversial aspects of speciation of African cichlid fishes. *Ibidem* 135: 569–578.
FRYER, G. & T. D. ILES, 1972. The cichlid fishes of the Great Lakes of Africa. T.F.H.Publications, Neptune City, 641 p.
GADD, K. G., L. C. NIXON, E. TAUBE & M. H. WEBSTER, 1962. The Lusitu tragedy. *Cent. Afr. J. Medicine* 8 (Suppl.): 491–508.
GALLAGHER, J., I. MACADAM, J. SAYER & L. VAN LAVIEREN, 1972. Pulmonary tuberculosis in free-living lechwe antelope in Zambia. *Trop. Anim. Hlth. Prod.* 4: 204–213.
GARROD, D. J. 1958. The growth of *Tilapia esculenta* Graham in Lake Victoria. *Hydrobiol.* 12: 268–298.
GERKING, S. D. 1962. Production and food utilization in a population of bluegill sunfish. *Ecol. Monogr.* 32: 31–78.
GILCHRIST, J. D. F. & W. W. THOMPSON, 1913. The freshwater fishes of South Africa. *Ann. S. Afr. Mus.* 11: 321–463.

GRAY, J. 1968. Animal locomotion. Weidenfeld and Nicolson, London. 479 p.
GRAY, R. W. & C. W. ANDREWS, 1971. Age and growth of the American eel (*Anguilla rostrata* (LeSueur)) in Newfoundland waters. *Can.J.Zool.* 49: 121–128.
GREENWOOD, P. H. 1958. Evolution and speciation in the *Haplochromis* fauna (Pisces, Cichlidae) of Lake Victoria. Proc. 15th Int. Congr. Zool.: 147–149.
GREENWOOD, P. H. 1966. The fishes of Uganda. The Uganda Society, Kampala, 131 p.
GREENWOOD, P. H., D. E. ROSEN, S. H. WEITZMAN & G. S. MYERS, 1966. Phyletic studies of teleostean fishes, with a provisional classification of living forms. *Bull. Am. Mus. Nat. Hist.* 131: 339–455.
GRZIMEK, M. and B. GRZIMEK, 1960. Census of plains animals in the Serengeti National Park, Tanganyika. *J. Wildlife Management* 24: 27–37.
GULLAND, J. A. 1969. Manual of methods for fish stock assessment. Part 1. Fish population analysis. *FAO Manuals in Fisheries Science* 4: 1–154.
HARDING, D. 1964. Hydrology and fisheries in Lake Kariba. *Verh. Internat. Verein. Limnol.* 15: 139–149.
HARDING, D. 1966. Lake Kariba. The hydrology and development of fisheries. In: Man-Made lakes, Symposia of the Institute of Biology 15: 7–20.
HAYES, F. R. & E. H. ANTHONY, 1964. Production capacity of North American lakes as related to the quantity and the trophic level of fish, the lake dimensions, and the water chemistry. *Trans. Amer. Fish. Soc.* 93: 53–57.
HENSEL, K. 1966. Age and growth rate of the brown bull-head (*Ictalurus nebulosus* Le Sueur, 1819) in backwaters of the innundation area of the River Elbe in Czechoslovakia. *ACTA F.R.N. Univ. Comen., 12-Zoologia* XIII: 171–191.
HICKLING, C. F. 1956. (Unpublished report on the potential of Lake Kariba following a visit to the Zambezi area). Looked up by COCHE, 1971.
HICKLING, D. F. 1961. Tropical Inland Fisheries. Longmans, London.
HILE, R. 1970. Body-scale relation and calculation of growth in fishes. *Trans. Amer. Fish. Soc.* 99(3): 468–474.
HOCHMAN, L. 1966. Zum Wachstum des Welses (*Silurus glanis* L.). Sbor. *ČSAZV, Živocišná výroba* 6: 295–302.
HOLČÍK, J. 1960. Age and growth of bitterling (*Rhodeus sericeus amarus*). *Rozpravy ČSAV. řada MPV* 70, 10: 30–11 (in Slovak).
HOLČÍK, J. 1966. The development and forming of the fish fauna in the Orava dam reservoir. *Biologické práce SAV* 12: 5–75 (in Slovak).
HOLČÍK, J. 1967a. Life history of the roach *Rutilus rutilus* (L) in the Klíčava reservoir. *Věst. Čs. spol. zool.* 31: 336–348.
HOLČÍK, J. 1967b. Annulus formation on the scales of six fish species from the Klíčava reservoir (Czechoslovakia). *Ibidem* 31: 159–161.
HOLČÍK, J. 1967c. Life history of the rudd – *Scardinius erythrophthalmus* (Linnaeus, 1758) in the Klíčava reservoir. *Ibidem* 4: 335–348.
HOLČÍK, J. 1969. On the phenomenon of Rosa Lee as exhibited by perch and roach in the Klíčava reservoir. *Zool. listy* 18: 93–97.
HOLČÍK, J. 1970a. The Klíčava Reservoir (An Ichthyological Study). *Biologické Práce SAV (Bratislava)* 15(3): 1–96.
HOLČÍK, J. 1970b. Standing crop, abundance, production and some ecological aspects of fish populations in some inland waters of Cuba. *Věst. Čs. spol. zool.* 34(3): 184–201.
HOLČÍK, J. 1970c. Abundance, ichthyomass and production of fish populations in three types of water-bodies in Czechoslovakia (man-made lake, trout lake, arm of the Danube River). *Acta Biologica Jugoslavica, Ichthyologia (Sarajevo)* 2(1): 37–52.
HOLČÍK, J. & V. MIŠÍK, 1962. A biological classification of streams of the southern slopes of Vihorlat and Blatska lowland. *Biologia (Bratislava)* 17: 517–522 (in Slovak).
HOLČÍK, J. & J. J. DUYVENÉ DE WIT, 1964. Systematic status of the bitterlings from Asia Minor and notes on the geographical variability of *Rhodeus sericeus* (Pallas, 1776) in the area of its distribution. *Zeitschrift für wissenschaftliche Zoologie* 169: 396–412.

HOLDEN, M. J. 1963. The populations of fish in dry season pools of the river Sokoto. *Colonial Office Fishery Publications* 19: 1–58.
HOOPER, F. F. 1949. Age analysis of a population of the Ameiurid fish *Schilbeodes mollis* (Hermann). *Copeia* 1949 (1): 34–38.
HORTON, P. A. 1961. The bionomics of brown trout in a Dartmoor stream. *J. Anim. Ecol.* 30: 311–338.
HORTON, R. E. 1945. Erosional development of streams and their drainage basins; hydrophysical approach to quantitative morphology. *Bull. Geol. Soc. Amer.* 56: 275–370.
HRBAČEK, J. 1969. Relations between some environmental parameters and the fish yield as a basis for a predictive model. *Verh. Internat. Verein. Limnol.* 17: 1069–1081.
HRBAČEK, J., V. HRUŠKA & O. OLIVA, 1952. On nourishment and growth of Vltava River wels. *Československý rybař:* 94–95 (in Czech).
HRUŠKA, V. & O. OLIVA, 1953. Additional remarks on nourishment and growth of Vltava River wels. *Československé rybářství:* 6 (in Czech).
HUET, M. 1946. Note préliminaire sur les relations entre la pente et les populations piscicoles des eaux courantes. Règle des pentes. 13e Biologische Jaarboek, Dodanaca, Bruxelles: 232–243.
HUET, M. 1954. Biologie, profils en long et en travers des eaux courantes. *Bul. franc. Pisciculture* 175: 41–53.
HUET, M. 1959. Profiles and biology of Western European streams as related to fish management. *Trans. Amer. Fish. Soc.* 88: 155–163.
HUET, M., A. LELEK, J. LIBOSVÁRSKY & M. PEŇAZ, 1969. Contribution à l'identification des zones piscicoles de quelques cours d'eau de Moravie (Tchécoslovaquie). *Verh. Internat. Verein. Limnol.* 17: 1103–1111.
HUET, M. & J. A. TIMMERMANS, 1963. La population piscicole de la Sémois inférieure, grosse rivière Belge du type supérieur de la zone à barbeau. Trav. St. R. Eaux et For. (Groenendaal) D 36.
HUET, M. & J. A. TIMMERMANS, 1966. La population piscicole de l'Ourthe. *Ibidem* 16: 1192–1203.
HURLEY, D. A. 1972. The American eel *(Anguilla rostrata)* in eastern Lake Ontario. *J. Fish. Res. Bd. Canada* 29: 535–543.
HUTCHINSON, G. E. 1969. Eutrophication, past and present. In: Eutrophication: causes, consequences, correctives, p. 17–26, Nat. Acad. Sc. Washington, D.C.
HYNES, H. B. N. 1970. The ecology of running waters. Univ. of Toronto Press. 555 p.
IVLEV, V. S. 1945. The biological productivity of waters. *Usp. sovrem. biol.* 19(1): 98–120 (in Russian). English version 1966, *J. Fish. Res. Bd. Canada* 23(11): 1727–1759.
JACKSON, P. B. N. 1960. The ecological effects of flooding by the Kariba Dam upon Middle Zambezi fishes. Proc. 1st Fed. Sci. Congress (Salisbury): 277–284.
JACKSON, P. B. N. 1961a. The fishes of Northern Rhodesia. A check list of indigenous species. The Government Printer, Lusaka, 140 p.
JACKSON, P. B. N. 1961b. Ichthyology. The fish of the Middle Zambezi. Kariba Studies, Manchester University Press, 36 p.
JACKSON, P. B. N. 1961c. The impact of predation, especially by the tigerfish (*Hydrocyon vittatus* Cast.) on African freshwater fishes. *Proc. zool. Soc. Lond.* 136: 603–622.
JACKSON, P. B. N. 1971. The African great lakes fisheries: past, present and future. *Afr. J. Trop. Hydrobiol. Fish.* 1: 35–49.
JARMAN, P. J. 1969. The effect of the creation of Lake Kariba upon the terrestrial ecology of the Middle Zambezi valley, with particular reference to the large mammals. The Nuffield Lake Kariba Research Station Sinamwenda Report 1962-1968: 26–33.
JENKINS, R. M. 1958. The standing crop of fish in Oklahoma ponds. *Proc. Okla. Acad. Sci.* 28: 157–172.
JENKINS, R. M. 1967. The influence of some environmental factors on standing crop and harvest of fishes in U.S. reservoirs. Reservoir Fishery Resources Symposium, American Fisheries Society, South Division: 298–321.
JENKINS, R. M. 1970. The influence if engineering design and operation and other environ-

mental factors on reservoir fishery resources. *Water Resources Bulletin* 6: 110–119.
JUBB, R. A. 1958. A preliminary report on the collection of freshwater fishes made by the Barnard Cap Expeditions to the Caprivi Strip, 1949, the lower Sabi River, 1950, and to Barotseland, 1952. *Occ. pap. Nat. Mus. S. Rhodesia* 22B: 177–189.
JUBB, R. A. 1960. Some Lake Kariba fish problems. *Piscator* 47: 112–119.
JUBB, R. A. 1961a. An illustrated guide to the freshwater fishes of the Zambezi River, Lake Kariba, Pungwe, Sabi, Lundi, and Limpopo Rivers. Stuart Manning, Bulawayo. 171 p.
JUBB, R. A. 1961b. The freshwater eels (*Anguilla* spp.) of Southern Africa. An introduction to their identification and biology. *Ann. Cape Prov. Mus.* 1: 15–48.
JUBB, R. A. 1964a. The eels of South African rivers and observations on their ecology. In: DAVIS, D. H. S. edit. Ecological Studies in Southern Africa. Monographiae Biologicae (Junk, The Hague) 14: 186–205.
JUBB, R. A. 1964b. Some fishes of the Victoria Falls region. The Victoria Falls. A Handbook to the Victoria Falls, the Batoka Gorge and part of the Upper Zambesi River (edit. B. M. FAGAN), Commission for the Preservation of Natural and Historical Monuments and Relicts, Northern Rhodesia. pp. 129–140.
JUBB, R. A. 1964c. A new species of *Barbus* (Pisces, Cyprinidae) from the Upper Zambezi River. *Ann. Mag. Nat. Hist.* 13, 7: 539–542.
JUBB, R. A. 1967. Freshwater fishes of Southern Africa. A. A. Balkema, Cape Town, Amsterdam, 248 p.
JUBB, R. A. & F. L. Farquharson, 1965. The fresh water fishes of the Orange River drainage basin. *S. Afr. Journ. Sc.* 61: 118–125.
KEAY, R. W. J. & A. AUBREVILLE, 1959. Vegetation map of Africa. L'Association pour L'Etude Taxonomique de la Flore d'Afrique Tropicale. Oxford Univ. Press.
KIMPE DE, P. 1964. Contribution à l'étude hydrobiologique du Luapula-Moero. *Annls. Mus. r. Afr. cent. (Tervuren), sc. zool.* 128: 1–238.
KIRKA, A. 1969. Age composition, growth, maturity and fecundity of *Salmo trutta* morpha *fario* (Linnaeus, 1758) and *Cottus poecilopus* (Heckel, 1836) in the Orava and Vah river basins. *Works of the Laboratory of Fishery Research (Bratislava)* 2: 189–198.
KOBAYASI, H. 1953. Comparative studies of the scales in Japanese freshwater fishes, with special reference to phylogeny and evolution. *Jap. J. Ichthyol.* 2: 246–260.
KRAEV, N. P. 1966. *Tilapia* in cooling ponds GRES. Rybnoe chozjajstvo, No. 4 (in Russian, not seen in original).
KRYZHANOVSKY, S. G. 1949. Ecomorphological rules of development of the cyprinoid, cobitid and siluroid fishes. *Trudy Instituta morfologii zhivotnykh* 1: 5–332 (in Russian).
KUEHNE, R. A. 1962. A classification of streams, illustrated by fish distribution in an eastern Kentucky creek. *Ecology* 43: 608–614.
LÁC, J. & M. ERTL, 1961. A course of fish elimination in the Danube dead arm by means of DDT emulsion. *Biologia (Bratislava)* 16(2): 103–109 (in Slovak).
LAGLER, K. F. ed. 1969. Man-made lakes planning and development. UNDP/FAO Rome, 71 p.
LAGLER, K. F., J. KAPETSKY & D. STEWART, 1971. The fisheries of the Kafue River Flats, Zambia, in relation to the Kafue Gorge Dam. Central Fisheries Research Institute, Chilanga, Zambia, Technical Report 1. 161 p.
LAGLER, K. F. & J. R. VALLENTYNE, 1956. Fish scales in a sediment core from Linsley Pond, Connecticut. *Science* 124: 368.
LAGUS, C. 1959. Operation Noah. W. Kimber, London, 176 p.
LAMPLUGH, G. W. 1907. The geology of the Zambezi basin around the Batoka Gorge. *Quart. J. Geol. Soc.* 63: 162–216.
LAPIN, J. J. & J. G. YUROVICKY, 1959. On intraspecific legalities of maturation and dynamic of fecundity in fish. *Zhurnal obshchey biologii* 20(6): 439–446 (in Russian).
LARKIN, P. A. 1956. Interspecific competition and population control in freshwater fish. *J. Fish. Res. Bd. Canada* 13(3): 327–342.
LE CREN, E. D. 1965. Some factors regulating the size of populations of freshwater fish. *Mitt. Intern. Verein. Limnol.* 13: 88–105.

LEE, R. 1920. A review of the methods of age and growth determination in fishes by means of scales. *Min. Agric. and Fish., Fish. Invest. (Ser. 2)* 4: 1–32.
LEGER, L. 1945. Economie biologique et productivité de nos rivières à Cyprinides. *Bul. franc. Pisciculture* 139: 49–69.
LE KUANG LONG, NGUYEN DINH ZAY, NGUYEN KUANG VINH. 1961. First results of investigation of the temperature influence on *Tilapia mossambica* Peters, acclimatised in Vietnam from 1961. *Z. obsc. biol.* 22(6) (in Russian, not seen in original).
LELEK, A. 1968. The vertical distribution of fishes in the Ebo Stream and notes to the fish occurrence in the Lake Bosumtwi, Ashanti, Ghana. *Zool. listy* 17: 245–252.
LE ROUX, P. J. 1956. Feeding habits of the young of four species of *Tilapia*. *South African Journal of Science* 53: 33–37.
LEWIS, W. M. 1949. The use of vertebrae as indicators of the age of the northern black bullhead *Ameiurus m. melas* (Rafinesque). *Iowa State Coll. Jour. Sci.* 23: 209–218.
LIVINGSTONE, D. 1857. Missionary Travels and Researches in South Africa. John Murray, London. 519 p.
LIVINGSTONE, D. & C. LIVINGSTONE, 1865. Narrative of an expedition to the Zambesi and its tributaries. H. Hamilton, London.
LIVINGSTONE, D. A. 1967. Postglacial vegetation of the Ruwenzori Mountains in equatorial Africa. *Ecol. Monographs* 37: 25–52.
LIVINGSTONE, D. A. 1971. Speculations on the climatic history of mankind. *Amer. Scientist* 59: 332–337.
LIVINGSTONE, D. A. & R. L. KENDALL, 1969. Stratigraphic studies of East African lakes. *Mitt. Internat. Verein. Limnol.* 17: 147–153.
LORANT, M. 1959. Will tilapia be space travellers' food? Food Manufacture and Distributor 34(12).
LOUBENS, G. 1969. Étude de certains peuplements ichthyologiques par des pêches au poison (1re note). *Cah. O.R.S.T.O.M., sér. Hydrobiol.* 3(2): 45–73.
MAAR, A. 1959. The fish potential of Lake Kariba. Fed. Min. Agric. Proc. First Fish. Day S. Rhod.: 50.
MAAR, A. 1960. Dams and drowned-out stream fisheries in Southern Rhodesia (Federation of Rhodesia and Nyasaland). – Athens Proceedings of the I.U.C.N. 7th Technical Meeting (1958) 4: 139–151.
MALAISSE, F. 1968. Ecologie et aménagement piscicole d'un cours d'eau tropical: la Luanza (Haut-Katanga, Rép. dém. Congo). *Trav. Ser. Sylv. Pisc. Univ. Off. Congo* 1: 3–8.
MALAISSE, F. 1969a. Les faciès d'un cours d'eau tropical: la Lunza (Haut-Katanga, Rép. dém. Congo). *Verh. Internat. Verein. Limnol.* 17: 936–940.
MALAISSE, F. 1969b. La pêche collective par empoisonnement au 'buba' (*Tephrosia vogelii* Hook. f.). Son utilisation dans l'étude des populations de poissons. *Les Naturalistes Belges* 50: 481–500.
MANN, H. 1963. Aal-Fang und Aal-Wirtschaft in Deutschland. *Fette, Seifen, Austrichtmittel, die Ernährungsindustrie* 65: 146–150.
MANN, K. H. 1965. Energy transformations by a population of fish in the river Thames. *J. Anim. Ecol.* 34: 253–275.
MANN, K. H. 1969. The dynamics of aquatic ecosystems. *Advan. Ecol. Res.* 6: 1–81.
MARGALEF, R. 1968. Perspectives in ecological theory. Univ. Chicago Press, 111 p.
MARGALEF, R. 1969. Diversity and stability: a practical proposal and a model of interdependence. In: Diversity and stability in ecological systems. *Brookhaven Symp. Biol.* 22: 25–37.
MARLIER, G. 1953. Etude biogéographique du bassin de la Ruzizi, basée sur la distribution des poissons. *Ann. Soc. r. zool. Belg.* 84: 175–224.
MARTIN, P. B. & H. E. WRIGHT, JR. 1967. Pleistocene Extinctions: The Search for a Cause. Yale University Press, New Haven.
MATHEWS, C. P. 1970. Estimates of production with reference to general surveys. *Oikos* 21: 129–133.
MATHEWS, C. P. 1971. Contribution of young fish to total production of fish in the River

Thames near Reading. *J. Fish. Biol.* 3: 157–180.

MATTHES, H. 1966. The food of *Tilapia mortimeri* Trewavas (syn. *Tilapia mossambica* Peters) in Lake Kariba. *Fish. Res. Bull. Zambia* 4: 47–49.

MATTHES, H. 1968a. The food and feeding habits of the tigerfish, *Hydrocyon vittatus* (Cast., 1861) in Lake Kariba. *Beaufortia* 15: 143–153.

MATTHES, H. 1968b. Preliminary Investigations into the Biology of the Lake Tanganyika Clupeidae. *Fish. Res. Bull. Zambia.* 4: 39–45.

MCCONNELL, W. J. 1963. Primary productivity and fish harvest in a small desert impoundment. *Trans. Amer. Fish. Soc.* 92: 1–12.

MCFADDEN, J. T. & E. L. COOPER, 1962. An ecological comparison of six populations of brown trout *(Salmo trutta)*. *Trans. Amer. Fish. Soc.* 91: 53–62.

MCKINLEY, D. 1964. The new mythology of 'man in nature'. *Perspectives in Biology and Medicine* 7: 93–105.

MCLACHLAN, A. J. 1969a. Substrate preferences and invasion behaviour exhibited by larvae of *Nilodorum brevibucca* Kieffer (Chironomidae) under experimental conditions. *Hydrobiologia* 33: 237–249.

MCLACHLAN, A. J. 1969b. The effect of aquatic macrophytes on the variety and abundance of benthic fauna in a newly created lake in the tropics (Lake Kariba). *Arch. Hydrobiol.* 16: 212–231.

MCLACHLAN, A. J. 1970. Submerged trees as a substrate for benthic fauna in the recently created Lake Kariba (Central Africa). *The Journal of Applied Ecology* 7(2): 253–266.

MCLACHLAN, S. M. 1971. The rate of nutrient release from grass and dung following immersion in lake water. *Hydrobiologia* (The Hague) 37(3–4): 521–530.

MENON, M. D. 1950. The use of bones, other than otoliths in determining the age and growth rate of fishes. *Cons. Internat. Explor. Mer., Jour. du Cons.* 16: 311–335.

MILLS, D. 1970. Preliminary observations on fish populations in some Tweed tributaries. Annual Report to the River Tweed Commissioners, Tweeddale Press, Hawick, 7 p.

MIRONOVA, N. V. 1969. On the biology of *Tilapia mossambica* Peters, in natural and laboratory conditions. *Voprosy ichtiologii* 9: 628–638 (in Russian).

MIŠIK, V. 1959. The ichthyofauna of the Kysuca River. *Biologicke prace SAV* 5 (4): 7–39 (in Slovak).

MORTIMER, M. A. E. 1960. Observations on the biology of *Tilapia andersoni* (Castelnau), (Pisces, Cichlidae), in Northern Rhodesia. *Joint Fish. Res. Organization Annual Report* 9: 42–67.

MORTIMER, M. A. E. 1965. Fish production from a stream in Northern Rhodesia. Proc. Central African Scientific and Medical Congress, Pergamon Press, Oxford, 405–414.

MUNTEMBA, M. 1970. The political and ritual sovereignty among the Kikuni Leya of Zambia. *Zambia Museums Journal* 1: 28–39.

MÜLLER, H. 1962. Die Aal-Wirtschaft in den Binnengewässern der Deutschen Demokratischen Republik. *Z. Fisch.* 10: 573.

MÜLLER, K. 1954. Die Fischbesiedlung und die regionale Einstufung der Fliessgewässer der nordschwedischen Waldregion. *Ber. limnol. Flussst. Freudenthal.* 6: 51–56.

NEEL, J. K. 1953. Certain limnological features of a polluted irrigation stream. *Trans. Am. Fish. Soc.* 72: 119–135.

NESS, J. & R. C. DUGDALE, 1959. Computation of production for populations of aquatic midge larvae. *Ecology* 40: 425–430.

NIKOLSKY, G. V. 1953a. On the biological basis for the rate of exploitation and methods of regulating the abundance of fish. Ocherki po obshchim voprosam ikhtiologii: 306–318 (in Russian).

NIKOLSKY, G. V. 1953b. On the theoretical backgrounds of surveys about the dynamics of fish number. *Trudy soveshtchaniy AN SSSR* 1: 77–93 (in Russian).

NIKOLSKY, G. V. 1963. Ekologiya ryb. Vysshaya Shkola Press, Moscow, 368 p. (English version 1963, The ecology of fishes, Academic Press, London and New York, 352 p.).

NIKOLSKY, G. V. 1965. Theory of fish populations as the biological background for rational exploitation and management of fishery resources. Izd. Nauka, Moscow, 382 p. (in Russian).

English version Oliver and Boyd, Edinburg, 323 p.
NORMAN, J. R. (edit. P. H. GREENWOOD) 1963. A History of Fishes. Ernest Benn, London, 398 p.
ODUM, E. P. 1969. The strategy of ecosystem development. *Science* 164: 262–270.
ODUM, H. T. 1971. Environment, power, and society. Willey-Interscience, New York, 331 p.
OLIFF, W. D. 1960. Hydrobiological studies on the Tugela River system. Part I. The main Tugela River. *Hydrobiologia* 14: 281–385.
OLIFF, W. D. 1963. Idem. Part III. The Buffalo River. *Ibidem* 21: 355–379.
OLIVA, O. 1955. The composition of fish populations in three natural ponds in the Elbe River region. *Universitas Carolina, Biologica* 1(1): 61–74 (in Czech).
OLIVA, O. 1960. Further contribution to the knowledge of fish populations composition in Elbe River backwaters. *Věst. Čs. spol. zool.* 24(1): 42–49 (in Czech).
PANTELOURIS, E. M., A. ARNASON & F.-W. TESCH, 1970. Genetic variation in the eel. II. Transferrins, haemoglobins and esterases in the eastern North Atlantic. Possible interpretations of phenotypic frequency differences. *Genet. Res.* 16: 277–284.
PANTULU, V. R. & V. D. SINGH, 1962. On the use of otoliths for the determination of age and growth of *Anguilla nebulosa nebulosa* McClelland. *Proc. Indian Acad. Sci. Sect. B* 55(5): 263–275.
PEŇÁZ, M. 1968. Some biological and technical aspects of the methods of back calculation of growth in fishes. Vol. invited lectures of 'The International Conference of Ageing and Growth of Fishes'. Smolenice, Czechoslovakia: 19–35 (mimeographed).
PETR, T. 1967. Fish population changes in the Volta Lake in Ghana during its first sixteen months. *Hydrobiologia* 30: 193–220.
PIKE, E. & T. CAREY, 1965. The Kafue floodplain. In: M. A. E. MORTIMER, ed. The fish and fisheries of Zambia. Falcon Press, Ndola. Zambia. 97 p.
PIVNIČKA, K. 1971. Fecundity, growth, mortality and production of fish populations in the Klíčava reservoir with regard to their density in years 1957–1970. Ph.D. thesis, Charles University, Prague (unpublished).
POLL, M. 1956. Poissons Cichlidae in: Exploration hydrobiologique du Lac Tánganika (1946–1947) 3: 619 p.
RADOVICH, J. 1973. Some observations on sustained yield in the sport fisheries, In: Items for Fishery Scientists from the Sport Fishing Institute, Washington, January–February, 1973 (and the following discussion in July and September issues).
RASMUSSEN, C. J. 1952. Size and age of the silver eel (*Anguilla anguilla* L.) in Esrum Lake. *Rep. Dan. Biol. St.* 54: 3–36.
REGIER, H. A. 1968. The potential misuse of exotic fish as introductions. In: A Symposium on Introductions of Exotic Species. *Ont. Dept. Lands and Forests Research Report* 82: 91–111.
REGIER, H. A. 1969. Ecological aspects of overcoming world hunger. *Canadian Audubon* 32: 12–16.
REGIER, H. A. 1970a. Community transformation – some lessons from large lakes. Preprint of a paper presented at the 50th year Anniversary Celebrations, University at Washington, College of Fisheries, Seattle, 13 p.
REGIER, H. A. 1970b. Current problems in assessing Lake Victoria's stocks. Fish Stock Assessment on African Inland Waters, FAO Working Paper 1: 1–9 (mimeographed).
REGIER, H. A., V. C. APPLEGATE & R. A. RYDER, 1969. Ecology and management of the walleye in western Lake Erie. *Tech. Rep. Gt. Lakes Fish. Comm.* 15: 1–101.
REGIER, H. A., A. J. CORDONE & R. A. RYDER, 1971. Total fish landings from fresh waters as a function of limnological variables, with special reference to lakes of East-Central Africa. Fish Stock Assessment on African Inland Waters, FAO Working Paper 3: 1–13 (mimeographed).
REGIER, H. A. & E. B. COWELL, 1972. Applications of ecosystem theory, succession, diversity, stability, stress and conservation. *Biol. Cons.* 4: 83–88.
RENSBURG VAN, H. 1968. Ecology and development. Part I of Ecology of the Kafue Flats. Vol. IV. In: Multipurpose Survey of the Kafue River Basin. Food and Agricultural

Organization of the United Nations, Rome, Italy. 138 p.
RENSBURG VAN, K. J. 1966. Growth of *Tilapia mossambica* (Peters) in De Hoop Vlei and Zeekoe Vlei. *Dept. Nat. Consv. Invest. Rep.* 9: 1–7.
REYNOLDS, B. 1968. The material culture of the peoples of the Gwembe Valley. Kariba Studies 3, Manchester Univ. Press, 262 p.
RICKER, W. E. 1946. Production and utilization of fish populations. *Ecol. Monogr.* 16: 374–391.
RICKER, W. E. 1958. Handbook of computations for biological statistics of fish populations. *Bull. Fish. Res. Bd. Canada* 119: 1–300.
RICKER, W. E. ed. 1968. Methods for assessment of fish production in fresh waters. Blackwell Scientific Publ. IBP Handbook 3, Oxford.
RICKER, W. E. 1969. Effect of size-selective mortality and sampling bias on estimates of growth, mortality, production, and yield. *J. Fish. Res. Bd. Canada* 26(3): 479–541.
RICKER, W. E. & R. E. FOERSTER, 1948. Computation of fish production. *Bull. Bingham Oceanogr. Coll.* 11: 173–211.
RINEY, T. 1963. The impact of man on the tropical environment. 9th Techn. Meeting, Nairobi, Kenya: 17–20.
ROBINS, E. & R. LEGGE, 1959. Animal Dunkirk. H. Jenkins, London, 188 p.
ROUX, P. 1961. Growth of *Tilapia mossambica* Peters in some Transvaal impoundments. *Hydrobiologia* 18(1–2): 165–175.
RUDENKO, G. P. 1962. Age composition of fishes, ichthyomass and fish production of perch lakes. *Nautchno-techn. biull. GosNIORCH* 16: 33–37 (in Russian).
RUPP, R. S. & S. E. DEROCHE, 1965. Standing crops of fishes in three small lakes compared with C^{14} estimates of net primary productivity. *Trans. Amer. Fish. Soc.* 94: 9–25.
RYDER, R. A. 1965. A method for estimating the potential fish production of north-temperate lakes. *Trans. Amer. Fish. Soc.* 94(3): 214–218.
SCHNABEL, Z. E. 1938. The estimation of the total fish population of a lake. *Amer. Math. Monthly* 45: 348–352.
SCHWASSMANN, H. O. 1971. Biological rhythms. In: HOAR, W. S. & RANDALL, D. J. edits., Fish Physiology, vol. 6. Academic Press, New York: 371–428.
SCUDDER, T. 1960. Fishermen on the Zambezi. *The Rhodes-Livingstone Institute Journal* 27: 41–49.
SCUDDER, T. 1962. The ecology of the Gwembe Tonga. Kariba Studies II (Manchester Univ. Press): 1–274.
SCUDDER, T. 1967. Lake Kariba fishermen. Preliminary FAO consultant's report, 14 p. (typescript).
SCUDDER, T. 1968. Social anthropology, man-made lakes and population relocation in Africa. *Anthropol. Quart.* 41: 168–175.
SCULLY, R. 1972. Physico-chemical parameters and gill net catches at four floodplain habitats on the Kafue Flats, Zambia. M.Sc. thesis, University of Idaho Graduate School. 59 p. (unpublished).
SHELFORD, V. E. 1911. Ecological succession. I. Stream fishes and the method of physiographic analysis. *Biol. Bull. Mar. Biol. Lab. Woods Hole* 21: 9–35.
SINHA, V. R. P. & J. W. JONES, 1967. On the age and growth of the freshwater eel (*Anguilla anguilla*). *J. Zool. (London)* 153: 99–117.
SMITH, J. J. 1971. Fish' climb 7 ft. dam wall. *Rhod. Herald*, 2 April 1971.
SNEED, K. E. 1951. A method for calculating the growth of channel catfish, *Ictalurus lacustris punctatus*. *Trans. Amer. Fish. Soc.* 80 (1950): 174–183.
SOMMANI, E. 1953. Il concetto di 'zona ittica' e il suo reale sinificato ecologico. *Boll. Pesca Piscic. Idrobiol.* 7: 61–71.
SOULSBY, J. J. 1963. Fishing survey at Kariba. *Rhod. Agric. Journal* 60: 135–138.
STARMACH, K. 1956. Fishery and biological characteristic of rivers. *Polskie Archiwum Hydrobiologii* 1: 89–135 (in Polish).
STRAŠKRABA, V., J. HOLČIK, M. LEGNER, J. KOMARKOVÁ, K. HOLČIKOVÁ, J. FOTT & M. PEREZ-EIRIZ, 1968. First contribution to the limnology of Cuban lakes and reservoirs. Mimeographed preprint of a paper to be published in Poeyana (Havana): 1–63.

SYCH, R. 1971. Some considerations on the theory of age determination of fish from their scales – finding proofs of reliability. FAO/EIFAC Technical Paper 13: 1–68.
TAIT, C. C. 1965. A preliminary investigation into the possibility of transferring Lake Tanganyika sardines to Lake Kariba. *Fish. Res. Bull. Zambia* 1: 7–8.
TAIT, C. C. 1967a. Kafue River and floodplain reserach: Hydrological data. *Fish. Res. Bull. Zambia* 3: 26–28.
TAIT, C. C. 1967b. Idem: Mass fish mortalities. *Ibidem* 3: 28–30.
TALBOT, L. M., W. J. A. PAYNE, H. P. LEDGER, L. D. VERDCOURT & M. H. TALBOT, 1965. The meat production potential of wild animals in Africa. A review of biological knowledge. *Commonwealth Bureau of Animal Breeding and Genetics Edinburgh Technical Communication* 16: vii + 1–42.
TAYLOR, C. R. 1970. Dehydration and heat: Effects on temperature regulations of East African ungulates. *Am. J. Physiol.* 219: 1136–1139.
TAYLOR, C. R. 1971. Ranching arid lands: physiology of wild and domestic ungulates in the desert. Botswana Notes and Records, Special Edition 1, Proceedings of the Conference on Sustained Production from Semi-Arid Areas, October 1971, Gaberone: 167–180 + 10 Fig. and 2 Tab.
THOMPSON, L. 1950. Science and the study of mankind. *Science* 3: 559–563.
TJURIN, P. V. 1962. Factor of natural mortality of fish and its significance in regulation of fisheries. *Voprosy ichtiologii* 2(3): 403–427 (in Russian).
TJURIN, P. V. 1963. Biological basis of fishery management on inland impoundments. Pischepromizdat, Moscow (in Russian).
TRAUTMAN, M. B. 1942. Fish distribution and abundance correlated with stream gradient as a consideration in stocking programs. Trans. Seventh North Am. Wild. Conf.: 211–223.
TREWAVAS, E. 1973. I On the cichlid fishes of the genus *Pelmatochromis* with proposal of a new genus for *P. congicus;* on the relationship between *Pelmatochromis* and *Tilapia* and the recognition of *Sarotherodon* as a distinct genus and II A new species of cichlid fishes of rivers Quanza and Bengo, Angola, with a list of the known Cichlidae of these rivers and a note on *Pseudocrenilabrus natalensis* Fowler. *Bull. British Mus. (Nat. Hist.), Zoology* 25: 1–37.
TURNER, W. R. 1960. Standing crops of fishes in Kentucky farm ponds. *Trans. Amer. Fish. Soc.* 89(4): 333–337.
VAN DER LINGEN, M. I. 1971. A case study of Lake Kariba. Prepublication copy presented at the Symposium on Man-Made Lakes, Tennessee, 135 p. mimeographed. Published 1973 under the title: Lake Kariba: Early history of south shore, in Man-Made Lakes, Their Problems and Environmental Effects. Geophys. Monogr. Ser. 17: xx–xx.
VAN OOSTEN, J. 1953. A modification in the technique of computing average lengths from the scales of fishes. *The Progressive Fish-Culturist* 15: 85–86.
VOGEL, J. O. 1970. The Kalomo Culture of Southern Zambia. Some notes toward a reassessment. *Zambia Museums Journal* 1: 77–88.
VOVK, F. I. 1956. On the methodology of reconstruction of the growth of fish from scale. *Trudy biol. st. 'Borok'* 2: 351–392 (in Russian).
WARREN, C. E., J. H. WALES, G. E. DAVIS & P. DOUDOROFF, 1964. Trout production in an experimental stream enriched with sucrose. *J. Wildl. Mgmt.* 28: 617–660.
WATERS, T. F. 1969. The turnover ratio in production ecology. *Amer. Natur.* 103: 173–185.
WHITEHEAD, P. J. & V. D. VAN SOMEREN, 1959. Records of young eels in Kenya rivers. *Nature* 183: 950.
WILLIAMS, N. 1960. A review of the Kafue River fishery. *Rhod. Agric. J.* 57: 1.
WILLIAMS, R. 1971. Fish ecology of the Kafue River and flood plain environment. *Fish. Res. Bull. Zambia* 5: 305–330.
WINBERG, G. C. ed. 1971. Symbols, units and conversion factors in studies of freshwater productivity. IBP/PF, Londo,n 24 p.
ZAWISZA, J. 1961. The growth of fishes in the lakes of Węgorzewo area. *Rocz. Nauk Rolniczych* 77-B-2: 681–748 (in Polish).

Annual Report, Department of Game and Fisheries, Zambia, for 1964, 1965. 38 p.
Department of Meteorology, Lusaka. Zambian Climatological Summary. Pub. monthly by the Government Printer, Lusaka, 7 p.
Fishery Statistics (Natural Waters), Ann. pub. of the Central Statistical Office, Lusaka, Zambia.
Multipurpose survey of the Kafue River Basin, 1968. Food and Agricultural Organization of the United Nations, Rome, Italy. Vol. 3. Climatology and hydrology. vii + 46 p.
SWECO (Swedish Consulting Group). 1967. Kafue Gorge hydroelectric power project. Project report (for) Republic of Zambia, Ministry of Transport, Power and Communications, Stockholm. 114 p.

*

Research on Lake Kariba now takes place mainly on the Rhodesian side and studies by Rhodesian authors have begun to appear after this Monograph went into print. I felt it would be useful to insert these last minute references with annotations in order to shed more light on the original text of this Monograph.

KENMUIR, D. H. S. 1973a. The ecology of the tigerfish, *Hydrocynus vittatus* Castelnau, in Lake Kariba. *Occ. Pap. natn. Mus. Rhod.* B5 (3): 115–170.

This is an important study covering many aspects of tigerfish ecology in the Sanyati Basin of Lake Kariba. Length measurements are, however, ambiguous: according to an explanation given in Methods, fork length was measured but later standard length is quoted. Furthermore, the author of the applied method of back-calculation of growth should be ROSA M. LEE (1920) and not GAIGHER (1967) or CHAEFFER (1965). Interesting and novel results are presented on habitats, movements and sizes of early juveniles, as well as on the formation of scales and teeth. In spite of the apparent difficulty in identification of annuli of the tigerfish studied (which was not my experience), the time of annulus inception and the back-calculated growth rates 'agree fairly closely with mine until the 4th year', and even thereafter the differences are negligible when possible standard-fork length exchange and scale reading difficulties are considered. It may be interesting to compare the possible relationship of changes in tooth type and food habit to juvenile marks on scales. The stomach content analyses made in 1969 on a good number of specimens revealed a clear shift in the diet of the tigerfish from *Alestes lateralis* (20.6%) and small cichlids (34%) to *Limnothrissa miodon* (41.4%) and even fewer *A. lateralis* (9.6%) and cichlids (15%) in 1970.

The fish succession in Lake Kariba is supplemented by several remarks. It is implied that a decline in density of *Micralestes acutidens* occurs simultaneously with the expansion of *Alestes lateralis*. *A. lateralis* is documented in the L.K.F.R.I. collection from January 1965; it is suspected that this fish arrived 'accidentally during stocking operations or naturally via the Victoria Falls'. The reason for the omission of other evidence of this phenomenon, published earlier, is not explained.

KENMUIR, D. H. S. 1973b. The commercial exploitation of stocks of tigerfish, *Hydrocynus vittatus* Castelnau, in Lake Kariba, Rhodesia, by means of small-meshed gill nets. *Rhod. J. agric. Res.* 11: 171–179.

A valuable article demonstrating the danger of conventional gear to the limited stock of tigerfish and its susceptibility to overfishing. About 7 kg/ha of tigerfish is estimated to form a sustainable yield which is a much lower figure than my original assessment. Of interest are also the incidental data on the appearance of *Limnothrissa miodon* on the Rhodesian side of the lake. 'The first sardine was found in the stomach of a tigerfish cought at Redcliffe Island in June, 1969'. The clupeid, however, was not found in the stomachs of tigerfish caught at the International Tigerfish Tournament held in October, 1969, whereas it constituted 70% of all items found in the stomachs of tigerfish at the same Tournament one year later, in August, 1970.

CAULTON, M. S. & B. J. HILL. 1973. The ability of *Tilapia mossambica* (Peters) to enter deep water. *J. Fish. Biol.* 5: 783–788.
It has been proved experimentally that the ability of pressure compensation of the *Sarotherodon mossambicus* is limited to certain depths. Whereas juveniles are more tolerant of the deep and in a matter of a few minutes are capable of quick compensation to a depth of 33 m, the ability of adult males to compensate is only to 20 m and females to 13 m and that at a rate of four to seven days. In some species the limits of fish distribution in Lake Kariba can, henceforth, be given by morphophysiological, endogenous factors.

LELEK, A. 1973. Sequence of changes in fish populations of the new tropical man-made lake, Kainji, Nigeria, West Africa. *Arch. Hydrobiol.* 71 (3): 381–420.
The conclusions on fish taxa succession in Lake Kainji, based on samples without determination of selectivity values of the gear used, are somewhat dubious. In spite of this the relative data show some trends also noted in Lake Kariba, i.e. invasion of *Alestes* sp. into the open lake and wide distribution of clupeids, decrease in the densities of citharinids and mudsuckers.

BRYLINSKY, M. & K. H. MANN. 1973. An analysis of factors governing productivity in lakes and reservoirs. *Limnology and oceanography* 18 (1): 1–14.
Input of solar energy has a greater influence on primary production that the nutrient load; both nutrients and morphological conditions displayed low correlation with phytoplankton production when applied to lakes within a wide latitude range. Nutrient load and morphological factors assumed greater importance 'within fairly homogenous ecological regions'. In the last meaning both Ryder's (1965) index = total dissolved solids: mean depth, or Schindler's (1971) index = surface area + drainage area: volume, have a similar pattern as HUET's (1946) gradient rule (see p. 464).

BOWMAKER, A. P. 1973. Hydrophyte dynamics in Mwenda Bay Lake Kariba. *Kariba Studies* 3: 42–59.
The succession of *Salvinia molesta* D.S. Mitchell and the emergent and submerged littoral hydrophytes described in this paper are complementary to the succession of fishes. Although introduced into the Upper Zambezi River as late as 1949, *Salvinia* invaded Lake Kariba in the first years of filling (1959–1962), reaching in 1963 the peak of 22% of the lake's surface. The invasion of other hydrophytes subsequent to 1966 was related more to the stabilization of the water level. 'During the filling phase (December 1958 to 1963), the lake level rose 31 m in 1959, 11 m in 1960 and 1961, 8 m in 1962 and 6 m in 1963. Each year gains occurred between February and August and between September and December. In 1963 the lake was allowed to fill (487,75 m a.m.s.l.) well above normal operating level (ca. 484 m a.m.s.l.) and was then rapidly spilled to excavate the stilling pool below the wall, so that between September 1963 and March 1964 the water level dropped at a rate of 1.2 m/month. Until this stage was reached, conditions were obviously not conducive to the establishment of shoreline vegetation,...'. The successful development of littoral plants, occurs only when the half-cycle amplitudes of fluctuation do not exceed 2 m and the rates of fluctuation are less than 0.6 per month. It is essential for the survival of some fish juveniles and expansion of grazing areas. Consequently, the years in which the water fluctuation exceeds the given values should also be the years of low fish production (1964, 1966, 1969 and 1972).

Of some interest to the contents of this monograph can also be a paper quoted by BOWMAKER which has not been seen by me:
BEGG, G. W. 1973. The biological consequences of discharge above and below Kariba dam. Onzième Congrès des Grands Barrages, Madrid: 421.

JUBB, R. A. 1974. The distribution of *Tilapia mossambicus* Peters, 1852, and *Tilapia mortimeri* Trewavas, 1966, in Rhodesian waters. *Arnoldia (Rhodesia)* 6 (25): 1–14.
The author believes in the existence of two separate species, *Sarotherodon mossambicus* and *S. mortimeri*. On the one hand he attempted to list additional characters for the separation

of these two taxa and on the other he implied that stocking by man could have changed the natural geographical separation of both forms and resulted in overlapping of most key characters. The reasons for combining *Tilapia* with a masculine gender *mossambicus* are not explained (see Art. 30 of ICZN). Puzzling are also other decisions combining masculine *Sarotherodon* with feminine *shirana* etc. All this has been published with the correct gender in the quoted paper of TREWAVAS (1973). Furthermore, why should *Tilapia andersoni* have been introduced into Lake Kariba when it could have invaded the lake naturally.

My intentions to give *S. mortimeri* only subspecific status were not understood; no thoughs are devoted to this by the author, though most of the gathered evidence points in this direction. If there are reasons at all for further separation within this species and the existence of *Tilapia placida* is confirmed, then, most probably, *Sarotherodon mossambicus mossambicus* and *S. mossambicus mortimeri* should be accompanied by a third subspecies *S. mossambicus placidus*.

MABAYE, A. B. E. 1973. The role of ecological studies in the rational management of fish stocks. *Afr. J. Trop. Hydrobiol. Fish.*, Special Issue 2: 143–160.

From the submission of my FAO Final Report on Lake Kariba in 1971 and the publication of this Monograph so much time has elapsed that some parts of the results are appearing in articles of other authors. In this case, to the advantage of our common goals, some of my results are presented in a slightly different context. This gives an interesting insight into different ways of reasoning by two independent persons; what this means is that various data in the present Monograph can yield other conclusions that those presented. It is best illustrated by the case of the Luapula *Labeo altivelis* which according to my judgement does not necessarily prove what the author says (see p. 555). In general this article places our Monograph very favourably in the whole context of the national economy, although I am tempted to add that my intentions were more than this.

KUDHONGANIA, W. A. 1973. Past trends and recent research on the fisheries of Lake Victoria in relation to possible future developments. *Ibidem:* 93–106.

This is in my opinion the first excellent paper on Lake Victoria fish stocks and, contrary to other similar papers, one that employs, with minor exceptions, correct terminology. It is also the first paper that does not pretend to have non-existent evidence. If my working papers and discussions in Jinja had anything to do with this result I would consider my effort a worth while contribution. The speculative 'potential yield' estimates (p. 100), however, seem to have dangerously inflated values, similar to those of Lake Kariba (see p. 443). The most valid results are probably those on the vertical distribution of fishes, while the generalizations on 'biomass, ichthyomass and yield' really mean values on the catchable part of stocks. The true potential of trawls for quantitative fish stock studies was nevertheless proved very well.

APPENDIX A

List of Symbols Used

a	adult period
A	total production
B	biomass, ichthyomass
B'	initial biomass
$\bar{B}$	mean biomass
B_t	standing stock
C_w	specific weight rate of growth
E	catch
E'	actual catch
$\bar{E}$	sustained catch
E_{max}	maximum sustained catch
E_{landed}	landed catch
F	fishing mortality
ϕC_w	index of stock weight growth intensity
ϕE	index of species commercial importance
ϕH	index of species average size
G	instantaneous growth rate (various growth seasons)
G_x	instantaneous growth rate (same growth season)
h	absolute increment
H	rate of increase in biomass
i	single age group
j	juvenile period
l	standard length
M	natural mortality
N	abundance
N'	density
P'	available production
$\bar{P}$	final production
s	senective period
t	time, number of days
$\bar{w}$	interpolated mean weight of an individual
Y	yield
$\bar{Y}$	sustained yield
Y_A	total yield
$Y_{P'}$	available yield
Y_{max}	maximum sustained yield
Z	total mortality (instantaneous rate)
$\bar{Z}$	mean daily rate of total mortality

APPENDIX B

Efficiency of Cove-Rotenone Samples

by GEORGE P. BAZIGOS

From a statistical point of view the aggregate of the survey units we are interested in i.e. fish in a lake, is a finite and delimited population. Sampling can be undertaken in order to estimate various characteristics of the population. Specifically in quantitative rotenone samples (qrs) interest centers most frequently on the following magnitudes of the population:
1. estimation of the abundance of fish and by species,
2. estimation of the mass of fish and by species, and
3. estimation of biological parameters describing the fish population. In this appendix we are dealing with the statistical efficiency of the sampling method used for the cove-piscicide samples at Lake Kariba to provide estimates of the abundance and mass of the fish population.

For our purposes it is reasonable to assume that, at a given time t_0 the area distribution of an individual species is determined by a number of factors (F). Specifically, the total number of factors F can be considered to consist of three groups of individual factors:

$$F = F_1 + F_2 + F_3,$$

where group F_1 equals unknown factors affecting the area distribution of a given species, group F_2 equals known factors which cannot be taken into account when designing a qrs, and group F_3 equals known factors which can be taken into account when designing a qrs.

In practice, factors of Group-F_3 are used as general 'control characteristics of stratification' in the design process of a large scale survey. The idea behind this operation is to divide the lake in which the fish population is living into strata which are more or less homogeneous within. If there is a significant correlation between the general control characteristics of stratification and the area distribution of the various species, a precise estimate of a stratum's magnitudes can be obtained from a small sample from that stratum.

The Figure B1 provides a graphical explanation of the domains of the factors determining the area distribution of a given species at the t_0. On a dynamic basis (over time) the behaviour of the above factors (F_1, F_2, F_3) is subject to change affecting the area distribution of a given species. The figure gives a graphical explanation of the established argument. From the above statement it is obvious that, in order to establish the true distribution pattern of the individual species, hundreds of variables would be involved. This procedure is, however, outside of the scope of this appendix.

For statistical purposes we can introduce the assumption that, at time t_0 'statistical equilibrium' exists between species as far as their area distribution is concerned. This equilibrium exists because of the relationship among the variables involved. This argument simplifies the statistical manipulation of the problem when dealing with many species.

The 'Area Typology' by main groups of species. – I propose in this study, to discuss the 'Area Typology' in terms of the main groups of species (1. economically preferred species, 2. secondary species, 3. accompanying species). At a later stage I shall discuss the problem in more depth by considering individual species.

To identify the 'Area Typology' of the lake as far as the main groups of species is concerned simple statistical tools were used. The idea to this lies in presenting by easily calculated coefficients the relationship existing among major groups of species in space (e.g. on a stratum basis). Here, the following three magnitudes receive consideration: 1. relative abundance of fish, 2. relative mass of fish, 3. average weight per fish. By using the standardized sample data of the quantitative rotenone samples (qrs) at Lake Kariba four diagrams were produced (Fig. B2–B5). In Fig. B2 the relationship in space of the relative abundance of fish among

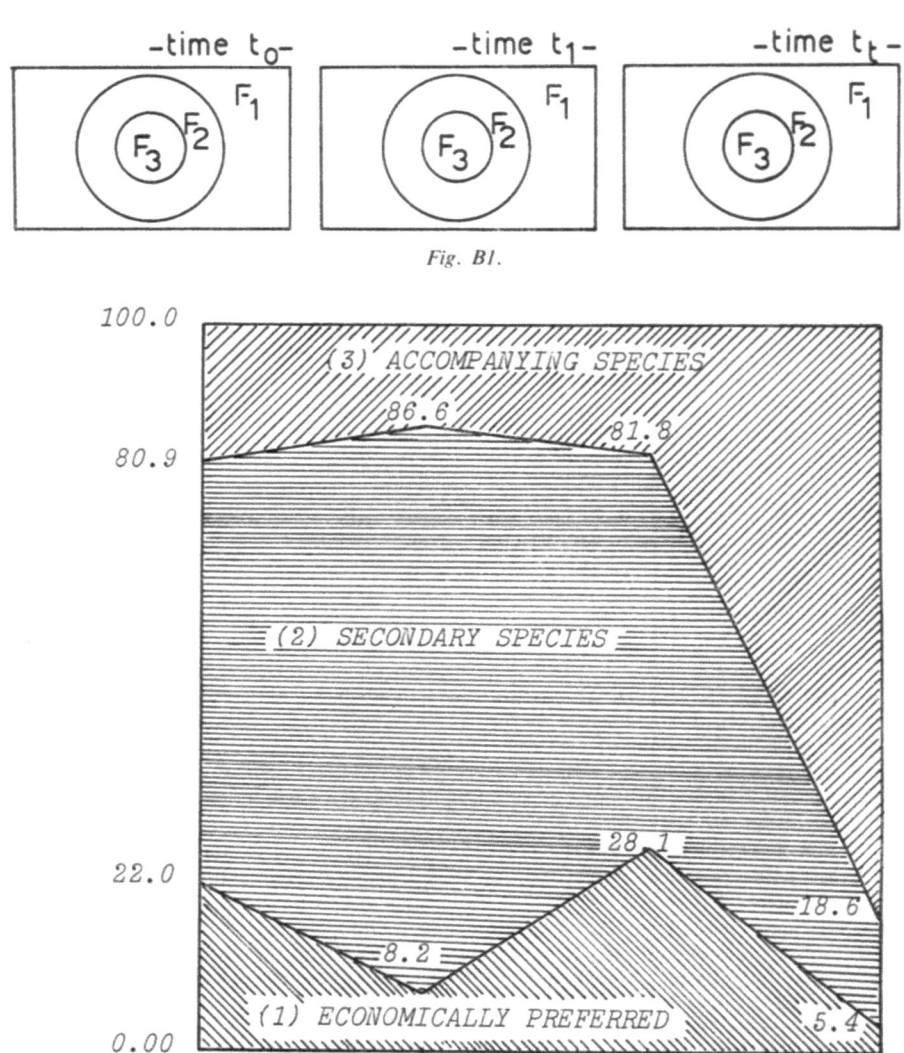

Fig. B1.

Fig. B2.

the three major groups of species is given. Specifically, for a better understanding of the 'Area Typology' of the lake the term 'Major groups of species concentration coefficient, MS-CC' was introduced; this coefficient was estimated for each individual stratum (see Table B8, B12 and corresponding text on p. 585) and for the population as a whole. The MS-CC expresses the average number of fish of a given major group of species per ha as a percentage of the total average number of fish per ha. The estimated MS-CC's are given in Table B1.

Table B1. Estimated MS-CC's by stratum and for the population as a whole.

Major groups of species	Stratum			
	1 + 2 + 3	1	2	3
1. Economically preferred	22.0	8.2	28.1	5.4
2. Secondary	58.9	78.4	53.7	13.2
3. Accompanying	19.1	13.4	18.2	81.4
Total	100.0	100.0	100.0	100.0

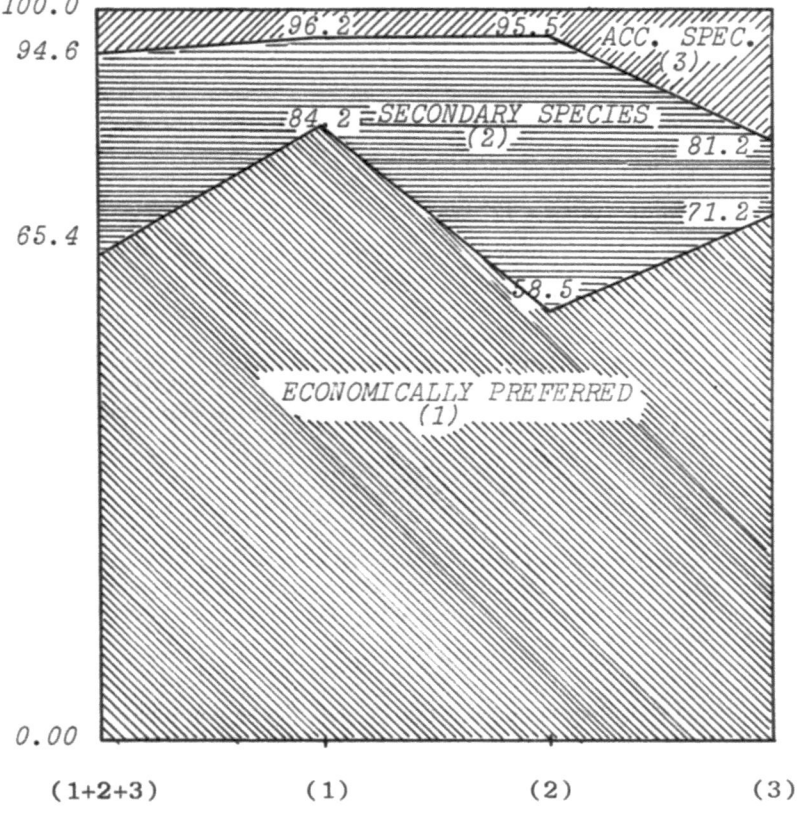

Fig. B3.

Table B2. Estimated MS-ICC's by stratum and for the population as a whole.

Major groups of species	Stratum			
	1 + 2 + 3	1	2	3
1. Economically preferred	65.4	84.2	58.5	71.2
2. Secondary	29.2	12.0	37.0	10.0
3. Accompanying	5.4	3.8	4.5	18.8
Total	100.0	100.0	100.0	100.0

From table B1 it is obvious that there are sound diversities in the 'typology' of the individual strata and this is reflected in the estimated MS-CC's. We found that there is a high participation of secondary species in strata 1 and 2 and a moderate one in stratum 3. There are sound differences in the participation of the accompanying species between strata 1 and 2 on one hand and stratum 3 on the other. Economically preferred species have a relatively high participation in stratum 2. The participation of these species in strata 1 and 3 can be considered as rather low.

The relationship in space among the major groups of species, as far as their abundance is concerned, is not attended by an equal relationship when considering the mass of the fish. Fig. B3 provides the existing relationship in space among major groups of species (in terms of weight). Also, in Table B2 the 'Ichthyomass Concentration Coefficients' were estimated (MS-ICC). The MS-ICC expresses the average weight of the total number of fish of a given major group of species per ha as a percentage of the average weight of the total number of fish per ha.

From Table B2 it is obvious that economically preferred species have the highest participation in all the strata in terms of weight. Secondary species have a relatively high participation in stratum 2 and a moderate one in strata 1 and 3. Accompanying species appear with a low participation in strata 1 and 2 and with a moderate one in stratum 3.

To facilitate attempts to describe the 'typology' of the various areas of the lake, I am suggesting the construction of an 'Index of Area Typology, $I^{(at)}$'. For the construction of the index[1] the values of the MS-CC's and MS-ICC's will be used. Specifically, the formula for the $I^{(at)}$ should be,

$$_1I_k^{(at)} = {_1W_k'} \frac{_1W_k}{\sum\limits_k {_1W_k}}$$

where, $_1W_k'$ = estimated value of the MS-CC for the kth major group of species within the 1th stratum; $_1W_k$ = estimated value of the MS-ICC for the kth major group of species within the 1th stratum. Further, in order to have data comparable in space, the estimated index should fulfil the condition,

$$\sum\limits_k {_iI_k^{(at)}} = 1.$$

In the Table B3 the estimated values of $I^{(at)}$ are given by stratum and for the population as a whole.

From the estimated indices one can see similarities in the 'typology' of stratum 1 and 2 of

---

[1] In the present study a simple $I^{(at)}$ was calculated. At a later stage I will try to construct a weighted $I^{(at)}$ by taking into account a proper weighting system.

Table B3. Estimated values of $I^{(at)}$ by stratum and for the population as a whole.

Major groups of species	Stratum			
	1 + 2 + 3	1	2	3
1. Economically preferred	44.1	41.1	44.3	18.8
2. Secondary	52.7	55.9	53.5	6.4
3. Accompanying	3.2	3.0	2.2	74.8
Total	100.0	100.0	100.0	100.0

the lake. Stratum 3 should be considered as one with a special 'typology'. The items of information on the 'typology' of the various areas of the lake are of interest in any attempt to establish efficient sampling designs for biological and statistical large scale surveys.

Finally, in order to complete our studies of the 'Area Typology' of the lake some other characteristic values were calculated. Figure B4 and B5 provide the estimated average weight per fish by major group of species, by stratum and for the population as a whole. From the produced histograms one can see striking differences in the average weight of fish between strata for almost all the major groups of species. It should be of interest to match these data with the results of the 'Catch Assessment Survey' at Lake Kariba and specifically with those data indicating the area distribution of the fishing effort and the level of fishing activities within strata. In such a case one can investigate whether fishing is responsible for the variability in the mean weight of fish between strata.

The principal component analysis. – In the previous section I stated that, at a given time t_0 the area distribution of an individual species is determined by a number of factors. Further, for statistical purposes we introduced the idea of 'statistical equilibrium', among species as far as their area distribution is concerned. In this section we move a step ahead by trying to answer the following two questions: 1. Is it possible to analyse the set of variables which determine the area distribution of the various species into a more fundamental group of components? 2. Which portion of the total variance can be accounted for by each component? The best method in this field is the Principal Component Analysis[2].

For this analysis the sample data (on a standardized area sampling unit basis, 1 ha) of the qrs at Lake Kariba were used. Further, in order to avoid lengthy calculations I have used data (totals) for each of the major groups of fish species: 1. economically preferred fish species; 2. secondary species; 3. accompanying species.

For calculation purposes the following notation was used: Variable $_{(1)}X$ = Number of fish of the economically preferred species; Variable $_{(2)}X$ = Number of fish of the secondary species; Variable $_{(3)}X$ = Number of fish of the accompanying species, (suffix k = 1,2,3) (r_{11} = estimated linear correlation coefficient between the variables $_1X$, $_1X$; r_{12} = estimated linear correlation coefficient between the variables $_1X$, $_2X$; r_{13} = estimated linear correlation coefficient between the variables $_1X$, $_3X$).

The method is based on the assumption that each variable $_{(k)}X$ consists of a systematic part, the mathematical expectation, and a random component. We want to find a linear function

$$U = a_{1(1)}X + a_{2(2)}X + a_{3(3)}X.$$

The function U ought to be chosen in such a fashion that, the principal components reproduce the original correlations between the variables $_{(k)}X$.

---

[2] KENDALL, M. G. 'A course in multivariate analysis', London (1957). TINTNER, G. 'Econometrics', John Wiley and Sons, Inc. New York (1954).

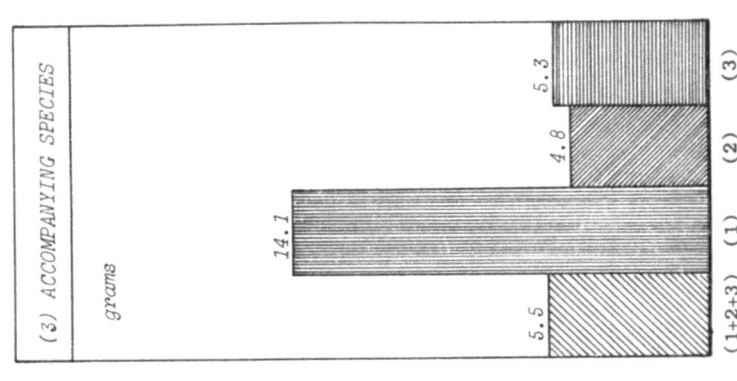
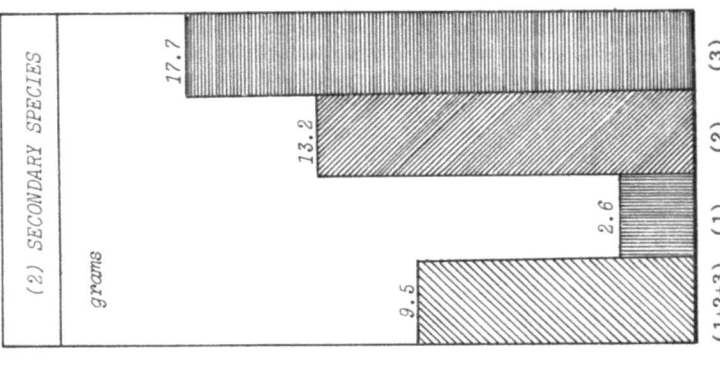
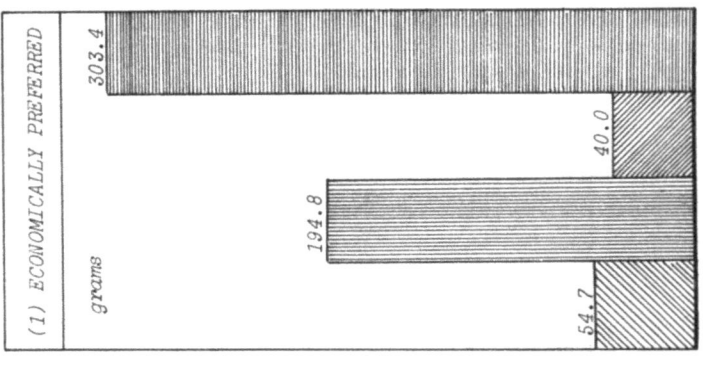

Fig. B4.

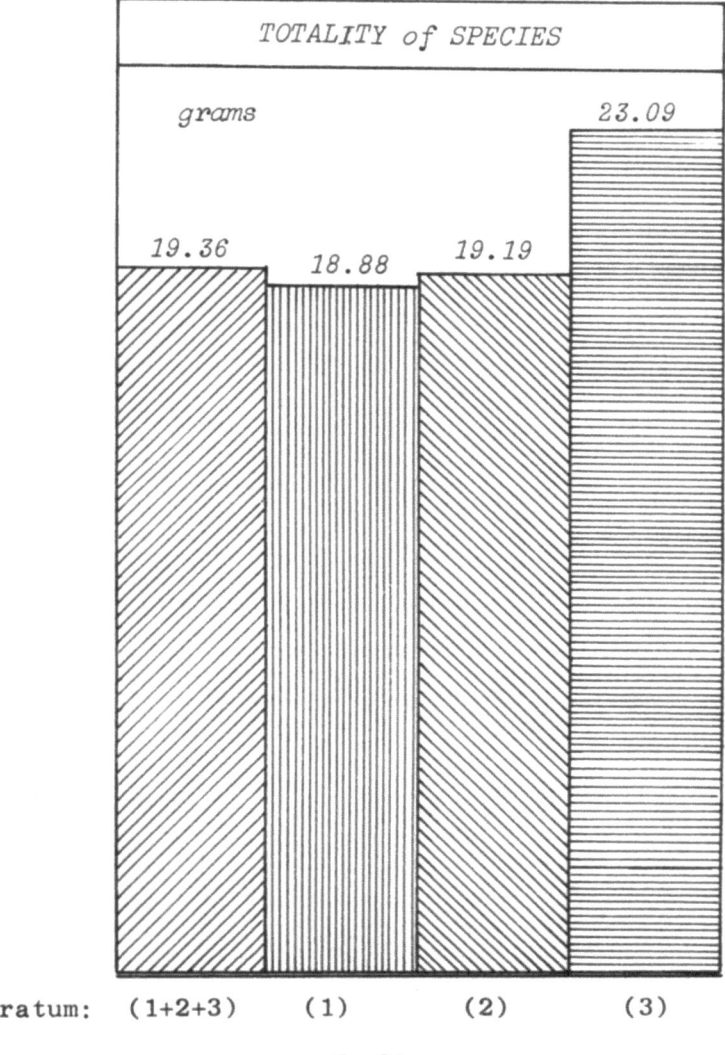

Fig. B5.

By using the sample data of the qrs the 'Coefficients of Correlation' were estimated for the pairs of the variables under study (the calculations were made under the assumption that the time is a fixed variable which does not affect the obtained results). The correlation matrix is given in Table B4.

The system of the linear equations which yield the coefficients of the first and largest principal component is:

$$r_{11} a_1 + r_{12} a_2 + r_{13} a_3 = \lambda a_1$$
$$r_{21} a_1 + r_{22} a_2 + r_{23} a_3 = \lambda a_2 \quad (1)$$
$$r_{31} a_1 + r_{32} a_2 + r_{33} a_3 = \lambda a_3$$

Table B4. Correlation matrix.

	$_{(1)}X$	$_{(2)}X$	$_{(3)}X$
$_{(1)}X$	$r_{11} = 1.000\ 000$	$r_{12} = 0.149\ 300$	$r_{13} = 0.755\ 530$
$_{(2)}X$		$r_{22} = 1.000\ 000$	$r_{23} = 0.423\ 070$
$_{(3)}X$			$r_{33} = 1.000\ 000$

where λ is Lagrange multiplier. Substituting the values of $r_{kk'}$ in the above system we get:

$$1.000\ 000\ a_1 + 0.149\ 300\ a_2 + 0.755\ 530\ a_3 = \lambda a_1$$
$$0.149\ 300\ a_1 + 1.000\ 000\ a_2 + 0.423\ 070\ a_3 = \lambda a_2 \quad (2)$$
$$0.755\ 530\ a_1 + 0.423\ 070\ a_2 + 1.000\ 000\ a_3 = \lambda a_3$$

The system (2) of equations can have only non-trivial solutions if its determinant becomes zero. The determinant equation becomes:

$$\begin{vmatrix} 1.000\ 000 - \lambda & 0.149\ 300 & 0.755\ 530 \\ 0.149\ 300 & 1.000\ 000 - \lambda & 0.423\ 070 \\ 0.755\ 530 & 0.423\ 070 & 1.000\ 000 - \lambda \end{vmatrix} = 0 \quad (3)$$

From this we have:
$$\lambda_1 = 0.20$$
$$\lambda_2 = 0.86$$
$$\lambda_3 = 1.96 \text{ (maximum)}.$$

Substituting λ_3 in system (2) the coefficients a_k of the equations of the system were calculated. The contribution of the first component to the variance of the standardized variables $_{(1)}X$, $_{(2)}X$, $_{(3)}X$ are the squares of a_1, a_2, a_3 under the condition that

$$a_1^2 + a_2^2 + a_3^2 = \lambda_3$$

The estimated values are:

$$a_1^2 = 0.685\ 077 \qquad a_1 = 0.827\ 693$$
$$a_2^2 = 0.396\ 180 \qquad a_2 = 0.629\ 428$$
$$a_3^2 = 0.878\ 743 \qquad a_3 = 0.937\ 412$$

Further, the first principal component is described by the following linear equation,

$$U = 0.345\ 659\ _{(1)}X + 0.262\ 860\ _{(2)}X + 0.391\ 481\ _{(3)}X. \quad (4)$$

An evaluation of the estimated values of a_k^2 leads to the following conclusions:
1. The first principal component explains about 70 per cent of the variance of $_{(1)}X$ (Economically preferred species), about 40 per cent of the variance of $_{(2)}X$ (Secondary species) and about 90 per cent of the variable $_{(3)}X$ (Accompanying species); 2. The total variance of the three standardized variables $_{(1)}X$, $_{(2)}X$, $_{(3)}X$, is evidently 3. Hence, since $\lambda_3 = 1.96$ it appears that the first principal component explains about 65 per cent of the total variance of the variables $_{(1)}X$, $_{(2)}X$, $_{(3)}X$.

One can argue that, there is a 'statistical equilibrium' in space among the fish species (major groups) and that their general trend in space can be explained by the first principal component which accounts for most of the variability. The first principal component is described by the equation

$$U = 0.345\ 659\ _{(1)}X + 0.262\ 860\ _{(2)}X + 0.391\ 481\ _{(3)}X.$$

The function U minimizes the error variances while its own variance is unity. It is interesting to note that accompanying species have the greatest weight.

Regressions in the area stratification. – In this section an attempt is made to study the relationship between the variable X (number of fish) and Y (weight of fish) in the established area strata of the qrs. Specifically, the following questions received consideration: Can the regression lines of Y on X in the different strata be regarded as the same? If not, in which respect do they differ? The method used for this study is the 'Analysis of Covariance'.

For the study, the sample data (on a standardized area sampling unit basis, 1 ha) of strata 2 and 3 were only used. Sample data of stratum 1 were considered as insufficient for this kind of analysis. The purpose was to examine whether the linear regressions of weight of fish on number of fish (Y on X) is the same in the above two strata (2,3). They may differ in slope, elevation or in the residual variances. In terms of the model, we have,

$$_1Y_h = {_1}a + {_1}B_1 X_h + {_1}e_h$$

where, suffix $1 = 2,3$ refers to the two strata we are dealing with and suffix h refers to the area sampling units within the strata.

Specifically, in order to simplify our calculations the method was applied to the totality of species and by major group of species (1. economically preferred fish species, 2. secondary species, 3. accompanying species). The following symbols were employed in our case: $_1x_h$ = total number of fish in the hth selected area sampling unit within stratum 1; $_1\bar{x}$ = estimated average number of fish (total) per area sampling unit within stratum 1; $\bar{x}$ = estimated average number of fish (total) per area sampling unit irrespective of stratum, etc. $_1C_{xx} = \underset{h}{s} (_1x_h - {_1}\bar{x})^2$; $_1C_{xy} = \underset{h}{s} (_1x_h - {_1}\bar{x})(_1y_h - {_1}\bar{y})$; $_1C_{yy} = \underset{h}{s} (_1y_h - {_1}\bar{y})^2$; $C_{xx} = \underset{1\ h}{s\ s} (_1x_h - \bar{x})^2$; $C_{xy} = \underset{1\ h}{s\ s} (_1x_h - \bar{x})(_1y_h - \bar{y})$; $C_{yy} = \underset{1\ h}{s\ s} (_1y_h - \bar{y})^2$, etc. For the variable Y the same notation was used as for the X variable.

The most convenient approach to this study is to estimate first the regression equations; secondly, to test the 'hypothesis of homogeneity' by comparing the estimated residual mean squares for each stratum. If the hypothesis of homogeneity is valid then we should proceed with the comparison of the slopes and elevation of the regression equations.

The estimated regression coefficients of the linear regression equations (Y on X) for each of the major groups of species within each stratum ($1 = 2,3$) and irrespective strata ($= 2 + 3$) were calculated by the usual formulae:

within strata:
    1. $_1b = {_1}C_{xy}/{_1}C_{xx}$
    2. $_1\hat{a} = {_1}\bar{y} - {_1}b_1\bar{x}$

Irrespective of strata:
    1. $b = C_{xy}/C_{xx}$
    2. $\hat{a} = \bar{y} - b\bar{x}$, (L = 2 + 3).

The estimated regression coefficients are given in table B5, the fitted straight lines are displayed in Fig. B6.

Residual mean squares in each stratum for the major groups of species were estimated by using the following formula:

$$_1S^2_{y.x} = \frac{1}{_1n - 2}\left(_1C_{yy} - \frac{_1C^2_{xy}}{_1C_{xx}}\right).$$

The Table B6 provides the obtained estimates.

The hypothesis of homogeneity of the residual mean squares was tested by using the two-tailed F-test. The Table B7 gives the calculated values of F and the corresponding theoretical values of F (a = 1%, 5%). It is evident that there are heterogeneous variances in all the four cases except the last (3. accompanying species). The rejection of the hypothesis of homogeneity for the economically preferred species, secondary species and total of species indicates that,

Table B5. Estimated regression coefficients (Y on X); x in thousands of fish, y in kg.

Major groups of species	Stratum					
	2 + 3		2		3	
	â	b	â	b	â	b
Total	29.94	13.19	62.56	12.00	16.16	2.30
1. Economically preferred species	140.16	6.86	287.93	0.72	23.10	120.48
2. Secondary species	−33.62	18.00	−101.94	20.48	3.45	6.52
3. Accompanying species	1.75	4.38	3.53	4.00	−0.65	5.66

Table B6. Estimated residual mean squares ($_1S^2{}_{y.x}$).

Major groups of species	Stratum	
	2	3
Total	55 212.8	632.2
1. Economically preferred species	144 874.1	1 438.8
2. Secondary species	7 599.1	69.6
3. Accompanying species	15.1	2.7

the relationship between the variables X and Y for the respective groups of species is not the same in the different strata (1 = 2,3).

The hypothesis of homogeneity was accepted for the residual mean squares of the accompanying species. We now compare the slopes or the regression coefficients of the estimated regression lines i.e. we test the hypothesis that the population regression lines are parallel. To make the test, the two-tailed F-test was used

$$F = \frac{V_2}{V_1}$$

where,

$$V_2 = \frac{1}{L-1} S_1 C_{xx} (_1b - \hat{b})^2, \quad \hat{b} = \frac{1}{S_1 C_{xx}} S_1 C_{xx\,1}b$$

and

$$V_1 = \frac{1}{n-2} S_1 \left(_1C_{yy} - \frac{_1C^2{}_{xy}}{_1C_{xx}} \right),$$

L = number of strata (L = 2)

$n = \overset{3}{\underset{1=2}{S}} {}_1n$: total number of sampling units in strata 2 and 3 (n = 8).

The calculated value of F (5.63) was compared with the theoretical value of F (5%, 1% − 7.70, 21.19). For these data the hypothesis that the slopes do not differ is accepted (Fig. B6).

Table B7. Estimated and theoretical values of F-test.

Major groups of species	Estimated values of F	Theoretical values of F	
		5%	1%
Total	873.6	19.00	99.00
1. Economically preferred	100.74	19.00	99.00
2. Secondary	109.15	19.00	99.00
3. Accompanying	5.64	19.00	99.00

Table B8. Allocation of the overall sample size of ASU's between strata.

Area stratification	Number of selected ASU's
Stratum: 1	2
Stratum: 2	4
Stratum: 3	4
Stratum: 1 + 2 + 3	10

The hypothesis of homogeneity of variances and the assumption of parallel lines for accompanying species were justified. It remains to test the null hypothesis $_{1=2}a = {_{1=3}}a$, i.e. the hypothesis that the population regression lines coincide; that is, since they are parallel, that they have the same elevation.

The difference, $_{1=2}a - {_{1=3}}a$, is estimated by the difference between the adjusted means

(1) $\bar{d} = [{_{1=2}}\bar{y} - b_w ({_{1=2}}\bar{x} - \bar{x})] - [{_{1=3}}\bar{y} - b_w ({_{1=3}}\bar{x} - \bar{x})]$.

Substituting the values of the respective magnitudes in (1) we get

$\bar{d} = [22.56 - 4.387 (4.75 - 3.32)] - [10.11 - 4.387 (1.90 - 3.32)]$
$\bar{d} \simeq 0$

This difference justifies the hypothesis that $_{1=2}a = {_{1=3}}a$ (the same conclusion was obtained by using the F-test).

From a sampling point of view, the sampling method of the qrs at Lake Kariba can be described as 'single stage stratified sample' with equal probabilities. Specifically, the lake under study was first divided into a number of sections (strata) by using general ecological criteria as control characteristics of stratification. Within each stratum a number of 'Area Sampling Units' (ASU's) was selected for the survey. The criteria used for the selection of the ASU's were those of accessibility and compatability with the available sampling gear. As a result of this procedure ASU's of different sizes were selected; the selected ASU's ranged in size from 0.021 ha to 5.273 ha. Also, the size of the selected samples of ASU's varies from stratum to stratum. The Table B8 provides the allocation of the overall sample size between strata. In order to ensure that there are at least two ASU's within each stratum we introduced some adjustment in the allocation of the original overall sample size between strata.

It has been noted that, for our purposes, the following variables received consideration: x = number of fish, y = weight of fish. Further, for the purpose of producing data suitable for statistical manipulation, the sample data of the qrs at Lake Kariba were expressed in terms of

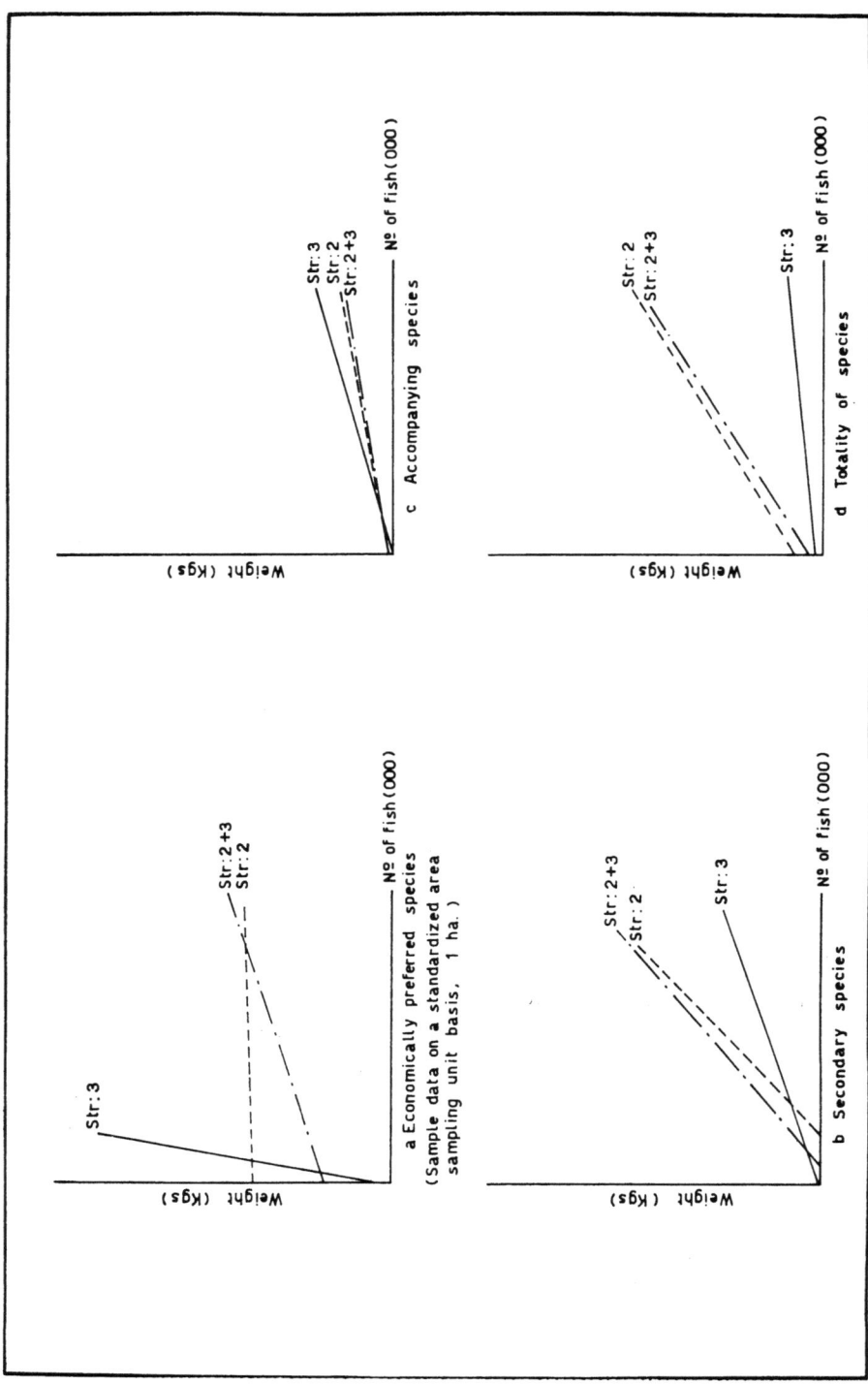

Fig. B6.

ASU's of fixed size: 'Standardized Area Sampling Units' of 1 ha (S-ASU's). For the statistical manipulation of the sample data the following suffixes and symbols were employed: suffix 1: refers to the stratum, suffix h: refers to the S-ASU, suffix k: refers to the major group of species, suffix e: refers to the minor group of species, and suffix i: refers to the species (genera, families), ($_1x_{hi}$: number of fish of species i^{th} in the hth S-ASU within the 1th stratum; $_1x_i = S\ _1x_{hi}$:
$\quad\quad\quad\quad\quad\quad\quad\quad\quad\quad\quad\quad\quad\quad\quad\quad\quad\quad\quad\quad\quad\quad\quad\quad\quad\quad\quad\quad\quad\quad\quad\quad\quad\quad\quad\quad\quad\quad\quad\quad\quad$ h

total number of fish of species i within the 1th stratum (sample data); $x_i = S\ S_1x_{hi}$: total
$\quad\quad\quad\quad\quad\quad\quad\quad\quad\quad\quad\quad\quad\quad\quad\quad\quad\quad\quad\quad\quad\quad\quad\quad\quad\quad\quad\quad\quad\quad\quad\quad\quad\quad$ 1 h

number of fish of species i in the sample, etc.). Symbols for the variable Y were established in a manner parallel to the symbols of the variable X.

In this section we consider some characteristic values describing the survey variables. Studies of this type should be based on the frequency distribution of the variables under study. Here the only data available were those obtained from the sample of the qrs and these data were used for our purposes. The following characteristic values were calculated (estimated): 1. average number of fish by species and groups of species per ha and by stratum and for the population as a whole (relative abundance); 2. relative variability per unit of the variable (X), CV(x) and 3. Personian measure of skewness ($S_{k(x)}$).

For the variable (Y) the estimated characteristic values were: relative mass of fish, relative variability per unit of the variable (Y), CV(y), and Personian measure of skewness ($S_{k(y)}$).

For estimation purposes the following formulae were employed:

$_1\bar{x}_i = \dfrac{1}{_1n} S\ _1x_{hi}$: Relative abundance of fish of the ith species in stratum 1.
$\quad\quad\quad\quad$ h

$\bar{x}_i = \dfrac{1}{n} S\ S\ _1x_{hi}$: Relative abundance of fish of the ith species in the population.
$\quad\quad\quad\quad$ 1 h

$_1\bar{x} = \dfrac{1}{_1n} S\ S_1x_{hi}$: Relative abundance of fish in the 1th stratum, etc.
$\quad\quad\quad\quad$ h i

$CV(_1x_i)$: coefficient of variation per unit of the Variable (X) for the ith species within the 1 stratum. Specifically,

$$CV(_1x_i) = \dfrac{\sqrt{_1S_i^2}}{_1\bar{x}_i}, \quad \text{where} \quad _1S_i^2 = \dfrac{_1C_{xxi}}{_1n - 1}, \quad \text{etc.}$$

$$S_{k(xi)} = \dfrac{_1\bar{x}_i - _1T_{oi}}{_1S_{xi}}, \quad \text{where To = Mode etc.}$$

The estimated characteristic values indicate that there is a high variability per unit of the variable X in nearly all strata as far as the individual species and major groups of species are concerned. In stratum 1 the variability per unit of the variable X for the major groups of species (economically preferred) can be considered as low. The variability per unit of the variable Y follows the same pattern as that of the variable X. Since the precision of a sample estimate depends on the one hand on the variability in the population and on the other hand on the size of the survey sample one can foresee that the expected precision of the estimated characteristics of the qrs will be low.

By using the standardized sample data of the qrs at Lake Kariba, the precision of the estimated characteristics were calculated, estimating parallel magnitudes, average number: weight of fish per standardized area sampling unit by species and major groups of species and the corresponding relative sampling errors. As was predicted the precision of the obtained estimates of the survey characteristics is generally low, the only exception being for the estimates in stratum 1 of economically preferred species and *Sarotherodon mossambicus*.

From a sampling point of view, the precision of the obtained estimates can be expressed as follows: with a probability 95 per cent the true population value of the survey characteristics would yield a value somewhere between $\bar{x} \pm 2S\bar{x}$, $\bar{y} \pm 2S\bar{y}$. The Tables B9 and B10 give the estimated lower and upper values of the established confidence intervals (a = 5%) for the major groups of species and irrespective of species.

Increasing the precision of the sample estimates. – As we have seen in previous sections under the established sampling conditions of the qrs the point estimates of the survey characteristics are generally of a low precision. This however does not imply that the survey was not successful or that the results of the survey cannot be used for comparisons in space or in time. It simply means that the obtained point estimates are not very close to the parameter(s) under investigation and that the sampling distribution of the estimates is widely dispersed. In this section I will discuss some of the factors which might affect the precision of the sample estimates. i.e. the effect of the length of the survey period, the effect of the size of the area sampling units, etc.

It has been noted that for the manipulation of the sample data we assumed that time is a fixed variable which does not affect the results of the survey. In order to obtain some indication of the effect of the length of the survey period on the obtained estimates the sample data of the qrs were used. First, the sampling area units of the qrs were regrouped according to the year when the survey took place i.e. 1968, 1969, 1971 and the respective magnitudes (relative abundance, ichthyomass), were calculated within the established classes. It is obvious that in this kind of classification of the S-ASU's the pre-stratification of the survey population was not taken into account. Further, there was no principle for the allocation of the overall sample size of units between classes. As it was noted above this exercise can be used to provide some indications of the effect of the length of the 'survey period' on the obtained sample estimates. In the Table B11 the calculated relative abundance/ichthyomass of fish for the years 1968, 69, 71 are given by major group of species and for the total of species. The fluctuation in time of the estimated magnitudes are shown in Figure B7. There are some indications that time might be one of the factors affecting the variance of the magnitudes.

From a sampling point of view, a population can be divided into units in various ways. In 'Area Sample Surveys' the sample units can be of various sizes and shapes, each of which creates a different subdivision of the survey population into sampling units. A change in the type of the unit usually affects both the cost of taking the sample and the precision obtained from it. In qrs the clustering of the survey population has the characteristics of a rather peculiar procedure. As has been noted, the criteria used for the selection of the ASU's are those of accessibility and compatibility with the available gear. As a result of this procedure area sampling units of different sizes and shapes were selected. The Table B12 gives an idea of the diversity in size of the sample units of the qrs at Lake Kariba.

By using the sample data of the qrs the average total number of fish and the average total weight of fish per ASU were calculated within the established size classes. These magnitudes were calculated for the total of species and for the major groups of species (Table B13). In this table one can see marked differences of the magnitudes under study between the established size classes. It is indicated that, small size ASU's have a high density of fish, medium size ASU's have a moderate density of fish and large size ASU's have a low density of fish. For a better understanding of the existing relationship of the magnitudes between the established size classes, the respective ratios were calculated (Table B14). For the calculation of the ratios the values of class-3 were taken as the 'base figures = 1.00'.

From the above analysis one can argue that, the different sizes of the selected 'Area Sampling Units' introduce another factor affecting the precision of the obtained estimates. Another interesting point in this analysis is the indication that estimates derived from samples based on small size ASU's are likely to differ significantly from the estimates obtained from samples based on medium or large size ASU's. In such a case one might be faced with the problem of biased sampling through the selection procedure of the ASU's.

A further analysis of the obtained results revealed that there is a lack of homogeneity between size classes as far as the composition of major groups of species is concerned. To test this hypothesis the χ^2 test was used. Specifically for our analysis the sample data of size class-1 and size class-2 were used. Sample data of size class-3 were considered as insufficient for this kind of analysis. Table B15, provides the calculation made for the χ^2 test.

Table B9. Estimated confidence intervals for some of the characteristics of the qrs at Lake Kariba (average number of fish per ha), in thousands.

Major groups of species	Stratum							
	1 + 2 + 3		1		2		3	
	Lower limit	Upper limit	Lower limit	Upper limit	Lower limit	Upper limit	Lower limit	Upper limit
Total	2.591	27.201	−10.591	45.843	1.371	50.817	−0.651	5.321
1. Econ. preferred	−1.869	8.409	1.292	1.586	−5.344	20.004	0.064	0.188
2. Secondary	0.382	17.196	−11.774	42.382	−1.710	29.734	0.015	0.601
3. Accompanying	0.440	5.234	−0.109	1.875	−0.385	9.889	−0.917	4.717

Table B10. Estimated confidence intervals for some of the characteristics of the qrs at Kariba Lake (average weight of fish in kg per ha).

Major groups of species	Stratum							
	1 + 2 + 3		1		2		3	
	Lower limit	Upper limit	Lower limit	Upper limit	Lower limit	Upper limit	Lower limit	Upper limit
Total	202.326	374.538	181.232	484.406	−143.120	1144.707	0.878	106.874
1. Econ. preferred	47.723	329.579	187.544	372.980	−17.722	604.099	6.447	70.168
2. Secondary	−48.628	217.060	−10.215	90.351	−144.745	514.831	−1.614	12.538
3. Accompanying	5.154	25.978	3.904	21.077	1.746	43.378	−5.891	26.106

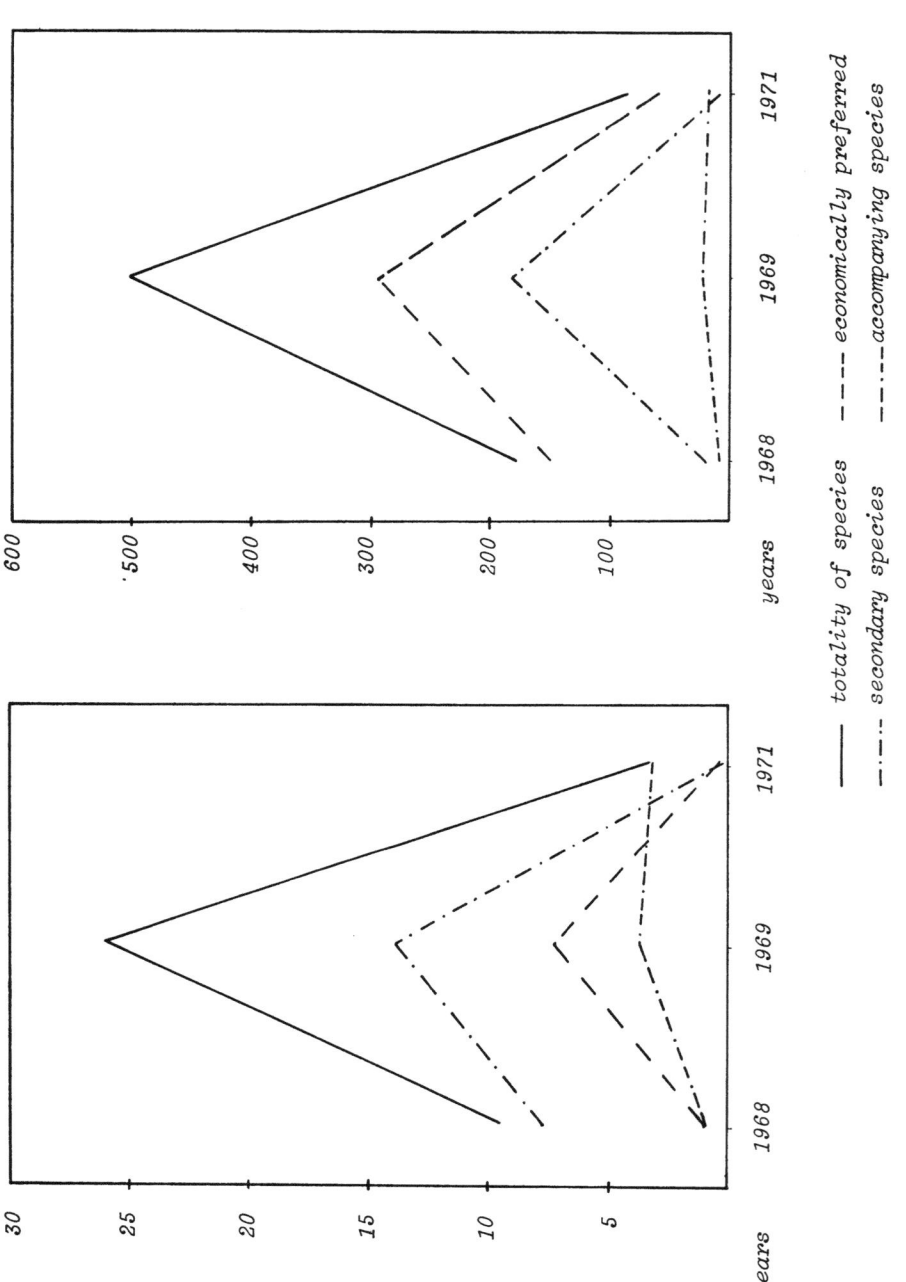

Fig. B7.

——— totality of species ---- economically preferred
–·–·– secondary species –··–··– accompanying species

Table B11. Calculated average number of fish and average weight of fish per ha within the established classes.

Major groups of species	Year 1968		Year 1969		Year 1971	
	Average No. of fish/ha (000)	Average weight of fish/ha (kgs)	Average No. of fish/ha (000)	Average weight of fish/ha (kgs)	Average No. of fish/ha (000)	Average weight of fish/ha (kgs)
Total	9.44	177.45	26.09	500.79	3.41	85.67
Economically preferred	0.78	148.95	7.33	293.19	0.12	58.98
Secondary	7.85	20.97	14.01	185.04	0.23	9.06
Accompanying	0.81	7.54	4.76	25.56	3.06	17.62

Table B14. Estimated ratios of the average total number of fish and average total weight of fish per ASU between the established size classes.

Size classes of the ASU's	Ratios	Total species		Economically preferred species		Secondary species		Accompanying species	
		Average No. of fish	Average weight of fish	Average No. of fish	Average weight of fish	Average No. of fish	Average weight of fish	Average No. of fish	Average weight of fish
Class-1	R_{13}	7.66	3.88	23.04	2.55	4.61	14.53	30.85	5.36
Class-2	R_{23}	3.32	1.70	2.83	1.58	2.81	1.99	13.82	3.94
Class-3	R_{33}	1.00	1.00	1.00	1.00	1.00	1.00	1.00	1.00

Table B12. Allocation of the overall sample size of ASU's in classes according to their sizes.

Size classes of the ASU's	No. of selected ASU's	Strata in which ASU's belong
Size class 1 = 0.01–0.50 ha	4	2,3
Size class 2 = 0.51–1.50 ha	4	1,3
Size class 3 = 3.00 ha and over	2	2,3
Total	10	

Table B13. Average total number of fish and average total weight of fish per ASU within the established size classes (number of fish in thousands, weight in kg).

Major group of species	Size class-1		Size class-2		Size class-3	
	No. of fish	weight of fish	No. of fish	weight of fish	No. of fish	weight of fish
0. Total	24.845	460.46	10.775	201.54	3.242	118.17
1. Economically pref.	7.211	259.49	0.885	161.22	0.313	101.83
2. Secondary	12.788	179.69	7.797	24.66	2.772	12.37
3. Accompanying	4.845	21.28	2.170	15.66	0.157	3.97

Table B15. Computation of χ^2 for the average total number of fish per ASU and average total weight of fish (size class 1,2).

Major groups of species	Average total number of fish per ASU			Average total weight of fish per ASU		
	o_1	o_1'	$\frac{(o_1 - o_1')^2}{o_1'}$	o_2	o_2'	$\frac{(o_2 - o_2')^2}{o_2'}$
1. Economically pref.	7,211	5,593	468,070	259.50	292.60	3.74
2. Secondary	12,788	14,358	171,670	179.70	142.16	9.91
3. Accompanying	4,846	2,170	470	21.30	25.74	0.77
		χ^2_1	640,210		χ^2_2	14.42

From the ditribution of χ^2 for a = 0.05 and a = 0.01 and 2 degrees of freedom, the corresponding tabulated values are 5.99 and 9.21 respectively. The χ^2 test tells us that the hypothesis of homogeneity is discredited.

From the above discussion it is obvious that, in the design process of a qrs the first question to be answered is the determination of the 'optimum size' of the ASU of the survey. The choice of the unit involves striking a balance between relative precision and relative cost.

It has been noted that the sample method used for the qrs was 'single stage stratified sample' with equal probabilities. For the determination of the sample size of the survey for a required precision of the mean a loose procedure was used; the method of 'simple random sample' with replacement was applied without any prior information or good guess about the value of the mean of the control variable of the survey and its variability per unit. As a result of this procedure the sample size of the qrs was estimated at ten ASU's which were arbitrarily allocated between strata. One can argue that another factor which affects the precision of the sample estimates of the survey is the sample size of the qrs.

If there is one thing that distinguishes sampling theory from general statistical theory, it is the degree of emphasis laid on the use of auxiliary information for improving the precision of estimates. In our case, for the sample estimates of the survey characteristics the estimator of the mean for single stage stratified sample was employed. It is important to see whether, by making use of auxiliary information, other estimators can be used to achieve higher precision. For example, it might be found that a relationship of a given type exists between the ichthyomass per ha qrs and catches per ha (Catch Assessment Survey). In such a case, it is natural to try to develop estimators based on the existing relationship of these two magnitudes; these estimators might be proved more efficient than those employed for the sample estimates of the qrs.

APPENDIX C

Tables of Mean Sizes of Individual Species, Life Intervals and Growth Intensity

C 1

Species	Absolute and relative growth values		0	1	2	3	4	5	6	7	8	9	10
Sarotherodon mossambicus mortimeri	Standard length in mm	l	55	55	92	127	161	(181)[1]	(196)				
	Wet weight in g	w̄	7		36	90	200	(280)	(350)				
	Absolute increment in mm	h		37	35		34	20	15				
	Index of the species average size	ΦH			4.02								
	Specific weight rate of growth	Cw	700	414	150	122		40	25				
	Index of stock weight growth intensity	ΦCw		j	346	a		s					
Mormyrops deliciosus	Standard length in mm	l	231	231	328	416	468	497	545	559			
	Wet weight in g	w̄	136		348	663	937	1089	1430	1500			
	Absolute increment in mm	h		97	88		52	29	48	14			
	Index of the species average size	ΦH					9.08						
	Specific weight rate of growth	Cw	13600	156	90	41		16	31	5			
	Index of stock weight growth intensity	ΦCw		j		2322	a			s			
Tilapia rendalli	Standard length in mm	l	66	66	109	143	180	207	(230)	(250)	(268)		
	Wet weight in g	w̄	14		69	127	312	520	(720)	(956)	(1180)		
	Absolute increment in mm	h		43	34		37	27	23	20	18		
	Index of the species average size	ΦH			4.50								
	Specific weight rate of growth	Cw	1400	393	84	146		67	38	33	23		
	Index of stock weight growth intensity	ΦCw		j	418		a		s				

Growth seasons

[1] In parentheses are length interpolated from the growth curve.

C 2

Species	Absolute and relative growth values		0	1	2	3	4	5	6	7	8	9	10
								Growth seasons					
Clarias gariepinus	Standard length in mm	l		82	131	183	254	319	383	443	505	548	610
	Wet weight in g	w̄		6.8	26	58	320	500	750	900	1630	1920	2700
	Absolute increment in mm	h	82	49	52	71	65	64	60	62	43	62	
	Index of the species average size	ΦH						6.10					
	Specific weight rate of growth	Cw	680	282	123	452	56	50	20	81	18	41	
	Index of stock weight growth intensity	ΦCw		j				180		a			
Hydrocynus vittatus	Standard length in mm	l		196	306	380	423	462	488	523			
	Wet weight in g	w̄		140	590	1050	1530	2000	2400	3000			
	Absolute increment in mm	h	196	110	74	43	39	26	35				
	Index of the species average size	ΦH					7.50						
	Specific weight rate of growth	Cw	14000	321	78	46	31	20	25				
	Index of stock weight growth intensity	ΦCw	j				a 2074						
Mormyrus longirostris	Standard length in mm	l		267	354	414	468	505	541	581			
	Wet weight in g	w̄		250	633	1023	1600	2008	2483	2900			
	Absolute increment in mm	h	267	87	60	54	37	36	40				
	Index of the species average size	ΦH				9.02							
	Specific weight rate of growth	Cw	25000	153	62	56	25	24	17				
	Index of stock weight growth intensity	ΦCw		j		4220	a			s			

Species	Absolute and relative growth values		\multicolumn{11}{c}{Growth seasons}											
			0	1	2	3	4	5	6	7	8	9	10
Heterobranchus longifilis	Standard length in mm	l	69	124	187	260	333	416	512	570	614	683	
	Wet weight in g	w	8	22	82	220	490	990	1870	2600	3200	4200	
	Absolute increment in mm	h		55	63	73	73	83	96	58	44	69	
	Index of the species average size	$\varnothing$H						6.83					
	Specific weight rate of growth	Cw	800	175	273	168	123	102	89	39	23	31	
	Index of stock weight growth intensity	$\varnothing$Cw				j		182			a		
Sargochromis codringtoni	Standard length in mm	l	87	130	179	216							
	Wet weight in g	w	26	90	240	440							
	Absolute increment in mm	h		43	49	37							
	Index of the species average size	$\varnothing$H			5.40								
	Specific weight rate of growth	Cw	2600	246	167	83							
	Index of stock weight growth intensity	$\varnothing$Cw			165								
Labeo altivelis	Standard length in mm	l	79	129	192	242	282	294	365	381	400		
	Wet weight in g	w	13	58	185	386	600	678	1260	1420	1630		
	Absolute increment in mm	h		50	63	50	40	12	71	16	19		
	Index of the species average size	$\varnothing$H					5.64						
	Specific weight rate of growth	Cw	1300	346	219	109	55	13	86	13	15		
	Index of stock weight growth intensity	$\varnothing$Cw		j			a		406	s			

C 4

Species	Absolute and relative growth values		0	1	2	3	4	Growth seasons 5	6	7	8	9	10
Hippopotamyrus discorhynchus	Standard length in mm	l		77	111	139	163	182					
	Wet weight in g	w̄		9	28	56	94	133					
	Absolute increment in mm	h	77	34	28	24	19						
	Index of the species average size	ΦH			3.64								
	Specific weight rate of growth	Cw	900	211	100	68	41						
	Index of stock weight growth intensity	ΦCw	j		105		a						
Malapterurus electricus	Standard length in mm	l		53	93	137	182	235	287	332	365	405	495
	Wet weight in g	w̄		4	21	62	150	310	530	820	1100	1500	2800
	Absolute increment in mm	h	53	40	44	45	53	52	45	33	40	90	
	Index of the species average size	ΦH						4.95					
	Specific weight rate of growth	Cw	400	425	195	142	107	71	55	34	36	87	
	Index of stock weight growth intensity	ΦCw		j				155			a		
Alestes lateralis	Standard length in mm	l		46	77	106							
	Wet weight in g	w̄		2	12	36							
	Absolute increment in mm	h	46	31	29								
	Index of the species average size	ΦH		3.53									
	Specific weight rate of growth	Cw	200	500	200								
	Index of stock weight growth intensity	ΦCw	j	a 300	s								

C 5

Species	Absolute and relative growth values		0	1	2	3	4	5	6	7	8	9	10
								Growth seasons					
Eutropius depressi-rostris	Standard length in mm	l		64	95	119	141	168	189				
	Wet weight in g	w̄		4	12	22	37	62	85				
	Absolute increment in mm	h	64	31	24	22	27	22					
	Index of the species average size	ΦH				3.36							
	Specific weight rate of growth	Cw	400	200	83	68	68	37					
	Index of stock weight growth intensity	ΦCw		j		a 164			s				
Marcusenius macro-lepidotus	Standard length in mm	l		95	130	160	180	218	232				
	Wet weight in g	w̄		13	35	66	98	178	222				
	Absolute increment in mm	h	95	35	30	20	38	14					
	Index of the species average size	ΦH				4.36							
	Specific weight rate of growth	Cw	1300	169	89	48	82	25					
	Index of stock weight growth intensity	ΦCw		j		a 338			s				
Schilbe mystus	Standard length in mm	l		52	77	102	123	141	164				
	Wet weight in g	w̄		2	7	15	26	37	53				
	Absolute increment in mm	h	52	25	25	21	18	23					
	Index of the species average size	ΦH				3.07							
	Specific weight rate of growth	Cw	200	250	114	73	42	43					
	Index of stock weight growth intensity	ΦCw		j 159		a		s					

C 6

Species	Absolute and relative growth values		0	1	2	3	4	5	6	7	8	9	10
Synodontis zambezensis	Standard length in mm	l		80	105	129	150	164	190	226	259		
	Wet weight in g	w̄		13	30	55	87	135	177	295	440		
	Absolute increment in mm	h	80	25	24	21	14	24	36	33			
	Index of the species average size	ϕH					3.21						
	Specific weight rate of growth	Cw	1300	131	83	58	55	31	67	49			
	Index of stock weight growth intensity	ϕCw	j				222	a					
Haplochromis darlingi	Standard length in mm	l		47	56	59							
	Wet weight in g	w̄		3.5	5.5	6.1							
	Absolute increment in mm	h	47	9	3								
	Index of the species average size	ϕH		2.80									
	Specific weight rate of growth	Cw	350	57	11								
	Index of stock weight growth intensity	ϕCw	j	203	a	s							
Synodontis nebulosus	Standard length in mm	l		44	64	82	100	115					
	Wet weight in g	w̄		2	6	13	23	35					
	Absolute increment in mm	h	44	20	18	18	15						
	Index of the species average size	ϕH			2.30								
	Specific weight rate of growth	Cw	200	200	117	77	52						
	Index of stock weight growth intensity	ϕCw	j		129	a							

C 7

Species	Absolute and relative growth values			Growth seasons										
			0	1	2	3	4	5	6	7	8	9	10
Brachyalestes imberi	Standard length in mm	l		59	116	130							
	Wet weight in g	w̄		5	48	70							
	Absolute increment in mm	h	59	57	14								
	Index of the species average size	ΦH		5.80									
	Specific weight rate of growth	Cw	500	860	46								
	Index of stock weight growth intensity	ΦCw	j	a | s									
				680									
Micralestes acutidens	Standard length in mm	l		43.3	56.5	63.5							
	Wet weight in g	w̄		1.14	3.70	5.00							
	Absolute increment in mm	h	43.3	13.2	7.0								
	Index of the species average size	ΦH		2.82									
	Specific weight rate of growth	Cw	114	225	35								
	Index of stock weight growth intensity	ΦCw	j	a | s									
				169									

APPENDIX D

Tables of Individual Species and Single Samples Production Computations

Table D1. Computation of production for *Mormyrus longirostris* stock from Lake Kariba (after data from locality D – Buffalo Island cove, 1968).

Growth seasons		0	1	2	3	4	5	6	7	8	Total Σ or m
Interpolated N'/ha at time of annulus formation	N'	(49)	30	19	12	7	4	3	2	1	78
Interpolated mean weight in g of 1 specimen	$\bar{w}$		250	633	1023	1600	2008	2483	2900	2910	
Initial biomass in kg/ha	B'		7.50	12.03	12.28	11.20	8.03	7.45	5.80	2.91	67.20
Rate of increase in biomass	H	+5.0308	+0.4724	+0.0204	−0.0917	−0.3325	−0.0753	−0.2507	−0.6892		
Instantaneous growth rate	G	5.5215	0.9290	0.4800	0.4473	0.2270	0.2125	0.1548	0.0039		0.9970
Instantaneous rate of total mortality	Z	0.4907	0.4566	0.4596	0.5390	0.5595	0.2878	0.4055	0.6931		0.4865
Mean biomass	$\bar{B}$	219.76	15.28	12.16	10.52	6.79	6.69	5.12	2.10		278.42
Total production (= G$\bar{B}$) in kg/ha/yr	A	1213.40	14.19	5.84	4.70	1.54	1.42	0.79	0.008		1241.89
Total yield in kg/ha/yr	Y_A			5.84	4.70	1.54	1.42	0.79	0.008		14.30
Available production in kg/ha/yr	P'	7.50	7.28	4.68	4.04	1.63	1.42	4.96	0.01		31.52
Available yield in kg/ha/yr	$Y_{P'}$			4.68	4.04	1.63	1.42	4.96	0.01		15.32

Table D2. Computation of production for *Hydrocynus vittatus* stock from Lake Kariba (after data from locality C – Chikanka Island cove, 1968).

Growth seasons		0	1	2	3	4	5	6	Total Σ or m
Interpolated N'/ha at time of annulus formation	N'	(50)	28	16	9	5	3	2	63
Interpolated mean weight in g of 1 specimen	$\bar{w}$		160	420	800	1150	2110	2300	
Initial biomass in kg/ha	B'		4.48	6.72	7.20	5.75	6.33	6.40	36.88
Rate of increase in biomass	H	+4.4954	+0.4053	+0.0689	−0.2246	+0.9600	−0.3193		
Instantaneous growth rate	G	5.0752	0.9651	0.6443	0.3630	0.6069	0.0862		1.2901
Instantaneous rate of total mortality	Z	0.5798	0.5598	0.5754	0.5876	0.5109	0.4055		0.5365
Mean biomass	$\bar{B}$	88.71	8.17	7.58	5.06	10.63	5.49		125.64
Total production (= G$\bar{B}$) in kg/ha/yr	A	450.22	7.88	4.88	1.84	6.45	0.47		471.74
Total yield in kg/ha/yr	Y_A			3	1.84	6.45	0.47		11.76
Available production in kg/ha/yr	P'	4.48	4.16	3.42	1.75	2.88	0.38		17.07
Available yield in kg/ha/yr	$Y_{P'}$			2	1.75	2.88	0.38		7.01

Table D3. Computation of production for *Mormyrops deliciosus* stock from Lake Kariba (after data from locality D – Buffalo Island cove, 1968).

Growth seasons		0	1	2	3	4	5	6	7	8	9	10	Total
Interpolated N'/ha at time of annulus formation	N'	(108)	69	44	28	18	11	7	4	3	2	1	187
Interpolated mean weight in g of 1 specimen	$\bar{w}$												
Initial biomass in kg/ha	B'		136 9.38	348 15.31	663 18.56	937 16.87	1089 11.98	1430 10.01	1500 6.00	3450 10.35			98.46
Rate of increase in biomass	H	$+4.4646$	$+0.4898$	$+0.1926$	-0.0960	-0.3418	-0.1799	-0.5117	$+0.5451$				
Instantaneous growth rate	G	4.9127	0.9395	0.6446	0.3459	0.1507	0.2721	0.0478	0.8329				1.0183
Instantaneous rate of total mortality	Z	0.4481	0.4497	0.4520	0.4419	0.4925	0.4520	0.5595	0.2878	0.4055	0.6931		0.4682
Mean biomass	$\bar{B}$	187.02	19.76	20.16	16.73	14.19	9.16	4.68	13.59				285.29
Total production ($= G\bar{B}$) in kg/ha/yr	A	918.77	18.56	12.99	5.79	2.14	2.49	0.22	11.39				972.35
Total yield in kg/ha/yr	Y_A			12.99	5.79	2.14	2.49	0.22	11.39				35.02
Available production in kg/ha/yr	P'	9.38	8.82	4.93	1.67	2.39	0.28	5.85					42.65
Available yield in kg/ha/yr	$Y_{P'}$		8.82	4.93	1.67	2.39	0.28	5.85					23.94

Table D4. Computation of production for *Mormyrops deliciosus* stock from Lake Kariba (after data from locality E, F, G – coves of Chete Island and Chipepo Bay, 1969).

Growth seasons		0	1	2	3	4	5	6	7	8	Total
Interpolated N'/ha at time of annulus formation	N'	(11)	8	5	4	3	2	1	1	1	25
Interpolated mean weight in g of 1 specimen ('x')	$\bar{w}$		63	206	447	741	1225	2070	2835	3450	
Initial biomass in kg/ha	B'		0.50	1.03	1.79	2.22	2.45	2.07	2.83	3.45	16.34
Rate of increase in biomass	H	$+3.8247$	$+0.7148$	$+0.5516$	$+0.2176$	$+0.3266$	-0.1678	$+0.3144$	$+0.1961$		
Instantaneous growth rate	G	4.1431	1.1848	0.7747	0.5054	0.5027	0.5246	0.3146	0.1963		1.0183
Instantaneous rate of total mortality	Z	0.3184	0.4700	0.2231	0.2878	0.1761	0.6924	0.0002	0.0002		0.2710
Mean biomass	$\bar{B}$	5.83	1.49	2.38	2.51	2.93	1.93	3.27	3.89		24.23
Total production ($= G\bar{B}$) in kg/ha/yr	A	24.24	1.76	1.84	1.27	1.47	1.01	1.03	0.76		33.38
Total yield in kg/ha/yr	Y_A				1.27	1.47	1.01	1.03	0.76		5.54
Available production in kg/ha/yr	P'	0.50	0.71	0.96	0.88	0.97	0.84	0.61			6.23
Available yield in kg/ha/yr	$Y_{P'}$				0.88	0.97	0.84	0.61			4.06

Table D5. Computation of production for *Sargochromis codringtoni* stock from Lake Kariba (after data from locality C – Chikanka Island cove, 1968).

Growth seasons		0	1	2	3	4	Total
Interpolated N'/ha at time of annulus formation	N'	(74)	34	14	6	3	57
Interpolated mean weight in g of 1 specimen	$\bar{w}$		56	128	250	355	
Initial biomass in kg/ha	B'		1.90	1.79	1.50	1.06	6.25
Rate of increase in biomass	H	+3.2478	−0.0607	−0.1777	−0.3427		
Instantaneous growth rate	G	4.0254	0.8267	0.6694	0.3506		1.4680
Instantaneous rate of total mortality	Z	0.7776	0.8874	0.8471	0.6933		0.8013
Mean biomass	$\bar{B}$	14.50	1.72	1.39	0.89		18.50
Total production ($= G\bar{B}$) in kg/ha/yr	A	58.37	1.42	0.93	0.31		61.03
Total yield in kg/ha/yr	Y_A			0.93	0.31		1.24
Available production in kg/ha/yr	P'	1.90	1.01	0.73	0.31		3.95
Available yield in kg/ha/yr	$Y_{P'}$			0.73	0.31		1.04

Table D6. Computation of production for *Sargochromis codringtoni* stock from Lake Kariba (after data from locality E – Chete Island cove, 1969).

Growth seasons		0	1	2	3	4	Total
Interpolated N'/ha at time of annulus formation	N'	(140)	46	18	7	3	74
Interpolated mean weight in g of 1 specimen	$\bar{w}$		26	98	213	355	
Initial biomass in kg/ha	B'		1.20	1.76	1.49	1.06	5.51
Rate of increase in biomass	H	+2.1453	+0.3886	−0.1682	−0.3366		
Instantaneous growth rate	G	3.2581	1.3269	0.7763	0.5108		1.4680
Instantaneous rate of total mortality	Z	1.1128	0.9383	0.9445	0.8474		0.9607
Mean biomass	$\bar{B}$	4.19	2.16	1.38	0.91		8.64
Total production ($= G\bar{B}$) in kg/ha/yr	A	13.65	2.87	1.07	0.46		18.05
Total yield in kg/ha/yr	Y_A			1.07	0.46		1.53
Available production in kg/ha/yr	P'	1.20	1.30	0.80	0.43		3.73
Available yield in kg/ha/yr	$Y_{P'}$			0.80	0.43		1.23

Table D7. Computation of production for *Sargochromis codringtoni* stock from Lake Kariba (after data from locality F – cove in Chipepo Bay, 1969).

Growth seasons		0	1	2	3	4	Total
Interpolated N'/ha at time of annulus formation	N'	(142)	88	59	42	27	216
Interpolated mean weight in g of 1 specimen	$\bar{w}$		43	51	235	355	
Initial biomass in kg/ha	B'		3.78	3.01	9.87	9.58	26.24
Rate of increase in biomass	H	+3.2827	−0.2291	+1.1877	−0.0291		
Instantaneous growth rate	G	3.7612	0.1706	1.5278	0.4125		1.4680
Instantaneous rate of total mortality	Z	0.4785	0.3997	0.3401	0.4416		0.4150
Mean biomass	$\bar{B}$	29.45	2.70	19.01	9.75		60.91
Total production ($= G\bar{B}$) in kg/ha/yr	A	110.77	0.46	29.04	4.02		144.29
Total yield in kg/ha/yr	Y_A			29.04	4.02		33.06
Available production in kg/ha/yr	P'	3.78	0.47	7.73	3.24		15.22
Available yield in kg/yr	$Y_{P'}$			7.73	3.24		10.97

Table D8. Computation of production for *Heterobranchus longifilis* stock from Lake Kariba (after data from locality C, D, F – Chikanka, Buffalo and Chipepo coves, 1968, 1969).

Growth seasons		0	1	2	3	4	5	6	7	8	Total
Interpolated N'/ha at time of annulus formation	N'	(7)	5	4	3	2	1	1	1	1	18
Interpolated mean weight in g of 1 specimen	$\bar{w}$		7.9	22	82	220	490	990	1870	2600	
Initial biomass in kg/ha	B'		0.04	0.09	0.25	0.44	0.49	0.99	1.87	2.60	6.77
Rate of increase in biomass	H	+1.7305	+0.8011	+1.0278	+0.5815	+0.1076	+0.7033	+0.6360	+0.3296		
Instantaneous growth rate	G	2.0669	1.0242	1.3156	0.9870	0.8007	0.7033	0.6360	0.3296		0.9829
Instantaneous rate of total mortality	Z	0.3364	0.2231	0.2878	0.4055	0.6931	0	0	0		0.3892
Mean biomass	$\bar{B}$	0.11	0.14	0.14	0.44	0.59	0.53	1.43	2.64	3.08	8.96
Total production ($= G\bar{B}$) in kg/ha/yr	A	0.23	0.14	0.18	0.58	0.42	1.01	1.01	1.68	1.01	5.65
Total yield in kg/ha/yr	Y_A									1.01	1.01
Available production in kg/ha/yr	P'	0.04	0.06	0.18	0.28	0.27	1.48		0.88	0.73	3.92
Available yield in kg/yr	$Y_{P'}$									0.73	0.73

Table D9. Computation of production for *Clarias gariepinus* stock from Lake Kariba (after data from locality C – Chikanka Island cove, 1968).

Growth seasons		0	1	2	3	4	5	6	7	8	9	10	11	Total Σ or m
Interpolated N'/ha at time of annulus formation	N'	(28)	22	17	14	11	9	7	5	4	3	3	2	97
Interpolated mean weight in g of 1 specimen	$\bar{w}$		6.8	26	58	320	500	750	900	1630	1920	2700	3010	
Initial biomass in kg/ha	B'	0.15	0.44	0.81	3.52	4.50	5.25	4.50	6.52	5.76	8.10	6.02	45.57	
Rate of increase in biomass	H	+1.6756	+1.0833	+0.6083	+1.4668	+0.2454	+0.1543	−0.1541	−0.3709	−0.1241	+0.3410	−0.2969		
Instantaneous growth rate	G	1.9169	1.3412	0.8024	1.7079	0.4462	0.4055	0.1823	0.5940	0.1637	0.3410	0.1086		0.7281
Instantaneous rate of total mortality	Z	0.2413	0.2579	0.1941	0.2411	0.2008	0.2512	0.3364	0.2231	0.2878	0	0.4055		0.2399
Mean biomass	$\bar{B}$	0.39	0.79	1.12	8.04	4.97	5.50	4.07	7.87	5.25	9.62	5.26		59.81
Total production (= G$\bar{B}$) in kg/ha/yr	A	0.75	1.06	0.90	13.73	2.22	2.23	0.74	4.67	0.86	3.28	0.57		31.01
Total yield in kg/ha/yr	Y_A									4.67	0.86	3.28	0.57	9.38
Available production in kg/ha/yr	P'	0.15	0.33	0.45	2.88	1.62	1.75	0.75	2.92	0.90	2.34	0.62		14.71
Available yield in kg/ha/yr	$Y_{P'}$									2.92	0.90	2.34	0.62	6.78

Table D10. Computation of production for *Sarotherodon mossambicus mortimeri* stock from Lake Kariba (after data from locality C – Chikanka Island cove, 1968).

Growth seasons		0	1	2	3	4	5	6	7	8	Total
Interpolated N'/ha at time of annulus formation	N'	(725)	340	160	76	36	17	8	4	2	643
Interpolated mean weight in g of 1 specimen	$\bar{w}$		6	25	87	217	1680	2100	2250	2300	
Initial biomass in kg/ha	B'		2.04	4.00	6.61	7.81	28.56	16.80	9.00	4.60	79.42
Rate of increase in biomass	H		+1.0347	+0.6732	+0.5026	+0.1668	+1.2963	−0.5305	−0.6241	−0.6713	
Instantaneous growth rate	G		1.7918	1.4271	1.2470	0.9140	2.0467	0.2231	0.0690	0.0220	0.9676
Instantaneous rate of total mortality	Z		0.7571	0.7539	0.7444	0.7472	0.7504	0.7536	0.6931	0.6933	0.7366
Mean biomass	$\bar{B}$		3.55	5.67	8.53	8.68	58.81	13.03	6.66	3.35	108.28
Total production (= G$\bar{B}$) in kg/ha/yr	A		6.36	8.09	10.64	7.93	120.37	2.91	0.46	0.07	156.83
Total yield in kg/ha/yr	Y_A					7.93	120.37	2.91	0.46	0.07	131.74
Available production in kg/ha/yr	P'		2.04	3.04	4.71	4.68	24.87	3.36	0.60	0.10	43.40
Available yield in kg/ha/yr	$Y_{P'}$					4.68	24.87	3.36	0.60	0.10	33.61

Table D11. Computation of production for *Sarotherodon mossambicus mortimeri* stock from Lake Kariba (after data from locality D – Buffalo Island cove, 1968).

Growth seasons		0	1	2	3	4	5	Total Σ or m
Interpolated N'/ha at time of annulus formation	N'	(1200)	278	78	22	6	2	386
Interpolated mean weight in g of 1 specimen	$\bar{w}$		9	50	95	170	1600	
Initial biomass in kg/ha	B'		2.50	3.90	2.09	1.02	3.20	12.71
Rate of increase in biomass	H		+0.7346	+0.4440	−0.6238	−0.7172	+1.1432	
Instantaneous growth rate	G		2.1972	1.7148	0.6419	0.5819	2.2420	1.4756
Instantaneous rate of total mortality	Z		1.4626	1.2708	1.2657	1.2991	1.0988	1.2794
Mean biomass	$\bar{B}$		3.66	4.85	1.55	0.73	5.95	16.74
Total production (= G$\bar{B}$) in kg/ha/yr	A		8.04	8.32	0.99	0.42	13.34	31.11
Total yield in kg/ha/yr	Y_A				0.99	0.42	13.34	14.75
Available production in kg/ha/yr	P'		2.50	3.20	0.99	0.45	2.86	10.00
Available yield in kg/ha/yr	$Y_{P'}$				0.99	0.45	2.86	4.30

Table D12. Computation of production for *Sarotherodon mossambicus mortimeri* stock from Lake Kariba (after data from locality E$_2$ – Chete intermittent stream cove, 1969).

Growth seasons		0	1	2	3	4	5	6	7	8	Total
Interpolated N'/ha at time of annulus formation	N'	(6200)	2860	1310	628	293	138	64	30	14	5337
Interpolated mean weight in g of 1 specimen	$\bar{w}$		6	32	65	168	1600	2150	2250	2300	
Initial biomass in kg/ha	B'		17.16	41.92	40.82	49.22	220.80	137.60	67.50	32.20	607.22
Rate of increase in biomass	H	+1.0181	+0.8932	−0.0266	+0.1872	+1.5008	−0.4729	−0.7124	−0.7402		0.9676
Instantaneous growth rate	G	1.7918	1.6740	0.7086	0.9496	2.2538	0.2955	0.0454	0.0220		
Instantaneous rate of total mortality	Z	0.7737	0.7808	0.7352	0.7624	0.7530	0.7684	0.7578	0.7622		0.7617
Mean biomass	$\bar{B}$	29.89	67.35	45.42	55.00	518.85	109.11	48.17	22.75		896.54
Total production (= G$\bar{B}$) in kg/ha/yr	A	53.56	112.74	32.18	52.23	1169.38	32.24	2.19	0.50		1455.02
Total yield in kg/ha/yr	Y$_A$				52.23	1169.38	32.24	2.19	0.50		1256.54
Available production in kg/ha/yr	P'	17.16	34.06	20.72	30.18	197.62	35.20	3.00	0.70		338.64
Available yield in kg/ha/yr	Y$_{P'}$				30.18	197.62	35.20	3.00	0.70		266.70

Table D13. Computation of production for *Labeo altivelis* stock from Lake Kariba (after data from locality D – Buffalo Island cove, 1968. All material).

Growth seasons		0	1	2	3	4	5	6	7	8	9	Total
Interpolated N'/ha at time of annulus formation	N'	78	45	24	14	8	4	3	2	1	0.3	102.3
Interpolated mean weight in g of 1 specimen	$\bar{w}$		13	58	185	386	600	678	1260	1420	1630	
Initial biomass in kg/ha	B'		0.58	1.39	2.59	3.09	2.40	2.03	2.52	1.42	0.49	16.51
Rate of increase in biomass	H	+2.0149	+0.8669	+0.6209	+0.1760	−0.2520	−0.1656	+0.2142	−0.5735	+0.1379		0.8218
Instantaneous growth rate	G	2.5650	1.4955	1.1599	0.7355	0.4411	0.1222	0.6197	0.1196	0.1379		
Instantaneous rate of total mortality	Z	0.5501	0.6286	0.5390	0.5595	0.6981	0.2878	0.4055	0.6931			0.5790
Mean biomass	$\bar{B}$	1.86	2.22	3.58	3.46	2.11	1.92	2.75	1.08	0.53		19.51
Total production (= G$\bar{B}$) in kg/ha/yr	A	4.77	3.32	4.15	2.54	0.93	0.23	1.70	0.13	0.07		17.84
Total yield in kg/ha/yr	Y$_A$					0.93	0.23	1.70	0.13	0.07		3.06
Available production in kg/ha/yr	P'	0.58	1.08	1.78	1.61	0.86	0.23	1.16	0.16	0.06		7.52
Available yield in kg/ha/yr	Y$_{P'}$					0.86	0.23	1.16	0.16	0.06		2.47

Table D14. Computation of production for *Tilapia rendalli* stock from Lake Kariba (after data from locality C – Chikanka Island cove, 1968).

Growth seasons		0	1	2	3	4	5	Total
Interpolated N'/ha at time of annulus formation	N'	(1120)	465	192	82	33	14	786
Interpolated mean weight in g of 1 specimen	$\bar{w}$		13	80	200	440	510	
Initial biomass in kg/ha	B'		6.04	15.36	16.40	14.52	7.14	59.46
Rate of increase in biomass	H	+1.6861	+0.9323	+0.0655	−0.1217	−0.7099		1.2469
Instantaneous growth rate	G	2.5650	1.8170	0.9163	0.7885	0.1476		
Instantaneous rate of total mortality	Z	0.8789	0.8847	0.8508	0.9102	0.8575		0.8764
Mean biomass	$\bar{B}$	15.83	25.28	15.47	13.49	5.11		75.18
Total production (= G$\bar{B}$) in kg/ha/yr	A	40.60	45.93	14.17	10.64	0.75		112.09
Total yield in kg/ha/yr	Y_A			14.17	10.64	0.75		25.56
Available production in kg/ha/yr	P'	6.04	12.86	9.84	7.92	0.98		37.64
Available yield in kg/ha/yr	$Y_{P'}$			9.84	7.92	0.98		18.74

Table D15. Computation of production for *Tilapia rendalli* stock from Lake Kariba (after data from locality F – cove in Chipepo Bay, 1969).

Growth seasons		0	1	2	3	4	Total
Interpolated N'/ha at time of annulus formation	N'	(3070)	880	252	75	22	1229
Interpolated mean weight in g of 1 specimen	$\bar{w}$		12	49	100	240	
Initial biomass in kg/ha	B'		10.65	12.35	7.50	5.28	35.69
Rate of increase in biomass	H	+1.2355	+0.1564	−0.4985	−0.3511		1.3702
Instantaneous growth rate	G	2.4849	1.4069	0.7134	0.8755		
Instantaneous rate of total mortality	Z	1.2494	1.2505	1.2119	1.2266		1.2346
Mean biomass	$\bar{B}$	20.69	13.70	5.92	4.44		44.75
Total production (= G$\bar{B}$) in kg/ha/yr	A	51.41	19.27	4.22	3.89		78.79
Total yield in kg/ha/yr	Y_A			4.22	3.89		8.11
Available production in kg/ha/yr	P'	10.56	9.32	3.82	3.08		26.78
Available yield in kg/ha/yr	$Y_{P'}$			2.00	3.08		5.08

Table D16. Computation of production for *Marcusenius macrolepidotus* stock from Lake Kariba (after data from locality E – Chete Island cove, 1969).

Growth seasons		0	1	2	3	4	5	6	Total
Interpolated N'/ha at time of annulus formation	N'	(718)	190	51	14	4	1	1	261
Interpolated mean weight in g of 1 specimen	w̄		13	35	66	98	178	222	
Initial biomass in kg/ha	B'		2.47	1.78	0.92	0.39	0.18	0.22	5.96
Rate of increase in biomass	H	+1.2357	−0.3248	−0.6586	−0.8571	−0.7898	+0.2209		0.9004
Instantaneous growth rate	G	2.5650	0.9904	0.6343	0.3955	0.5966	0.2209		1.3153
Instantaneous rate of total mortality	Z	1.3293	1.3152	1.2929	1.2526	1.3864	0.0000		7.70
Mean biomass	B̄	4.91	1.50	0.67	0.26	0.12	0.24		14.72
Total production (= GB̄) in kg/ha/yr	A	12.59	1.49	0.42	0.10	0.07	0.05		14.72
Total yield in kg/ha/yr	Y$_A$			0.42	0.10	0.07	0.05		0.64
Available production in kg/ha/yr	P'	2.47	1.12	0.43	0.13	0.08	0.04		4.27
Available yield in kg/ha/yr	Y$_{P'}$			0.43	0.13	0.08	0.04		0.68

Table D17. Computation of production for *Malapterurus electricus* stock from Lake Kariba (after data from locality C, E, F, G – 1968, 1969; Z = 0.29, t = 190).

Growth seasons		0	1	2	3	4	5	6	7	8	9	10	Total
Interpolated N'/ha at time of annulus formation	N'	(13178)	4612	1582	527	179	61	21	7	3	1	1	6994
Interpolated mean weight in g of 1 specimen	w̄		3.9	21	62	150	310	530	820	1100	1500	2800	
Initial biomass in kg/ha	B'		17.99	33.22	32.67	26.85	18.91	11.13	5.74	3.30	1.50	2.80	154.11
Rate of increase in biomass	H	+0.3110	+0.6137	−0.0169	−0.1961	−0.3508	−0.5300	−0.6622	−0.5536	−0.7884	+0.6241		0.7937
Instantaneous growth rate	G	1.3610	1.6835	1.0826	0.8836	0.7259	0.5363	0.4364	0.2938	0.3102	0.6241		
Instantaneous rate of total mortality	Z	1.0500	1.0698	1.0995	1.0797	1.0767	1.0663	1.0986	0.8474	1.0986	0.0000		1.0541
Mean biomass	B̄	21.02	45.51	38.28	24.82	15.92	8.64	4.19	2.52	1.04	3.85		165.79
Total production (= GB̄) in kg/ha/yr	A	28.61	76.62	41.44	21.93	11.56	4.63	1.83	0.74	0.32	2.40		190.08
Total yield in kg/ha/yr	Y$_A$				21.93	11.56	4.63	1.83	0.74	0.32	2.40		9.92
Available production in kg/ha/yr	P'	17.99	27.05	21.61	15.75	9.76	4.62	2.03	0.84	0.40	1.30		101.35
Available yield in kg/ha/yr	Y$_{P'}$				15.75	9.76	4.62	2.03	0.84	0.40	1.30		9.19

Table D18. Computation of production for *Synodontis zambezensis* stock from Lake Kariba (after data from locality C – Chikanka Island cove, 1968).

Growth seasons		0	1	2	3	4	5	6	7	8	Total
Interpolated N'/ha at time of annulus formation	N'	(145)	78	43	28	12	7	3	2	1	174
Interpolated mean weight in g of 1 specimen (same growth seasons)	$\bar{w}_x$		11	20	39	100	115	174	230	440	
Initial biomass in kg/ha	B'		0.86	0.86	1.09	1.20	0.80	0.52	0.46	0.44	6.23
Rate of increase in biomass	H	+1.7778	+0.0024	+0.2389	+0.0943	−0.3992	−0.4333	−0.1265	−0.0444		
Instantaneous 'X' growth rate (for the same growth season)	G_x	2.3979	0.5978	0.6679	0.9416	0.1398	0.4141	0.2790	0.6487		0.7608
Instantaneous rate of total mortality	Z	0.6201	0.5954	0.4290	0.8473	0.5390	0.8474	0.4055	0.6931		0.6221
Mean biomass	$\bar{B}$	2.38	0.36	1.24	1.20	0.66	0.42	0.44	0.39		7.09
Total production (= $G\bar{B}$) in kg/ha/yr	A	5.71	0.21	0.83	1.13	0.09	0.17	0.12	0.25		8.51
Total yield in kg/ha/yr	Y_A				1.13	0.09	0.17	0.12	0.25		1.76
Available production in kg/ha/yr	P'	0.86	0.39	0.53	0.73	0.10	0.18	0.11	0.21		3.11
Available yield in kg/ha/yr	$Y_{P'}$				0.73	0.10	0.18	0.11	0.21		1.33

Table D19. Computation of production for *Synodontis zambezensis* stock from Lake Kariba (after data from locality G – Chipepo Bay second cove, 1969).

Growth seasons		0	1	2	3	4	5	6	7	Total
Interpolated N'/ha at time of annulus formation	N'	(663)	274	115	48	19	8	3	1	468
Interpolted mean weight in g of 1 specimen (same growth season)	$\bar{w}_x$		8	22	52	100	160	130	230	
Initial biomass in kg/ha	B'		2.19	2.53	2.50	1.90	1.28	0.39	0.23	11.02
Rate of increase in biomass	H	+1.1959	+0.1434	−0.0136	−0.2727	−0.3951	−1.1885	−0.5281		
Instantaneous 'X' growth rate (for the same growth season)	G_x	2.0794	1.0117	0.8602	0.6539	0.4700	0.2076	0.5705		0.7769
Instantaneous rate of total mortality	Z	0.8835	0.8683	0.8738	0.9266	0.8651	0.9809	1.0986		0.9281
Mean biomass	$\bar{B}$	4.25	2.65	1.84	1.65	1.05	0.23	0.18		11.85
Total production (= $G\bar{B}$) in kg/ha/yr	A	8.84	2.68	1.58	1.08	0.49	0.05	0.10		14.82
Total yield in kg/ha/yr	Y_A				1.08	0.49	0.05	0.10		1.72
Available production in kg/ha/yr	P'	2.19	1.61	1.44	0.91	0.48	0.09	0.10		6.82
Available yield in kg/ha/yr	$Y_{P'}$				0.91	0.48	0.09	0.10		1.58

Table D20. Computation of production for *Hippopotamyrus discorhynchus* stock from Lake Kariba (after data from locality D – Buffalo Island cove, 1968).

Growth seasons		0	1	2	3	4	5	6	Total
Interpolated N′/ha at time of annulus formation	N′	(100,000)	15,300	2320	372	58	9	1	18060
Interpolated mean weight in g of 1 specimen	$\bar{w}$		6	22	49	80	132	170	
Initial biomass in kg/ha	B′		91.80	51.04	18.23	4.64	1.19	0.17	167.07
Rate of increase in biomass	H	−0.0855	−0.5870	−1.7505	−1.3682	−1.3625	−1.9441		0.8560
Instantaneous growth rate	G	1.7918	1.2993	0.8007	0.4902	0.5008	0.2530		
Instantaneous rate of total mortality	Z	1.8773	1.8863	1.8306	1.8584	1.8633	2.1971		1.9188
Mean biomass	$\bar{B}$	62.49	38.75	8.60	2.53	0.65	0.07		113.09
Total production (= $G\bar{B}$) in kg/ha/yr	A	111.97	50.35	6.89	1.24	0.32	0.02		170.79
Total yield in kg/ha/yr	Y_A			6.89	1.24	0.32	0.02		8.47
Available production in kg/ha/yr	P′	91.80	37.12	10.04	1.80	0.47	0.04		141.27
Available yield in kg/ha/yr	$Y_{P'}$			10.04	1.80	0.47	0.04		12.35

Table D21. Computation of production for *Eutropius depressirostris* stock from Lake Kariba (after data from locality C – Chikanka Island cove, 1968).

Growth seasons		0	1	2	3	4	5	6	Total
Interpolated N′/ha at time of annulus formation	N′	(900)	360	145	58	23	9	4	599
Interpolated mean weight in g of 1 specimen	$\bar{w}$		3.4	13	25	40	58	95	
Initial biomass in kg/ha	B′		1.22	1.88	1.45	0.92	0.52	0.38	6.37
Rate of increase in biomass	H	+0.3076	+0.4319	−0.2625	−0.4550	−0.5667	−0.3073		0.7606
Instantaneous growth rate	G	1.2238	1.3412	0.6539	0.4700	0.3716	0.5034		0.9026
Instantaneous rate of total mortality	Z	0.9162	0.9093	0.9164	0.9250	0.9383	0.8107		
Mean biomass	$\bar{B}$	1.44	2.34	1.26	0.73	0.40	0.33		6.50
Total production (= $G\bar{B}$) in kg/ha/yr	A	1.76	3.14	0.82	0.34	0.15	0.17		6.38
Total yield in kg/ha/yr	Y_A				0.34	0.15	0.17		0.66
Available production in kg/ha/yr	P′	1.22	1.39	0.70	0.34	0.16	0.15		3.96
Available yield in kg/ha/yr	$Y_{P'}$				0.34	0.16	0.15		0.65

Table D22. Computation of production for *Eutropius depressirostris* stock from Lake Kariba (after data from locality D – Buffalo Island cove, 1968).

Growth seasons		0	1	2	3	4	5	6	Total
Interpolated N'/ha at time of annulus formation	N'	(260)	122	60	28	13	6	3	232
Interpolated mean weight in g of 1 specimen	$\bar{w}$		3.8	11	22	26	58	95	
Initial biomass in kg/ha	B'		0.46	0.66	0.62	0.47	0.35	0.28	2.84
Rate of increase in biomass	H	+0.5784	+0.3523	−0.0690	−0.2751	−0.2960	−0.1999		0.7590
Instantaneous growth rate	G	1.3350	1.0629	0.6932	0.4924	0.4770	0.4934		0.7439
Instantaneous rate of total mortality	Z	0.7566	0.7106	0.7622	0.7675	0.7730	0.6933		
Mean biomass	$\bar{B}$	0.62	0.78	0.61	0.40	0.31	0.25		2.97
Total production (= G$\bar{B}$) in kg/ha/yr	A	0.83	0.83	0.42	0.20	0.15	0.12		2.55
Total yield in kg/ha/yr	Y_A				0.20	0.15	0.12		0.47
Available production in kg/ha/yr	P'	0.46	0.43	0.31	0.18	0.13	0.11		1.62
Available yield in kg/ha/yr	$Y_{P'}$				0.18	0.13	0.11		0.42

Table D23. Computation of production for *Eutropius depressirostris* stock from Lake Kariba (after data from locality F – cove in Chipepo Bay, 1969).

Growth seasons		0	1	2	3	4	5	6	Total
Interpolated N'/ha at time of annulus formation	N'	(850)	370	164	73	32	14	6	659
Interpolated mean weight in g of 1 specimen	$\bar{w}$		4	12	20	34	58	80	
Initial biomass in kg/ha	B'		1.48	1.97	1.46	1.09	0.8	0.5	7.30
Rate of increase in biomass	H	+0.5546	+0.2849	−0.2986	−0.2941	−0.2925	−0.5256		0.7303
Instantaneous growth rate	G	1.3863	1.0986	0.5108	0.5307	0.5341	0.3215		0.8255
Instantaneous rate of total mortality	Z	0.8317	0.8137	0.8094	0.8248	0.8266	0.8471		
Mean biomass	$\bar{B}$	1.96	2.23	1.27	0.93	0.69	0.39		7.47
Total production (= G$\bar{B}$) in kg/ha/yr	A	2.72	2.45	0.65	0.49	0.37	0.12		6.80
Total yield in kg/ha/yr	Y_A				0.49	0.37	0.12		0.98
Available production in kg/ha/yr	P'	1.48	1.31	0.58	0.45	0.34	0.13		4.29
Available yield in kg/ha/yr	$Y_{P'}$				0.45	0.34	0.13		0.92

Table D24. Computation of production for Schilbe mystus stock from Lake Kariba (after data from locality D – Buffalo Island cove, 1968; $\bar{Z} = 0.16$, $t = 40$).

Growth seasons		0	1	2	3	4	5	6	7	8	Total
Interpolated N'/ha at time of annulus formation	N'	(120)	66	37	19	11	5	3	2	1	144
Interpolated mean weight in g of 1 specimen	$\bar{w}$		2.1	6.5	15	26	37	53	73	98	
Initial biomass in kg/ha	B'		0.14	0.24	0.28	0.29	0.18	0.16	0.15	0.10	1.54
Rate of increase in biomass	H	+0.1439	+0.5513	+0.1699	+0.0034	−0.4356	−0.1515	−0.0853	−0.3986		
Instantaneous growth rate	G	0.7419	1.1299	0.8363	0.5500	0.3528	0.3594	0.3202	0.2945		0.5731
Instantaneous rate of total mortality	Z	0.5980	0.5786	0.6664	0.5466	0.7884	0.5109	0.4055	0.6931		0.5984
Mean biomass	$\bar{B}$	0.15	0.32	0.30	0.00	0.15	0.15	0.13	0.08		1.28
Total production (= G$\bar{B}$) in kg/ha/yr	A	0.11	0.36	0.25	0.00	0.05	0.05	0.04	0.02		0.88
Total yield in kg/ha/yr	Y_A					0.05	0.05	0.04	0.02		0.16
Available production in kg/ha/yr	P'	0.14	0.16	0.16	0.12	0.05	0.05	0.04	0.02		0.74
Available yield in kg/ha/yr	$Y_{P'}$					0.05	0.05	0.04	0.02		0.16

Table D25. Computation of production for Brachyalestes imberi imberi stock from Lake Kariba (after data from locality D – Buffalo Island cove, 1968).

Growth seasons		0	1	2	3	4	Total
Interpolated N'/ha at time of annulus formation[1]	N'	(20)	10	5	3	1	19
Interpolated mean weight in g of 1 specimen	$\bar{w}$		3.5	10	18	30	
Initial biomass in kg/ha	B'		0.035	0.005	0.054	0.030	0.124
Rate of increase in biomass	H	+0.5597	+0.3567	+0.0769	−0.5878		
Instantaneous growth rate	G	1.2528	1.0498	0.5878	0.5108		0.8503
Instantaneous rate of total mortality	Z	0.6931	0.6931	0.5109	1.0986		0.7489
Mean biomass	$\bar{B}$	0.047	0.006	0.058	0.023		0.134
Total production (= G$\bar{B}$) in kg/ha/yr	A	0.059	0.006	0.034	0.012		0.111
Available production in kg/ha/yr	P'	0.035	0.032	0.024	0.012		0.103

[1] $\bar{Z} = 0.30$, $t = 40$; N_0' 3, N_1' 11, N_2' 5, N_3' 3, N_4' 1.

Table D26. Computation of production for *Brachyalestes imberi imberi* stock from Lake Kariba (after data from locality F – cove in Chipepo Bay, 1969).

Growth seasons		0	1	2	3	Total
Interpolated N'/ha at time of annulus formation[1]	N'	(1000)	85	7	1	93
Interpolated mean weight in g of 1 specimen	$\bar{w}$		3.5	10	18	
Initial biomass in kg/ha	B'		0.30	0.07	0.02	0.39
Rate of increase in biomass	H	−1.2124	−1.4474	−1.3581		
Instantaneous growth rate	G	1.2528	1.0498	0.5878		0.9635
Instantaneous rate of total mortality	Z	2.4652	2.4972	1.9459		2.3028
Mean biomass	$\bar{B}$	0.174	0.037	0.011		0.222
Total production (= G$\bar{B}$) in kg/ha/yr	A	0.218	0.039	0.006		0.263
Available production in kg/ha/yr	P'	0.297	0.045	0.008		0.350

[1] $\bar{Z} = 0.16$, $t = 200$; N_0' 155, N_1' 85, N_2' 7.

Table D27. Computation of production for *Alestes lateralis* stock from Lake Kariba (after data from locality C – Chikanka Island, 1968).

Growth seasons		0	1	2	Total
Interpolated N'/ha at time of annulus formation[1]	N'	(5000)	402	31	433
Interpolated mean weight in g of 1 specimen	$\bar{w}$		4.5	35	
Initial biomass in kg/ha	B'		1.81	1.08	2.89
Rate of increase in biomass	H	−1.0168	−0.5110		
Instantaneous growth rate	G	1.5041	2.0513		1.7777
Instantaneous rate of total mortality	Z	2.5209	2.5623		2.5416
Mean biomass	$\bar{B}$	1.14	0.84		1.98
Total production (= G$\bar{B}$) in kg/ha/yr	A	1.71	1.72		3.43
Available production in kg/ha/yr	P'	1.81	0.94		2.75

[1] $\bar{Z} = 0.51$, $t = 200$ gives N_0' (2406), N_1' 402, N_2' 31.

Table D28. Computation of production for *Alestes lateralis* stock from Lake Kariba (after data from locality D – Buffalo Island, 1968).

Growth seasons		0	1	2	3	4	Total
Interpolated N'/ha at time of annulus formation[1]	N'	(26000)	1882	137	10	1	2030
Interpolated mean weight in g of 1 specimen	w̄		4.5	35	52	60	
Initial biomass in kg/ha	B'		8.47	4.79	0.52	0.06	13.84
Rate of increase in biomass	H	−1.1216	−0.5691	−2.2215	−2.1595		
Instantaneous growth rate	G	1.5041	2.0513	0.3959	0.1431		1.0236
Instantaneous rate of total mortality	Z	2.6257	2.6204	2.6174	2.3026		2.5415
Mean biomass	B̄	5.09	3.66	0.21	0.02		8.98
Total production (= GB̄) in kg/ha/yr	A	7.66	7.51	0.08	0.003		15.25
Available production in kg/ha/yr	P'	8.47	4.18	0.17	0.008		12.83

[1] Z̄ = 0.53, t = 40 gives N₀' 30694, N₁' 1882, N₂' 137.

Table D29. Computation of production for *Alestes lateralis* stock from Lake Kariba (after data from locality E – Chete Island cove, 1969).

Growth seasons		0	1	2	3	Total
Interpolated N'/ha at time of annulus formation[1]	N'	(780,000)	6688	53	1	6742
Interpolated mean weight in g of 1 specimen	w̄		4.5	35	52	
Initial biomass in kg/ha	B'		30.10	1.85	0.05	32.00
Rate of increase in biomass	H	−3.2549	−2.7865	−3.5745		
Instantaneous growth rate	G	1.5041	2.0513	0.3959		1.3171
Instantaneous rate of total mortality	Z	4.7590	4.8378	3.9704		4.5224
Mean biomass	B̄	8.89	0.62	0.02		9.53
Total production (= GB̄) in kg/ha/yr	A	13.37	1.27	0.008		14.65
Available production in kg/ha/yr	P'	30.10	1.62	0.02		31.74

[1] Z̄ = 0.71, t = 140; N₀' 7028, N₁' 6688, N₂' 53, N₃' 1.

Table D30. Computation of production for *Alestes lateralis* stock from Lake Kariba (after data from locality F – cove in Chipepo Bay, 1969).

Growth season		0	1	2	3	4	Total
Interpolated N'/ha at time of annulus formation[1]	N'	(5400000)	106408	2160	41	1	108610
Interpolated mean weight in g of 1 specimen	$\bar{w}$		4.5	35	52	60	
Initial biomass in kg/ha	B'		478.84	75.60	2.13	0.06	556.63
Rate of increase in biomass	H	−2.4228	−1.8459	−3.5685	−3.5705		1.0236
Instantaneous growth rate	G	1.5041	2.0513	0.3959	0.1431		
Instantaneous rate of total mortality	Z	3.9269	3.8912	3.9644	3.7136		3.8755
Mean biomass	$\bar{B}$	180.07	34.52	0.58	0.02		215.19
Total production (= G$\bar{B}$) in kg/ha/yr	A	270.84	70.81	0.23	0.003		341.88
Available production in kg/ha/yr	P'	478.84	65.88	0.70	0.008		545.43

[1] $\bar{Z} = 0.74$, t = 200; N_0' 131,713, N_1' 106,408, N_2' 686, N_3' 41.

Table D31. Computation of production for *Haplochromis darlingi* stock from Lake Kariba (after data from locality A – cove near Siavonga, 1968).

Growth seasons		0	1	2	3	Total
Interpolated N'/ha at time of annulus formation	N'	(609)	107	19	4	130
Interpolated mean weight in g of 1 specimen	$\bar{w}$		4.3	7.2	8.5	
Initial biomass in kg/ha	B'		0.4601	0.1368	0.034	0.6309
Rate of increase in biomass	H	−1.0340	−3.4643	−3.4219		
Instantaneous growth rate	G	0.6701	0.5155	0.1660		0.4505
Instantaneous rate of total mortality	Z	1.7014	3.9798	3.5879		3.0897
Mean biomass	$\bar{B}$	0.2861	0.0382	0.0096		0.3339
Total production (= G$\bar{B}$) in kg/ha/yr	A	0.1917	0.0197	0.0016		0.2130
Available production in kg/ha/yr	P'	0.2247	0.0551	0.0052		0.2850
Available yield in kg/ha/yr	$Y_{P'}$		0.0551	0.0052		0.0603

Table D32. Computation of production for *Haplochromis darlingi* stock from Lake Kariba (after data from locality C – Chikanka Island cove, 1968).

Growth seasons		0	1	2	Total
Interpolated N'/ha at time of annulus formation	N'	(306)	37	4	41
Interpolated mean weight in g of 1 specimen	$\bar{w}$		2.3	3.8	
Initial biomass in kg/ha	B'		0.0851	0.0152	0.1003
Rate of increase in biomass	H	−3.8236	−4.6203		0.7716
Instantaneous growth rate	G	1.0411	0.5021		0.7716
Instantaneous rate of total mortality	Z	4.8647	5.1224		4.9935
Mean biomass	$\bar{B}$	0.0218	0.4487		0.4705
Total production (= G$\bar{B}$) in kg/ha/yr	A	0.0227	0.2253		0.2480
Available production in kg/ha/yr	P'	0.0629	0.0060		0.0689
Available yield in kg/ha/yr	$Y_{P'}$		0.0060		0.0060

Table D33. Computation of production for *Haplochromis darlingi* stock from Lake Kariba (after data from locality D – Buffalo Island cove, 1968).

Growth seasons		0	1	2	3	Total
Interpolated N'/ha at time of annulus formation	N'	(536)	98	18	3	119
Interpolated mean weight in g of 1 specimen	$\bar{w}$		2.9	5.15	4.7	
Initial biomass in kg/ha	B'		0.2842	0.0927	0.0141	0.3910
Rate of increase in biomass	H	−3.4836	−3.3276	−4.0344		
Instantaneous growth rate	G	0.6592	0.5743	−0.0914		0.3807
Instantaneous rate of total mortality	Z	4.1428	3.9019	4.1258		4.0568
Mean biomass	$\bar{B}$	0.079	0.02	0.19		0.30
Total production (= G$\bar{B}$) in kg/ha/yr	A	0.0521	0.0154	0.0177		0.0852
Available production in kg/ha/yr	P'	0.1372	0.0405	0.0014		0.1791
Available yield in kg/ha/yr	$Y_{P'}$		0.0405	0.0014		0.0419

Table D34. Computation of production for *Haplochromis darlingi* stock from Lake Kariba (after data from locality E – Chete Island cove, 1969).

Growth seasons		0	1	2	3	Total
Interpolated N'/ha at time of annulus formation	N'	(1119)	331	122	15	468
Interpolated mean weight in g of 1 specimen	$\bar{w}$		1.5	4.0	7.6	
Initial biomass in kg/ha	B'		0.4965	0.488	0.114	1.0985
Rate of increase in biomass	H	−1.7338	−1.3171	−4.1844		
Instantaneous growth rate	G	1.0705	0.9808	0.6418		0.8977
Instantaneous rate of total mortality	Z	2.8043	2.2979	4.8262		3.3095
Mean biomass	$\bar{B}$	0.24	0.27	1.76		2.27
Total production (= G$\bar{B}$) in kg/ha/yr	A	0.2522	0.2663	1.1302		1.6487
Available production in kg/ha/yr	P'	0.3707	0.3050	0.0540		0.7297
Available yield in kg/ha/yr	$Y_{P'}$		0.3050	0.0540		0.359

Table D35. Computation of production for *Haplochromis darlingi* stock from Lake Kariba (after data from locality F – cove I in Chipepo Bay, 1969).

Growth seasons		0	1	2	Total
Interpolated N'/ha at time of annulus formation	N'	(24068)	2043	160	2203
Interpolated mean weight in g of 1 specimen	$\bar{w}$		5.6	7.5	
Initial biomass in kg/ha	B'		1144.8	1200	2344.8
Rate of increase in biomass	H	−2.2184	−6.5726		
Instantaneous growth rate	G	3.4610	0.2921		1.8765
Instantaneous rate of total mortality	Z	5.6794	5.8647		5.7720
Mean biomass	$\bar{B}$	4596.83	3923.88		8520.71
Total production (= G$\bar{B}$) in kg/ha/yr	A	15.9096	1.1462		17.0558
Available production in kg/ha/yr	P'	9.4795	0.3040		9.7835
Available yield in kg/ha/yr	$Y_{P'}$		0.3040		0.3040

Table D36. Computation of production for *Haplochromis darlingi* stock from Lake Kariba (after data from locality G – cove II in Chipepo Bay, 1969).

Growth seasons		0	1	2	3	4	Total
Interpolated N'/ha at time of annulus formation	N'	(4624)	1183	313	82	22	
Interpolated mean weight in g of 1 specimen	$\bar{w}$		3.3	4.7			
Initial biomass in kg/ha	B'		3903.9	1471.1			5375.0
Rate of increase in biomass	H		−2.2671	−2.7081			
Instantaneous growth rate	G		0.8718	0.3537			0.6127
Instantaneous rate of total mortality	Z		3.1389	3.0618			3.1003
Mean biomass	$\bar{B}$		1542.20	493.80			2036.01
Total production (= G$\bar{B}$) in kg/ha/yr	A		1.3440	0.1746			1.5186
Available production in kg/ha/yr	P'		2.2714	0.4382			2.7096
Available yield in kg/ha/yr	$Y_{P'}$			0.4382			0.4382

Table D37. Computation of production for *Synodontis nebulosus* stock from Lake Kariba (after data from locality F – Chipepo Bay cove, 1969).

Growth seasons		0	1	2	3	4	5	6	7	Total
Interpolated N'/ha at time of annulus formation	N'	(1463)	537	222	89	34	14	6	2	
Interpolated mean weight in g of 1 specimen	$\bar{w}$		1.9	5.8	12	22	35	48	61	
Initial biomass in kg/ha	B'		1.02	1.29	1.07	0.75	0.49	0.29	0.12	904
Rate of increase in biomass	H	−0.3604	+0.2327	−0.1871	−0.3561	−0.4310	−0.5313	−0.8591		
Instantaneous growth rate (average)	G	0.6419	1.1160	0.7270	0.6062	0.4643	0.3158	0.2397		0.5873
Instantaneous rate of total mortality	Z	1.0023	0.8833	0.9141	0.9623	0.8874	0.8471	1.0988		0.9422
Mean biomass	$\bar{B}$	0.86	1.43	0.99	0.64	0.40	0.22	0.08		4.62
Total production (= G$\bar{B}$) in kg/ha/yr	A	0.55	1.60	0.72	0.39	0.18	0.07	0.02		3.53
Total yield in kg/ha/yr	Y_A						0.07	0.02		0.09
Available production in kg/ha/yr	P'	1.02	0.87	0.55	0.34	0.18	0.08	0.03		3.07
Available yield in kg/ha/yr	$Y_{P'}$						0.08	0.03		0.11

Table D38. Computation of production for *Micralestes acutidens* stock from Lake Kariba (after data from locality D – Buffalo Island cove, 1968).

Growth seasons		Eggs	0	1	2	3	Total
Interpolated N'/ha at time of annulus formation	N'	46786	(1400)	120	11	1	132
Interpolated mean weight in g of 1 specimen	w̄	0.0002112		1.16	2.50	5.00	
Initial biomass in kg/ha	B'	0.00988		0.139	0.027	0.005	0.18
Rate of increase in biomass	H		−5.9517	−1.6217	−1.7048		
Instantaneous growth rate	G		0.0141	0.7679	0.6931		0.4917
Instantaneous rate of total mortality	Z	3.5092	(5.9658)	2.3896	2.3979		2.6883
Mean biomass	B̄		0.023	0.013	0.002		0.038
Total production (= GB̄) in kg/ha/yr	A		0.00032	0.00998	0.00139		0.01169
Available production in kg/ha/yr	P'		0.139	0.015	0.002		0.156

APPENDIX E

Morphometry of Sampling Sites and Standing Crop Tables

Cove at Siavonga Harbour Peninsula

(Experiment A. 20th. – 22nd. March, 1968)

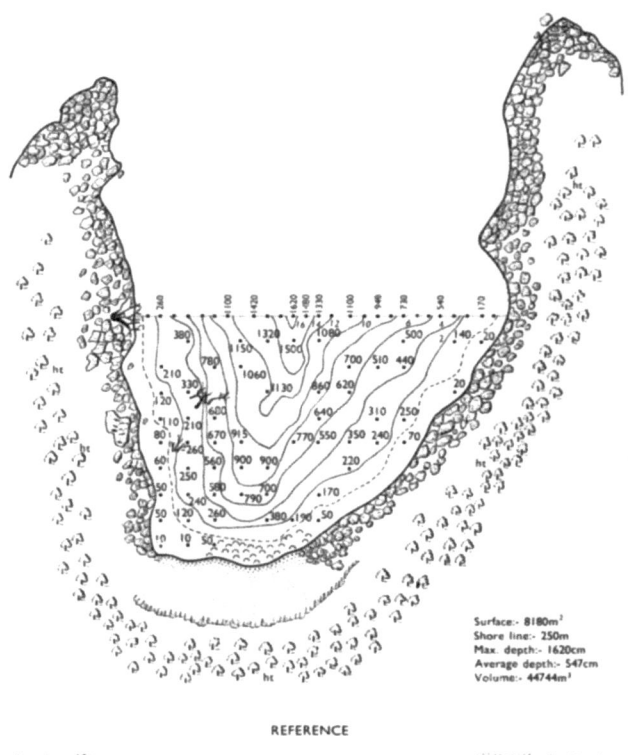

Fig. E1.

Table E1. Abundance and standing crop of fishes in cove near Siavonga (surface 8180 m^2, shoreline 250 m).

Sample A 14 species	Standard length ranges in mm	Average length of 1 fish in mm	Number of specimens	Relative		Average weight of 1 fish in g	Total weight in g	Relative	
				number of fish per 1 ha	number of fish per 100 m shoreline			weights in kg per 1 ha	weights in kg per 100 m shoreline
Synodontis zambezensis	39–205	95	116	142	46	29.9	3470	4.24	1.39
Hydrocynus vittatus	206–270	240	7	9	28	277.8	1945	2.38	0.78
Hippopotamyrus discorhynchus	11–157	104	71	87	3	27.3	1942	2.37	0.78
Haplochromis darlingi	11– 60	36	398	486	159	1.8	754	0.92	0.30
Sarotherodon mossambicus mortimeri	20–125	40	93	114	37	4.7	445	0.54	0.18
Tilapia rendalli	31–125	66	17	21	7	24.6	419	0.51	0.17
Heterobranchus longifilis	127–166	153	5	6	2	44	220	0.27	0.09
Pseudocrenilabrus philander	17– 48	31	100	122	40	0.9	97	0.12	0.04
Barbus unitaeniatus	42– 61	52	32	39	13	2.9	96	0.12	0.04
Micralestes acutidens	30– 52	46	40	49	16	1.8	72	0.09	0.03
Alestes lateralis	13– 53	43	35	43	14	1.8	66	0.08	0.03
Sarotherodon andersoni	13– 70	37	15	18	6	3.8	57	0.07	0.02
Mormyrops deliciosus	68–132	100	2	2	0.8	12	24	0.03	0.01
Aplocheilichthys johnstoni	21	21	1	1	0.4	0.15	0.15	0.00018	0.00006
Total			932	1139	372	10	9607	11.74	3.86

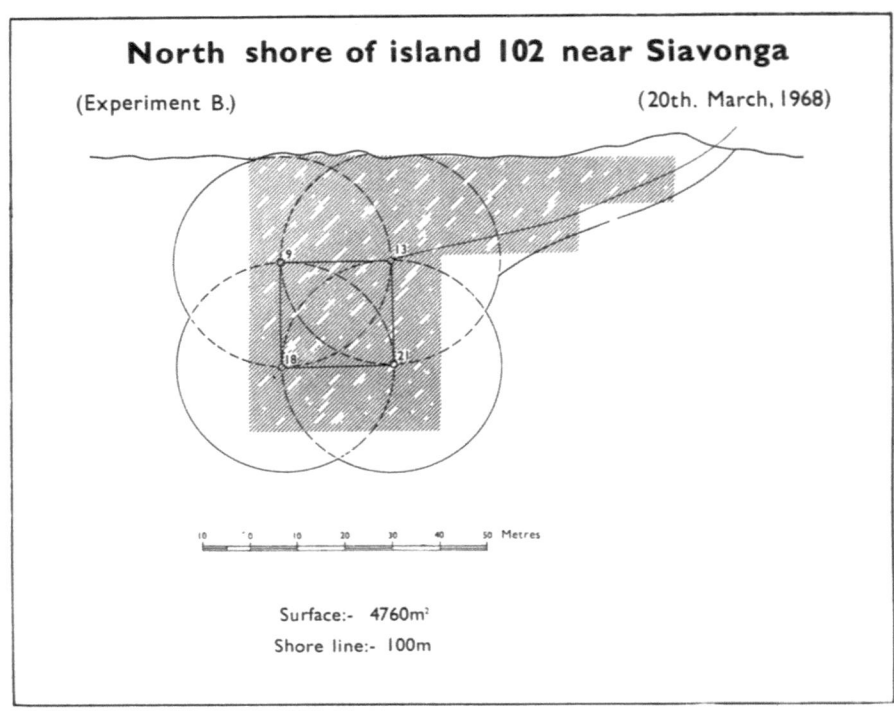

Fig. E2.

Table E2. Abundance and standing crop of explosion sensitive fishes in rocky shore area of island N 102 near Siavonga (surface affected 4760 m², shoreline 100 m).

Sample B 9 species	Standard length ranges in mm	Average length of 1 fish in mm	Number of specimens	Relative		Average weight of 1 fish in g	Total weight in g	Relative	
				number of fish per 1 ha	number of fish per 100 m shoreline			weights in kg per 1 ha	weights in kg per 100 m shoreline
Hydrocynus vittatus	90–321	204	31	65	31	217.0	6730	14.14	6.73
Sarotherodon mossambicus mortimeri	50–345	142	7	15	7	564.7	3953	8.30	3.95
Tilapia rendalli	205–280	233	5	10	5	786.0	3930	8.26	3.93
Sarotherodon andersoni	294	294	1	2	1	1950	1950	4.10	1.95
Alestes lateralis	15– 60	44	307	645	307	2.2	669	1.40	0.67
Haplochromis darlingi	16– 80	44	185	389	185	3.0	560	1.18	0.56
Pseudocrenilabrus philander	15– 51	30	117	246	117	1.0	114	0.24	0.11
Barbus unitaeniatus	40– 57	48	10	21	10	2.4	24	0.05	0.02
Micralestes acutidens	45– 51	49	3	6	3	2.3	7	0.01	0.007
Total			666	1399	666	26.93	17937	37.68	17.93

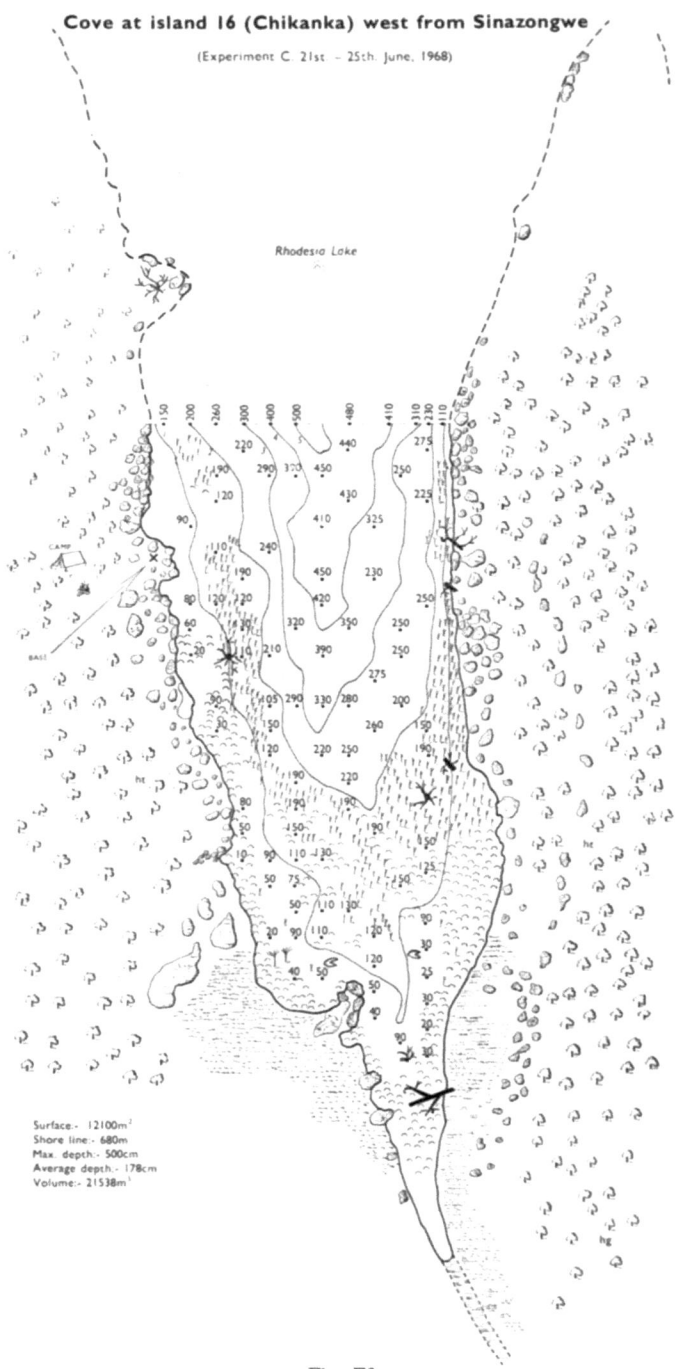

Fig. E3.

Table E3. Abundance and standing crop of fishes from cove of Chikanka Island (surface 12100 m^2, shoreline 680 m).

Sample C 29 species	Standard length ranges in mm	Average length of 1 fish in mm	Number of specimens	Relative number of fish per 1 ha	Relative number of fish per 100 m shoreline	Average weight of 1 fish in g	Total weight in g	Relative weights in kg per 1 ha	Relative weights in kg per 100 m shoreline
Tilapia rendalli	37–425	126	582	481	85	193	112517	93.00	16.55
Clarias gariepinus	30–710	334	89	73	13	850.7	75712	62.57	11.13
Sarotherodon mossambicus mortimeri	35–299	101	479	396	70	79.7	38192	31.56	5.62
Hydrocynus vittatus	48–420	131	296	245	43	128	37886	31.31	5.57
Heterobranchus longifilis	103–963	400	7	6	1	2584.9	18094	14.95	2.66
Hippopotamyrus discorhynchus	33–185	125	329	272	48	44.6	14678	12.13	2.16
Sargochromis codringtoni	72–270	157	45	37	7	187	8417	6.96	1.24
Synodontis zambezensis	26–278	158	60	49	9	130.5	7828	6.47	1.15
Labeo altivelis	57–315	210	10	8	1	453	4530	3.74	0.67
Distichodus schenga	63–230	191	28	23	4	237.1	6638	5.48	0.98
Alestes lateralis	22– 60	40	1759	1451	259	1.5	2699	2.23	0.40
Haplochromis carlottae	200–250	225	4	3	0.5	548.7	2195	1.81	0.32
Malapterurus electricus	380	380	1	0.8	0.1	1240	1240	1.02	0.18
Synodontis nebulosus	34–121	94	34	28	5	20	678	0.56	0.10
Eutropius depressirostris	18–179	74	30	23	4	13.3	398	0.33	0.06
Haplochromis darlingi	24– 88	42	165	136	24	2.4	396	0.33	0.06
Brachyalestes imberi imberi	60–150	103	11	9	2	35.3	388	0.32	0.06
Serranochromis robustus jallae	167–255	196	2	1.5	0.2	129.2	258	0.21	0.04
Serranochromis macrocephalus	162–204	183	2	1.5	0.2	101.5	203	0.17	0.03
Marcusenius macrolepidotus	165–191	178	2	2	0.3	97	194	0.16	0.03
Schilbe mystus	75–139	100	13	11	2	14.8	192	0.16	0.03
Barbus unitaeniatus	35– 58	47	79	65	12	2.3	186	0.15	0.03
Micralestes acutidens	36– 53	44	44	36	6	1.6	72	0.06	0.01
Sargochromis giardi	109	109	1	0.8	0.1	64	64	0.05	0.009
Pseudocrenilabrus philander	24– 58	42	22	18	3	2.4	53	0.04	0.008
Barbus fasciolatus	28– 35	31	52	43	8	0.7	37	0.03	0.005
Labeo lunatus	92	92	1	0.8	0.1	14	14	0.02	0.003
Labeo congoro	70	70	1	0.8	0.1	9.5	9.5	0.008	0.001
Barbus poechii	53	53	1	0.8	0.1	3.2	3.2	0.002	0.0005
Total			4149	3424	607.7	805.09	3340309	275.83	49.106

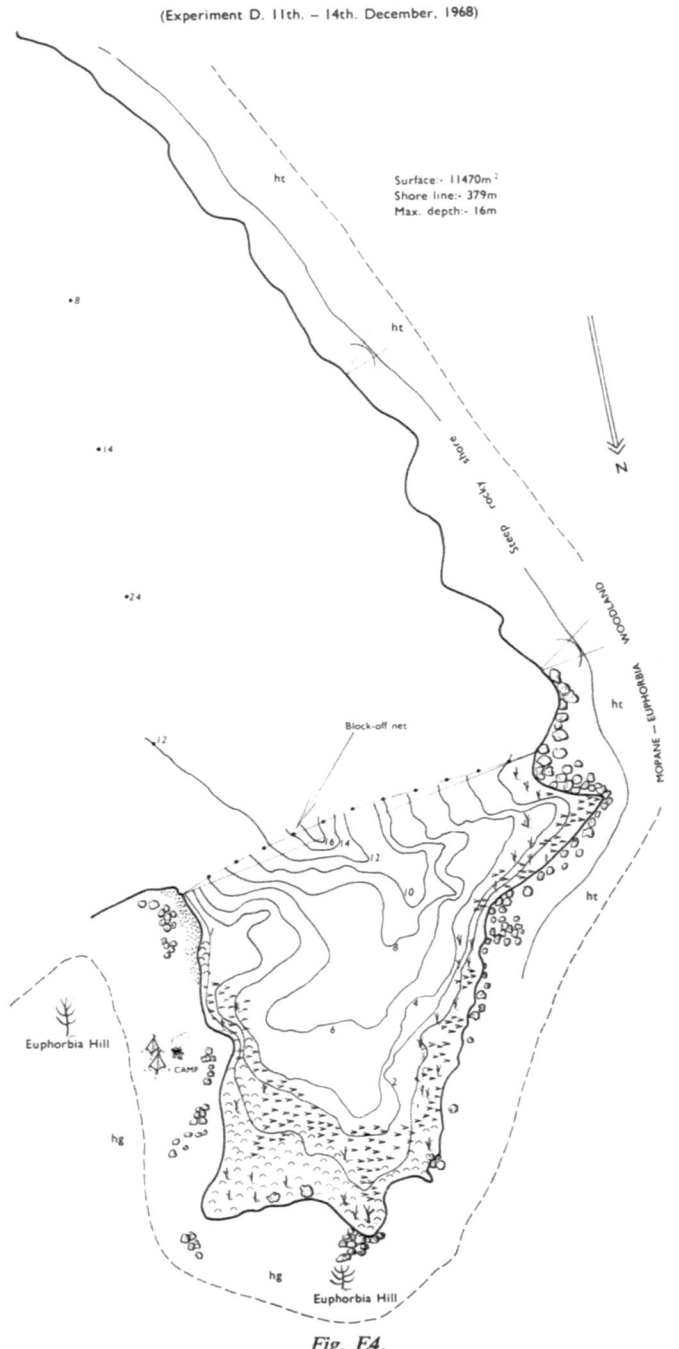

Fig. E4.

Table E4. Abundance and standing crop of fishes in south cove at Island N 20 (surface 11470 m², shoreline 379 m).

Sample D	Standard length ranges in mm	Average length of 1 fish in mm	Number of specimens	Relative number of fish per 1 ha	Relative number of fish per 100 m shoreline	Average weight of 1 fish in g	Total weight in g	Relative weights in kg per 1 ha	Relative weights in kg per 100 m shoreline
29 species									
Mormyrops deliciosus	17–825	291	297	259	70	550	163465	142.51	43.13
Mormyrus longirostris	215–741	431	53	46	14	1319	69905	60.95	18.44
Hippopotamyrus discorhynchus	26–576	137	810	706	213	67	51848	45.20	13.68
Hydrocynus vittatus	35–521	160	385	336	88	129	49763	43.38	13.13
Clarias gariepinus	110–676	417	34	30	9	1083	36839	32.12	9.72
Tilapia rendalli	65–273	135	187	163	49	168	31499	27.46	8.31
Alestes lateralis	10–80	35	32319	28177	8505	0.9	29129	25.40	7.69
Sarotherodon mossambicus mortimeri	24–310	91	467	407	123	54	25346	22.11	6.67
Synodontis zambezensis	33–283	129	253	221	67	91	22917	20.00	6.05
Labeo altivelis	30–334	145	102	89	27	139	14175	12.36	3.74
Eutropius depressirostris	68–282	138	298	260	78	44	13061	11.39	3.45
Sargochromis codringtoni	37–215	114	137	119	36	78	10669	9.30	2.81
Sargochromis giardi	46–223	117	126	111	33	83	10451	9.11	2.76
Malapterurus electricus	255–431	315	13	11	3	750	9750	8.50	2.57
Heterobranchus longifilis	120–393	253	13	11	3	292	3797	3.31	1.00
Labeo congoro	90–435	262	2	2	0.5	1260	2521	2.20	0.66
Haplochromis darlingi	12–100	48	607	529	160	70	2460	2.14	0.64
Schilbe mystus	86–203	140	35	30	9	46	1618	1.41	0.43
Marcusenius macrolepidotus	147–230	197	9	8	2	136	1229	1.07	0.32
Brachyalestes imberi imberi	46–147	97	24	21	6	29	703	0.61	0.18
Synodontis nebulosus	80–235	105	29	25	8	23	673	0.59	0.18
Distichodus schenga	72–127	94	30	26	8	21	637	0.55	0.17
Barbus unitaeniatus	30–69	40	202	176	53	3	554	0.48	0.15
Micralestes acutidens	30–62	46	187	163	49	2	369	0.32	0.10
Distichodus mossambicus	91–130	103	11	10	3	27	297	0.26	0.06
Barbus fasciolatus	12–39	26	272	237	72	2	131	0.11	0.03
Barbus poechii	68–91	78	8	7	2	12	94	0.08	0.02
Labeo cylindricus	105	105	1	1	0.3	29	29	0.02	0.008
Aplocheilichthys johnstoni	18–30	23	43	37	11	0.2	10	0.009	0.003
Total			36954	32218	9710	14.93	551941	482.949	146.101

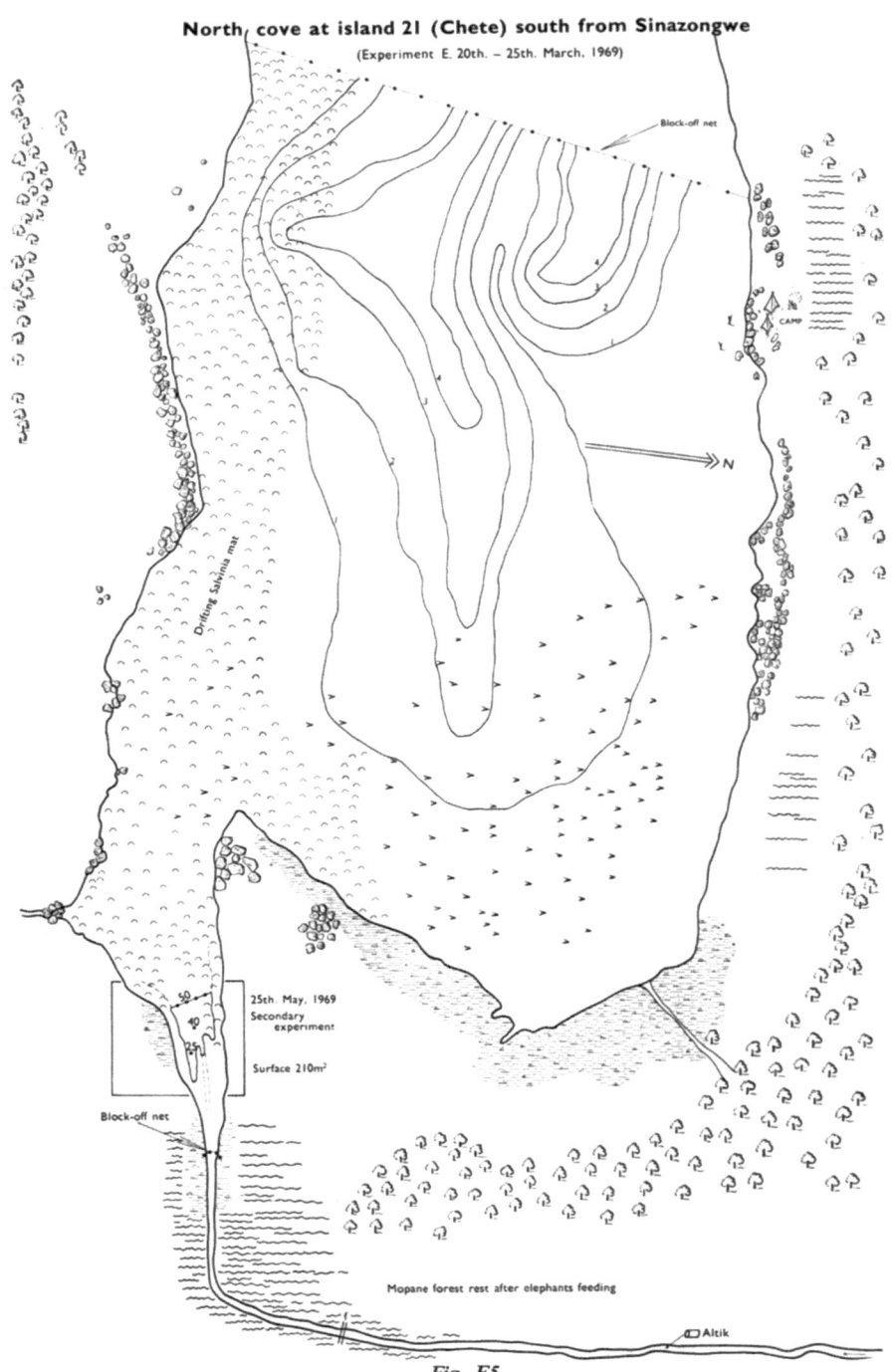

Fig. E5.

Table E5a. Abundance and standing crop of fishes from north cove of Chete Island (surface 52730 m^2, shoreline 1065 m).

Sample E$_1$ 28 species	Standard length ranges in mm	Average length of 1 fish in mm	Number of specimens	Relative number of fish per 1 ha	Relative number of fish per 100 m shoreline	Average weight of 1 fish in g	Total weight in g	Relative weights in kg per 1 ha	Relative weights in kg per 100 m shoreline
Tilapia rendalli	27– 455	138	1838	349	173	196	360449	68.36	33.84
Clarias gariepinus	85– 780	499	194	37	18	1593	309119	58.62	29.02
Sarotherodon mossambicus mortimeri	20– 565	120	2290	434	216	124	296567	56.24	27.85
Hippopotamyrus discorhynchus	50– 355	147	3387	642	319	71	241103	45.72	22.64
Hydrocynus vittatus	30– 610	158	806	153	76	226	174967	33.18	16.43
Heterobranchus longifilis	105–1170	788	11	2	1	9155	100707	19.10	9.46
Malapterurus electricus	246– 540	417	42	8	4	1717	72100	13.67	6.77
Mormyrops deliciosus	299–1005	478	57	11	5	1166	66460	12.60	6.24
Alestes lateralis	15– 87	39	28612	5426	2699	1.3	37385	7.09	3.51
Synodontis zambezensis	65– 310	158	264	50	25	122	32196	6.11	3.02
Sargochromis codringtoni	35– 245	140	217	41	20	146	31676	6.01	2.97
Labeo altivelis + L. rubropunctatus	70– 325	189	71	13	7	242	17153	3.25	1.61
Marcusenius macrolepidotus	109– 250	158	155	29	15	70	10838	2.05	1.02
Haplochromis darlingi	18– 73	40	2257	428	213	2.4	5419	1.03	0.51
Distichodus schenga	112– 263	155	22	4	2	161	3552	0.67	0.33
Eutropius depressirostris	100– 260	152	55	10	5	55	3005	0.57	0.28
Barbus unitaeniatus	25– 87	45	1424	270	134	2.04	2903	0.55	0.27
Brachyalestes imberi imberi	60– 155	113	23	4	2	45	1038	0.20	0.10
Synodontis nebulosus	70– 175	116	27	5	2	37	1014	0.19	0.09
Mormyrus longirostris	397	397	1	0.2	0.09	850	850	0.16	0.08
Pseudocrenilabrus philander	20– 65	35	459	87	43	1.5	682	0.13	0.06
Barbus fasciolatus	18– 54	29	377	71	36	0.70	252	0.05	0.02
Schilbe mystus	155– 221	180	3	0.6	0.3	64	237	0.04	0.02
Labeo cylindricus	100	100	1	0.2	0.09	25	25	0.005	0.002
Micralestes acutidens	25– 52	43	12	2	1	137	20	0.004	0.002
Barbus lineomaculatus	34	34	2	0.4	0.20	7.5	15	0.003	0.001
Aplocheilichthys johnstoni	25	25	1	0.2	0.09	0.30	0.27	0.00005	0.0002
Total			42608	8080	4017	41.53	1769732	335.60	166.14

Table E5b. Abundance and standing crop of fishes in the cove of intermittent Elephant Stream (surface 210 m^2).

Sample E$_3$	Standard length ranges in mm	Average length of 1 fish in mm	Number of specimens	Relative number of fish per 1 ha	Average weight of 1 fish in g	Total weight in g	Relative weights in kg per 1 ha
9 species							
Sarotherodon mossambicus mortimeri	17–195	43	390	18571	12.5	4875	232.14
Alestes lateralis	33– 55	47	236	11238	2.4	569	27.09
Clarias gariepinus	49–128	75	84	4000	5.4	456	21.71
Haplochromis darlingi	35– 60	48	132	6286	3.3	441	21.00
Tilapia rendalli	18–130	41	23	1095	11.4	262	12.48
Barbus unitaeniatus	40– 57	49	39	1857	2.5	97	4.62
Hydrocynus vittatus	40– 81	59	15	714	3.8	57	2.71
Sargochromis codringtoni	25– 36	31	40	1905	0.9	35	1.67
Pseudocrenilabrus philander	36– 46	41	8	381	2.1	17	0.81
Total			967	46047	7.0	6809	324.23

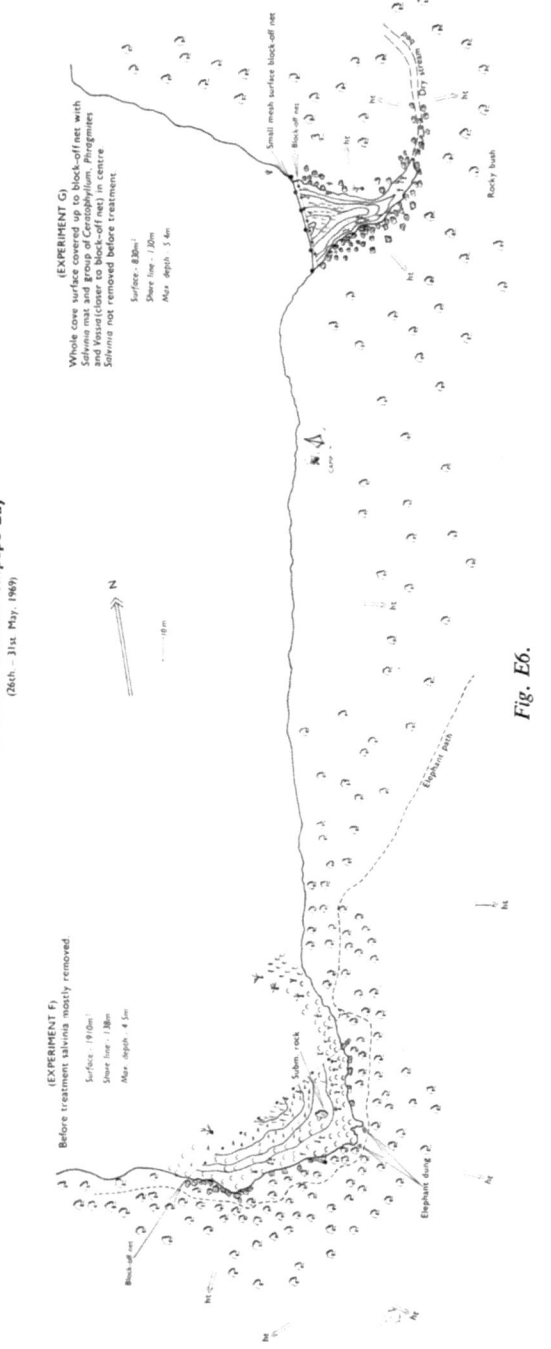

Fig. E6.

Table E6a. Abundance and standing crop of fishes in a cove of the east shore of Chipepo Bay (surface 1910 m^2, shoreline 138 m).

Sample F 23 species	Standard length ranges in mm	Average length of 1 fish in mm	Number of specimens	Relative number of fish per 1 ha	Relative number of fish per 100 m shoreline	Average weight of 1 fish in g	Total weight in g	Relative weights in kg per 1 ha	Relative weights in kg per 100 m shoreline
Hippopotamyrus discorhynchus	22–200	102	981	5136	701	25	24733	129.49	17.92
Malapterurus electricus	170–485	329	19	99	14	968	18400	96.33	13.33
Alestes lateralis	12– 95	39	11100	58115	7929	1.2	13757	72.03	9.97
Synodontis zambezensis	25–245	113	191	1000	136	50	9564	50.07	6.93
Hydrocynus vittatus	30–430	81	262	1372	187	34	8895	46.57	6.45
Clarias gariepinus	45–540	120	56	293	40	95.6	5352	28.02	3.88
Tilapia rendalli	50–190	121	44	230	31	117	5152	26.97	3.73
Sargochromis codringtoni	42–260	119	41	215	29	118	4842	25.35	3.51
Heterobranchus longifilis	63–380	194	12	63	9	269	3232	16.92	2.34
Haplochromis darlingi	17– 84	45	653	3419	466	3	2167	11.34	1.57
Eutropius depressirostris	40–205	106	93	487	66	22	2112	11.05	1.53
Sarotherodon mossambicus mortimeri	24–175	81	50	262	36	38	1920	10.05	1.39
Mormyrops deliciosus	35–470	102	31	162	22	47	1481	7.75	1.07
Barbus fasciolatus	15– 42	30	1776	9298	1269	0.6	1169	6.12	0.85
Synodontis nebulosus	55–115	86	77	403	55	14	1098	5.75	0.80
Marcusenius macrolepidotus	55–210	154	12	63	9	71	855	4.48	0.62
Pseudocrenilabrus philander	23– 59	35	443	2319	316	1.5	650	3.40	0.47
Mormyrus longirostris	160–280	220	2	10	1	163	326	1.71	0.24
Brachyalestes imberi imberi	32– 90	53	34	178	24	4	137	0.72	0.10
Barbus unitaeniatus	24– 55	42	76	398	54	1.7	134	0.70	0.10
Labeo altivelis	140	140	1	5	0.7	73	73	0.38	0.05
Micralestes acutidens	42	42	1	5	0.7	1.1	1.36	0.007	0.010
Aplocheilichthys johnstoni	29	29	1	5	0.7	0.4	0.43	0.002	0.0003
Total			15956	83537	11396.1	6.65	106051	555.21	76.86

Table E6b. Abundance and standing crop of fishes in a cove of the east shore of Chipepo Bay (surface 830 m², shoreline 130 m).

Sample G 25 species	Standard length ranges in mm	Average length of 1 fish in mm	Number of specimens	Relative number of fish per 1 ha	Relative number of fish per 100 m shoreline	Average weight of 1 fish in g	Total weight in g	Relative weights in kg per 1 ha	Relative weights in kg per 100 m shoreline
Hippopotamyrus discorhynchus	45–250	114	602	7253	463	34.4	20713	249.55	15.93
Sarotherodon mossambicus mortimeri	25–405	135	46	554	35	415.6	19118	230.34	14.71
Mormyrops deliciosus	127–642	348	30	361	23	566.7	17002	204.84	13.08
Malapterurus electricus	335–505	415	6	72	5	1693.5	10160	122.41	7.81
Clarias gariepinus	41–560	331	10	120	8	914	9140	110.12	7.03
Tilapia rendalli	175–305	248	9	108	7	1015.5	9085	109.46	6.99
Alestes lateralis	18–70	44	2379	28663	1830	1.4	3334	40.17	2.56
Sargochromis codringtoni	75–225	150	13	157	10	192.7	2505	30.18	1.93
Synodontis zambezensis	60–232	116	44	530	34	53.7	2364	28.48	1.82
Hydrocynus vittatus	30–305	73	69	831	53	33.1	2285	27.53	1.76
Eutropius depressirostris	26–231	116	60	723	46	31	1858	22.38	1.43
Heterobranchus longifilis	40–350	187	6	72	5	204.7	1228	14.79	0.94
Labeo altivelis	110–275	160	8	96	6	107.9	863	10.40	0.66
Mormyrus longirostris	175–260	226	5	60	4	152.4	762	9.18	0.59
Haplochromis darlingi	27–70	46	101	1217	78	3.3	338	4.07	0.26
Barbus fasciolatus	23–40	30	360	4337	277	0.7	244	2.94	0.19
Labeo congoro	190	190	1	12	0.8	220	220	2.65	0.17
Synodontis nebulosus	58–95	80	9	108	7	12.3	111	1.34	0.08
Marcusenius macrolepidotus	135–160	147	2	24	1	53.5	107	1.29	0.08
Brachyalestes imberi imberi	30–85	50	30	361	23	3.4	102	1.23	0.08
Pseudocrenilabrus philander	26–42	34	45	542	35	1.3	57	0.69	0.04
Barbus unitaeniatus	37–60	45	24	289	18	2	48	0.58	0.04
Labeo cylindricus	95	95	1	12	0.8	21	21	0.25	0.02
Barbus lineomaculatus	36	36	1	12	0.8	9	9	0.11	0.007
Aplocheilichthys johnstoni	28–29	28	2	24	1	0.2	0.4	0.0048	0.0003
Total			3863	46538	2971.4	26.3	101674.4	1224.98	78.21

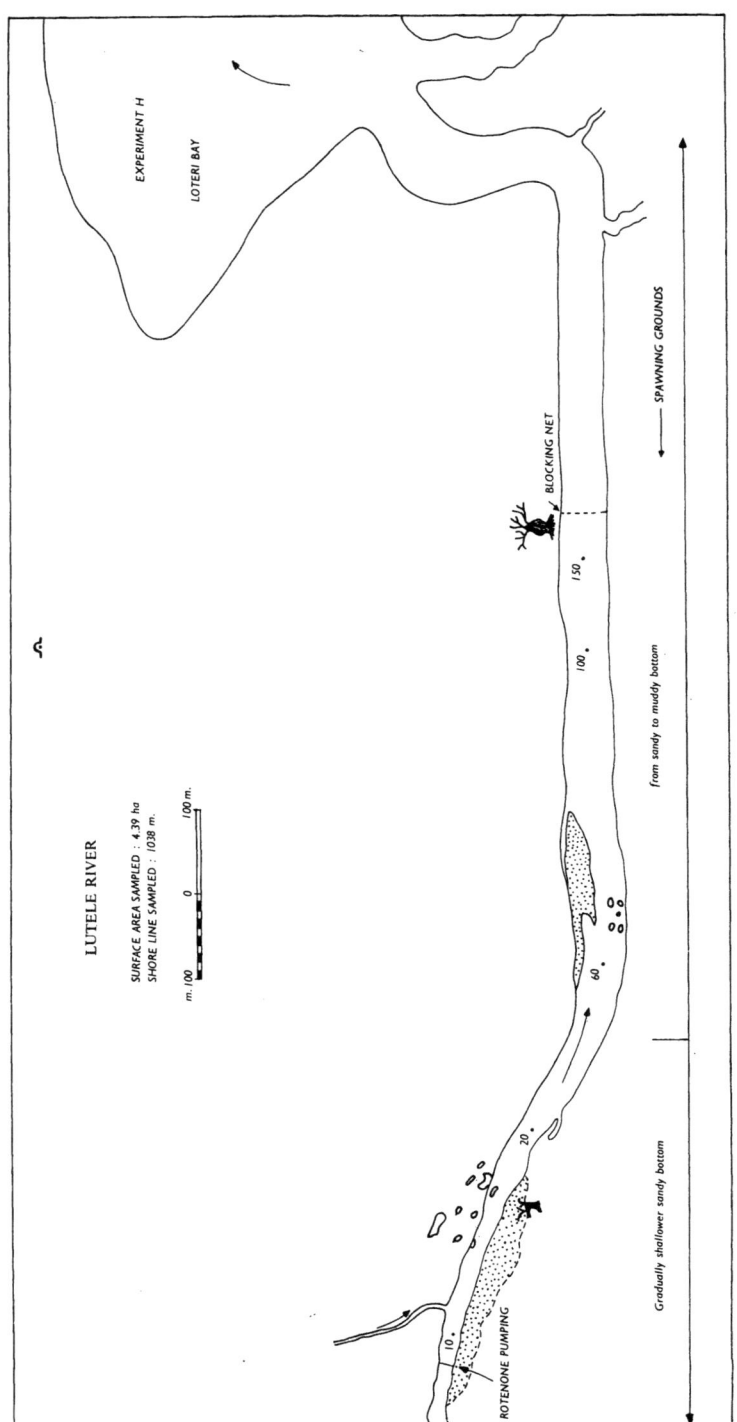

Fig. E7.

Table E7a. Abundance and standing crop of fishes in inshore area of Loteri Bay (surface 1 ha, shoreline 100 m).

Sample H 21 species	Standard length ranges in mm	Average length of 1 specimen in mm	Number of specimens	Average weight of 1 specimen in g	Total weight in kg
Synodontis zambezensis	64–270	146	240	104	24.97
Mormyrus longirostris	300–702	507	8	2382	19.06
Hydrocynus vittatus	151–625	226	46	351	16.16
Tilapia rendalli	115–305	240	15	954	14.32
Clarias gariepinus	255–590	406	12	831	9.97
Barbus unitaeniatus	21– 64	39	5591	1.0	8.10
Sargochromis codringtoni	40–227	101	78	86	6.72
Mormyrops deliciosus	244–845	460	3	2219	6.66
Malapterurus electricus	230–520	337	5	1196	5.98
Haplochromis carlottae	120–210	176	20	268	5.36
Eutropius depressirostris	21–242	164	85	62	5.25
Hippopotamyrus discorhynchus	39–265	106	124	33	4.10
Heterobranchus longifilis	370–450	410	2	920	1.84
Haplochromis darlingi	14– 79	34	526	2	0.84
Sarotherodon mossambicus mortimeri	140–190	161	3	205	0.62
Marcusenius macrolepidotus	122–230	179	5	107	0.54
Alestes lateralis	19–122	37	231	1	0.03
Brachyalestes imberi imberi	65– 85	75	2	11	0.02
Labeo cylindricus	70	70	1	7	0.007
Barbus lineomaculatus	36– 43	40	4	1	0.005
Limnothrissa miodon	20– 25	22	2	0.2	0.0003
Total			7003	9741.2	130.552

Table E7b. Abundance and standing crop of fishes in the Lutele River estuary (surface 4.39 ha, shoreline 1038 m).

Sample I 12 species	Standard length ranges in mm	Average length of 1 fish in mm	Number of specimens	Relative		Average weight of 1 fish in g	Total weight in g	Relative	
				number of fish per 1 ha	number of fish per 100 m shoreline			weights in kg per 1 ha	weights in kg per 100 m shoreline
Tilapia rendalli	268–305	256	101	23	10	1056	106667	24.29	10.28
Hydrocynus vittatus	110–550	235	92	21	9	310	28551	6.50	2.75
Sarotherodon mossambicus mortimeri	129–340	213	49	11	5	576	28208	6.42	2.72
Malapterurus electricus	246–540	348	7	1	0.7	1225	8575	1.95	0.83
Synodontis zambezensis	80–190	114	86	20	8	44	3748	0.85	0.36
Brachyalestes imberi imberi	75–195	106	37	8	3	38	1359	0.31	0.13
Clarias gariepinus	160–260	205	4	1	0.4	106	425	0.09	0.04
Hippopotamyrus discorhynchus	120–165	130	6	1	1	47	285	0.06	0.03
Labeo cylindricus	125–195	160	2	0.4	0.2	83	167	0.04	0.02
Marcusenius macrolepidotus	90–140	125	4	1	0.4	32	128	0.03	0.01
Eutropius depressirostris	100–140	112	4	1	0.4	20	80	0.02	0.007
Haplochromis darlingi	32–40	35	3	0.7	0.3	1	4	0.0009	0.0004
Total			395	89.1	38.4	3538	178197	40.561	17.177

APPENDIX F

Time of Annulus Inception: A Pond Experiment

by EUGENE K. BALON & E. MICHAEL CHADWICK

On January 29, 1969, two fish species of cichlid, *Sarotherodon andersoni* and *Pseudocrenilabrus philander* were placed into two concrete ponds of the Department of Wildlife, Fisheries and National Parks in Chilanga, Zambia. At monthly intervals, a sub-sample of ten fish was netted from each pond and the standard length, weight, and three key scales were taken. This procedure was repeated at least twelve times. The scales were removed from the first row above the lateral line beginning behind the origin of the dorsal fin and after cleaning in water were mounted between two microscope slides. The slides were examined at a magnification of 46× on a Bausch & Lomb microprojector.

From a total of 229 scale sets for *S. andersoni* and 197 sets for *P. philander*, 87 and 51 respectively, were regenerated and could not be used for further analysis; 34 and 20 sets had indistinct or undefinable annuli and were also not used.

Difficulty arises as to what is an annulus. EL ZARKA (1961) considered a separation of sclerites on scales from *Tilapia zilli* as false marks, while it was these distinct gaps between the circuli that van RENSBURG (1966) counted as annuli for *S. mossambicus*. Here only continuous cutting over of old sclerites by a new sclerite has been considered as an annulus (BALON, 1962b; BERG & GRIMALDI, 1967; SYCH, 1971). The annulus could easily be distinguished in this fashion and was remarkably clear on the oral and lateral sections of the scale as a dark line of cut over circuli, often followed by a series of lagunae, especially in older fish. The annulus was almost always discernable throughout its entire perimeter.

Some smaller fish, notably *P. philander*, had juvenile marks which could be dismissed after a back calculation revealed the fish to be too small to be in their second year of growth. These juvenile marks also lacked the distinct cutting over found in true annuli.

The oral radius was measured from the scale center to the tip of the longest lobe nearest to the longitudinal axis. The distance to each annulus was also measured. The number of circuli from the final annulus to the scale edge was then counted (Table F1).

The number of circuli beyond the final annulus was averaged for each monthly sub-sample and plotted in Fig. F1. For *S. andersoni* the lowest circuli count was in March for 1969 and 1970, suggesting that this month is the time of annulus inception (i.e. when the number of circuli beyond the final annulus equals zero) for the bulk of the population in this species.

The situation of *P. philander* was not as clear cut and three possible periods of annulus inception – in April, August and January – were manifested (Fig. F2). We have no explanation for this phenomenon. Speculatively it can be inferred that a multiple spawning may have occurred in stable pond habitats and that spawning marks are mistaken for an overlap with the annuli. A fish with only a few circuli beyond the final annulus was arbitrarily designated as having a newly formed annulus (fewer than 6 for *S. andersoni*; fewer than 4 for *P. philander*).

The most conclusive illustration gives the relationship of average radius length (in mm) from the final annulus to the scale edge versus the total oral radius length in *S. andersoni* (Fig. F3). The period of the smallest scale increment (March 1969 and April 1970), should correspond to the time of annulus inception. The largest scale increments were measured for January.

GARROD (1958) postulated that two rings are incepted yearly on scales from *T. esculenta* corresponding to breeding periods associated with the two equinoctial rains. In our Chilanga case only one ring is formed annually for both species, otherwise Table F1 would present more age groups and a different allocation of age group percentages. Those in group one after

Table F1. Percentage of fish having respective numbers of annuli.

	Annuli	Before April 29	After April 29
S. andersoni	1	77.3	10.4
	2	17.0	62.5
	3	3.8	23.0
	4	1.9	4.1
P. philander	1	76.4	27.1
	2	23.6	58.6
	3	0	14.3

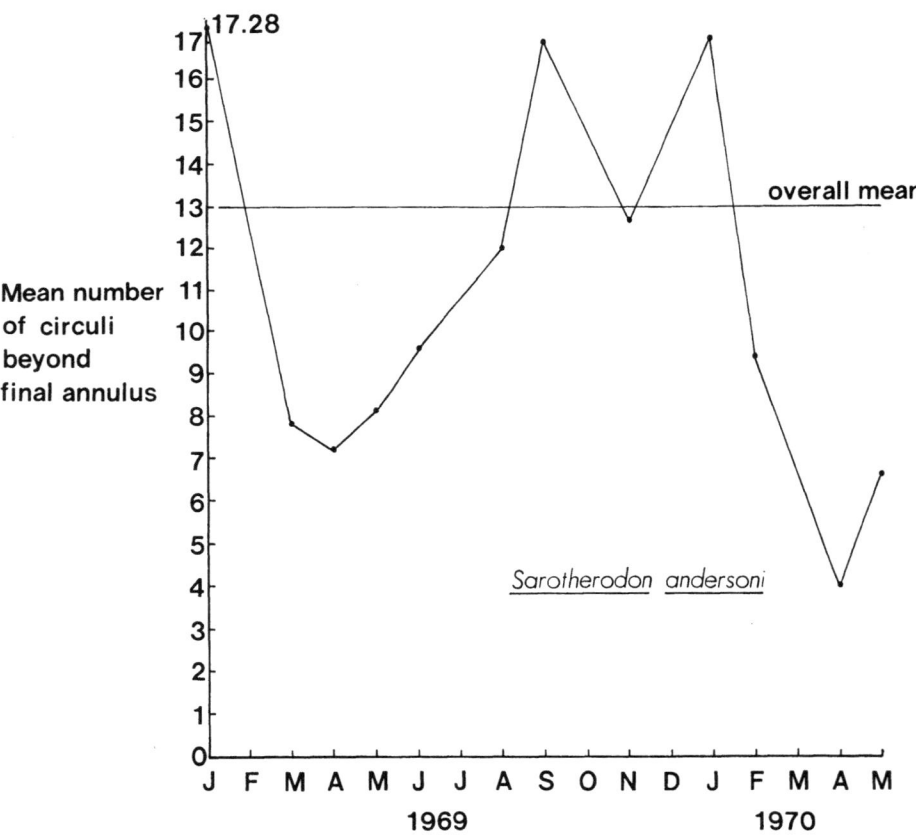

Fig. F1.

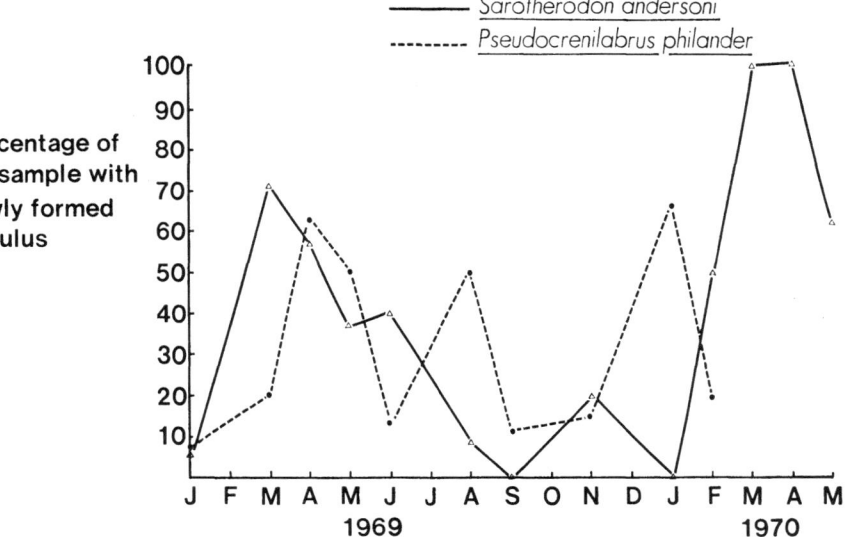

Fig. F2.

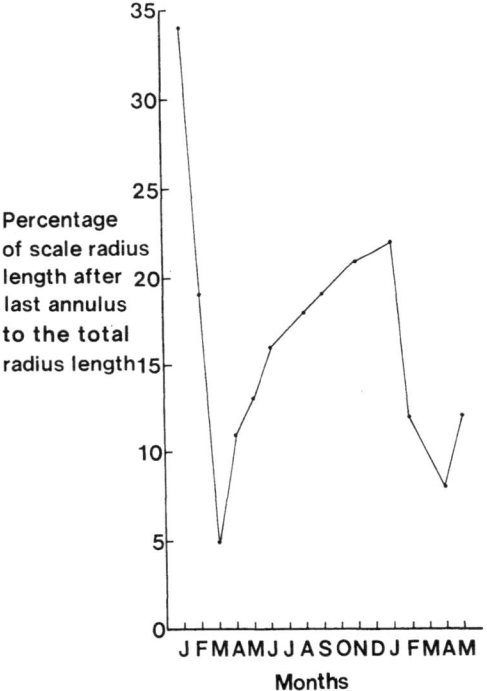

Fig. F3.

April 29 are fish which form their annulus later in the year. Perhaps as DE BONT (1967) suggests, because environmental conditions are nearly constant in tropical regions there is a much greater variation in the time of annulus inception than in temperate and sub-tropical regions. Possibly, when annulus overlaps with a spawning mark, variations in spawning times, induced by hydrological conditions, stock density and harvest are also responsible.

Why do the pond fishes in Chilanga form annuli differently from the Lake Kariba and Kafue Floodplain fishes (see sections 3, 5 and 6.4)? In the latter cases all the evidence places the annulus inception into the period of October to January.

For the moment scant evidence is available for a reliable explanation of this discrepancy in annulus inception. The theorem given in the chapter on eels (section 5) can be extended here. The annulus inception is determined genetically as an expression of endogenous circaannual rhythm and is timed by environmental factors. Certain fish species may carry an annulus inception rhythm from the original area of distribution, where it was regulated at different time sequences of environmental factors. In a newly invaded area annulus inception will consequently be shifted somewhere between the original rhythm and new regulating factors timed several months earlier or later than the regulating factors in original area (SCHWASSMANN, 1971).

In Lake Kariba and Kafue Floodplain fishes multiple regulating factors probably timed the annulus inception. The temperature and hydrological regimes are easy to enumerate; both were different in the Chilanga ponds. Well and stream water supply in these ponds influenced the water temperature and rains had a limited or a different effect on the water level of ponds than it had in natural habitats. In the absence of the usual timing factors the annulus inception may have been postponed in a similar manner as spawning activities in ponds evidently differ in time from those taking place in natural habitats. One is tempted to explain these phenomena in alternative ways, however, until more evidence is gathered rein cannot be given to mere speculation.

APPENDIX G

Lepidological Study: Key Scales of Lake Kariba Fishes

Thirty-four fish species found in Lake Kariba have bodies covered with scales. Seven are scaleless. Scales were used for age determination and for back-calculations of growth. For this purpose three scales of each specimen were collected, always from the same place, on the left side of the body. These are the key scales (Fig. G1).

Since scales from other parts may differ significantly from the key scales presented here (CHU, 1935), the conclusions are valid solely for key scales. Fishes could be identified from preserved scales found in cultural layers of excavations or in the refuse of temporary settlements (LAGLER & VALLENTYNE, 1956; BALON, 1963a). It is however hoped that even within archaeological or anthropological material scales close to the key scales in character could then be selected. The atlas of Lake Kariba fish scales presented here later also serve as background for phylletic studies (CHU, 1935; KOBAYASI, 1953).

For the sake of simplicity only maximum lengths of scales (Fig. G2a) in longitudinal axis of the fish body and maximum scale heights (Fig. G2b) in dorso-ventral axis were measured. The caudal part (posterior) is the area from the scale center (focus) towards the tail of the fish. This is usually the visible part covered by epidermis when removed from the body of the fish. The oral part (anterior) is the opposite area from the caudal part embedded deep in the skin of the fish; the lateral parts are the two scale areas situated dorsally and ventrally from the center and the diagonal parts are the transverse lobes separating oral and lateral parts. Circuli (striae) are the concentric, or other, ridges on the scale and the channels (radii) are the various radial, longitudinal, transverse or reticulated 'cracks' on the scale. Sometimes part of the scale's caudal edge is covered with spines, ctenii of ctenoid scales (Fig. G2). The scale sizes are expressed in percentages of standard length.

The scale photographs are arranged in phylletic order corresponding to the list given with Figure G1. The identification keys are arranged according to the scale types and characters. Each species is represented with an average key scale, selected out of three, each from 10 specimens of different sizes of the same species.

The shape of scales usually differs with growth, especially the a to b ratio. For example, the proportions in ten individuals of *Mormyrops deliciosus* were for scale lengths (a) from 2.08 to 1.03%, and scale heights (b) from 0.75 to 1.43% of standard length. The extremes were not necessarily characteristic of the smallest (Fig. G3, a = 1.83%, b = 1.17%) or largest (Fig. G4, a = 1.39%, b = 1.05%) individuals.

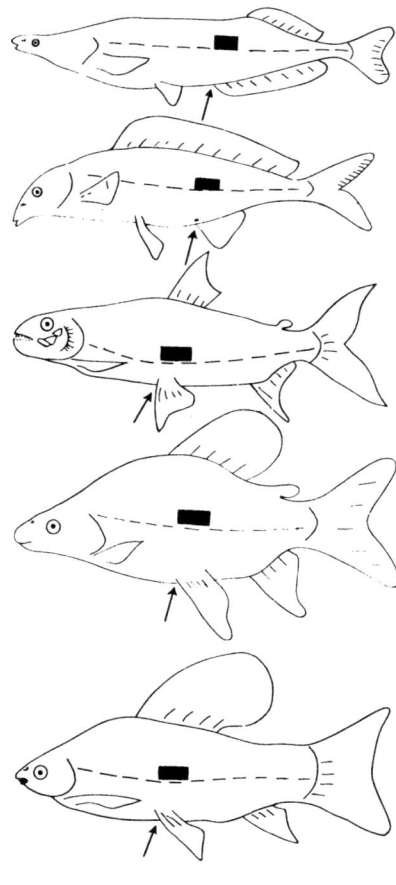

Anguilla nebulosa labiata
Limnothrissa miodon
Mormyrops deliciosus

Hippopotamyrus discorhynchus
Marcusenius macrolepidotus
Mormyrus longirostris

Hydrocynus vittatus
Alestes lateralis
Brachyalestes imberi imberi
Micralestes acutidens

Distichodus mossambicus
Distichodus schenga

Barbus poechii
Barbus paludinosus
Barbus unitaeniatus
Barbus lineomaculatus
Barbus fasciolatus
Labeo cylindricus
Labeo lunatus
Labeo congoro
Labeo altivelis
Barilius zambezensis
Aplocheilichthys johnstoni

Sarotherodon andersoni
Sarotherodon mossambicus mortimeri
Tilapia rendalli
Sargochromis giardi
Sargochromis codringtoni
Serranochromis macrocephalus
Haplochromis carlottae
Haplochromis darlingi
Pseudocrenilabrus philander

Fig. G1.

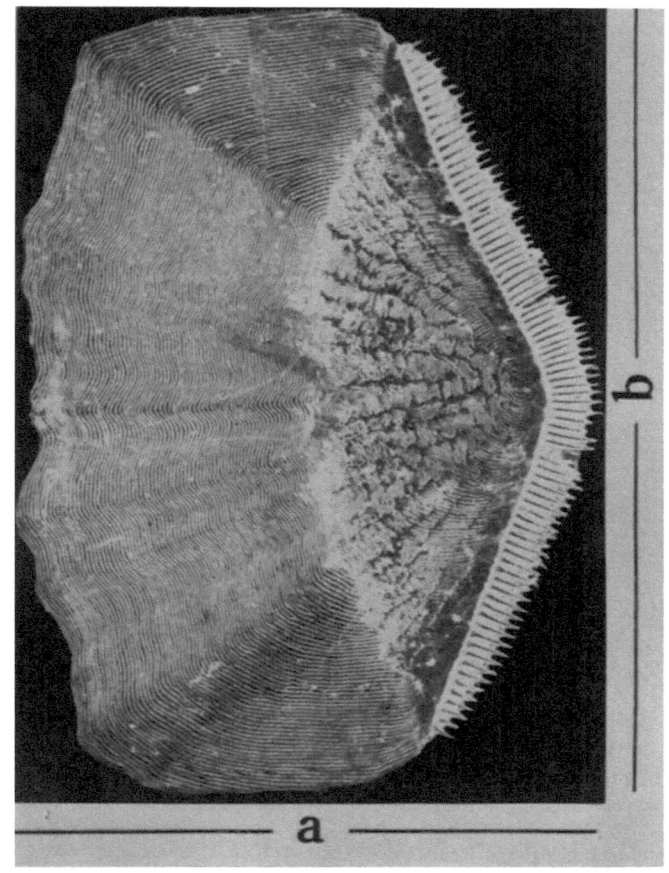

Fig. G2.

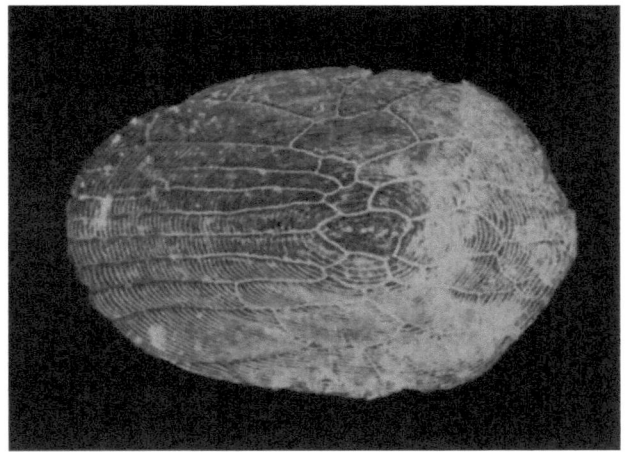

Fig. G3.

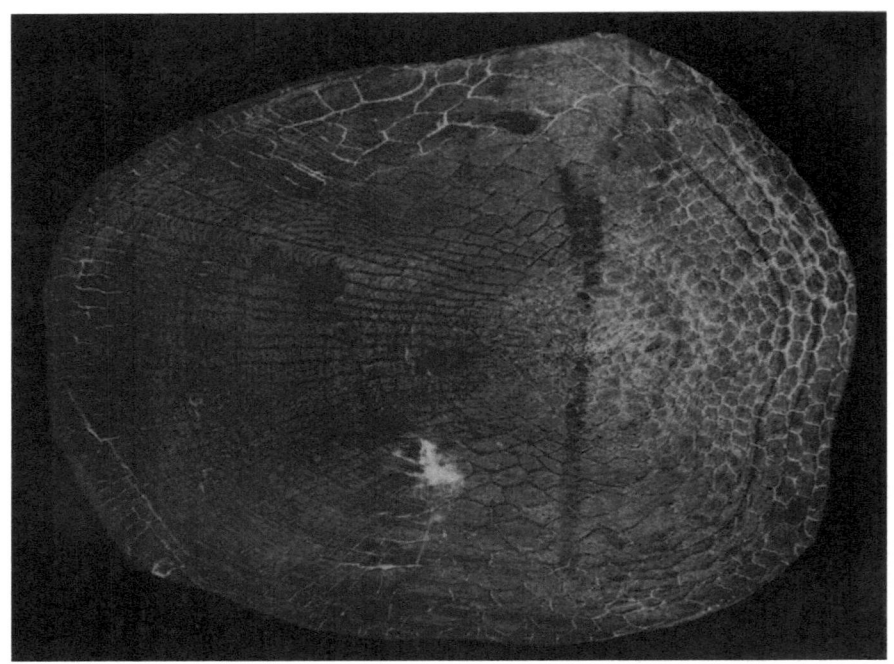

Fig. G4.

1	(3)	Circuli form a chain pattern	2
2		Scales extremely small (less than 0.5% SL) and elongated	
		(1) *Anguilla nebulosa labiata*	
3	(5)	Circuli arranged dorso-ventrally	4
4		Scales circular and large (3.9% SL) . . . . (2) *Limnothrissa miodon*	
5	(42)	Circuli in continuous concentric arrangement	6
6	(12)	Scales with channels forming reticulated pattern, esp. in the caudal part . . .	
		MORMYRIDAE	
7	(11)	Scale length larger than scale neight	8
8	(9)	Oval shaped scale, with center shifted toward caudal edge (a = 2.5% SL) . .	
		(3) *Mormyrops deliciosus*	
9	(51)	Center approximately in the middle of the scale (a = 1.5% SL)	
		(6) *Mormyrus longirostris*	
10	(11)	Scale of circular shape (a = 3% SL) . . . . (5) *Marcusenius macrolepidotus*	
11		Scale height the same or larger than scale length (a = 1.8 cf. b = 2% SL) . . .	
		(4) *Hippopotamyrus discorhynchus*	
12	(50)	Scales with channels arranged radially	13
13	(42)	Channels from the center	14
14	(28)	With the edge of oral part irregularly lobed	15
15	(27)	With radial channels in the oral as well as caudal part.	16
16	(17)	With one radial channel in the oral part, four or more channels in the caudal part (a = 3.5% SL) . . . . (7) *Hydrocynus vittatus*	
17	(18)	About six radial channels (a = 5.2% SL, b = 6.1% SL)	
		(10) *Micralestes acutidens*	
18	(19)	About eight radial channels (a = 4.8, b = 5.6% SL) . . (8) *Alestes lateralis*	

19 (23)	About eight primary and some secondary[1] radial channels		20
20 (21)	Medium large scales (5.2% SL)	(13) *Barbus poechii*	
21 (22)	Very large scales (9% SL)	(15) *Barbus unitaeniatus*	
22	In the caudal part some channels branched (9% SL) . . . . (9) *Brachyalestes imberi imberi*		
23 (29)	More than 8 primary and secondary channels		24
24 (25)	Primary channels only in the caudal part, more than 10 secondary in the oral part (a = 5.4, b = 6.8% SL)	(17) *Barbus fasciolatus*	
25 (41)	Primary and secondary channels in the oral and caudal part		26
26 (27)	a = 6.9, b = 7.7% SL	(14) *Barbus paludinosus*	
	a = 5.7, b = 6.2% SL	(16) *Barbus lineomaculatus*	
27 (41)	With radial channels in the oral part only		28
28 (41)	With the edge of oral part regularly lobed between the channel terminals . . .		29
29 (33)	About 12 to 15 channels and only slightly marked lobes between them		30
30 (31)	a = 3.2, b = 4.2% SL	(24) *Sarotherodon andersoni*	
31 (32)	a = 5.2, b = 6.4% SL	(25) *Sarotherodon mossambicus mortimeri*	
32 (33)	a = 2.9, b = 3.6% SL	(27) *Sargochromis giardi*	
33 (39)	More than 14 channels and well marked lobes between them		34
34 (35)	a = 4.0, b = 5.3% SL	(28) *Sargochromis codringtoni*	
35 (36)	a = 5.2, b = 6.3% SL	(26) *Tilapia rendalli*	
36 (37)	a = 3.4, b = 4.1% SL	(29) *Serranochromis macrocephalus*	
37 (38)	a = 3.3, b = 4.0% SL	(30) *Haplochromis carlottae*	
38 (39)	a = 3.5, b = 4.5% SL	(32) *Pseudocrenilabrus philander*	
39 (30)	Seventeen channels (a = 4.4, b = 6.7% SL) .	(23) *Aplocheilichthys johnstoni*	
40 (41)	Ten channels only, and ctenii (a = 3.7, b = 5.0% SL) . . . . (31) *Haplochromis darlingi*		
41 (42)	Radial channels in caudal part only (a = 4.8, b = 5.8% SL) . . . . (22) *Barilius zambezensis*		
42 (43)	Circuli open toward the caudal edge		43
43 (46)	Square diagonal edges		44
44 (45)	Scale elongated (a = 6.0, b = 4.5% SL)	(18) *Labeo cylindricus*	
45 (46)	Scale more or less circular (a = 5.0, b = 4.5)	(20) *Labeo congoro*	
46	Scales with rounded diagonal edges		47
47 (48)	Scale elongated (a = 5.3, b = 4.7% SL)	(19) *Labeo lunatus*	
48 (49)	Scale with rounded diagonal edges (a = 4.3, b = 4.2% SL)	(21) *Labeo altivelis*	
50 (51)	Scales with no channels		51
51 (52)	Center close to caudal edge		52
52 (53)	One row of long ctenii along caudal edge		53
53 (54)	Higher than long (a = 2.0, b = 2.7)	(11) *Distichodus mossambicus*	
54	Scale square edged but of the same length as height (a = 2.8, b = 2.8% SL) . . (12) *Distichodus schenga*		

---

[1] Primary radial channels run regularly from the center, secondary are only on the edge of scales.

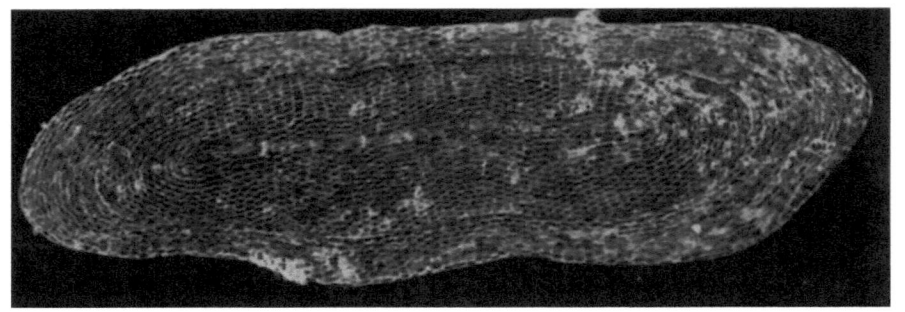

1. *Anguilla nebulosa labiata*

2. *Limnothrissa miodon*

3. *Mormyrops deliciosus*

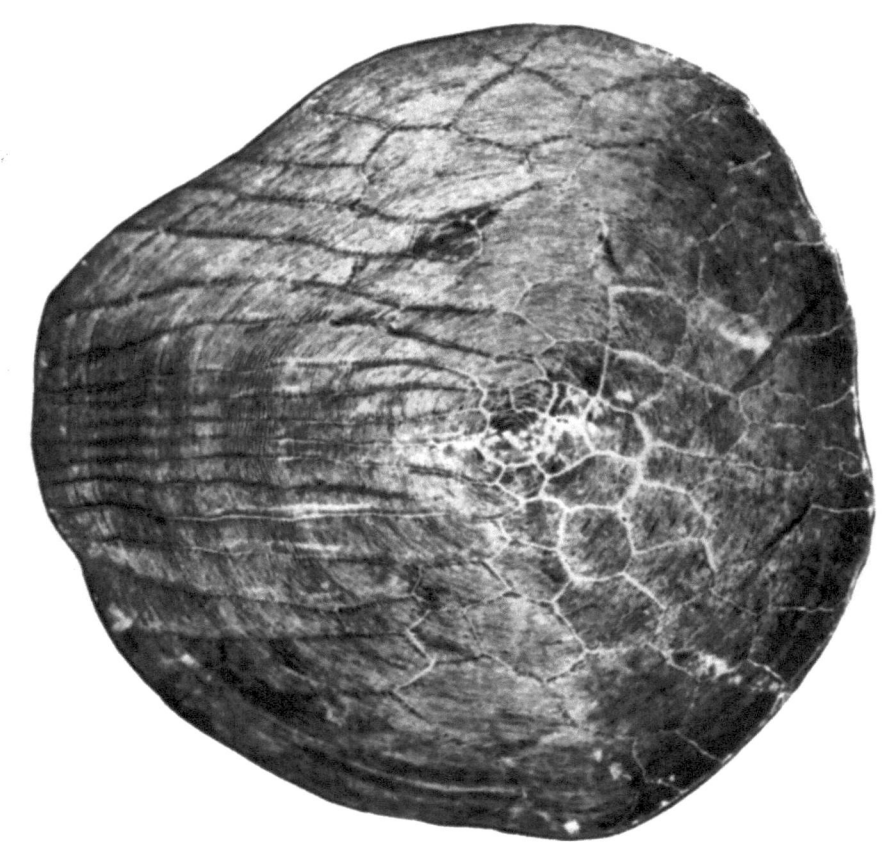

4. *Hippopotamyrus discorhynchus*

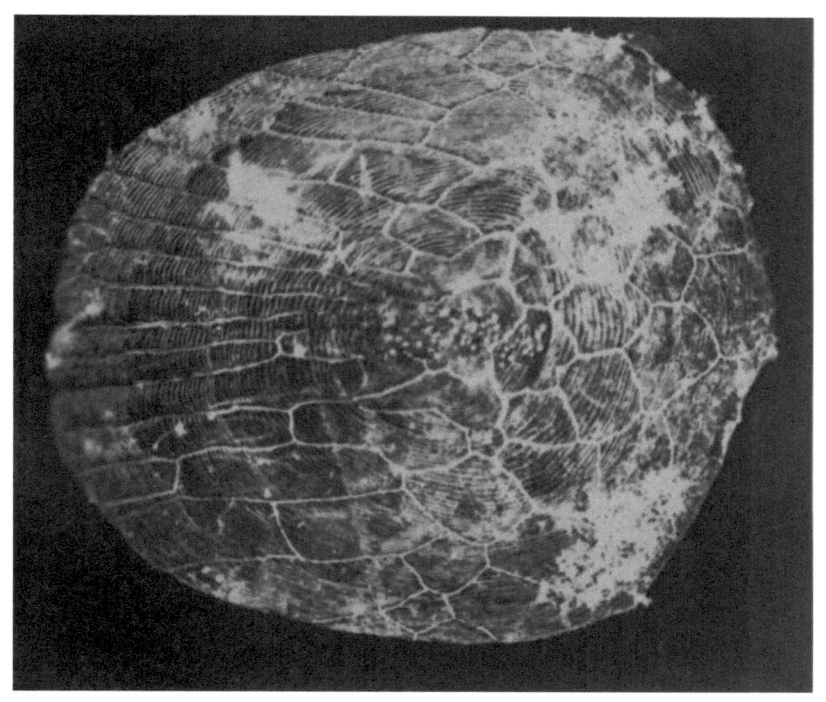

5. *Marcusenius macrolepidotus*

6. *Mormyrus longirostris*

7. *Hydrocynus vittatus*

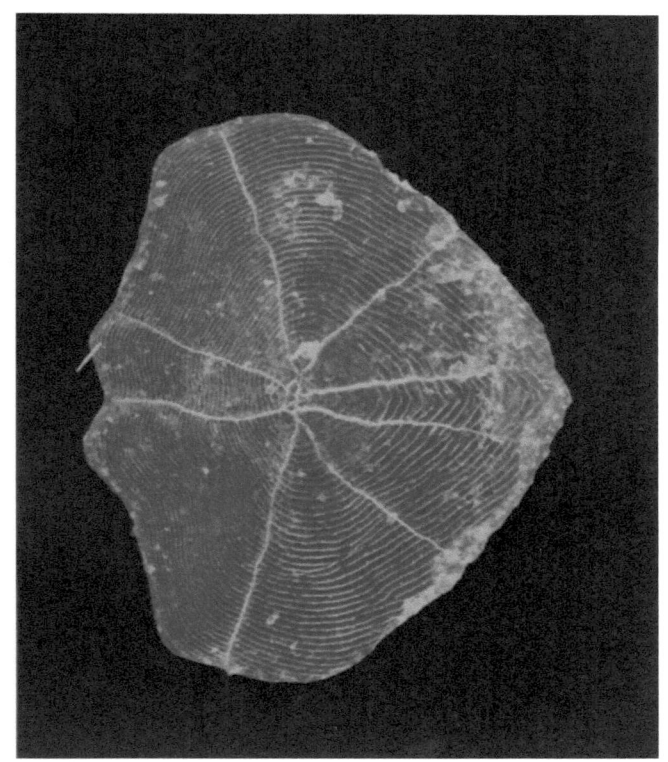

8. *Alestes lateralis*

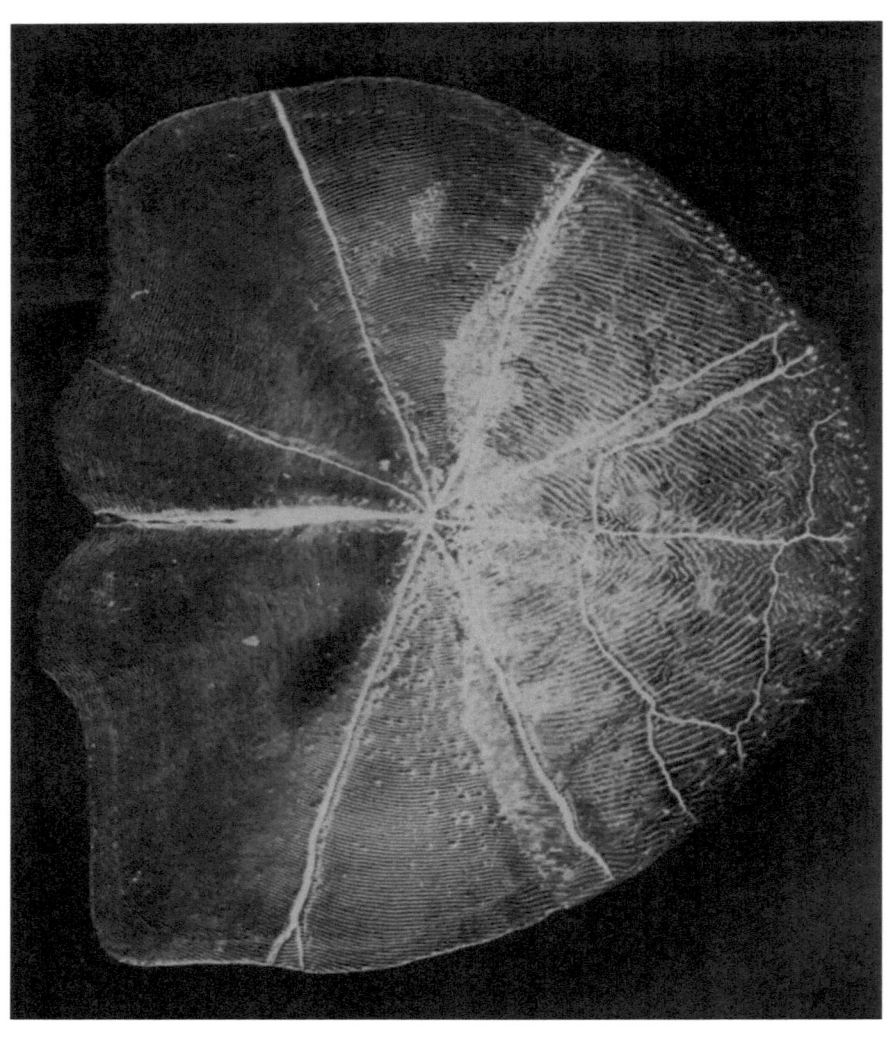

9. *Brachyalestes imberi imberi*

10. *Micralestes acutidens*

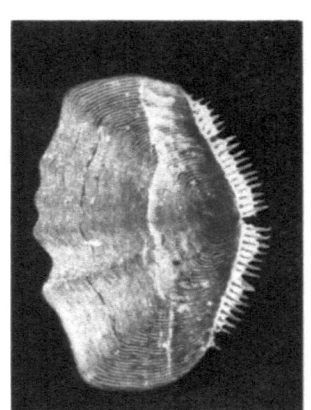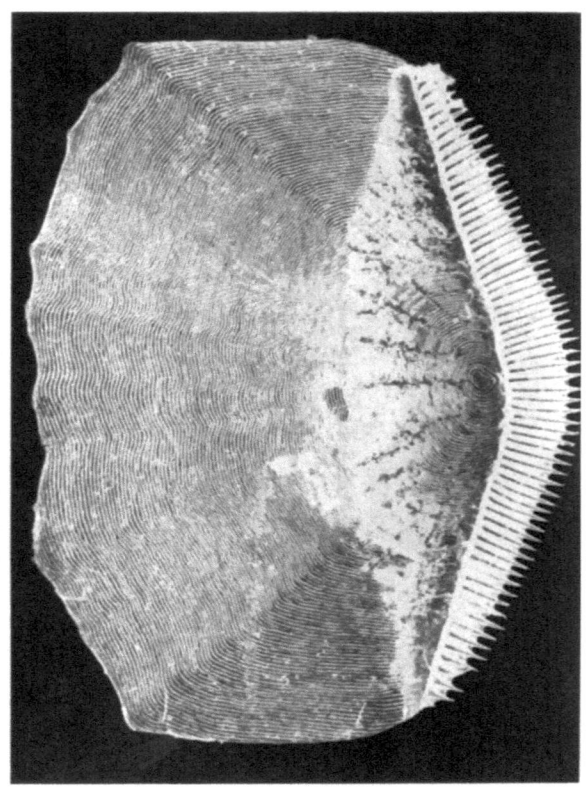

11. *Distichodus mossambicus*

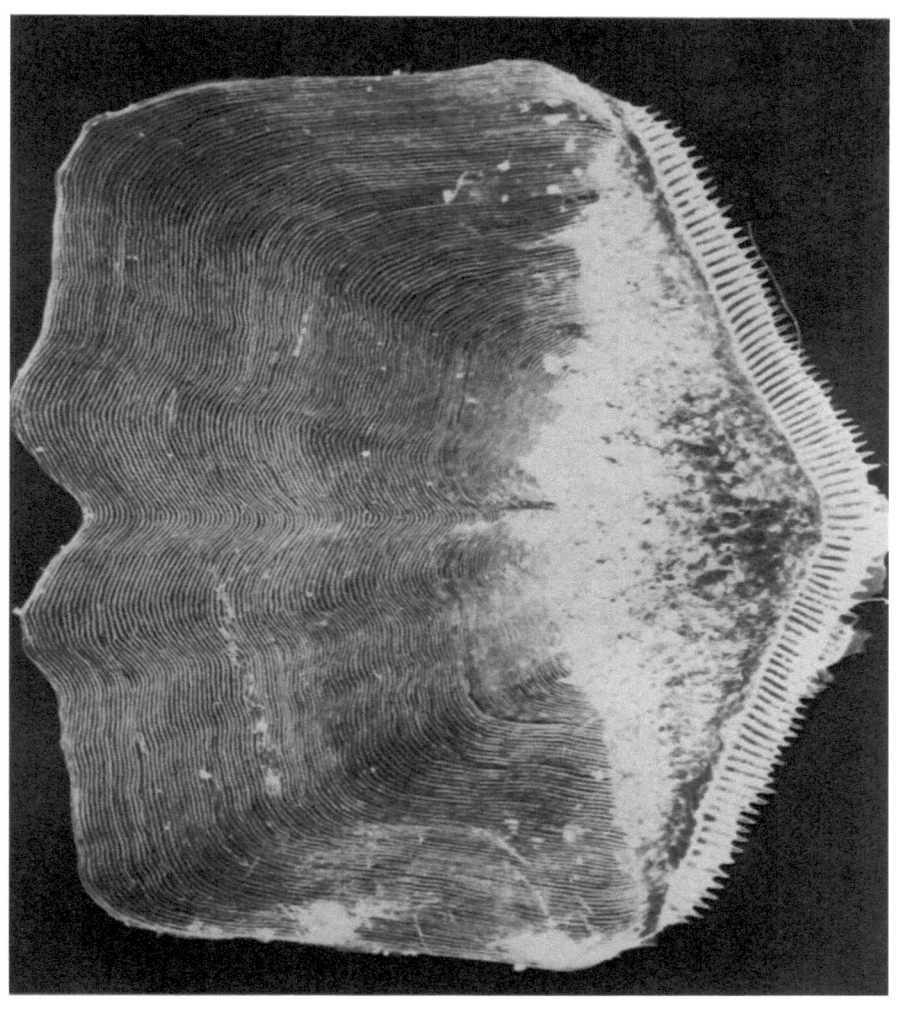

12. *Distichodus schenga*

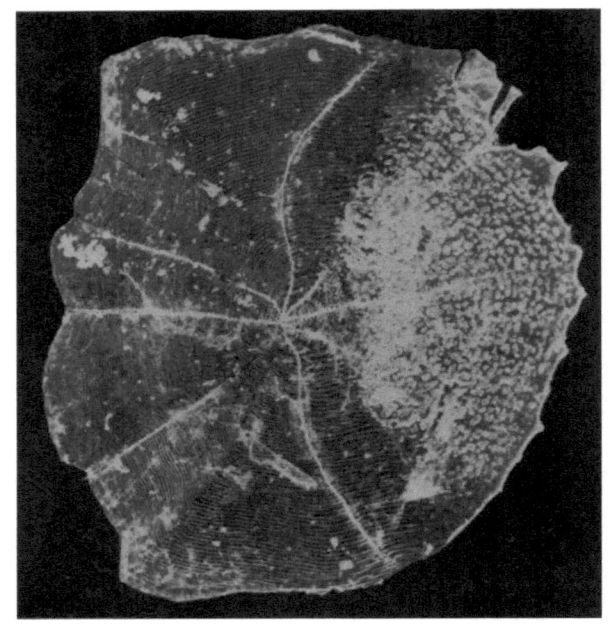

13. *Barbus poechii*

14. *Barbus paludinosus*

15. *Barbus unitaeniatus*

16. *Barbus lineomaculatus*

17. *Barbus fasciolatus*

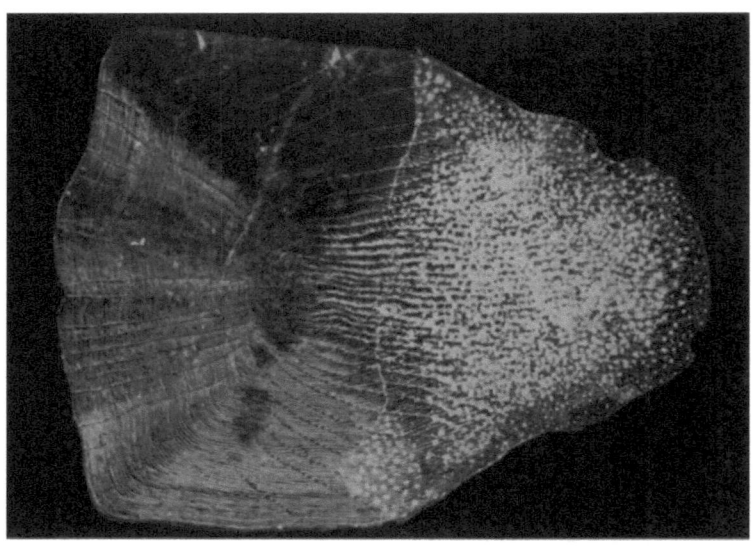

18. *Labeo cylindricus*

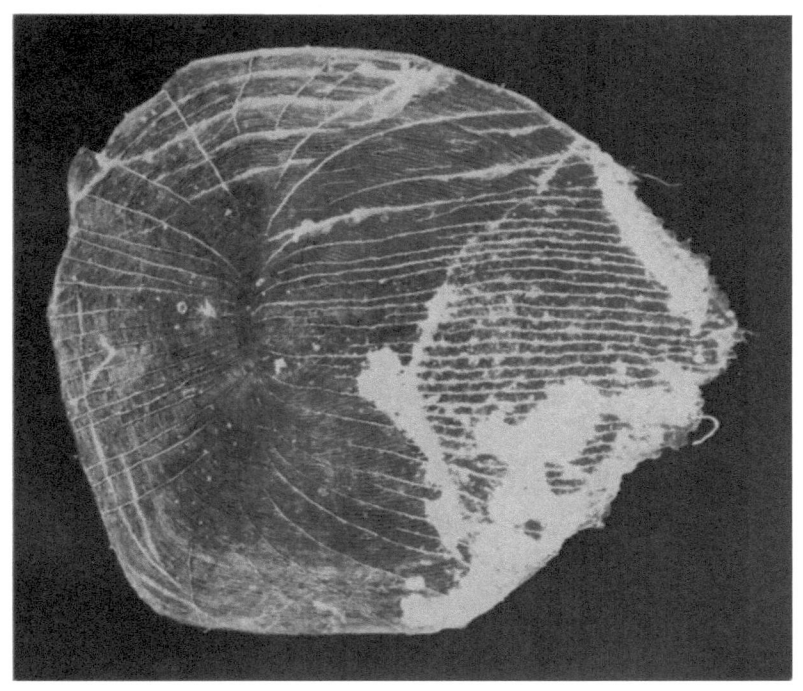

19. *Labeo lunatus*

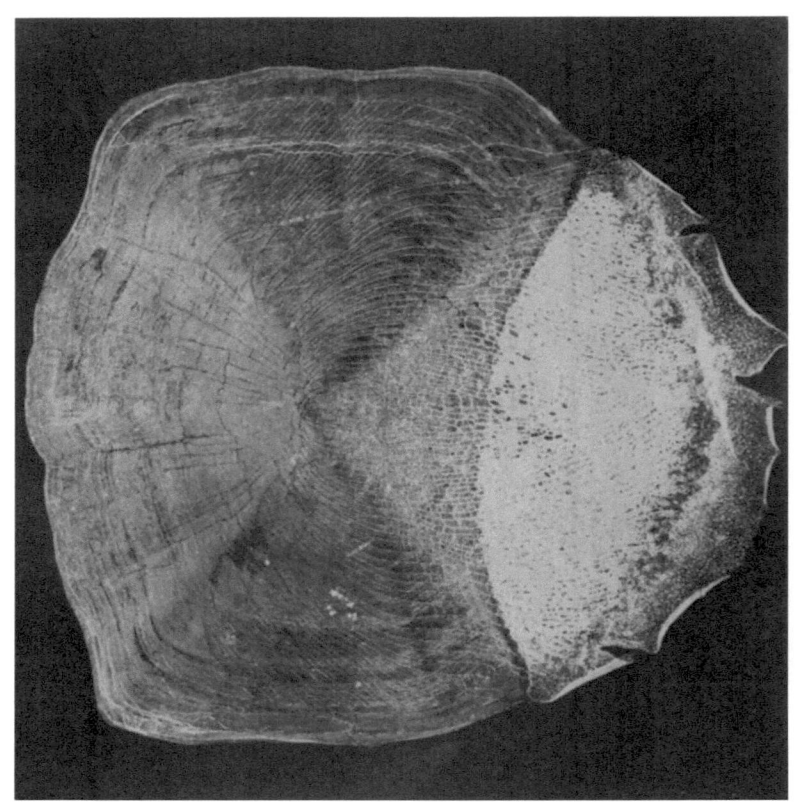

20. *Labeo congoro*

21. *Labeo altivelis*

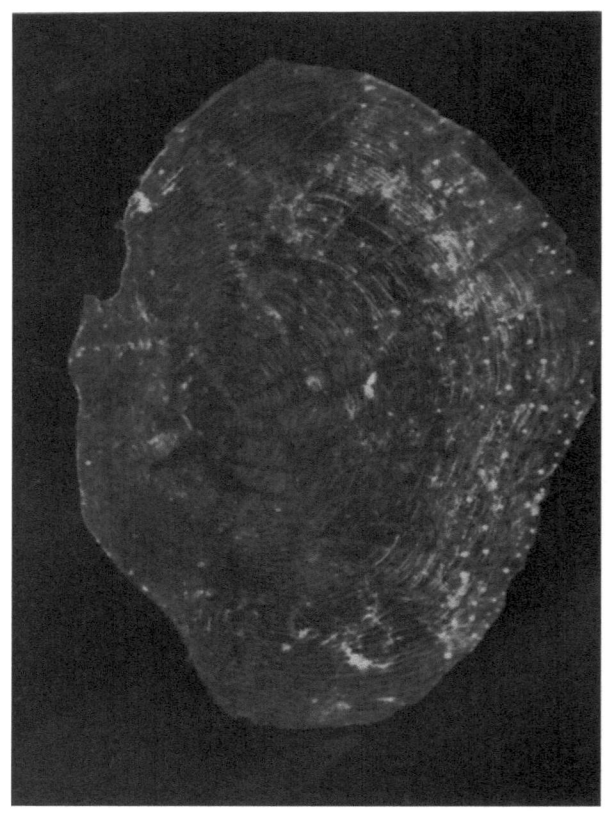

22. *Barilius zambezensis*

23. *Aplocheilichthys johnstoni*

24. *Sarotherodon andersoni*

25. *Sarotherodon mossambicus mortimeri*

26. *Tilapia rendalli*

27. *Sargochromis giardi*

28. *Sargochromis codringtoni*

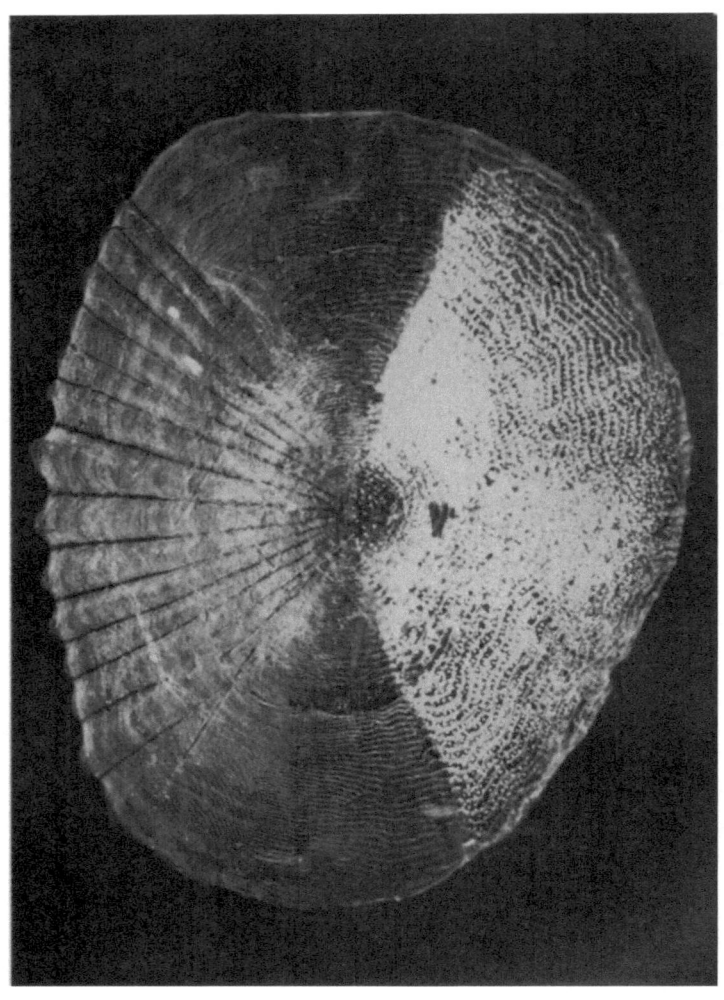

29. *Serranochromis macrocephalus*

30. *Haplochromis carlottae*

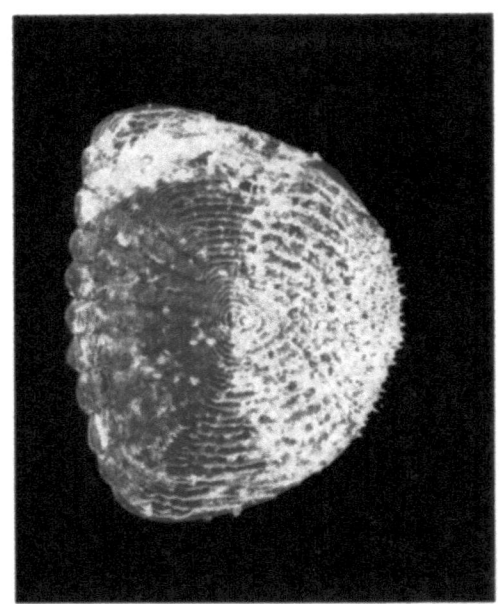

31. *Haplochromis darlingi*

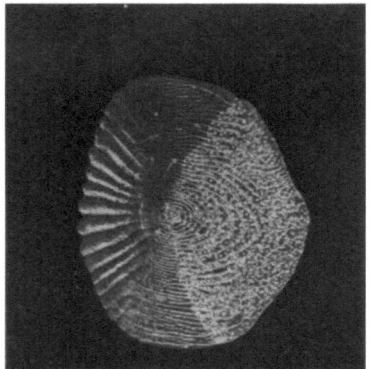

32. *Pseudocrenilabrus philander*

PART III

Plates

An Annotated Photographic Summary of the
Lake Kariba Ecosystem

by

EUGENE K. BALON

... So the more we dam the rivers, then the sooner we are damned.
(...)
There are benefits, of course, which may be countable, but which
Have a tendency to fall into the pockets of the rich,
While the cost are apt to fall upon the shoulders of the poor.
So cost-benefit analysis is nearly always sure,
To justify the building of a solid concrete fact,
While the Ecologic Truth is left behind in the Abstract.

>KENNETH E. BOULDING (1972)
>in "A Ballad of Ecological Awareness"
>accompanying "The Careless Technology"

Plate 1. Kariba Dam with one spill-gate open: Rhodesian power station on the left; cofferdam for construction of another power station now being built in Zambia to the right, and boom, a buoy barrier to keep weeds from engulfing the dam site at the entrance into the gorges (12 January, 1971). All aerial photographs in this part were taken from a Cessna aircraft flying at an altitude of 260 m.

Plate 2. Steep shore of clean washed rocks is bordered by a partially submerged forest. It is estimated that the trees will remain in this condition for 50 or more years (cove at north side of the entry into Kota Kota narrows, 4 April, 1968).

Plate 3. The rising water of the lake found its way into remote secondary valleys: here through a narrow passage (arrow) the lake formed Loteri Bay which backs up the new entry of the Lutele River. In less than one year after introduction the anchoveta invaded this landlocked part of the lake. Nesting grounds of the kurper and redbreasted bream were discovered in the river above its new entry (12 January, 1971).

Plate 4. Exposed shores, especially those of islands, were heavily eroded as a result of strong wave action (island east of Sinazongwe, the site of anchoveta releases, 24 April, 1970).

Plate 5. The first cove selected for quantitative sampling was near the dam at Siavonga West.

Plate 6. In the cove of Chikanka Island (sample C) began the series of quantitative sampling of Basin III (12 January, 1971).

Plate 7. Toxaphene was distributed from a running boat and mixed by the action of the engine propeller (Chikanka, 23 June, 1968).

Plate 8. Cove of Buffalo Island situated near the old submerged Zambezi River bed, was treated next (12 January, 1971).

Plate 9. The north cove of Chete Island yielded the largest quantitative sample (E); it was entered by the separately sampled Elephant Stream (es) (12 January, 1971).

Plate 10. Elephants had to be chased away from the shores of Chete (23 March, 1969).

Plate 11. Silt deposits along Elephant Stream bear tracks of our nighttime visitor, the laughing hyena (24 March, 1969).

Plate 12. Tigerfish, *Hydrocynus vittatus*, from the Lutele River.

Plate 13. The tigerfish had uncomfortably sharp teeth.

Plate 14. This *Labeo rubropunctatus* was originally identified as *L. altivelis*. It came from sample E at Chete Island (21 March, 1969).

Plate 15. Only one senile tigerfish (28 kg) had visibly wornout teeth.

Plate 16. A most common catch: the bottlenose, bream and tigerfish from an overnight gill-net set at Kota Kota (4 April, 1968).

Plate 17. Distichodus schenga from sample E.

Plate 18. Silver catfish, *Eutropius depressirostris*, from sample E.

Plate 19. A tailless silver or butter catfish was found in sample G from Kota Kota (29 May, 1969).

Plate 20. Female (above) and male parrotfish, *Hippopotamyrus discorhynchus*, from sample E.

Plate 21. The bull-dogfish, *Marcusenius macrolepidotus*, from sample E.

Plate 22. The rednose mudsucker, *Labeo altivelis*, from sample E.

Plate 23. The purple mudsucker, *Labeo congoro*, from sample G.

Plate 24. The DARLING's dwarf bream, *Haplochromis darlingi* (above) and the MOFFAT's dwarf bream, *Pseudocrenilabrus philander*, the smallest cichlids in Lake Kariba from sample E.

Plate 25. The shore line of a small island below the Kota Kota narrows; dead inundated trees and eroded rocky shores. Wave action exposes fine agates, amethysts and other gem stones; cloudy water area marks distribution of piscicide at collecting site during preliminary qualitative survey (17 July, 1968).

Plate 26. Some steep shores, remnants of the Gwembe Valley hillocks, are broken up into dangerous ravines (second island near the anchoveta release, 24 April, 1970).

Plate 27. When the soil was washed away the remaining rocks became an occasional resting place for cormorants. Sooner or later these rocks will also be eroded (island of anchoveta release, see Fig. 68).

Plate 28. In the rainy season access to lake shores was not easy and attempts sometimes ended disagreeably.

Plate 29. A portable Furuno F-860 echosounder was at first used to locate areas inhabited by fish and for construction of bathymetric maps (cove F at Chipepo Bay, 27 May 1969).

Plate 30. Later, for long-run traces, a more sensitive ELAC echosounder was installed in a catamaran (25 April, 1970).

Plate 31. For detailed mapping the ELAC echosounder was mounted on smaller crafts (photographed by A. G. COCHE).

Plate 32. Two sets of block nets had to be prepared (Chikanka cove, 22 June, 1968).

Plate 33. A blocking net with a large size mesh was set first and later, inside, along the measuring boom, a small size mesh net was placed into position (22 June, 1968).

Plate 34. Even the smallest juveniles, mixed in with floating weeds and submerged plants had to be sampled.

Plate 35. To prevent birds from creating a heavy bias in our estimates sampling had to proceed quickly.

Plate 36. Most fishes, specially the larger ones were measured on the spot and then separated into species (24 March, 1969).

Plate 37. In the end the parched and rather odiforous site was abandoned by most of the hired hands; only my assistants remained to finish the job.

Plate 38. Had they known that these hippos were nearby, they would surely have left as well.

Plate 39. After blocking-off the cove at the Kota Kota foothills (sample F) we attempted to remove floating salvinia in order to facilitate subsequent recovery of poisoned fish (27 May, 1969).

Plate 40. Blocked-off 0.5 ha square areas were quantitatively sampled in the lagoons of the Kafue floodplain (near Lochinvar, 3 June, 1970).

Plate 41. The head of a Namazambwe vundu.

Plate 42. Around the large Namazambwe Bay (Basin II) the Tonga people still adhere to their pre-impoundment farming practices. The bay later yielded an abundant eel harvest.

Plate 43. The Siengwazi Fall on the Kalomo River in the rainy season (12 January, 1971).

Plate 44. The African mottled eel, *Anguilla nebulosa labiata*, from the pool below Siengwazi Fall was first evidence of eels surmounting Kariba Dam (2 October, 1969).

Plate 45. A long-line with 175 hooks baited with beef and fish chunks was set at different depths across Lake Kariba (near the island N 162 at 42 fathoms, 29 January, 1971).

Plate 46. There was an abundance of large eels at depths of 25 to 40 m (near the Siavonga Peninsula, 28 January, 1971).

Plate 47. Special trap-nets were constructed. When baited and set at proper depths they proved to be a most efficient eel catching gear (Namazambwe, 20 February, 1971).

Plate 48. The drainage area of Lake Kariba is very diverse, numerous rivers and intermittent streams enter the lake forming habitats which have been harvested throughout the ages by local inhabitants (Lutele River estuary, 12 January, 1971).

Plate 49. From ancient times to the present simple weirs have been effective in harvesting fish that come into the streams as the rainy season progresses (Lusito River at Chipangula, 22 April, 1969).

Plate 50. Utilizing blocking nets or natural barriers, streams were quantitatively sampled with toxaphene or rotenone (Lusito River at Chipangula, 22 April, 1969).

Plate 51. The golden mudsucker, *Labeo cylindricus*, was one of the most common fishes in the streams and lotic parts of the lake (Kalomo River near boma, 28 April, 1969).

Plate 52. The Elephant Stream entry retained back-up water from the lake also in the dry season. Two months earlier all fish were eradicated here and now fish were back again in a higher biomass but lower diversity (25 May, 1969).

Plate 53. In the Kafue floodplains roofs of huts serve as platforms for the drying of fish. The fish are subsequently sold to itinerant traders.

Plate 54. The fish production on the floodplain may be substantially affected by the feeding and dunging of hippos. The herds remain relatively stable as they are feared by local hunters (photograph by J. M. KAPETSKY).

Plate 55. Lusenga nets employed by local fishermen to harvest clupeid schools at Lake Tanganyika (Mpulungu, 1 November, 1970).

Plate 56. Lusenga nets were employed at Lake Kariba after *Limnothrissa miodon* became well established there. They proved ineffective (Sinazongwe, near President Island, 14 October, 1969).

Plate 57. More successful than the Lusenga net, the frame lift net was employed to catch clupeids in the same place the following night.

Plate 58. Mixed *A. lateralis* and *L. miodon* catches were obtained in these early days (15 October, 1969).

Plate 59. Elephants tore off and ate the bark and wood of this baobab tree (south cove at Chete, 8 January, 1969).

Plate 60. In spite of this the baobabs remain alive and bloom every year, and contribute to a change in diet for passing herds of ungulates.

Plate 61. Every day at least once, the elephants come to the lake and deposit part of their undigested food directly into the water (south beach of Chete, 8 February, 1969).

Plate 62. The rest of the manure will also eventually find its way into the lake via rain run-off.

Plate 63. Attempts to cut through the fern mats with special high powered, closed cooling system vessels nearly ended in tragedy when after several hundred meters the mats closed and locked in the boats (Devils Gorge, 20 January, 1970).

Plate 64. Hunting, gathering and subsistence farming was the way of life in pre-Kariba Lake times (Namazambwe, 19 March, 1969).

Plate 65. The President of Zambia is deeply concerned about his people's well being.

Plate 66. His Natural Resources Development College in Lusaka is instructing its students in the management of the inherited wealth of western technology (at N.R.D.C., 2 April, 1969).

Plate 67. Promises of enormous fish harvests from Lake Kariba proved empty (Chikanka fishing camp, 23 June, 1968).

Plate 68. The catches, to be dryed for transportation to remote villages, decreased drastically and so did the temporary fishing camps. The following year the Chikanka Island camp was abandoned.

Plate 69. People and animals had to adapt to a new way of life; used tin cans and gasoline barrels replace clay pottery (Sinachukuju, 26 September, 1968).

Plate 70. The Tonga knew their country intimately and utilized a wide range of plants and animals; resettlement changed all this (Lufua River estuary, 20 October, 1968).

Plate 71. The elephant herds were capable of browsing a mopane forest down to bushlike remnants, fertilizing the forest floor with their dung at the same time (Chete Island, 8 February, 1970).

Plate 72. The area around the headwaters of Elephant Stream was an example of such a forest destroyed. But was it destruction? The herds moved on to other parts of the island and returned when the forest regenerated (25 March, 1969).

Plate 73. Sample B was taken with plastic explosives linked by an explosive cord, fired electrically from the shore.

Plate 74. Collection with rotenone at the edge of Victoria Falls, at time of lowest water level yielded excellent samples (3 November, 1971).

Plate 75. Immediately after the water drops down the lip of Victoria Falls it enters deep gorges which recapitulate the headwater topography of the pre-Middle Zambezi River (3 November, 1971).

Plate 76. Large *Barbus marequensis* locked in lotic refuge below Siengwazi Fall.

Plate 77. To explain the invasion of Lake Kariba by Upper Zambezi fishes samples from the edge of Victoria Falls were needed. First attempt with electrofisher yielded poor results (25 November, 1970).

Plate 78. Blister beetle *Mylabris dicinta* seasonally invaded the gardens of the resettled Tonga.

Plate 79. Baobab flowers bloom in October.

Plate 80. Sorghum harvest transported to Sinazeze Village by Tonga women.

Plate 81. These usually well hidden beauties give nightly concerts; one member of this ensemble is the green reed-frog *(Hyperolius tuberilinguis)*.

Plate 82. The lechwe antelopes roaming the Kaufe floodplain in the hundreds of thousands have to be protected (Chunga, 25 June, 1970).

Plate 83. The Zambezi River below Kariba Dam is a new habitat as is the lake above. Will the ecosystem adapt to the new flow regime?

Plate 84. For the time being a Tonga mother takes all this as inevitable (Lusito, 31 January, 1969).

Images taken with the multi-spectral scanner (MSS) in a wavelength 0.7 to 0.8 micrometers (upper red – lower infrared) from the ERTS-1 satellite orbiting approximately 1054 km above the Earth's surface. Each image covers an area of about 213 by 213 kilometers and is reproduced from the scale 1:1 000 000. Obtained from EROS Data Center, Geological Survey of the United States Department of the Interior and selected by E. K. BALON.

Image 1. Zambezi River and the upper part of Lake Kariba during a relatively low cloud cover on 17 January, 1973. Notice the different character and river width above and below the Victoria Falls (marked as 1), Basins I (marked 2) and II (3), and Namazambwe Bay (4).

Image 2. A major portion of Lake Kariba with a dense cloud cover abscuring the Kota Kota (5) Narrows (29 December, 1972). Notice the large island which is Chete (6); Chikanka Island (7) and the group of small islands where most of the anchoveta larvae flown from Lake Tanganyika were released (8).

Image 3. Major portion of Basin IV, Zambezi River below the Dam (9) and the Kafue River (23 November, 1972). Kafue floodplain (10), Kafue Reservoir (11) and the Kafue River entry into the Zambezi (12).

Image 4. The lowest portion of Lake Kariba and the Zambezi River below (22 November, 1972). Kariba Dam (9), Zambezi and Kafue River (14) and Luangwa River (15) confluences.

Images taken with the multi-spectral scanner (MSS) in a wavelength 0.7 to 0.8 micrometers (upper red – lower infrared) from the ERTS-1 satellite orbiting approximately 1054 km above the Earth's surface. Each image covers an area of about 213 by 213 kilometers and is reproduced from the scale 1:1 000 000. Obtained from EROS Data Center, Geological Survey of the United States Department of the Interior and selected by E. K. BALON.

1 ▼

▼ 2

3 ▼

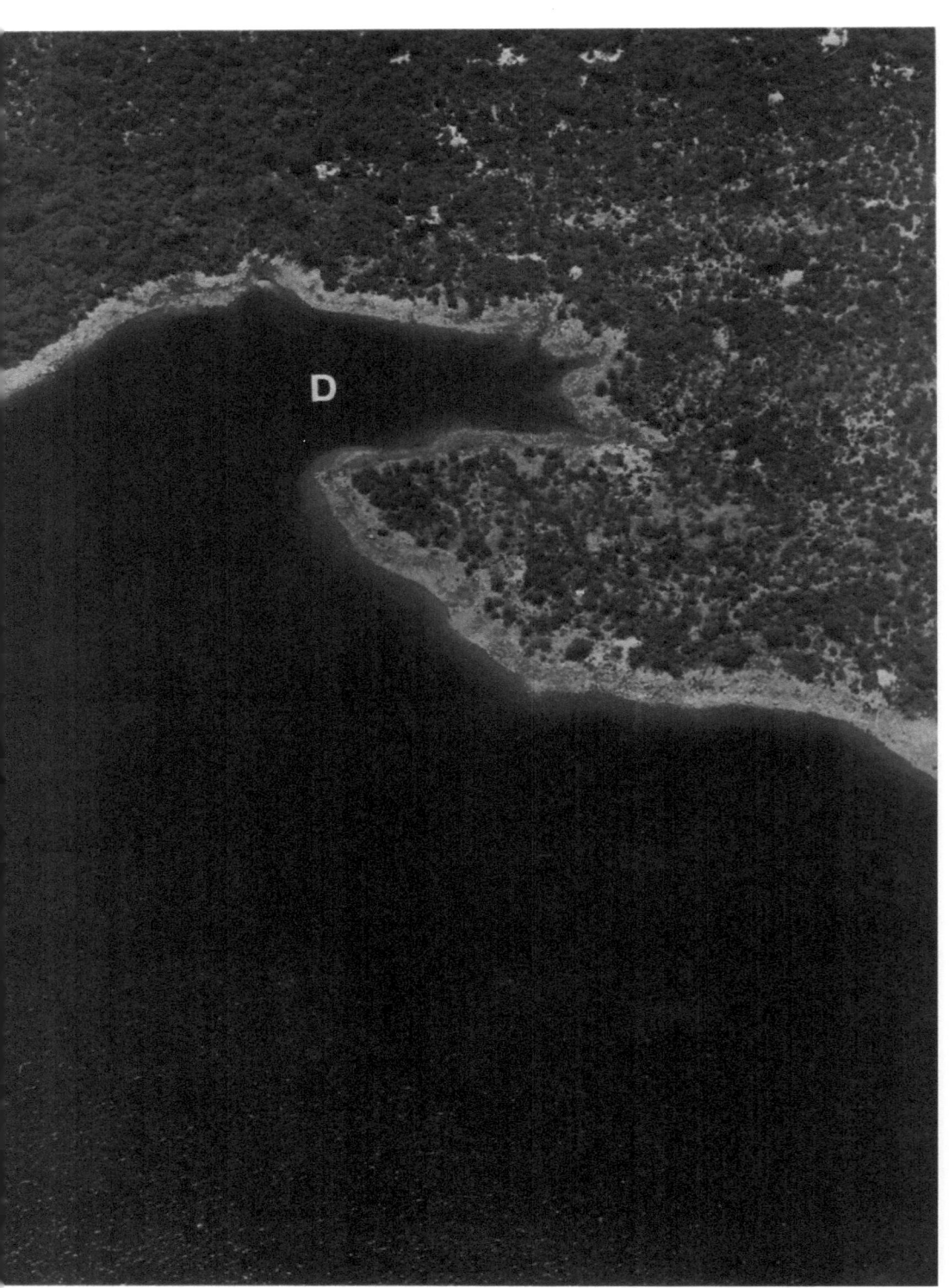

▲ 9

13▲

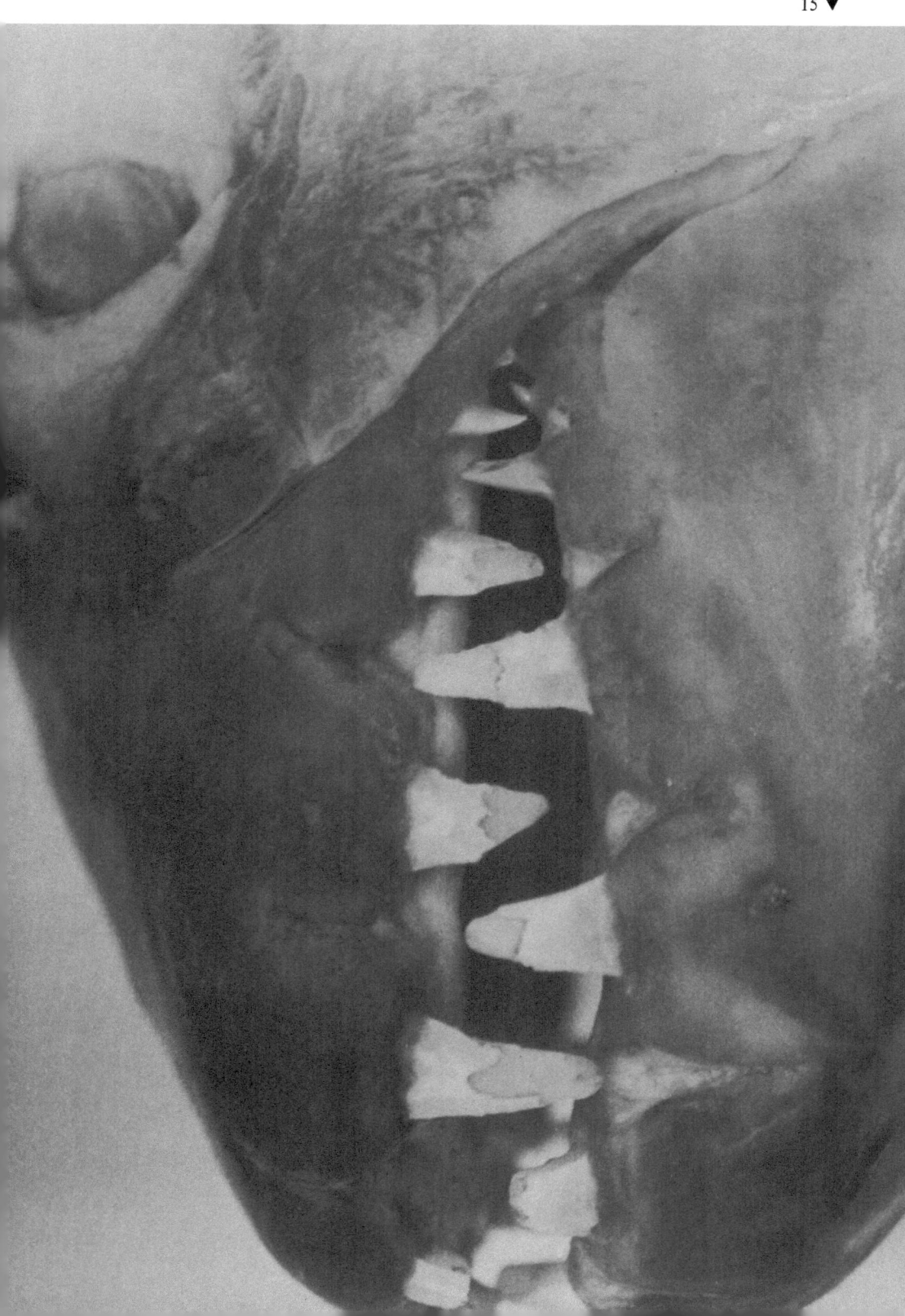

17 ▲   18 ▲

◀ 23

24 ▼

▲ 27

28 ▼

29 ▲

◀ 30

31 ▲

32 ▲

▲ 33

34 ▶

35 ▲

36 ▼

37 ▲

38 ▼

39 ▲

40 ▼

42 ▼

44 ▼

▼ 45

48 ▲

▲ 50

▲ 51

52

53 ▲

54 ▼

55 ▲

56 ▼

▲ 57

▼ 58

59 ▲

61 ▲

▼ 68

69 ▼

70 ▶
71 ▶▶
72 ▶▶▶

◀ 73   74 ▼

75 ▼

76 ▼

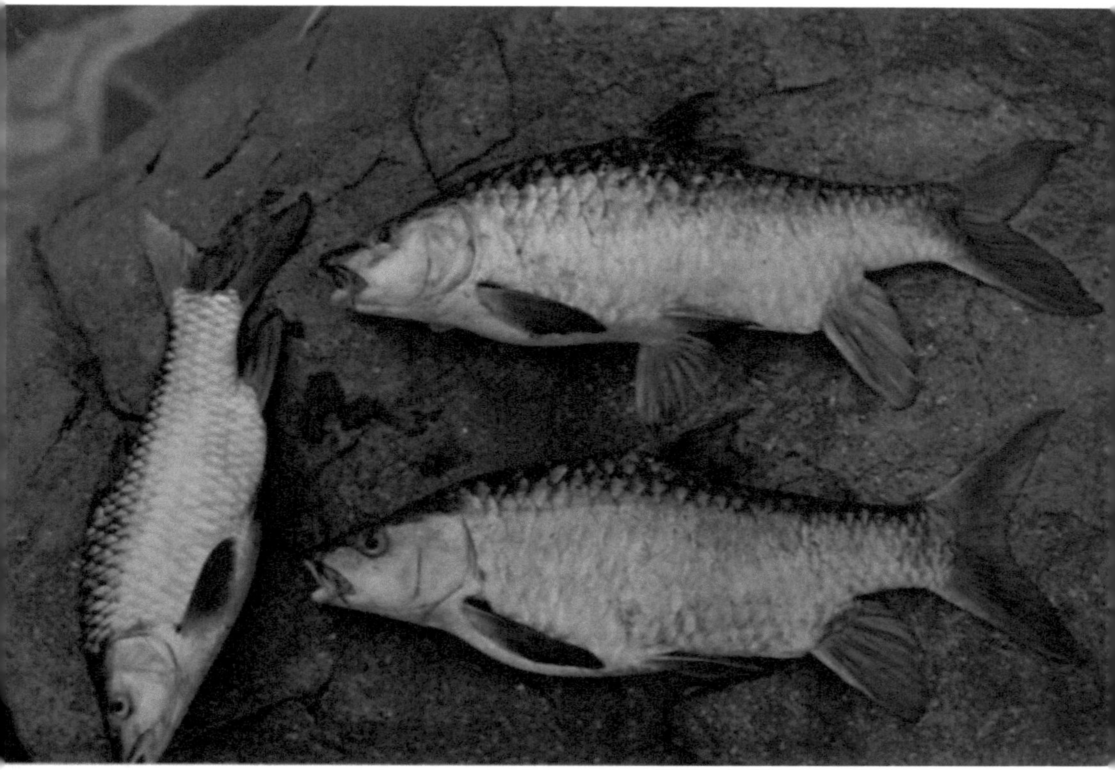

77 ▼

78 ▼

79 ▼

81 ▼

82 ▼

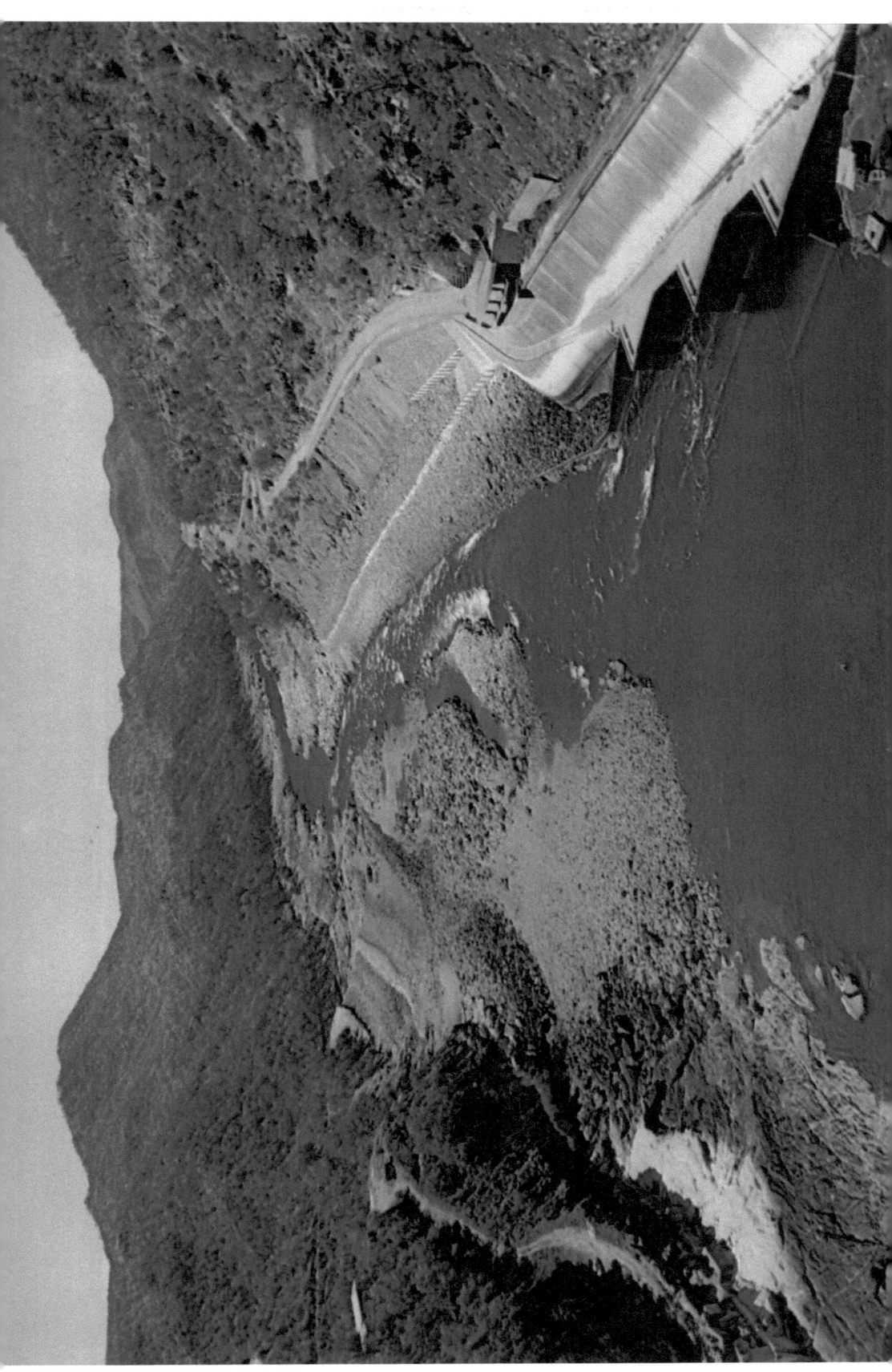

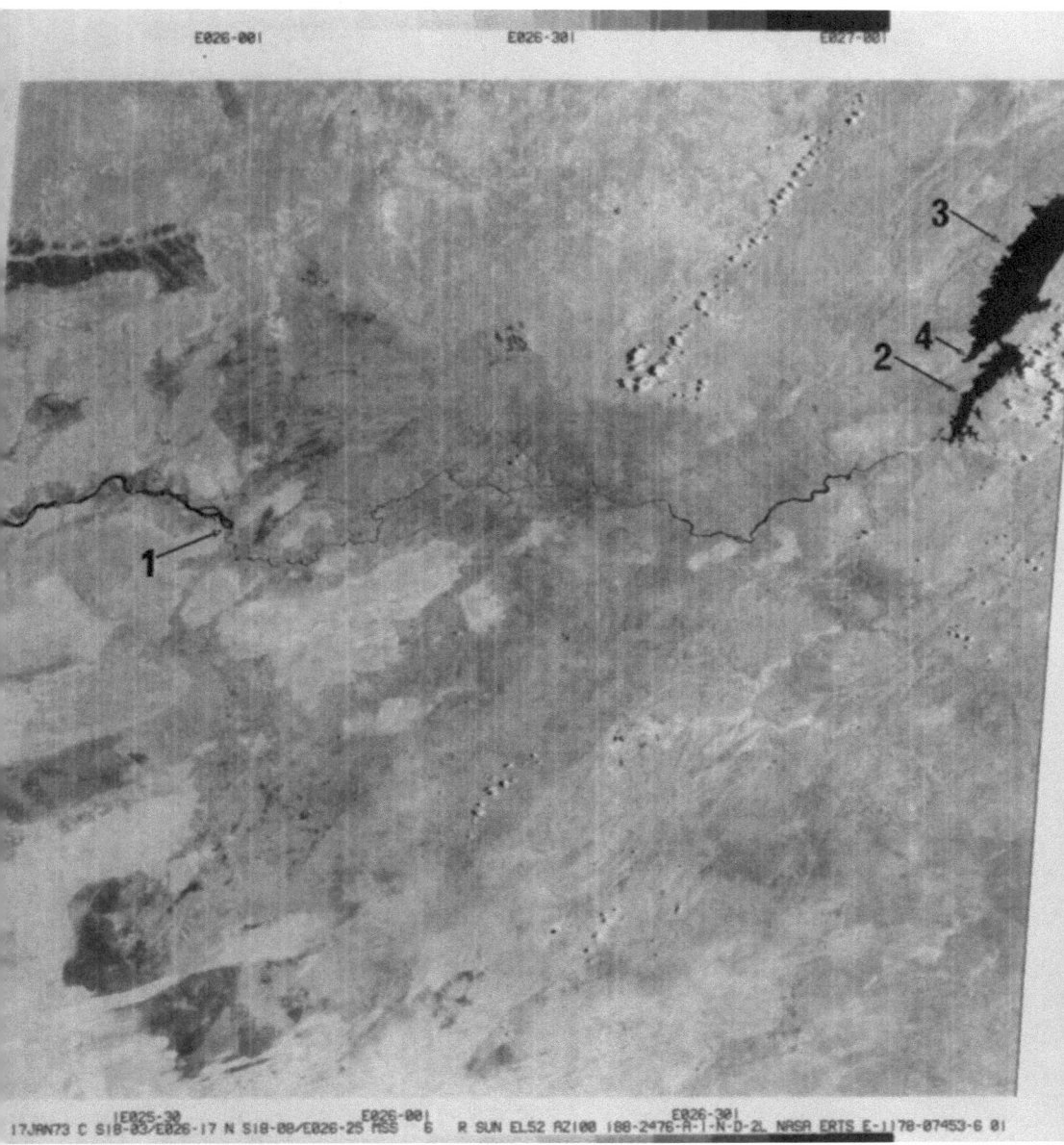

Image 1.

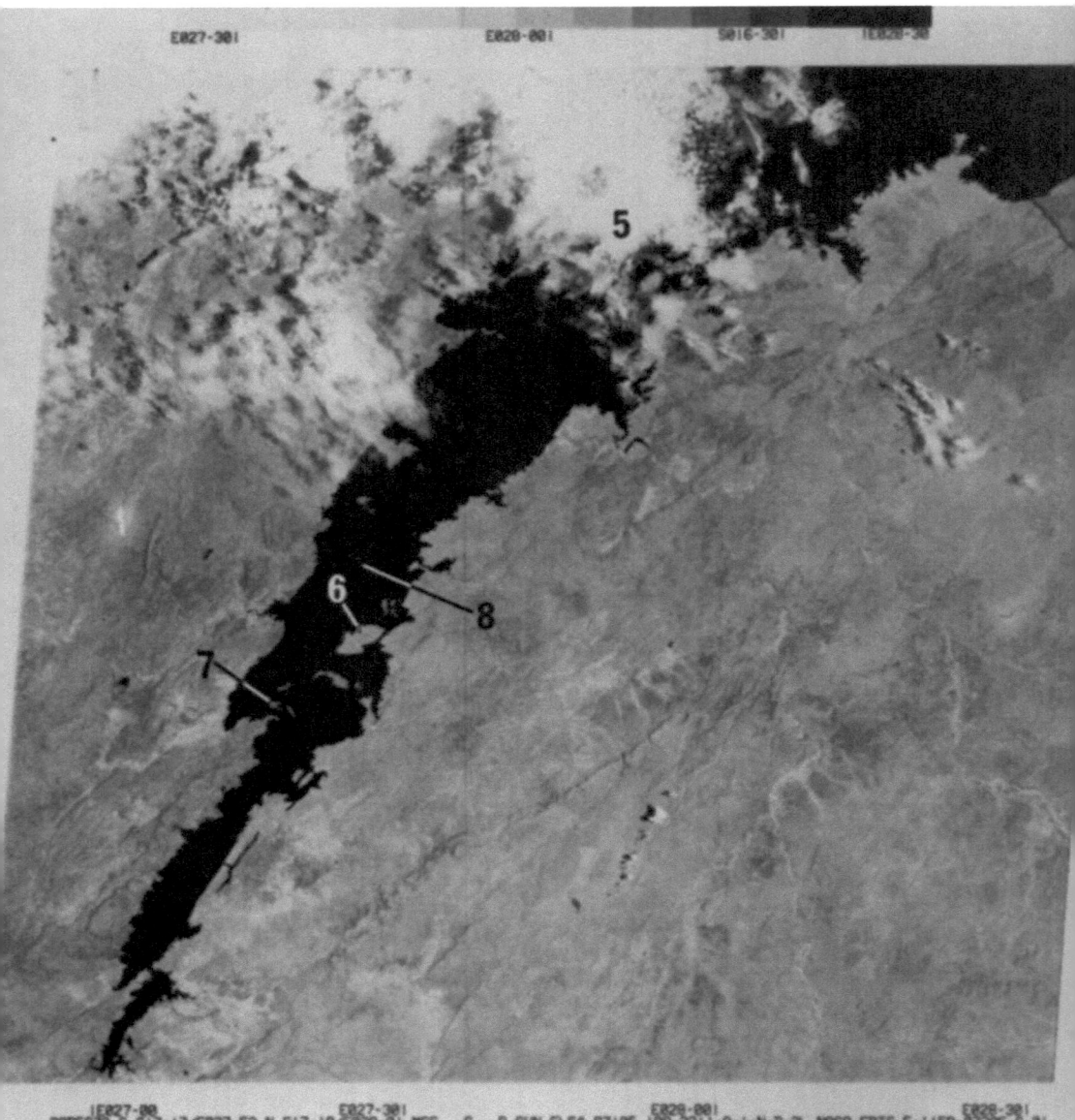

Image 2.

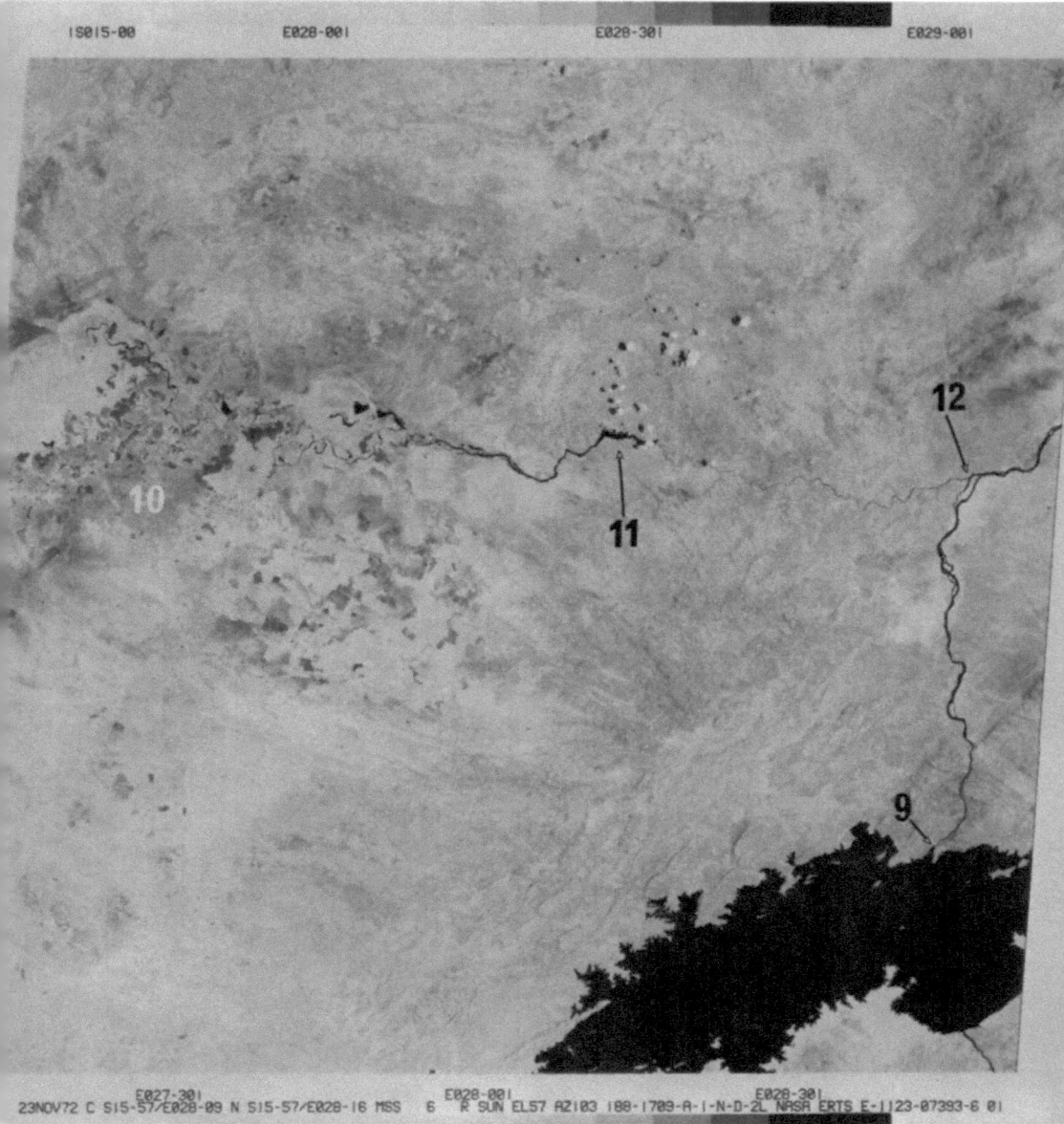

Image 3.

Image 4.

GENERAL INDEX

A

Abundance 278, 282, 576, 590, 591
Acacia 44
   albida 42
   nigriscens, ichtyotoxine 471
Acknowledgements 8, 264, 502
Actual catch 551
   trophic status 235
Adansonia digitata, baobab 41
Ademium obesum, ichthyotoxine 471
African mottled eel, *Anguilla nebulosa labiata* 709
Air, humidity 27
Akosombo Dam 7
Albert Lake 119
   alkalinity 230
   conductivity 230
   light penetration 129
   light transmission 120, 128
   light vertical attenuation 128
   productivity indices 234
Alestes lateralis 262, 264, 266, 436, 480, 492
   abundance 406, 438, 475, 477, 479, 480, 509, 535, 627, 629, 631, 633, 635, 636, 638–641
   behaviour 535
   biomass 438, 440, 543
   density 437, 439
   distribution 478, 481, 493, 529
   fecundity 480, 482
   growth 480, 599
   ichthyomass 480
   key scales 657
   life span 428
   meristic characters 482
   morphometric characters 482
   mortality rates 439
   production 438, 441, 480, 617–619
      available 441
   reproduction 483
   size & production values 444
   standing crop 477, 627, 629, 631, 633–641
      stock 509
   stock composition 269, 270
Algal vegetation 45
Alkalinity 191, 204
   total 56–59, 190, 199–202, 209, 210, 222, 227–233
Alkali metals 197
Alluvium 21, 44
   lower-terrace 23

Altitude, correction factor 164
Ameiurus melas melas 307
Amphilius platychir 463, 492
   abundance 466, 469
   distribution 494
   numbers sampled 490, 491
   standing crop 468, 469
Anacystis 46
Anchoveta, Tanganyican, *Limnothrissa miodon*
   abundance 536, 537
   behaviour 529, 538, 540
   density 440
   distribution 526–529, 535, 536
   fecundity 533
   fishing technique 526, 527, 536
   gonads 532
   growth 530, 533
   length 531
      frequency 532
   maturity 533, 534
Anemometer 27
Angola belt 27
Animal rescue 'Operation Noah' 46, 255, 556
   sanctuaries 46
Anionic composition 202
Annual heat budget 162
   index 162
Annulus 278, 643
   inception 643, 646
Anguilla mossambica 448, 449
Anguilla nebulosa labiata, African mottled eel 262, 264, 446–449, 463, 709
   abundance 466, 469
   density 439
   distribution 450, 493
   juvenile migration 464
   key scale 652
   lengths 449, 450
   life span 428
   standing crop 466–469
Aplocheilichthys johnstoni, abundance 477, 627, 633–639
   distribution 494
   key scale 669
   standing crop 477, 627, 633–639
      katangae, distribution 494
      hutereani, distribution 494
Aquatic vegetation 43
Area 87
   cleared 43
   depth zones 96

sampling units 588
typology 575, 576, 578, 579
Aswan High Dam 7
Atitlan Lake, stability 161
  tropicality index 160
  work budget 160
Auchenoglanis ngamensis, distribution 494
Available, production/biomass ratio 342
  yield 282, 342

B

Balovale 51
Ballad, ecological 735
Bandama River 7
  water quality 61, 62
Bangweulu Lake 191
  alkalinity 230
  chemical composition 224–226
  conductivity 185, 230
  euphotic zone 126
  ionic composition 195
  light penetration 129
  light transmission 120, 127
  mineral content 224, 228
  productivity indices 234
  silicate 224
  sulphate 201
  total solids 185
Bangweulu-Luapula-Mweru catchment 217
Baobab, *Adamsonia digitata* 41, 717
  flowers 730
Barbus afrohamiltoni, distribution 493
  afrovernayi, distribution 493
  annectens 463, 492
    abundance 466, 469, 493
    standing crop 468, 469
  barnardi, distribution 493
  barotseensis 463, 492
    abundance 466, 469
    distribution 493
    standing crop 468, 469
  belcrossi, distribution 493
  eutaenia, distribution 493
  fasciolatus 492
    abundance 477, 631–639
    distribution 493
    key scale 664
    standing crop 477, 631–639
  haasianus, distribution 493
  lineomaculatus 547
    abundance 635, 639, 641
    distribution 493
    key scale 663
    standing crop 635, 639, 641

  marequensis 463, 729
    abundance 466, 469
    standing crop 468, 469
  marequensis codringtoni 492
    abundance 491
    distribution 493
    numbers sampled 490, 491
  marequensis marequensis, distribution 493
  manicensis, distribution 493
  multilineatus, distribution 493
  neefi, distribution 493
  paludinosus 465, 475, 492, 547
    abundance 466–469, 477
    distribution 494
    key scale 662
    standing crop 466–469, 477
  poechii 463, 475, 492, 547
    abundance 466–469, 477, 631–633
    distribution 494
    key scale 662
    standing crop 466–469, 477, 631, 633
  puellus, distribution 494
  radiatus 463, 492
    abundance 466, 469
    distribution 494
    standing crop 466–469
  thamalakensis, distribution 494
  tangandensis, distribution 494
  trimaculatus, distribution 494
  unitaeniatus 492, 547
    abundance 477, 627–641
    distribution 494
    key scale 663
    standing crop 477, 627–641
  viviparus 463
    abundance 466, 469
    distribution 494
    standing crop 468, 469
Barilius zambezensis 463
  abundance 466, 469
  distribution 494
  key scale 668
  numbers sampled 490, 491
  standing crop 468, 469
Barotse Flood Plain 13, 18, 51
Basalts, Batoka 20
Basement Complex, rocks 99
Basins 84
Bathymetry 88, 93
Bathythermograph 105
Batoka Gorge 18, 20
  basalt 18
Belgian streams, standing stock 543
Benue River 7
Benthos 46, 47

Bicarbonate 199–205, 210, 213, 217, 218, 227, 228
Binga Township 28, 83
Biochemical activity 166
   nutrients 229
Biological production, volume 142
Birgean heat budget 147
Biomass 275, 278, 282, 438, 440, 544
   ichthyomass 278
   initial 278
   mean 278
   submerged tree fauna 41
Blister beetle, *Mylabris dicinta* 730
Block nets 702
Bolsena Lake 149
   annual heat index 153
   BAHB 153
   maximum heat content 153, 156, 157
   residual heat 153
   tropicality index 153, 154
Boom 679
Bosmina longirostris 47
Bottlenose, *Mormyrus longirostris* 349
Bottom deposits 23
Bourget Lake 153
Brachyalestes imberi imberi, spot-tail 264, 436, 480
   abundance 438, 477, 479, 631–642
   biomass 438, 439, 543
   density 437
   distribution 478, 481, 493
   fecundity 480, 482
   growth 602
   index of average size 429
   key scale 658
   length increment 480
   life span 428
   mortality rate 439
   production 438, 441, 480, 616, 617
   replacement 483
   reproduction 483
   size and production values 444
   standing crop 477, 631–642
Brachionus falcatus 47
Brachystegia 42
Breeze, land-sea 29
Budget, annual heat 162
   chemical 214
Buffalo Island, cove D 685
Bulawayo 16
Buleya-Malima Plain 23, 28
Bull-dogfish, *Marcusenius macrolepidotus* 696
Bumi 20
Bush, clearing 41
Bush-cleared fishing grounds 43, 99

Butter catfish, *Schilbe mystus*

C

Cabora Bassa 7, 101
   morphometry 100
   rapids 13, 18
Calcium 56, 57, 198, 209–218, 222, 228
CAPCOR, Central African Power Corporation 27, 77
Carbonate 59, 199–202, 205, 210, 211, 224, 227, 228
Carrying capacity 278, 553
Catch 278, 529
   actual 278, 443
   assessment survey 579
   clupeids 715, 716
   determining factor 530
   landed 278
   maximum sustained 278
Catfish tailless 695
Cation concentration 197, 198
Cationic ratio 59
Cavally River 7
Central African Federation 8
Central Fisheries Research Institute (CFRI) 8
Central Plains 25
Ceratophyllum demersum 45
Chad Lake 100
Characidae 406, 428
Chemical budget 213–219, 227, 228
   composition 210, 216, 217, 224–228
      evolution 222, 223
      flood water 212
      rainfall 212
      Zambezi River 58
   constituents 207, 208
   cycle 204–206
   evolution 219–222, 228
   quality 234
Chemocline 209
Chemocycle 227
Chete Gorge 180
Chete Island 98, 424, 472, 474, 686
   elephants 687
   sampling 474
   wildlife 474
Chezya River 67
Chilanga 8
Chiloglanis neumanni 463, 495
   abundance 466, 469
   distribution 494
   standing crop 468, 469
   swierstrai, distribution 494

Chimene River 83
  drainage area 67
  water quality 73
Chipepo Bay 426, 637
  evaporation 79
  harbour 28
Chirundu 23, 28
Chisi Lake 230
Chironomidae 41, 46, 47
Chloride 56, 57, 200–205, 213–218, 224, 227
Chlorophyta 46
Chobe 13, 18, 51
Chabbo Boma Mission 28
Charara River 20
Chrysophyta 46
Cichlasoma tetracanthus 295
Circulation 142, 181, 209
Clariallabes platyprospos, distribution 494
Clarias gariepinus, Zambezi barbel 264, 318, 342, 433, 463, 495
  abundance 318, 438, 466, 469, 475, 477, 509, 631, 633, 635–639, 641, 642
  age 318, 320
  body-vertebra relationship 319
  biomass 322, 438, 440, 544
  density 320, 437
  distribution 494
  growth 320, 321, 597
  index of average size 429
  length increments 323
  life span 428
  maturity 428
  mortality 320, 439
  minimum harvestable size 318, 320, 442
  production 320, 438, 441, 608
  size and production values 444
  standing crop 466, 468, 469, 477, 631–642
  standing stock 318–320, 509
  weight increments 320
  yield 322, 443, 445
  ngamensis 463
    abundance 466, 469, 509
    distribution 494
    standing crop 468, 469, 509
  submarginatus 492, 494
  theodorae 463, 492
    abundance 466, 469
    distribution 494
    numbers sampled 490, 491
    standing crop 468, 469
Clearing, bush 41
Climate 25, 27, 35, 214, 215
  annual cycle 35
  Gwembe Valley 27, 36
  humid mesothermal 25

Köppen 26
  normal 27
  Binga 30
  Chipepo 31
  Kariba Airport 32
  relative 40
  stations 35
  steppe 25
Climatographs 29, 33
Cladocera 47
Closterium kuetzingii 45
Clupeid, behaviour 538, 540
  density 529
  harvest 715
  introduction 524, 525
  sampling localities 528
  schools 540
Coelenterata 47
Coffer-dam 679
Colophospermum mopane 41
Colonization, plants 45
Commiphora 41, 42
Communities, explosive development 43
  sudd 45
Como Lake 149, 153, 156
Comoe River 7
Conductivity 56, 57, 183–192, 200, 209, 219, 221, 222, 227, 229, 230, 233
  electrical 63
  equivalent 189, 190
Congo River, water quality 61, 62
Canopy, grass 44
  woody 44
Convergence zone, intertropical 25
Co-ordinates, geographical 84
Coptostomabarbus wittei, distribution 494
Cornish-jack, *Mormyrops deliciosus* 349
Correlation matrix 582
Cove A 682
  C 683
  D 685
  E 686
  F 705
Ctenopoma ctenotis, distribution 495
  multispinis 495
Cuban lagunas, available production 543
Cultivation, semi-permanent 23
Cultural tribes 554
Cycle, rainfall 51
  runoff 51
Cyanophyta 46

D

Danube River 255, 256, 542, 543

Darling's dwarf bream, *Haplochromis darlingi* 419, 697
Dartmoor stream, production 543
Deer Creek, production 543
Deka River 84
Dendritic shores 86
Density 278
  current 131, 132, 214
Deoxygenation 173, 175, 177, 178, 180
Deposits, Quaternary 21
Depth 92
  maximum 169
  mean 88
  mixed layer 142
  contours 87, 91, 92
  of visibility 109
Development, biological communities 43
  explosive 43
  hydrophytes 43
Devil's Gorge 83, 84, 98, 102
Diptera 46, 47
Discharge, monthly 68
Dissolved, gases 164
  inorganic matter 217
  minerals 216
  oxygen 164, 167, 168, 172, 178–181
  solids 556
Distichodus mossambicus, sikapapa, abundance 633
    distribution 493
    key scale 660
    standing crop 633
  schenga, schenga 694
    abundance 477, 631, 633, 635
    distribution 493
    key scale 661
    standing crop 477, 633, 635
Distribution, controlling factors 575
Downwarping 20
Drawdown level 82

E

Ebo Stream 465
Echinochloa 498
Echosounder 701
Echosounding 83
Ecological history 255, 256
Ecosystem, influences 213
Edward Lake 119, 120, 128, 129, 230, 234
Eels 446, 466, 709
  abundance 467
  age 452
  catch 454, 455
  density 440, 455–458

distribution 446
growth 448, 450, 453
history 449
juveniles 450, 463
limits of distribution 449
otolith 447, 448
production 456, 458
sampling 447
standing crop 467
stock size 456
taxonomic status 448, 449
yield 456
Eichornia 7
ELAC Echograph Castor 537
Electric catfish, *Malapterurus electricus* 325
Electrofishing 729
Elephants 687, 717
  browsing 725
  manure 718
Elephant Stream 472, 476, 477, 686, 688, 713, 726
  fish biomass 475
  fish density 475
  standing crop 475
Elimination 279
Eudorina elegans 45
Energy 163
  budget 549
  content 146
Engraulicypris brevianalis, distribution 494
Epilimnion 139, 159, 160, 163, 170, 183
  dissolved oxygen 171
Erosion 20, 83, 99
Escarpment 16
  grit 20
Eunotia garusica 45
  pectinalis 45
Euphorbia 41
  candelabrum, ichthyotoxine 471
Euphotic zone, depth 110, 126, 164, 176
European eels, length-weight relationship 450, 451
Eutrophy 46, 549
Eutropius depressirostris, silver catfish 264, 333, 434, 694
  abundance 336–338, 477, 631, 633, 635, 638–642
  age 333
  annual increments 337
  biomass 336–342, 438, 440, 543
  density 333, 336, 437
  distribution 494
  growth 336, 600
  length-vertebra relationship 334, 335
  life span 428

minimum harvestable size 336, 342
mortality rate 439
production 339-342, 438, 441, 614, 615
   available 336-338, 342, 441
size and production values 444
standing crop 477, 631, 633, 635, 638-642
standing stock 336, 342
yield 339-342
   available 336, 338, 342, 445
   total 443
yangambianus, distribution 494
Evaporation 29, 33, 66, 79, 80, 215
Evapo-transpiration 33
Explosives, sampling with 727

F

Faulting 20
Fauna 41
   of submerged trees 41
Fern, explosive growth 43
   mats 719
   Salvinia auriculata 43, 181
Fish, abundance 438, 592
   age determination 265
   annual increments 265
   area inhabited 259
   density 266, 268, 542
   distribution 234, 265
      limits 182
   drying 714
   growth 262, 265, 266, 268, 272
   harvest 235, 722
   invasion 547
   life intervals 266
   resource 546
   mortality 265
   oxygen requirements 176
   population 262
   potential 542
   production 82, 183, 188, 255-259, 459, 549, 551, 553, 556
   resource 545, 552
   size 266
   standing stock 258
   succession 262
   surplus 546, 547
   total density 439
Fishing, bush-cleared grounds 43
   camps 722
   methods 529, 531
Fisheries Training Center 47
Filling phase 80
Flood 51
   base 65, 67

peaks 51
season 77
storage level 82
water, chemical composition 212
Flora 41
Flow, character 56, 57
Zambezi 51
Food and Agriculture Organization of the UN (FAO) 8
Food production 556

G

Gache-Gache River 21
Gale 29
Game reserves 46
Gases, dissolved 164
Gastropods 47
Geneva Lake 149, 153-156
Geographical co-ordinates 84
Geology 213, 214
George Lake 129, 230, 234, 543
Global radiation, see radiation
Glossary 278
Gneisses 20, 21, 23
Golden mudsucker, *Labeo cylindricus* 712
Gradient, thermal 163
Gravity center, depth 161
Great Slave Lake 111
Green bream, *Sargochromis codringtoni* 280
Growth, intensity 430
   rate 271
Gwaai River 16, 21, 67
Gwelo 16
Gwembe Valley 16, 18, 20, 21, 25, 174
   cross section 22
   cultivable land 23
   fauna 46
   human population 47
   profile 65
   Tonga 47
   vegetation 41

H

Haplochromis carlottae, rainbow bream 492, 547
   abundance 420, 631, 641
   distribution 495
   key scales 675
   standing crop 631, 641
   darlingi, Darling's dwarf bream 264, 436, 492, 697
      abundance 419, 420, 438, 475, 477, 627, 629, 631, 633, 635, 636, 638-642

age 421
biomass 422, 438, 440, 544
body-scale relationship 424
density 437, 549
distribution 495
food 426, 427
growth 421, 422, 425–428, 601
habitat 420–422, 426
ichthyomass 420
juvenile marks 422
key scales 421, 676
life span 428
mortality rate 421, 422, 439
numbers sampled 490, 491
production 422, 424, 427, 438, 441, 549, 551, 619–622
standing crop 477, 627, 629, 630, 633, 635–642
frederici, distribution 495
Hardness 56, 57, 198, 199, 202
Harvest 278
Heat, budget 157, 556
annual 162, 163
Birgean 147
content 150, 156, 163
annual cycle 154
maximum 152, 153, 156, 162, 163
seasonal variation 155
gain 160
index 149
annual 153, 162, 163
maximum 153, 157, 162, 163
incomes 154
residual 152, 162, 163
Hemichromis fasciatus 492
distribution 495
numbers sampled 490, 491
Hemigrammocharax machadoi, distribution 493
multifasciatus, distribution 493
Hepsetus odoe, abundance 509
distribution 493
standing crop 509
Heterobranchus longifilis, vundu 262, 324, 433
abundance 438, 477, 627, 631, 633, 635, 638–641
biomass 438, 440, 543
body-vertebra relationship 327
density 437
distribution 494
growth 325, 328, 329, 332, 598
life span 428
maturity 428
minimum harvestable size 442
mortality 439

production 438, 441, 607, 441, 444
standing crop 477, 627, 631, 633, 635, 638–641
yield 443, 445
Hippopotamus 704, 714
Hippopotamyrus discorhynchus, parrotfish 264, 380, 435, 492, 695
abundance 438, 477, 627, 631, 633, 638–642
age 381, 386, 387
biomass 382, 392, 394, 438, 440, 543
density 437
distribution 493
growth 390, 394, 599
key scales 654
life span 428
minimum harvestable size 394, 442
mortality 394, 439
production 392, 438, 441, 614
available 382, 441
scales 381–383
standing crop 477, 627, 631, 633, 635, 638–642
yield 392
available 382, 445
total 443
Horokiwi Stream, production 543
standing stock 543
Human, interference 42
population 41
Humidity, relative 29, 30, 38
Hunter-food gatherers 47
Hydrocynus vittatus, tigerfish 69, 264, 267, 268, 318, 432, 463, 492, 536, 689
abundance 406, 438, 466, 469, 477, 627, 629, 631, 633, 635–642
annulus 285
biomass 267, 438, 440, 543
density 437
distribution 493, 529, 597
index of average size 429
key scales 656
life span 428
minimum harvestable size 442
mortality 439
production 438, 440, 604
available 441
size and production values 44
standing crop 468, 469, 477, 627, 629, 631, 633, 635–642
yield 443, 445
Hydro-electric, development 7
scheme 77, 87
Hydrogen-ion concentration 183
Hydrogen sulphide 7, 175, 180, 215
Hydrology 77

755

Hydrophytes, attached 45
  emergent 45
  free-floating 43
  submerged 45
  vascular 43
Hyena 688
Hypolimnion 61, 159, 160, 165, 166, 169, 170, 179, 180, 183
  dissolved oxygen 171
  oxygen content 174
Hypolimnetic, areal deficit 170, 181
    temperate oligotrophy 170
    tropical oligotrophy 170
  oxygen depletion 170, 171, 173, 181
Hypopanchax jubbi, distribution 495
Hypsographic curves 89

I

Ichthyomass 298
Ictalurus lacustris punctatus 307
Identification key, scales 650, 651
Index, of average size 429
  of commercial importance 431
  of species average size 429
  of stock weight growth intensity 429
Indian Ocean 13
Inga Dam 7
Intervals in life history 275
Islands 98
Ions, inorganic 191
Ionic, composition 59, 191, 192, 195, 200, 201, 205, 209, 210, 227
  concentration 189, 190, 197
  contribution, rainfall 211
Isohyets 27

J

Jebel-Aulia Lake 129
Jellyfish, *Limnocnida rhodesiae* 47
Johnston Falls 555
Joint Fisheries Research Organization 8, 51

K

Kabompo River 13
Kafue 13
  lagoons, fish abundance 508
    fish standing stock 508
  River 13
    dissolved solids 500
    fish, abundance 508
    fish, fauna 499
    phosphates 500
    nutrient concentration 500
    standing crop 467
    standing stock 508
    water composition 60
    water quality 61, 62
  floodplain 497
    ecological fluctuations 521, 522
    fish abundance 509
    fish density 514
    fish production 498, 520
    fish standing stocks 509
    fishery 499, 520
    flooding effects 521, 522
    geography 498
    lechwe 732
    physical and chemical parameters 501
    production 519, 521
    primary production 500
    sampled localities 508, 510
    stock density 510, 512
    total production 512, 514, 518, 521
    vegetation 498
    wildlife 498, 499, 519, 520
    yield 514, 518
Kainji Dam 7
Kainji Lake 101
  chemical composition 225
  circulation 180
  dissolved oxygen 179
  general characteristics 244, 245
  light transmission 120
  mineral content 224, 228
  morphometry 100
  thermal cycle 146
Kalahari 18
  sands 98
  sediments 18
Kalomo, Culture 470
  River, archeological history 470
    biomass 467
    drainage 462
    eels 446
    fish abundance 466, 469
    fish distribution 465
    fish standing crop 465, 469
    fishing 470, 471
    geographical characteristics 465, 470
    gradient 461, 465
    highveld plateau 463
    history 460, 461, 470
    localities sampled 462
    local people and fish resource 467, 470
    lowveld (gorge and valley floor) 463
    middleveld (escarpment) 463
    population density 467

sources 460, 461
species composition 472
stream gradient 459, 460
turnover ratio 467
Kaolinite 23
Kariba, airport 28
   Dam 7, 16, 18, 27, 219, 221, 228, 262, 419, 446, 497, 679
      annual discharges 66
      annual flow regime 54
      catchment 13
      closed 8
      purpose 555
      spilling flow 61
      sluice gates 61
      turbine flow 61
      water discharges 77
      water level 84, 61
   eels, length-weight relationship 450, 451
   Gorge 83, 180, 184
      flow regime 61
   Hills 23
   Lake, actual catch 551
      actual trophic status 235
      area of depth zones 96
      basin depth contours 87
      bathymetry 93
      biological development 235
      breadths 86
      chemical budget 214, 227, 228
      chemical composition 225–228
      chemical cycle 204–206
      chemical evolution 219–222, 228
      depth 92
      depth contours 87, 91, 92
      ecological history 255, 256
      energy budget 146, 549
      fish, area inhabited 259
      fish invasion 484, 497, 547
      fish potential 542
      fish succession 262
      floor 18
      future trophic status 235
      general characteristics 244, 245
      geology 213, 214
      islands 98
      length 86
      light penetration 123, 125, 530
      mean slopes of depth contours 92
      mineral content 213
      monthly discharges 64, 68
      morphometry 85, 99, 100, 235
      nutrient content 235, 548
      nutrient load 262, 550, 551
      nutrient turnover 549
      optical properties 108
      physiography 213
      planimetry 87
      shore-line development 86
      stability 548–550, 557
      surface area 86, 87, 90
      thermal energy 146
      total catch 442
      total dissolved solids 184–186, 220, 233, 551
      trophic status 233
      tributary rivers 67
      vegetation 484, 552
      visibility 110
      volume 87, 88, 90, 92–94, 216
      water budget 67
      water capacity 93
      water fluctuations 523
      water level 262
   tragedy 556
   weed 181
Karro, floor deposition 20
   rocks 20
   sequence 21
   soils 98
Katanga System 18
Kentucky ponds, standing stock 543
Kirkia 42
Kličava Reservoir 543
Kneria auriculata 492, 493
   polli 493
Konkoure River 7
Köppen 25
Kossou, Dam 7
   Lake 101, 181
      dissolved oxygen 179
      general characteristics 244, 245
      morphometry 100
Kota Kota 20
Kurper bream, *Sarotherodon mossambicus mortimeri* 343
Kyoga Lake 230, 234

L

Labeo 69
   altivelis, rednose mudsucker 246, 432, 696
      abundance 394, 396, 438, 477, 631, 633, 638, 639
      age 396, 398, 399, 406
      biomass 438, 440, 443
      density 400, 437
      distribution 494
      growth 400–403, 598
      juvenile marks 401

key scales 395, 667
length composition 398
life span 428
maturity 397–401, 428
minimum harvestable size 404, 406, 442
mortality 439
production 404, 405, 438, 441, 610
scales 396, 397
standing crop 631, 633, 638, 639
yield 406, 443, 445
congoro, purple mudsucker 697
  abundance 631, 633, 639
  distribution 494
  key scales 666
  standing crop 631, 633, 639
cylindricus, golden mudsucker 463, 547, 712
  abundance 466, 467, 469, 477, 633, 635, 639, 641, 642
  distribution 494
  key scales 664
  standing crop 466–469, 477, 633, 635, 639, 641, 642
cylindricus annectens 490, 491, 494
lunatus, sailfin mudsucker 492, 547
  abundance 631
  distribution 494
  key scales 665
  standing crop 631
rubropunctatus 691
  abundance 635
  distribution 494
  standing crop 635
Lagarosiphon ilicifolius 45
Lake Kariba, drainage 13
  Fishery Research Institute (LKFRI) 8
  geological map 20–21
  seismic activity 24
Lake Kossou, Ivory Coast 99
Lakes, man-made 7
Lechwe, antelopes 732
Lamellibranchs 47
Lepomis macrochirus 295
Leptoglanis rotundiceps 492, 494
Life intervals 275
Light, attenuation 112, 118, 119, 129, 130
  green, penetration 121, 124
  incident 112
  intensity 105, 111, 118
  optical code 119
  optical depth 125
  optical properties 112, 121
  penetration 116, 121, 123, 125, 127, 129, 530
  radiation 149

relative intensities 112, 114, 115
spectral composition 118
total penetration 114, 115
transmission 120, 127
vertical attenuation coefficient 116
vertical illumination 112
visible 111, 112
wavelength penetration 117
Limestone 21
Limnocnida rhodesiae, jellyfish 47
Limnological cycle 137
Limnothrissa miodon, Tanganyikan anchoveta 262, 446
  abundance 536, 537, 641
  behaviour 529, 538, 540
  density 439
  distribution 493, 526–529, 535, 536
  fecundity 533
  fishing methods 526, 527, 536
  gonads 532
  growth 530, 533
  introduction 524, 547
  key scales 652
  length-frequency 531, 532
  life span 428
  maturity 533, 534
  standing crop 641
Livingstone 51
  monthly discharge 68
  Pump Station 51
    annual discharges 66
    evaporation 66
  river discharges 65
Localities, sampled 261
Lonchocarpus capassa, ichthyotoxine 471
Long-lines 709
Loteri Bay 681
Lower Zambezi River 13
Luapula River 191
  water composition 60, 62
Luangwa 13, 27
  Valley 21
Luanza River, standing crop 467
Lucerne Lake 149, 153
Ludwigia stolonifera 45
Lufua River 70
  chemistry 72
  conductivity 70
  drainage area 67
Lugamo Lake 153, 230
Lukulu 13
Lukunzu River 83
Lungwebungu River 13
Luputa Gorge 18
Lusaka 42

Lusaka-Choma ridge 65
Lusenga nets 715
Lusitu 23
Lusito River 711, 712
Lutele River 640, 681, 711
Luvua River 54
Luwani 42

M

Madumabisa mudstones 20, 23
Magnesium 56–58, 198, 209–211, 217, 218, 222
Malapterurus electricus, electric catfish 264, 324, 435
  abundance 438, 477, 631, 633, 635, 638, 641, 642
  biomass 438, 440, 543
  body-vertebra relationship 326
  density 437
  distribution 494
  growth 325, 328–331, 599
  life span 428
  maturity 428
  minimum harvestable size 442
  mortality 439
  production 438, 440, 612
    available 441
  standing crop 477, 631, 633, 635, 638–642
  yield 443, 445
Malawi Lake 100, 230
Mammals 46
Man in Gwembe Valley 47
Mana-Pools 46
Manica Platform 18
Manure 718
Map of sampling sites 473
Marcusenius ansorgi 490–493
  castelnaui 493
  macrolepidotus, bull-dogfish 264, 380, 381, 434, 492, 547, 696
    abundance 438, 477, 509, 631, 633, 635, 638–642
    age 381, 389
    biomass 383, 394, 438, 440, 543
    density 394, 437
    distribution 493
    growth 391, 394, 600
    key scales 655
    life span 428
    minimum harvestable size 442
    mortality 394, 439
    production 393, 438, 441, 612
    scales 382, 383
    standing crop 477, 631, 633, 635, 638, 639, 641, 642
    standing stock 509
    yield 382, 393, 394, 443, 445
Mastacembelus mellandi 490–492, 495
Masumo River 228
Mat, *Salvinia* 45
Matabele 16
Matusadona Range 16
Maximum, heat content 154, 162
  sustained yield 555
Mazoe Dam 230
Mbilizi 83
McIlwaine Lake 230
Mead Lake 149, 153, 157
Mean depths 88
Mechanical energy 147
Melosira granulata 45
Mesotrophy 46
Metalimnion 139, 159, 160, 163
Metals, alkali 197
Meteorological data 35
Micaschist 23
Micralestes acutidens, silver robber 262, 318, 492
  abundance 406, 412, 416, 418, 438, 477, 627, 629, 631, 633, 635, 638
  age 407, 410, 411
  annulus 411
  biomass 416–419, 438, 439, 544
  body-scale relationship 406
  density 417, 419, 437
  distribution 493
  fecundity 412–415
  growth 407, 412, 414, 417, 602
  habitats 419
  key scales 659
  life span 428
  maturity 412
  mortality 407, 411, 412, 415, 417, 439
  production 438, 441, 623
    available 407, 417, 419, 441
  reproduction 419
  scales 207, 407, 409
  sex composition 410
  standing crop 477, 627, 629, 631, 633, 635, 638
Microcystis flos-aquae 45
Middle Zambezi River 13, 25, 84, 492
  fish invasion 496
Mid-Zambezi Valley 18, 20, 200
Mineral 202
  contents 59, 183, 186, 191, 192, 205, 209, 213, 219, 224, 227, 228
  export 217, 218
  import 216

total 200
Minimum harvestable size 278
Miscellaneous rivers 70
Moffat's dwarf bream, *Pseudocrenilabrus philander* 697
Mollusca 47
Molteno 20
  sandstones 99
Mongu 27
Monomictic reservoir 142
Mopane 41, 42, 44
  woodland 41
Morar Lake 153, 154
Mormyrops deliciosus, cornish-jack 264, 349, 433
  abundance 349, 438, 477, 627, 633, 635, 638–641
  age 358, 360, 366–369
  biomass 378, 438, 440, 543
  density 370, 437
  distribution 493
  growth 372, 373, 378, 379, 596
  index of average size 429
  juvenile marks 378
  key scales 361, 363, 652
  life span 428
  minimum harvestable size 376, 378, 379, 442
  mortality 439
  production 438, 440, 605
    available 378, 441
    total 378
  reproduction 367
  standing crop 477, 627, 633, 635, 638–641
  stock density 367
  yield 378, 443, 445
Mormyrus ellenbergeri 493
  lacerda 493
  longirostris, bottlenose 349, 433, 463
    abundance 349, 438, 466, 469, 477, 633, 635, 638–641
    age 359, 370
    biomass 379, 438, 440, 543
    density 370, 437
    distribution 493
    growth 374, 375, 378, 380, 597
    juvenile marks 378
    key scales 362, 655
    life span 428
    minimum harvestable size 377, 380, 442
    mortality 439
    production 438, 440, 604
    reproduction 367
    standing crop 468, 469, 477, 633, 635, 638, 639, 641

  yield 379, 443, 445
Morphoedaphic index 233
Morphology 99
Morphometry 84, 85, 100, 235
Mortality 278, 279
Mottled eel 449
Mozambique Plain 18
Mpedele-Mutulanganga 23
Mulolo 20
Mweru Lake 54, 186, 188, 195, 201, 203, 224–233
Mweru-Luapula 120, 126–129
Mweru Wantipa Lake 230, 234
Mylabris dicinta, blister beetle 730

N

Najas 45
Namazambwe Bay 707
  eels density 456
Nannocharax macropterus 493
Naodza River 20, 21
Nasser Lake 126, 146, 181, 224, 225
Nasser-Nubia Lake 100, 101, 179, 244, 245
Natural tribes 554
Ness Lake 153
Nesting grounds 681
Ngami depression 18
Ngonye Falls 27
Niger River 7, 61, 62
Niles 61, 62
Nitrates 203, 209, 235
Nitrate-nitrogen 201–203, 209–215, 224, 227
Nitrogenous organic matter 201
North American streams, standing stock 543
North bank scheme 77
Northern Highlands 25
Nutrient, concentration 209
  content 235, 548
  dissolved 233
  load 262, 551
  retention by *Salvinia* 45
  turnover 549

O

Okavango 13
Oklahoma ponds, standing stock 543
Oligochaeta 47
Oligomesotrophic 108
Oligotrophy 170
Operation Noah, animal rescue 46, 255, 556
Optical properties 108
Orava Lake 430
Orange River 7

Orta Lake 153
Oryza barthi 498
Oxycline 209
Oxycycle 165, 171, 173-178, 181
Oxygen, concentrations 165
   deficit 167, 169, 170
   dissolved 164, 174-179
   distribution 165, 166, 174, 181
   fish distribution limits 182
   ratios 172
   saturation 165

P

Panicum repens 45
Papyrus swamps 121
Parrotfish, *Hippopotamyrus discorhynchus* 380, 695
Paulinia acuminata 45
Pebbly arkose 20
Pelagic water, mineral content 183
Pellonula afzelius 535, 536
Pelmatochromis ruweti 495
Perca fluviatilis 320, 325
Periods of life 279
Petersius rhodesiensis 463, 466-469
Petrocephalus catostoma 490-493
pH 56, 57, 59
Phases, biological communities 43
Phosphate 209, 213, 227, 235
   -phosphorus 201-203, 209-211, 224, 227
Photographs 677
Physiography 213
Phytoplankton 45
Pistia stratiotes 7, 45, 219
Plain eel 449
Plain squeaker, *Synodontis zambezensis* 298
Planimetry 86, 87
Plants, sudd 45
Plateau tract 18
Plates 677
Population 279
Post-filling phase 82
Potamogeton pusillus 45
   sweinfurthii 45
Potassium 56-58, 198, 209-211, 213, 215, 217, 218, 222
Potential energy 147
Potential fish harvest 233
Pre-Cambrian rocks 18
Precipitation 29
President of Zambia 721
Primary production 142, 146
Primary productivity 111
Processing 703, 704

Production 257, 382, 426, 438, 548
   available 265, 275, 278, 282, 345, 441
   biological 229, 234
   ecological 278, 521
   final 278
   harvestable 275
   organic 178, 180, 215
   terrestrial 215
   total 259, 275, 278, 441, 543, 544
Productivity 203, 229, 235
   biological 188, 201, 209
   organic 170, 182
   primary 234
Protopterus annectens brieni 493
Protozoa 47
Psammite 21
Pseudocrenilabrus philander, Moffat's dwarf bream 463, 466, 547, 697
   abundance 420, 469, 477, 627, 629, 631, 635-639
   distribution 495
   juvenile marks 643
   key scales 676
   numbers of annuli 644
   pond experiment 643
   standing crop 468, 469, 477, 627, 629, 631, 635-639
Pterocarpus 42
Purple mudsucker, *Labeo congoro* 697
Pweto 54
Pyramid Lake 160, 161
Pyranometer 29
Pyrrophyta 46

R

Radiant energy 105
Radiation, global 29, 30, 35, 39
   solar 112
Rain belt 27
Rainfall 27, 29, 35, 38, 79, 211
   chemical composition 212
   mineral composition 228
Rainy season 25
Rate of weight increase 430
Ratio of fish production to primary production 545
Red-breasted bream, *Tilapia rendalli* 311
Rednose mudsucker, *Labeo altivelis* 696
Reed-frog 732
Relative humidity, see humidity
Reoxygenation 179
Replacement of *B. imberi* 480
Resettle, local population 23
Residual heat 149, 151, 162
Rhabdalestes rhodesiensis 493

Rhodesia 7, 99
  islands 98
  shoreline length 86
Rhodesian Meteorological Department 25
  Highlands 25
  Plateau 16
Rivers, discharges 65
  secondary, chemistry 211
    ionic content 211
Riverine characteristics 142
Roads 700
Rotifera 47
Runoff 79
Rutilus rutilus 320, 325

S

Sake Lake 230
Salinity 54, 59, 183, 186–188, 192, 195, 203, 211, 219–221, 227, 228
Salisbury 16
Salts 203, 205
Salvinia auriculata 7, 43, 215, 219, 235, 551
  autecology 45
  control 45
  coverage 43
  mat 45, 70, 99, 449
  weed 181
Sampled localities 260, 261
Sampling 702, 703, 705
  edge of Victoria Falls 728
  with plastic explosives 727
  statistical efficiency 575
  streams 712
  water, chemical properties 107
    optical properties 102
    thermal properties 105
  methodology 102
Sands, residual 21
Sandstones 20, 23
Sanyati Gorge 21, 83, 180
  River 20, 25, 67, 131, 173, 178, 180, 189, 211
    catchment area 67
    chemistry 71
    drainage area 69
    flooding 69
    ionic composition 70, 73
    mineral content 73
    water quality 69
Sargochromis codringtoni, green bream 262, 337, 432, 492
  abundance 296, 420, 438, 477, 631, 633, 635–641
  age 280, 287–289, 298
  annulus 285, 345
  average size 293
  biomass 295, 296, 298, 438, 440, 543
  body-scale relationship 281
  density 348, 437
  distribution 495
  growth 280, 281, 290, 291, 294–296, 348, 598
  habitat 280
  initial biomass 282
  juvenile marks 285, 286
  key scales 673
  life span 428
  minimum harvestable size 283, 298, 442
  mortality 295, 439
  production 281, 282, 438, 441, 607, 608
  production available 282, 297, 298, 441
  scales 282, 284–286
  size and production values 444
  standing crop 477, 631, 633, 635, 638, 639, 641
  standing stock 280, 295, 296
  survival rate 289
  yield 297
  yield available 282, 445
  giardi, pink bream 492, 547
    abundance 631, 633, 420, 495
    key scale 672
    standing crop 631, 633
Sarotherodon 99
  hybrid 467
  andersoni, three-spot bream 463, 492, 547
    abundance 469, 509, 627, 629
    annulus formation 504
    biomass 513
    density 519
    distribution 495
    growth 502–506, 514
    juvenile marks 503
    key scales 669
    minimum harvestable size 514
    pond experiment 643
    reproduction 514
    size 507
    standing crop 466–469, 627, 629
    standing stock 509
    weight-length relationship 502
  macrochir, green-headed bream 463
    abundance 509
    annulus formation 504
    biomass 513
    density 513
    distribution 495
    growth 502, 503, 506
    juvenile marks 503
    body-scale relationship 502

size 507
standard length-total length relationship 502
standing stock 509
weight-length relationship 502
mossambicus mortimeri, kurper bream 264, 433
  abundance 343, 438, 475, 477, 627, 629, 631, 633, 635, 636, 638, 639, 641, 642
  age 343, 347, 348
  annulus 345, 349
  biomass 345, 347, 355–357, 440, 438, 543
  body-scale relationship 343, 344
  density 437
  distribution 343, 495
  food habits 348
  growth 347–351, 596
  harvest 349
  key scales 670
  length increments 354
  life span 428
  minimum harvestable size 349, 442
  mortality 343, 347, 439
  production 438, 440, 551, 552, 609, 610
    available 343, 349, 441
  scales 345, 346
  standing crop 475, 477, 627, 629, 631, 633, 635–642
  stock 349
  survival rate 343, 348
  weight increments 354
  yield 343, 349, 443
niloticus 311
Sassandra River 7
Sauberer's nomograph 113
Savannah woodland 41
Savory Dam 230
Scales, circuli 647
  identification key 650, 651
  location of key scales 648
  maximum height 647
  maximum length 649
Schilbe mystus, butter catfish 262, 324, 435, 492, 547
  abundance 438, 477, 509, 631, 633, 635
  biomass 438, 439, 543
  body-vertebra relationship 329
  density 437, 491
  distribution 494
  growth 324, 328–330, 600
  life span 428
  minimum harvestable size 442
  mortality 439
  numbers sampled 490, 491
  production 438, 441, 616

  available 441
  size and production values 444
  standing crop 477, 631, 633, 635, 509
  yield 445
Schist 21
Seasonal variation 82
Sebungwe 83
  River 21
  water quality 73
Secchi disc, visibility 58
Secondary rivers, catchment areas 67
  discharge 65, 66, 77
  drainage area 67
  dry season 67
  inflow 66
  ionic composition 73
  monthly discharges 67, 68
  water quality 67
Sediments, ion exchange 23
  Karroo 20
Sedimentation 82
Seismography 24
Senanga 13
Senegal River 7
Sengwa 23
  River 67
Serranochromis 547
  angusticeps, abundance 509
    distribution 495
    standing stock 509
  longimanus 495
  macrocephalus, purple-faced bream 311, 492
    abundance 631
    distribution 495
    key scales 672
    numbers sampled 490
    standing crop 629
  robustus jallae, olive bream 463, 492
    abundance 466, 469, 631
    distribution 495
    standing crop 466–469, 631
  thumbergi 490–492, 495
Shand 27
Shores 680, 698, 699
  eroded 681
  rocks 700
Shoreline, composition 20
  development 86
  length 86
  Rhodesian 20
  types 86
Sibilobilo 20, 98
  Narrows 73
Siengwazi Fall 708, 709

eels 463
Sikolwinzola 83
Silica 201, 203, 227
Silicate 216–218, 224, 228
Silt load, visibility 58
  Zambezi River 58
Siluroid fishes, age determination 328
Silver catfish, *Eutropius depressirostris* 333, 694
Silver robber, *Micralestes acutidens* 406
Simpson's formula 88
Sinakatenge 23
Sinamwenda River 23, 83
Sinazongwe 23, 99
Single stage stratified sampling 585
Sodium 56, 57, 59, 197, 198, 209, 210, 213, 215, 218, 222, 228
Soils 21
  lime-accumulating 21
  pedocal 21
  type 42
Sokoto River 542
Solids 192
  total 56, 57, 183, 184, 205, 220, 227, 234
Sorghum 731
South Equatorial Divide 13
Species 279
  diversity 550
Specific weight rate of growth 275
Spillage 80
Spotted squeaker, *Synodontis nebulosus* 325
Springs, mineral 21
  thermal 21
Stability 160–163, 548, 549, 557
  oscillations 550
Stabilization 179
Stagnation 170, 183
Standing crop 233, 279
  stock 257, 279, 420, 543
Sterculia 41, 42
Stock 279
  total 443
Stormberg Age, see Upper Karroo 18
Strata volumes 88
Stratification, stability 147
  thermal 165, 215
Submerged, forest 680
  woodland 41
Sudd, plants 45
Surface areas 86, 87, 90
  of islands 99
Sulphate 56, 57, 59, 200, 201, 203, 217, 218, 224, 227
Summer heat incomes 154
Sunshine 38

Binga 30
Symbols, a list 574
Synodontis
  leopardinus 490–492, 494
  macrostigma 509
  nebulosus, spotted squeaker 262, 324, 436, 492
    abundance 438, 477, 631, 633, 635, 638, 639
    biomass 438, 439, 544
    body-vertebra relationship 328
    density 437
    distribution 494
    growth 328–330, 601
    life span 428
    mortality 439
    production 438, 441, 622
      available 441
    size and production values 444
    standing crop 477, 631, 633, 635, 638, 639
  nigromaculatus 490–494
  zambezensis, plain squeaker 262, 298, 300, 342, 434
    abundance 438, 477, 627, 631, 633, 635, 638–642
    age 298, 306–308
    annuli 300, 308
    biomass 438, 440, 543
    density 437
    distribution 494
    growth 298, 299, 601
    length 306, 309, 310
    life span 428
    minimum harvestable size 306, 442
    mortality 439
    production 274, 438, 441, 613
      available 441
    standing crop 477, 627, 631, 633, 635, 638–642
    yield 306, 445

T

Tanganyika Lake 439
  alkalinity 230
  anchoveta, behaviour 540
    food 524
    growth 530
    length 531
    maturity 533
  clupeid, abundance 541
    distribution 541
    length 524
    life span 524

conductivity 230
Limnothrissa miodon 524, 530, 531, 533, 540
morphometry 100
Stolothrissa tanganicae 524, 540
Taxocene 279
Taxon 279
Tchad Lake 542
Temperature 56, 57
 air 27, 29, 30, 35, 37
 curves 134, 135, 140
 vertical distribution 132, 133
Tephrosia vogelii, ichthyotoxine 471
Terminalia 42
Terms, usage 278
Thames River 542–544
Thermal, characteristics 163
 cycles 131, 133, 137, 139, 142, 144, 146, 165, 181, 227, 235
 energy 146
 gradient 131–133, 136, 145, 163
 profiles 105, 132
 properties 131, 162
 stratification 165, 215, 227
Thermocline 131–133, 136–139, 145, 163, 174, 181, 183, 209
Thicket, londe, lusaka, luumpa 41
Thornthwaite's method 35
Thun Lake 149, 153
Thunderstorms 29
Tigerfish, *Hydrocynus vittatus* 689
 senile 692
 teeth 690
Tilapia 99
 rendalli, red-breasted bream 262, 268, 311, 432, 492, 499
  abundance 438, 477, 509, 627, 629, 631, 633, 635–642
  age 312, 315
  annulus 345, 504
  biomass 438, 440, 513, 543
  body-scale relationship 311, 314, 502
  density 311, 318, 348, 437, 475, 513
  distribution 495
  growth 316, 318, 348, 502, 503, 506, 507, 596
  juvenile marks 503
  key scales 671
  length increments 317
  life span 428
  minimum harvestable size 318, 442
  mortality 312, 315, 347, 439
  production 276, 277, 311, 438, 441, 611
   available 441
  scales 312, 313

  size 507
  size and production values 444
  standard length-total length relationship 502
  standing crop 475, 477, 627, 629, 631, 633, 635–642
  standing stock 311, 509
  stock composition 271, 273
  stock weight growth intensity 507
  weight increments 317
  weight-length relationship 502
  yield 443, 445
 melanopleura = *Tilapia rendalli*
 sparrmani 463, 466–469, 509
  distribution 494
  standing crop 466–469
 tholloni 311
Time of annulus inception 643
Tribe, cultural 554
 natural 554
Tributary rivers 67
Total, dissolved solids 551
 mortality 275
 standing stock 440
 solids 185, 186, 191, 233
Tonga 734
 origin 47
 population density 47
 resettlement 47
 subsistence cultivators 47
Toxaphene, distribution 684
Trap-nets 710
Trees, submerged 41
Trophic status 233
Trophogenic zone 164, 178, 182, 234
Tropicality index 151, 153, 162, 163
Tropodiaptomus kraepelini 47
Turbines, corrosion of 7
 discharge 82
 flow 80
Turnover ratio 279, 543

U

Umniati River 67
United Nations Development Programme (UNDP) 8
Uplift 20
Upper, Karroo 18
 Zambezi River 13
Usage of terms 278

V

Valley Tonga 47

Vallisneria aethiopica 45
Vegetation, aquatic 41
  bush-cleared, regeneration 44
  canopy 44
  Gwembe Valley 41, 42
  semi-aquatic 43
  shoreline 54
  terrestrial 41
  types 42
Victoria Falls 13, 18, 43, 262, 420, 491
  edge sampling 728
  fish, barrier 484, 486, 496
    density 491, 492
    distribution 492, 495, 496
  geography 484, 486, 489
  geology 486, 489
  history, formation 486
  map 485
  species composition 491
  sampling localities 489
Victoria Lake 100, 119, 120, 126–129, 147, 149, 151, 153, 154, 156, 157, 195, 201, 224–228, 230, 233, 420
Visibility 58
  depth 103, 108–111
Volta Lake 100, 101, 119, 120, 126, 128, 129, 146, 163, 165, 179, 224–227, 244, 245, 535, 536
  River 7, 61, 62
Volume 87, 88, 90, 92–94, 216
Vossia cuspidata 498
Vundu, *Heterobranchus longifilis* 325

W

Water, balance components 80
  budget 77, 79
  calcico-carbonate 59
  capacity 93
  composition 59
  deficiency 35
  fern, *Salvinia auriculata* 43
  fluctuations 523
  gains 78–80
  level, fluctuation 43, 80–82, 88
    rising 262
  losses 78, 79
  movements 83
  operating level 63
  quality 54, 56, 57, 61, 62
  total annual loss 33
Weir 711
Whistler's Bend Reservoir 543
Wind 29, 157, 164
  characteristics 34

congo air 25
  katabatic 29
  northeast monsoon 25
  southeast trade 25
  speed 27, 30
  warming effect 157
  work 160–163
  work-curve 159
Work, budget 160
  constant 160
  of wind 160–163
  total 162
Wyland Lake 543

X

Xanthidium subtrilobum 45
Xyloborus torquatus 46

Y

Yield 279, 320, 441, 442
  available 275, 279, 282, 298, 320, 345, 382, 442, 443, 545
  maximum, potential 443
    sustained 279, 443, 546
  sustained 279
  total 275, 279, 442, 544, 545

Z

Zambezi, barbel *Clarias gariepinus* 318
  catchment 18
    geology 18, 19
  channel 165
  climate 25
  flood 119
  flow regime 51
  inflow 54
  Plain 16
  River 7, 13, 51, 121–123, 125, 131–133, 137, 139, 142, 147, 149, 154, 156, 166, 189, 191, 195, 198–200, 203, 205, 209–211, 215–218, 222, 227, 228, 420, 446, 449, 492, 550
    alkalinity, $CaCO_3$ 56, 57
    annual flow 54
    annual regime 53
    calcium 56, 57
    catchment above Kariba Dam 11
    channel 21
    chemical composition 58, 212
    chloride 56, 57
    conductivity 56, 57
    discharge 13, 52, 77

    drainage divide 13
    fish invasion routes 487, 488
    flow regime 55
    general physiography 14
    hardness, $CaCO_3$ 56, 57
    hydrographs 52
    ionic composition 73, 193, 194, 213
    longitudinal profile 13, 14
    magnesium 56, 57
    pH 56, 57
    physical features 15
    potassium 56, 57
    salinity 54, 56, 57, 193
    silt load 58, 82
    sodium 56, 57
    sulphate 56, 57
    temperature 56, 57
    total solids 56, 57
    water quality 55, 56, 59
    zones 51
  Valley 20
Zambia 8, 164, 180
  islands 98
  president of 721
  shoreline length 86
Zambian High Plateau 25
Zhemtchuzhnoe Lake 542
Zongwe River 67, 73
Zooplankton 47

MIX
Papier aus verantwortungsvollen Quellen
Paper from responsible sources
FSC® C105338

If you have any concerns about our products,
you can contact us on
ProductSafety@springernature.com

In case Publisher is established outside the EU,
the EU authorized representative is:
**Springer Nature Customer Service Center GmbH
Europaplatz 3, 69115 Heidelberg, Germany**

Printed by Libri Plureos GmbH
in Hamburg, Germany